AIRCRAFT ELECTRICITY AND ELECTRONICS

AIRCRAFT ELECTRICITY AND ELECTRONICS

Third Edition

Ralph D. Bent
James L. McKinley

Gregg Division
McGraw-Hill Book Company

New York Atlanta Dallas St. Louis San Francisco Auckland Bogotá Guatemala Hamburg Johannesburg Lisbon London Madrid Mexico Montreal New Delhi Panama Paris San Juan São Paulo Singapore Sydney Tokyo Toronto

AVIATION TECHNOLOGY SERIES

Aircraft Powerplants, Bent/McKinley
Aircraft Maintenance and Repair, Bent/McKinley
Aircraft Basic Science, Bent/McKinley
Aircraft Electricity and Electronics, Bent/McKinley

Library of Congress Cataloging in Publication Data

Bent, Ralph D
Aircraft electricity and electronics.

(Aviation technology series)
First-2d editions by Northrop Institute of Technology, Inglewood, Calif. published under title: Electricity and electronics for aerospace vehicles.
Includes index.
1. Airplanes—Electric equipment. 2. Airplanes—Electronic equipment. 3. Guided missiles—Guidance systems. 4. Astronautics—Communication systems. I. McKinley, James L., joint author. II. Northrop Institute of Technology, Inglewood, Calif. Electricity and electronics for aerospace vehicles. III. Title. IV. Series.
TL690.B36 1981 629.1 80-23602

ISBN 0-07-004793-6

This text has been produced with the assistance and collaboration of the Institute of Technology of Northrop University.

Aircraft Electricity and Electronics, Third Edition

1 2 3 4 5 6 7 8 9 0 SMSM 8 9 8 7 6 5 4 3 2 1

ISBN 0-07-004793-6

The editors for this book were D. E. Gilmore and Mitsy Kovacs, the designers were Charles A. Carson and Caryl Spinka the art supervisor was George T. Resch, and the production supervisor was Priscilla Taguer. Cover photo: Courtesy of General Dynamics

CONTENTS

ACKNOWLEDGMENTS

The authors wish to express appreciation to the following organizations for their generous assistance in providing illustrations and technical information for this text.

AiResearch Manufacturing Company, Division of the Garrett Corporation, Los Angeles, California

AMP Incorporated, Harrisburg, Pennsylvania

Beech Aircraft Company, Wichita, Kansas

The Bendix Corporation
Aerospace Electronics Group, Arlington, Virginia
Avionics Division, Fort Lauderdale, Florida
Communications Division, Baltimore, Maryland
Electric Power Facility, Eatontown, New Jersey
Electrical Components Division, Sydney, New York
Flight Systems Division, Teterboro, New Jersey

The Boeing Company, Renton, Washington

Cessna Aircraft Company, Wichita, Kansas

Collins Air Transport Division, Rockwell International

Collins General Aviation Division, Rockwell International

Dayton Aircraft Products Division, Dayton International, Inc. Fort Lauderdale, Florida

Delco Remy Division, AC Spark Plug Division, General Motors Corporation, Anderson, Indiana

Federal Aviation Administration, Washington, D.C.

General Electric Company, Battery Division, Gainesville, Florida

Global Navigation, Inc., Torrance, California

Government Electronics Division, Motorola, Inc., Scottsdale, Arizona

International Rectifier Company, El Segundo, California

King Radio Corporation, Olathe, Kansas

Lockheed California Company, Burbank, California

Marathon Battery Company, Division of Marathon Manufacturing Company, Waco, Texas

McDonnell Douglas Corporation, Douglas Aircraft Division, Long Beach, California

Narco Avionics Division, Narco Scientific Industries, Fort Washington, Pennsylvania

Northrop University, Inglewood, California

Piper Aircraft Corporation, Lock Haven, Pennsylvania

Prestolite Electrical, Division of the Eltra Corporation, Toledo, Ohio

Prestolite Wire, Division of the Eltra Corporation, Port Huron, Michigan

Radio Corporation of America, Van Nuys, California

Ryan Stormscope, Columbus, Ohio

Saft America, Incorporated, Valdosta, Georgia

Sperry Flight Systems, Avionics Division, Sperry Rand Corporation, Phoenix, Arizona

Sundstrand Aviation Electric Power, Sundstrand Corporation, Rockford, Illinois

Symbolic Displays, Incorporated, Irvine, California

Teledyne Battery Products, Redlands, California

Weston Instruments Division, Sangamo Weston, Incorporated, Newark, New Jersey

PREFACE

Aircraft Electricity and Electronics is one of the four texts in the Aviation Technology Series. This edition has been revised to include material which, when used in connection with classroom discussions, lectures, demonstrations, and practical application, should assist the student in attaining the proficiency levels defined in current Federal Aviation regulations.

In preparing this edition, the authors have reviewed at length *A National Study of the Aviation Mechanics Occupation* together with FAR Part 147 and FAA Advisory Circulars 65-2C and 43.23-1A. In addition, numerous suggestions and recommendations were solicited and received from aviation instructors, aircraft manufacturers, aviation operators, and maintenance specialists. Both the new material and the revised sequence of chapters reflect the composite result of all these sources.

This textbook provides thorough coverage of electrical and electronic theory at a level which may be easily understood by the student who does not have a knowledge of advanced mathematics. Following the chapters explaining fundamental theory, the applications to electrical and electronic systems are described. Although a detailed study of electronic systems and solid-state devices as applied to microelectronics is beyond the scope of this text, the last several chapters provide descriptions of these systems and devices as applied to modern aircraft needs. Electronics as applied to aircraft is usually termed *avionics* which is a contraction of *aviation electronics.*

In the earlier sections of the text, specific information is given concerning typical aircraft electrical equipment, power systems, and basic electronic circuits. A thorough study of these portions of the text will give the technician a solid foundation on which to build for more advanced work in electric and electronic technology. The later sections of the text provide general information on the application of avionics in aircraft.

For the person who is not an electrical or avionics specialist but who is assigned to work on equipment in which electrical and avionic systems are installed, the information contained in this text will provide an increased appreciation of the systems and their application to aircraft.

Each topic in the *Aviation Technology Series* has been explained in a logical sequence so that the student may advance step by step and build a solid foundation for increased learning. Numerous pictures, charts, and drawings should give students an enhanced understanding of the explanations and descriptions included in each text.

The subjects in the *Aviation Technology Series* have been organized to provide a wealth of classroom material for instructors in public and private technical schools, training departments of aircraft manufacturing companies, vocational schools, high schools, and technical departments of colleges. The series should also be of substantial value to those who seek self development in aviation technology.

Ralph D. Bent
James L. McKinley

CHAPTER 1

Fundamentals of Electricity

This present period in history may well be called *the age of electronics* because electricity and electronics have become vital in every facet of modern technology. This is particularly true in the aviation and aerospace fields because all modern aircraft and spacecraft are very largely dependent upon electronics and electricity for communications, navigation, and control. **Electronics** is merely a special application of electricity wherein precise manipulation of electrons is employed to control electric power for a vast number of functions.

Airframe and powerplant maintenance technicians are not usually required to have an extensive knowledge of electronic phenomena; however, they should understand the basic principles of electricity and electronics and be able to perform a variety of service operations involved in the installation of electric and electronic equipment on an airplane. The repair, overhaul, and testing of electronic equipment is usually performed by *avionic* specialists who have had extensive training in this type of work.

Previous to the last century, little was known concerning the nature of electricity. Its manifestation in the form of lightning was considered by many to be a demonstration of divine displeasure. During the last century, the causes of electrical phenomena have been accurately determined, and we are now able to employ electricity to perform a multitude of tasks.

Today electricity is so common that we take it for granted. Without it there would be no modern automobiles, refrigerators, electric irons, electric lights, streetcars, airplanes, missiles, spacecraft, radios, x-ray, telephones, or television. Life, in the modern sense, could not continue; we would soon revert to the horse-and-buggy era.

One function of electricity in an airplane is to ignite the fuel-air charge in the engine. Electricity for this purpose is supplied by magnetos coupled to the engine. In the case of gas-turbine engines such as turbojets or turboprops, electrical ignition is needed only at the time of starting the engines. In addition to providing engine ignition, electricity supplies light, heat, and power. For example, it operates position lights, identification lights, landing lights, cabin lights, instrument lights, heaters, retractable landing gear, wing flaps, engine cowl flaps, radio, instruments, and navigation equipment. Airliners contain many miles of electric wiring and hundreds of electric and electronic components; hence it is obvious that any person engaged in the servicing, operation, maintenance, or design of such aircraft must have a thorough understanding of electrical principles. This applies to pilots, aircraft and powerplant technicians, instrument technicians, flight engineers, design engineers, maintenance engineers, and many others interested in the technical aspects of aircraft operation and maintenance.

THE ELECTRON THEORY

Many persons who are unfamiliar with electricity believe that an understanding of the subject is extremely difficult to attain and that only a few individuals of superior intelligence can hope to learn much about it. This is not true. A few hours of study will enable almost anyone with sufficient interest to understand the basic principles. These principles are Ohm's law, magnetism, electromagnetic induction and inductance, capacitance, and the nature of direct and alternating currents. These fundamentals are not difficult to master, and almost all electrical applications and phenomena may be explained in terms of these principles.

MOLECULES AND ATOMS

Matter is defined as anything that occupies space; hence everything that we can see and feel constitutes matter. It is now universally accepted that matter is composed of molecules, which, in turn, are composed of atoms. If a quantity of a common substance, such as water, is divided in half, and the half is then divided, and the resulting quarter divided, and so on, a point will be reached where any further division will change the nature of the water and turn it into something else. The smallest particle into which any compound can be divided and still retain its identity is called a **molecule.**

If a molecule of a substance is divided, it will be found to consist of particles called **atoms.** An atom is the smallest possible particle of an element, and until recently it was considered impossible to divide or destroy an atom.

There are more than 100 recognized elements, several of which have been artificially created from various radioactive elements. An **element** is a substance that cannot be separated into different substances except by nuclear disintegration. Common elements are iron, oxygen, aluminum, hydrogen, copper, lead, gold, silver, and so on. The smallest division of any of these elements will still have the properties of that element.

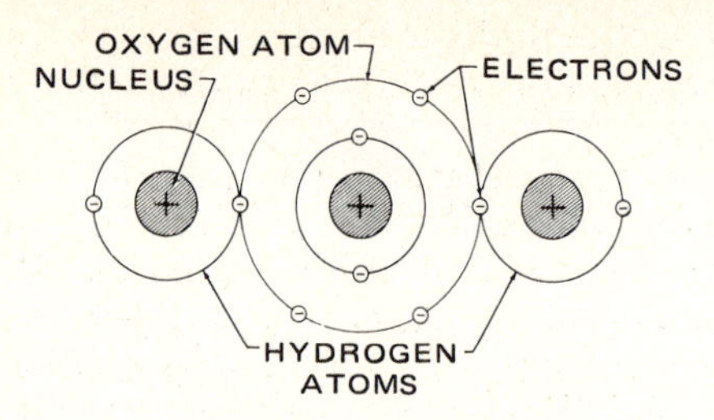

Fig. 1-1 Drawing to illustrate a water molecule.

A **compound** is a chemical combination of two or more different elements, and the smallest possible particle of a compound is a molecule. For example, a molecule of water (H_2O) consists of two atoms of hydrogen and one atom of oxygen. A diagram representing a water molecule is shown in Fig. 1-1.

ELECTRONS, PROTONS, AND NEUTRONS

Many discoveries have been made that greatly facilitate the study of electricity and provide new concepts concerning the nature of matter. One of the most important of these discoveries has dealt with the structure of the atom. It has been found that an atom consists of infinitesimal particles of energy known as electrons, protons, and neutrons. All matter consists of one or more of these basic components. The simplest atom is that of hydrogen, which has one electron and one proton as represented in the diagram of Fig. 1-2*a*. The structure of an oxygen atom is indicated in Fig. 1-2*b*. This atom has eight protons, eight neutrons, and eight electrons. The protons and neutrons form the **nucleus** of the atom; electrons revolve around the nucleus in orbits varying in shape from an ellipse to a circle and may be compared to the planets as they move around the sun. A **positive** charge is carried by each proton, **no** charge is carried by the neutrons, and a **negative** charge is carried by each electron. The charges carried by the electron and the proton are equal but opposite in nature; thus an atom which has an equal number of protons and electrons is electrically neutral. The charge carried by the electrons is balanced by the charge carried by the protons.

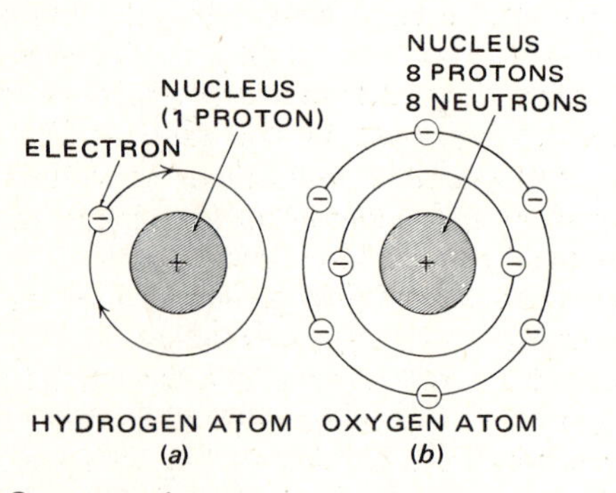

Fig. 1-2 Structure of atoms.

Through research on the weight of atomic particles, scientists have found that a proton weighs approximately 1845 times as much as an electron and that a neutron has the same weight as a proton. It is obvious, then, that the weight of an atom is determined by the number of protons and neutrons contained in the nucleus.

It has been explained that an atom carries two opposite charges: a positive charge in the nucleus and a negative charge in each electron. When the charge of the nucleus is equal to the combined charges of the electrons, the atom is neutral; but if the atom has a shortage of electrons, it will be **positively charged.** Conversely, if the atom has an excess of electrons, it will be **negatively charged.** A positively charged atom is called a **positive ion,** and a negatively charged atom is called a **negative ion.** Charged molecules are also called ions.

ATOMIC STRUCTURE AND FREE ELECTRONS

The path of an electron around the nucleus of an atom describes an imaginary sphere or shell. Hydrogen and helium atoms have only one shell, but the more complex atoms have numerous shells. When an atom has more than two electrons, it must have more than one shell, since the first shell will accommodate only two electrons. This is shown in Fig. 1-2*b*. The number of shells in an atom depends upon the total number of electrons surrounding the nucleus.

The atomic structure of a substance is of interest to the electrician because it determines how well the substance can conduct an electric current. Certain elements, chiefly metals, are known as **conductors** because an electric current will flow through them easily. The atoms of these elements give up electrons or receive electrons in the outer orbits with little difficulty. The electrons that move from one atom to another are called **free electrons.** The movement of free electrons from one atom to another is indicated by the diagram in Fig. 1-3, and it will be noted that they pass from the outer shell of one atom to the outer shell of the next. The only electrons shown in the diagram are those in the outer orbits.

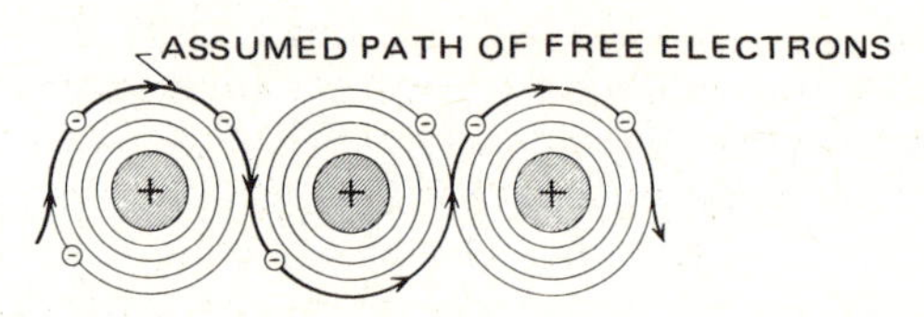

Fig. 1-3 Assumed movement of free electrons.

An element is a conductor, nonconductor (insulator), or semiconductor, depending upon the number of electrons in the outer orbit of the atom. If an atom has less than four electrons in the outer orbit, it is a conductor. If it has more than four atoms in the outer orbit, it is an **insulator.** A **semiconductor** material such as germanium or silicon has four electrons in the outer orbit of its atoms. These materials have a very high resistance to current flow when in the pure state; however, when measured amounts of other elements are added, the material can be made to carry current. The nature and use of semiconductors are discussed in a later chapter.

To cause electrons to move through a conductor, a force is required, and this force is supplied in part by the electrons themselves. When two electrons are near each other and are not acted upon by a positive charge, they repel each other with a relatively tremendous force. It is said that if two electrons could be magnified to the size of peas and were placed 100 ft apart, they would repel each other with tons of force. It is this force which is utilized to cause electrons to move through a conductor.

Electrons cluster around a nucleus because of the neutralizing positive force exerted by the protons in the nucleus and also because of an unexplained phenomenon called the **nuclear binding force.** If the binding force were suddenly removed, there would be an explosion like that of the atomic bomb. The force of the atomic-bomb explosion is the result of an almost infinite number of atoms disintegrating simultaneously.

The movement of electrons through a conductor is due, not to the disintegration of atoms, but to the repelling force which the electrons exert upon one another. When an extra electron enters the outer orbit of an atom, the repelling force immediately causes another electron to move out of the orbit of that atom and into the orbit of another. If the material is a conductor, the electrons move easily from one atom to another.

We are all familiar with the results of passing a hard rubber or plastic comb through the hair. When the hair is dry, a faint crackling sound may be heard and the hair will stand up and attempt to follow the comb. As the comb moves through the hair, some of the electrons in the hair are dislodged and picked up by the comb. The reason for the transfer is probably that the outer orbits of the atoms of the material in the comb are not filled; they therefore attract electrons from the hair. When the hair is agitated by the comb, the unbalanced condition existing between the atoms of the comb and of the hair causes the electrons to transfer. The hair now becomes positively charged because it loses electrons, and the comb becomes negatively charged because it gains electrons.

When the hair is thus charged, it will tend to stand up, and the single strands will repel one another because each has a similar charge. If the comb is then brought near the hair, the hair will be attracted by the comb because the hair and the comb have unlike charges. The attraction is the result of the electrons on the comb being attracted by the positive charge of the hair.

Static charging by friction between two or more dissimilar materials is called **triboelectric** charging. This type of charging is an important factor in the design and installation of electric and electronic equipment in aircraft or space vehicles.

A charged body, such as a comb or plastic rod, may be used to charge other bodies. For example, if two pith balls are suspended near each other on fine threads, as in Fig. 1-4*a*, and each ball is then touched with a charged plastic rod, a part of the charge is conveyed to the balls. Since the balls will now have a similar charge, they will repel each other as in Fig. 1-4*b*. If the rod is rubbed with a piece of fur, it will become negatively charged and the fur positively charged. By touching one of the balls with the rod and the other with the fur, the balls are given opposite charges. They will then attract each other as shown in Fig. 1-4*c*.

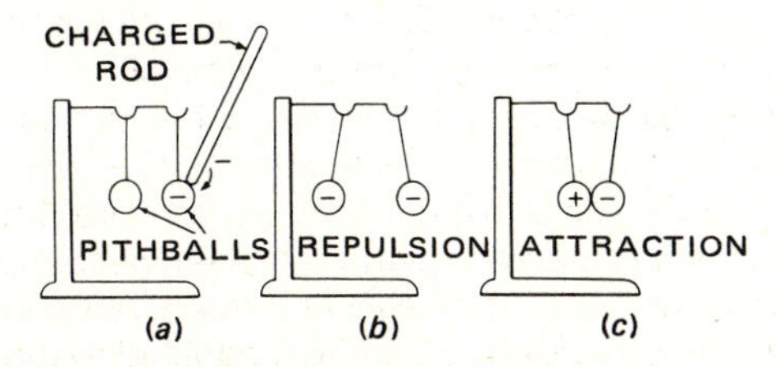

Fig. 1-4 Reaction of like and unlike charges.

The behavior of a charged body indicates that it is surrounded by an invisible field of force. This field is assumed to consist of lines of force extending in all directions and terminating at a point where there is an equal and opposite charge. A field of this type is called an **electrostatic field.** When two oppositely charged bodies are in close proximity, the electrostatic field is relatively strong. If the two bodies are joined by a conductor, the electrons from the negatively charged body flow along the conductor to the positively charged body, and the charges are neutralized. When the charges are neutral, there is no electrostatic field.

DIRECTION OF CURRENT FLOW

It has been shown that an electric current is the result of the movement of electrons through a conductor. Since a negatively charged body has an excess of electrons and a positively charged body a deficiency of electrons, it is obvious that the electron flow will be **from** the negatively charged body **to** the positively charged body when the two are connected by a conductor. It is therefore clear that electricity flows from negative to positive.

Until recently, however, it was assumed that electric current flowed from positive to negative. This was because the polarities of electric charges were arbitrarily assigned names without the true nature of electric current being known. The study of radio and other electronic devices has made it necessary to consider the true direction of current flow, but for all ordinary electrical applications, the direction of flow may be considered to be in either direction so long as the theory is used consistently. Even though there are still some texts which adhere to the old conventional theory that current flows from positive to negative, it is the purpose of this text to consider all current flow as moving from negative to positive. Electrical rules and diagrams are arranged to conform to this principle in order to prevent confusion and to give the student a true concept of electrical phenomena.

The student will sometimes read or hear the statement "electron flow is from negative to positive, and current flow is from positive to negative." This statement is a fallacy because current flow consists of electrons moving through a conductor, and the movement is from negative to positive as explained in this section. The student should fix this principle firmly in mind to avoid being confused when encountering an application of the old "conventional" current-flow theory.

It is expected that eventually all writers and teachers will teach the principle as it actually is; however, it often takes many years to correct a false idea, and the student is warned to exercise care in the study of electricity. Particular care must be paid to rules dealing with current flow and its effects.

STATIC ELECTRICITY

ELECTROSTATICS

The study of the behavior of static electricity is called **electrostatics.** The word **static** means stationary or at rest, and electric charges which are at rest are called **static electricity.** In the previous section it was shown that static electric charges may be produced by rubbing various dissimilar substances together and triboelectric charging takes place. The nature of the charge produced is determined by the types of substances. The following list of substances is called the **electric series,** and the list is so arranged that each substance is positive in relation to any which follow it, when the two are in contact:

1. Fur	**6.** Cotton	**11.** Metals
2. Flannel	**7.** Silk	**12.** Sealing wax
3. Ivory	**8.** Leather	**13.** Resins
4. Crystals	**9.** The body	**14.** Gutta percha
5. Glass	**10.** Wood	**15.** Guncotton

If, for example, a glass rod is rubbed with fur, the rod becomes negatively charged, but if it is rubbed with silk it becomes positively charged.

When a nonconductor is charged by rubbing it with a dissimilar material, the charge remains at the points where the friction occurs because the electrons cannot move through the material; however, when a conductor is charged, it must be insulated from other conductors or the charge will be lost.

An electric charge may be produced in a conductor by induction if the conductor is properly insulated. Imagine that the insulated metal sphere shown in Fig. 1-5 is charged negatively and brought near one end of a metal rod which is also insulated from other conductors. The electrons constituting the negative charge in the sphere repel the electrons in the rod and drive them to the opposite end of the rod. The rod then has a positive charge in the end nearest the charged sphere and a negative charge in the opposite end. This may be shown by suspending pith balls in pairs from the middle and ends of the rod by means of conducting threads. At the ends of the rod, the pith balls separate as the charged sphere is brought near one end; but the balls near the center do not separate because the center is neutral. As the charged sphere is moved away from the rod, the balls fall to their original positions, thus indicating that the charges in the rod have become neutralized.

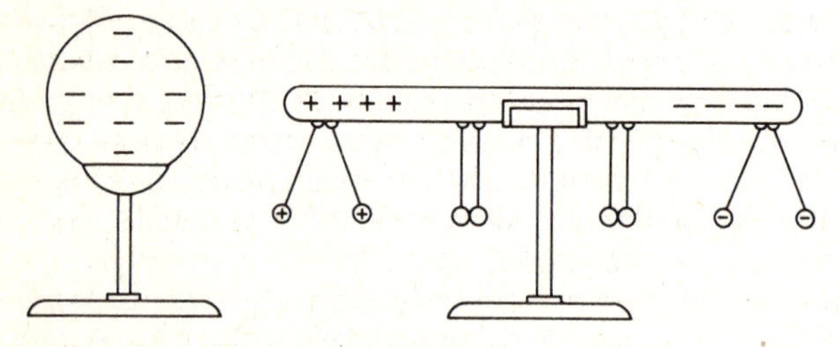

Fig. 1-5 Charging by induction.

LIGHTNING

The familiar flash of lightning is nothing but an enormous spark caused by the discharge of static electricity from a highly charged cloud. Clouds become charged because of friction between their many minute particles of water, air, and dust. Lightning is most commonly found in cumulus and cumulonimbus clouds. These latter are the towering, billowy clouds frequently seen in the summer; they are caused by warm moist air moving up into colder areas where condensation takes place. Such clouds have air currents moving up through their centers at speeds which are sometimes in excess of 100 miles per hour (mph) [161 kilometers per hour (km/h)]. The turbulence caused by these updrafts is largely responsible for the electric charges which cause lightning.

Although serious damage to an aircraft as the result of lightning is rare, studies have been made to establish safe procedures when lightning may be encountered. Such studies have indicated that a positive charge develops in the forward portion of the cloud, where the updrafts are more pronounced. Thus it seems that the rising air currents are removing electrons from that portion of the cloud. The negative charge develops in the rear portion of the cloud and is separated from the positive charge by a neutral area. When the difference between the charges becomes great enough, a flash of lightning occurs and the cloud becomes neutral for a time in that particular area.

The use of weather radar in modern airliners has helped pilots to avoid flying through thunderstorms where the danger of lightning would be greatest. Danger areas show up clearly on the radar **scopes** at a sufficient distance for the pilot to have adequate time to fly around them.

STATIC ELECTRICITY AND THE AIRPLANE

As mentioned previously, the effects of static electricity are of considerable importance in the design, operation, and maintenance of aircraft. This is particularly true because modern airplanes are equipped with radio and other electronic equipment. The pop and crackle of static is familiar to everyone who has listened to a radio receiver when static conditions are prevalent. An airplane in flight picks up static charges because of contact with rain, snow, clouds, dust, and other particles in the air. The charges thus produced in the aircraft structure result in **precipitation static** (p static). The charges flow about the metal structure of the airplane as they tend to equalize, and if any part of the airplane is partially insulated from another part, the static electricity causes minute sparks as it jumps across the insulated joints. Every spark causes p-static noise in the radio communication equipment and also causes disturbances in other electronic systems. For this reason, the parts of an airplane are **bonded** so that electric charges may move throughout the airplane structure without causing sparks. Bonding the parts of an airplane simply means establishing a good electrical contact between them. Movable parts, such as ailerons, flaps, and rudders, are connected to the main structure of the airplane with flexible woven-metal leads called **bonding braid.**

The **shielding** of electronic devices and wiring is also necessary to help eliminate the effects of p static on electric equipment in the airplane. Shields consist of metal coverings which intercept undesirable waves and prevent them from affecting sensitive electronic systems.

An airplane in flight often accumulates very high electric charges, not only from precipitation, but also from the high-velocity jet-engine exhaust as it flows through the tailpipe. When the airplance charge becomes sufficiently high, electrons will be discharged into the surrounding air from sharp or pointed sections of the airplane. The level at which this begins is called the **corona threshold.** Corona discharge is often visible at night, emanating from wing tips, tail sections, and other sharply pointed sections of a plane. The visible discharge is called *Saint Elmo's fire.*

Corona discharge occurs as short pulses at very high frequencies, thus producing energy fields which couple with radio antenna fields to cause severe interference. The solution to the problem is to cause the charge on the airplane to be partially dissipated in a controlled manner so that the energy level of the discharge will be reduced and the effects of the discharge will cause a minimum of interference. In the past, static-discharge **wicks** were used to reduce the charge on the airplane. See Fig. 1-6.

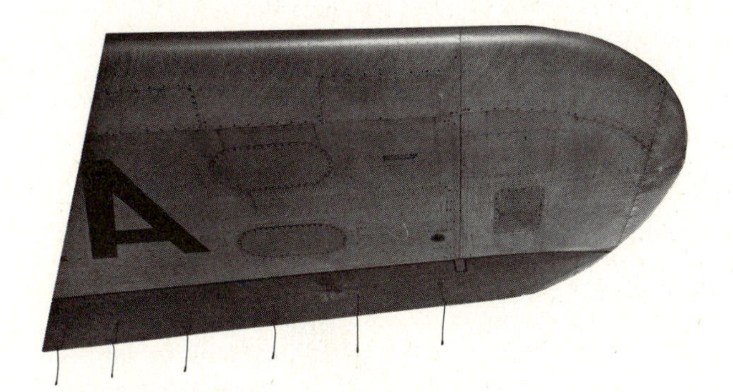

Fig. 1-6 Static discharge wicks.

Because of the high speeds of jet aircraft and the fact that they are powered by gas-turbine engines that tend to increase static charges, it became necessary to develop static-discharge devices more effective than the wicks formerly used. A type of discharger that has proved most successful is called a **null field discharger.** These dischargers are mounted at the trailing edges of outer ailerons, vertical stabilizers, and other points where high discharges tend to occur. They produce a discharge field which has minimum coupling with radio antennas. Typical installations are shown in Fig. 1-7.

Static charges must be taken into consideration when an airplane is being refueled. Gasoline or jet fuel flowing through the hose into the airplane will usually cause a static charge to develop at the nozzle of the hose unless a means is provided whereby the charge may bleed off. If the nozzle of the fuel hose should become sufficiently charged, a spark could occur and cause a disastrous fire. To prevent such an occurrence, the nozzle of the fuel hose is connected electrically to the aircraft by means of a grounding cable or other device, and the aircraft is grounded to the earth. In this way, the fuel nozzle and the aircraft are kept neutral with the earth, and no charges can develop sufficient to create a spark.

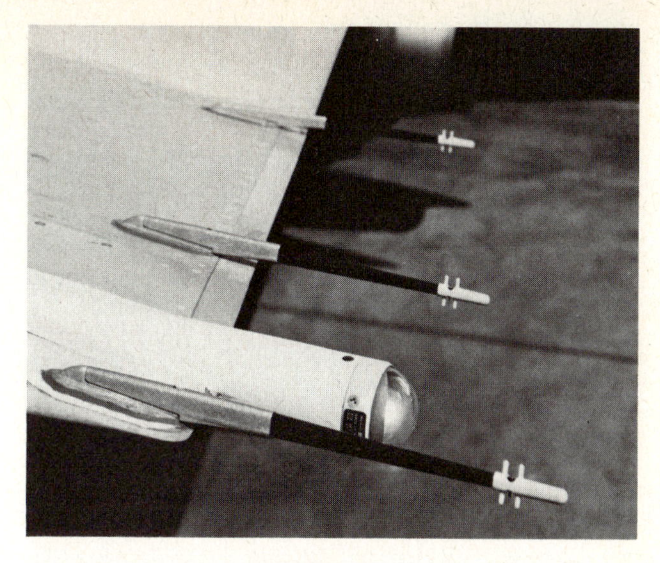

Fig. 1-7 Installation of Null Field Dischargers. (Dayton Aircraft Products Div., Dayton International)

THE ELECTRIC CURRENT

DEFINITION

An electric **current** is defined as a flow of electrons through a conductor. In an earlier part of this chapter it was shown that the free electrons of a conducting material move from atom to atom as the result of the attraction of unlike charges and the repulsion of like charges. If the terminals of a battery are connected to the ends of a wire conductor, the negative terminal forces electrons into the wire and the positive terminal takes electrons from the wire; hence as long as the battery is connected, there is a continuous flow of current through the wire until the battery becomes discharged.

It is said that an electric current travels at more than 186 000 miles per second (mps) [299 000 km/s]. Actually, it would be more correct to say that the effect, or force, of electricity travels at this speed. Individual electrons move at a comparatively slow rate from atom to atom in a conductor, but the influence of a charge is "felt" through the entire length of a conductor instantaneously. A simple illustration will explain this phenomenon. If we completely fill a tube with tennis balls, as shown in Fig. 1-8, and then push an additional ball into one end of the tube, one ball will fall out the other end. This is similar to the effect of electrons as they are forced into a conductor. When electrical pressure is applied to one end of the conductor, it is immediately effective at the other end. It must be remembered, however, that under most conditions, electrons must have a conducting path before they will leave the conductor.

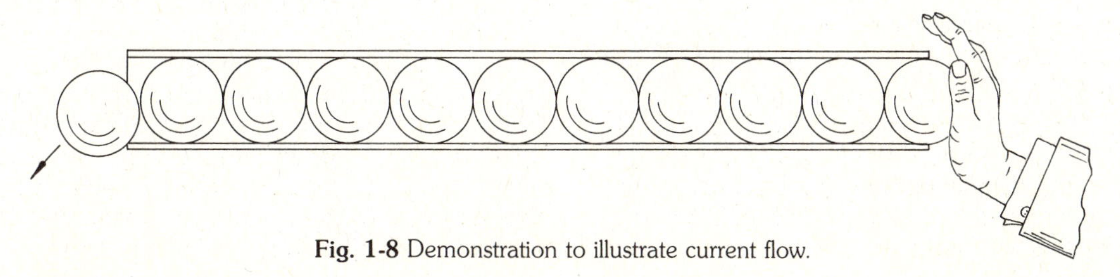

Fig. 1-8 Demonstration to illustrate current flow.

POTENTIAL DIFFERENCE AND ELECTROMOTIVE FORCE

Just as water flows in a pipe when there is a difference of pressure at the ends of the pipe, an electric current flows in a conductor because of a difference in electrical pressure at the ends of the conductor. If two tanks containing water at different levels are connected by a pipe with a valve, as shown in Fig. 1-9, water flows from the tank with the higher level to the other tank when the valve is open. The difference in water pressure is due to the higher water level in one tank.

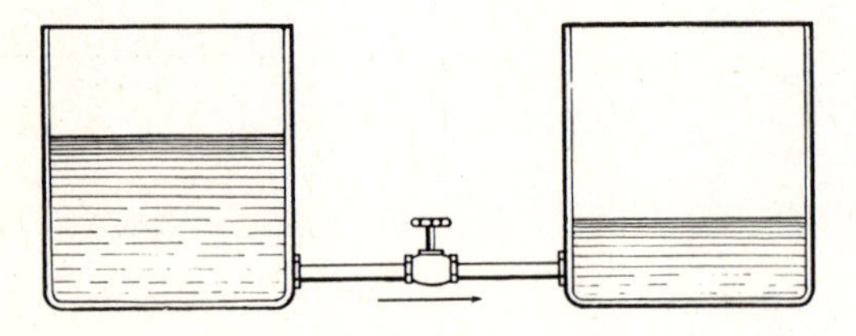

Fig. 1-9 Difference of pressure.

It may be stated that in an electric circuit, a large number of electrons at one point will cause a current to flow to another point where there is a small number of electrons if the two points are connected by a conductor. In other words, when the electron level is higher at one point than at another point, there is a *difference of potential* between the points. When the points are connected by a conductor, electrons flow from the point of high potential to the point of low potential. There are numerous simple analogies which may be used to illustrate potential difference. For example, when an automobile tire is inflated, there exists a difference of potential (pressure) between the inside of the tire and the outside. When the valve is opened, the air rushes out. If the tip of an old-fashioned light bulb is broken off, air rushes into the bulb because the inside of the bulb is at a lower pressure than the atmosphere. In this case the bulb represents a positive charge and the atmosphere a negative charge.

For all practical purposes, the earth is considered to be electrically neutral; that is, it has no charge. Therefore, if a positively charged object is connected to the earth, electrons flow from the earth to the object; and if a negatively charged object is connected, the electrons flow from the object to the earth.

The force which causes electrons to flow through a conductor is called **electromotive force,** abbreviated emf, or *electron-moving* force. The practical unit for the measurement of emf or potential difference is the **volt** (V). The word volt is derived from the name of the famous electrical experimenter, Alessandro Volta (1745-1827), of Italy, who made many contributions to the knowledge of electricity.

Electromotive force and potential difference may be considered the same for all practical purposes. When there is a potential difference, or difference of electrical pressure, between two points, it simply means that a field of force exists which tends to move electrons from one point to the other. If the points are connected by a conductor, electrons will flow as long as the potential difference exists.

With reference to Fig. 1-9, it may be stated that a difference of potential exists between the two water tanks because the weight of the water in one tank exerts a greater pressure than the weight of the water in the other tank. We may compare the difference in pressure at the ends of the connecting pipe with emf. If the water in one tank exerts a pressure of 10 pounds per square inch (psi) [68.95 kilopascals (k Pa)] at the end of the pipe, and the water in the other tank exerts a pressure of 5 psi [34.48 k Pa], there is a difference of 5 psi between the ends of the pipe. In like manner, we may say that there is an emf of 5 V between two electric terminals.

Since potential difference and emf are measured in volts, the word **voltage** is commonly used instead of longer terms. For example, we may say that the voltage of an aircraft storage battery is 24. This means that there is a potential difference of 24 V between the terminals. In simple terms, **1 volt is the emf required to cause current to flow at the rate of 1 ampere through a resistance of 1 ohm.** The terms *ampere* and *ohm* will be clarified in the study of Ohm's law.

RESISTANCE

Resistance is that property of a conductor which tends to hold back, or restrict, the flow of an electric current; it is encountered in every circuit. Resistance may be termed *electrical friction* because it affects the movement of electricity in a manner similar to the effect of friction on mechanical objects. For example, if the interior of a water pipe is very rough because of rust or some other material, a smaller stream of water will flow through the pipe at a given pressure than would flow if the interior of the pipe were clean and smooth.

The unit used in electricity to measure resistance is the **ohm.** The ohm is named for the German physicist Georg S. Ohm (1789–1854), who discovered the relationship between electrical quantities known as Ohm's law. The practical value of the ohm will be discussed in the study of this law. The symbol for ohm or ohms is the Greek letter omega (Ω).

It has been explained that materials which have a relatively large number of free electrons are conductors. When an emf is not acting on a conductor, it is assumed that the free electrons are moving at random from atom to atom and filling the gaps in outer orbits of atoms deficient in electrons. When an emf is applied to a conductor, the free electrons begin to move in a definite direction through the material, provided that there is a complete circuit through which the current can flow. The greater the emf applied to a given circuit, the greater the current flow.

The best conductors of electricity in the order of their conductivity are silver, copper, gold, and aluminum, but the use of gold or silver for conductors is limited because of the cost. The resistance of a copper wire of a given diameter and length is lower than that of an aluminum wire of the same size; but for a given weight of each material, aluminum has the lower resistance. For this reason aluminum wire may be used to advantage where the weight factor is important.

Gold is used extensively in modern electronic equipment to provide corrosion-free contacts for *plug-in* modules and other units which can be removed and replaced for service or repair. The many *black boxes* containing complex electronic circuitry can be quickly and easily repaired merely by removing a circuit module and plugging in another. The gold at the contacts provides positive electrical connections whenever a change is made.

The resistance of a standard length and cross-sectional area of a material is called its **resistivity.** For example, the resistivity of copper wire is 10.4 Ω per circular-mil-foot (cmil·ft). This means that 1 foot (ft) [30.48 centimeters (cm)] of copper wire having a cross-sectional area of 1 cmil, or 0.001 in [0.0254 millimeters (mm)], diameter will have 10.4 Ω resistance. For aluminum, the resistivity is 19.3 Ω/cmil·ft.

Insulators are materials which have relatively few free electrons. There are no perfect insulators, but many substances have such high resistance that for practical purposes they may be said to prevent the flow of current. Substances having good insulating qualities are dry air, glass, mica, porcelain, rubber, plastic, asbestos, and fiber compositions. The resistance of these substances varies to some extent, but they may all be said effectively to block the flow of current.

According to the electron theory, the atoms of an insulator do not give up electrons easily. When an emf is applied to such a substance, the outer electron orbits are distorted; but as soon as the emf is removed, the electrons return to their normal positions. If, however, the emf applied is so strong that it strains the atomic structure beyond its elastic limit, the atoms lose electrons and the material becomes a conductor. When this occurs, the material is said to be ruptured.

The resistance of a wire varies inversely with the area of the cross section. For example, if the area of the cross section of one wire is twice the cross-sectional area of another wire of the same length in material, the larger wire has one-half the resistance of the smaller wire. When the cross-sectional area of a wire remains constant, the resistance increases in proportion to length. For example, a wire 2 ft [60 cm] long has twice the resistance of a similar wire 1 ft [30 cm] long.

Temperature is another factor which affects the resistance of a wire. Usually, the resistance of a wire increases with an increase in temperature. However, some substances such as carbon, decrease in resistance as the temperature increases. The degree of resistance change due to temperature variation is not constant but depends upon the material. Some materials have a greater variation of resistance as a result of a given temperature change than other materials.

The general rule for the resistance of a conductor is as follows: **The resistance of a given conductor varies directly as its length, and inversely as the area of its cross section, when the temperature remains constant.** This may be expressed as a formula:

$$R = \frac{KL}{S}$$

K is a constant which depends upon the resistivity of the material; for example, copper has a resistivity of 10.4 Ω at 68°F [20°C]. In the formula, L is the length of the wire in feet, and S is the cross-sectional area in circular mils. To find the resistance of 300 ft [9.14 m] of copper wire having a cross-sectional area of 100 cmil [0.0507 mm²], the formula is applied as follows:

$$R = \frac{10.4 \times 300}{100} \qquad R = 31.2\ \Omega$$

As indicated in the previous paragraph, the cross-sectional area of a wire is measured in circular mils. One mil is one-thousandth of an inch. One circular mil is the area of a circle having a diameter of 1 mil, or 0.001 in. The area of a square having sides equal to 1 mil is 1 square mil (mil²) [0.000645 mm²]. These areas are illustrated in Fig. 1-10.

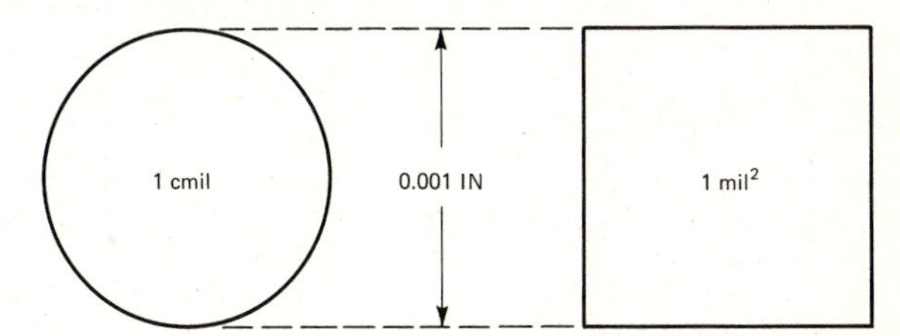

Fig. 1-10 The circular mil.

The formula for the area of a circle is

$$\frac{\pi d^2}{4} \qquad \text{or} \qquad 0.7845d^2$$

If a circle has a diameter (d) of 1 mil, the area in square mils is 0.7854×1^2, or 0.7854 mil². Since a circular mil is defined as the area of a circle having a diameter of 1 mil, then 1 cmil is equal to 0.7854 mil², and

$$1 \text{ mil}^2 = \frac{1}{0.7854} \text{ cmil}$$

The formula, A (area) $= 0.7854d^2$, gives the area of a circle in square mils when the diameter is in mils. Since

$$1 \text{ mil}^2 = \frac{1}{0.7854} \text{ cmil}$$

the area of a circle in circular mils may be given as

$$A \text{ (cmil)} = \frac{0.7854d^2}{0.7854} = d^2$$

Hence, when we wish to know the area of a circle in circular mils, we merely square the diameter.

Resistance in electric circuits produces heat just as mechanical friction produces heat. This is called the **heat of resistance.** Normally the heat of resistance is dissipated as fast as it is produced, and the wire of the circuit may become only slightly warm. However, if the current flowing in the wire is so great that it generates heat faster than the heat can be carried away by the surrounding air or insulation, the wire will eventually overheat. This may lead to the burning of the insulation and a possible fire. Tables are available which give the current-carrying capacity of copper wire according to size. For continuous-duty circuits, these limits must not be ex-

TABLE 1-1 Current-carrying Capacities for AN-S-C-48 Electric Cable

AWG wire size	Continuous rating, A: In bundles or conduit	Continuous rating, A: Single cable in free air	Intermittent rating, A
20	7	10	15
18	10	15	20
16	13	20	25
14	18	30	35
12	24	40	48
10	32	55	67
8	44	70	90
6	60	95	115
4	80	125	155
2	110	170	210
1	125	190	240
1/0	150	230	300
2/0	175	260	340
3/0	190	310	410
4/0	225	375	500

ceeded. Table 1-1 gives the current-carrying capacities of commonly used sizes of aircraft electric wire.

Two sections of wire having the same resistance generate the same amount of heat when they carry equal currents; but if one wire has a greater surface, it can carry more current without damage because it can dissipate the heat faster than the other. For example, if one section of copper wire has a length of 1 in [2.54 cm] and a cross-sectional area of 10 cmil [0.00507 mm^2], and another section of copper wire is 2 in [3.08 cm] long and has a cross-sectional area of 20 cmil [0.01014 mm^2], the resistance of the two sections of wire is the same. However, the larger wire can carry more current because it can dissipate heat more rapidly.

CURRENT

When it is necessary to measure the flow of a liquid through a pipe, the rate of flow is often measured in **gallons per minute.** The gallon is a definite quantity of liquid and may be called a unit of quantity. The unit of quantity for electricity is the **coulomb** (C), named for Charles A. Coulomb (1736-1806), a French physicist who conducted many experiments with electric charges. One coulomb is the amount of electricity which, when passed through a standard silver nitrate solution, will cause 0.001118 gram (g) of silver to be deposited upon one electrode. (An electrode is a terminal, or pole, of an electric circuit.) A coulomb is also defined as 6.28×10^{18} electrons, that is, 6.28 billion billion electrons.

The rate of flow for an electric current is measured by the number of coulombs per second passing a given point in a circuit. Instead of designating the rate of flow in coulombs per second, a unit called the **ampere** (A) is used. **One ampere is the rate of flow of 1 coulomb per second.** The ampere was named in honor of the French scientist André M. Ampère (1775-1836).

The flow of electricity through a conductor is called a **current.** Hence, when current is mentioned, it indicates a flow of electricity measurable in amperes.

OHM'S LAW

In mathematical problems, emf is expressed in **volts** and the symbol E is used to indicate the emf until the actual number of volts is determined. R is the symbol for resistance in ohms, and I is the symbol for current, or amperage. The letter I may be said to represent the **intensity** of current. The letter symbols E, R, and I have an exact relationship in electricity given by Ohm's law. This law may be stated as follows: **The current in an electric circuit is directly proportional to the emf (voltage) and inversely proportional to the resistance.** Ohm's law is further expressed by the statement: **1 volt causes 1 ampere to flow through a resistance of 1 ohm.** The equation for Ohm's law is

$$I = \frac{E}{R}$$

which indicates that the current in a given circuit is equal to the voltage divided by the resistance.

An equation is defined as a proposition expressing equality between two values. It may take as many forms as those shown for Ohm's law in Fig. 1-11. The different

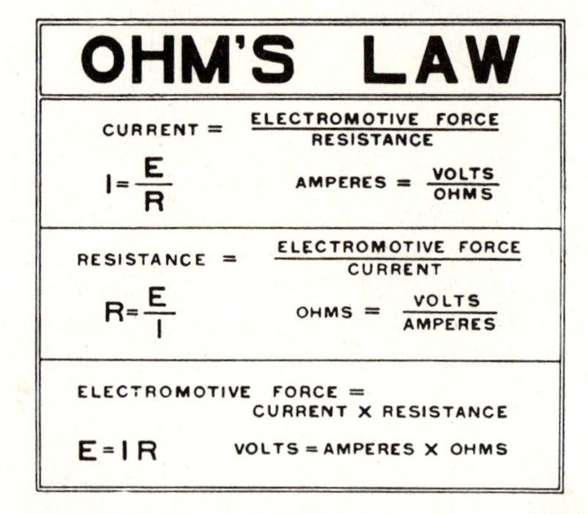

Fig. 1-11 Equations for Ohm's law.

forms for the Ohm's law equation are derived by either multiplication or division. For example,

$$R(I) = R\left(\frac{E}{R}\right) \qquad \text{becomes} \qquad RI = \frac{RE}{R}$$

Then

$$RI = E \qquad \text{or} \qquad E = IR$$

In a similar manner, if both sides of the equation $E = IR$ are divided by I, we arrive at the form

$$R = \frac{E}{I}$$

Thus we find it simple to determine any one of the three values if the other two are known. Ohm's law may be used to solve any common direct-current (dc) circuit problem because any such circuit, when operating, has voltage, amperage, and resistance. To solve alternating-current (ac) circuit problems, other values must be taken into consideration. These will be discussed in the section on alternating current.

From the study of Ohm's law, it has been seen that the current flowing in a circuit is directly proportional to the voltage and inversely proportional to the resistance. If the voltage applied to a given circuit is doubled, the current will double. If the resistance is doubled and the voltage remains the same, the current will be reduced by one-half (see Fig. 1-12). The circuit symbol for a battery which is the power source for these circuits and the circuit symbol for a resistor or resistance are indicated in the illustration.

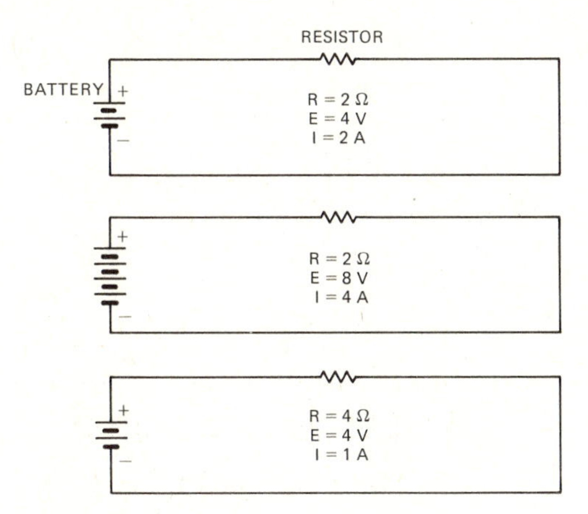

Fig. 1-12 Effects of resistance and voltage.

The equations of Ohm's law are easily remembered by using the simple diagram shown in Fig. 1-13. By covering the symbol of the unknown quantity in the diagram with the hand or a piece of paper, the known quantities are found to be in their correct mathematical arrangement.

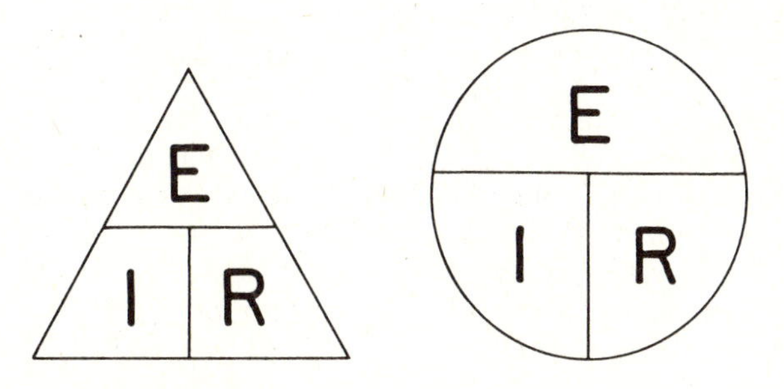

Fig. 1-13 Diagrams for Ohm's law.

For example, if it is desired to find the total resistance of a circuit in which the voltage is 10 and the amperage is 5, cover the letter R in the diagram. This leaves the letter E over the letter I; then

$$R = \frac{E}{I} = \frac{10}{5} \quad \text{or} \quad R = 2\ \Omega$$

If it is desired to find the voltage in a circuit when the resistance and the amperage are known, cover the E in the diagram. This leaves I and R adjacent to each other; they are therefore to be multiplied according to the equation form $E = IR$.

It is important for electricians or technicians who are to perform electrical work on an airplane to achieve a thorough understanding of Ohm's law, because this knowledge will enable them to determine the correct size and length of wire to be used in a circuit, the proper sizes of fuses and circuit breakers, and many other details of a circuit and its components. Further study of the use of Ohm's law is made in the next section of this chapter.

ELECTRIC POWER AND WORK

Power means the rate of doing work. One horse-power (hp) [746 watts (W)] is required to raise 550 pounds (lb) [249.5 kilograms (kg)] a distance of 1 ft [30.48 cm] in 1 s. When 1 lb [0.4536 kg] is moved through a distance of 1 ft, 1 foot-pound (ft·lb) [13.82 cm·kg] of work has been performed; hence, 1 hp is the power required to do 550 ft·lb [7601 cm·kg] of work per second. The unit of power in electricity and in the SI metric system is the **watt** (W), which is equal to 0.00134 hp. Conversely, 1 hp is equal to 746 W. In electrical terms, **1 watt is the power expended when 1 volt moves 1 coulomb per second through a conductor; that is, 1 volt at 1 ampere produces 1 watt of power.** The formula for electric power is

$$W = EI \quad \text{or} \quad \text{Watts} = \text{volts} \times \text{amperes}$$

Another unit used in connection with electrical work is the **joule** (J) named for James Prescott Joule (1818-1889), an English physicist. **The joule is a unit of work, or energy, and represents the work done by 1 watt in 1 second.** This is equal to 0.7376 ft·lb. To apply this principle, let us assume that we wish to determine how much work in joules is done when a weight of 1 ton is raised 50 ft. First we multiply 2000 by 50 and find that 100 000 ft·lb of work is done. Then, when we divide 100 000 by 0.7376, we determine that approximately 135 575 J of work, or energy, were used to raise the weight.

A joule is also defined as 10^7 ergs. An **erg** can be defined as a dyne-centimeter (dyn·cm), that is, the energy expended when a force of 1 dyne (dyn) is expended through a distance of 1 cm in the direction of the force. A **dyne** is the force required to impart an acceleration of 1 cm/s^2 to a mass of 1 g. One newton (N) is equal to 10^5 dyn, which is the force necessary to move a mass of 1 kg at an acceleration rate of 1 m/s^2. This is equal to 0.2248 lb.

It is wise for the technician to understand and have a good concept of the joule because this is the unit designated by the SI metric system for the measurement of work or energy. Other units convertible to joules are the British thermal unit (Btu), calorie (cal), foot-pound, and watthour (Wh). All these units represent a specific amount of work performed.

Electric power expended in a circuit is manifested in the form of heat or motion. In the case of electric lamps, electric irons, electric cooking ranges, etc., power is expended in the form of heat. In an electric motor or electromagnet, the power is expended in the form of motion, and work is done. An electric current flowing through a wire will always produce heat, although in many cases the rise in temperature is not noticeable. The heat generated in a given circuit is proportional to the square of the current, as shown by the following formulas:

$$W = EI \quad \text{and} \quad E = IR$$

By substitution

$$W = IR \times I \qquad \text{or} \qquad W = I^2 R$$

When energy is lost in an electric circuit in the form of heat, it is called an $I^2 R$ loss because $I^2 R$ represents the heat energy lost, measured in watts.

Since we know the relationship between power and electrical units, it is simple to calculate the approximate amperage to operate a given motor when the efficiency and operating voltage of the motor are known. For example, if it is desired to install a 3-hp [2.238 kilowatts (kW)] motor in a 24-V system and the efficiency of the motor is 75 percent, we proceed as follows:

$$1 \text{ hp} = 746 \text{ W}$$

$$W = 3 \times 746 = 2238 \text{ W}$$

$$I = \frac{2238}{24} = 93.25 \text{ A}$$

Since the motor is only 75 percent efficient, we must divide 93.25 by 0.75 to find that approximately 124.33 A is required to operate the motor at rated load. Thus, in a motor that is 75 percent efficient, 2984 W of power is required to produce 2.238 W of power at the output.

DC CIRCUITS

TYPES OF CIRCUITS

To cause a current to flow in a conductor, a difference of potential must be maintained between the ends of the conductor. In an electric circuit this difference of potential is normally produced by a battery or a generator; so it is obvious that both ends of the conductor must be connected to the terminals of the source of emf.

Figure 1-14 shows the components of a simple circuit with a battery as the source of power. One end of the circuit is connected to the positive terminal of the battery and the other to the negative terminal. A switch is incorporated in the circuit to connect the electric power to the load unit, which may be an electric lamp, bell, relay, or any other electric device that could be operated in such a circuit. When the switch in the circuit is closed, current from the battery flows through the switch and load and then back to the battery. Remember that the direction of current flow is from the negative terminal to the positive terminal of the battery. The circuit will operate only when there is a continuous path through which the current may flow from one terminal to the other. When the switch is opened (turned off), the path for the current is broken and the operation of the circuit ceases.

One of the most common difficulties encountered in electric systems is the *open* circuit. This means simply that there is a break somewhere in the circuit and that no current can flow. An open circuit is shown in Fig. 1-15. When the circuit is complete and the current can flow, it is called a *closed* circuit. The circuit in Fig. 1-14 is a closed circuit when the switch is closed.

Another common cause of circuit failure is called a *short circuit.* A short circuit exists when an accidental contact between conductors allows the current to return to the battery through a short, low-resistance path, as shown in Fig. 1-16. This failure is prevented by making sure that all insulation on the wires is in good condition and strong enough to withstand the voltage of the power source. Furthermore, all wiring should be properly secured with insulated clamps or other devices so that it cannot rub against any structure and wear through the insulation.

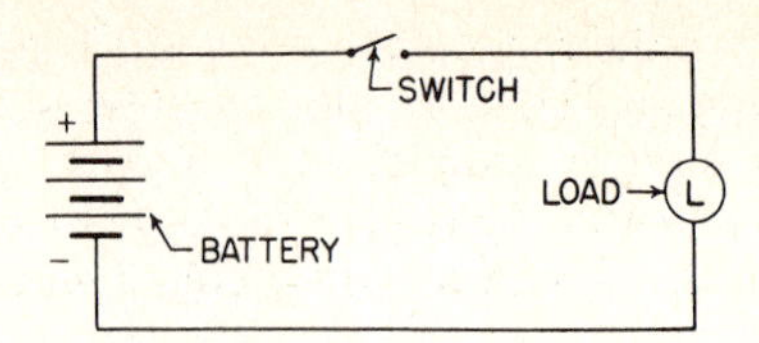

Fig. 1-14 A simple circuit.

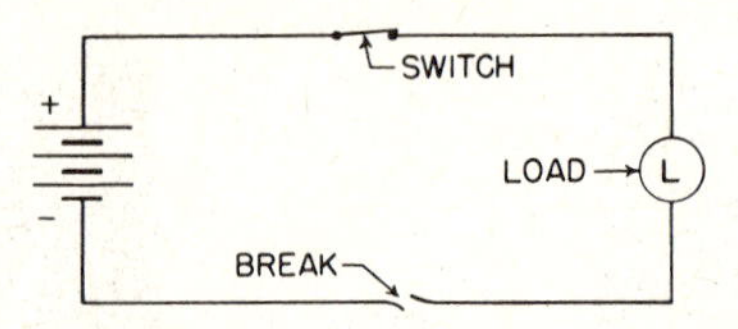

Fig. 1-15 An open circuit.

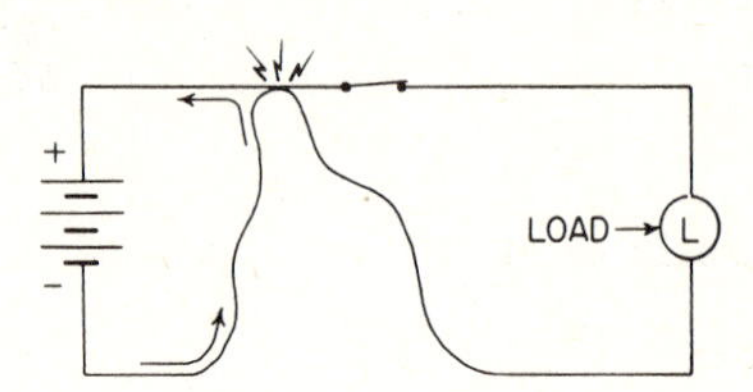

Fig. 1-16 A short circuit.

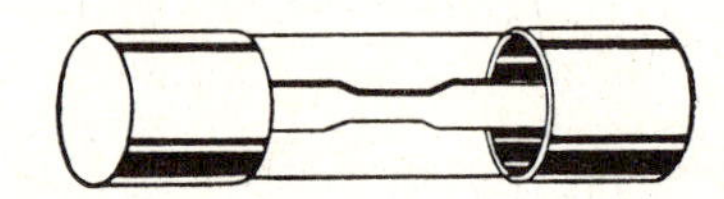
Fig. 1-17 A simple fuse.

The danger in a short circuit is that an excessive amount of current may flow through limited portions of the circuit, causing wires to overheat and burn off the insulation. If the short circuit is not discovered immediately, the wiring is likely to become red hot and may melt. Many fires are caused by short circuits, but the danger is largely overcome by the installation of protective devices, such as fuses or circuit breakers.

A **fuse** is a portion of a circuit composed of a metal or alloy with a low melting point. If the current in the circuit becomes too great, the fuse will melt and open the circuit. A simple fuse is shown in Fig. 1-17.

The **circuit breaker** is a mechanical device designed to open a circuit when the current flow exceeds a safe limit. Usually the circuit breaker contains an element which reacts to heat. The heat causes the metal to expand, and the expansion trips the contact points to an open position.

Heavy-duty commercial circuit breakers are usually operated by the magnetic force created by the current flow. An overload of current will give the electromagnet sufficient strength to open the circuit by means of a spring-loaded switching device. Magnetism and electromagnetism are explained in a later section of this text.

The circuit breakers employed in aircraft systems are usually of the thermal type; that is, they react to heat as explained above. Typical aircraft circuit breakers are illustrated in Fig. 1-18.

Fig. 1-18 Circuit breakers.

Since airplanes are usually constructed of metal, the airplane structure may be used as an electric conductor. In the circuit in Fig. 1-14, if one terminal of the battery and one terminal of the load are connected to the metal structure of the airplane, the circuit will operate just as well as with two wire conductors. A diagram of such a circuit is shown in Fig. 1-19. When a system of this type is used in an airplane, it is called a **grounded** or **single-wire** system. The ground circuit is that part of the complete circuit in which current passes through the airplane structure. Any unit connected electrically to the metal structure of the airplane is said to be grounded. When an airplane employs a single-wire electric system, it is important that all parts of the airplane be well bonded to provide a free and unrestricted flow of current throughout the structure. This is particularly important for aircraft in which sections are joined by adhesive bonding.

There are two general methods for connecting units in an electric system. These are illustrated in Fig. 1-20. The first diagram shows four lamps connected in **series.** In a circuit of this type, all the current flowing in the circuit must pass through each unit in the circuit. If one of the lamps should burn out, the circuit is broken and the other lamps in the circuit will stop burning. A familiar example of such a circuit is a set of series Christmas-tree lights.

In a **parallel** circuit there are two or more paths for the current, and if the path through one of the units is broken, the other units will continue to function. The units of an aircraft electric system are usually connected in parallel; hence, the failure of one unit will not impair the operation of the remainder of the units in the system. A simple parallel circuit is illustrated in the second diagram of Fig. 1-20. A circuit which has some of the units connected in series and the others connected in parallel is called a **series-parallel** circuit (see Fig. 1-21).

Ohm's law may be used to determine the electrical values in any common circuit even though it may contain a number of different load units. In order to solve such a circuit, it is necessary to know whether the units are con-

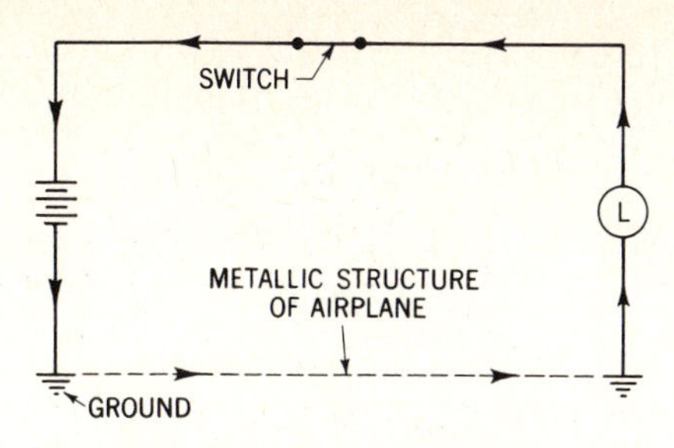

Fig. 1-19 Drawing to illustrate the single-wire system.

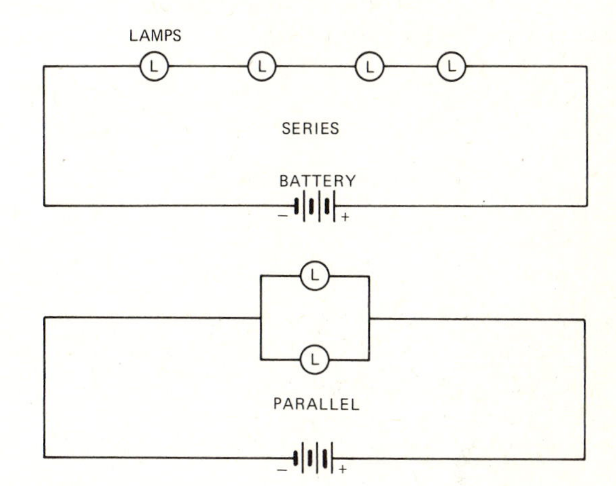

Fig. 1-20 Series and parallel circuits.

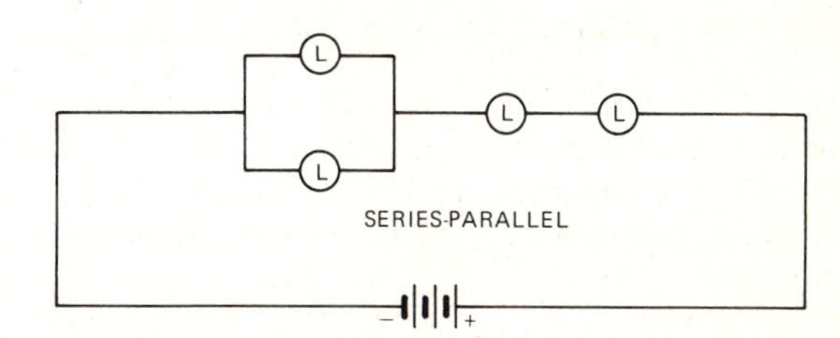

Fig. 1-21 A series-parallel circuit.

nected in series, parallel, or in a combination of the two methods. When the type of circuit is determined, the proper formula may be applied.

When a current flows through a resistance, the voltage across the resistance is equal to the product of the current and the resistance. The voltage is known as the **IR drop.** The *IR* drop in a complete circuit is equal to the voltage of the supply. This is shown in the water analogy in Fig. 1-22. Assume that the internal resistance of the battery is 0.1 Ω, the resistance of each lamp is 25 Ω, the resistance in the circuit is 100 Ω, and the battery voltage is 24 V. When load resistances are connected in series, we add them to find the total and then divide the total resistance into the voltage of the source to find the current, thus:

$$I = \frac{24}{0.1 + 25 + 25 + 100} = \frac{24}{150.1}$$
$$= 0.1599 \text{ A}$$

The voltage drop across any unit of the circuit may be found easily, because the current is the same through each unit. The voltage drop in the battery is found by

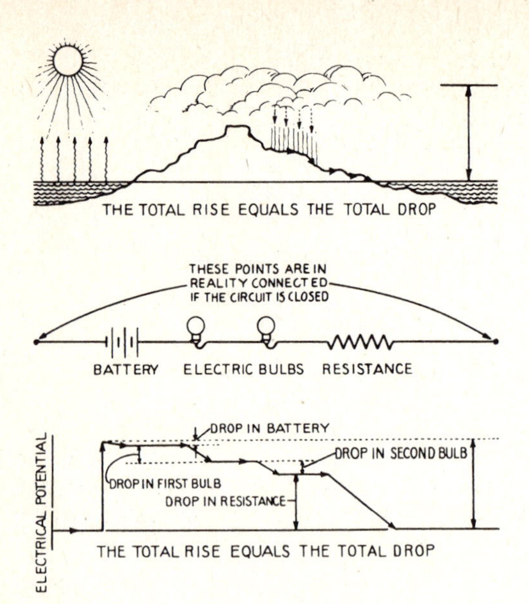

Fig. 1-22 Water analogy of voltage drop.

multiplying 0.1599 by 0.1. This gives a drop of 0.01 599 V in the battery. In like manner, the drop across each lamp is found to be 3.9975 V, and the drop across the circuit resistance is found to be 15.99 V. We may check the accuracy of the calculations by adding all the *IR* drops in the circuit; the sum is found to be 24 V, which is the same as the source.

For ordinary aircraft circuits, it is not necessary to consider the internal resistance of the battery because it is negligible. In the circuit of the foregoing example, it will be noted that the *IR* drop in the battery is very small compared with the *IR* drops in the other parts of the circuit.

SOLVING SERIES CIRCUITS

As explained previously, two or more units are connected in series when the quantity of electrons (current) flowing through each unit of a circuit in a given length of time is the same for each unit. This applies, not only in quantity per unit of time (amperes), but also in the strict sense of the word *same*. Two or more units do not have to be adjacent to each other in a circuit to be in series. In the circuit of Fig. 1-23, it can be seen that the current flow through each unit in the circuit must be the same, regardless of the direction of current flow. If we replace the load resistor R_2 with an electronic system or device contained in a *black box* as shown in Fig. 1-24, the current flow in each resistor will still be the same, provided that the total resistance of the black-box load is the same as it was for R_2. In this case, we regard the black box as a single unit rather than concern ourselves with the separate components within the black box. Thus we see that there is only one path for current flow in a series circuit; however, an individual load unit may consist of more than one component within itself. Note that the black box in Fig. 1-24 is shown with several resistances connected in a network within the box. In the series circuit under consideration, we are only concerned with the total resistance of the black-box unit.

The load units adjacent to each other in a circuit are connected in series if there are no electrical junctions between the two units. This is illustrated in Fig. 1-25. In

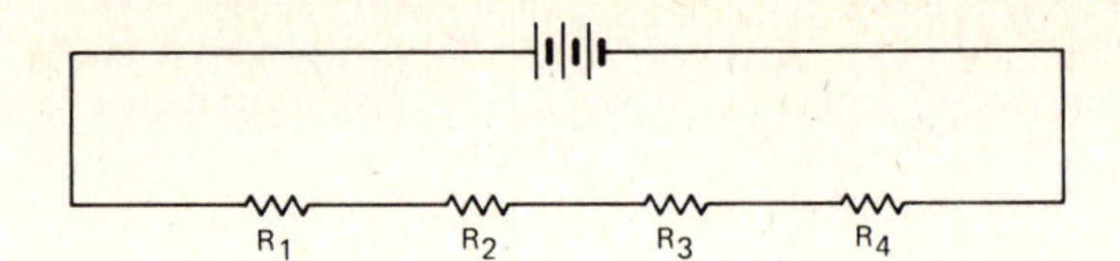

Fig. 1-23 A series circuit with loads indicated as single resistors.

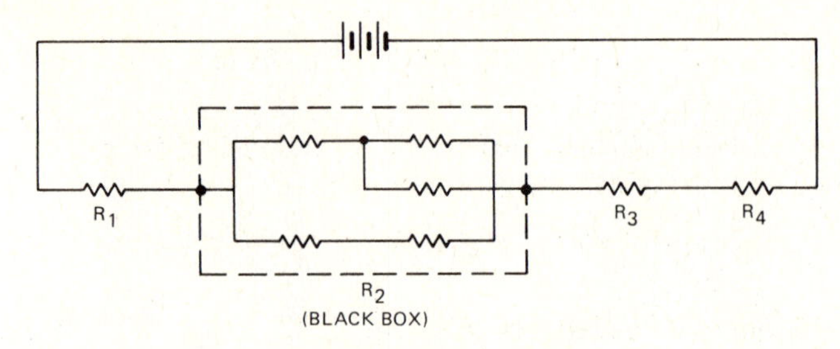

Fig. 1-24 Series circuit diagram to show how an individual load may be other than a simple resistor.

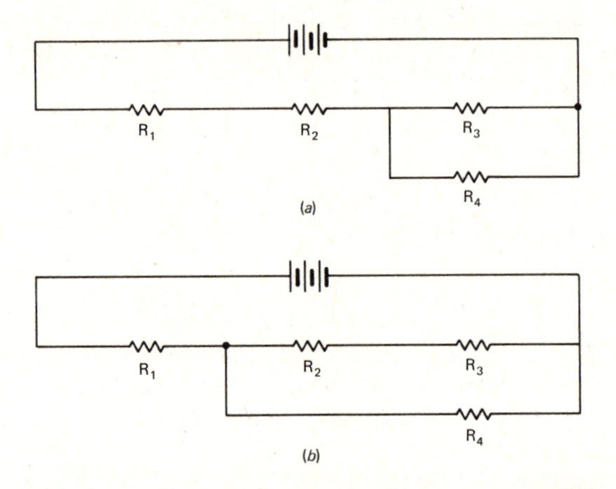

Fig. 1-25 Circuit diagram showing load units connected both in parallel and in series with variations in arrangement.

circuit *a*, R_1 and R_2 are connected in series because there is no electrical junction between them to take a part of the current, and all the current flowing through R_1 must also pass through R_2. In circuit *b*, R_1 and R_2 are not connected in series because the current which flows through R_1 is divided between R_2 and R_4. Note, however, that R_2 and R_3 are in series because the same current must pass through both of them.

Examine the circuit of Fig. 1-26 in which R_1, R_2, and R_3 are connected in series, not only to each other, but also to the power source. The electrons flow from negative to positive in the circuit and from positive to negative in the power source. The same flow, however, exists in every

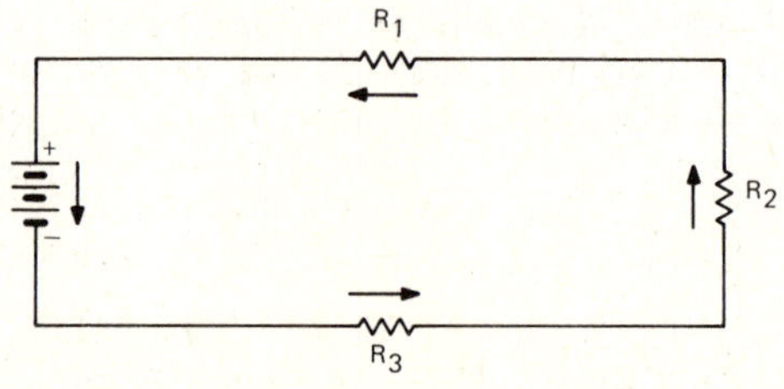

Fig. 1-26 Current flow is the same through each load unit (resistor).

part of the circuit, because there is only one path for current flow. Since the current is the same in all parts of the circuit,

$$I_t = I_1 = I_2 = I_3$$

that is, the total current is equal to the current through R_1, R_2, or R_3.

RESISTANCE AND VOLTAGE IN A SERIES CIRCUIT

In a series circuit, the total resistance is equal to the sum of all the resistances in the circuit; hence,

$$R_t = R_1 + R_2 + R_3 + \cdots$$

This principle was illustrated in Fig. 1-22.

The voltage (potential difference) measured between any two points in a series circuit depends upon the resistance between the points and the current flowing in the circuit. Figure 1-27 shows a circuit with three resistances connected in series. The difference in potential maintained by the battery between the ends of the circuit is 24 V.

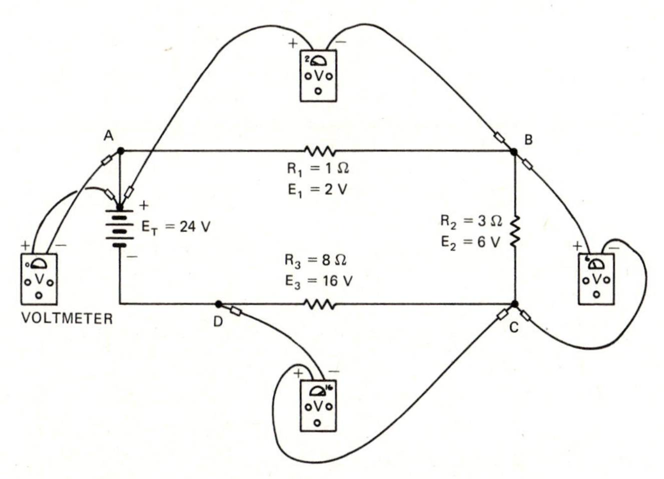

Fig. 1-27 Individual voltages when added are equal to the total voltage applied to the circuit.

As previously explained in the discussion of Ohm's law, the voltage between any two points in a circuit can be determined by the equation

$$E = IR$$

that is, the voltage is equal to the current multiplied by the resistance. In the circuit of Fig. 1-27, we have given a value of 1 Ω to R_1, 3 Ω to R_2, and 8 Ω to R_3. According to our previous discussion, the total resistance of the circuit is expressed by

$$R_t = R_1 + R_2 + R_3$$

or

$$R_t = 1 + 3 + 8$$
$$= 12\ \Omega$$

Since the total voltage E_t for the circuit is given as 24, we can determine the current in the circuit by Ohm's law, using the form

$$I = \frac{E}{R}$$

Then

$$I_t = \frac{24}{12}$$
$$= 2\ \text{A}$$

Since we know that the current in the circuit is 2 A, it is easy to determine the voltage across each load resistor. Since $R_1 = 1\ \Omega$, we can substitute this value in Ohm's law to find the voltage difference across R_1.

$$E_1 = 2 \times 1$$
$$= 2\ \text{V}$$

In like manner,

$$E_2 = 2 \times 3$$
$$= 6\ \text{V}$$

and

$$E_3 = 2 \times 8$$
$$= 16\ \text{V}$$

When we add the voltages in the circuit, we find

$$E_t = E_1 + E_2 + E_3$$
$$= 2 + 6 + 16$$
$$= 24\ \text{V}$$

We have determined by Ohm's law that the total of the voltages (voltage drops) across units in a series circuit is equal to the voltage applied by the power source, in this case the 24-V battery.

In a practical experiment, we can connect a voltmeter (voltage-measuring instrument) from the positive terminal of the battery in a circuit such as that shown in Fig. 1-27 to the point *A*, and the reading will be zero. This is because there is no appreciable resistance between these points. When we connect the voltmeter between the positive terminal of the battery and point *B*, the instrument will give a reading of 2 V. By similar use of the voltmeter, we measure between points *B* and *C* and obtain a reading of 6 V, and between points *C* and *D* for a reading of 16 V. In a circuit such as that shown, we can assume that the resistance of the wires connecting the resistors is negligible. If the wires were quite long, it would be necessary to consider their resistances in analyzing the circuit.

As we have shown, in a series circuit, the voltage drop across each resistor (load unit) is directly proportional to the value of the resistor. Since the current through each unit of the circuit is the same, it is obvious that it will take a higher electrical pressure (voltage) to push the current through a higher resistance and it will require a lower pressure to push the same current through a lower resistance.

The voltage across a load resistor is a measure of the work required to move a unit charge (given quantity of electricity) through the resistor. Electric energy is consumed as current flows through a resistor and the electric energy is converted to heat energy. As long as the power source produces electric energy as rapidly as it is consumed, the voltage across a given resistor will remain constant.

Remember that electric power is measured in watts and can be converted to horsepower, because 746 W is equal to 1 hp. Power is the rate of doing work, and work can be measured in foot-pounds. If 55 lb [24.95 kg] is raised 10 ft [304.8 cm], 550 ft·lb [746 J] of work has been done. If this work is done in 1 s, 1 hp [746 W] has been employed, because 1 hp is required to do 550 ft·lb of work in 1 s. If we apply a 24-V source to a load having 0.77 Ω resistance, the current flow will be approximately 31.2 A. Then, since power (watts) is equal to voltage multiplied by current ($W = EI$), we can find the power being produced in the circuit. That is,

$$W = 24 \times 31.2$$
$$= 748.8 \text{ W} \quad \text{or} \quad 1.003 \text{ hp}$$

Students who have mastered Ohm's law and the three fundamental formulas for series circuits can apply their knowledge to the solution of any series circuit where sufficient information is given. The following examples are shown to illustrate the techniques for solution:

Example *A*—Fig. 1-28:

$$E_t = 12 \text{ V}$$
$$I_1 = 3 \text{ A}$$
$$R_2 = 2 \ \Omega$$
$$R_3 = 1 \ \Omega$$

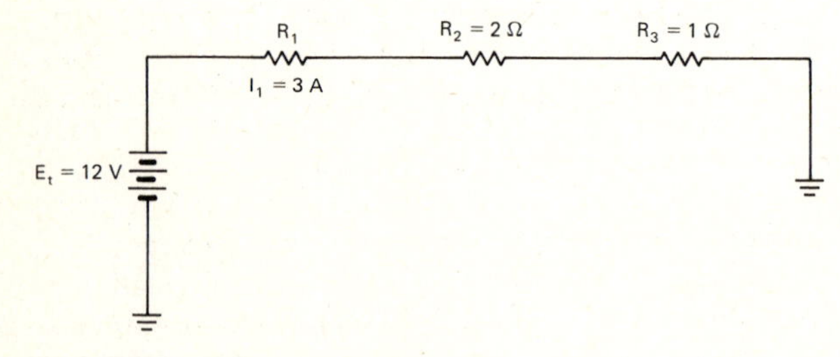

Fig. 1-28 Series circuit for Example A.

Since I_1 is given as 3 A, it follows that I_t, I_2, and I_3 are also equal to 3 A. Then

$$R_t = \frac{E_t}{I_t}$$
$$R_t = \frac{12}{3}$$
$$= 4 \ \Omega$$

$$E_2 = 2 \times 3$$
$$= 6 \text{ V}$$
$$E_3 = 1 \times 3$$
$$= 3 \text{ V}$$

Since $R_1 + R_2 + R_3 = R_t$, we can easily determine that $R_1 = 1 \ \Omega$. By using the formula $I = ER$, we find that $E_1 = 3$ V.

The solved problem may then be expressed as follows:

$E_t = 12$ V	$I_t = 3$ A	$R_t = 4 \ \Omega$
$E_1 = 3$ V	$I_1 = 3$ A	$R_1 = 1 \ \Omega$
$E_2 = 6$ V	$I_2 = 3$ A	$R_2 = 2 \ \Omega$
$E_3 = 3$ V	$I_3 = 3$ A	$R_3 = 1 \ \Omega$

Example *B*—Fig. 1-29:

$$E_t = 24 \text{ V}$$
$$R_1 = 30 \ \Omega$$
$$R_2 = 10 \ \Omega$$
$$R_3 = 8 \ \Omega$$

Then

$$R_t = 30 + 10 + 8$$
$$= 48 \ \Omega$$
$$I_t = \frac{E_t}{R_t}$$
$$= \frac{24}{48}$$
$$= \tfrac{1}{2} \text{ A}$$

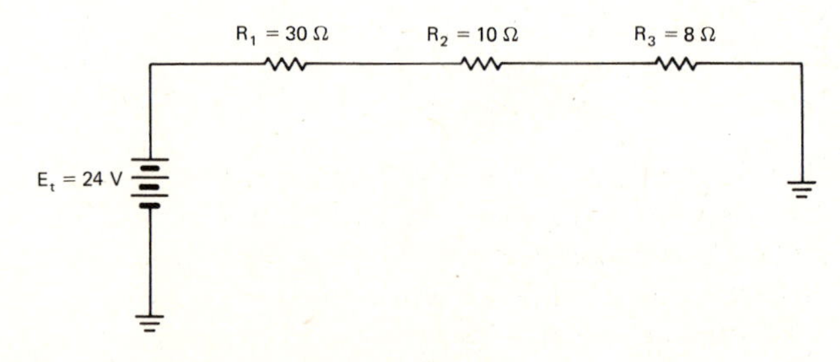

Fig. 1-29 Series circuit for Example B.

E_1, E_2, and E_3 are determined by multiplying each resistance value by $\frac{1}{2}$, the current value of the circuit. The solved circuit is shown in Fig. 1-30.

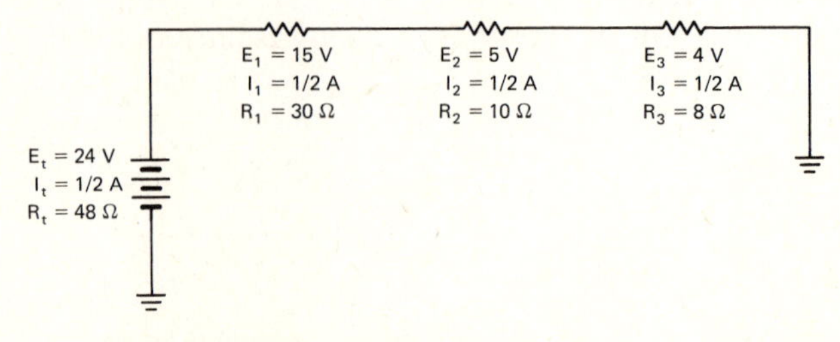

Fig. 1-30 Solved circuit for Example B.

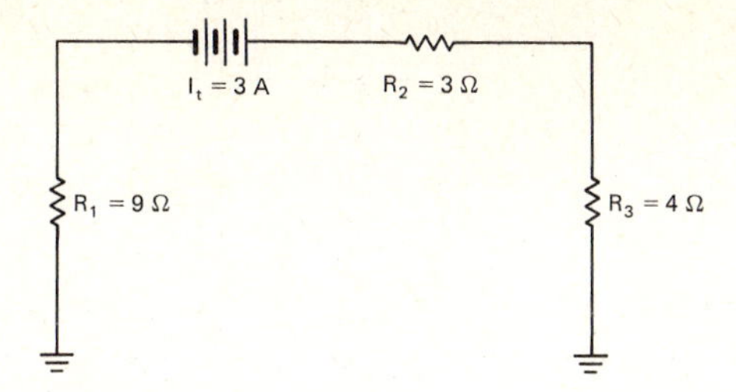

Fig. 1-31 Series circuit for Example C.

Example *C*—Fig. 1-31:

This circuit presents the case where current and resistance are known, and it is required to find the individual and total voltages. The known circuit values are as follows:

$$I_t = 3\ \text{A}$$
$$R_1 = 9\ \Omega$$
$$R_2 = 3\ \Omega$$
$$R_3 = 4\ \Omega$$

From the values given, we can easily determine that the total resistance is 16 Ω. The voltages can then be determined by Ohm's law:

$$E = IR$$
$$E_t = I_t \times R_t$$
$$= 3 \times 16$$
$$= 48\ \text{V}$$

The values of the solved circuit are then as shown below:

$E_t = 48$ V	$I_t = 3$ A	$R_t = 16\ \Omega$
$E_1 = 27$ V	$I_1 = 3$ A	$R_1 = 9\ \Omega$
$E_2 = 9$ V	$I_2 = 3$ A	$R_2 = 3\ \Omega$
$E_3 = 12$ V	$I_3 = 3$ A	$R_3 = 4\ \Omega$

It will be noted in all the circuits presented thus far that the values are always in accordance with Ohm's law formulas. It is recommended that the student check the problems given to verify the results.

Example *D*—Fig. 1-32:

The values for the circuit shown are indicated in the illustration. It is left up to the student to work out the solution. Remember that the total resistance for a series circuit is equal to the sum of the individual resistances.

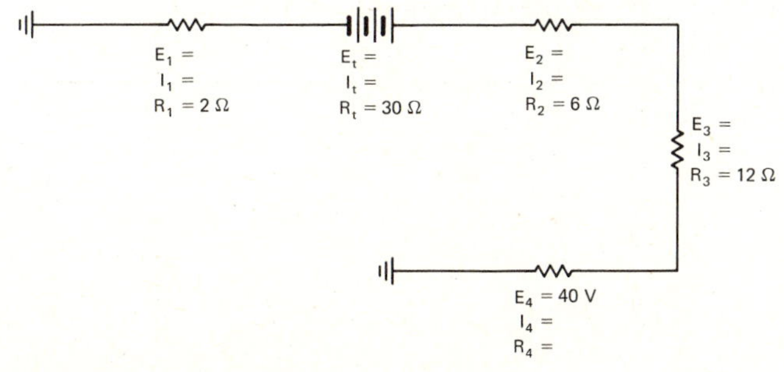

Fig. 1-32 Series circuit for Example D.

PARALLEL CIRCUITS

The connection of load units in **parallel** may be defined in a number of ways. One suitable definition is as follows: **Two or more units are connected in parallel if they have common negative connections and common positive connections to the power source.** This definition cannot apply to ac circuits, however, because the power source rapidly changes polarity. In this case, we can state that the units are connected in parallel if they have common or direct connections to the power source. A parallel circuit is shown in Fig. 1.33. In the diagram, positive (+) signs and negative (−) signs are placed at the ends of the load units merely to show the polarity of the individual connections. The lines between the battery and the load units indicate electrical connections and do not indicate any particular size or type of electric wire. When we analyze a circuit of this type, we assume that the resistance of the wire is negligible and that the source of power has no internal resistance.

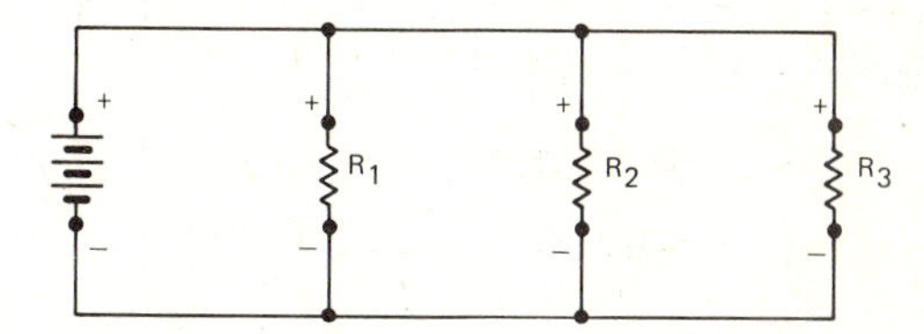

Fig. 1-33 A simple parallel circuit.

A typical arrangement of parallel circuitry is found in the electric circuit or system for a private-dwelling house. All the power outlets and the lighting fixtures are connected in parallel. This is true even though there may be a number of separate circuits feeding through separate fuses or circuit breakers. In home power systems, large appliances such as water heaters, electric furnaces, ovens, etc., are usually connected to a separate power source with higher voltage and, therefore, cannot be in parallel with the lights and normal power outlets.

The resistors (load units) do not need to be arranged as in Fig. 1-33 to be connected in parallel. The three circuits of Fig. 1-34 show loads connected in parallel. Circuits *a* and *b* are identical to the circuit of Fig. 1-33, and circuit *c*

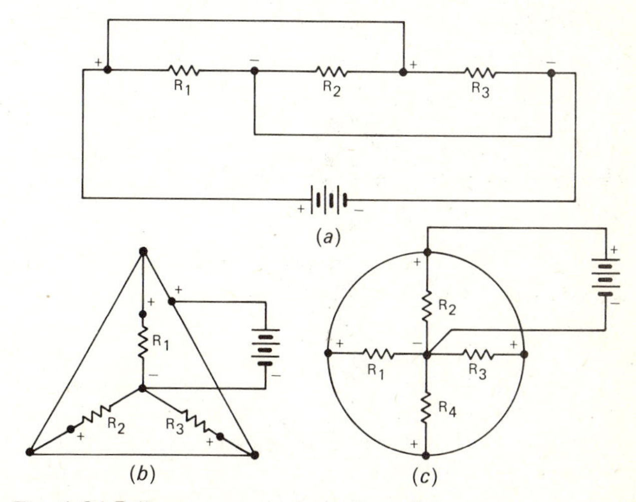

Fig. 1-34 Different arrangements of parallel circuits.

has an additional load unit connected in parallel. A careful examination of the circuits will reveal that the connections are in common for each side of the power source. There is a direct connection (current path) without resistance from any one negative terminal of a load unit to the negative terminal of any other load unit and to the negative terminal of the power source. The same condition is true with respect to all positive terminals.

There may be some junctions between two or more resistors connected in parallel, but these junctions do not change the fact that the resistances are still connected in parallel. It will be noted in Fig. 1-35 that the three resistances, R_1, R_2, and R_3, have common terminals with

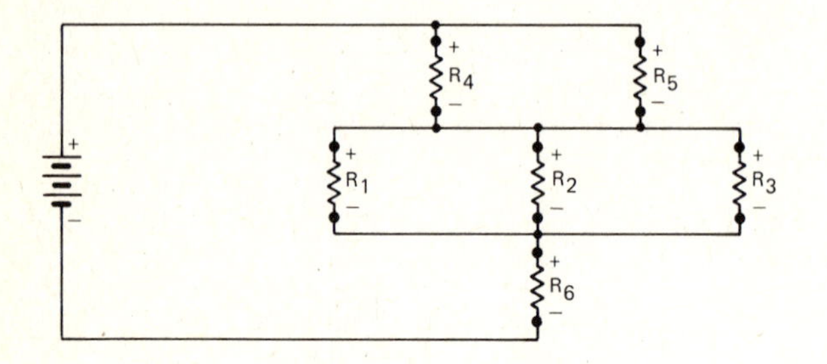

Fig. 1-35 Parallel grouping of resistors in a circuit.

one another even though there are other resistances connected between their common terminals and the power source. It will further be noted that R_4 and R_5 are connected in parallel because they have positive terminals connected together and negative terminals connected together. The resistance R_6 is in series, not with any other single resistance, but with the parallel groups.

The voltage across any resistance in a parallel group is equal to the voltage across any other resistance in the group. Note in Fig. 1-36 that the voltage of the source of 12 V. Since the terminals of the source are connected

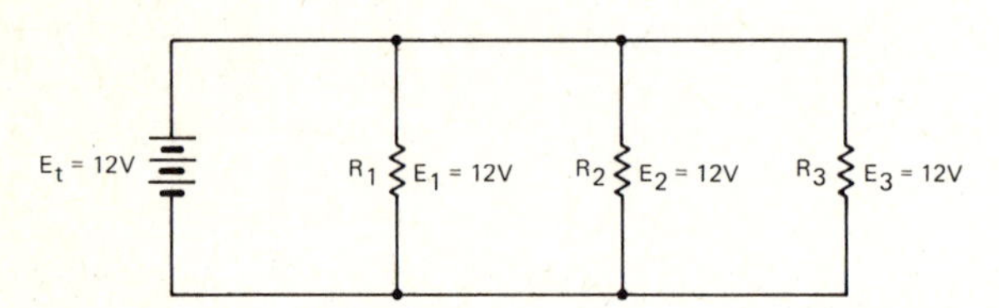

Fig. 1-36 Voltages in a parallel circuit.

directly to the terminals of the resistances, the difference in potential across each resistance is the same as that of the battery or source. By testing with a voltmeter, it would be found that the potential difference across each resistance in the circuit would be 12 V. The formula for voltage in a parallel circuit is

$$E_t = E_1 = E_2 = E_3 = \cdots$$

Remember that electric current or electron flow is expressed in amperes and that 1 A indicates a rate of flow of 1 C/s. In like manner, liquid flow is often expressed in gallons or liters per minute.

In the circuit of Fig. 1-37, the current through R_1 is given as 4 A, the current through R_2 is 2 A, and the current through R_3 is 6 A. To supply this current flow through the

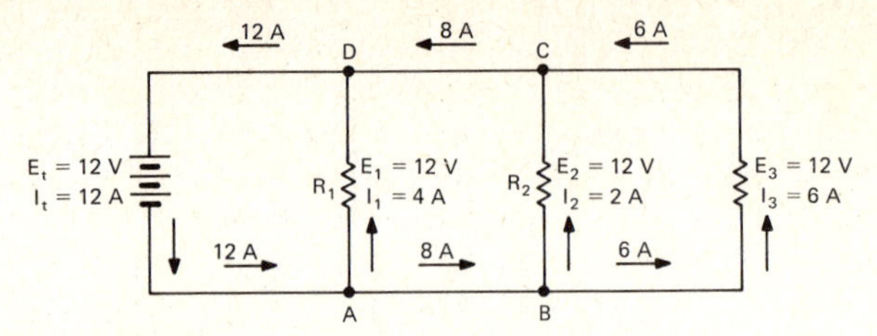

Fig. 1-37 Current flow in a parallel circuit.

three resistances, the power source must supply 4 + 2 + 6, or a total of 12 A to the circuit. It must be remembered that the power source does not actually manufacture electrons, but it does apply the pressure to move them. All the electrons that leave the battery to flow through the circuit must return to the battery. The power source for a circuit can be compared to a pump which moves liquid through a pipe.

An examination of the circuit in Fig. 1-37 reveals that a flow of 12 A comes from the negative terminal of the battery, and at a point A, the flow divides to supply 4 A for R_1 and 8 A for the other two resistors. At point B the 8 A divides to provide 2 A for R_2 and 6 A for R_3. On the positive side of the circuit, 6 A joins 2 A at point C, and the resulting 8 A joins 4 A at point D before returning to the battery. The formula for current in a parallel circuit is then seen to be

$$I_t = I_1 + I_2 + I_3 + \cdots$$

Since the current flow and voltage are given for each resistor in Fig. 1-37, it is easy to determine the value of each resistance by means of Ohm's law; that is,

$$R = \frac{E}{I}$$

Then $R_1 = \frac{12}{4} = 3\ \Omega$, $R_2 = \frac{12}{2} = 6\ \Omega$, $R_3 = \frac{12}{6} = 2\ \Omega$, and $R_t = \frac{12}{12} = 1\ \Omega$.

It will be noted that the total resistance for the group of resistances is less than the resistance of any single resistance in the circuit. At first this seems to be illogical, but when it is remembered that resistance is the opposition to current flow and that each resistance in a parallel circuit provides another path for current flow, then it is easy to see that the more resistances there are connected in parallel in a circuit, the more paths there are for current to flow. It is obvious, then, that the total resistance would be less than for one resistance. Resistance to current flow is determined by the material and the physical dimensions of a conductor ($R = KL/S$), and we shall, for the purpose of explanation, assume that the material and length of the resistances, or resistors, in the circuit of Fig. 1-37 are the same. The difference in resistance for each resistor will, therefore, be determined by the cross-sectional area. This is illustrated in Fig. 1-38.

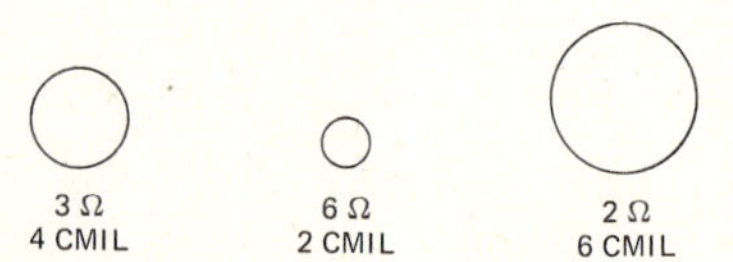

Fig. 1-38 Illustration to show effect of cross-section area on the resistance of a conductor.

In the illustration it can be seen that the resistance decreases as the cross-sectional area increases. A resistor having a cross-sectional area equal to the total of the three resistors would have a cross-sectional area of 12 cmil [0.00608 mm^2]. Since the resistance of a conductor or resistor is inversely proportional to the cross-sectional area, other factors being equal, it is easily determined that the resistor having a cross-sectional area of 12 cmil would have a resistance of 1 Ω.

The formula for the total resistance in a parallel circuit can be derived by use of Ohm's law and the formulas for total voltage and total current. Since

$$I_t = I_1 + I_2 + I_3$$

and

$$I = \frac{E}{R}$$

we can replace all the values in the preceding formula for current with their equivalent values in terms of voltage and resistance. Thus we arrive at the equation

$$\frac{E_t}{R_t} = \frac{E_1}{R_1} + \frac{E_2}{R_2} + \frac{E_3}{R_3}$$

In a parallel circuit, $E_t = E_1 = E_2 = E_3$. Therefore, we can divide all the terms in the previous equation by E and arrive at the formula

$$\frac{1}{R_t} = \frac{1}{R_1} + \frac{1}{R_2} + \frac{1}{R_3}$$

Solving for R_t, the equation becomes

$$R_t = \frac{1}{1/R_1 + 1/R_2 + 1/R_3}$$

This equation can be expressed verbally as follows: **The total resistance in a parallel circuit is equal to the reciprocal of the sum of the reciprocals of the resistances.**

The **reciprocal** of a number is the quantity 1 divided by that number. For example, the reciprocal of 3 is $\frac{1}{3}$. When the reciprocal of a number is multiplied by that number, the product is always 1.

If the formula for total resistance in a parallel circuit is applied to the circuit problem of Fig. 1-37, we find

$$R_t = \frac{1}{\frac{1}{3} + \frac{1}{6} + \frac{1}{2}}$$

$$R_t = \frac{1}{\frac{2}{6} + \frac{1}{6} + \frac{3}{6}}$$

$$= \frac{1}{\frac{6}{6}}$$

$$= 1\ \Omega$$

If some or all the resistances in a parallel circuit are of the same value, the resistance value of one can be divided by the number of equal-value resistances to obtain the total resistance value. For example, if a circuit has four 12 Ω resistors connected in parallel, the value 12 can be divided by the number 4 to obtain the total resistance value of 3 Ω for the four resistances.

When two resistances are connected in parallel, we can use a formula derived from the general formula for R_t to determine the total resistance. The formula is as follows:

$$R_t = \frac{1}{1/R_1 + 1/R_2}$$

Inverting,

$$\frac{1}{R_t} = \frac{1}{R_1} + \frac{1}{R_2}$$

Using a common denominator,

$$\frac{1}{R_t} = \frac{R_2}{R_1 \times R_2} + \frac{R_1}{R_1 \times R_2}$$

Combining,

$$\frac{1}{R_t} = \frac{R_1 + R_2}{R_1 \times R_2}$$

Inverting,

$$R_t = \frac{R_1 \times R_2}{R_1 + R_2}$$

From the foregoing formula we find that when two resistors are connected in parallel, the total resistance is equal to the product of the two resistances values divided by their sum. If a 5 Ω resistance is connected in parallel with a 6 Ω resistence, we apply the formula thus:

$$R_t = \frac{5 \times 6}{5 + 6}$$

$$= \frac{30}{11}$$

$$= 2\tfrac{8}{11}\ \Omega$$

Another interesting fact we can observe from combining parallel resistances is that when two resistances are connected in parallel and one has twice the value of the other, the total resistance is equal to one-third the value of the larger resistance. If a 12-Ω resistance is connected in parallel with a 6-Ω resistance, the total resistance is $\frac{12}{3} = 4\ \Omega$.

The formula for two resistances connected in parallel can be used to solve a circuit where more than two resistors are connected in parallel. Consider a circuit such as that shown in Fig. 1-39. The solution to the circuit can be derived in steps.

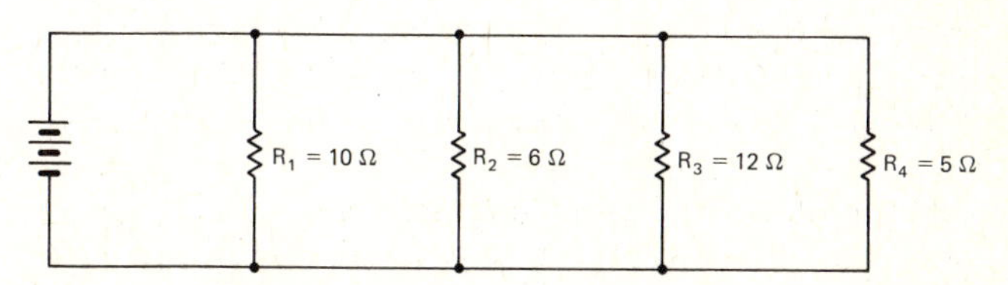

Fig. 1-39 Parallel circuit with resistance values.

1. Combine R_1 and R_4.

$$R_t = \frac{10 \times 5}{10 + 5}$$

$$= \frac{50}{15}$$

$$= \frac{10}{3}\ \Omega$$

2. Combine R_2 and R_3.

$$R_t = \frac{6 \times 12}{6 + 12}$$

$$= \frac{72}{18}$$

$$= 4\ \Omega$$

3. Combine the two totals.

$$R_t = \frac{\frac{10}{3} \times 4}{\frac{10}{3} + 4}$$

$$= \frac{\frac{40}{3}}{\frac{22}{3}}$$

$$= 1\tfrac{9}{11}\ \Omega$$

We can also use the conventional formula as previously given to solve this same problem.

$$R_t = \frac{1}{\frac{1}{10} + \frac{1}{6} + \frac{1}{12} + \frac{1}{5}}$$

$$R_t = \frac{1}{\frac{6}{60} + \frac{10}{60} + \frac{5}{60} + \frac{12}{60}}$$

$$R_t = \frac{1}{\frac{33}{60}}$$

$$= \tfrac{60}{33}$$

$$= 1\tfrac{9}{11}\ \Omega$$

With practice, the student can learn to simplify parallel circuits, first combining the resistances of equal value, then combining pairs with a ratio of 2:1. A final combination of the results will then give the total.

A parallel circuit having some resistances unknown in value, but having a current value given with a known resistance value, can be solved through the use of Ohm's law and the formula for total resistance. See Fig. 1-40.

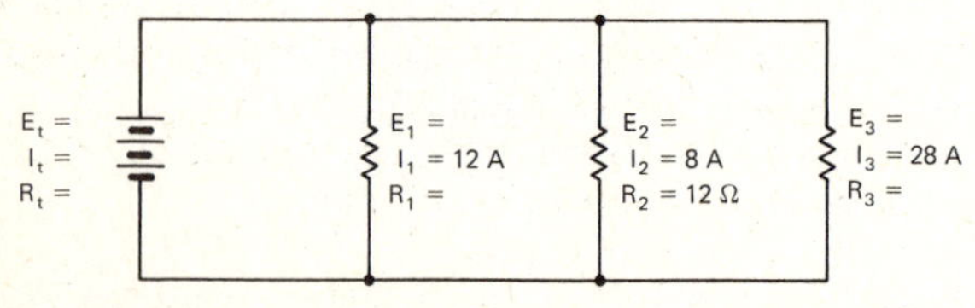

Fig. 1-40 Diagram to show a method for solving a parallel circuit.

An examination of this circuit reveals that $I_2 = 8$ A and $R_2 = 12\ \Omega$. With these values it is apparent that the voltage across R_2 is equal to 96 V. That is,

$$E_2 = I_2 \times R_2$$

$$= 8 \times 12$$

$$= 96\ \text{V}$$

Since the same voltage exists across all the load resistors in a parallel circuit, we know that E_t, E_1, and E_3 are all equal to 96 V. We can then proceed to find that $R_1 = \frac{96}{12}$ or 8 Ω and $R_3 = \frac{96}{28}$ or $3\frac{3}{7}$ Ω. Since total current is equal to the sum of the current values, $I_t = 12 + 8 + 28$ or 48 A. The total resistance is then $\frac{96}{48} = 2\ \Omega$.

In any circuit where a number of load units are connected in parallel or in series, it is usually possible to simplify the circuit in steps and derive an equivalent circuit. A sample circuit with its equivalent is illustrated in Fig. 1-41. To simplify this circuit, we can start by combining the current values through R_1 and R_3 to arrive at $10 + 8 = 18$ A. Hence, $I_{1,3} = 18$ A. The known resistances in the circuit can be combined all at once using the standard formula, or they can be combined in steps. Since we find that R_2 is twice the value of R_4, we know that the total for the two resistances is one-third the value of the larger resistance or 24 Ω. Hence, $R_{2,4} = 24\ \Omega$. This value can be combined in a similar manner with R_5. The total value of the three resistances $R_{2,4,5}$ is then $\frac{24}{3}$ or 8 Ω.

The total current value for the three known resistances is 30 − 18 or 12 A; that is, $I_{2,4,5} = 12$ A. The voltage can then be found by Ohm's law:

$$E = I_{2,4,5} \times R_{2,4,5} = 12 \times 8 = 96\ \text{V}$$

Since the voltage is the same for all units in a parallel circuit, we can compute the balance of the unknown values for the complete solution thus:

$E_t = 96$ V	$I_t = 30$ A	$R_t = 3.2\ \Omega$
$E_1 = 96$ V	$I_1 = 10$ A	$R_1 = 9.6\ \Omega$
$E_2 = 96$ V	$I_2 = 1\frac{1}{3}$ A	$R_2 = 72\ \Omega$
$E_3 = 96$ V	$I_3 = 8$ A	$R_3 = 12\ \Omega$
$E_4 = 96$ V	$I_4 = 2\frac{2}{3}$ A	$R_4 = 36\ \Omega$
$E_5 = 96$ V	$I_5 = 8$ A	$R_5 = 12\ \Omega$

In a parallel circuit where total resistance is known and all the resistance values except one are known, a formula derived from $R_t = (R_1 \times R_2)/(R_1 + R_2)$ can be used. If it is assumed that the unknown resistance is R_1, the procedure is as follows:

Multiply both sides of the formula by $R_1 + R_2$.

$$R_t\,(R_1 + R_2) = R_1 \times R_2$$

or

$$R_t \times R_1 + R_t \times R_2 = R_1 \times R_2$$

Divide both sides of the equation by R_1.

$$\frac{R_t \times R_1 + R_t \times R_2}{R_1} = \frac{R_1 \times R_2}{R_1}$$

Then

$$R_t + \frac{R_t \times R_2}{R_1} = R_2$$

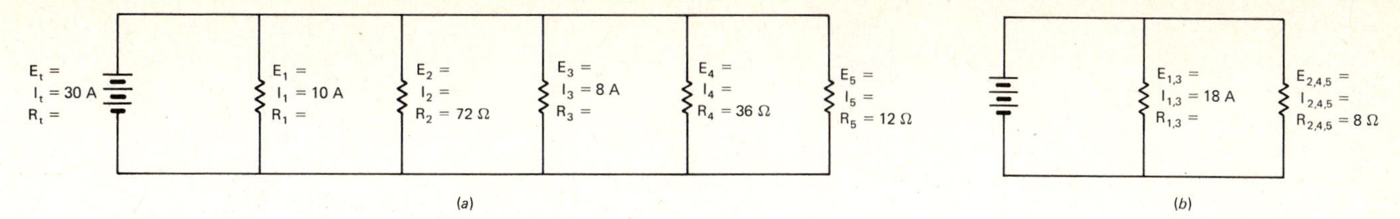

Fig. 1-41 Development of an equivalent circuit from an original circuit.

Subtract R_t from both sides of the equation.

$$\frac{R_t \times R_2}{R_1} = R_2 - R_t$$

To solve for R_1, divide both sides of the equation by $R_t \times R_2$ and invert. Then

$$R_1 = \frac{R_t \times R_2}{R_t - R_2}$$

The formula above can be made applicable to any problem if the unknown resistance is called R_x and the known resistance is called R_k.

If a parallel circuit contains two resistances with $R_k = 12\ \Omega$ and $R_t = 4\ \Omega$, we apply the formula thus:

$$R_x = \frac{12 \times 4}{12 - 4}$$

$$= \frac{48}{8}$$

$$= 6\ \Omega$$

If a circuit contains more than two resistances connected in parallel, all the known resistances are combined to give R_k for the formula.

SERIES-PARALLEL CIRCUITS

As the name implies, a series-parallel circuit is one in which some load units are connected in series and some are connected in parallel. Such a circuit is shown in Fig. 1-42. In this circuit it is quickly apparent that the resistances R_1 and R_2 are connected in series and the resistances R_3 and R_4 are connected in parallel. When the two parallel resistances are combined according to the parallel formula, one resistance, $R_{3,4}$, is found and this value is in series with R_1 and R_2 as shown in Fig. 1-43. The total resistance R_t is then equal to the sum of R_1, R_2, and $R_{3,4}$.

If certain values are assigned to some of the load units in the circuit of Fig. 1-42, we can solve for the unknown values and arrive at a complete solution for the circuit. For the purposes of this problem, the following are known:

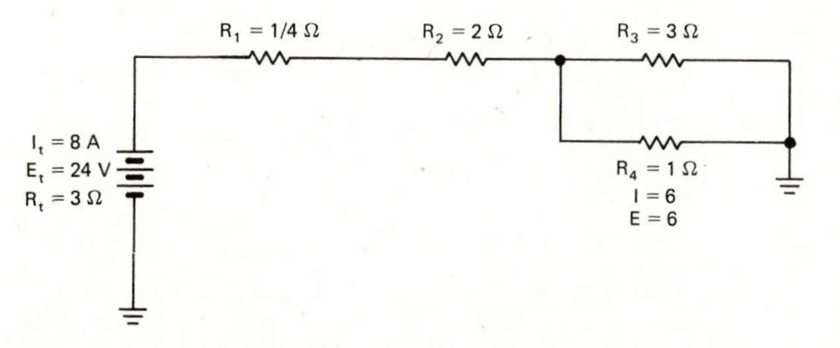

Fig. 1-42 A simple series-parallel circuit.

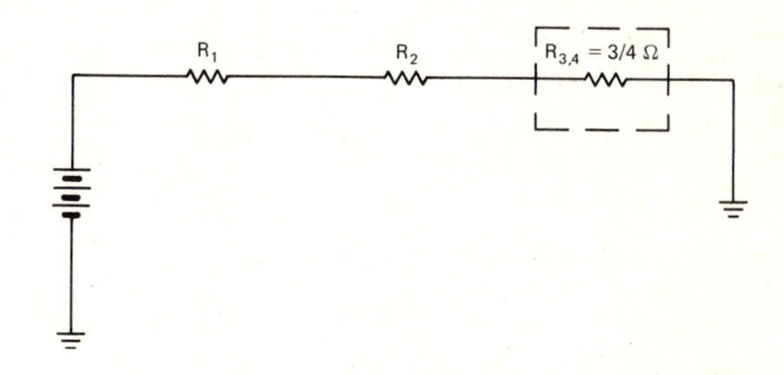

Fig. 1-43 Series equivalent of the preceding circuit.

$$E_t = 24\ \text{V}$$

$$I_1 = 8\ \text{A}$$

$$R_2 = 2\ \Omega$$

$$I_3 = 2\ \text{A}$$

$$E_4 = 6\ \text{V}$$

Since R_3 and R_4 are connected in parallel, the voltage across each has to be the same. It is apparent then that E_3 will equal 6 V. By Ohm's law, we see that R_3 must be 3 Ω.

The current through R_1 is given as 8 A; hence the current through R_2 is also 8 A, since the two load units are connected in series. The 8 A must then be divided between R_3 and R_4. Since $I_3 = 2$ A, I_4 must be 6 A, and R_3 must be 1 Ω. We can combine R_3 and R_4 by means of the formula for two resistances connected in parallel.

$$R_{3,4} = \frac{R_3 \times R_4}{R_3 + R_4}$$

$$R_{3,4} = \frac{3 \times 1}{3 + 1}$$

$$= \tfrac{3}{4}\ \Omega$$

Since E_t for the circuit is 24 V, and since the current flow through R_1 and R_2 is 8 A, we know that the total resistance for the circuit must be 3 Ω ($R = E/I$).

The additional values are easily determined by Ohm's law, and the complete solution becomes:

$E_t = 24$ V	$I_t = 8$ A	$R_t = 3\ \Omega$
$E_1 = 2$ V	$I_1 = 8$ A	$R_1 = \frac{1}{4}\ \Omega$
$E_2 = 16$ V	$I_2 = 8$ A	$R_2 = 2\ \Omega$
$E_3 = 6$ V	$I_3 = 2$ A	$R_3 = 3\ \Omega$
$E_4 = 6$ V	$I_4 = 6$ A	$R_4 = 1\ \Omega$

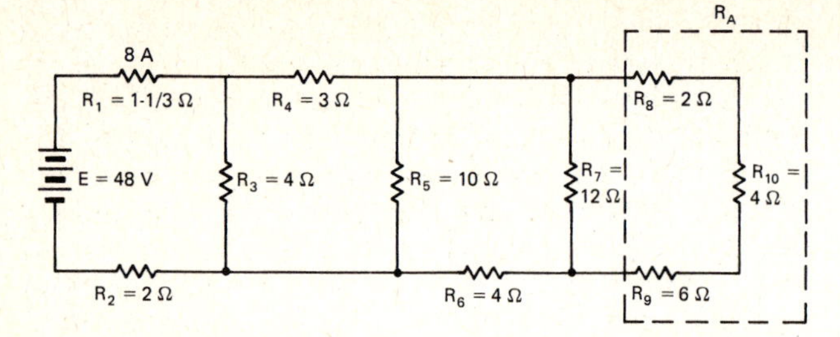

Fig. 1-44 Series parallel circuit to illustrate solution procedure.

The solution of a series-parallel circuit such as that shown in Fig. 1-44 is not difficult provided that the load-unit (resistance) values are kept in their correct relationships. To determine all the values for the circuit shown, we must start with R_8, R_9, and R_{10}. Since these resistances are connected in series with each other, their total value is $2 + 4 + 6 = 12\ \Omega$. We shall call this total R_A; that is, $R_A = 12\ \Omega$. The circuit can then be drawn as in Fig. 1-45, which is the equivalent of the original circuit.

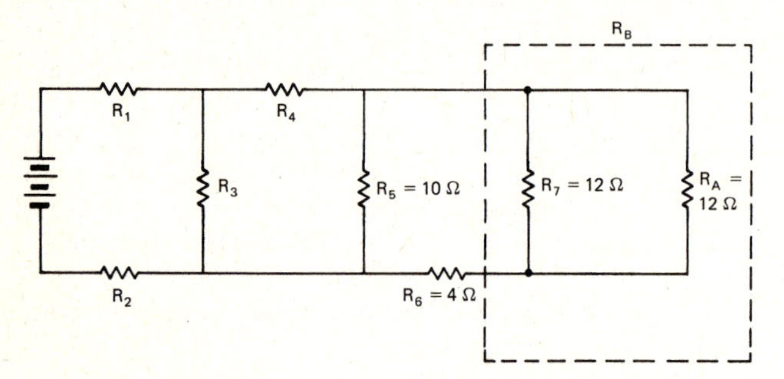

Fig. 1-45 First equivalent circuit for preceding circuit.

In the circuit of Fig. 1-45 it can be seen that R_7 and R_A are connected in parallel. The formula for two parallel resistances can be used to determine the resistance of the combination. We shall call this combination R_B. Then

$$R_B = \frac{R_7 \times R_A}{R_7 + R_A}$$

$$= \frac{12 \times 12}{12 + 12}$$

$$= \frac{144}{24}$$

$$R_B = 6\ \Omega$$

Now an equivalent circuit can be drawn as in Fig. 1-46 to further simplify the solution. In this circuit we combine the two series resistances, R_B and R_6, to obtain a value of 10 Ω for R_C. The equivalent circuit is then drawn as in Fig. 1-47.

Since the new equivalent circuit shows that R_5 and R_C are connected in parallel and that each has a value of 10 Ω, we know that the combined value is 5 Ω. We designate this new value as R_D and draw the circuit as in Fig. 1-48. R_D is connected in series with R_4; hence, the total of the two resistances is 8 Ω. This is designated R_E for the equivalent circuit of Fig. 1-49. In this circuit we solve the

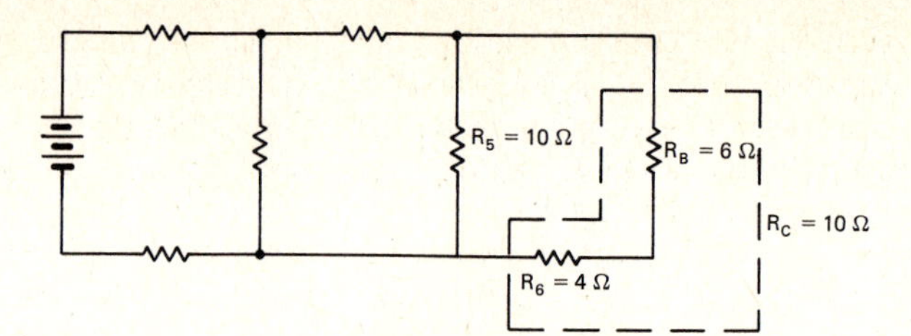

Fig. 1-46 Second equivalent circuit.

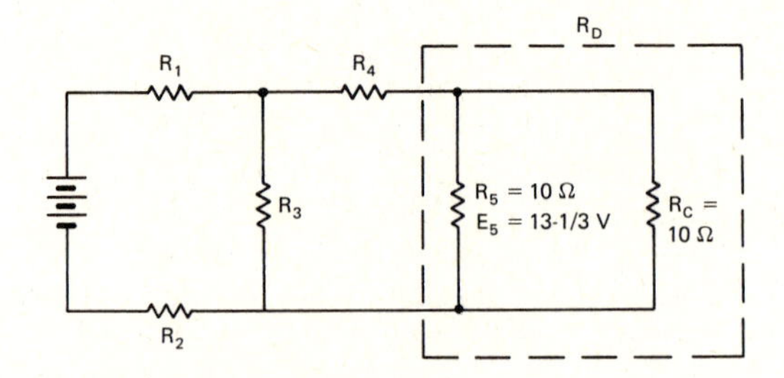

Fig. 1-47 Third equivalent circuit.

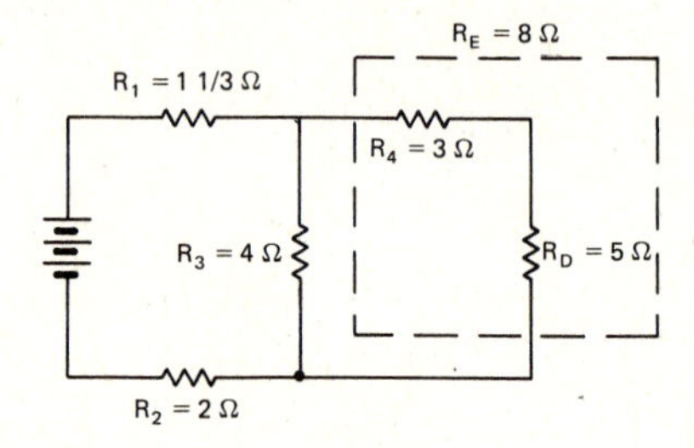

Fig. 1-48 Fourth equivalent circuit.

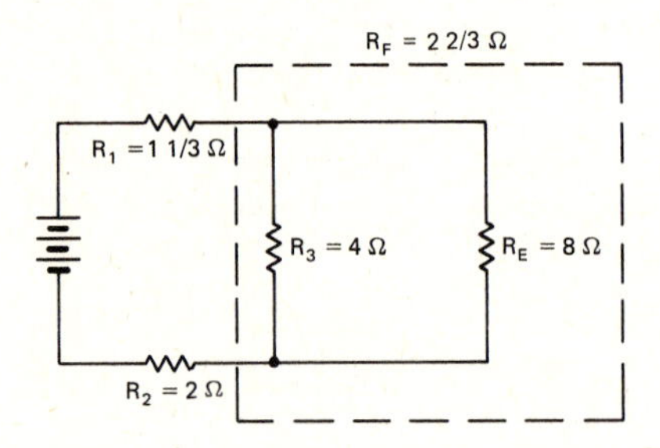

Fig. 1-49 Fifth equivalent circuit.

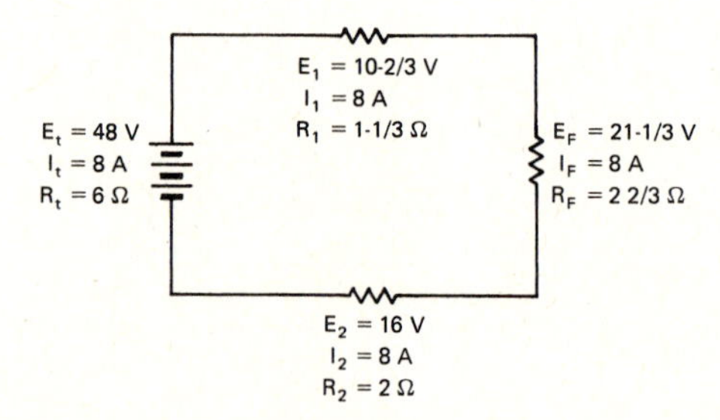

Fig. 1-50 Final equivalent circuit.

parallel combination of R_3 and R_E to obtain the value of $2\frac{2}{3}\ \Omega$ for R_F. The final equivalent circuit is shown in Fig. 1-50 with R_1, R_F, and R_2 connected in series. These resistance values are added to find the total resistance for the circuit.

$$R_t = 1\tfrac{1}{3} + 2\tfrac{2}{3} + 2 = 6\ \Omega$$

With the total resistance known and E_t given as 48 V, it is apparent that $I_t = 8$ A ($I_t = E_t/R_t$). The values for the entire circuit can be computed using Ohm's law and proceeding in a reverse sequence from that used in determining total resistance.

First, since $I_t = 8$ A, I_1, I_F, and I_2 must each be 8 A because the resistances are shown to be connected in series in Fig. 1-50. By Ohm's law ($E = IR$) we find that $E_1 = 10\frac{2}{3}$ V, $E_F = 21\frac{1}{3}$ V, and $E_2 = 16$ V. Referring to Fig. 1-49, it can be seen that $21\frac{1}{3}$ V exist across R_3 and R_E. This makes it possible to determine that $I_3 = 5\frac{1}{3}$ A and $I_E = 2\frac{2}{3}$ A. In Fig. 1-48 we note that I_4 and I_D must both be $2\frac{2}{3}$ A because the two resistances are connected in series. Then $E_4 = 8$ V and $E_D = 13\frac{1}{3}$ V. Since E_D is the voltage across R_5 and R_C in the circuit of Fig. 1-47, it is easily found that $I_5 = 1\frac{1}{3}$ A and $I_C = 1\frac{1}{3}$ A. In the circuit of Fig. 1-46 it is apparent that $1\frac{1}{3}$ A must flow through both R_B and R_6 because they are connected in series and we have already noted that $I_C = 1\frac{1}{3}$ A. Then $E_B = 8$ V and $E_5 = 13\frac{1}{2}$ V.

Since $E_B = 8$ V, we can apply this voltage to the circuits as shown in Figs. 1-44 and 1-45 and note that both E_7 and E_A are 8 V. Then $I_7 = \frac{8}{12}$ or $\frac{2}{3}$ A and $I_A = \frac{2}{3}$ A. Since R_8, R_9, and R_{10} are connected in series and the same current, $\frac{2}{3}$ A, flows through each, $E_8 = 1\frac{1}{3}$ V, $E_9 = 4$ V, and $E_{10} = 2\frac{2}{3}$ V.

The completely solved circuit is shown in Fig. 1-51. A check of all the values given will reveal that they comply with the requirements of Ohm's law.

INTERNAL RESISTANCE OF A POWER SOURCE

In our previous discussions of electric circuits, we have assumed that the source of power had infinite capacity and no internal resistance. This is not possible in actual circuits because all materials offer some resistance to current flow. In a battery, for example, the internal resistance is determined by the materials of which the plates are constructed, the type of electrolyte, the physical dimensions of the plates, and the temperature. In a generator, the internal resistance depends upon the size and length of conductors in the armature, the temperature, and the materials of which the electric conducting components are made. It is, therefore, logical to assume that a 100-A generator is much larger than a 30-A generator of the same voltage. It must be explained, however, that the output of a generator can be increased by designing it with a special cooling system. A generator provided with an efficient cooling system can produce much more power than a generator of the same size without adequate cooling.

In the design of an actual electric system, the internal resistance of the power source must be considered as well as the resistance of the wiring. Such resistance can be shown as an additional load unit in a drawing of the circuit. This resistance is drawn to be connected in series with the power source as shown in Fig. 1-52.

The open-circuit (no-load) voltage of a lead-acid storage battery rated at 12 V is usually about 13.2 V or 2.2 V per cell. In the circuit of Fig. 1-52, two electric lamps are connected in parallel with a motor. All the units in the circuit are controlled by switches so that they can be turned on or off at will. When both the lights are on and the motor is off, the total resistance of the circuit R_t is $\frac{2.6}{2} + 0.02$ or 1.32 Ω. I_t is then $\frac{13.2}{1.32}$ or 10 A. The voltage drop across the battery under these conditions is 10×0.02 or 0.2 V, and the voltage applied to the lamps is 13 V. Each lamp is then being supplied with 5 A.

If the motor switch is turned on, a much larger current will be flowing from the battery, and the effect of the internal resistance becomes apparent. In this case we compute the total resistance as follows:

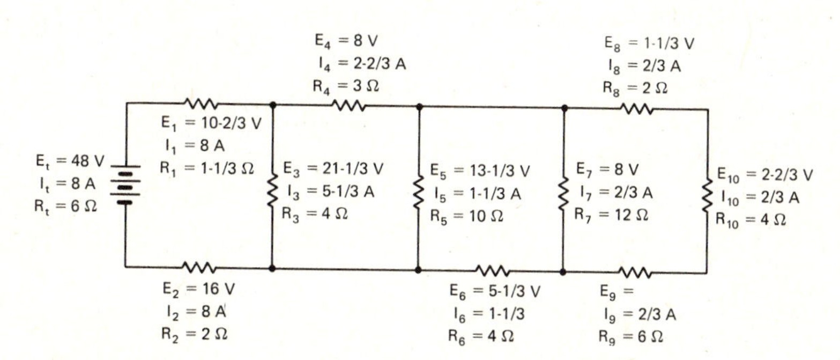

Fig. 1-51 Completely solved circuit.

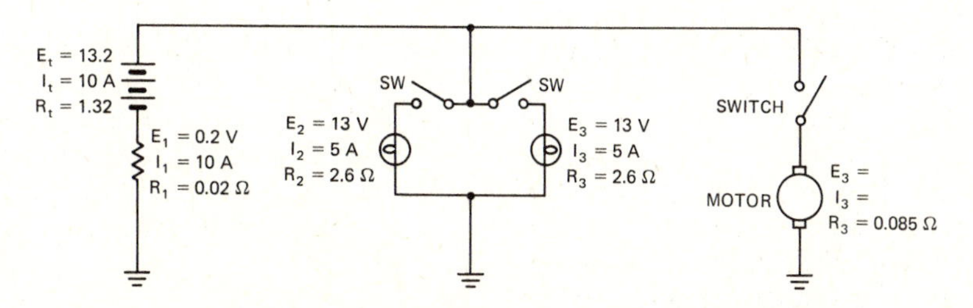

Fig. 1-52 Circuit to demonstrate effects of internal resistance.

$$R_t = \frac{1.3 \times 0.085}{1.3 + 0.085} + 0.02$$
$$= 0.08 + 0.02$$
$$= 0.1\ \Omega$$

Then

$$I_t = \frac{13.2}{0.1}$$
$$= 132\ \text{A}$$

The voltage drop across the internal resistance then becomes 132 × 0.02 or 2.64 V. Subtracting the voltage drop from the total leaves 10.56 V across the load units. With this voltage, the current through the lamps will be 4 A, and the lights will not burn as brightly as before the motor switch was turned on. This effect is very commonly observed when a person starts an automobile engine with the lights turned on. If the battery is low in charge, the lights may go completely out when the starter switch is turned on. It is most important, therefore, that the battery in either an automobile or an airplane be kept at a high level of charge.

The generators or alternators used to supply power for the operation of electric systems in aircraft are controlled by voltage regulators. These regulators compensate for voltage drops, due to internal resistance, by increasing the generator field strength, thus maintaining a constant voltage for all normal load conditions.

KIRCHHOFF'S LAWS

The circuits in this chapter are all solvable by means of Ohm's law as demonstrated. There are, however, many circuits which are more complex and which cannot be solved by Ohm's law. For these circuits, Kirchhoff's laws may provide the necessary techniques and procedures.

Kirchhoff's laws were discovered by Gustav Robert Kirchhoff, a German physicist of the nineteenth century. The two laws may be stated as follows:

Law No. 1. The algebraic sum of the voltages rises (drops) encountered in tracing around any closed-loop circuit is zero.

Law No. 2. The algebraic sum of the currents arriving at or leaving any junction point in a circuit is zero.

In the simple series circuit of Fig. 1-53, all the voltages measured around the circuit, not including the power source, are equal to 200 V when added. We note that the polarities of the load units are opposite in direction to the polarity of the battery. If we give the battery voltage a positive value, then each load-unit voltage will have a negative value. The algebraic sum may be expressed thus:

$$E_t - E_1 - E_2 - E_3 - E_4 = 0$$

Figure 1-54 shows a circuit to illustrate the principle of Kirchhoff's second law. In this circuit, it will be noted that I_1 and I_2 are flowing to point A and that I_3, I_4, and I_5 are

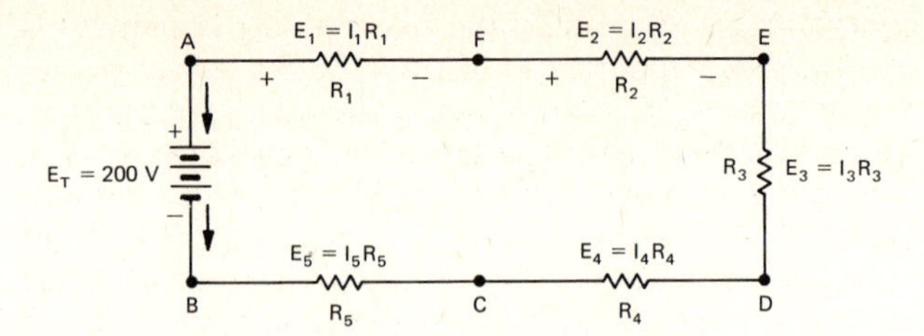

Fig. 1-53 Circuit to illustrate Kirchhoff's first law.

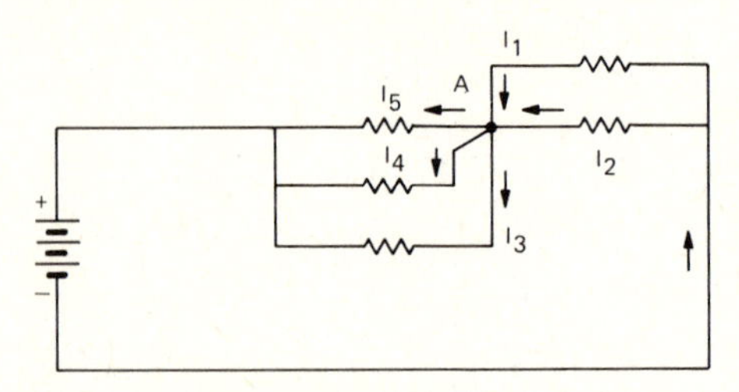

Fig. 1-54 Circuit to illustrate Kirchhoff's second law.

flowing away from point A. If we give a positive value to the currents flowing *to* the point and a negative value to the currents flowing *from* the point, we can express the principle thus:

$$I_1 + I_2 - I_3 - I_4 - I_5 = 0$$

Since the solution of complex circuits by means of Kirchhoff's laws is not an essential function of the aviation technician, we shall not attempt to explain the mathematical applications of the laws in this text.

THE RHEOSTAT

In practical electric circuits it is often necessary to insert resistances which may be adjusted to reduce the voltage applied to a load. For example, in a lighting circuit it may be desirable to provide for operating the lights at varying degrees of brilliance. This is usually accomplished by means of variable resistors called **rheostats.** A drawing of a rheostat is shown in Fig. 1-55.

A rheostat is usually constructed by winding high-resistance wire radially on a circular form made of a nonconducting material. A sliding contact arm is mounted on a shaft located in the center of the circular resistance, with one end of the contact arm resting on the bare wire. The

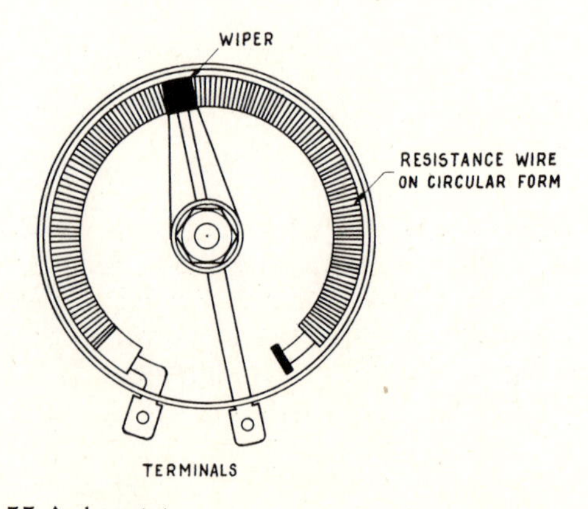

Fig. 1-55 A rheostat.

contact arm is often called the **wiper.** When one terminal of a circuit is connected to one end of the resistance and the other terminal is connected to the sliding contact arm, it is possible to vary the resistance of the circuit by rotating the shaft and moving the contact arm along the resistance. When the arm is moved in one direction, it places additional coils of resistance wire in the circuit, thus increasing the resistance. If the sliding arm is moved in the opposite direction, it removes a part of the resistance from the circuit. A circuit with a rheostat is illustrated in Fig. 1-56.

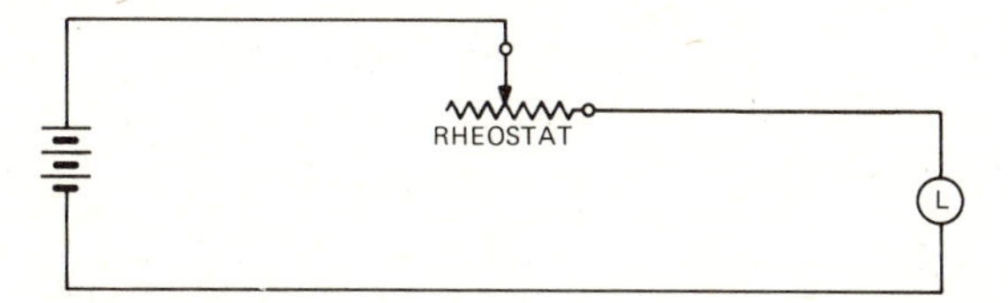

Fig. 1-56 Diagram showing a rheostat connected in a circuit.

A rheostat connected in series with a lamp circuit may be used to reduce the voltage as necessary to produce the desired brilliance. Ohm's law may be used to determine the resistance value of the rheostat required for any particular circuit. It must be noted that both the resistance and the current-carrying capacity of the rheostat must be correct for the circuit and voltage involved.

SOLUTION OF A RESISTANCE BRIDGE CIRCUIT

When resistances are connected in a bridge circuit as shown in Fig. 1-57*a*, it will be noted that two Δ (delta) circuits are formed. These circuits share the resistance R_5 in common. Because of this, it is not possible to solve the circuit by the methods we have explained previously. A mathematical method has been devised whereby the circuit can be solved by converting one of the Δ circuits to an equivalent Y circuit.

Figure 1.57*b* represents an equivalent circuit where the Δ circuit *ABD* of Fig. 1-57*a* has been converted to the equivalent Y circuit *ABD* in Fig. 1-57*b*. This conversion is accomplished with formulas as follows:

$$R_a = \frac{R_1 \times R_5}{R_1 + R_4 + R_5}$$

$$R_b = \frac{R_1 \times R_4}{R_1 + R_4 + R_5}$$

$$R_c = \frac{R_4 \times R_5}{R_1 + R_4 + R_5}$$

The circuit of Fig. 1-57*b* is a simple series-parallel type and can be solved as we have explained previously.

For an example of how the circuit of Fig. 1-57 can be solved, we shall first assign resistance values to the resistors in Fig. 1-57*a*. $R_1 = 2\ \Omega$, $R_2 = 8\ \Omega$, $R_3 = 4\ \Omega$, $R_4 = 4\ \Omega$, and $R_5 = 10\ \Omega$. Then

$$R_a = \frac{2 \times 10}{2 + 4 + 10} = \frac{20}{16} = 1.25\ \Omega$$

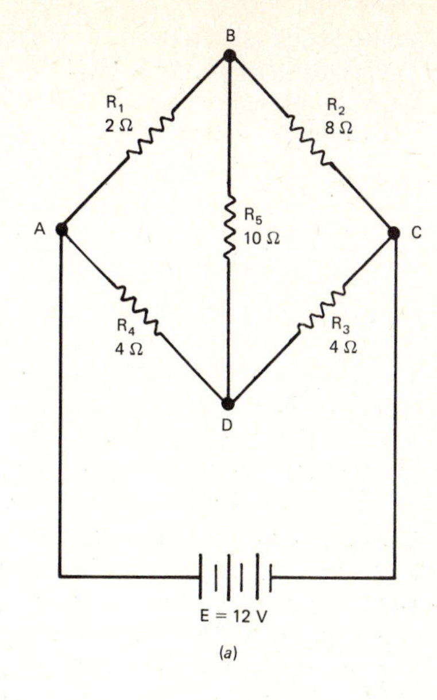

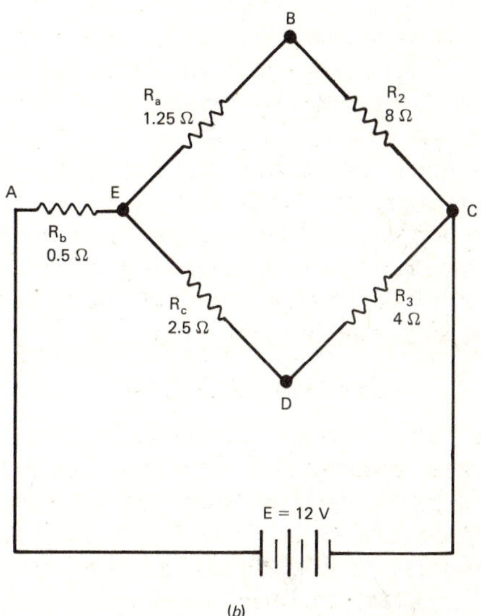

Fig. 1-57 Circuit diagrams to illustrate the conversion of a delta circuit to an equivalent Y circuit and the solution of a resistance-bridge circuit.

$$R_b = \frac{2 \times 4}{2 + 4 + 10} = \frac{8}{16} = 0.5\ \Omega$$

$$R_c = \frac{4 \times 10}{2 + 4 + 10} = \frac{40}{16} = 2.5\ \Omega$$

In the circuit of Fig. 1-57*b*, R_a and R_2 are in series, and R_c is connected in series with R_3. Since series circuit values are added to determine the value of the total, we add the series resistances in this case. Then $R_a + R_2 = 1.25 + 8 = 9.25\ \Omega$, and $R_c + R_3 = 2.5 + 4 = 6.5\ \Omega$. The combination of $R_a + R_2$ is in parallel with the combination of $R_c + R_3$; hence we use the parallel formula for two resistances to determine the equivalent value.

$$R_t = \frac{6.5 \times 9.25}{6.5 + 9.25} = \frac{60.125}{15.75} = 3.82\ \Omega$$

Since the parallel circuit is in series with R_b, we add the total of the parallel resistances (3.82 Ω) to R_b (0.5 Ω) to obtain the combined equivalent resistance for the circuit; that is,

$$0.5 + 3.82 = 4.32\ \Omega$$

Since 12 V is applied to the bridge circuit, the current through the circuit is 12/4.32 = 2.78 A.

The method described for the solution of bridge circuits is useful when it is desired to determine values in circuits utilizing the Wheatstone bridge, such as those described in Chapter 15 for certain instruments.

REVIEW QUESTIONS

1. For what purposes is electricity used in aircraft?
2. Discuss the need for the aviation maintenance technician to have a good understanding of electrical principles.
3. Define *molecule* and *atom.*
4. What particles are found in an atom?
5. What is an *element* in matter?
6. What is another name for a charged atom?
7. What makes some substances *conductors, nonconductors,* or *semiconductors?*
8. What force is required to cause electrons to move through a conductor?
9. Explain the nature of *static* charges.
10. What charge will be developed on a plastic rod which is rubbed with a piece of fur?
11. Explain the electrostatic field.
12. What is meant by *triboelectric* charging?
13. If a positively charged terminal is connected to a negatively charged terminal, in which direction will the electrons flow?
14. What undesirable effects are caused by static electricity during the operation of an airplane?
15. By what means are the effects of static reduced in an airplane?
16. What is meant by *precipitation static?*
17. Explain the functions of *bonding* and *shielding.*
18. What is a *null field discharger?*
19. What is an *electric current?*
20. What name is given to the unit of *electromotive force?*
21. To what physical force may electromotive force be compared?
22. What is the unit of electric current flow?
23. What is the unit electrical quantity?
24. Define *resistance* and give the unit of resistance.
25. What factors determine the resistance of a conductor?
26. Give the formula for the general rule of resistance.
27. Explain *resistivity.*
28. Define *circular mil.*
29. Give the three forms for the formula of *Ohm's law.*
30. Define *watt.*
31. Compare watts with *horsepower.*
32. What horsepower is expended in a circuit in which the voltage is 110 and the current is 204 A?
33. Show that the power expended in a given circuit is proportional to the square of the voltage.
34. What amperage is required to drive a 5-hp motor in a 110-V circuit when the motor has an efficiency of 60 percent?
35. What is the efficiency of an electric motor which delivers 10 hp in a 208-V circuit and draws a current of 48 A?
36. Explain *open circuit, closed circuit,* and *short circuit.*
37. What is meant by a *single-wire* power system?
38. Explain the difference between *series* circuits and *parallel* circuits.
39. What is the total resistance when resistances of 3, 4, 6, and 8 Ω are connected in parallel? In series?
40. What resistance would have to be connected in series with a 3-V lamp in a 28-V circuit when the operating current of the lamp is 0.5 A? What is the operating resistance of the lamp?
41. Determine the total resistance of the accompanying circuit and give the current flow for each resistance.

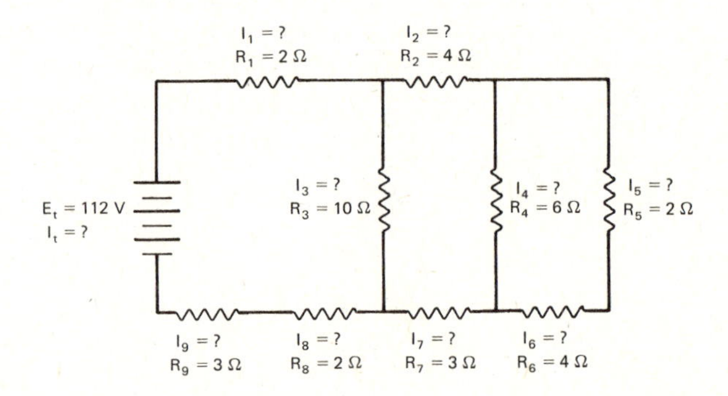

CHAPTER 2

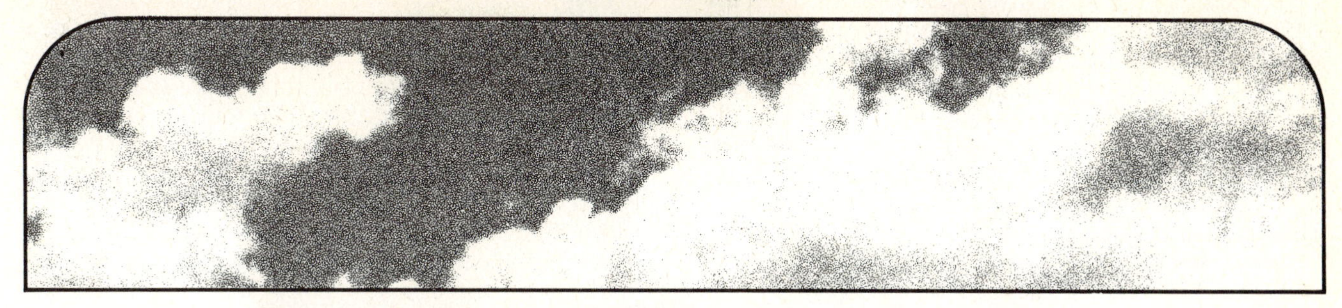

Magnetism and Electromagnetism

Most people have very little comprehension of the prominent influence which magnetism and its effects have on their daily lives. When we enumerate a few of the modern and not so modern conveniences which depend upon magnetism, we can see at once that civilization would regress about one hundred years if the force of magnetism were not available to us. Among the conveniences dependent upon magnetism for their operation are electric generators, the telegraph, the telephone, electric motors, radio, television, missile-guidance systems, radar, navigation equipment, and numerous other devices.

Magnetism is one of the fundamental forces involved in the use of electricity; hence it is essential that the mechanic or technician obtain a good understanding of the subject. In this chapter are explained magnetic theory and some of the uses of magnetic force in electric devices and systems.

THEORY OF MAGNETISM

THE MAGNET

Almost everyone has witnessed the effects of magnetism, and many have owned simple permanent magnets such as that illustrated in Fig. 2-1. A **magnet** may be defined as an object which attracts such magnetic substances as iron or steel. It produces a magnetic field external to itself which reacts with magnetic substances. A **permanent magnet** is one which maintains an almost constant magnetic field without the application of any magnetizing force. Some magnetized substances show practically no loss of magnetic strength over a period of several years.

A **magnetic field** is assumed to consist of invisible lines of force which leave the **north** pole of a magnet and enter the **south** pole. The direction of this force is assumed only in order to establish rules and references for operation. Whether there is any actual movement of force from the north pole to the south pole of a magnet is not known, but it is known that the force acts in a definite direction. This is indicated by the fact that a north pole will repel another north pole but will be attracted by a south pole.

A **natural magnet** is one found in nature; it is called a **lodestone,** or *leading stone.* The natural magnet received this name because it was used by early navigators to determine direction. The lodestone is composed of an oxide of iron called magnetite.

When first discovered, the lodestone was found to have peculiar properties. When it was freely suspended, one end always pointed in a northerly direction. For this reason, one end of the lodestone was called the *north-seeking* and the other the *south-seeking* end. These terms have been shortened to *north* and *south,* respectively. The reason that a freely suspended magnet assumes a north-south position is that the earth is a large magnet and the earth's magnetic field exists over the entire surface. In accordance with the assigned direction of a magnetic field, the magnetic lines of force leave the earth at a point near the south pole and enter near the north pole, as shown in Fig. 2-2. Since the north pole of a

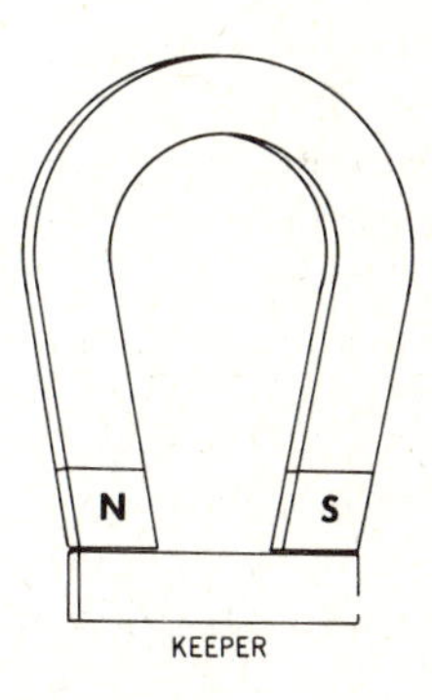

Fig. 2-1 A permanent magnet.

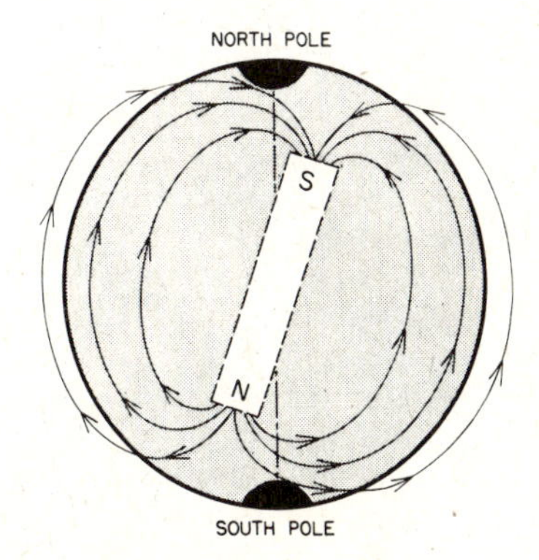

Fig. 2-2 Earth's magnetic field.

magnet is attracted to the south pole of another magnet and repels another north pole, we know that the magnetic pole near the geographic north pole of the earth is actually a south magnetic pole, and that the magnetic pole near the geographic south pole of the earth is actually a north magnetic pole (see Fig. 2-2).

The magnetic poles of the earth are not at the geographic poles. The magnetic pole in the Northern Hemisphere is located at a point north and west of Hudson Bay in Baffinland, and the magnetic pole in the Southern Hemisphere is in Victoria Land. For this reason, a magnetic compass points to true north only when it is in a location where the magnetic pole is approximately in line with the geographic pole. The line on the earth's surface where the compass points to true north is called the **agonic line.** The agonic line and **isogonic lines** are illustrated in Fig. 2-3.

The true nature of magnetism is not clearly understood, although its effects are well known. One theory which seems to provide a logical explanation of magnetism assumes that atoms or molecules of magnetic substances are in reality small magnets. It is known that electrons moving through a conductor produce a magnetic field around the conductor. Because of this fact, it is reasoned that electrons moving around the nucleus of an atom create minute magnetic fields. In magnetic substances such as iron it is assumed that most of the electrons are moving in one general direction around the nuclei; hence these electrons produce a noticeable magnetic field in each atom, and each atom or molecule becomes a tiny magnet. When the substance is not magnetized, the molecules lie in all positions in the material, as shown in Fig. 2-4*a*, and their fields tend to cancel one another. When the substance is placed in a magnetic field, the molecules align themselves with the field, and the fields of the molecules add to the strength of the magnetizing field. A diagram of a magnetized substance is shown in Fig. 2-4*b*.

When a piece of soft iron is placed in a magnetic field, almost all the molecules in the iron align themselves with the field; but as soon as the magnetizing field is removed, most of the molecules return to their random positions, and the substance is no longer magnetized. Because some of the molecules tend to remain in the aligned position, every magnetic substance retains a slight amount of magnetism after having been magnetized. This produces a small amount of **residual magnetism** in the substance.

Certain substances, such as hard steel, are more difficult to magnetize than soft iron because of the internal friction among the molecules. If such a substance is placed in a strong magnetic field and is struck several blows with a hammer, the molecules become aligned with the field. When the substance is removed from the magnetic field, it will retain its magnetism; hence it is called a **permanent magnet.** Hard steel and certain metallic alloys, such as Alnico, an alloy containing nickel, aluminum, and cobalt, which have the ability to retain magnetism are able to do so for the same reason that they are difficult to magnetize; that is, the molecules do not shift their positions easily. When the molecules are aligned, all the north poles of the molecules point in the same direction and produce the north pole of the magnet. In like manner, the south poles of the molecules produce the south pole of the magnet.

Many substances have no appreciable magnetic properties. The atoms of these substances apparently have their electron orbits in positions such that their fields cancel one another. Among these substances are copper, silver, gold, and lead.

MAGNETIC DOMAINS

A theory of magnetism similar to that explained in the foregoing discussion treats magnetic materials as having magnetic domains. According to this concept, there are groups of billions of atoms in magnetic substances, each group having its atoms aligned to give the group a polarity. These groups are called **domains** and are apparently arranged in accordance with the crystalline structure of the material. When a substance is not magnetized, the magnetic domains are arranged so that their polarities cancel one another. Under the influence of a magnetic field, either all the domains align themselves with the applied field or those already aligned or nearly in alignment gain in magnetic strength while the misaligned domains weaken. The substance is completely magnetized when all the domains are aligned with the magnetizing field.

THE MAGNETIC CIRCUIT

The field of force existing between the poles of a magnet is called a **magnetic field.** The pattern of this field may be seen by placing a stiff paper over a magnet and sprinkling iron filings on the paper. As shown in Fig. 2-5, the iron filings will line up with the lines of magnetic force. It will be noted that the lines directly between the poles are straight

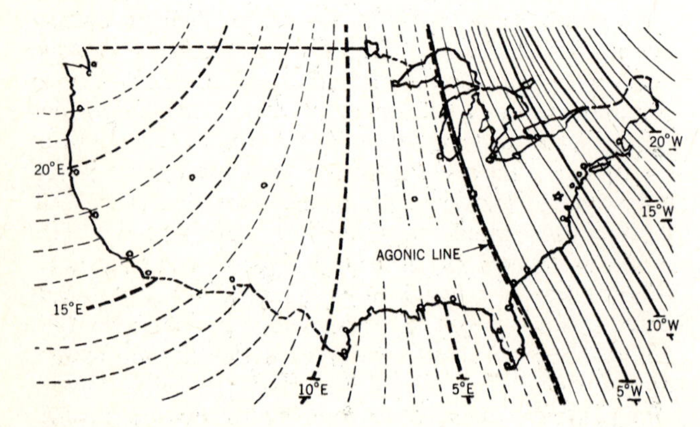

Fig. 2-3 Isogonic lines.

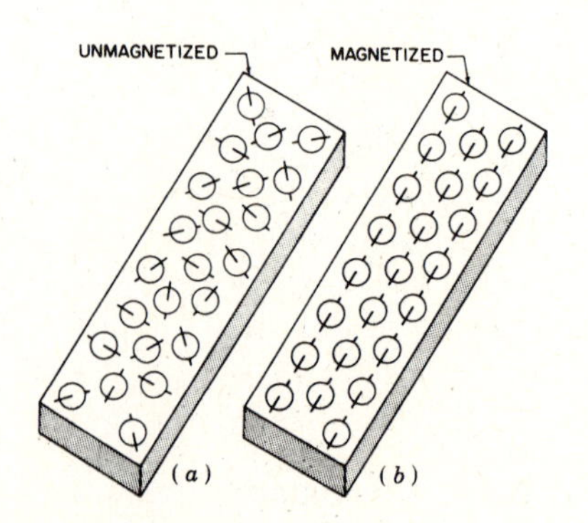

Fig. 2-4 Theory of magnetism.

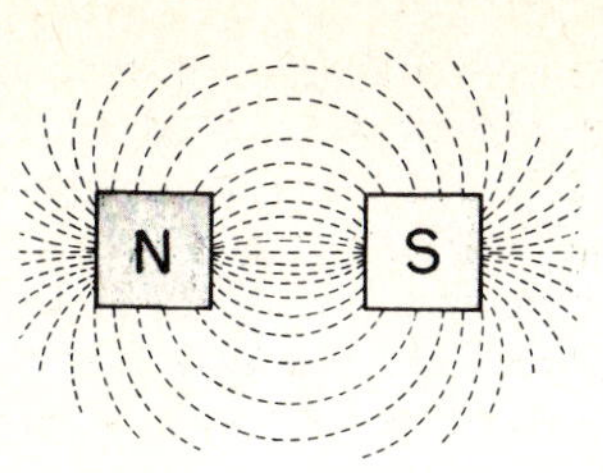

Fig. 2-5 A magnetic field.

but the lines farther from the direct path are curved. This curving is due to the repulsion of lines traveling in the same direction. If iron filings are sprinkled on a paper placed over two north poles, the field will have the pattern shown in Fig. 2-6. Here the lines of force from the two poles come out and curve away from one another.

Magnetic force, which is also called **magnetic flux,** is said to travel from north to south in invisible lines. We cannot say that this is literally true, but by assuming a direction, we provide a reference by which calculations can be made and effects determined. Since iron filings in a magnetic field arrange themselves in lines, it is logical to say that magnetic force exists in lines.

The space or substance traversed by magnetic lines of force is called the **magnetic circuit.** If a soft-iron bar is placed across the poles of a magnet, almost all the magnetic lines of force (flux) go through the bar, and the external field will be very weak. It is possible to magnetize a steel ring, as shown in Fig. 2-7, but when this is done, there is no apparent external field because the magnetic circuit is entirely within the ring. Such a ring is called a **Gramme ring.** Now, if one side of the ring is cut through, a magnetic force immediately becomes apparent, and there is a very strong magnetic field across the gap.

The external field of a magnet is distorted when any magnetic substance is placed in that field because it is

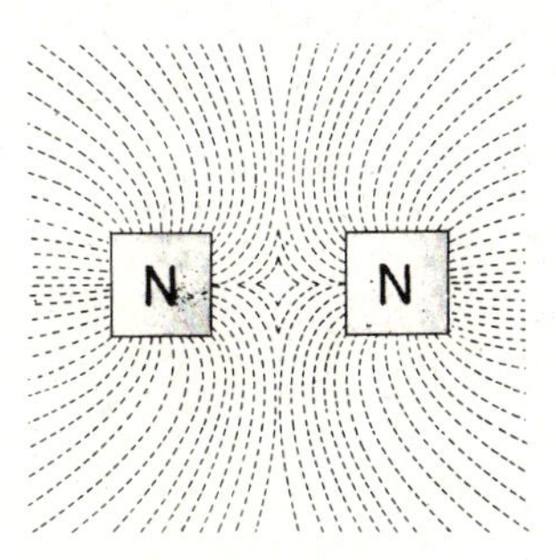

Fig. 2-6 Field between like magnetic poles.

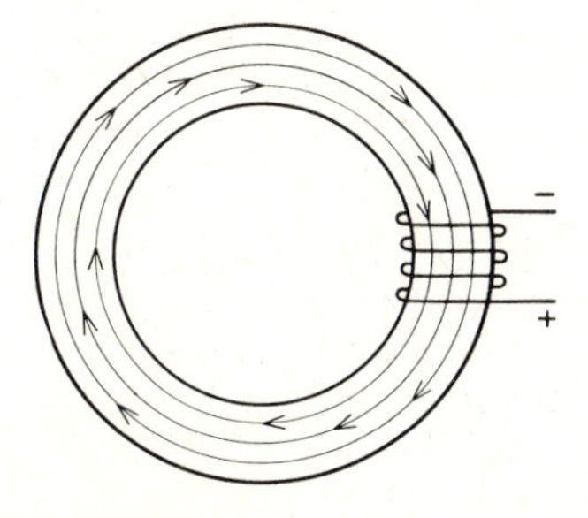

Fig. 2-7 A magnetized ring.

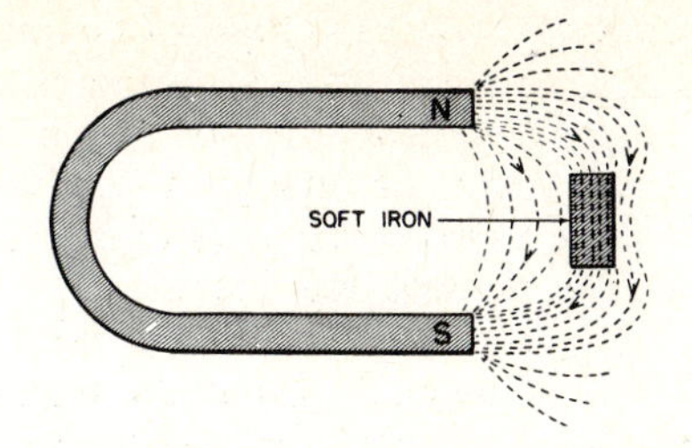

Fig. 2-8 Field distorted by a magnetic substance.

easier for the lines of force to travel through the magnetic substance than through the air (see Fig. 2-8). The ease with which a substance carries magnetic lines of force is called its **permeability.** The permeability of iron and its alloys is comparatively high but varies according to the nature of the alloy. The permeability of nonmagnetic materials is approximately the same as that of air and is given a value of unity (1). The opposition of a material to magnetic flux is called **reluctance** and compares to resistance in an electric circuit. The symbol for reluctance is $\mathcal{R}$ and the unit is **rel.**

The law for magnetic circuits is similar to Ohm's law for electric circuits. The force which causes magnetic flux to flow through a substance is called **magnetomotive force,** abbreviated mmf, and may be compared to emf. Mmf is measured in **gilberts** (Gb) or **ampere-turns,** and the symbol is F. The intensity of the flux is represented by the Greek letter phi (ϕ). The law for magnetic circuits may be stated as follows: **1 gilbert establishes a flux of 1 line of force through a material when the reluctance of the material is 1.**

The **maxwell** (Mx) is the unit of magnetic flux and represents 1 line of force. **When there is a magnetic flux of 1 maxwell per square centimeter (cm^2) of cross-sectional area, the flux density is 1 gauss.**

In the SI metric system, magnetomotive force is measured in ampere-turns. One ampere-turn is equal to 1.257 Gb. The unit of magnetic flux is the **weber** (Wb), which is equal to 10^8 Mx or 10^8 lines of force. The unit of flux density is the **tesla** (T), which is equal to 10 000 gauss (G) or a density of 10 000 lines of force per square centimeter of cross-sectional area of the material magnetized.

MAGNETIC DEVICES

ELECTROMAGNETS

It was noted previously in this section that an electric current flowing through a conductor creates a magnetic field around the conductor. In Fig. 2-9, the shaded circle represents a cross section of a conductor with current flowing in toward the paper. The current is flowing from negative to positive. When the current flows as indicated, the magnetic field is in a counterclockwise direction. This is easily determined by the use of the left-hand rule, which is based upon the true direction of current flow. When a wire is grasped in the left hand with the thumb pointing from negative to positive, the magnetic field around the conductor is in the direction that the fingers are pointing.

If a current-carrying wire is bent into a loop, the loop assumes the properties of a magnet; that is, one side of the

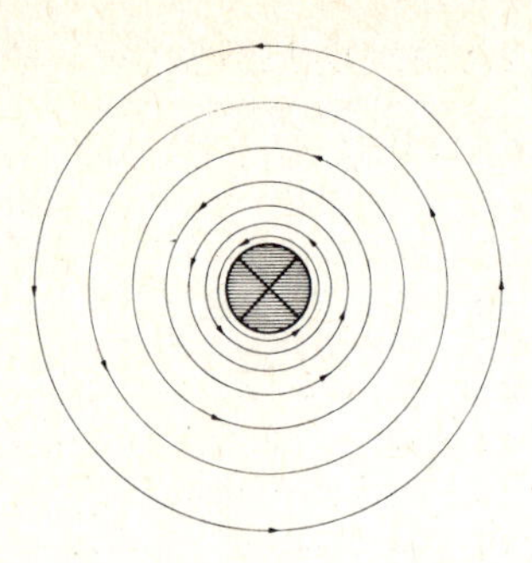

Fig. 2-9 Magnetic field around a conductor.

loop will be a north pole and the other side will be a south pole. If a soft-iron core is placed in the loop, the magnetic lines of force will traverse the iron core and it becomes a magnet. When a wire is made into a coil and connected to a source of power, the fields of the separate turns join and thread through the entire coil as shown in Fig. 2-10*a*. Figure 2-10*b* shows a cross section of the same coil. Note that the lines of force produced by one turn of the coil combine with the lines of force from the other turns and thread through the coil, thus giving the coil a magnetic polarity. The polarity of the coil is easily determined by the use of the **left-hand rule for coils: When a coil is grasped in the left hand with the fingers pointing in the direction of current flow, that is, from negative to positive, the thumb will point toward the north pole of the coil.**

When a soft-iron core is placed in a coil, an electromagnet is produced. Of course, the wire in the coil must be insulated so that there can be no short circuit between the turns of the coil. A typical electromagnet is made by winding many turns of insulated wire on a soft-iron core which has been wrapped with an insulating material. The turns of wire are placed as close together as possible to help prevent magnetic lines of force from passing between the turns. Figure 2-11 is a cross-sectional drawing of an electromagnet.

The strength of an electromagnet is proportional to the product of the current passing through the coil and the number of turns in the coil. This value is usually expressed in ampere-turns. If a current of 5 A is flowing in a coil of an electromagnet and there are 300 turns of wire in the coil, the coil will have an mmf of 1500 ampere-turns. Since the gilbert is also a measure of mmf and 1 ampere-turn is equal to 1.26 Gb, the mmf may also be given as 1890 Gb. The ultimate strength of the magnet also depends upon the permeability of the core material.

Electromagnets are used for many purposes, some of which were mentioned in the introduction to this chapter. The application of electromagnets will be discussed in detail as the various components of electric systems are described in later sections of this text.

The force exerted upon a magnetic material by an electromagnet is inversely proportional to the square of the distance between the pole of the magnet and the material. For example, if a magnet exerts a pull of 1 lb [0.4536 kg] upon an iron bar when the bar is $\frac{1}{2}$ in [1.27 cm] from the magnet, then the pull will only be $\frac{1}{4}$ lb [0.1134 kg] when the bar is 1 in [2.54 cm] from the magnet. For this reason, the design of electric equipment using electromagnetic actuation requires careful consideration of the distance through which the magnetic force must act. This is especially important in voltage regulators and relays.

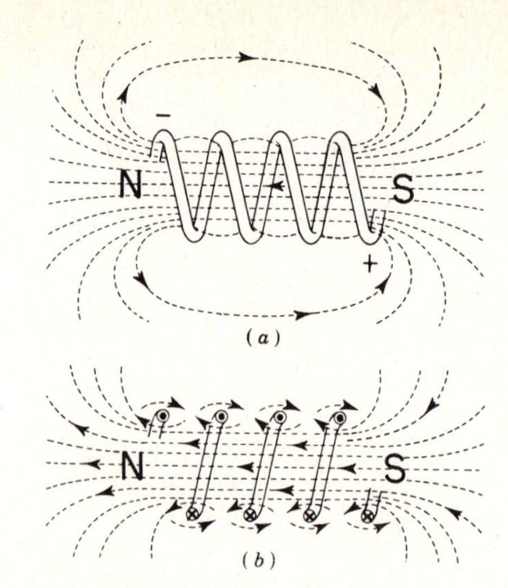

Fig. 2-10 The field of a coil.

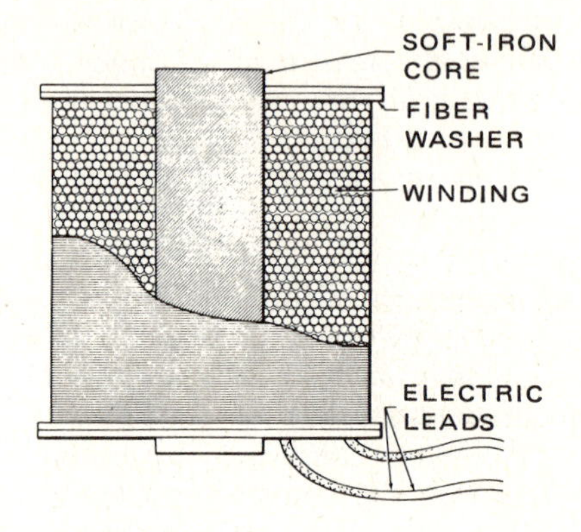

Fig. 2-11 An electromagnet.

SOLENOIDS

It has been explained that a coil of wire, when carrying a current, will have the properties of a magnet. Such coils are frequently used to actuate various types of mechanisms. If a soft-iron bar is placed in the field of a current-carrying coil, the bar will be magnetized and will be drawn toward the center of the coil, thus becoming the core of an electromagnet. By means of suitable attaching linkage, the movable core may be used to perform many mechanical functions. An electromagnet with a movable core, or without any core, is called a **solenoid.** The present practice among electricians is to call an electromagnetic device an electromagnet when it has a fixed core and a solenoid when it has a movable core or no core.

A solenoid is usually made with a split core. One part of the core is fixed permanently inside the coil and the other is left free to move (see Fig. 2-12). The two sections of the core are normally held apart by a spring; but when the coil is energized, the fixed core has a polarity opposite to that of the adjacent face of the movable core, and hence the movable core is attracted to the fixed core. This imparts motion through a connecting rod to the mechanical linkage.

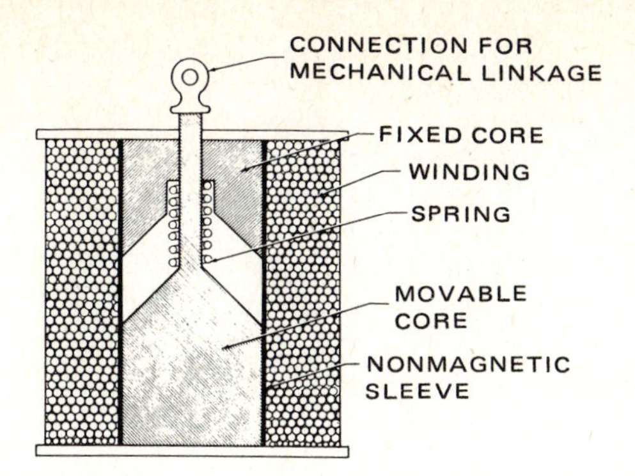

Fig. 2-12 A solenoid.

Solenoids are commonly used to operate switches, valves, circuit breakers, and several types of mechanical devices. The chief advantage of solenoids is that they can be placed almost anywhere in an airplane and can be controlled remotely by small switches or electronic control units. Although the use of solenoids is limited to operations where only a small amount of movement is required, they have a much greater range of movement than fixed-core electromagnets.

RELAYS

Electrically operated switches are often called **relays.** These may be simple switches of the single-throw type or more complex switches of the double-throw type, controlling several circuits. Figure 2-13 illustrates a typical single-throw relay. It consists of an electromagnet arranged so that its force may be used to close or open electric contacts. Because the contact points are closed when the electromagnet is energized, it is called a **normally open** relay.

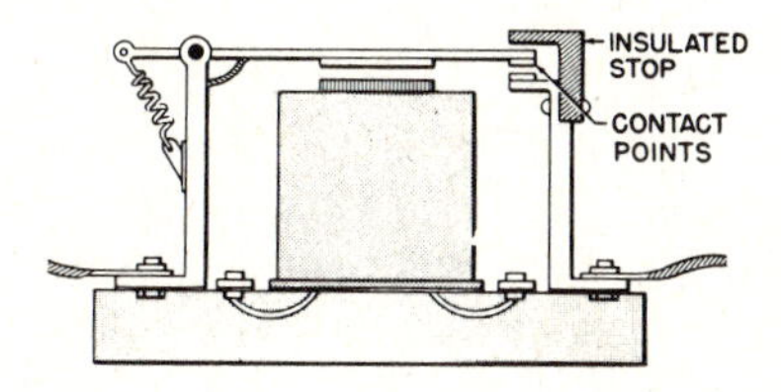

Fig. 2-13 A relay.

The part of the relay attracted by the electromagnet to close the contact points is called the **armature.** There are several types of armatures in electrical work, but in every case it will be found that an armature consists, in part, of a bar or core of material which may be acted upon by a magnetic field. In a relay, the armature is attracted to the electromagnet, and the movement of the armature either closes or opens the contact points. In some cases, the electromagnet operates several sets of contact points simultaneously.

Some heavy-duty relays are operated by solenoids. In such a relay, the movable core is connected to a contact bar which opens or closes an electric circuit. Such relays are used in circuits carrying high amperage. Starter systems in large airplanes frequently employ solenoid-type relays to control the power supply to the starter motor.

ELECTROMAGNETIC INDUCTION

BASIC PRINCIPLES

The transfer of electric energy from one circuit to another without the aid of electric connections is called **induction.** When electric energy is transferred by means of a magnetic field, it is called **electromagnetic induction.** This type of induction is universally employed in the generation of electric power. Electromagnetic induction is also the principle which makes possible the operation of electric transformers and the transmission of radio signals through air or outer space.

Electromagnetic induction occurs whenever there is a relative movement between a conductor and a magnetic field, provided that the conductor is cutting across (linking with) magnetic lines of force and is not moving parallel to them. The relative movement may be caused by a stationary conductor and a moving field or by a moving conductor with a stationary field. A moving field may be provided by a moving magnet or by changing the value of the current in an electromagnet.

The two general classifications of electromagnetic induction are **generator action** and **transformer action.** Both actions are the same electrically, but the methods of operation are different.

GENERATOR ACTION

The basic principle of generator action is shown in Fig. 2-14. As the conductor is moved through the field, a voltage is induced in it. The same action takes place if the conductor is stationary and the magnetic field is moved. The direction of the induced voltage depends on the direction of the field and may be determined by using the **left-hand rule for generators: Extend the thumb, forefinger, and middle finger of the left hand so that they are at right angles to one another, as shown in Fig. 2-15. Turn the hand so that the index finger points in the direction of the magnetic field and the thumb points in the direction of conductor movement. Then the middle finger will be pointing in the direction of the induced voltage.**

Another method for determining the direction of induced voltage is easily applied. Assume that the conductor moving through the field breaks off sections of the

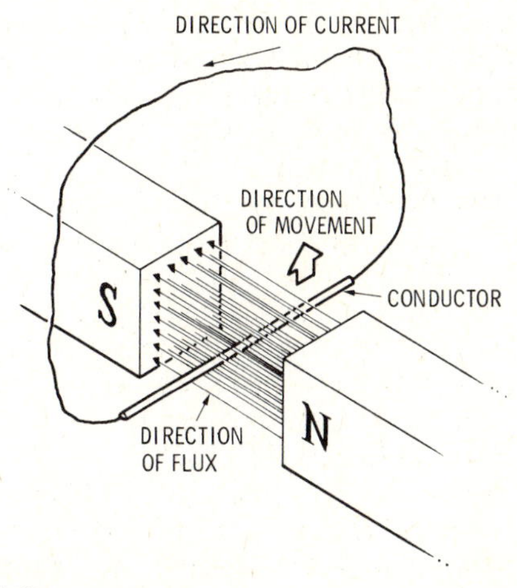

Fig. 2-14 Generator action.

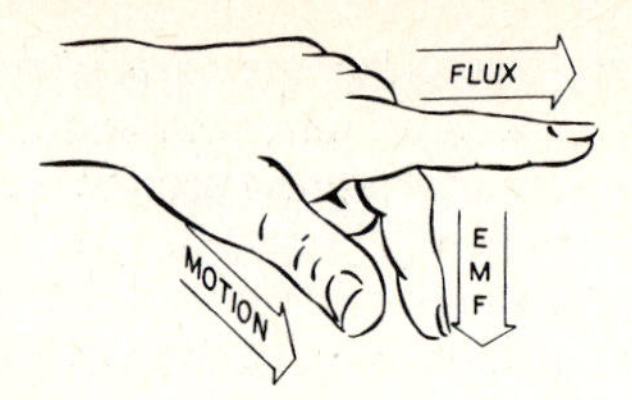

Fig. 2-15 Left-hand rule.

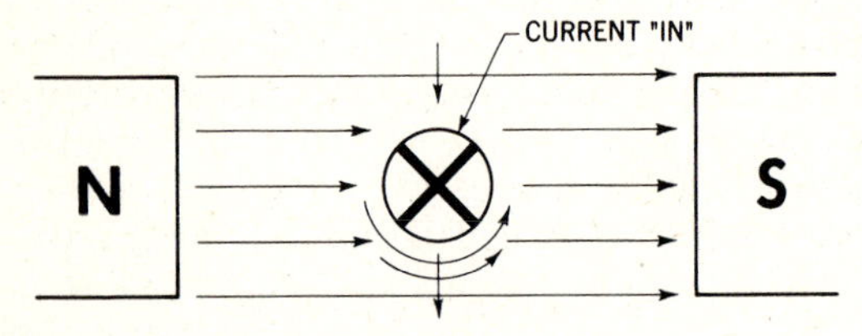

Fig. 2-16 A conductor cutting a field.

lines of force and causes them to wrap around the conductor as illustrated in Fig. 2-16. By applying the left-hand rule for conductors, it can be seen that if the lines of force curl around the conductor in a counterclockwise direction, the positive end of the conductor is toward the observer. Since the current travels from negative to positive, it is going into the paper in the illustration.

The left-hand rule for conductors may be stated as follows: **When a conductor is held in the left hand with the thumb pointing in the direction of current flow, the fingers will be pointing in the direction of the magnetic field. Conversely, if a conductor is held in the left hand with the fingers around the conductor in the direction of the magnetic field, the extended thumb will be pointing in the direction of the current flow.**

These rules are based upon the fact that current flows from negative to positive.

Figure 2-17 illustrates another kind of generator action. Here a bar magnet is pushed into a coil of wire (solenoid). A sensitive meter connected to the leads from the coil shows that a current flows in a certain direction as the magnet moves into the coil. As soon as the magnet stops moving, the current flow stops. When the magnet is withdrawn, the meter shows that the current is flowing in the opposite direction. The current induced in the coil is caused by the field of the magnet as it cuts across (links with) the turns of wire in the coil.

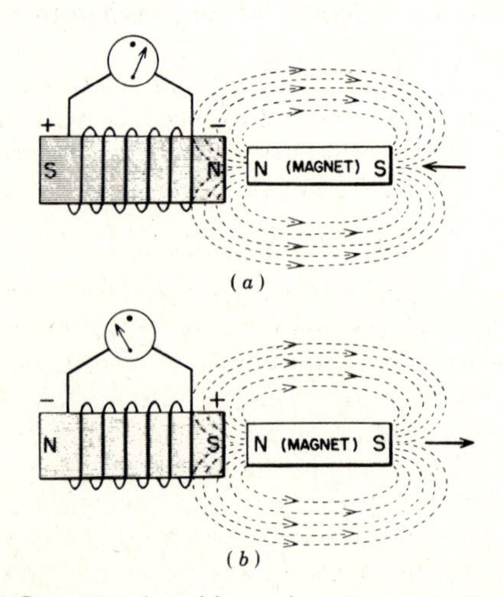

Fig. 2-17 Current induced by a changing magnetic field.

The characteristics of an induced current are stated in Lenz's law as follows: **An induced current is always in such a direction that its field opposes any change in the existing field.**

In Fig. 2-17*a* it will be seen that the north pole of the coil is adjacent to the north pole of the bar magnet; hence it opposes the insertion of the magnet into the coil. At the instant that the magnet begins to move out of the coil, current induced in the coil changes to the opposite direction; hence the field of the coil is reversed. The south pole of the coil field is now adjacent to the north pole of the bar magnet and opposes the withdrawal of the magnet (see Fig. 2-17*b*).

TRANSFORMER ACTION

In transformer action, the coil which produces the magnetic field and the coil in which the voltage is induced are both stationary. The movement of the magnetic field is produced by varying the strength of the current in the coil which produces the field.

If a battery is connected to the ends of one of two adjacent conductors, as in Fig. 2-18*a*, an expanding magnetic field is produced around the current-carrying conductor. We shall call this the primary conductor. As this field expands, the lines of force cut across the adjacent conductor, which we shall call the secondary. This induces a voltage in the secondary, and according to Lenz's law, this voltage will be in a direction which opposes the field of the primary conductor. The induced voltage is indicated for only a fraction of a second because the field becomes stationary as soon as the maximum current is flowing. Remember that there must be a current flowing in the conductor to produce a field and that there must be a change in current flow to cause a change in the strength of the field. When the switch in the primary circuit is opened, the field collapses and the lines of force cut across the secondary conductor in a direction opposite to

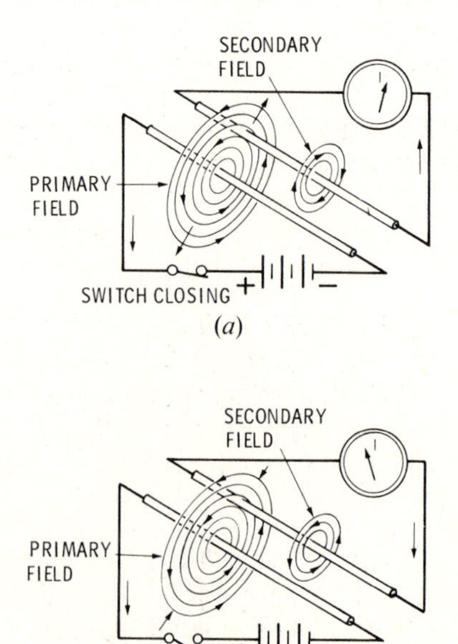

Fig. 2-18 Transformer action.

that which occurred when the battery was connected (see Fig. 2-18*b*). In either case, when the field is set up in the primary or when it collapses, the field of the secondary is in a direction which opposes the change.

When conductors are placed so that the field produced by one of the conductors induces a voltage in the other, the conductors are said to have **mutual inductance.**

The principle of mutual inductance described in the previous paragraphs applies to coils as well as to single conductors. Almost all practical applications of mutual inductance and transformer action involve the use of coils, because by this method the effects produced are much greater than they are by the use of single conductors.

The second part of Lenz's law states that **the value of the induced emf, in volts, is equal to one hundred-millionth volt for each line of magnetic force cut per second by the circuit in which the emf is produced.** In other words, when 100 million lines of force are being cut by a conductor each second, an emf of 1 V will be induced. It is therefore apparent that the greater the number of turns of wire in a coil, the greater the induced voltage for a given change in the magnetic field.

The formula for voltage developed by electromagnetic induction, in the SI metric system of measurements, is

$$E = \beta l v \sin \phi$$

where E = voltage, V
β = strength of the magnetic field, Wb/m^2
l = length of the conductor, m
v = linear velocity of the conductor through the magnetic field, m/s
ϕ = the angle of the conductor motion with respect to the direction of a magnetic field

The **weber** represents a magnetic-field strength of 10^8 lines of force.

The sine of angle ϕ (phi) can be determined from a table of trigonometric functions when the angle is known. The **sine** of an angle is defined in the chapter on alternating current.

An elementary transformer may be produced by placing one coil on a common core with another coil as shown in Fig. 2-19. The diagrams of Fig. 2-19 indicate the magnetic and electrical actions which take place. When a battery is connected to the primary coil, a field is set up in the iron core. Since the secondary winding is on this same core, a voltage is induced in the secondary. This voltage causes a current to flow in a direction opposite to that in the primary coil (see Fig. 2-19*a*). The induced current exists only during the time that the field is building up, which is but a fraction of a second. When the switch is opened, the primary field collapses and induces a current in the secondary coil, but this time the induced current is in the same direction as that of the primary coil. The resulting field opposes the collapse of the primary field as shown in Fig. 2-19*b*.

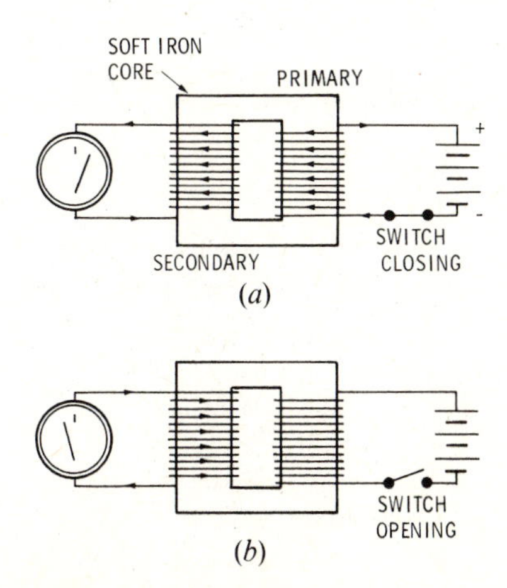

Fig. 2-19 A simple transformer.

A further description of transformers and their uses will be given in the section concerning alternating current.

SELF-INDUCTANCE

The magnetic field produced when a source of emf is connected to a coil induces an emf in that coil as well as in any other coil which is cut by the field. This occurs because the expanding field of the coil cuts across its own windings. The induced emf is in a direction which opposes the applied emf, thus satisfying the conditions of Lenz's law. When the circuit through the coil is broken, the field collapses and induces an emf which tends to maintain the existing field. This emf causes the current to arc at the point where the circuit is broken. For this reason, inductance coils are sometimes used to create a spark in the cylinder of an engine to ignite the fuel mixture. Modern engines use high-tension coils or magnetos with spark plugs for ignition.

The magnetic property of a coil or wire which causes it to induce an opposing emf within the same circuit is called **self-inductance.**

As previously stated, the voltage induced in a coil is proportional to the rate of cutting lines of force. When a circuit through a coil is broken, the field starts to collapse very rapidly. This causes the lines of force (magnetic flux) to cut across the windings at a rapid rate, with the result that a relatively high voltage is induced in the windings of the coil. This voltage is in a direction which opposes the collapse of the field; hence it tends to keep the current flowing. This action results in an arc across the switch points. The voltage induced in a coil when the circuit is broken is frequently many times as great as the applied voltage. A person may therefore receive a severe shock from a coil when the coil is disconnected from a relatively low-voltage source.

Because of the rapidly changing values of emf and current, the effects of self-inductance are very important where alternating current is used. These effects will be discussed in the section dealing with alternating current.

INDUCTION COILS

Induction coils, often called spark coils, are used extensively to produce the spark necessary for the ignition of fuel in gasoline engines. Such a coil is actually a special type of transformer. By means of the induction coil, the low voltage of a battery is stepped up to several thousand volts so that it will jump the gap of a spark plug and create a spark. The primary winding of an induction coil consists of relatively few turns of large wire and will carry sufficient current to set up a strong magnetic field through the soft-iron core. The core consists of thin iron laminations or insulated iron wires bound together in the size and shape desired. This type of construction reduces the effects of the **eddy currents** induced in the core. In a solid core, eddy currents are quite strong and cause heating as well as loss of power.

The secondary winding in an induction coil, placed concentrically around the primary winding, consists of many turns of fine wire. The number of turns in the secondary winding depends on the voltage to be produced by the coil.

When a magnetic field is built up by the primary coil, the magnetic flux links with the many turns of wire in the secondary coil and induces a relatively high voltage in the secondary. The voltage induced in the secondary coil is in approximately the same ratio to the voltage in the primary coil as the ratio of the number of turns in the secondary coil is to the number of turns in the primary coil. This relationship may be expressed by the following equation:

$$\frac{E_p}{E_s} = \frac{N_p}{N_s}$$

where E_p and E_s = voltages
N_p and N_s = number of turns of wire in primary and secondary coils

In an ignition system, the current through the primary coil is interrupted at the instant when a spark is desired in the cylinder of the engine. This is accomplished by means of a set of cam-actuated breaker points. The cam is driven by the engine and is timed to open the breaker points at the correct point for the firing of the fuel charge in the cylinder.

Figure 2-20 shows a schematic circuit for a typical induction coil used in an ignition system. It will be noted that a capacitor (condenser) is connected in parallel with the breaker points. The purpose of this capacitor is to cause the magnetic field to collapse more rapidly and to reduce the arcing at the breaker points. The capacitor accomplishes these functions by absorbing the voltage induced in the primary when the magnetic field collapses. The theory of capacitors is explained in another section of this text.

When the breaker points close in the primary circuit of an induction coil, the buildup of current is gradual, even though it appears to be almost instantaneous. Since the current buildup is gradual, so also is the buildup of the magnetic field. The full current flow does not take place instantaneously because of the effects of self-inductance as defined in Lenz's law. That is, as the current builds up, an opposing voltage is being induced in the circuit for as long as the current flow is increasing. The effect of the gradual increase in field strength is to reduce the induced voltage in the secondary winding.

As shown in Fig. 2-20, a capacitor is connected across the breaker points in the primary circuit. This capacitor counteracts the effects of self-inductance when the breaker points are opened; hence at this time there is an instantaneous collapse of the current flow and magnetic field. This very rapid collapse of the field results in a great increase in the voltage induced in the secondary coil. It is this high voltage which produces the spark at the spark gap.

Figure 2-21 illustrates graphically the approximate rate of current increase compared with current decrease in the primary winding of an induction coil as the breaker points are closed and then opened.

Some induction coils are provided with self-actuated breaker points. In coils of this type the breaker points are opened by the magnetic field of the primary coil. Figure 2-22 is a schematic diagram of an induction coil utilizing a vibrator mechanism as a circuit interrupter to produce the spark. The breaker points are normally held in the closed position by spring tension. When the switch is closed, the magnetic field of the core attracts the armature and opens the breaker points. This action breaks the primary circuit and causes the field to collapse. The collapse of the field induces a high voltage in the secondary circuit, and this voltage produces the spark across the spark gap. The collapse of the field also releases the contact points, which are immediately closed by spring tension. Again the field builds up in the primary, and the cycle is repeated. This action continues for as long as the switch is closed, and a steady shower of sparks occurs at the spark gap.

Coils of the type described in the foregoing paragraphs are often used as boosters in aircraft ignition systems.

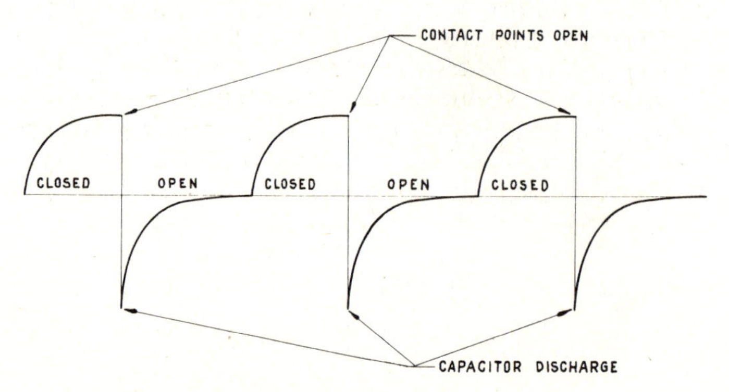

Fig. 2-21 Primary current in an induction coil.

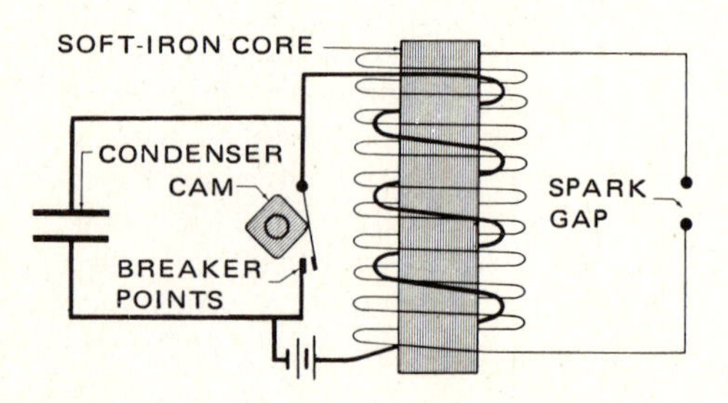

Fig. 2-20 Induction coil with a mechanical circuit breaker.

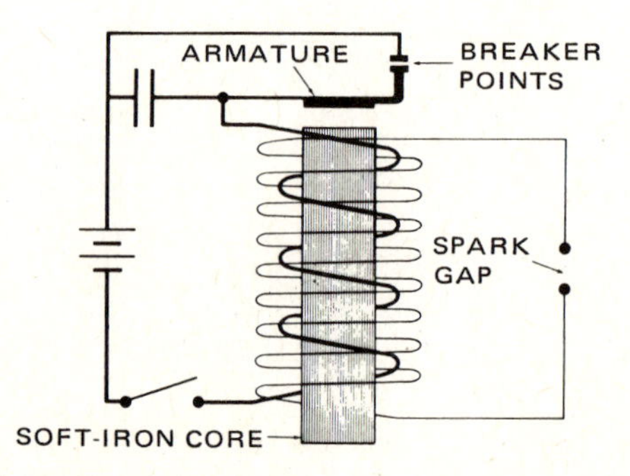

Fig. 2-22 Spark-coil diagram.

Such a coil supplies a spark during the starting of the engine, when the magneto is not turning fast enough to produce satisfactory ignition.

THE MAGNETO

A magneto is an excellent example of the application of electromagnetic induction. It has been explained that electromagnetic induction takes place whenever there is a relative movement between a conductor and a magnetic field. In a magneto, the magnetic field is provided by a permanent magnet. A coil of wire is located in such a position that it will be linked by the field of the permanent magnet whenever the field is set in motion. In some magnetos, the permanent magnet is rotated to produce the required motion of the field, and in others, the magnetic flux is carried by rotating, soft-iron inductors.

The coil of a high-tension magneto contains two separate windings. The primary winding consists of relatively large copper wire and the secondary winding of many thousands of turns of very fine wire. Thus the coil assembly is actually a high-ratio step-up transformer. Only the primary winding is included in the magneto for a low-tension ignition system.

The coil assembly is wound on a laminated soft-iron core which is mounted between the pole shoes of the magneto. The magnetic flux passes from one pole of the permanent magnet into one of the pole shoes and through the core of the coil. It returns to the opposite pole of the magnet through the other pole shoe, thus completing the magnetic circuit. When the permanent magnet rotates between the pole shoes, it causes a rapid reversal of the flux through the core of the coil, thus inducing a voltage in the coil.

Figure 2-23 is a diagram of a small high-tension ignition system for a six-cylinder opposed engine. Reference to this diagram will help the student to gain an understanding of the magneto operation.

The purpose of a magneto is to develop voltage of sufficient strength to jump the gap of a spark plug in the cylinder of an engine. This voltage is necessary at the instant when the fuel-air charge in the engine is to be ignited, and the magneto is designed to produce the maximum voltage only when it is required.

Since the value of an induced voltage depends upon the rate of current change in a coil, **breaker points** are incorporated in the primary circuit of the magneto to cause an instantaneous collapse of the primary current and magnetic field. During the time that the breaker points are closed, the primary current increases because of the flux change produced by the rotating permanent magnet. As the magnet rotates, the magnetic flux reaches maximum strength through the core of the primary coil and then begins to decrease. The primary current increases in a direction which sets up a field to oppose the decrease in the magnet field. When the maximum opposition exists between the rotating-magnet field and the primary-current field, the breaker points are opened by a **cam** and there occurs an instantaneous reversal of the magnetic field. This rapid flux reversal produces a very high voltage in the secondary winding of the coil.

Because of the opposition between the rotating-magnet field and the primary-current field, a high stress exists immediately before the breaker points open. This is because the flux from the rotating magnet has dropped to zero and has begun to exert force in the opposite direction. The primary field at this time is maintained entirely by the current flowing in the primary circuit. Hence, it can be seen that there are two magnetic forces acting against each other. The mmf of the primary coil is maintaining the field while the force of the rotating magnet is attempting to reverse the field. The stress thus produced is comparable to the tension of a spring. When the breaker points open, the action may be compared to the releasing of a trigger which permits the spring to snap forward violently. This, of course, represents the rapid reversal of the magnetic field.

A capacitor is connected in parallel with the breaker points to absorb the voltage induced in the primary coil by a sudden change in the magnetic field. This absorption results in a more rapid collapse of the field and prevents the current from arcing across the breaker points. If an arc were permitted at the breaker points, the points would

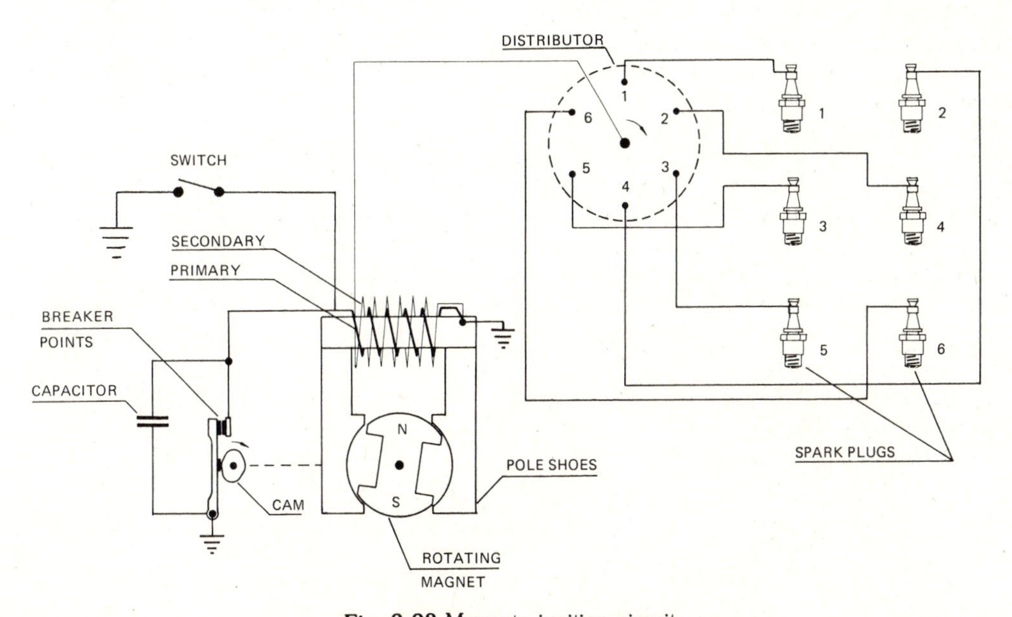

Fig. 2-23 Magneto ignition circuit.

burn and the primary current would decrease gradually instead of instantaneously.

As shown in Fig. 2-23, the secondary winding is wound upon the core with the primary coil. This arrangement provides for the maximum effect when the field collapses, and a very high voltage is consequently produced in the secondary coil. The fact that the secondary winding consists of many thousands of turns of fine wire is one of the principal reasons why it is possible to develop an emf of several thousand volts.

The secondary winding is connected through a **distributor** to the spark plugs of the engine, and the magneto is timed to the engine so that the high voltage is produced at the instant that it is needed to ignite the fuel-air charge in each cylinder. Note particularly in Fig. 2-23 that the distributor terminals are not connected to the correspondingly numbered spark plugs; engine cylinders do not fire in numerical sequence. The firing order of a typical six-cylinder opposed engine, such as that represented in the diagram, is 1-4-5-2-3-6; hence, the distributor terminals must be connected as follows: 1 to 1, 2 to 4, 3 to 5, 4 to 2, 5 to 3, and 6 to 6.

If the primary circuit of a magneto were closed continuously, the magneto would generate a relatively high alternating voltage, which would not, however, be high enough to jump the gap in a spark plug. The breaker points increase the voltage induced in the secondary coil because they produce an instantaneous collapse of the magnetic field. The breaker points also make it possible to time the spark for the instant at which it is needed in the cylinder.

There are many types of magnetos designed for aircraft engines, but it is beyond the scope of this text to describe in detail their construction and operation. The foregoing description of magneto theory is intended only to show the student how electromagnetic induction is applied in the operation of magnetos.

REVIEW QUESTIONS

1. Describe the properties of a permanent magnet.
2. What is the difference between substances required for permanent magnets and those used for temporary magnets?
3. Define *permeability; reluctance.*
4. Define *gilbert, maxwell, gauss, ampere-turn, weber,* and *tesla.*
5. When the direction of current flow through a coil is known, how do you determine the polarity of the coil?
6. What is the strength of an electromagnet when the current flow through the coil is 10 A and the coil has 200 turns? What is the strength in gilberts?
7. How does a pull of a magnet on a piece of steel at 1 in distance compare with the pull at 2 in distance?
8. Compare a *solenoid* with an *electromagnet.*
9. Describe a *relay.*
10. What conditions are necessary to produce electromagnetic induction?
11. Compare *generator action* with *transformer action.*
12. How do you determine the direction of current flow induced in a conductor?
13. Give Lenz's law.
14. How many lines of force must be cut per second to induce an emf of 1 V?
15. Define *self-inductance.*
16. Why are laminated iron cores used in transformers, induction coils, and other devices in which there are rapidly changing magnetic fields?
17. Describe an *induction coil.*
18. Describe the *coil* of a typical magneto.
19. What device is used to collapse the field in a magneto?
20. Describe briefly the operation of a magneto.

CHAPTER 3

Capacitors and Inductors

The purpose of this chapter is to explain the operation of capacitors and inductors, particularly as they function in a dc circuit. Although the operation of these units is also especially important in ac circuits, their functions in these circuits will be discussed in the chapter on alternating current.

Because they are identical, a capacitor may also be called a condenser. The term **capacitor** originated because the device is used to introduce capacitance into ac circuits. For the purposes of this chapter the word *capacitor* will be used, but it must be understood that everything said of a capacitor is also true of a condenser.

The **inductor,** or inductance coil, is merely a coil of insulated wire. Its characteristics were partially described in the discussion on electromagnetic induction, but will be more thoroughly explained in this section.

CAPACITORS

CAPACITOR THEORY

A capacitor consists of two conductors separated by an insulating medium which is capable of holding an electric charge. A simple capacitor, consisting of two metal plates separated by air, is shown in Fig. 3-1. The air, or other insulating material, between the plates of a capacitor is called the **dielectric.** When the plates of a capacitor are connected to a source of emf, the capacitor becomes charged. This charge consists of an excess of electrons on the negative plate and a corresponding deficiency of electrons on the positive plate. If the capacitor is disconnected from the source of emf, the charge will remain in the capacitor for a length of time depending upon the nature of the dielectric.

Unless there is a complete vacuum between the plates, the dielectric between the plates of a capacitor consists of a large number of atoms. This holds true whether the dielectric gaseous, liquid, or solid. Since the dielectric is an insulator, it contains very few free electrons; hence it takes a very high voltage to cause the electrons to break away from the dielectric's atoms and move through the material. When the capacitor is charged, an emf exists between the plates and acts upon the dielectric. Although the emf is not great enough to cause the electrons in the dielectric to break away from the atoms, it does cause them to shift a small distance in their orbits. This shifting of the electrons toward the positive plate of the capacitor creates what is known as a **dielectric stress,** which may be compared to a stretched rubber band. When the plates of a charged capacitor are connected by a conductor, the electrons flow from the negative plate to the positive plate, thus neutralizing the charge and relieving the dielectric stress. Then the capacitor is said to be *discharged.*

A clear understanding of the operation of a capacitor may be had by studying the hydraulic analogy shown in Fig. 3-2. The capacitor is represented by a chamber separated into two equal sections by an elastic diaphragm representing the dielectric. These chambers are connected to a centrifugal pump by means of pipes. The pump represents the generator in an electric circuit, and the valve in one of the pipes represents a switch. When the pump ro-

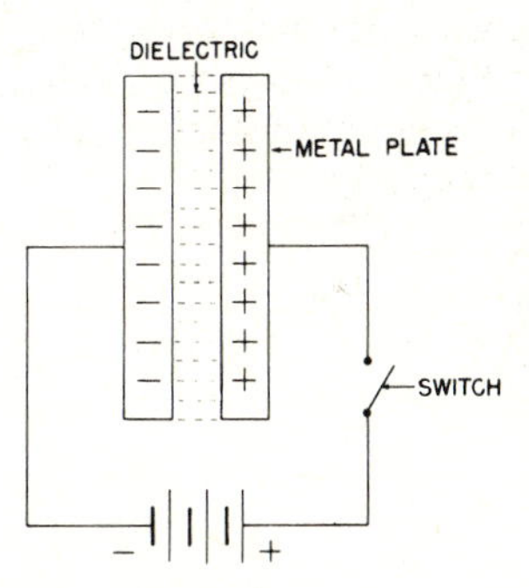

Fig. 3-1 Simple capacitor circuit.

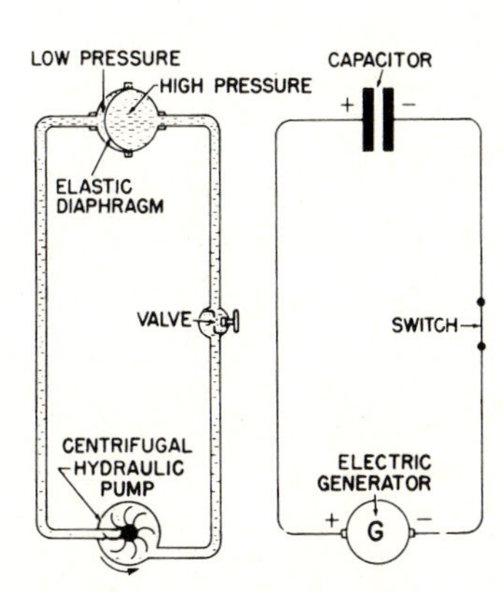

Fig. 3-2 Hydraulic analogy for a capacitor.

tates, it forces water into one of the chambers and causes the diaphragm to stretch. Water from the other chamber then flows out toward the pump. One of the chambers contains more water than the other, and the diaphragm, being stretched, maintains a pressure differential between the chambers. When the diaphragm pressure is equal to the pump pressure, the water will stop flowing, and the chamber may be said to be *charged.* If the valve is then closed, the diaphragm will maintain the differential of pressure between the sections of the chamber.

In the corresponding electric circuit, when the generator is running, electrons are forced into one plate of the capacitor and withdrawn from the other plate. When the potential difference between the plates is equal to the emf of the generator, the current flow will stop and the capacitor is charged. Now the switch may be opened and the charge will remain in the capacitor.

In the hydraulic "circuit," when the pump is stopped and the valve opened, the water will immediately flow from the section which has the higher pressure to the section which has the lower pressure. As soon as the pressures are equal, the flow will stop. In like manner, if the generator in the electric circuit is stopped and the switch is closed, the electrons will flow from the negatively charged plate, through the generator, and back to the positive plate.

An interesting experiment may be performed to demonstrate the operation of a capacitor. Obtain a sensitive ammeter with a center zero and connect it momentarily in series with a battery and a capacitor as shown in Fig. 3-3*a*. At the moment that the connection is made, there will be a movement of the ammeter needle to the right or left and then back to zero, if the capacitor is in good condition. Now remove the connections from the battery and touch the ends of the conductors together as shown in Fig. 3-3*b*. It will be noted that the ammeter needle now moves in a direction opposite to that in which it moved when the circuit was connected to the battery. In the first case, the capacitor was charging; in the second case, it was discharging. This is the reason that the needle of the ammeter moved in opposite directions. This experiment shows that electricity is stored in the capacitor when a voltage is applied and that the electricity will flow out of the capacitor when a circuit is provided for the electrons to flow from one plate to the other.

CAPACITANCE

The effect of a capacitor, that is, its ability to store an electric charge, is called **capacitance.** The unit of capacitance is the **farad** (F), which is the capacitance present when 1 V will store 1 C of electric energy in the capacitor. The farad is much too large a unit for practical purposes, and so a smaller unit called the **microfarad** (μf) is generally used. One microfarad is one-millionth of a farad. μ is the Greek letter mu. It is of interest to note that a capacitor with a capacitance of 1 F would probably weigh several thousand pounds.

Some capacitors have such small capacitance that even the microfarad is too large a unit for convenient expression of the value. In such cases the **picofarad** (*pf*) is used. One picofarad is equal to one-trillionth of a farad. This value may also be expressed $1\ pf = 10^{-12}$ F. The symbol for capacitance is the letter *C*.

The capacitance of a capacitor depends upon three principal factors: the area of the plates, the thickness of the dielectric, and the material of which the dielectric is composed. It will be readily apparent that two capacitors of the same size may differ considerably in capacitance because of a difference in the dielectric material.

To measure the dielectric characteristics of a material, a **dielectric constant** is used. Air is given a dielectric constant of 1 and is used as a reference for establishing the dielectric constants of other materials. Mica, which is commonly used as a dielectric in capacitors, has a dielectric constant of 5.8. This means that a capacitor having

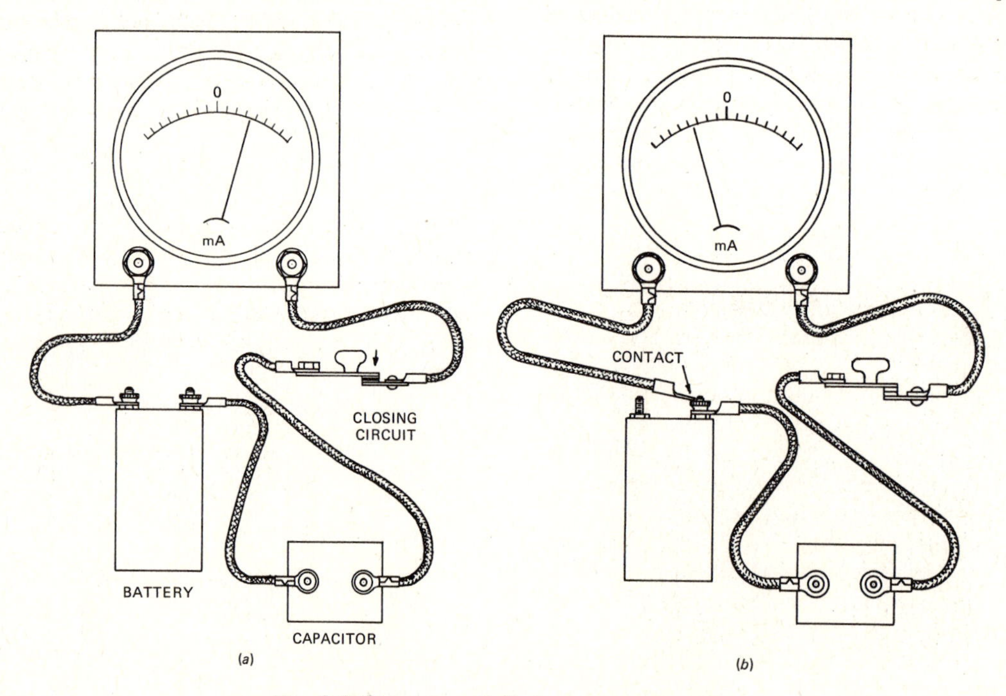

Fig. 3-3 Charging and discharging a capacitor.

mica as a dielectric has 5.8 times the capacitance of a similar capacitor having air as the dielectric.

In addition to the dielectric constant of a material, its insulating qualities must be considered. The insulating quality of a material is called its **dielectric strength** and is measured in terms of the voltage required to rupture (break down) a given thickness of the material. In selecting a capacitor for any purpose, it is important that the capacitance be correct and that the breakdown voltage of the capacitor be greater than the voltage to which the capacitor will be subjected when in use.

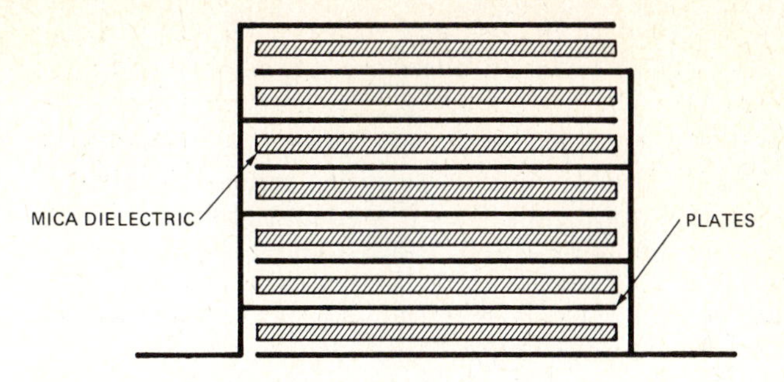

Fig. 3-5 Arrangement of a fixed capacitor.

TYPES OF CAPACITORS

There are two general types of capacitors: **fixed** and **variable.** The fixed capacitor is constructed with the plates and dielectric placed firmly together and covered with a protecting material such as waxed paper, plastic, ceramic material, or an insulated metal case. Because of its construction, the capacitance of a fixed capacitor cannot be changed.

Variable capacitors normally have fixed plates and movable plates arranged in such a manner that the dielectric effect between the plates may be changed by varying the distance between the plates or by moving one set of plates into or out of the other set. The construction of a typical variable tuning capacitor is shown in Fig. 3-4. Variable capacitors are used in radio and other electronic devices where it is necessary to change the capacitance to meet the requirements of a given circuit. The dielectric material in a variable capacitor is usually air.

Although the conducting elements of a capacitor are called plates, in a fixed capacitor they frequently consist of long strips of foil insulated with waxed paper and rolled together. The rolled plates are then covered with an insulating material and may then be placed in a protective case. The leads from the plates may be brought out at one end or both ends of the case, depending upon the design of the capacitor. Fixed capacitors of the mica type are often constructed as shown in Fig. 3-5. The plates are connected to form two groups with mica sheets separating the alternate plates.

When a relatively high capacitance is desired in a small capacitor, an **electrolytic** type is used. In a capacitor of this type, the dielectric is a liquid or paste known as an electrolyte. The electrolyte forms an oxide on one of the plates, which effectively insulates it from the other plate. The dielectric constant of the electrolyte is much greater than that of the commonly used dry materials; hence the capacitor has a considerably higher capacitance than the capacitors using dry materials. Electrolytic capacitors must be connected in a circuit with the correct polarity, because such a capacitor will allow current to flow through it in one direction. If the current flows through an electrolytic capacitor, the capacitance is lost and the plates will decompose. Precautions must be taken to ensure that electrolytic capacitors are not connected in reverse and that they are not overloaded. Often these capacitors will overheat and burst if they are not connected and used properly, and they may thus create a safety hazard.

Fig. 3-4 A variable capacitor.

MULTIPLE CAPACITOR CIRCUITS

When capacitors are connected in parallel (Fig. 3-6*a*), **the combined capacitance is equal to the sum of the capacitances.** The effect is the same as if one capacitor were used having a dielectric area equal to the total dielectric area of all the capacitors in the parallel circuit. Any multiple-plate capacitor is actually a group of capacitors connected in parallel. Since the capacitance varies directly as the area of the dielectric, it is apparent that two capacitors having the same dielectric area and connected in parallel have twice the capacitance of one, because the two capacitors have twice the dielectric area of the one.

The formula for capacitors connected in parallel is

$$C_t = C_1 + C_2 + C_3 \cdots$$

Capacitors in series present an unusual problem; the formula is similar to that used for resistances in parallel. **When capacitors are connected in series, the total capacitance is equal to the reciprocal of the sum of the reciprocals of the capacitances.** The formula is

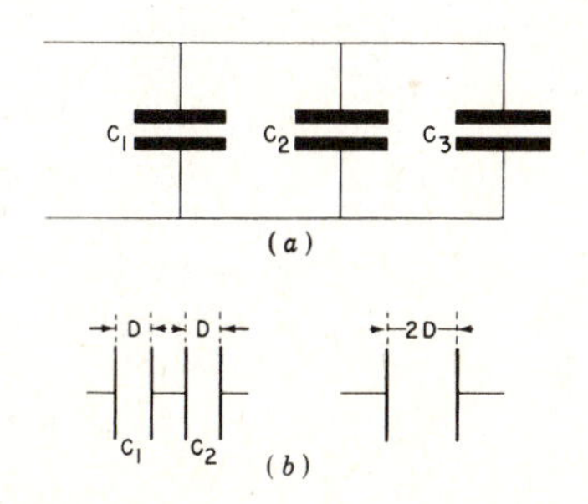

Fig. 3-6 Capacitors connected in parallel and series.

$$C_t = \frac{1}{1/C_1 + 1/C_2 + 1/C_3} \cdots$$

This rule for capacitances may also be stated as follows: **When capacitors are connected in series, the reciprocal of the total capacitance is equal to the sum of the reciprocals of the capacitances.** The formula is then

$$\frac{1}{C_t} = \frac{1}{C_1} + \frac{1}{C_2} + \frac{1}{C_3} \cdots$$

From the foregoing formulas, it will be found that the total capacitance, when capacitors are connected in series, is less than the capacitance of the lowest-rated capacitor in the series. The reason for this may be understood by observing a circuit where two capacitors of equal rating are connected in series (Fig. 3-6*b*). The two center plates will not contribute to the capacitance because their charges are opposite and will neutralize each other. The effect is that of two outside plates acting through a dielectric which has twice the thickness of the dielectric of one of the capacitors. Therefore, the total capacitance of the two capacitors is equal to one-half the capacitance of one of the capacitors. Remember that the capacitance of a capacitor varies inversely as the thickness of the dielectric.

EFFECTS AND USES OF CAPACITORS IN ELECTRIC CIRCUITS

When a capacitor is connected in series in a dc circuit, no current can flow because of the insulating qualities of the dielectric. When an emf is applied to such a circuit, there is a momentary flow of electrons into the negative plate of the capacitor and a corresponding flow out of the positive plate. As soon as the dielectric stress is equal to the applied emf, the flow of electrons stops. If the emf is removed, the charge remains in the capacitor until a path is provided through which the electrons may flow from the negative plate back to the positive plate. These effects have been explained previously and are illustrated in Fig. 3-3.

The charging and discharging of capacitors connected in series is illustrated in Fig. 3-7. Three capacitors are connected in series and to a battery through a double-throw switch. The switch is connected in a manner which makes it possible to connect the battery to the capacitors as shown in Fig. 3-7*a* or to disconnect the battery and short-circuit the capacitors as in Fig. 3-7*b*.

When the battery is connected to the capacitors, the electrons will flow from the negative terminal of the battery to the right-hand plate of capacitor 1. The electron charge in this plate will repel the electrons in the left-hand plate and cause these electrons to flow to the right-hand plate of capacitor 2. This, in turn, causes the electrons in the left-hand plate of capacitor 2 to be repelled, and they flow to capacitor 3. The positive charge of the battery attracts the electrons from the left-hand plate of capacitor 3. These same electrons are also repelled by the negative charge on the right-hand plate. The results are shown in Fig. 3-7*a*. All the right-hand plates of the capacitors have negative charges, and all the left-hand plates have positive charges; that is, they have few free electrons.

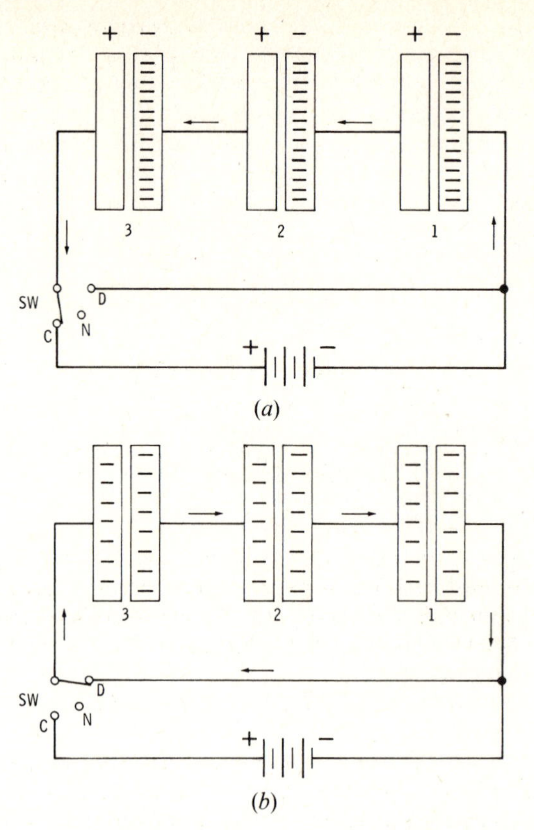

Fig. 3-7 Charging and discharging of capacitors connected in series.

If the switch is placed in the neutral position, the capacitors will hold their charges. However, when the switch is placed in the D position, the capacitors are short-circuited and the electrons flow in the direction shown to neutralize, or discharge, the plates. When no voltage is applied to the capacitors, the number of electrons on each of the plates for all capacitors will be approximately equal, assuming that the capacitors are identical in size and construction.

TIME CONSTANT

When a capacitor is connected to a voltage source, it takes a certain length of time for the capacitor to become fully charged. If a high resistance is connected in series with the capacitor, the time for charging is increased. For any given circuit containing capacitance and resistance only, the time in seconds required to charge the capacitor to 63.2 percent of its full charge is called the **time constant** for that circuit. This same time constant applies when the capacitor is discharged through the same resistance and is the time required for the capacitor to lose 63.2 percent of its charge.

The charging and discharging of a capacitor in terms of time constants is illustrated in the graph of Fig. 3-8. It will be noted that it takes six time constants to charge the capacitor to 99.8 percent of full charge. The discharge curve is the exact reverse of the charge curve. When the capacitor is short-circuited, it will lose 63.2 percent of its charge in one time constant and almost 99.8 percent of its charge in six time constants.

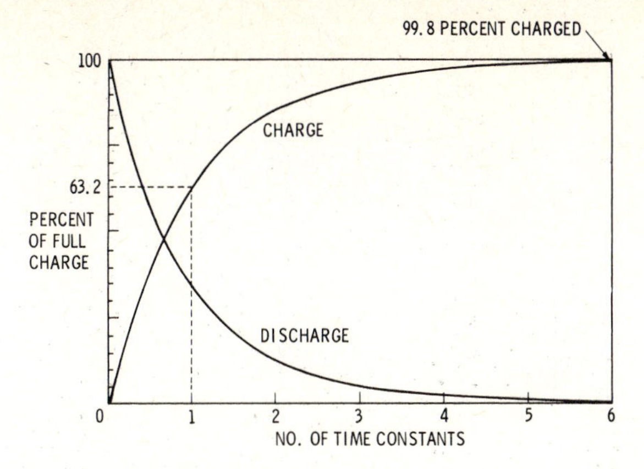

Fig. 3-8 Curves showing the charge and discharge of a capacitor according to the time constant.

To determine the length of a time constant in seconds for any particular capacitor-resistance circuit, it is necessary to multiply the capacitance in microfarads by the resistance in megohms (MΩ); that is,

$$T = CR$$

As an example of how the time constant may be used in determining the performance of a capacitance-resistance circuit, we shall assume that a 20-μf capacitor is connected in series with a 10 000-Ω resistor and that 110 V is applied to the circuit at intervals of $\frac{1}{2}$ s.

The time constant is equal to 20 × 0.01 or 0.2 s. (Note that 10 000 Ω is equal to 0.01 MΩ.) The time interval is given as $\frac{1}{2}$ s; hence, the number of time constants is 2.5. If we examine a time-constant chart or graph, we find that the voltage at 2.5 time constants will be approximately 92 percent of full voltage. Applying this to our problem shows us that 92 percent of 110 V is approximately 101 V. Thus, we find that the capacitor in this problem will charge to approximately 101 V.

SOME USES FOR CAPACITORS

When a capacitor is connected in parallel in a dc circuit, it opposes any change in the circuit voltage; for as voltage from the source rises, current flows into the capacitor and thus slows the voltage rise. If the voltage of the source remains at a higher level, the capacitor will charge to that level and will have no further effect on the circuit as long as the voltage remains constant. If the voltage from the source drops, the capacitor discharges into the circuit and holds the circuit voltage above that of the source for a short time. The property of capacitors to oppose changes in voltage is utilized in dc circuits to reduce or eliminate voltage pulsations. The voltage from a dc generator pulsates; that is, it varies slightly above and below the average value. When a capacitor of sufficient capacitance is connected in parallel with the generator, voltage pulsations are largely eliminated and a steady direct current is delivered. This is discussed further in the section dealing with generators.

Another use for capacitors has been mentioned in the discussion of induction coils and magnetos. In these instances, the capacitor is used to reduce the arcing at the breaker points. When the magnetic field in an induction coil or a magneto coil collapses, a relatively high voltage is induced in the primary winding. This voltage causes a spark to jump across the breaker points; the points burn and the magnetic field slowly collapses. When a capacitor is connected in parallel with the points, the voltage is absorbed by the capacitor and thus prevents arcing at the points.

When airplanes are equipped with radio, as most of them are, capacitors are used to reduce radio interference. Fluctuating voltages and currents in electric circuits cause the emanation of electromagnetic waves. These waves induce currents in the radio circuits and interfere with normal operation. Capacitors are connected in the electric circuits at points where they will be most effective in absorbing the momentary fluctuations of voltage; in this way they reduce the emanation of electromagnetic waves.

THE INDUCTANCE COIL

The terms **mutual inductance** and **self-inductance** have been defined in a previous section of this text. In the study of inductance coils, we are chiefly concerned with the principle of self-inductance. Any electric conductor possesses the property of inductance when there is a change of current flow in it. Hence, an inductor may be a straight piece of wire or a coil with thousands of turns of wire. Figure 3-9 illustrates a variety of inductance coils.

An understanding of the effect of inductance in a single conductor may be obtained by studying Fig. 3-10. The large circle represents the cross section of a conductor. For the purposes of this explanation, the sector *AOB* will be assumed to consist of many small conductors instead of being part of one large conductor. When the conductor is connected to a source of power so that a current flows outward toward the observer, a magnetic flux develops in the direction indicated in the drawing. The lines of flux (magnetic force) will move outward from the center as shown. In the sector *AOB* of the conductor, the lines of flux will then be moving upward across the many small conductors which we are assuming make up that sector of the large conductor. Since the lines of force are moving upward at this point, the small conductors are moving downward in relation to the lines of force.

The left-hand rule for generators may now be applied to determine the effect of the induced voltage. Observe that the direction of flux at the top of the conductor is to the right, and remember that the conductor is moving downward with respect to the flux lines. By pointing the index finger to the right, as in Fig. 3-11, and pointing the thumb downward, we find that the direction of induced voltage is into the paper. Thus we see that the voltage induced by the lines of flux cutting the conductor is in a direction opposite to the applied voltage. This is in accordance with Lenz's law (page 30).

The effect of inductance in the conductor is to make the current-flow increase gradual instead of instantaneous. The curves of Fig. 3-12 illustrate what takes place when a switch is closed in a dc circuit and again when the voltage is removed. When the switch is closed, the current at first rises rapidly and then increases more slowly until the level of the applied voltage is reached. If the switch remains

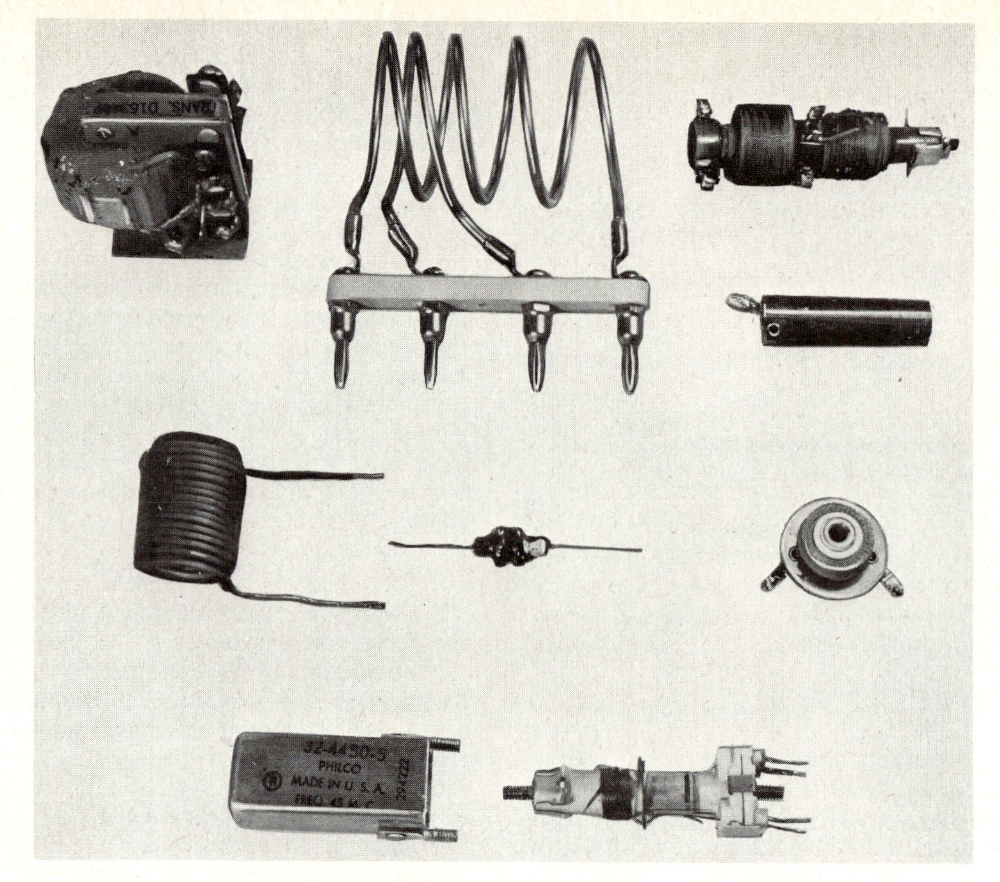

Fig. 3-9 Inductance coils.

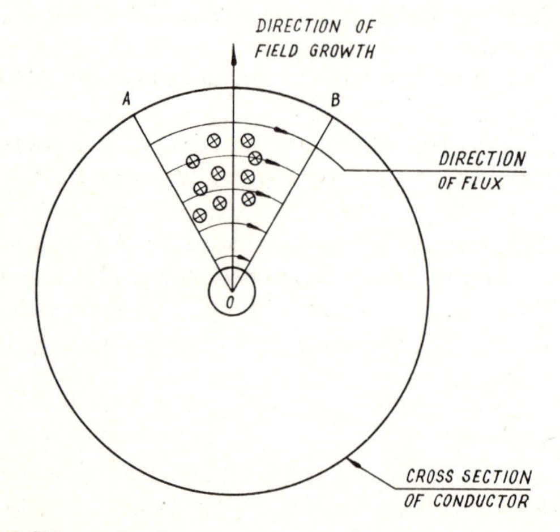

Fig. 3-10 Effect of inductance in a single conductor.

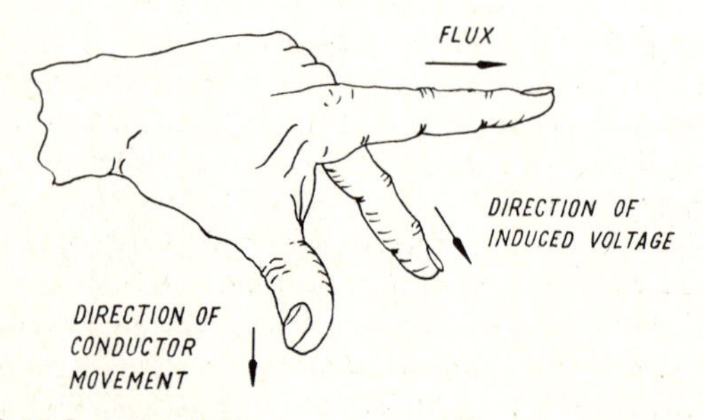

Fig. 3-11 Left-hand rule.

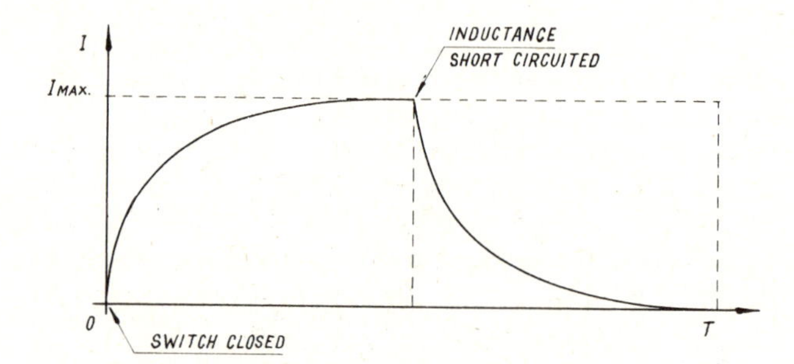

Fig. 3-12 Curve showing the effect of inductance on the rise and fall of current.

closed, the current value will continue steady. When the applied voltage is cut off, the current decay will be not instantaneous but gradual, as shown in the drawing. This is because the induced voltage at cutoff is in a direction which tends to keep the current flowing. To obtain this effect the voltage is removed by short-circuiting the inductance.

It has been stated in a previous section of this text that the field strength of an electromagnet depends upon the number of turns of wire in the coil, the current flowing in the coil, and the material in the core. Actually, an electromagnet and an inductance coil are essentially the same; hence, the effect of an inductance coil in a circuit also depends upon the number of turns of wire in the coil, the current flowing in the coil, and the material used in the core. Inductance coils are made with soft-iron cores when a high inductive effect is desired. When a low inductive effect is desired, the inductance coil has no core; that is, the core is made of air.

The inductance of a coil is measured in a unit called the **henry** (H), named for Joseph Henry (1797–1878), an American physicist. **One henry is the inductance of a coil when a change of current of 1 ampere per second will induce an emf of 1 volt.** The symbol for inductance is the letter L. The henry is too large a unit for most applications, and so a smaller unit called the **millihenry** (mH) is used. One millihenry is one-thousandth of a henry.

As in the case with capacitance in a circuit with resistance, a time constant is applied in a circuit containing inductance in series with a resistance. Referring to Fig. 3-8, the curves shown apply to an inductive circuit as well as a capacitive circuit. The curves for a circuit with inductance and resistance only are also shown in Fig. 3-12.

In the case of inductance, the maximum current flow in a circuit is delayed for a short time after the inductance coil is connected to a power source. The time constant is the time in seconds that is required for the current flow to reach 63.2 percent of maximum after the circuit is connected to the power source. The time constant for a decaying current is the time in seconds required for the current flow to fall to 36.8 percent of maximum. This is the same time as is required for the increase to 63.2 percent of maximum.

To determine the time constant for a circuit containing only inductance and resistance, it is necessary to divide the inductance (L) in henrys by the resistance (R) in ohms. Hence,

$$T = \frac{L}{R}$$

If a 10-H inductance coil is connected in series with a 200-Ω resistance, the time constant is $\frac{10}{200}$ or 0.05 s. With this combination in a circuit, the current will reach maximum in about $\frac{1}{3}$ s.

As discussed briefly in Chapter 2, inductance is utilized in spark coils and magnetos to produce the high-tension current needed for developing the spark at the electrodes of a spark plug. Remember that when current flow changes in a conductor, the changing magnetic field around the conductor induces a voltage which opposes the change in the magnetic field. This is in accordance with Lenz's law. When an inductance oil is disconnected from a source of current, the collapsing magnetic field induces an opposing voltage which is higher in value than the original applied voltage. The level of the induced voltage depends upon the inductance of the coil and the rate of current decay (time constant). A voltmeter designed to show voltage in either direction can be connected across the terminals of an induction coil to show the surge of reverse voltage at the time that the coil is disconnected from a power source such as a battery. Because of the **transient voltages** developed in circuits where inductances or induction coils are connected and disconnected, safety devices must be incorporated to protect circuit elements which may be damaged by the sudden high-voltage "spikes." Diodes are often used for this purpose and are connected in the circuit to block high reverse voltages or to bypass them so they will not be applied to transistors or other diodes which are not designed to withstand the voltages.

The principal uses of inductance coils are in ac systems, radios, and electronics. Inductance coils are discussed more fully in sections of this text dealing with these subjects.

REVIEW QUESTIONS

1. Define *capacitor.*
2. What name is given the insulating material between the plates of a capacitor?
3. What conditions exist in the plates of a capacitor when it is charged?
4. When a capacitor is connected to a source of voltage, what action takes place in the circuit?
5. Define *farad.*
6. What unit is normally used to indicate the capacitance of a capacitor?
7. Give the factors which determine the capacitance of a capacitor.
8. What is meant by *dielectric constant*?
9. Describe the construction of a typical fixed capacitor. Describe a variable capacitor.
10. How does the capacitance of an electrolytic capacitor compare with the capacitance of a dry capacitor of similar size?
11. What precaution must be taken in connecting an electrolytic capacitor in a circuit?
12. Give the formula for determining the total capacitance when capacitors are connected in parallel; in series.
13. What is the effect of a capacitor connected in parallel in a dc circuit?
14. Give three uses for capacitors in an aircraft electric system.
15. What is the total capacitance of a circuit when capacitors of 100, 200, and 300 μf are connected in parallel?
16. What would be the capacitance if the capacitors in the foregoing question were connected in series?
17. Define *time constant* as applied to a circuit containing capacitance and resistance only connected in series.
18. How is the time constant for a capacitive circuit determined?
19. What is the effect of an inductance coil in a circuit?
20. What law describes the principle of inductance?
21. What factors determine the inductance value of an inductance coil?
22. Compare the inductance of a coil having an iron core with that of a similar coil having an air core.
23. What unit is used to measure inductance?
24. Give the symbol for inductance.
25. How does the time constant for a circuit with inductance and resistance only compare with the time constant for a circuit containing capacitance and resistance only?
26. What is the time constant for a circuit having an inductance of 5 H and a resistance of 100 Ω?
27. Discuss *transient voltages* in an inductive circuit and explain how they are caused.
28. What protecting devices may be used to reduce the possibility of transient-voltage damage?

CHAPTER 4

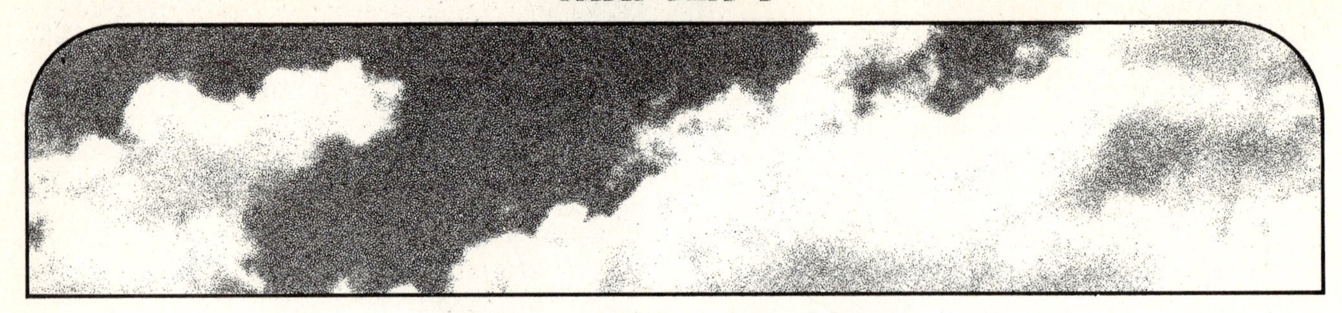

Alternating Current

A thorough understanding of alternating current is becoming increasingly important to aviation maintenance technicians, because modern jet aircraft utilize this type of power for both flight and ground operations. Those who wish to be classified as master maintenance technicians, those who expect to become electrical and electronics specialists, and others who seek supervisory positions in the electrical fields should study this section most carefully. Every principle discussed should be completely understood before the next is studied.

Until recently, the main electric systems in airplanes were powered with direct current. Many airplanes, particularly large military aircraft and modern jet airliners, are now equipped with ac systems supplying power for practically all purposes. Three-phase ac systems are used on large aircraft because they provide a great saving in weight compared with dc systems.

On airplanes having dc power systems, circuits requiring alternating current are supplied by means of inverters. An inverter is a device which changes direct current to alternating current. It may be a combination of a dc motor and an ac generator, a vibrator and transformer combination or a solid-state device called a **static inverter.**

Some of the units operated by alternating current in airplanes are instruments, fluorescent lights, radio equipment, electric motors, navigation equipment, and automatic pilot. This list does not include all the devices which are or may be operated by alternating current, nor is it intended to indicate that all types of the above named devices require alternating current. Airplanes not carrying the equipment listed or any other equipment requiring alternating current are very common. Some small airplanes have practically no electric equipment, but their utility is limited to daylight contact flight.

A good knowledge of the principles of alternating current is essential for the understanding of various electric devices. This is especially true of ac electric motors, generators, and transformers. This section explains the nature of alternating current and many of its characteristics and uses.

DEFINITION AND CHARACTERISTICS

Alternating current is defined as current which periodically changes direction and continuously changes in magnitude. The current starts at zero and builds up to a maximum in one direction, then falls back to zero, builds up to a maximum in the opposite direction, and returns to zero. In like manner, the voltage attains a maximum in one direction, drops to zero, rises to a maximum in the opposite direction, and then returns to zero.

It is difficult for some students to visualize the nature of alternating current, but there are many common devices which may be used to illustrate this principle. First, consider reciprocating (moving back and forth) devices such as a carpenter's saw, a connecting rod in an engine, or the pendulum in a clock. Each of these devices performs useful work with a reciprocating motion. Figure 4-1 shows a hydraulic analogy of an ac circuit performing work. The pump forces the fluid back and forth in the pipes and causes the working piston to move back and forth. This piston is connected to a crankshaft which converts the reciprocating motion of the piston to the rotary motion of the flywheel.

Values of alternating current and voltage are indicated by a **sine curve.** In Fig. 4-2, this curve represents a definite voltage or current value for a certain degree of rotation through the alternating cycle. One cycle begins at 0° and ends at 360°. The value of the alternating current is zero at 0°, maximum in one direction at 90°, zero at 180°, maximum in the opposite direction at 270°, and zero at 360°, as shown in the sine curve.

The sine curve is so named because it represents the ratio of a side of a right triangle to the hypotenuse. The vertical side is the side opposite the given angle and is the line BC in Fig. 4-3. The sine of the angle BAC is BC/AB. This is expressed mathematically as $\sin BAC = BC/AB$.

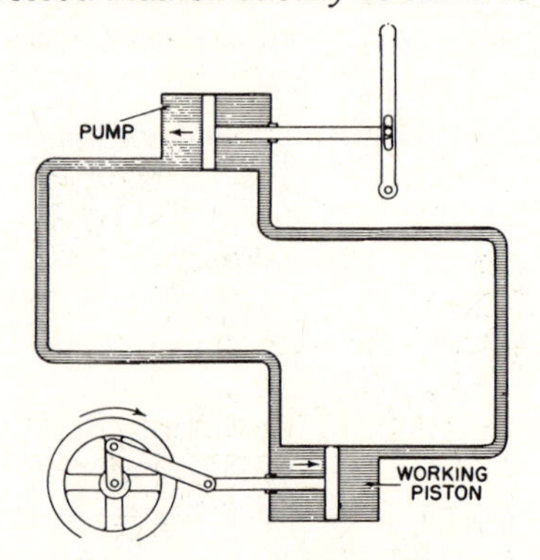

Fig. 4-1 Hydraulic analogy of alternating current performing work.

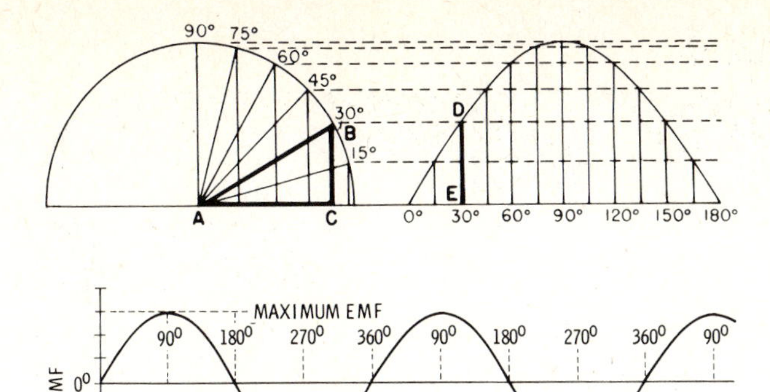

Fig. 4-2 A sine curve.

In the triangle ADE, the sine of the angle DAE is DE/AD. This is also the sine of the angle CAD. The sine curve shown in Fig. 4-2 represents the sines of all angles formed as the radius of the circle is rotated from 0° counterclockwise through 360°. Two positions of the radius are shown as AB and AD in Fig. 4-3. In quadrant I of the circle, the sine increases from 0 to 1 in value. In quadrant II, the sine decreases from 1 to 0. In quadrant III, the sine increases in a negative direction from 0 to -1; and in quadrant IV, the sine value changes from -1 to 0.

For practical purposes, the values of an alternating current may be considered to follow the sine curve. This may be understood by considering the generation of alternating current by a simple generator (see Fig. 4-4). A single loop of wire is placed so that it may be rotated in a magnetic field. As the loop is turned, the sides of the loop cut through the lines of force and an emf is induced in the sides of the loop. Since the side AB is moving up through the field, and the side CD is moving down through the field, the voltage induced in AB causes current to flow from A to B, and the voltage induced in CD causes current to flow from C to D. These voltages add together and cause current to flow in the direction $ABCD$. This is true for as long as AB is moving up and CD is moving down. When the coil is in a vertical position, the sides are moving horizontally and parallel to the lines of force. In this position, no voltage is induced. As the loop continues to rotate, the side AB moves down through the field and the side CD moves up. Then the direction of the current in the loop is reversed.

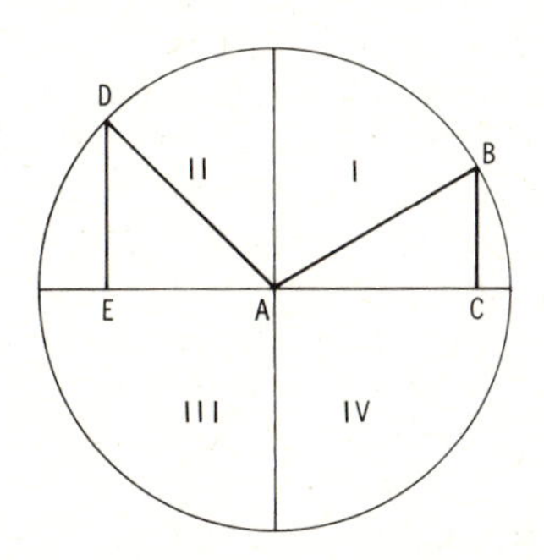

Fig. 4-3 Triangles to illustrate the sine of an angle.

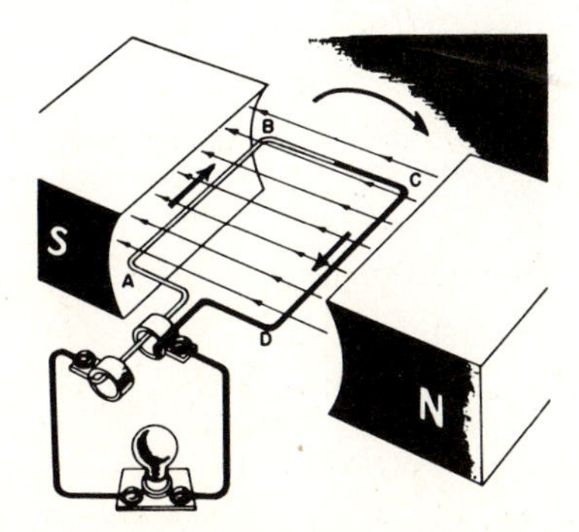

Fig. 4-4 A simple ac generator.

When the current is carried to an external circuit by means of slip rings, it travels in one direction while the loop moves to 180° and in the other direction while the loop moves from 180 to 360°. When it is vertical, the loop is in either the 0 or 180° position, and no voltage is induced. When the loop is in a horizontal position, the maximum voltage is induced because at this time the sides are cutting the greatest number of lines of force.

It has been found that the instantaneous value of the voltage induced in a loop as it rotates in a magnetic field is proportional to the sine of the angle through which the loop has rotated from 0°. Hence we use the sine curve to represent the values from 0 to 360°. The value of either the voltage or the amperage may be represented in this manner.

RMS, OF EFFECTIVE, VALUES

In order to determine the amount of power available from an alternating current, we must arrive at its effective value. It is obvious that effective value does not equal maximum value, because this value is attained only twice in the cycle. Even though the current during one half-cycle is equal and opposite in direction to that during the other half-cycle, the currents do not cancel each other; work is done whether the current is moving in one direction or the other. Therefore, the effective value must lie somewhere between the zero value and the maximum value.

The effective value of an alternating current is calculated by comparing it with direct current. The comparison is based on the amount of heat produced by each current under identical conditions. Since the heat produced by a current is proportional to the square of the current, it is necessary to find the square root of the mean square of a number of instantaneous values. The resultant value is called the **root-mean-square** (rms) current, and it may be obtained from the formula $I = I_m/\sqrt{2}$ where I is the effective value and I_m is the maximum. If the maximum current has a value of 1, then the rms value is equal to 0.707, or $I = 0.707I_m$. In like manner, the effective value of an alternating emf is found to be 0.707 multiplied by the maximum emf.

In all practical applications of alternating current, the values of voltage, or current, are stated according to their effective values rather than the maximum values. For example, when the voltage is given as 110, the maximum value of the voltage is $(1/0.707) \times 110 = 155.6$ V, approximately. Keeping this in mind, technicians should always make certain that any instrument or equipment that they use with a nominal voltage rating in alternating current has a safety factor sufficient to handle the maximum voltage.

FREQUENCY

It has been explained that one cycle of alternating current covers a period in which the current value increases from zero to maximum in one direction, returns to zero, increases to maximum in the opposite direction, and then returns to zero. The number of cycles occurring per second is the **frequency** of the current and is measured in a unit called the **hertz,** named for Heinrich Rudolph Hertz, a German physicist of the late nineteenth century who made a number of important discoveries and valuable contributions to electrical science. One hertz (Hz) is equal to 1 cycle per second [1 c/s.]. One kilohertz is equal to 1000 cycles per second [1000 c/s]. City lighting and power systems in the United States generally operate at a frequency of 60 Hz. Alternating currents in airplane circuits usually have a frequency of 400 Hz. This frequency is commonly used for modern jet aircraft as well as for a number of other applications. The word *alternation* is frequently used in discussing alternating current, and it means one half-cycle. It is apparent, therefore, that there are 120 alternations in a 60-Hz current.

The frequency of an alternating current has considerable effect on the operation of a circuit, for many units of electric equipment operate only on current of a certain frequency. Wherever such equipment is used, it is important to make sure it is designed for the frequency of the current in the circuit in which it is to be used. Units such as synchronous motors operate at speeds proportional to the frequency of the current even though the voltage is somewhat lower or higher than the rated voltage of the machine. It is also important to remember that a circuit designed for a given frequency may be easily overloaded by using a current of a different frequency, even though the voltage may remain the same. This is because of effects of inductive and capacitive reactance which will be explained later in this section.

PHASE

The phase of an alternating current or a voltage is the angular distance it has moved from 0° in a positive direction. The **phase angle** is the difference in degrees of rotation between two alternating currents or voltages, or between a voltage and a current. For example, when one voltage reaches maximum value 25° later than another, there is a phase angle of 25° between the voltages. Also, if the current in an ac system reaches maximum 20° after the voltage reaches maximum, there is a phase angle of 20° between the voltage and the current. The phase angle in electrical equations is usually represented by the Greek letter theta (θ).

It is very common in ac systems for the current to lag or lead the voltage. This is caused by inductance or capacitance in the circuit and will be explained as we proceed.

Figure 4-5 shows sine curves representing a current lagging the voltage and a current leading the voltage. In circuits where the currents and voltage do not reach maximum at the same time, they are said to be out of phase. In the top diagram of Fig. 4-5, notice that the heavy current line crosses the zero line after the light voltage line has crossed it. This means that the current reaches zero after the voltage. In like manner, the peak value of current occurs after the peak value of voltage. For this reason, we know that the current is lagging the voltage by several degrees. In Fig. 4-5*b*, it will be seen that the voltage is approximately 90° out of phase with the current. That is, the voltage follows the current by approximately 90°.

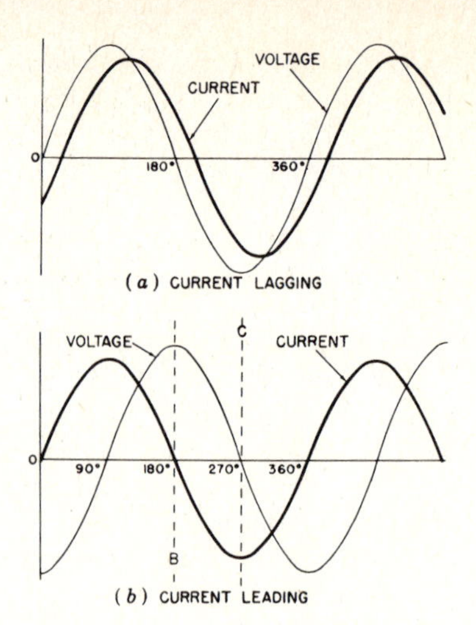

Fig. 4-5 Voltage and current out of phase.

CAPACITANCE IN AC CIRCUITS

When a capacitor is connected in series in an ac circuit, it appears that the alternating current is passing through the capacitor. In reality electrons are stored first on one side of the capacitor and then on the other, thus permitting the alternating current to flow back and forth in the circuit without actually passing through the capacitor.

A hydraulic analogy may be used to explain the operation of a capacitor in a circuit (see Fig. 4-6). The capacitor is represented by a chamber separated into two sections by an elastic diaphragm. The ac generator is represented by the piston-type pump. As the piston moves in one direction, it forces fluid into one section of the chamber and draws it out of the other section. The fluid flow represents the flow of electrons in an electric circuit. Thus it can be seen that there is an alternating flow of fluid in the lines and that work is done as the fluid moves back and forth, first filling one side of the chamber and then the other.

The operation of a capacitor in an ac circuit is for all practical purposes identical to the operation of the chamber just described. The electrons build up on one plate of the capacitor and flow out of the other plate. This

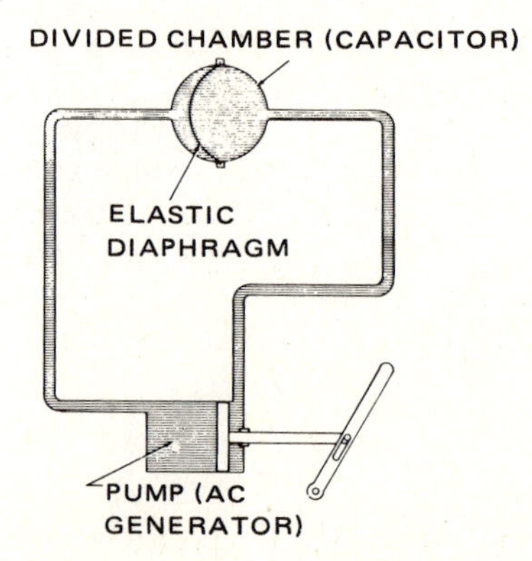

Fig. 4-6 Hydraulic analogy of an ac circuit with a capacitor in series.

establishes an electrostatic field which creates a dielectric stress causing the current to flow in the opposite direction. As soon as the voltage from the source begins to drop, the current starts to flow out of one plate of the capacitor and into the other.

The effect of a capacitor in an ac circuit is to cause the current to lead the voltage. If it were possible to have a circuit without any resistance, the current would lead the voltage by 90°. It is easy to understand this by referring to the current and voltage curves in Fig. 4-5*b* and considering the movement of the electrons through one alternation. As the voltage rises, the current begins to drop because of the dielectric stress in the capacitor. This, of course, means that opposition to the flow of current is developing. By the time the voltage has reached maximum, the capacitor is completely charged, and hence no current can flow. At this point (*B*) the current has a value of zero. As the voltage begins to drop, the current flows out of the capacitor in the opposite direction because the potential of the capacitor is higher than the potential on the line. By the time the voltage has dropped to zero, the current is flowing at a maximum rate because there is no opposition. This point on the curve is represented by the letter *C*.

It must be remembered that the above action takes place only when there is no resistance in the circuit. Since this is impossible, a circuit in which the current leads the voltage by as much as 90° does not exist. However, the study of such a circuit gives the student a clear understanding of the effect of capacitance.

The effects of capacitance in ac circuits are most pronounced at higher frequencies. Modern electronic circuits often produce frequencies of many millions of cycles per second (Hz). For this reason, special types of electronic and electric devices and equipment have been designed to reduce the effects of capacitance where these effects are detrimental to the operation of the circuit. The study of capacitive reactance in the next section will help to explain why capacitance may be troublesome at high frequencies.

CAPACITIVE REACTANCE

The effect of capacitance in an ac circuit is similar in some respects to that of resistance; it is called **capacitive reactance.** Since it opposes the flow of current in the circuit, it is measured in ohms.

The capacitive reactance in a circuit is inversely proportional to the capacitance and the ac frequency. This is because a large-capacity capacitor will take a greater charge than a low-capacity capacitor; hence it will allow more current to flow in the circuit. If the frequency increases, the capacitor charges and discharges more times per second; hence more current flows in the circuit. From this it can be seen why the reactance will decrease as capacitance or frequency increases.

The formula for capacitive reactance is

$$X_C = \frac{1}{2\pi fC}$$

where X_C = capacitive reactance, Ω
f = frequency, Hz
C = capacitance, F

To determine the capacitive reactance in a circuit in which the frequency is 60 Hz and the capacitance 100 μf, substitute the known values in the formula. Then

$$X_C = \frac{1}{2\pi \times 60 \times 100/1\,000\,000}$$

Remember that 1 μf is one-millionth of a farad; hence 100 μf is equal to 100/1 000 000 F. Therefore,

$$X_C = \frac{1}{6.283 \times 0.006} = \frac{1}{0.037\,698} = 26.5\ \Omega$$

INDUCTANCE IN AC CIRCUITS

The effect of inductance in ac circuits is exactly opposite to that of capacitance. Capacitance causes the current to lead the voltage, and inductance causes the current to lag. Figure 4-7 shows the voltage and current curves for a purely inductive circuit. According to Lenz's law, whenever a current change takes place in an inductance coil, an emf is induced which opposes the change in current. The emf induced will then be maximum when the rate of current change is the greatest. Since the current change is most rapid in an ac circuit when it is passing through the zero point, the induced voltage will be maximum at this same point, marked *A* in the Fig. 4-7. When the current reaches maximum, there is momentarily no current change, and hence the emf is zero at this point (*B*). Remember that to induce a voltage there must be a current change causing a change in magnetic flux. Hence, at point *B*, where there is no current change, there will be no voltage. Thus we find that the current lags the voltage by 90° in a purely inductive circuit. But since a purely inductive circuit is impossible because there is always resistance present, the current will never be as much as 90° behind the voltage.

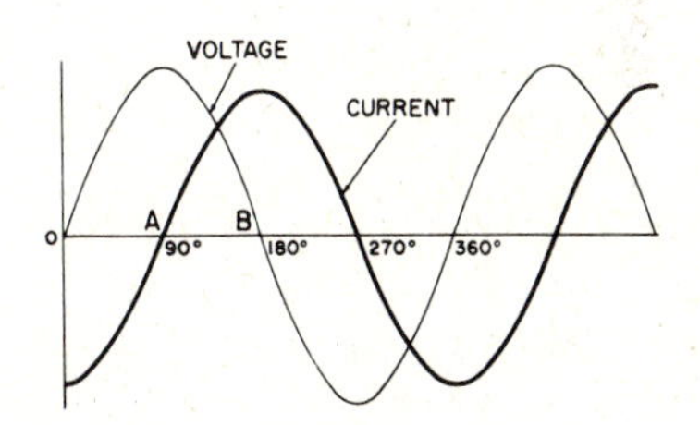

Fig. 4-7 Current lagging the voltage.

INDUCTIVE REACTANCE

The effect of inductance in an ac circuit is called **inductive reactance** and is measured in ohms because it impedes the flow of current in the circuit.

The inductive reactance in a circuit is proportional to the inductance of the circuit and the frequency of the alternating current. As the inductance is increased, the induced emf which opposes the applied emf is increased; hence, the current flow is reduced. Likewise, when the frequency of the current in the circuit is increased, the rate of current change in the inductance coil is also increased; hence, the induced opposing emf is higher and the current flow is again reduced.

Now we can clearly see that the effects of capacitance and inductance are opposite, since inductive reactance increases as the frequency increases and capacitive reactance decreases as the frequency increases. The formula for inductive reactance is

$$X_L = 2\pi f L$$

where X_L = inductive reactance, Ω
f = frequency, Hz
L = inductance, H

Let us assume that an inductance coil of 7 H is connected in a 60-Hz circuit and it is necessary to find the inductive reactance. By substituting the known values in the formula,

$$X_L = 2 \times 3.1416 \times 60 \times 7 = 2638.94\ \Omega$$

IMPEDANCE

In the study of Ohm's law for dc circuits, it was found that the current in a circuit was equal to the voltage divided by the resistance. In ac circuits it is necessary to consider capacitive reactance and inductive reactance before the net current in such a circuit can be determined. The combination of resistance, capacitive reactance, and inductive reactance is called **impedance,** and the formula symbol is Z.

It might appear that we could add the capacitive reactance, inductive reactance, and resistance to find the impedance, but this is not true. Remember that capacitive reactance and inductive reactance have opposite effects in an ac circuit. For this reason, to find the total reactance we use the difference in the reactances. If we consider inductive reactance as positive, because inductance causes the voltage to lead the current, and capacitive reactance as negative, because it causes the voltage to lag, then we can add the two algebraically; that is,

$$X_L + (-X_C) = X_t \qquad \text{or total reactance}$$

Now it might appear that we could add this result to the resistance to find the impedance, but again we must consider the effect of resistance in the circuit. We know that resistance in a circuit does not cause the current to lead or lag, and for this reason its effect is 90° ahead of inductance and 90° behind capacitance. Therefore, it is necessary to add resistance and reactance vectorially.

A **vector** is a line which represents a certain force exerted in a given direction. The length of the line represents the value of the force. Figure 4-8 shows how reactance and resistance are added vectorially to find the impedance in a circuit. Starting at the point A in the Fig. 4-8*a*, the line **AD** represents the resistance. **AB,** which is at right angles to **AD,** represents the inductive reactance. Now we form a parallelogram by drawing the line **BC** parallel to **AD** and the line **CD** parallel to **AB.** The line **AC** then represents the impedance. If **AB** (X_L) is 3 Ω and **AD** (R) is 5 Ω, then **AC** (Z) is 5.83 Ω as found by measuring.

Since **AC** is the hypotenuse of a right triangle, we can use the formula for a right triangle to find the value of **AC** when the other values are known. Then

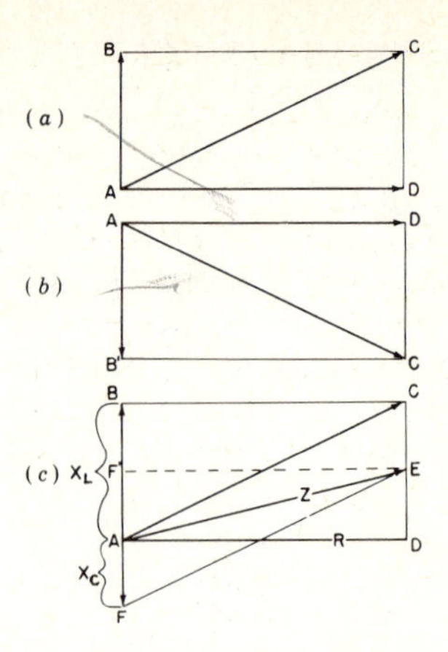

Fig. 4-8 Vector diagrams for combining resistance and reactance.

$$(\mathbf{AC})^2 = (\mathbf{AB})^2 + (\mathbf{AD})^2$$

or

$$Z^2 = X_L^2 + R^2$$

By applying the formula to the above problem it will be found that the answer stated is correct. To add capacitive reactance and resistance, vectorially, a triangle is formed as shown in Fig. 4-8*b*. The results obtained with this triangle will be the same as for inductive reactance, and the right-triangle formula may be used in either case.

When a circuit contains both capacitive and inductive reactances, the impedance may be found by combining the vectors as shown in Fig. 4-8*c*. First the line **AC** is drawn to show the vector sum of X_L and R. **AC** is then combined vectorially with **AF** (X_C). This forms the parallelogram *ACEF* with the diagonal **AE,** which represents the impedance. Another simple method for finding the impedance in this problem is to lay off **BF′** on **AB,** making **BF′** equal to **AF.** This makes the side **AF′** equal to the difference between X_L and X_C. Then draw **EF′** parallel to **AD** and draw **ED** parallel to **AF′**. The diagonal **AE** then represents the value of the impedance.

It will be noted in the foregoing problems that the resultant of the resistance and reactance vectors is the hypotenuse of a right triangle. The resistance vector is the base of the triangle, and the reactance vector is the vertical side. The triangle formed in this manner is called the **impedance triangle** and may be used to solve problems when it is necessary to combine reactance and resistance.

If we wish to use an algebraic formula in the above problem, the formula for a right triangle may be used after combining the capacitive reactance and the inductive reactance. This may be shown as follows:

$$Z^2 = (X_C - X_L)^2 + R^2$$

Of course it must be remembered that the smaller reactance must always be subtracted from the larger reactance to obtain the total reactance.

After the impedance is found in an ac circuit, the other values may be found by Ohm's law for alternating current. In this formula we merely substitute the symbol Z, meaning impedance, for the normal symbol R, meaning resistance. The formula then reads $I = E/Z$.

RESONANT CIRCUITS

Since capacitance and inductance have opposite effects in an ac circuit, the effect of one may be used to cancel the effect of the other. It will be remembered that inductive reactance increases in proportion to the frequency and that capacitive reactance is inversely proportional to the frequency. For this reason, a frequency may be found which will balance any capacitance with any inductance; that is, the capacitive reactance of any capacitor may be equal to the inductive reactance of any inductance coil at the proper frequency.

Consider a 5-μf capacitor and a 10-H inductance coil. What is the frequency at which the effect of one will cancel the effect of the other?

Since X_C must equal X_L, then

$$2\pi fL = \frac{1}{2\pi fC}$$

By transposing and simplifying we find that this formula reduces to

$$f = \frac{1}{2\pi\sqrt{LC}}$$

Substituting the values in the problem under consideration in this formula, the formula becomes

$$f = \frac{1}{6.28 \times \sqrt{10 \times 0.000\,005}}$$

or

$$f = 22.52 \text{ Hz}$$

This shows that the frequency at which a 5-μf capacitor and a 10-H inductance coil are equal in reactance is approximately 22.5 Hz. This is known as the resonant frequency for these two values and is expressed by the following rule: **When the capacitive reactance in a circuit is equal to the inductive reactance, the circuit is said to be resonant.**

It is obvious from the above problem that resonance can occur only at one frequency when the inductance and capacitance are fixed. It can also be seen that a circuit may be designed which will operate only through a very narrow range of frequencies. This is one of the basic principles of radio tuning. In radio, variable capacitors are used to change the resonant frequency of the circuit in order to select stations of different frequencies.

Figure 4-9 is a diagram of the resonant circuit discussed in the foregoing problem. Maximum current will flow in the circuit at a frequency of approximately 22.5 Hz; but the current will drop rapidly if the frequency is increased or decreased. It will be noted that the capacitor and inductance coil in the foregoing problem are connected in series. When connected in this manner, a circuit will pass maximum current at the resonant frequency.

Figure 4-10 shows a circuit diagram in which the capacitor and inductance coil are connected in parallel. A combination of this type is often called a *tank* circuit. The effect of a parallel resonant circuit is opposite to that of a series circuit. That is, the impedance furnished by the parallel combination is greatest at the resonant frequency. In the circuit diagram of Fig. 4-10, current will flow between points A and B when the frequency is not the same as the resonant frequency of the parallel circuit. Current of the resonant frequency is *trapped* in the parallel circuit.

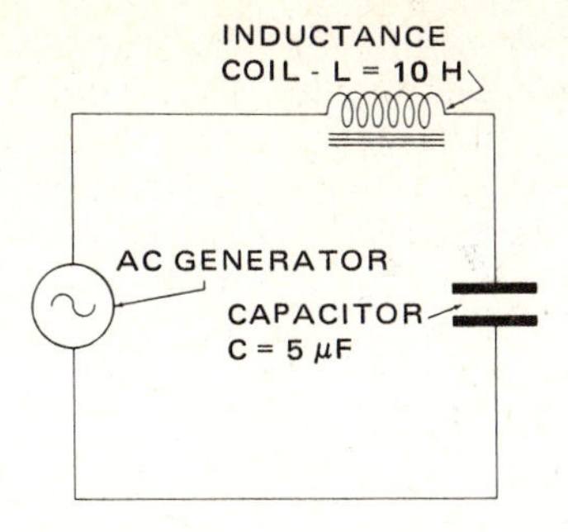

Fig. 4-9 Series resonant circuit.

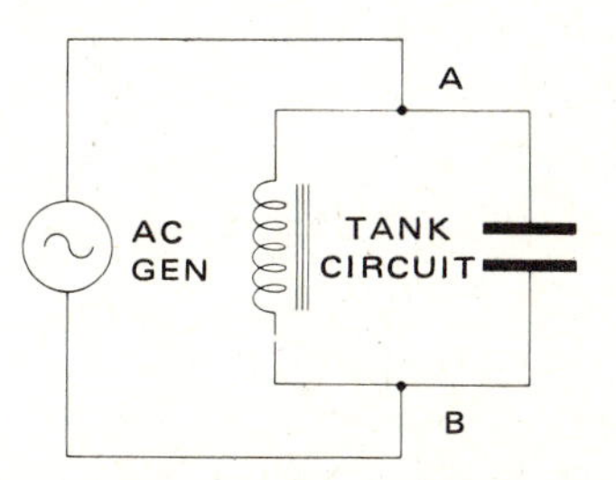

Fig. 4-10 Parallel resonant circuit.

The action of a parallel-resonant, or tank, circuit may be explained briefly as follows: In the circuit of Fig. 4-10, if current flows in from the generator at point A, it will immediately go to the capacitor, because the voltage of the capacitor is low and the inductive reactance of the inductance coil opposes flow through the coil. Therefore, during the time that current is flowing in at A, it first charges the capacitor. This, in effect, builds up a back pressure, which then opposes the current flow-in at A. While the capacitor is charging, a small amount of current flow begins building up through the inductance coil. Then, as voltage from the generator falls off at A, the inductance coil, because of inductance, will continue to direct the flow from A to B and to the opposite side of the capacitor. The capacitor will be charging at the same time that the electron flow increases at point B. The result is that there will be a back pressure at B which tends to stop the flow of electrons from the generator into the tank circuit. This cycle continues first in one direction and then the other, with the electric energy being stored first by the capacitor and then by the magnetic field of the inductance coil. At resonant frequency, the back pressure at A and B in the tank circuit is almost equivalent to the applied voltage and in time with the applied voltage; it therefore holds the current flow back.

If there were no resistance in the tank circuit, no current would flow through the tank at the resonant frequency because the back pressure would be equal to the applied pressure; but there would be relatively high current flowing back and forth through the inductance coil and the capacitor. In practice, since no circuit can be completely free of resistance, a very small current will flow in the circuit outside the tank at the resonant frequency.

This current flow will tend to keep the tank currents at maximum levels.

Parallel-resonant, or tank, circuits are used in radio to block unwanted frequencies and to provide high impedance at certain frequencies. These circuits are frequently used for tuning, and in such cases the capacitor in the tank circuit is variable so that the resonant frequency of the circuit may be changed.

RESISTANCE IN AC CIRCUITS

Resistance in an ac circuit has definite effects when connected with capacitance or inductance. It has been explained that in a purely capacitive or inductive circuit, current either leads or lags a voltage by 90°. If resistance is connected in series with such circuits, it causes the angle of lead or lag to decrease. The resistance decreases the flow of current, and the capacitor cannot become fully charged by the time the voltage reaches maximum. Therefore, when the voltage is maximum in a positive direction, a current continues to flow in a positive direction to charge the capacitor. This current flows for a time after the voltage begins to drop, but of course it cannot continue until the voltage drops to zero.

In a purely capacitive circuit, the current drops to zero as soon as the voltage reaches maximum, because the capacitor is fully charged at this time; but with resistance in the circuit, the current continues to flow until the charge in the capacitor (dielectric stress) is equal to the voltage. At this time, the current is zero. As the voltage continues to drop, the current begins to flow in the opposite direction, until the capacitor charge balances the applied voltage. This action is illustrated by the curves of Fig. 4-11.

A vector diagram (see Fig. 4-12) may be used to determine by how much the current leads the voltage. The horizontal line *AD* represents the resistance in the circuit, and the vertical line *AB* represents the capacitive reactance. By drawing a parallelogram *ABCD* and then drawing the diagonal *AC*, we establish the line representing the impedance. The angle between lines *AD* and *AC* is then the angle by which the current leads the voltage. If the resistance is 4 Ω and the capacitive reactance is 3 Ω, then $Z = \sqrt{4^2 + 3^2} = 5\ \Omega$. By measuring the angle *CAD* with a protractor, it is found to be approximately 37°. This angle is the phase angle between the current and the voltage and is represented by the Greek letter theta (θ).

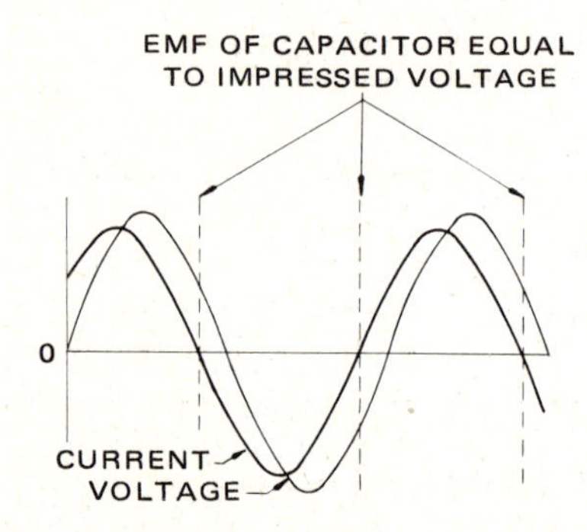

Fig. 4-11 Effect of resistance in a capacitive circuit.

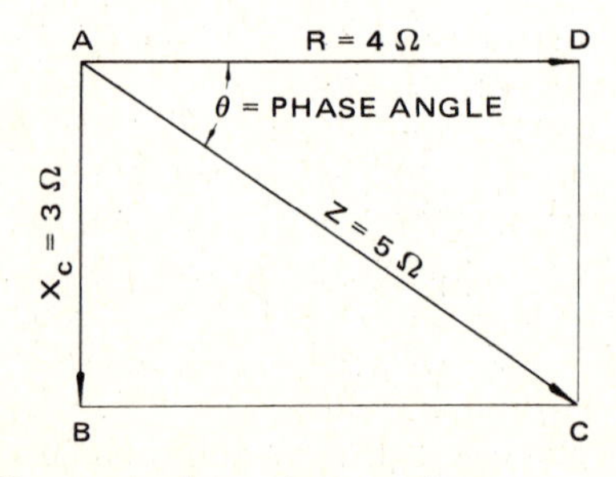

Fig. 4-12 Diagram to show phase angle.

When resistance is connected in series within an inductance, current is held back, and this reduces the opposing emf of the inductance coil. Remember that according to Lenz's law the strength of the opposing emf of an inductance coil is proportional to the rate of current change. Therefore, when the rate of current change is decreased because of the resistance in the circuit, there is a smaller opposing emf and the current may follow the voltage more closely than it would if there were no resistance. This results in a smaller angle of current lag.

The phase angle for an inductive circuit may be found in the same manner as it is for a capacitive circuit. In an inductive circuit, the angle represents current lag; in a capacitive circuit, the angle represents current leading the voltage. When the circuit contains both inductance and capacitance, the reactance which has the greater value determines whether the current lags or leads the voltage.

POWER IN AC CIRCUITS

In a dc circuit the power is equal to the product of the current and the voltage. This is true in an ac circuit only when the current and voltage are in phase. The curves in Fig. 4-13 may be used to explain this fact. Since power is developed whether the current is moving in one direction or the other, the power curves are positive. This may be proved algebraically because the product of like numbers is always positive. Therefore, when a negative current value is multiplied by a negative voltage value, the product is positive. Now it can be seen that the power taken at any instant is equal to the product of the voltage at that instant and the current at that instant. This is true whether the current and voltage are in phase or not.

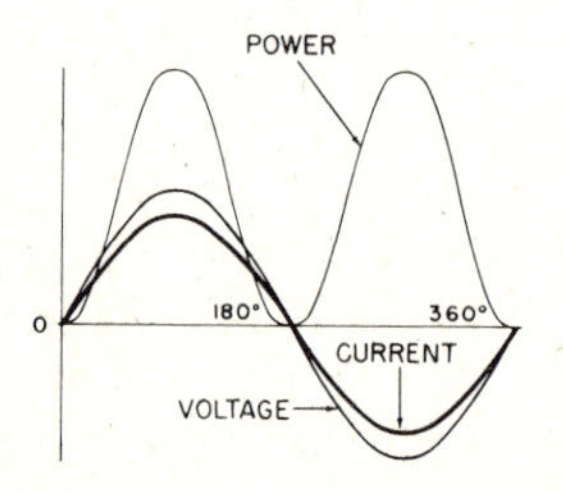

Fig. 4-13 Power curve when voltage and current are in phase.

Figure 4-14 shows the power curves for a circuit in which the current and voltage are out of phase. At the point where the current and voltage have opposite signs, the power is negative, as represented by the portion of the power curve below the zero line. The net power in such a circuit is not equal to the product of the effective voltage and the effective amperage, because a part of the power has a negative value.

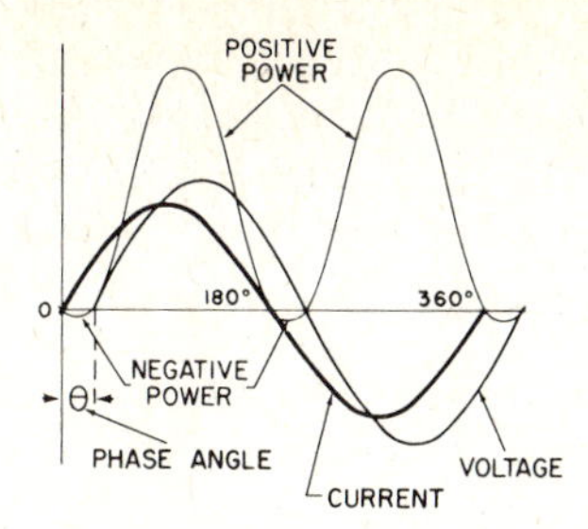

Fig. 4-14 Power curve when voltage and current are out of phase.

Power is measured in watts, and in a circuit where the voltage and current are in phase, it is equal to the product of the current and the voltage. When the current and the voltage are out of phase, the product of the voltage and amperage is designated as volt-amperes. To find the power in watts, it is necessary to multiply the volt-amperes by the cosine of the angle by which the current and voltage are out of phase, that is, the phase angle θ.

The cosine of an angle is the ratio of the side of a right triangle adjacent to the angle to the hypotenuse of the triangle. In Fig. 4-15 AC/AB is the cosine of angle A. The value of the cosine of any angle may be found in a table of trigonometric functions.

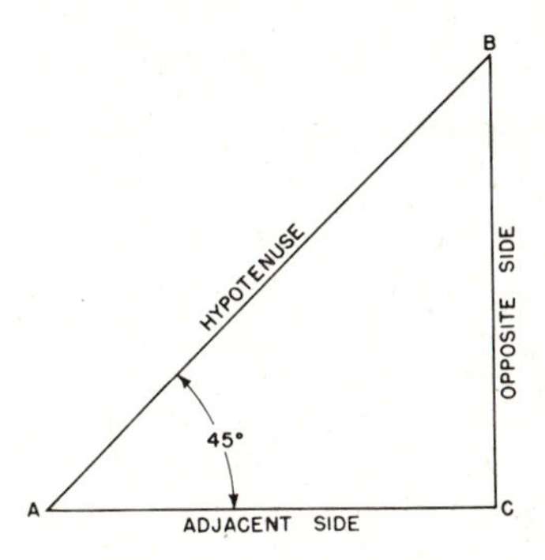

Fig. 4-15 Right triangle for demonstrating the cosine of an angle.

Since the cosine of the phase angle θ is used to determine the power of an ac circuit when the voltage and current are out of phase, it is called the **power factor** of the circuit. When the voltage and current are in phase, the value of the power factor is 1, or unity. Otherwise, the value of the power factor is always less than 1.

Power factor is very important in ac motor circuits, and in many cases, it is necessary to insert resistance or capacitance in a circuit to improve it. Induction motors contribute a great amount of inductance to a circuit, thereby decreasing the power factor. For this reason, high-resistance windings are frequently used in such motors to improve the power factor.

TRANSFORMERS

One of the chief advantages of alternating current is that it can be transmitted at a high voltage with a low power loss; the voltage can then be reduced to any desired value by means of transformers. We therefore find transformers frequently used in ac systems.

A schematic diagram of a transformer is shown in Fig. 4-16. It has been previously explained in the section on

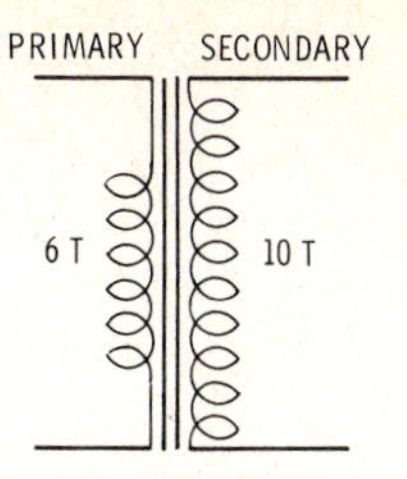

Fig. 4-16 Schematic diagram of a transformer.

electromagnetic induction that every conductor of an electric current has a magnetic field. If alternating current is flowing in a conductor, the magnetic field around the conductor expands and collapses rapidly as the current changes in magnitude and direction. This makes it possible to change the voltage of an alternating current by the means of mutual induction coil (transformer) without the aid of a circuit-breaking device.

A transformer consists of a primary winding and a secondary winding on either a laminated soft-iron or an annealed sheet-steel core. The secondary coil may be wound on the primary coil or on a separate section of the same core. This is illustrated in Fig. 4-17. The laminated core reduces the effect of **eddy currents** which otherwise would cause considerable heat and a loss of power.

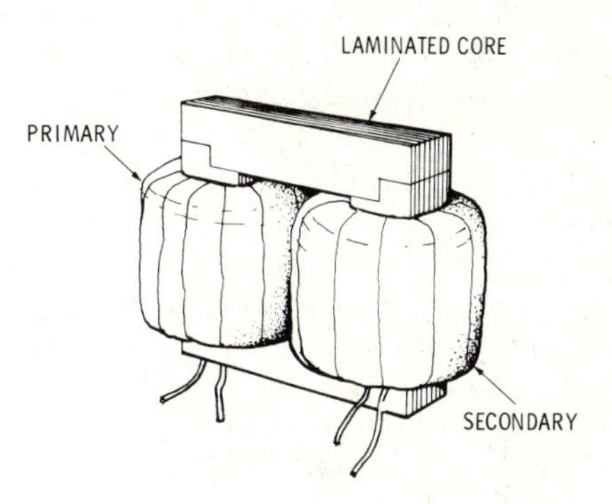

Fig. 4-17 Drawing of a transformer.

The theory of transformer operation is the same as that of an induction (booster) coil. The expanding field of the primary coil induces a voltage in the secondary coil, and if the terminals of the secondary windings are connected together or to a load, the voltage produces a current which sets up an opposing field in the secondary coil. As the current in the primary coil begins to decrease, the current in the secondary reverses and sets up a field which is in the same direction as the primary field. This conforms to Lenz's law because it opposes the collapse of the primary field. Remember that an induced emf (voltage) is always proportional to the rate of change in the magnetic field and that it opposes the change.

The emf induced in the secondary coil of a transformer is out of phase with the emf in the primary by nearly 180°. This is because the primary current is nearly 90° out of phase with the primary emf, owing to the inductance of the primary winding, and the emf of the secondary coil is 90° out of phase with the primary. In theory, the secondary emf of a circuit with no resistance would be exactly 180° out of phase with the primary, but since no circuit can be free of resistance, the two voltages cannot be 180°

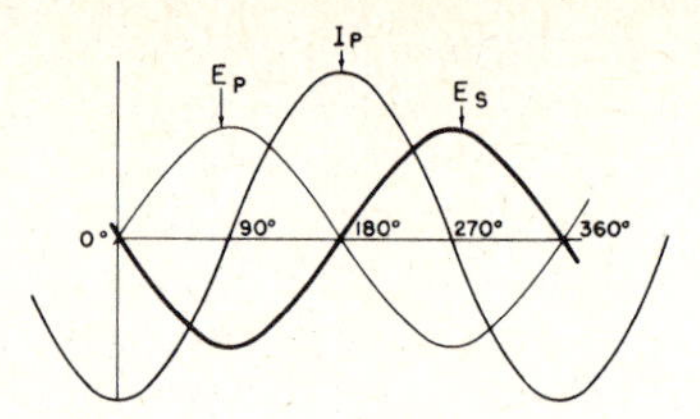

Fig. 4-18 Voltage and current in the coils of a transformer.

out of phase but will be somewhat less than 180°, depending on the resistance of the circuit.

A study of Fig. 4-18 will help the student to understand the phase relations in a transformer circuit. The curve E_p represents the emf applied to the primary coil of the transformer. I_p is the current in the primary, which lags behind the primary emf by almost 90° because of the inductance of the primary winding. Since the current change is greatest as it reverses direction, a maximum emf (E_S) is induced in the secondary at this point. When the current reaches a maximum value at 180° on the curve, there is an instant when there is no current change; hence at this point there is no induced emf in the secondary. As the current value decreases, the rate of change increases, and the secondary emf increases to oppose this change.

One of the most important features of a transformer is that the primary coil may be left connected to the line and will consume very little power unless the secondary circuit is closed. This is because of the inductive reactance of the primary winding. The primary current sets up a field which induces an opposing emf in the primary coil. This opposing emf is called **counter emf** and is almost equal to the emf applied to the coil; hence, only a very small current will flow in the coil.

We may consider that the field is a reservoir of power and that when the secondary circuit is closed, power is being drawn from the reservoir. Then current will flow in the primary circuit sufficient to maintain the field flux at a maximum value. If the secondary circuit is disconnected, no more power will be drawn from the field; hence, very little current will be necessary to maintain the field strength. From this we can see that the strength of the field remains almost constant as long as the load does not exceed the ultimate capacity of the transformer.

When the primary and secondary coils of a transformer are wound upon the same core, they are both affected by the same magnetic field. It will be remembered that the emf induced in a coil depends upon the lines of force being cut per second. Since both the primary and secondary coils are being cut by the same magnetic field, the ratio of the primary emf to the secondary emf is proportional to the ratio of the number of turns of wire in the primary to the number of turns in the secondary. For example, if the primary coil has 100 turns of wire and the secondary has 200 turns, then the emf of the secondary will have twice the value of the emf in the primary. The formula for these values is

$$\frac{E_P}{E_S} = \frac{N_P}{N_S}$$

where E_P = voltage in primary
E_S = voltage induced in secondary
N_P = number of turns in primary winding
N_S = number of turns in secondary winding

It is obvious that the output of a transformer cannot be greater than the power input. Since the power in a transformer is approximately equal to the voltage times the amperage, we can see that if the voltage in the secondary is higher than the voltage in the primary, then the amperage in the secondary must be lower than the amperage in the primary. In a transformer that is 100 percent efficient, the ratio of the amperage in the primary to the amperage in the secondary is inversely proportional to the ratio of the voltages. The formula for this relationship is

$$\frac{E_P}{E_S} = \frac{I_S}{I_P} \quad \text{or} \quad E_P I_P = E_S I_S$$

When the secondary of a transformer has more turns of wire than the primary and is used to increase voltage, it is called a **step-up** transformer. When the transformer is used to reduce the voltage, it is called a **step-down** transformer. In many cases, the same transformer may be used as either a step-up or a step-down transformer. The coil connected to the input voltage is called the primary, and the coil connected to the load is called the secondary. When a transformer is in use, the voltage capacity of its primary winding, which may usually be ascertained from the name or data plate, must not be exceeded.

If it becomes necessary to use more than one transformer in a circuit, with the transformers connected either in series or in parallel, it is most important that they be properly *phased.* Figure 4-19 illustrates a simplified circuit for two transformers connected in series. Note that the primary terminals P_1 of the first transformer and P_1 of the second transformer are connected to the same line of the power supply and that the P_2 terminals of the transformers are likewise connected to the same line of the power supply. With the primary circuits connected in this manner, the secondary terminals S_1 will be positive at the same time and negative at the same time. Therefore, to connect the two secondary circuits in series to obtain maximum voltage, S_2 of one transformer should be connected to S_1 of the other transformer and the opposite terminals S_1 and S_2 then used as output terminals. With this arrangement the voltages are additive and the total output will be 220 V if the individual secondary windings produce 110 V each. If the two secondary windings in this series were connected so that S_2 of one transformer were connected to S_2 of the other, then there would be no output from the two S_1 terminals, because the voltages would be working in opposite directions.

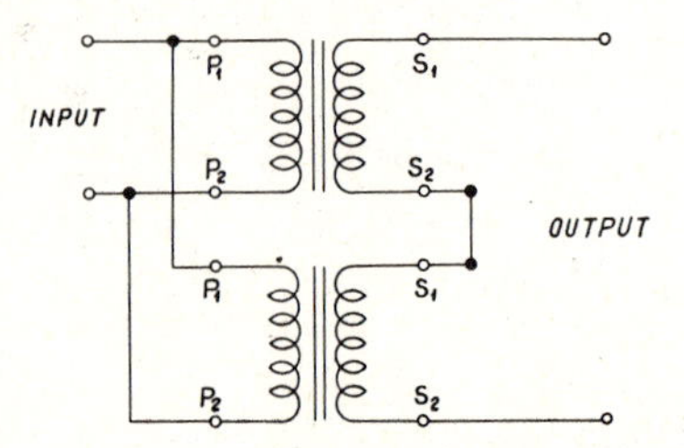

Fig. 4-19 Transformers connected properly in series.

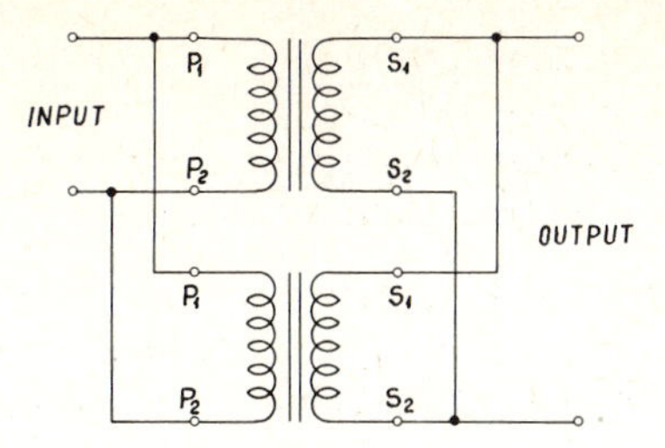

Fig. 4-20 Transformers connected in parallel.

In Fig. 4-20, transformers are shown connected in parallel. The primary windings are connected in the same manner as those in the circuit of Fig. 4-19. To connect the secondary windings in parallel, the two terminals S_1 and the two terminals S_2 are connected to the same line. The output between these lines will then have the same voltage as each individual winding. If the connection for one of the secondary windings is reversed, a short circuit will be created between the two secondary windings, and the transformers will be burned out, or the circuit breaker in the power supply will be opened.

POLYPHASE AC CIRCUITS

A polyphase ac circuit consists of two or more circuits which are usually interconnected and so energized that the currents through the separate conductors and the voltages between them have exactly equal periods but differ in phase. A difference in phase means that the voltages or currents do not reach peak positive or peak negative values at the same time, but the corresponding values of current or voltage are usually separated by an equal number of degrees. For example, in a three-phase ac system, No. 1 phase will reach a peak voltage in a positive direction 120° before the No. 2 phase, the No. 2 phase will reach the maximum positive voltage 120° before No. 3 phase, and so on. Thus the three phases are separated by an angle of 120°.

Modern jet aircraft of all types employ three-phase power systems because of the efficient transmission of power in such systems. Because of the great electric power requirements on large aircraft, a dc power system would add hundreds of pounds of weight in comparison with a three-phase ac system.

Figure 4-21 shows the schematic diagram of a delta-connected alternator stator. This alternator, which also may be called an ac generator, supplies three separate voltages spaced 120° apart. It is called a **delta-connected** alternator because the diagram is in the form of the Greek letter delta (Δ). With a delta connection, the voltage between any two terminals of an alternator is equal to

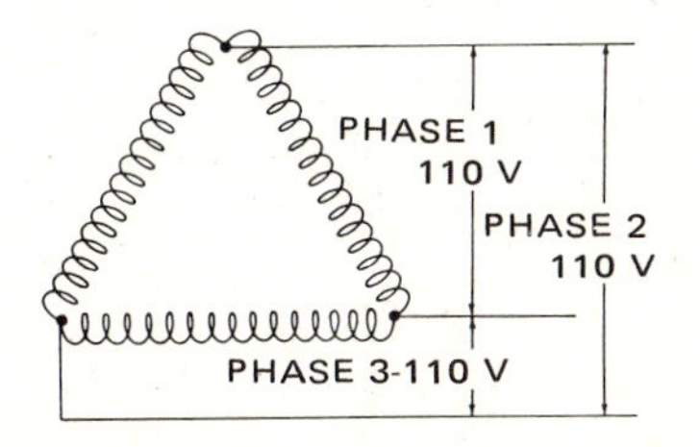

Fig. 4-21 Schematic diagram of a delta-wound alternator stator.

the voltage across one phase winding. The current through any terminal is equal to the vectorial sum of the current flowing in two of the phase windings. The vectorial sum of the currents in two of the phase windings is equal to 1.73 times the current flowing in one of the windings. It has been explained previously that a vector represents a certain force in a given direction. When two such forces are added, it is necessary to consider their direction as well as their magnitude.

Another method for connecting the phase windings of a three-phase system is illustrated in Fig. 4-22. This is known as a **Y connection.** An alternator of this type may have three or four terminals. When there are three terminals, the voltages between any two of the terminals are equal but 120° apart in phase. To operate single-phase equipment, any two of the terminals are used. When the alternator has four terminals, the fourth is common to all windings and is called the **neutral wire.** This makes it

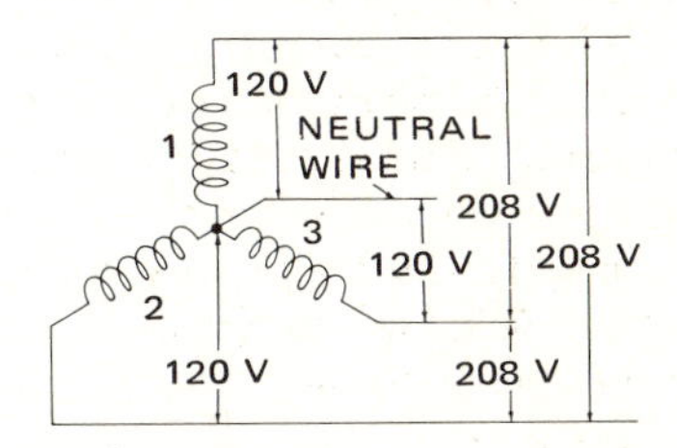

Fig. 4-22 Schematic diagram of a Y-wound alternator stator.

possible to obtain two different voltages from one machine. In the ac power system of a modern jet airliner, the neutral wire is grounded and the three phase connections, which may be *A*, *B*, and *C* or 1, 2, and 3, are connected to the power system of the airplane. In all cases, the separate phase terminals must be properly identified.

The voltage between any two of the three phase windings of the Y system is equal to the vectorial sum of the voltages of two of the phase windings. For example, if the emf across one winding is 120 V, the emf between two of the three phase terminals is 1.73 times 120, or 208 V. An arrangement of this kind is convenient because the 120-V circuit may be used for operating lights and other small loads and the three-phase 208 V circuit used to operate larger power equipment. On an airplane in which the neutral wire is grounded, a single-wire system may be used for all single-phase 120-V circuits. The 208-V three-phase power may be directed to a three-phase motor or other device requiring this type of power. Where it is necessary to obtain direct current for certain power needs, the three-phase current can be directed through a three-phase, full-wave rectifier. Rectifiers are explained in the next section of this chapter.

Single-phase equipment and such items as lights are operated by connecting one terminal of the unit to one of the three-phase conductors and the other terminal to the metal structure of the airplane (ground). Thus, the single-phase circuits make use of the voltage from one phase winding, and three-phase equipment is connected to the three separate phases. The single-phase voltage equals 120 V from ground to one of the phase terminals, single-phase 208 V from one phase terminal to another phase terminal, and 208 V for the three-phase system.

RECTIFIERS

Although alternating current performs many functions just as well as direct current, some systems, such as those for battery charging and various segments of electronic systems, require direct current. Direct current is obtained from alternating-current sources by means of **rectifiers.** A rectifier is a device which allows current to flow in one direction, but will oppose or stop the flow in the opposite direction. A rectifier may be compared to a check valve in a hydraulic system.

A variety of rectifiers or rectifying devices and systems have been invented and used, and many of these are still being used. A number of types of rectifiers are falling into disuse because **solid-state** technology has developed to a point where **solid-state rectifiers** are found to be most dependable and efficient for a wide range of dc power requirements. The term *solid-state* refers to devices in which a solid material is used to control electric currents through the manipulation of electrons within the material.

In this section we shall describe a few of the types of rectifiers which have been used and may still be used for some applications.

COPPER OXIDE RECTIFIER

When a plate of copper is coated with a layer of copper oxide, electrons will flow easily from the copper to the oxide, but they will not flow easily from the oxide to the copper. If one terminal of the alternator is connected to the copper and one to the copper oxide, as shown in Fig. 4-23, a current will flow during one half-cycle but not during the other half-cycle. This results in an intermittent direct current in the circuit, as shown in Fig. 4-24. When a single rectifier unit is placed in series in an ac circuit, the result is called **half-wave** rectification because only one-half the available current can be used in the dc circuit.

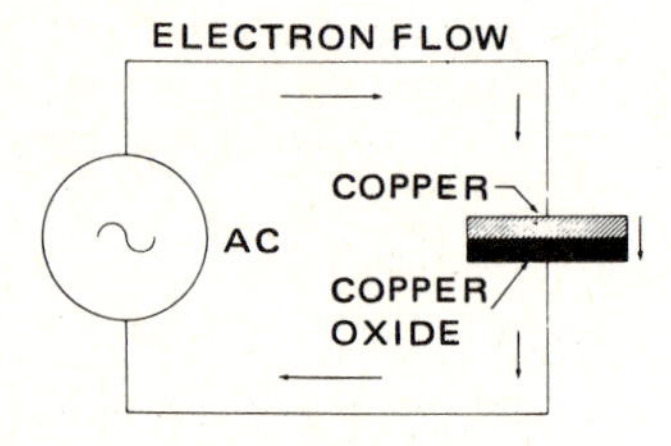

Fig. 4-23 Half-wave copper-oxide rectifier.

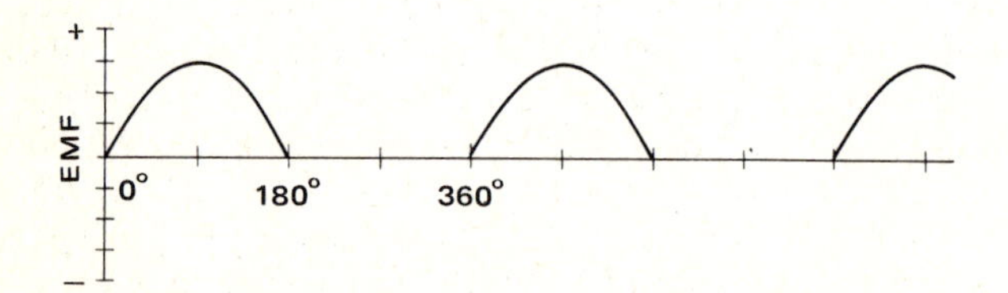

Fig. 4-24 Curve for half-wave rectified current.

FULL-WAVE RECTIFIER

In order to utilize all the available power from an ac supply, a group of rectifiers is connected in a bridge circuit as shown in Fig. 4-25. The points *A* and *D* are connected to the ac circuit. In the half-cycle during which point *A* is negative, electrons flow from *A* to *B*, through the battery to *C*, and out of the circuit at *D*. During the other half-cycle, electrons flow from *D* to *B*, through the battery to *C*, and out of the circuit through *A*. Since the full cycle of the alternating current is thus used, this system is called a **full-wave** rectifier. A graph of the output from a full-wave rectifier is shown in Fig. 4-26. The circuit of Fig. 4-25 represents a battery-charging circuit; a similar circuit, however, can be used wherever direct current is required. The terminal *B* in Fig. 4-25 is the negative terminal, and the terminal *C* is the positive terminal.

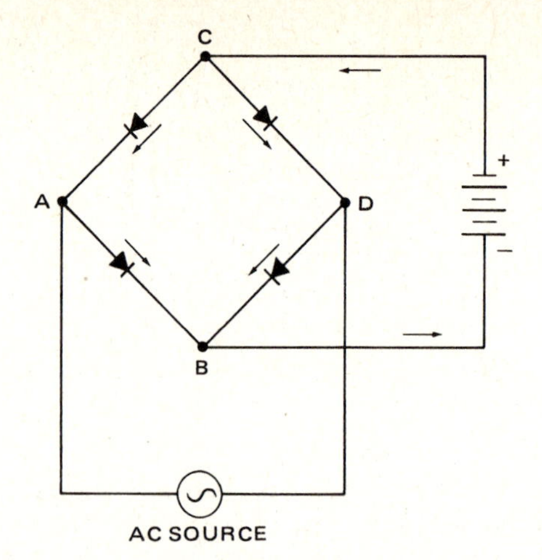

Fig. 4-25 A full-wave rectifier circuit.

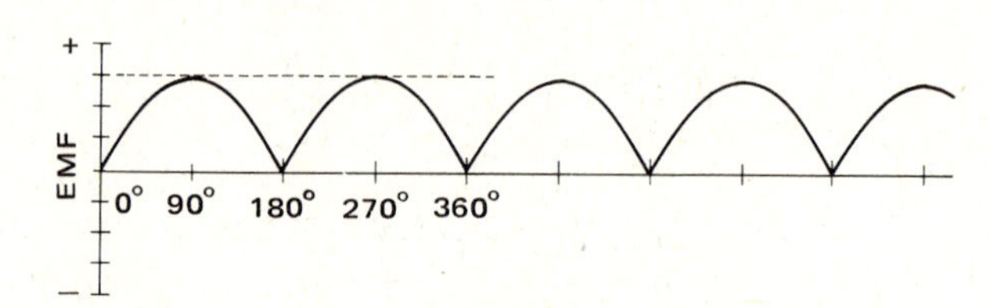

Fig. 4-26 Curve for full-wave rectified current.

SEMICONDUCTOR RECTIFIERS

To understand the principles of rectification as performed by a semiconductor rectifier, it is necessary to gain a concept of what actually takes place in the material of the rectifier. We shall, therefore, give a brief description of the structure of semiconductor materials and the electronic activity within such materials. Semiconductors are commonly called **solid-state** devices because they are solid and contain no loose or moving parts.

The principal semiconductor materials used for rectifiers are silicon and germanium. It was explained in the first chapter of this text that a semiconductor element has four electrons in the outer orbit or shell of each atom. Silicon has a total of 14 electrons in the atom, 4 of these being in the outer shell. Germanium atoms have 32 electrons with 4 in the outer shell of each. In the pure state, neither of these materials will conduct an electric current easily. This is because the atoms have a strong **valence bond** formed as the electrons in the outer shell of each atom pair with the atoms in adjacent atoms. This is shown in Fig. 4-27. The illustration is a two-dimensional concept of the **crystal lattice** for germanium. Actually, the electrons are in spherical shells rather than rings, and they rotate about the nuclei of the atoms. However, they still form energy

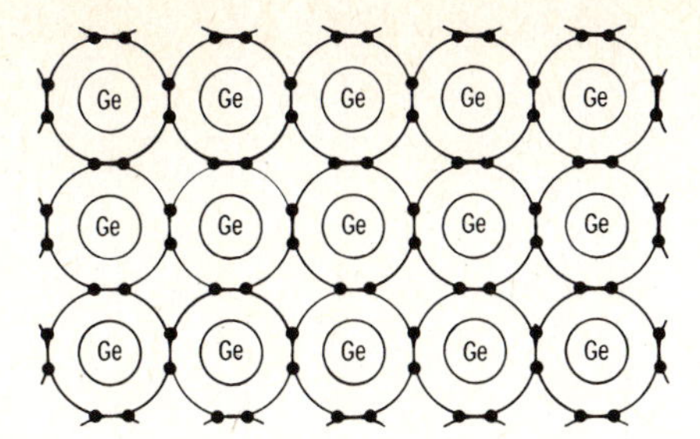

Fig. 4-27 Two-dimensional diagram of the crystal lattice for germanium.

bonds in the outer shells, and they are not easily moved from one atom to another. The only way this can happen is when a very high voltage is applied across the material, and the valence bonds are broken. It can be stated that pure germanium and silicon do not have free electrons to serve as current carriers.

To make germanium or silicon capable of carrying a current, a small amount of another element (impurity) is added. This is called **doping.** The element **antimony,** having the chemical symbol Sb, has five electrons in the outer shell of each atom. When this material is added to germanium, the germanium becomes conductive. The reason for this is that the fifth electron from the Sb atom cannot bond with the germanium electrons and is left free in the material. This is shown in Fig. 4-28. Remember that the germanium atoms have four electrons in the outer shell of each atom; hence, only four of the Sb electrons can become paired in the valence bonds.

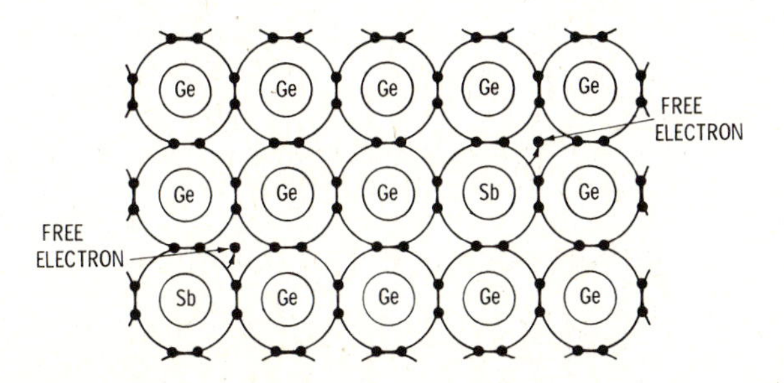

Fig. 4-28 Effect of adding antimony to germanium to form n-type material.

When germanium is treated with antimony, the resulting material is called *n*-type germanium, because it contains extra electrons which constitute negative charges. It must be remembered, however, that the material is still electrically neutral, because the total number of electrons in the material is balanced by the same number of protons. The Sb atom has 51 protons, and their positive charge balances the negative charge of the 51 electrons in each atom. One of 51 electrons is forced out of the outer shell of the Sb atom, and this becomes a free electron. The Sb is called a **donor,** because it *donates* electrons to the material.

When the element indium (In) is added to germanium, vacant spaces are left in the valence bonds, because indium atoms have only three electrons in the outer shell. The vacant spaces are called **holes.** The holes can be filled by electrons which break away from the valence bonds. When this occurs, another hole is left where the electron previously was situated. Thus the holes appear to move through the material.

The hole represents a net positive charge, because a balanced condition requires that a pair of electrons occupy each bond. When one of the electrons is missing, the bond lacks the normal negative charge; hence, it is positive and attracts electrons. An illustration of *p*-type germanium is shown in Fig. 4-29. The holes can be seen adjacent to the indium atoms. Indium added to germanium is called an **acceptor,** because it *accepts* electrons from other atoms.

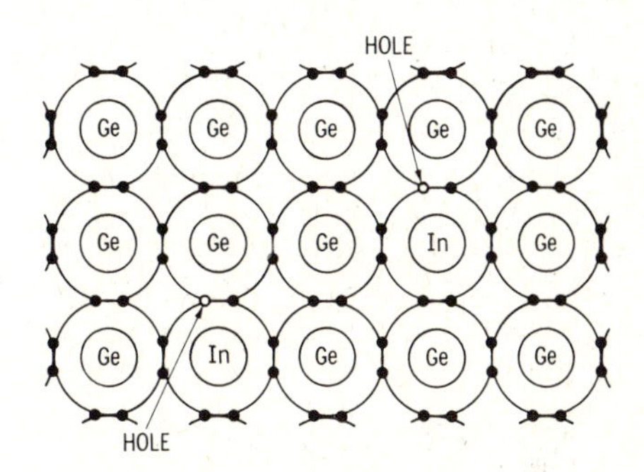

Fig. 4-29 Addition of indium to form p-type material.

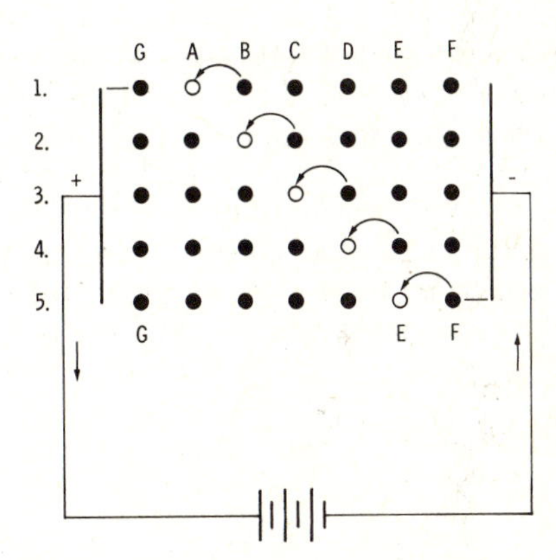

Fig. 4-30 Diagram to illustrate travel of a "hole" in p-type germanium.

In *p*-type germanium, the holes appear to act as current carriers, because they drift through the material toward a negative charge. This can be explained as shown in the drawing of Fig. 4-30. Assume that there is a hole at *A* in line 1 of the illustration. When a voltage is applied, the electron at *B* moves to occupy the hole at *A*, and the hole then appears at *B*. This action continues, and the electrons move from negative to positive as the hole moves from *A* to *E*. At this point, the hole is filled by a new electron from the outside circuit. The positive charge of the voltage source attracts the electron from the point *G*, and a new hole is formed to start its journey toward the negative terminal. In the external circuit, electrons flow from the negative terminal of the battery through the crystal and back to the positive terminal of the battery.

When a piece of *n*-type germanium forms a junction with a piece of *p*-type germanium, an interesting phenomenon takes place. Since there are holes (positive charges) in the *p*-type germanium and electrons (negative charges) in the *n*-type germanium, there is a drift of holes and electrons toward the junction. The holes are attracted by the negative charge of the electrons in the *n*-type material, and the electrons are attracted by the positive charge of the holes in the *p*-type material. Some of the electrons diffuse across the junction to fill holes on the positive side. This movement of charges leaves a large number of negative ions in the *p*-type material farthest from the junction and a large number of positive ions in the *n*-type material farthest from the junction. Remember that the material is electrically neutral, as a whole, before the junction is made, because the number of electrons is balanced by the number of protons. The material is still electrically neutral as a whole after the junction is made, but portions have negative charges and other portions have positive charges.

The stationary ions on each side of the junction provide charges which stop the movement of electrons across the junction. These charges result in a **potential barrier** with a voltage of 0.3 for germanium. Figure 4-31 illustrates the condition which exists when a junction of two different types of germanium is made. Note that holes move toward the junction from the *p*-type material, and electrons move toward the junction from the *n*-type material until the charges are balanced.

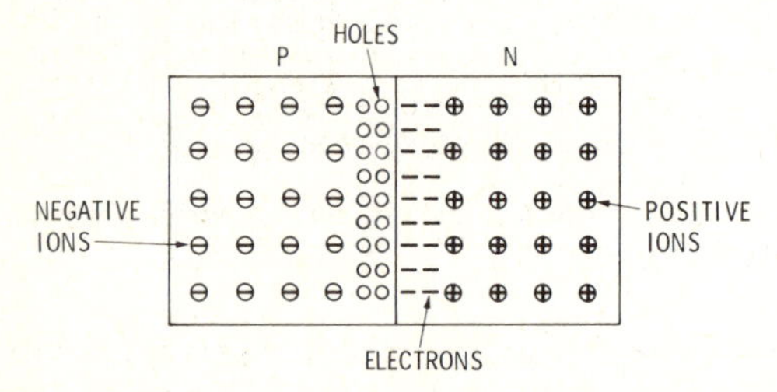

Fig. 4-31 Junction of p- and n-type materials to form a potential barrier.

When two types of germanium or silicon are joined as described in the previous paragraphs, a **diode** is formed. The word *diode* means *two electrodes.* The two parts of the material, *p*-type and *n*-type, comprise the two electrodes. If we connect a battery or other power source to the *p-n* diode, we find that current will flow through it in one direction, but not in the other direction. The diode therefore becomes a rectifier. This is explained in the illustrations of Figs. 4-32 and 4-33.

In the illustration of Fig. 4-32, the battery is connected with the negative terminal of the battery joined to the *n* side of the diode. In this way, the electrons flowing from the negative side of the battery neutralize the effect of the positive ions which would otherwise affect the current flow. This makes it possible for the electrons to flow across the barrier (junction) to occupy the holes and flow on toward the positive terminal. Thus the diode has become a good conductor in one direction, that is, from *n* to *p*.

In Fig. 4-33 we observe the condition when the battery is connected in the opposite direction, which is called **reverse bias.** Here the positive terminal of the battery is connected to the *n* side of the diode. The free electrons are drawn toward the positive charge until the potential

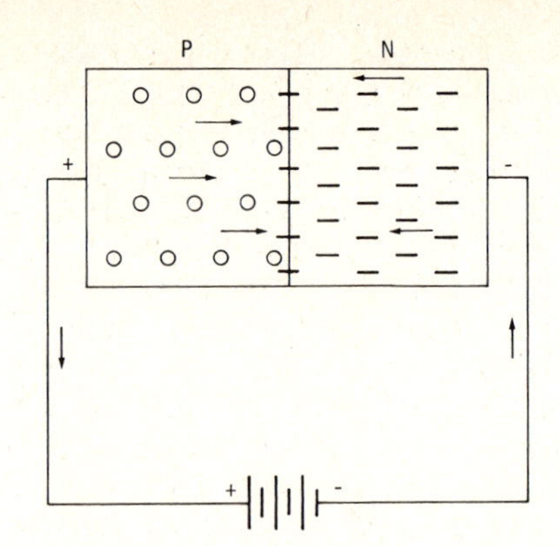

Fig. 4-32 P-n diode connected to provide forward bias.

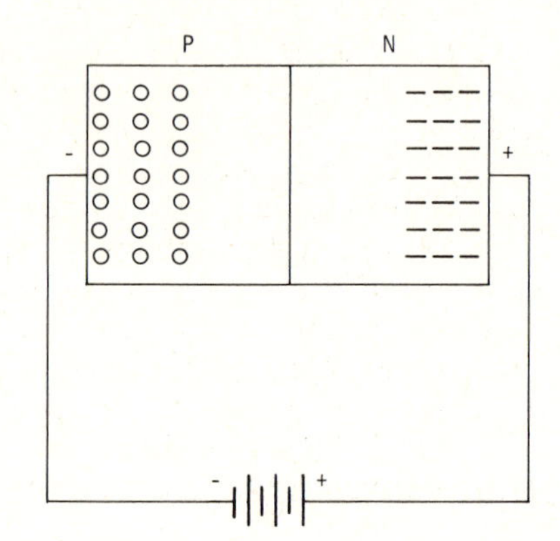

Fig. 4-33 P-n diode connected to provide reverse bias.

balances. The holes in the *p* side of the diode move toward the negative charge so there can be no movement of electrons across the junction. Under these conditions no current can flow.

From the foregoing explanations it can be seen that a single *p-n* diode (crystal diode) can serve as a half-wave rectifier, and four diodes connected in a bridge circuit can serve as a full-wave rectifier (see Fig. 4-25).

Figure 4-34 is a photograph of the components of an early silicon diode rectifier manufactured by the International Rectifier Corporation. As explained previously, the word *diode* means that the device has two electrodes, or terminals. In the illustration the units of the diode are arranged as follows, from left to right: terminal wire ("pigtail"), base, special soldering alloy, silicon wafer, aluminum wire, case with insulator and anode terminal, and terminal wire. During the assembly of this diode, the base terminal wire is first welded to the base. This assembly is then placed in a jig with the terminal wire pointing downward. A small disk of solder is then placed in the circular

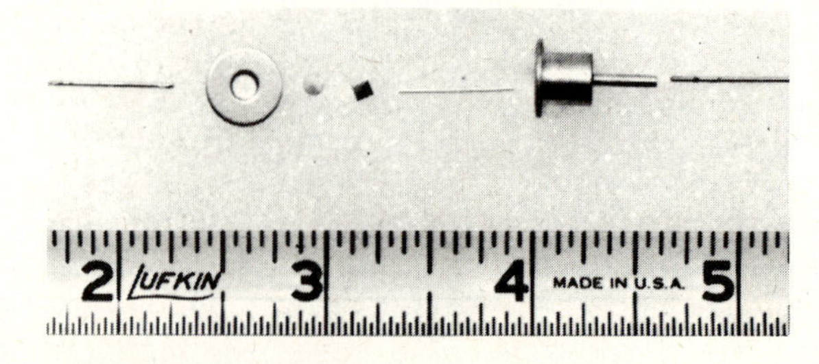

Fig. 4-34 Components of a silicon diode rectifier.

depression in the base, and the silicon wafer is placed on the solder. Finally, the pure aluminum wire is placed in a vertical position with its end bearing against the center of the silicon wafer. Still held in the jig, the unit is then passed through a furnace with carefully controlled temperatures. The heat fuses all parts together and creates an alloy junction at the point where the aluminum wire is joined to the silicon. This junction establishes the barrier which makes the unit a rectifier. Upon completion of the initial assembly and testing for performance, a case is placed over the unit and the flange of the case is resistance-welded to the base. Thereafter, the unit is placed in a heated vacuum chamber to remove all air and moisture and is then sealed by compressing the terminal stem. Several additional tests and inspections are performed before the unit is ready for the customer.

The performance of semiconductor rectifiers (crystal diodes) has made possible many new circuits of high performance and light weight for use in aircraft, space vehicles, and many industrial applications. It is possible to get the same performance from a few ounces of circuitry today as was formerly obtained from many pounds. This is in part accomplished by utilizing crystal diodes and transistors in place of electron tubes. High-power rectifiers have also been greatly improved by the use of semiconductors. For example, a germanium diode capable of handling more than 100 A weighs but a small fraction of the weight of a copper-oxide rectifier of the same capacity.

Silicon diode rectifiers that have a current-carrying capacity of more than 2500 A have been developed. These make it possible to convert ac power at high levels into dc power. One application of this capability is the variable-speed constant-frequency power supplies for large aircraft. In the past it has been necessary to employ a constant-speed drive (CSD) for each alternator to keep the 400-Hz outputs in phase (synchronized). It is now possible to allow the speed and output frequency to vary and still produce a 400-Hz ac power output for the aircraft electrical requirements. This is accomplished by changing the alternating current from the alternators to direct current by means of silicon diode rectifiers. The direct current is then changed to 400-Hz alternating current by means of static inverters. Variable-speed constant-frequency power systems are described later in this text.

A high-capacity silicon diode rectifier is shown in Fig. 4-35. This rectifier is constructed by employing a high-temperature solder to join a copper terminal to a wafer of *n*-type silicon that has been treated with an impurity such as boron to create a *p*-type material on the surface. The wafer may be from a few thousandths of an inch to 0.0625 in [1.59 mm] thick. The thickness depends upon the voltage that must be handled by the diode. The diameter of the wafer is determined by the current-carrying capacity required. The wafer in a 500-A diode may be 1.25 in [3.18 cm] in diameter or larger.

Since electric-power systems for large aircraft are almost entirely of the 400-Hz ac type, it is necessary that many rectifiers be used to provide direct current in the circuits where it is required. For this reason we find literally hundreds of diode rectifiers connected singly and in full-wave, three-phase rectifier systems on modern jet airplanes. If it were necessary to use electron tubes or older types of rectification methods for these circuits, hundreds of pounds would be added to the weight of the electric systems.

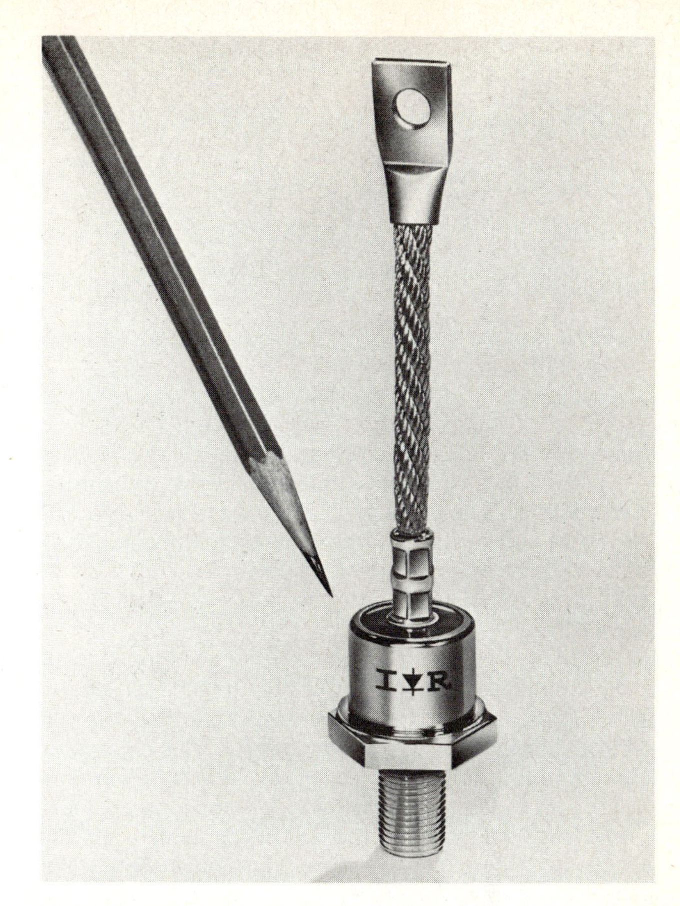

Fig. 4-35 A high-capacity silicon diode rectifier. (International Rectifier Company.)

THREE-PHASE RECTIFIER

It is often necessary to obtain direct current from three-phase power systems in aircraft; hence, three-phase rectifier units are employed. It would be possible to use a single-phase full-wave rectifier in one leg of a three-phase system; however, it is more efficient to use a rectifier system which utilizes the power from all three legs of the three-phase circuit. The output of a three-phase alternator is indicated in Fig. 4-36. It will be noted in the diagram that the voltages reach maximum 120° apart. A rectifier consisting of six diodes is connected in a manner

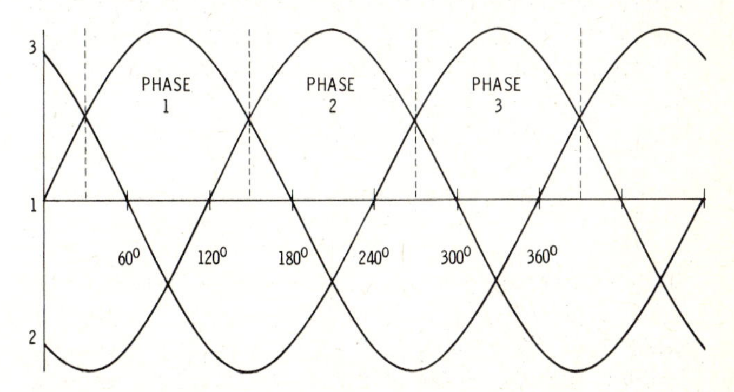

Fig. 4-36 Voltage output of a three-phase alternator.

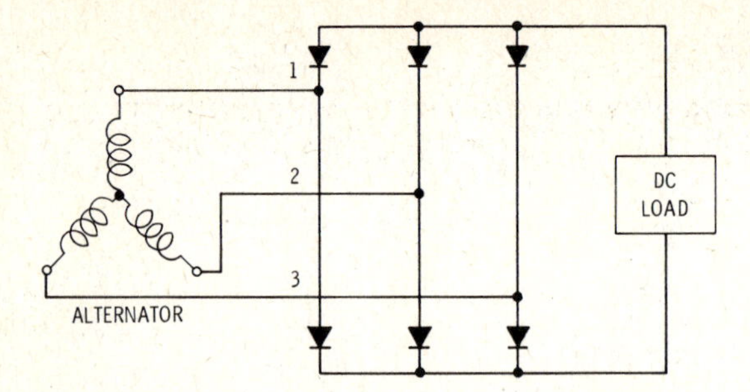

Fig. 4-37 Diagram showing six diodes connected as a full-wave rectifier for three-phase ac.

to provide one-way paths for the ac output as shown in Fig. 4-37. It can be seen in the diagram that the current flowing in each section of the three-phase circuit will always be flowing in the same direction in the output side of the rectifier, even though it reverses direction in the ac side of the circuit.

FILTERS

The output of a rectifier is a pulsating direct current unless some means is provided to level off the peaks and fill in the valleys of voltage and current. This smoothing process is accomplished by means of a filter which consists of an inductance or *choke* coil and one or more capacitors. The inductance coil is placed in series in the circuit, and the capacitors are connected in parallel or *across* the output. The method for connecting a filter is shown in Fig. 4-38.

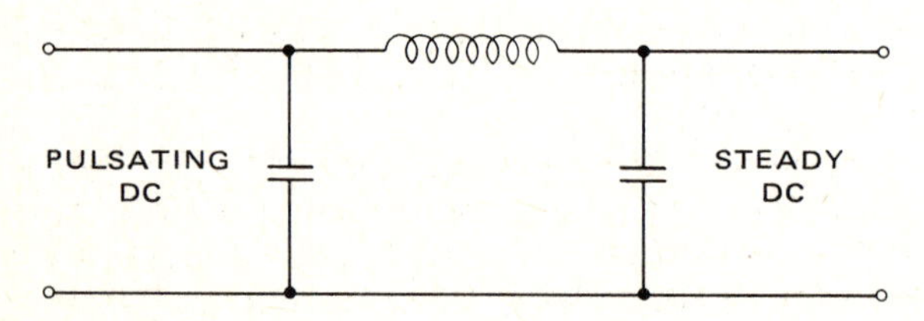

Fig. 4-38 A filter circuit.

The effect of a capacitor connected across a pulsating dc circuit is to oppose changes in voltage. The capacitor charges as the voltage rises and discharges back into the circuit as the voltage drops. The inductance coil generates an opposing voltage as the current increases, thus reducing the rate of current rise. When the current flow decreases, the magnetic field of the inductance coil begins to collapse and induces a voltage which tends to keep the current flowing. The effect of the combination of the inductance coil and capacitors, therefore, is to smooth both the voltage and current variations and produce a steady direct current.

ELECTRON-TUBE RECTIFIERS

Even though solid-state (semiconductor) rectifiers are now commonly used in most applications, electron-tube rectifiers are still used and should be understood. For the purpose of charging batteries in ground stations and supplying direct current for a variety of uses, electron-tube rectifiers may be employed.

There are a number of types of electron-tube rectifiers, one of the most common being the **tungar** rectifier. This name is derived from the fact that the rectifying element is a glass tube containing two tungsten electrodes, the space within the tube between the electrodes being filled with argon gas. The electrodes are called the **filament** and the **plate.** The filament may be heated to incandescence by means of an electric current and in this condition emits electrons from its surface. These electrons form a cloud around the filament but do not leave the vicinity of the filament unless there is a positively charged body near enough to cause them to travel across the intervening space. This is the function of the plate. When the plate is positively charged, electrons flow from the filament to the plate. When the plate is negatively charged, no electrons flow. This property of the tube, which permits current to flow in one direction but not in the other, makes it ideal for rectifying purposes.

A schematic diagram of a tungar rectifying circuit is shown in Fig. 4-39. In this circuit, only one-half of the incoming alternating current is utilized; hence it is a half-wave rectifier. A tungar rectifier circuit for full-wave rectification is shown in Fig. 4-40. In a circuit of this type both the negative and positive components of the alternating current are utilized in the dc circuit.

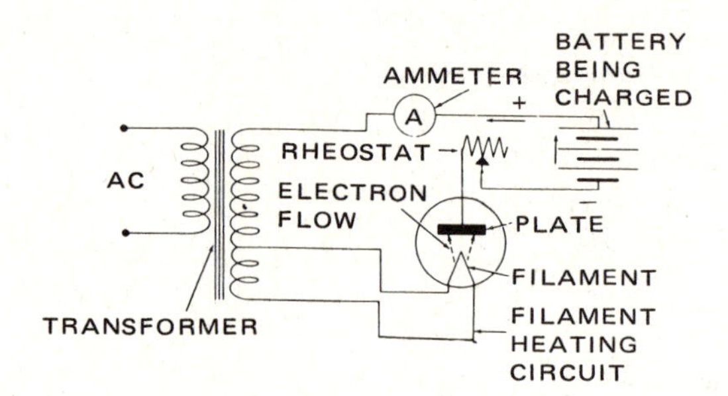

Fig. 4-39 Tungar half-wave rectifier circuit.

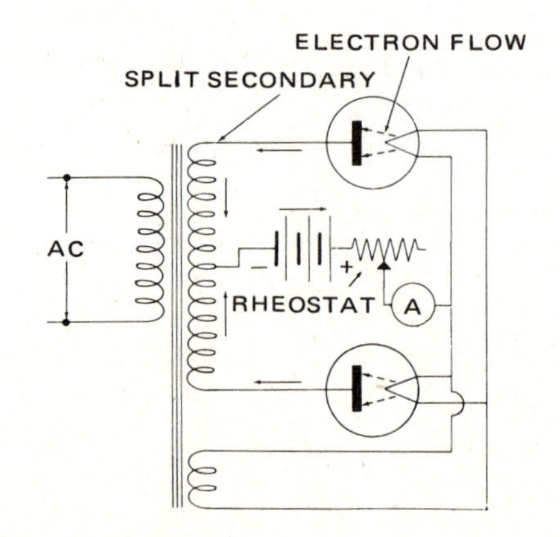

Fig. 4-40 Tungar full-wave rectifier circuit.

When it is necessary to obtain a higher value of direct current than is possible with vacuum-tube rectifiers, mercury-arc rectifiers are sometimes used. A rectifier of this type utilizes mercury vapor in a tube to increase the conductivity of the space between the electrodes. This is possible because the mercury vapor becomes ionized and supplies additional free electrons, thus permitting an increase in the flow of current.

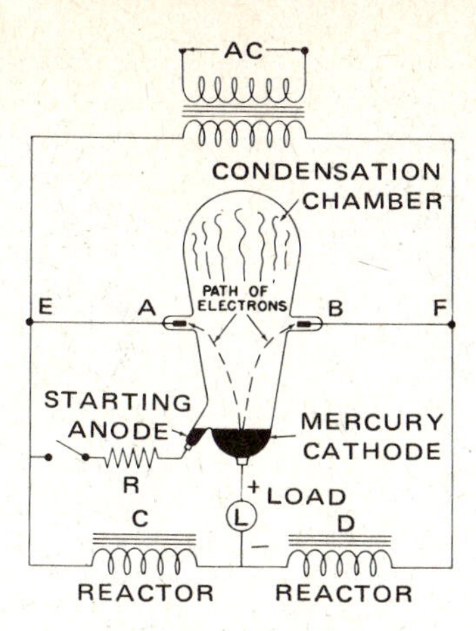

Fig. 4-41 Mercury-arc rectifier circuit.

A schematic diagram of a mercury-arc rectifying circuit is shown in Fig. 4-41. The cathode (negative electrode) is a pool of mercury at the bottom of a glass tube. The running anodes (positive electrodes) *A* and *B* are located in recesses on each side of the tube above the mercury pool. An anode required for starting the operation in the tube is near the mercury pool at the bottom. The upper part of the tube is a condensation chamber which permits the mercury vapor to condense and flow back to the cathode pool. When an ac supply is connected to the circuit, as shown in Fig. 4-41, the anode *A* is positive during one half-cycle and the anode *B* is positive during the other half-cycle. The operation of the rectifier is started by tilting the tube until some of the mercury flows over from the cathode pool and makes contact with the starting anode. This permits a flow of current which is limited by the resistance *R*. The tube is then returned to a level position. The mercury contact breaks and causes a spark at the surface, thus liberating electrons, which are immediately attracted to the positive anode. The flow of electrons thus started maintains a hot spot on the surface of the mercury cathode which causes it to emit still more electrons. Since the flow of electrons is always from the cathode to one of the two anodes, the direction of current is always the same in the circuit connected to the cathode.

In a mercury-arc rectifier, it is necessary to maintain a continuous flow of electrons from the mercury pool; otherwise, the operation will stop. The reactors in the circuit maintain this flow even though the alternating voltage drops to zero twice during each cycle. This may be clearly understood by considering the sequence of events as the voltage at the anodes passes through the zero point. Assume that the anode *A* is positive. The electronic flow will then be from *F*, through the reactor *D* and the load *L* to the cathode, and from the cathode to the anode *A*. As the ac voltage drops to zero, the inductance of the reactor *D* opposes the drop in current and sets up a local circuit in the direction *DLB*. This will maintain a flow of electrons through the tube until the point *F* becomes sufficiently positive to attract the electrons. At this time the electrons will flow from the point *E*, which is now negative, through the reactor *C* and the load *L*, to the anode *B*.

From the foregoing discussion it can be seen that the mercury-arc rectifier makes use of both half-waves of the alternating current; hence, it is a full-wave rectifier. Because of the reactors in the circuit, the current never falls to zero.

REVIEW QUESTIONS

1. Define *alternating current.*
2. What are the advantages of alternating current in large transport aircraft?
3. Explain the *sine curve.*
4. What is meant by *rms,* or *effective values,* of alternating current?
5. Explain *frequency.*
6. What is meant by *phase* in speaking of alternating current?
7. Why is it that alternating current appears to flow through a capacitor?
8. What is the effect of capacitance in an ac circuit?
9. Give the formula for *capacitive reactance.*
10. What is the capacitive reactance in a circuit when the capacitance is 1 μf and the frequency is 60 Hz?
11. How does frequency affect the capacitive reactance in a given circuit?
12. What is the effect of inductance in an ac circuit?
13. Give the formula for *inductive reactance.*
14. Compute the inductive reactance in a circuit where the frequency is 1000 kHz and the inductance is 20 mH.
15. Explain *impedance.*
16. Draw a diagram to show the combination of 15 Ω inductive reactance, 10 Ω capacitive reactance, and 4 Ω resistance.
17. Compute the impedance in an ac circuit which has the following values: $f = 1400$ kHz, $L = 5$ mH, $C = 2$ μf, $R = 600$ Ω.
18. What is the resonant frequency in an ac circuit when the capacitance is 2 μf and the inductance is 50 mH?
19. Of what does a *tank* circuit consist?
20. What purpose does a tank circuit serve?
21. How may we determine the phase angle between voltage and current in an ac circuit when the reactances and resistance are known?
22. How is the phase angle utilized in determining the power in an ac circuit?
23. What is meant by *power factor* in an ac circuit?
24. Why can ac voltage be changed by means of a transformer?
25. Why is it that a transformer will not draw an appreciable amount of power when no load is connected to the secondary, even though the primary is connected?
26. Compare the number of windings in the primary coil of a step-up transformer with the number of windings in the secondary coil.
27. Give the formula for expressing the voltage values in the circuits of a transformer with respect to the number of turns in the primary and secondary windings.
28. Give the formula for maximum current values in the primary and secondary windings of a transformer.
29. What is meant by *phasing* when connecting more than one transformer in the same circuit?
30. Explain a three-phase ac circuit. Show the difference between the delta and Y circuits.

31. In a Y-connected three-phase circuit with the neutral wire grounded, if the voltage between phase 1 and ground is 110 V, what will the voltage be between phase 1 and phase 2?
32. How is it possible to operate single-phase equipment from a three-phase power system?
33. Explain the operation of a rectifier.
34. What is the difference between a half-wave rectifier and a full-wave rectifier?
35. How is full-wave rectification accomplished?
36. What is meant by a *solid-state* device?
37. Explain the principle of a semiconductor.
38. What materials are used in crystal diodes?
39. What is the difference between an *n*-type material and a *p*-type material in a semiconductor unit?
40. What is a *hole* in a semiconductor?
41. Why are solid-state devices more effective than electron tubes in electronic circuits?
42. Describe a high-capacity silicon diode rectifier.
43. Explain the operation of an electron-tube rectifier.
44. Why are filters used in the output side of rectifier circuits?

CHAPTER 5

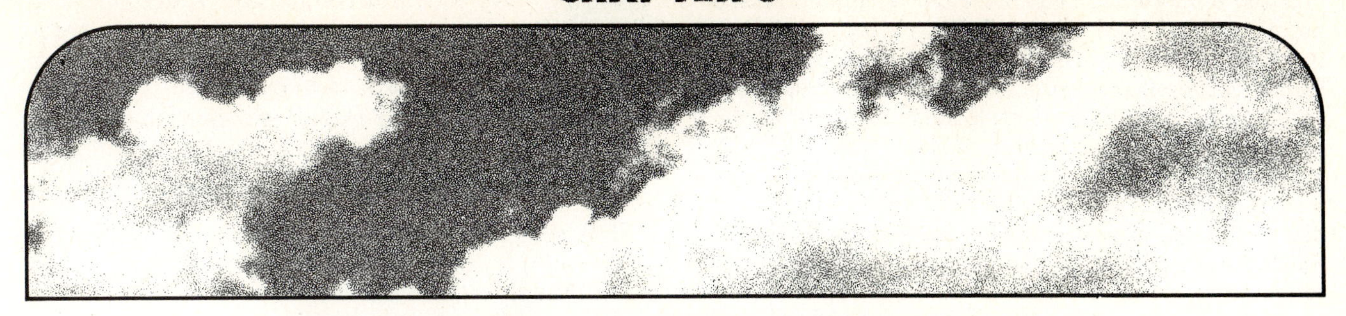

Electric Measuring Instruments

The fundamental units of electrical measurement are the ampere, volt, ohm, and watt. To measure electrical values in terms of these units, certain instruments are required. We are all familiar with such common measuring devices as scales for measuring weight, rules for length, thermometers for temperature, and speedometers for speed. The common electric measuring instruments are the **ammeter, voltmeter, ohmmeter,** and **wattmeter.** The unit measured by each of these instruments is clearly indicated by its name. There are many electric measuring instruments in addition to those mentioned above, but for the purposes of this chapter, a discussion of these basic instruments is considered sufficient.

Electric measuring instruments may be divided into three general classifications. These are the **indicating** instruments, from which the electrical value is read directly as a needle moves across a dial; the **nonindicating** instruments, from which a reading is obtained by adjusting dials, switches, etc., and then noting their positions; and **recording** instruments, which make a continuous record of required values over a period of time. In this study of electrical fundamentals, we are primarily interested in the indicating instruments. Other types will be discussed in appropriate sections of the text.

METER MOVEMENTS

The basic principle of many electric instruments is that of the **galvanometer.** This is a device which reacts to minute electromagnetic influences caused within itself by the flow of a small amount of current. A simple galvanometer is shown in Fig. 5-1. It consists of a magnetized needle suspended within a coil of wire. When a current is passed through the wire, a magnetic field is produced and the magnetized needle attempts to align itself with this field. Practical galvanometers cannot be constructed as simply as the one described above, but they all operate because of the reaction between magnetic and electromagnetic forces.

Any device designed to indicate a flow of current, particularly a very small current, and which operates on the principle of two interacting magnetic fields, may be called a galvanometer. A permanent magnet pivoted so that it can turn in response to the influence of a current-carrying coil, or a current-carrying coil placed in a magnetic field and pivoted so that it can turn in response to the field produced by a current flow, may be used as a galvanometer. In any event, the rotating part must be balanced by a spring which will tend to hold it in the zero position when there is no current flow.

The most common types of electric measuring instruments employ a moving coil and a permanent magnet. This arrangement is known as the **d'Arsonval** or **Weston** movement and is illustrated in Fig. 5-2. The coil, consisting of fine wire, is pivoted and mounted so that it may

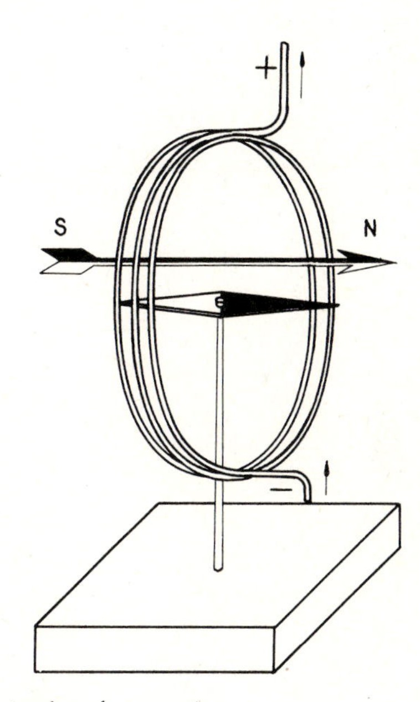

Fig. 5-1 A simple galvanometer.

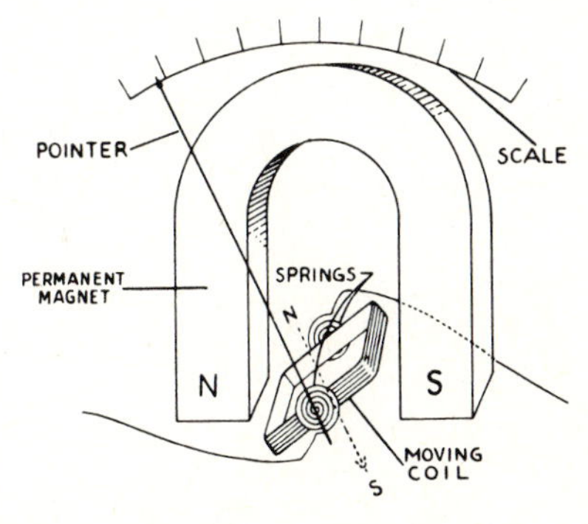

Fig. 5-2 D'Arsonval or Weston meter movement.

rotate in the magnetic field of the permanent magnet's poles. When a current flows in the coil, a magnetic field is produced. The north pole of this field is repelled by the north pole of the permanent magnet and attracted by its south pole. As shown in Fig. 5-2, this will cause the coil to rotate to the right. The magnetic force causing the rotation is proportional to the current flowing in the coil and is balanced against a coil spring. The result is that the distance of rotation will increase as the current flow in the coil increases. The needle attached to the coil and moving along a scale will then indicate the amount of current flowing in the coil.

It is quite apparent that the Weston movement, used alone, is not suitable for the measurement of alternating current. Such current would produce rapid reversals of polarity in the moving coil that would cause the needle only to vibrate. Under these conditions, no indication could be obtained.

A movement similar to the Weston movement, but suitable for ac measurements, employs an electromagnet in place of the permanent magnet. This is called a **dynamometer** movement (see Fig. 5-3). The moving coil can be connected in either series or parallel with the electromagnet circuit. When a movement of this kind is used, the indicating needle will always move in the same direction regardless of the direction of the current through the instrument. This is because the polarity of both the moving coil and the electromagnet changes when the current direction changes; hence, the direction of torque (twisting force) remains the same. The movement will therefore operate with alternating current.

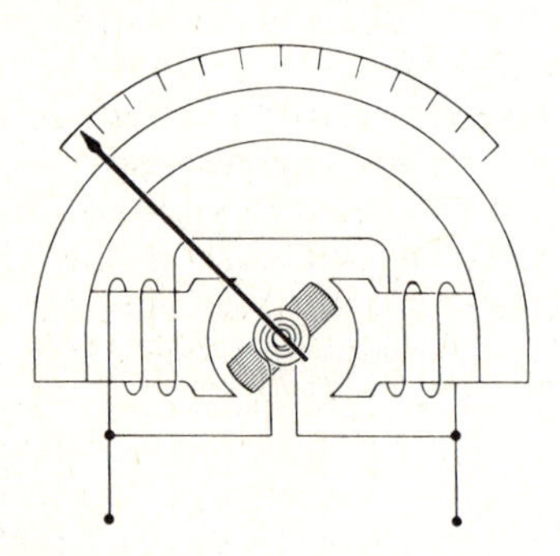

Fig. 5-3 Dynamometer movement.

IRON-VANE MOVEMENT

Another type of movement used with alternating current is illustrated in Fig. 5-4. This is called an iron-vane mechanism and, as the name implies, it employs an iron vane through which electromagnetic forces act to move the indicating needle. The iron vane is attached to a pivoted shaft and is free to move into the coil whenever the coil is energized. Also mounted on the shaft is the indicating needle; hence, the vane and the needle move together in response to a current flow in the coil. The movement of the vane-and-needle assembly is balanced by a coil spring which holds the needle in the zero position when no current is flowing.

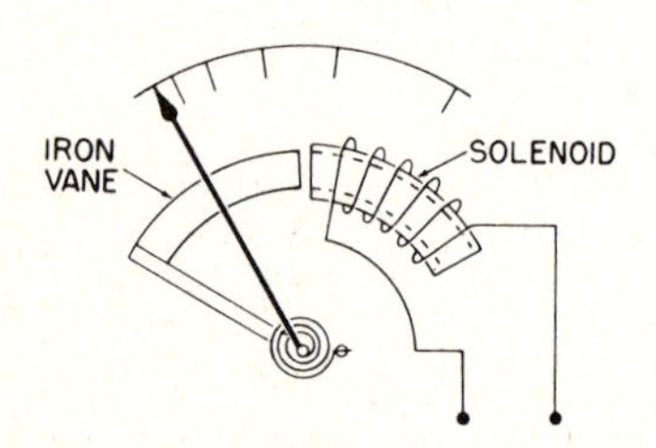

Fig. 5-4 Iron-vane movement.

CONSTRUCTION FEATURES

Electric meters must be constructed with the utmost care and precision. This is so because some meter movements must respond to currents as small as a few millionths of an ampere. Some of the moving parts are more accurately machined and finished than the works of an expensive watch. For this reason, such instruments must be handled with great care to prevent shock or vibration damage which would result in a loss of accuracy.

Because of the sensitivity required of electric-meter movements, it is necessary that the pivot-shaft bearings be as nearly frictionless as possible. This is accomplished by using jewel bearings similar to those used in fine watches for over two centuries. Figure 5-5 shows three different types of jewel bearings used in watches, clocks, and electric instruments. The ring-and-end-stone jewel bearing is used in watches, the ring being required to keep the tiny teeth of the watch gears constantly in mesh.

For electric instruments the V-jewel bearing is used because the friction of the ring bearing is much too great for the sensitivity required. The pivot may have a radius at its tip ranging from 0.0005 in [0.00127 cm] to as high as 0.003 in [0.0076 cm], depending on the weight of the mechanism and the vibration it will encounter. The radius of the pit in the jewel is somewhat greater; contact is therefore made with a circle a fraction of a thousandth of an inch across. The design shown in the illustration has the least friction of any practical type of instrument bearing.

Although the moving elements of instruments are designed to be of the lowest possible weight, the extremely small contact area between the pivot and jewel results in the large stresses for which the bearing must be designed. For example, a moving element weighing 300 milligrams (mg) [0.00066 lb], resting on the area of a circle of

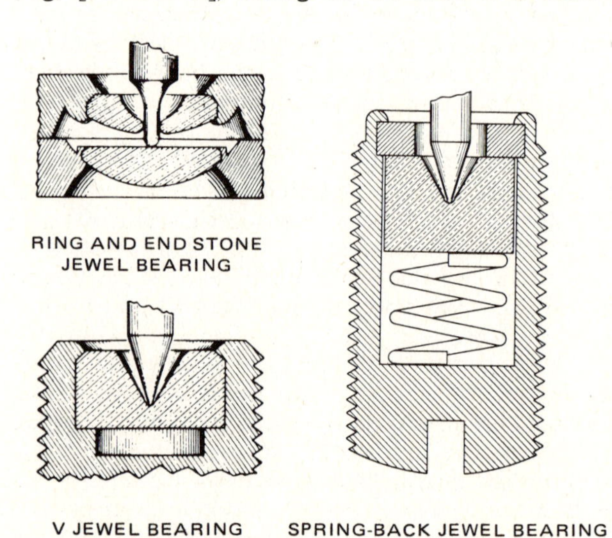

Fig. 5-5 Jeweled bearings.

0.0002-in [0.0005-cm] diameter, produces a force of about 10 tons per in² [1406 kg/cm²]. From this it can be seen that if an instrument is dropped or jarred, the bearing stresses can easily be increased to the level where permanent damage is done.

Some instruments are designed to withstand rather severe shocks, and these are supplied with spring-back jewel bearings such as that shown in Fig. 5-5. This construction permits the pivot shaft to move axially when it is subjected to shock, with the result that the stresses are greatly reduced.

TAUT-BAND MOVEMENT

A rather ingenious development in instrument movements has largely eliminated the friction problems and the need for pivoted bearings. In this instrument movement, the moving coil is suspended on a taut platinum-iridium band held by spring tension in the instrument frame. This type of unit is called a **taut-band movement.** Figure 5-6 illustrates the construction of an instrument utilizing the taut-band suspension for the moving coil.

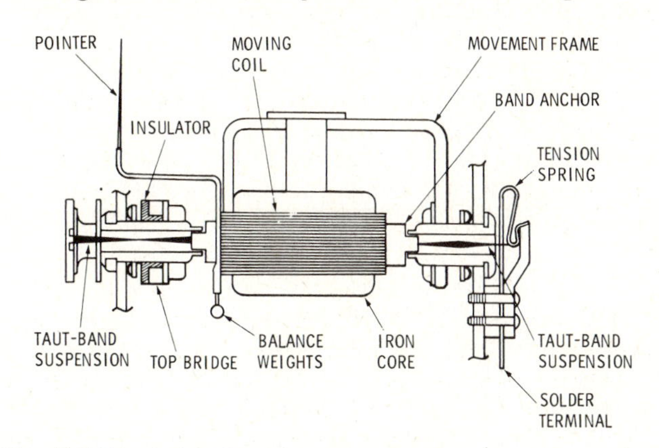

Fig. 5-6 Taught-band instrument movement.

It can be seen that the taut-band instrument is not nearly as sensitive to shock as the type having jewel bearings, because shocks are taken up by the elasticity of the taut band. The band is not subject to corrosion because of the material from which it is made. Since there are no parts rubbing against one another as in a bearing, the friction is eliminated, and the movement can respond to extremely small magnetic influences. For this reason, the movement can be designed for very high sensitivity.

DESIGN FOR UNIFORM SCALE

Since the magnetic force acting upon a magnetic substance is inversely proportional to the distance between the magnet and the substance acted upon, instruments using the magnetic principle do not have a uniform scale unless special construction features are incorporated. In the Weston meter movement, a uniform scale is obtained by placing a cylindrical iron core inside the moving coil (see Fig. 5-7). This arrangement results in a uniform magnetic field in the air space between the core and the poles. The coil rotates in the cylindrical space a distance proportional to the amount of current flowing in the coil windings.

It is apparent from a study of a typical instrument diagram that there is a limit to the range through which the

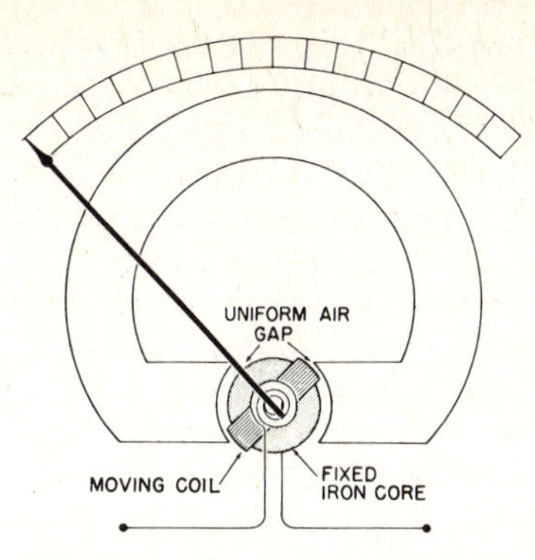

Fig. 5-7 Iron core to provide a uniform field.

indicating needle can act. In a conventional meter movement, such as that in Fig. 5-7, this range is approximately 100°. Some meters require a greater range and must be specially constructed.

SENSITIVITY

As previously stated, some meters must be constructed with a high degree of sensitivity. The sensitivity is determined by the amount of current required to produce a full-scale deflection of the indicating needle. Very sensitive movements may require as little as 0.000 05 A to produce a full-scale deflection. This value is commonly called *20 000 Ω/V*, because it requires 20 000 Ω to limit the current to 0.000 05 A when an emf of 1 V is applied. Movements having a sensitivity of 1000 Ω/V are commonly used by electricians when the power consumed by the instrument is of no consequence. In electronic work, where very small currents and voltages must be measured, instruments of very high sensitivity are required. Electronic measuring instruments, such as the vacuum-tube voltmeter (VTVM) or the solid-state voltmeter (SSVM), are normally used for the measurement of currents and voltages in electronic circuits. These instruments are designed to isolate the measuring circuit from the circuit being measured, hence very little loading is applied to the circuit being measured.

To understand the importance of sensitivity in an instrument for testing certain values where current flow is very small, it is well to consider a specific example. In the circuit in Fig. 5-8 a 100-V battery is connected across two resistors in series. Each resistor has a value of 100 000 Ω, making the total resistance of the circuit 200 000 Ω. Since the two resistors are equal in value, it is obvious that the voltage across each will be 50 V. If we wish to test this voltage by means of a voltmeter which has a

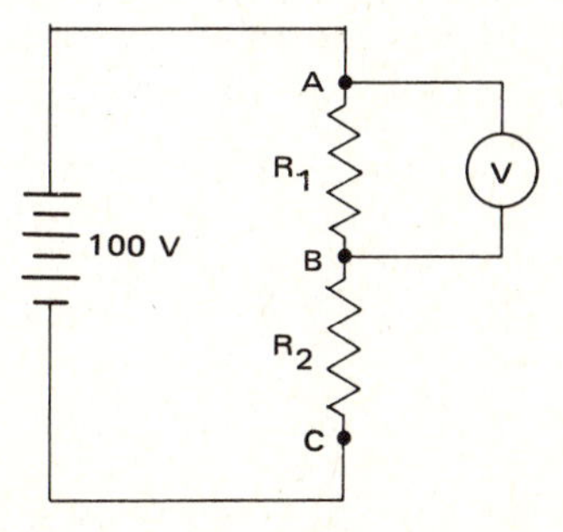

Fig. 5-8 Demonstration to show need for high sensitivity in a voltmeter.

1000-Ω/V sensitivity, we will discover that a large error is introduced into the reading.

Assume that the voltmeter has a range of 100 V and that it is connected across R_1 between the points A and B. Since the voltmeter has a sensitivity of 1000 Ω/V, its total resistance will be 100 000 Ω. When this is connected in parallel with R_1, the resistance of the parallel combination becomes 50 000 Ω, and the total resistance of the circuit is now 150 000 instead of 200 000 Ω. With the resistance between A and B 50 000 Ω and the resistance between B and C 100 000 Ω, the voltage drop will be 33.3 V between A and B and 66.7 V between B and C. It is apparent then that the voltmeter used would not be satisfactory for this test.

If we connect a voltmeter with 20 000 Ω/V sensitivity across R_1, we will obtain a much more accurate indication of the operating voltage. The voltmeter has an internal resistance of 2 000 000 Ω, and this resistance, combined in parallel with R_1, will produce a resistance of 95 238 Ω. This resistance in series with the 100 000 Ω of R_2 will produce a voltage drop of approximately 48.7 V across R_1 and 51.3 V across R_2. The reading of the voltmeter is then 48.7 V, which is probably as accurate as necessary for normal purposes.

THE AMMETER

Most of the electric measuring instruments in common use employ one of the meter movements described in the foregoing sections of this chapter. These movements are adapted to the desired purpose by the use of parallel or series resistances in the instrument circuit. Usually the resistances are enclosed in an instrument case, but in some of the higher-capacity ammeters a shunt resistance is connected as a part of the external circuit.

When a resistance is connected in parallel with the terminals of a meter, it is called a **shunt** resistance. A shunt resistance, also called an *instrument shunt,* may be defined as a particular type of resistor designed to be connected in parallel with a meter to extend the current range beyond some particular value for which the instrument is already competent. In general, the word *shunt* means *connected in parallel.* For example, a shunt circuit is an electrical design in which two or more circuits are connected in parallel.

A simple ammeter with low capacity may be constructed by using relatively large wire in the moving coil. If the wire is large enough to carry the full amperage of the circuit in which the meter is used, it is not necessary to incorporate a shunt resistance. The shunt resistance is employed in most ammeters because it makes it possible to use the same meter movement for a wide range of current measurement.

A typical ammeter circuit with a shunt resistance is shown in Fig. 5-9. If we assume that a current of 0.01 A causes a full-scale deflection of the indicating needle and that the resistance of the movement is 5 Ω, we can calculate the voltage required to produce a full-scale deflection. By applying Ohm's law we find this to be 0.05 V. This instrument can be made to measure almost any current value by using a shunt resistance of the correct value. Suppose that it is necessary to use the ammeter where the current range is from 0 to 30 A. Since 30 A must flow through the parallel combination of the meter and the shunt resistance and only 0.01 A can flow through the meter, then 30 − 0.01, or 29.99, A must flow through the shunt resistance. We know that 0.05 V across the meter provides a current of 0.01 A; hence we must find a resistance that will cause a voltage drop of 0.05 V when a current of 29.99 A is flowing through it. By Ohm's law,

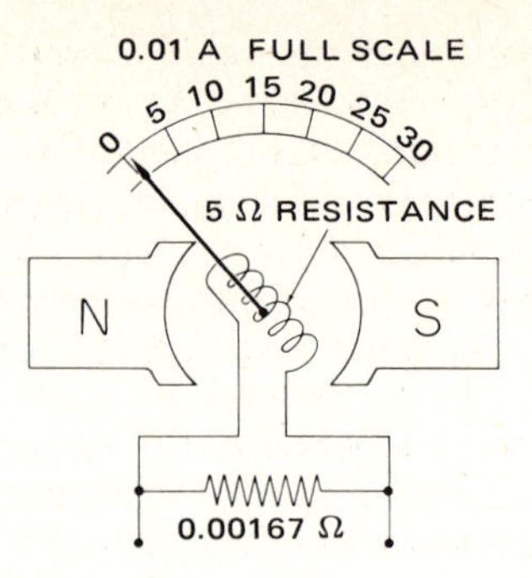

Fig. 5-9 Ammeter circuit.

$$R = \frac{0.05}{29.99} = 0.001\ 67\ \Omega$$

If we wish to use the same meter movement for a range of 500 A, the value of the shunt resistance may be determined as for a 30-A range. Since 0.01 A will flow through the instrument at full-scale deflection, 499.99 A must flow through the shunt. The resistance of the shunt must be such that 499.99 A will cause a voltage drop of 0.05 V. This value is obtained by dividing 0.05 by 499.99. The required resistance is found to be approximately 0.0001 Ω. It is very difficult to construct a resistance of exactly 0.0001 Ω, and temperature changes also cause some variations; hence, practical ammeters for high amperage employ a movement less sensitive than that described above.

Greater accuracy may be obtained with a sensitive movement by incorporating a series resistance into it. For example, if a resistance of 995 Ω is connected in series with the movement, then the value of the shunt resistance may be increased to approximately 0.02 Ω. This will provide much greater accuracy because the higher resistance of the shunt reduces the error factor. Typical ammeter shunts are shown in Fig. 5-10.

Milliammeters, which are used to measure current values in thousandths of amperes, do not necessarily require shunt resistances. If the instrument has a sensitivity of 100 Ω/V, it has a range of 0 to 10 milliamperes (mA) without a series or shunt resistance. A shunt resistance

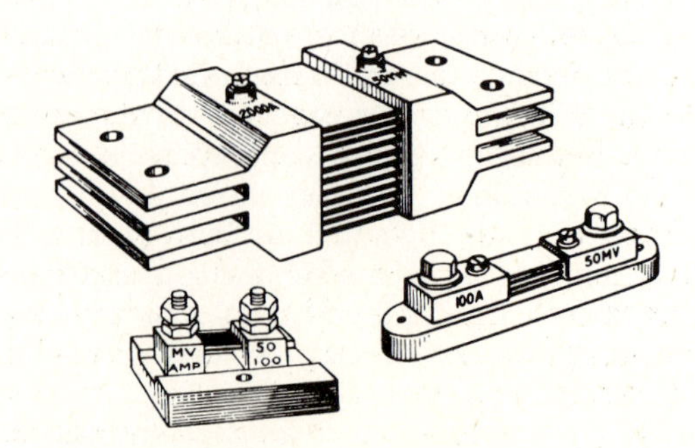

Fig. 5-10 Ammeter shunts.

may be incorporated to increase the range to any desired value. To measure current in units smaller than the milliampere, a **microammeter** is used. One microampere (μa) is one-millionth of an ampere. An ammeter with a sensitivity of 20 000 Ω/V has a full-scale deflection of the indicating needle with a current of 50 μa through the movement. Such a meter may be used to measure current in a range of 0 to 50 μa.

The proper method for connecting an ammeter in a circuit is shown in Fig. 5-11. Note that the ammeter and shunt are in parallel with each other and in series with the load. An ammeter of the proper type uses a negligible amount of power for its operation; hence, it will not interfere with the operation of the load. The instrument must *not* be connected in parallel with the source of power. The ammeter and its shunt are designed to offer as little resistance as possible in a circuit; hence, if it is connected in parallel with the power source, it will act as a direct short circuit. This will not only prevent the operation of the circuit, but also in most cases, cause irreparable damage to the instrument and to the power source.

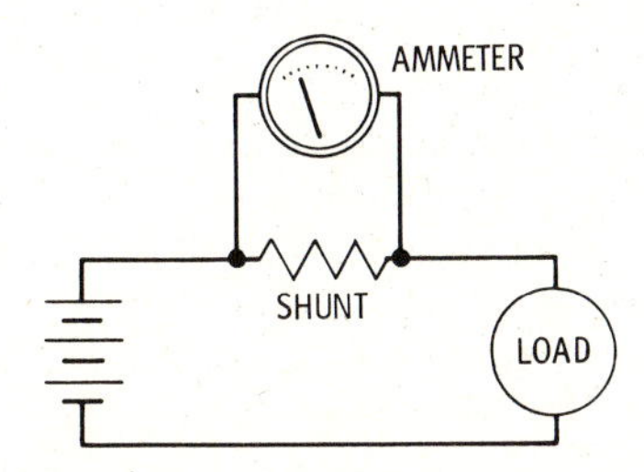

Fig. 5-11 Ammeter connected in a circuit.

An exception to the rule for connecting an ammeter in a circuit occurs when the ammeter is used for testing a dry-cell battery. The maximum current from a No. 6 dry cell is between 20 and 30 A; hence, an ammeter with a 30-A range may be safely connected across the terminals of the cell. The internal resistance of the cell prevents an excessive current from flowing even though the cell is short-circuited. Although this test with an ammeter is used to determine the condition of the dry cell, the same test should *never* be performed on a storage cell.

An ammeter of the correct capacity and type may be used to determine how much current a particular load in an aircraft will draw. If it is desired to find out how much current flows in a starter circuit, it is merely necessary to disconnect one of the power cables to the starter motor and connect the ammeter between the starter cable and the terminal on the starter motor. Care must be taken to see that the ammeter is connected with the correct polarity. In a system with a negative ground, the positive (+) terminal of the ammeter is connected to the power cable from the starter relay and the negative (−) terminal of the ammeter is connected to the power terminal of the starter motor.

The ammeter used to check the current in a starter system should have a range up to 500 A because the initial current flow is very high. After the ammeter is properly connected with cable as large as the starter cable, the starter switch may be closed. It will be noted that there is a very high surge of current at first and this rapidly falls off to a much lower value.

THE VOLTMETER

A voltmeter of the moving-coil type actually measures the current flow through the instrument; but since the current flow is proportional to the voltage, the instrument dial may be marked in volts. The meter movement is adapted to the measurement of voltage by the use of series resistances.

Figure 5-12 shows a schematic diagram of a voltmeter circuit. Assuming that the meter movement has a full-scale deflection at a current of 0.001 A and an internal resistance of 10 Ω, it is easily determined that 0.01 V is the maximum that can be applied to the instrument without the addition of a series resistance. If we wish to give the instrument a range of 0 to 30 V, we use Ohm's law to find the required series resistance. Since the current through the instrument must be 0.001 A for a 30-V reading, we proceed as follows:

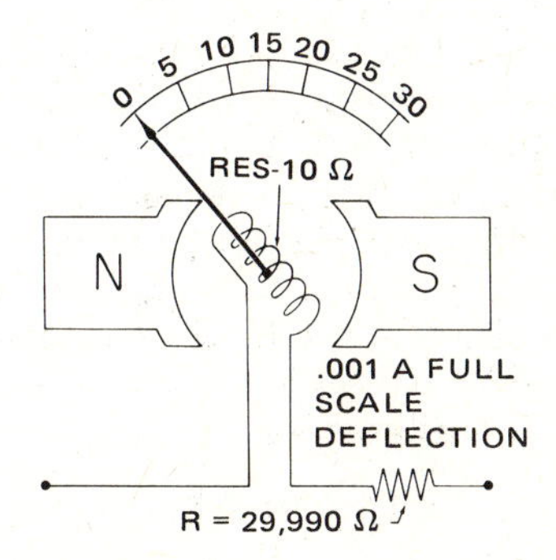

Fig. 5-12 Voltmeter circuit.

$$R = \frac{30}{0.001} = 30\ 000\ \Omega$$

Since the internal resistance of the movement is 10 Ω, this value must be subtracted from the total required resistance. The series resistance required is then 29 990 Ω.

Voltmeters usually have the necessary series resistance built into the instrument itself. The range of such an instrument may be increased by the use of additional series resistances called **multipliers.** Resistances of this type are used with test instruments when the instrument must be capable of measuring a wide range of voltages. For example, to double the range of a voltmeter, it is necessary merely to add a series resistance equal to the total resistance of the instrument. The resistance of a voltmeter may be determined by testing it with an ohmmeter, provided that the range of the voltmeter is greater than the voltage employed for operation within the ohmmeter.

The proper method for connecting a voltmeter in a circuit is shown in Fig. 5-13. Instead of being connected in

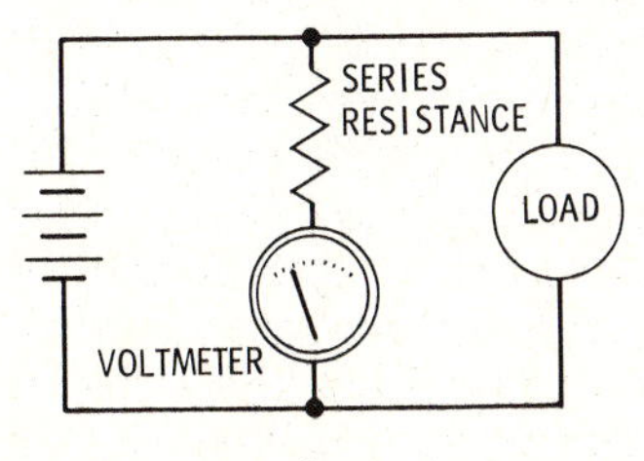

Fig. 5-13 Voltmeter connected in a circuit.

series, like the ammeter, the voltmeter is connected in parallel with the power supply. This is possible because of the high resistance of the instrument. If it were connected in series, the resistance would block the current flow and the circuit could not operate. Furthermore, a series arrangement could not measure the voltage of the source of power, even if the circuit could operate.

If the resistance of a voltmeter is too low, it disturbs the conditions of the circuit, and an accurate reading cannot be obtained. This is particularly true in circuits having a low current flow such as electronic circuits. In such circuits it is necessary to use very sensitive (high-resistance) instruments. As explained previously, it is common practice to use electronic instruments in such cases.

Since a voltmeter has sufficient resistance so that it may be connected in parallel with the power supply, it causes no damage to the instrument if it is connected in series with the load. The only effect is to prevent the operation of the circuit.

Voltmeters are used in airplanes so that the pilot, or another member of the crew, may be kept informed concerning the operation of the electric system. They are not usually installed in small airplanes with single-generator systems, but where it is necessary to operate two or more generators in parallel the voltmeter is essential to aid in balancing the output of the generators.

The voltmeter is a valuable instrument for troubleshooting and checking electric and electronic circuits. In every case, the technician must be sure that the voltmeter being used is of the correct range and that it is the proper type for the current in the circuit, whether alternating or direct current.

If a particular electric unit is not functioning, the first step is to determine whether electric power is being delivered to the unit. This is quickly accomplished by testing with the voltmeter. In a system with negative (−) ground, the positive prod or alligator clip connected to the voltmeter is touched to the terminal of the unit being checked and the negative prod is touched to the metal of the airplane. If power is reaching the test point, the voltmeter should read system voltage.

Circuits can be tested while "hot," that is, with power on, or they may be tested by connecting the voltmeter with the power off, then turning the power on and observing the response of the voltmeter. In testing "hot" circuits, care must be taken to avoid causing short circuits by allowing some metal object to bridge between the terminal being tested and the metal of the aircraft. The test leads of the voltmeter should be well insulated, and the test prods or alligator clips should be insulated except at the points where electrical contact is to be made.

On large aircraft where 208-V 400-Hz circuits are to be tested, great care must be taken to avoid contacting "hot" parts of circuits with the bare hands or any other part of the body. Severe shocks will occur at any time a circuit is completed through the body. Working procedures established by the company operating large aircraft are usually such that the danger of shock is minimized.

For testing the normal operating circuits of an aircraft electric system, the sensitivity of the test voltmeter need not be high. A voltmeter with a sensitivity of 1000 Ω/V will usually suffice. For testing electronic circuits where current flow is very low, the test voltmeter should be of the vacuum-tube or solid-state type. As explained previously, these instruments isolate the test function from the circuit so no load is applied to the circuit.

THE OHMMETER

The ohmmeter, as its name implies, is an instrument for measuring resistance. The type most commonly used by aircraft mechanics and electronics technicians employs a moving-coil galvanometer similar to those described for ammeters and voltmeters. To make the movement capable of measuring resistance directly, it is necessary merely to provide a source of electric power and a suitable resistance.

Figure 5-14 is a schematic diagram of a simple ohmmeter circuit. The principle of operation follows Ohm's law, and a 3-V battery provides the power necessary for operation. The meter movement has a sensitivity of 1000 Ω/V and an internal resistance of 10 Ω. The total series resistance throughout the circuit must be 3000 Ω to provide a full-scale deflection of the indicating needle when the emf of the power source is 3 V. This is attained by placing a 2500-Ω fixed resistance and a 490-Ω variable resistance in series with the battery and the meter movement. The variable resistance makes it possible to compensate for the lowering of the battery voltage over a period of time.

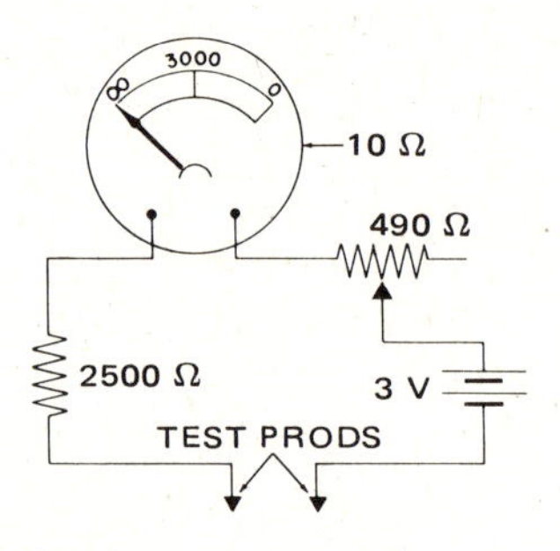

Fig. 5-14 A simple ohmmeter circuit.

When the test prods are in contact with each other, the indicating needle moves to the full-scale position. This point is marked zero because it indicates that there is zero resistance between the test prods. If a resistance of 3000 Ω is placed between the test prods, the needle travels halfway across the scale. This point on the scale is marked 3000 Ω. If the test prods are separated, that is, placed to measure the resistance of the air between them, the indicating needle remains at the extreme left side of the scale. This point is marked infinity (∞) because the resistance of the air is so great that no measurable current passes through the circuit; hence, for practical purposes, the resistance is infinite.

The range of resistance readings on the scale of the ohmmeter described above is from zero to infinity, but the practical range is from approximately 100 to 30 000 Ω. The scale divisions for the very high resistances are so close together that the probability of error increases tremendously as the value of the reading becomes higher. The basic range of the ohmmeter may be changed by the use of resistances as multipliers. For higher resistances, it is necessary to use a higher voltage or a more sensitive movement. In either case, the current-limiting resistance must be increased.

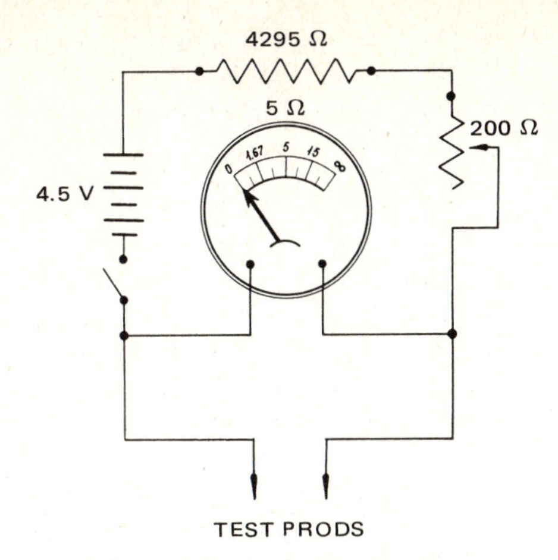

Fig. 5-15 Ohmmeter for testing low resistances.

An ohmmeter designed for the measurement of very low resistances must be connected so that the resistance to be measured acts as a shunt resistance across the test prods. A circuit for this type of ohmmeter is shown in Fig. 5-15. The meter movement is connected in series with a battery, a switch, a fixed resistor, and a variable resistor. If we assume that the meter has an internal resistance of 5 Ω and that 1 mA will produce a full-scale deflection, then a total circuit resistance of 4500 Ω will provide for a full-scale deflection when a 4.5-V battery is used to power the instrument. To test a resistance, the switch is closed and the indicating needle moves to the extreme right of the scale; that is, it indicates infinite resistance. If we place a 5-Ω resistor between the test prods, the needle will take a position at half scale. This is because 0.5 mA is now passing through the meter and 0.5 mA through the resistor. If a 15-Ω resistor is placed between the prods, the indicating needle will take a position three-fourths the distance from the zero end of the scale. This is because $\frac{3}{4}$ mA will be flowing through the meter and $\frac{1}{4}$ mA through the resistor.

The ohmmeter is a most useful instrument for testing or checking electric circuits and appliances. It not only measures resistance but also is an excellent continuity tester.

To measure the resistance of any electric item such as a resistor, coil, or length of wire, it is necessary merely to connect the test prods of the ohmmeter to the terminals of the item being tested. Before the test is made, the ohmmeter should be adjusted so the scale reading is zero when the prods are connected together and infinity when the prods are separated. The range of the ohmmeter used should be such that the indicating needle will move to a point in the center two-thirds of the scale when the unit being tested is connected to the test prods. With some instruments, the manufacturer may suggest a different portion of the scale as providing the most accurate results.

In testing units connected in a circuit, all electric power in the circuit must be turned off. It is good practice to disconnect or remove power sources from a circuit or unit when testing portions of the circuit with an ohmmeter. Electric power in a circuit being tested with an ohmmeter will usually cause damage to the instrument and will certainly prevent a correct reading of resistance values. It is often necessary to disconnect one side of a unit from a circuit to prevent other parts of the circuit from affecting the reading.

To use an ohmmeter as a continuity tester, it is necessary merely to contact the terminals of the circuit being tested with the test prods of the ohmmeter. This is illustrated in the drawing of Fig. 5-16, which shows how a section of circuit wiring can be checked. Usually the wiring for a number of circuits is included in a bundle called a **harness** and it is not possible to inspect the individual wires visually. The particular circuit to be tested can be identified at each end by means of coding on the wire and in the circuit diagram provided in the aircraft maintenance manual. After the wire to be tested is identified, the prods of the ohmmeter are touched to the two terminals of the wire simultaneously. If the wire is in good condition, the ohmmeter will read near the zero resistance point.

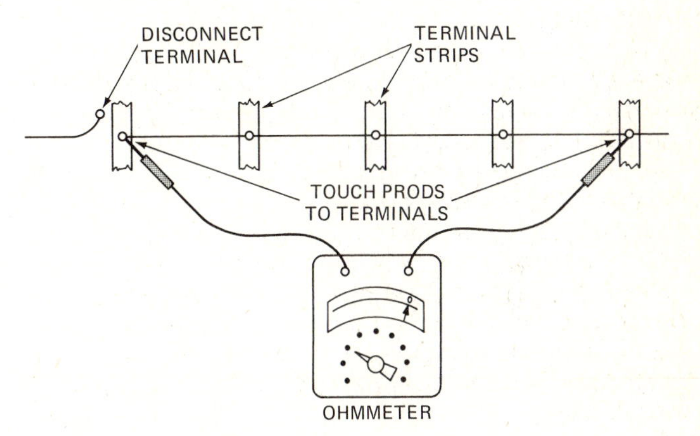

Fig. 5-16 Use of an ohmmeter as a continuity tester.

AC MEASURING INSTRUMENTS

The principal types of ac instrument movements are classified as follows: iron-vane meters, hot-wire meters, thermocouple meters, dynamometers, and meters using rectifiers. The iron-vane movement and the dynamometer have been discussed in a previous section of this chapter. These movements are satisfactory for alternating current of relatively low frequencies within a range of 15 to 1000 Hz. For frequencies of over 150 Hz, the dynamometer-type movements must have corrections applied.

The **hot-wire meter** was one of the first movements used for HF ac measurements which was not adversely affected by the frequency of the current. The principle of this meter is illustrated in Fig. 5-17. The actuating element is a wire which expands as its temperature in-

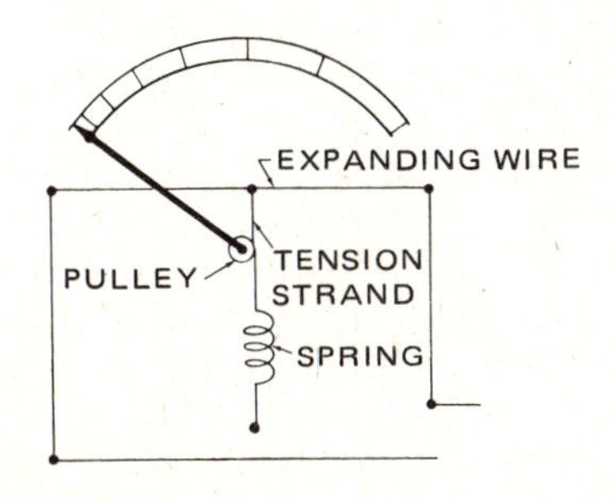

Fig. 5-17 A hot-wire meter.

creases. This thermal-sensitive wire usually consists of platinum silver or platinum iridium. The ends of this wire are connected to the source of power. As the current flows through the wire, it heats and expands, and this expansion allows the tension spring to move the tension strand which is wrapped around a small pulley on the pointer shaft. As the tension strand moves, the pointer shaft is rotated and the pointer moves across the face of the dial. Since the expansion of the wire is proportional to its temperature and the temperature increases as the square of the current flowing through the wire, the scale is marked according to the square law.

Because of several undesirable features limiting its usefulness, the hot-wire movement has been superseded by the thermocouple meter for the measurement of HF currents. The accuracy of the hot-wire instrument is impaired by changes in temperature, and in operation it consumes a relatively large amount of power. Because of these features, it is not suitable for use in circuits in which the voltage and current values are very small or in which substantial changes in temperature are encountered.

A satisfactory movement for the measurement of HF ac values is the **thermocouple meter.** A thermocouple circuit consists of two junctions of two dissimilar metals in one circuit. When one junction is heated an emf develops, and a current flows in the circuit. This action is illustrated in Fig. 5-18. The emf developed by a thermocouple depends on the nature of the metals and the difference in temperature between the heated junction (hot junction) and the cold junction.

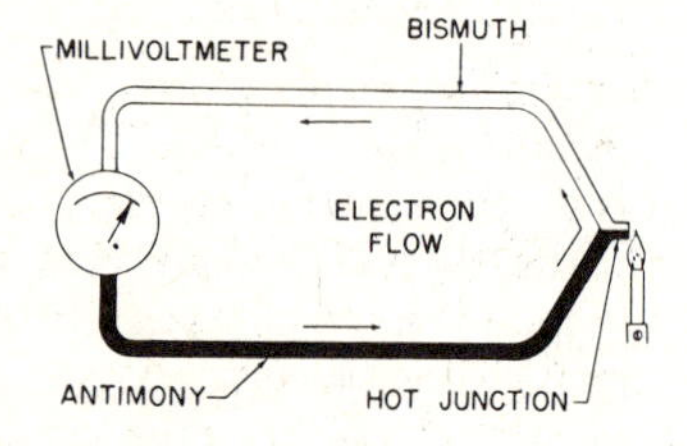

Fig. 5-18 Thermocouple principle.

The **thermoelectric effect** (production of an emf by the application of heat) of a thermocouple is caused by the atomic structure of the metals employed. The thermocouple of electric measuring instruments usually consists of an alloy of antimony with an alloy of bismuth, because these metals will produce a greater emf per degree of difference in temperature between the hot junction and the cold junction than will other combinations of metals.

When two dissimilar metals are placed in contact, there is a momentary flow of electrons from one metal to the other. This is because one metal gives up electrons more easily than the other. The electron flow will not continue, however, because the emf on one metal soon balances the emf on the other.

If conductors of antimony and bismuth are connected at two points, as in Fig. 5-18, there is a momentary flow of electrons from the antimony to the bismuth at both junctions. Thereafter, however, if the junctions have the same temperature, there is no flow of current through the circuit because the difference of potential at one junction is equal to the difference of potential at the other junction. But when heat is applied at one of the junctions, the potential difference will increase at that point because the heat increases the tendency of the electrons to flow from the antimony to the bismuth. Since the emf at the hot junction is greater than the emf at the cold junction, a current will continue to flow for as long as there is a difference in temperature between the two junctions. If a sensitive meter is inserted in the circuit at the cold junction, the current flow in the circuit can be measured.

A schematic diagram of a thermocouple-meter circuit is shown in Fig. 5-19. The terminals of the source of current are connected to the instrument at *A* and *B*. Current then flows through the heating element and causes a rise in temperature at the thermocouple. The emf produced by the thermocouple causes a current to flow through the sensitive meter movement, and the indicating needle gives a reading proportional to the current produced by the thermocouple.

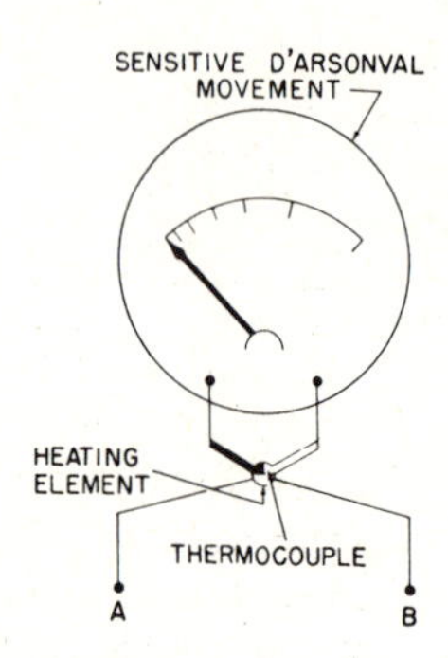

Fig. 5-19 Thermocouple meter.

It must be noted that the heating effect at the thermocouple is proportional to the square of the current flowing through the heating element and that the current produced by the thermocouple is directly proportional to the difference in temperature between the hot junction and the cold junction. For this reason, the movement of the indicating needle is proportional to the square of the current, and the scale must be marked according to the square law.

The direction of current through the heating element does not influence the direction of the current produced by the thermocouple; hence, the instrument is satisfactory for measuring alternating current. A thermocouple scale marked for direct current also indicates the effective value for alternating current because this effective value is determined by comparing the heating effect of alternating current with the heating effect of direct current.

Thermocouple meters are constructed in such a manner that the heating effect of an electric current is almost immediately apparent; hence, a reading is obtained very quickly. Because of the sensitivity of the heating element, the meter does not have a large safety factor. For this reason the capacity of the meter must not be exceeded at any time. Even very careful electricians burn out meters occasionally; so it is necessary to be sure that the current supplied by the circuit being tested does not exceed the capacity of the instrument.

As previously explained, the principal use of thermocouple meters is in the measurement of HF ac values. By the application of proper calibration, thermocouple instruments may be used to measure current or voltages up to frequencies of 100 megahertz (MHz) (100 000 000 Hz).

One of the most common instruments for use with relatively low-frequency alternating currents is a dc meter movement connected in a circuit with a full-wave rectifier as shown in Fig. 5-20. An instrument such as this is often called a **rectifier instrument.** A full-wave rectifier is used to change the alternating current to direct current. When such a meter is used for both ac and dc measurements, two scales are usually provided. This is because the movement of the needle, when measuring ac values, is proportional to the average value of the current rather than the effective value. If the dc scale only is used, the reading on the dc scale must be multiplied by 1.11 to obtain the ac effective values.

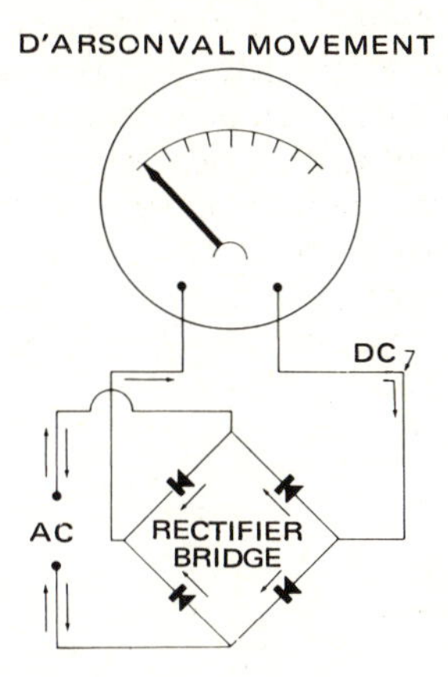

Fig. 5-20 Rectifier-type meter.

When using a rectifier-type meter with HF alternating currents, the error increases in proportion to the frequency. This error is caused by the capacitive effect of the rectifier elements. At extremely high frequencies, the rectifier will pass a substantial amount of current in both directions; hence, the current will flow through the instrument in both directions. The result is that the current in one direction is reduced by the current in the opposite direction.

THE WATTMETER

Wattmeters are not frequently used by the aircraft maintenance technician, but a short discussion of the principles of such meters will aid in the understanding of the measurement of electric power.

The unit for the measurement of electric power is the watt. One watt is the power expended when a current of 1 A is flowing under the pressure of 1 V. In an electric circuit, the power in watts is equal to the product of the voltage and the amperage. This is true in dc circuits and in ac circuits when the voltage and current are in phase.

Since electric power involves both amperage and voltage, the wattmeter must be capable of multiplying these values. A schematic diagram of a wattmeter connected in a circuit is shown in Fig. 5-21. The instrument is constructed in a manner similar to a dynamometer movement, but the circuits for the magnetic field and the moving coil are separate. One of the windings must provide a field proportional to the current in the circuit, and the other must produce a field proportional to the voltage. Since the current windings must be heavy enough to carry the current of the circuit, the current coil is stationary. The moving coil carries the voltage winding because this winding must have a high resistance and is relatively light in weight.

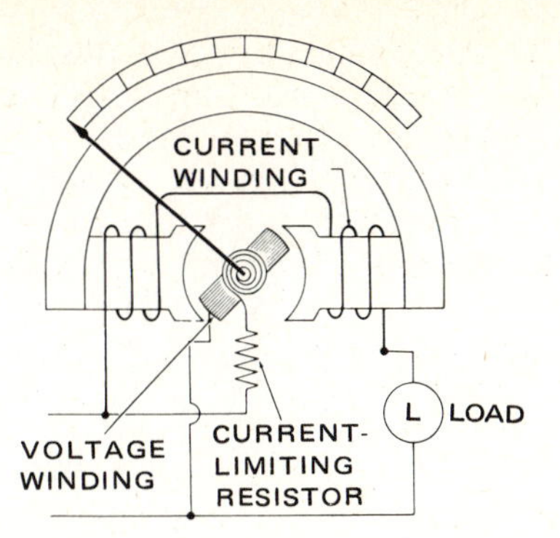

Fig. 5-21 Wattmeter circuit.

The current circuit is connected in series with the load circuit, and the voltage circuit is connected in parallel with the load circuit. A current-limiting resistance is connected in series with the voltage coil. When the circuits are connected in this manner, the strength of the stationary field is proportional to the load current, and the strength of the moving-coil field is proportional to the voltage across the load. The indicating needle moves a distance proportional to the product of the voltage and the amperage; hence, the scale may be marked directly in watts.

THE MULTIMETER

In practice, the functions of a voltmeter, ohmmeter, and ammeter (or milliammeter) are usually combined in an instrument called a **multimeter** or **voltohmmeter** (VOM). These combination instruments make it possible to take a wide range of electrical measurements with just one basic instrument. One type of multimeter is shown in Fig. 5-22. This instrument will measure ac and dc voltage up to 5000 V, resistance from zero to infinity, and dc current from zero to 10 A. One meter movement is used to provide all the indications by means of the various scales

Fig. 5-22 A typical multimeter.

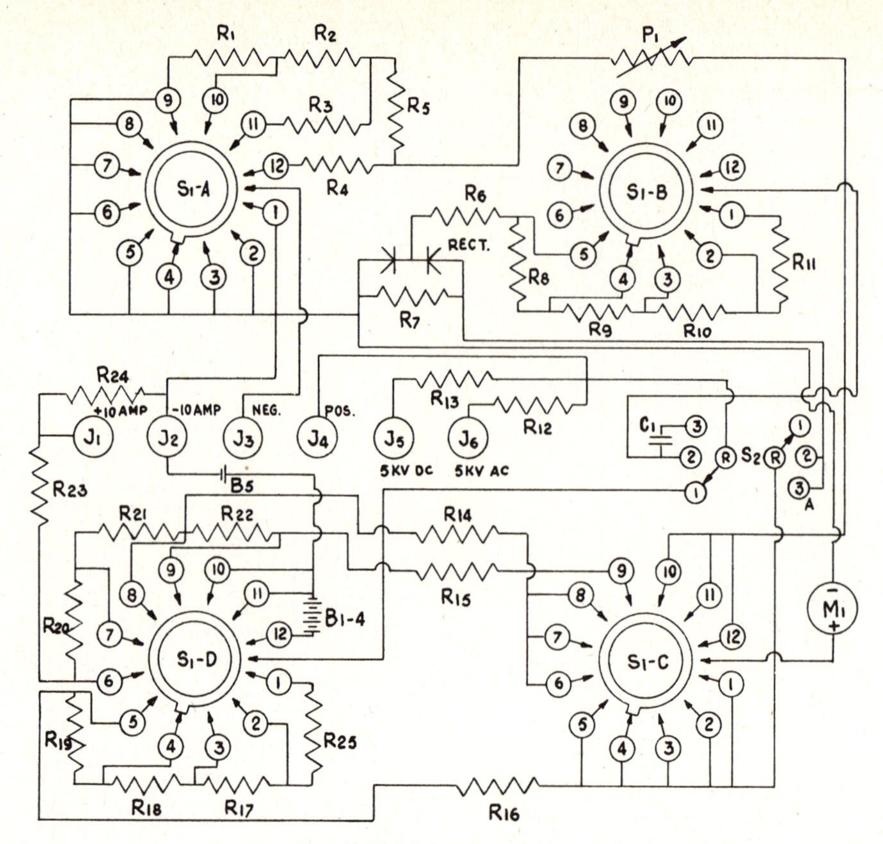

Fig. 5-23 Multimeter circuit.

on the dial. For each type of indication, the instrument is adjusted by means of a rotary, multiwafer switch and by plugging the test leads into the proper jacks.

The circuit for the multimeter in Fig. 5-22 is shown in Fig. 5-23. It will be noted that the four wafers of the rotary switch have 12 positions. In the circuit diagram the switch is set for position 4. In this position, the meter is set to measure in the 5000-V dc range. The test leads are plugged into the common jack (J_3) and the 5000-V jack (J_5). A careful study of the circuit will reveal the many combinations available.

Another, more sophisticated, multimeter is shown in Fig. 5-24. This is a solid-state VOM that provides the sensitivity necessary for testing voltage in low-current electric or electronic circuits. This is accomplished by incorporating a dual field-effect transistor (FET) which isolates the indicating circuit from the circuit being tested. This is the same principle involved in vacuum-tube voltmeters (VTVM) in that the test circuit does not impose any load on the circuit being tested. This assures an accurate reading for all voltage measurements.

The multimeter shown in Fig. 5-24 provides for the measurement of resistance from zero ohms to infinity, ac and dc voltages from zero to 1000 V, or 1 kilovolt (kV), and current from zero to 10 A dc. In measuring current above 300 mA, one test lead is plugged into the dc 10-A jack. The measuring circuits are powered by a 9-V battery and a 1.5 V dry cell. Adjustments are provided so the meter can be adjusted to zero before making tests.

When using a multimeter, the technician must follow the instructions provided in the instruction manual for the instrument. In testing an unknown voltage or amperage, it is important that the meter range be set above the highest level likely to be encountered. It is good practice to set the highest range available to start, then adjust downward until the reading falls in the upper one-third of the scale.

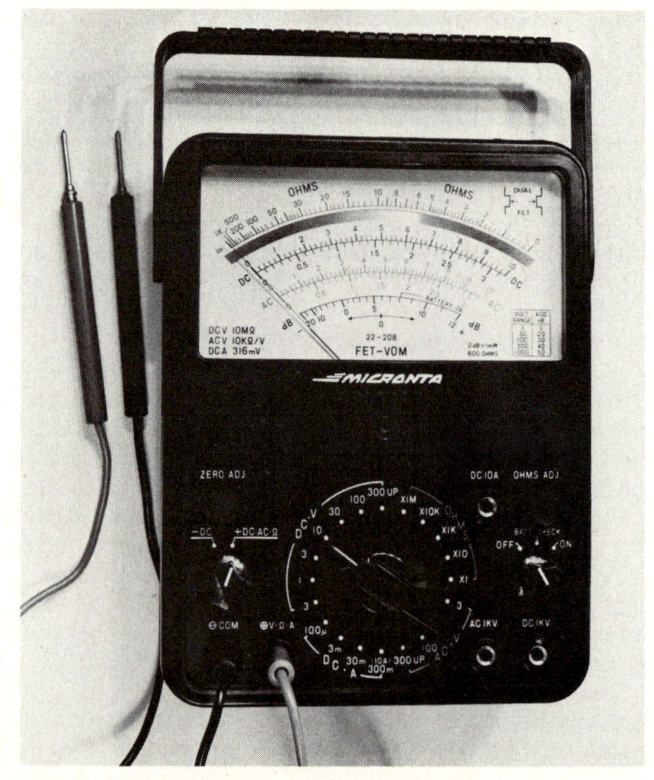

Fig. 5-24 A solid-state multimeter.

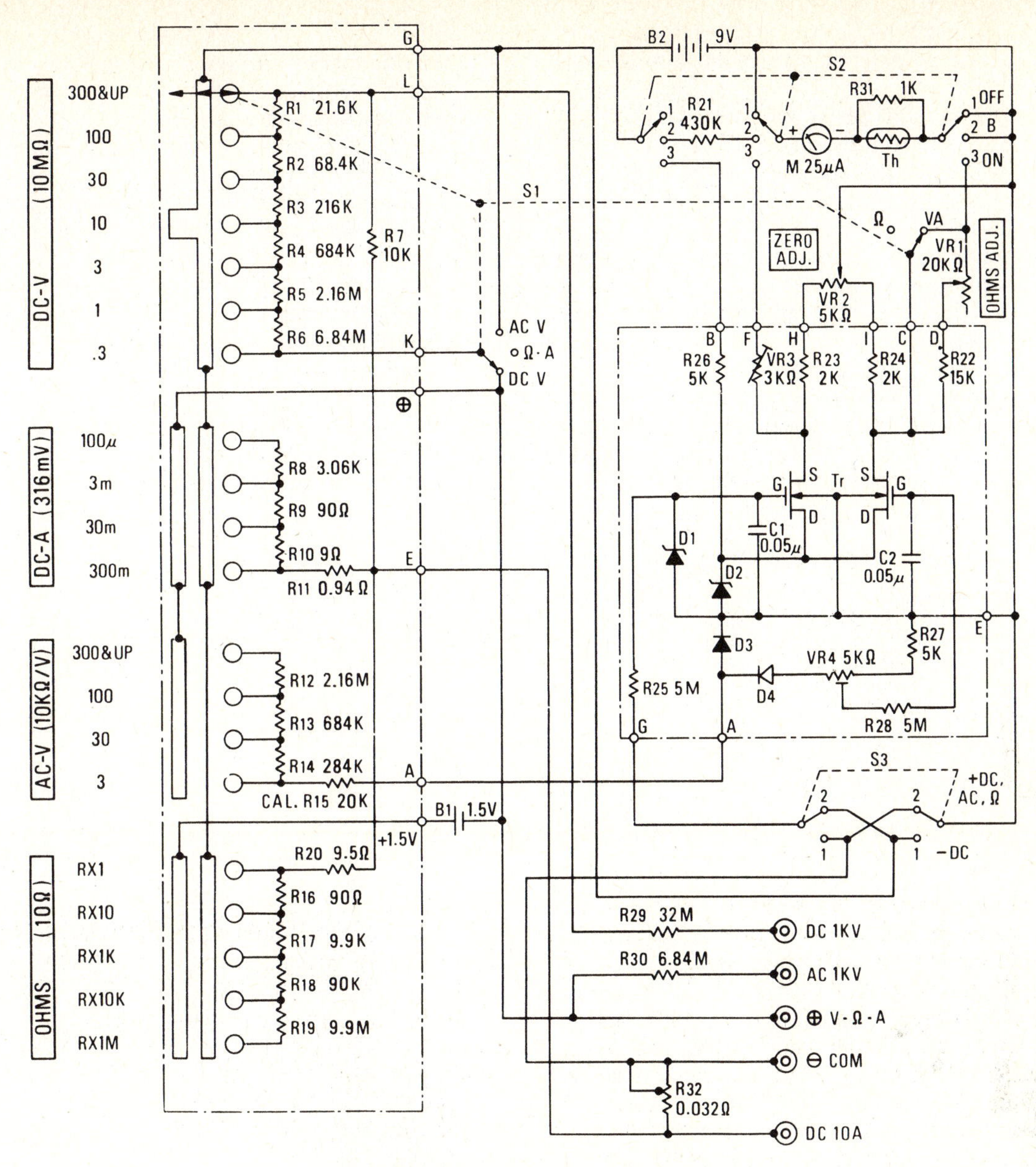

Fig. 5-25 Solid-state multimeter circuit.

This practice will avoid damage to the instrument from overload and provide for maximum accuracy. The meter movement is provided with automatic protection against overload. When voltages above the 30-V range are tested, care must be taken to avoid shock. This is particularly important in the 300- to 1000-V range.

The circuit for the multimeter in Fig. 5-24 is shown in Fig. 5-25. The range switch is a rotary type even though it is shown as straight-line contacts in the schematic diagram. As the switch is rotated, the circular contact shown in each position is electrically connected to the contact strips. Note that only one set of contacts is made for each position of the switch.

REVIEW QUESTIONS

1. Give three general classifications of electric measuring instruments.
2. Explain the operation of a galvanometer.
3. Describe the *d'Arsonval* or *Weston* meter movement.
4. Compare the *Weston* movement with the *dynamometer* movement.
5. For what type of current measurement is the iron-vane movement most suitable.
6. Describe the type of pivot bearing most commonly used in electrical-instrument movements.
7. What is meant by a *taut-band* movement?
8. Explain a meter sensitivity of 20 000 Ω/V.
9. What is the reason for using a very sensitive instrument to test voltages in a circuit?
10. Why is a vacuum-tube voltmeter or a solid-state voltmeter needed for testing electronic circuits?
11. Show the difference between a meter circuit connected as an ammeter and one connected as a voltmeter.
12. What important precaution must be observed when connecting an ammeter in a circuit?

13. How must a voltmeter be connected with respect to a load in order to determine the voltage applied to the load?
14. A meter movement has a full-scale deflection at 1 mA and an internal resistance of 15 Ω. In this case, what value should the multiplier resistor have if the movement is to be used in a voltmeter which has a range of 10 V?
15. If you wish to use the movement described in the foregoing question for an ammeter with a range of 100 A, what should be the value of the shunt resistance?
16. A meter movement having an internal resistance of 10 Ω and a full-scale deflection of 1 mA is used in an ohmmeter. The power supply for the ohmmeter is a 3-V dry cell. What is the required series resistance, and what will be the midscale reading?
17. Show how a meter movement must be connected to provide for the indication of very low resistances.
18. Name five types of instrument movements used for the measurement of ac values.
19. What is the most satisfactory movement for measuring HF ac values?
20. Explain the operation of a thermocouple.
21. Draw a schematic circuit for a rectifier-type instrument.
22. What is the function of a wattmeter?
23. Why is the wattmeter capable of indicating power consumption?
24. Describe a *multimeter.*
25. List and explain the important operations to be performed before connecting a multimeter for a test.
26. What are the advantages of a *solid-state* multimeter or *voltohmmeter* (VOM)?

CHAPTER 6

Batteries

In the past, the number of different types of electric cells and batteries has been comparatively low. The average person was not likely to encounter more than two types: the carbon-zinc dry cell and the lead-acid storage battery. A few other types of cells were made for special purposes; however, these were not generally known.

Today, hundreds of types and sizes of cells and batteries are in use. This is because of the many electric and electronic devices invented for the needs of the average man, as well as for science, technology, medicine, and other fields. Tiny mercury cells are used in hearing aids, cameras, electric watches, and other devices utilizing miniaturized circuits. Cells are made in many sizes, shapes, and types to meet the requirements of a wide variety of electrical loads from heavy-duty to low-drain and intermittent service.

We shall not attempt to consider all types of cells and batteries in this chapter, but will examine the theory and construction of a sufficient number of types so that technicians will understand the use and care of those they may come into contact with on or off the job. These will include those described as carbon-zinc, mercury, lead-acid, and nickel-cadmium.

DRY CELLS AND BATTERIES

VOLTAIC CELLS

In an earlier portion of this text, it was explained that various dissimilar substances have opposite polarities with respect to one another and that when two such substances are rubbed together, one will have a positive charge and the other a negative charge. Dissimilar metals also have this property, and when two such metals are placed in contact with each other, there will be a momentary flow of electrons from the one having a negative characteristic to the one having a positive characteristic. If two plates of dissimilar metals are placed in a chemical solution called an **electrolyte,** opposite electric charges will be established on the two plates.

An electrolyte is technically defined as a compound which, when molten or in solution, conducts electric current and is decomposed by it. In simple terms, an electrolyte is a solution of water and a chemical compound which will conduct an electric current. The electrolyte in a standard storage battery consists of sulfuric acid and water. Various salts dissolved in water will also form electrolytes.

An electrolyte will conduct an electric current because it contains positive and negative ions. When a chemical compound is dissolved in water, it separates into its component parts. Some of these parts carry a positive charge, and others carry a negative charge.

The action of an electrolyte will be clear if a specific case is considered. When a rod of carbon and a plate of zinc are placed in a solution of ammonium chloride, the result is an elementary voltaic cell (see Fig. 6-1). The carbon and zinc elements are called **electrodes.** The carbon, which is the positively charged electrode, is called the **anode,** and the zinc plate is called the **cathode.**

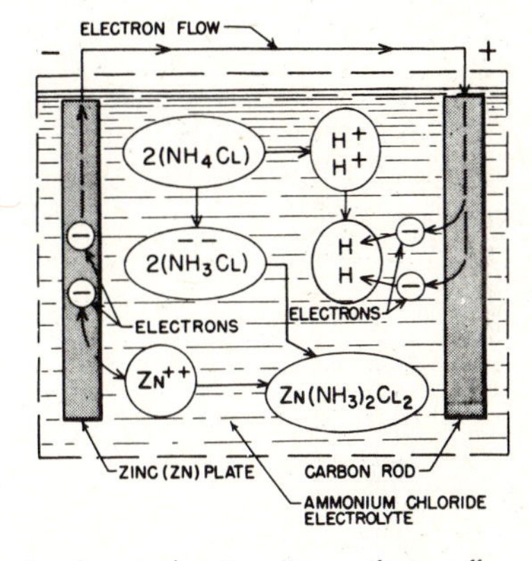

Fig. 6-1 Electrochemical action in a voltaic cell.

As soon as the zinc (Zn) plate is placed in the electrolyte, zinc atoms begin to go into solution as ions, each leaving two electrons at the plate. This causes the zinc plate to become negatively charged. The zinc ions in the solution are positive because each one lacks the two electrons left at the plate. This positive charge causes the zinc ions to remain near the zinc plate because the plate has become negative. The effect of the zinc ions gathered near the plate is to stop the decomposition of the zinc plate for as long as the negative charge of the plate is balanced by the positive charge of the zinc ions in solution.

The ammonium chloride in solution in the electrolyte apparently separates into positive hydrogen ions and a combination of ammonium and chlorine which is negatively charged. When the two electrodes are connected by an external conductor, the free electrons from the zinc plate flow to the carbon rod; and the hydrogen ions move to the carbon rod, where each ion picks up one electron and becomes a neutral hydrogen atom. The positive zinc ions combine with the negative ammonium chloride to take the place of the hydrogen ions released into solution. The effect of these chemical actions is to remove electrons from the carbon rod and to liberate free electrons at the zinc plate. This results in a continuous supply of electrons available at the negative (zinc) electrode. When the two electrodes are connected, the electrons will flow to the carbon rod, where the hydrogen ions become hydrogen atoms as the result of their neutralization by the electrons. Eventually, hydrogen gas bubbles form on the carbon rod and insulate it from the solution. This is called **polarization** and will cause the current flow to stop until the hydrogen is removed. For practical voltaic cells, it is necessary to employ a method of depolarization.

The standard **dry cell** used in flashlights and for other purposes for which a low-voltage dc supply is desired employs a compound called manganese dioxide (MnO_2) to prevent the accumulation of hydrogen at the positive electrode in the cell. Figure 6-2 is a drawing of this type of cell. A dry cell is so called because the electrolyte is in the form of a paste; the cell may therefore be handled without the danger of spillage. The zinc can is the negative electrode, and the paste electrolyte is held in close contact with the zinc by means of a porous liner. The space between the carbon rod and the zinc can is filled with manganese dioxide saturated with electrolyte. Graphite is mixed with the manganese dioxide to reduce the internal resistance of the cell. The top of the cell is sealed with a wax compound to prevent leaking and drying of the electrolyte. Many cells are encased in a tin-plated steel can to make them more durable; a layer of insulating material is then placed between the inner zinc can and the outer can to prevent short circuiting.

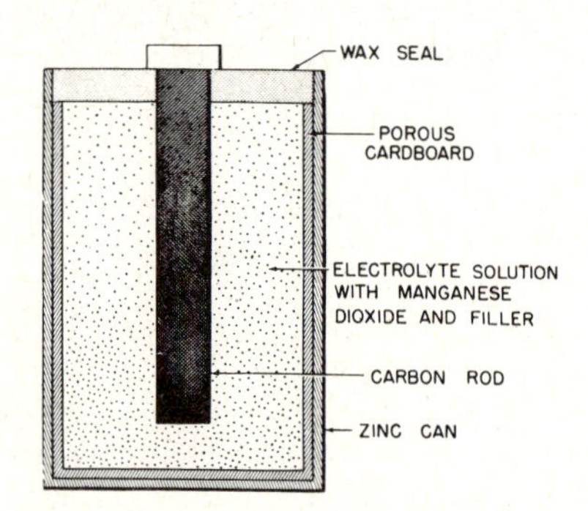

Fig. 6-2 Construction of a simple dry cell.

The emf developed by a zinc-carbon cell is approximately 1.5 V. The voltage of any cell depends upon the materials used as electrodes. As previously stated, dissimilar metals always have a definite polarity with respect to one another. For example, if nickel and aluminum are placed in an electrolyte, the nickel will be positive and the aluminum negative. However, if nickel and silver are acted upon by the same electrolyte, the nickel will be negative and the silver positive. The more active a metal is chemically, the greater its negative characteristic. A lead-acid secondary cell, such as those employed in storage batteries, develops an emf of 2.1 V. The electrodes (plates) are composed of lead for the negative and lead peroxide for the positive.

In a **secondary cell,** the chemical action which produces the electric current can be reversed. This is accomplished by applying a voltage higher than that of the cell to the cell terminals; this causes a current to flow through the cell in a direction opposite to that in which the current normally flows. The positive terminal of the charging source is connected to the positive terminal of the cell, and the negative terminal of the charging source is connected to the negative terminal of the cell. Since the voltage of the charger is higher than that of the cell, electrons flow into the negative plate and out of the positive plate. This causes a chemical action to take place which is the reverse of that which occurs during operation of the cell; the elements of the cell return to their original composition. At this time, the cell is said to be *charged.* Secondary cells may be charged and discharged many times before they deteriorate to the point at which they must be discarded.

A cell which cannot be recharged satisfactorily is called a **primary cell.** The elementary voltaic cell described previously in this section is a primary cell. Some of the elements deteriorate as the cell produces current; hence, the cell cannot be restored to its original condition by charging. The common flashlight cell is a familiar example of a primary cell.

The negative plate of a primary cell deteriorates because the material goes into solution with the electrolyte. In the secondary cell, the material of the plates does not go into solution but remains in the plates, where it undergoes a chemical change during operation.

BATTERIES

A **battery** consists of a number of primary or secondary cells connected in series to obtain a desired voltage. Figure 6-3 is an illustration of the arrangement of an early type of the 22.5-V B battery. The cells in the illustration are connected in series. When cells are connected in this

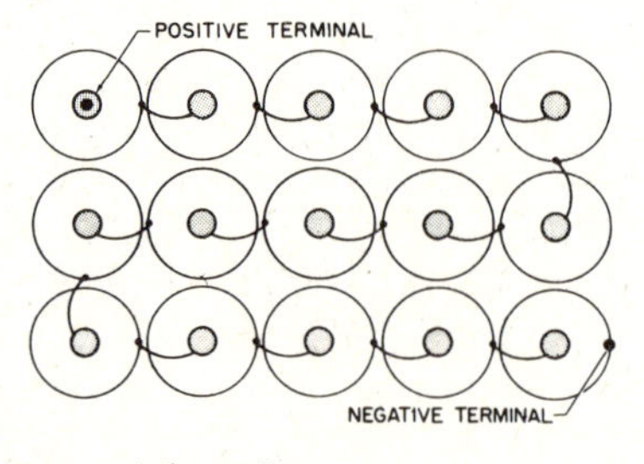

Fig. 6-3 A battery of dry cells.

manner, the total voltage of the battery is equal to the sum of the voltages of the cells. The cells supply 1.5 V each; hence, the total voltage here is equal to 1.5×15, or 22.5, V.

When cells are connected in series, the positive terminal of one cell must be connected to the negative terminal of the next cell, and so on through the complete series of cells. The open positive terminal of the cell at one end of the series is the positive terminal of the battery, and the open negative terminal at the other end is the negative terminal of the battery.

When small, compact batteries are desired for lightweight electronic devices, the individual cells are usually made in a flat, rectangular shape. The cells are then stacked together so that there is no waste space between them. In this manner, a fairly high-voltage battery can be built for use in a small space. Such batteries have a very low, but adequate, current capacity, inasmuch as the type of electronic equipment for which the battery is designed usually requires very little amperage. The construction of a typical battery pack is shown in Fig. 6-4.

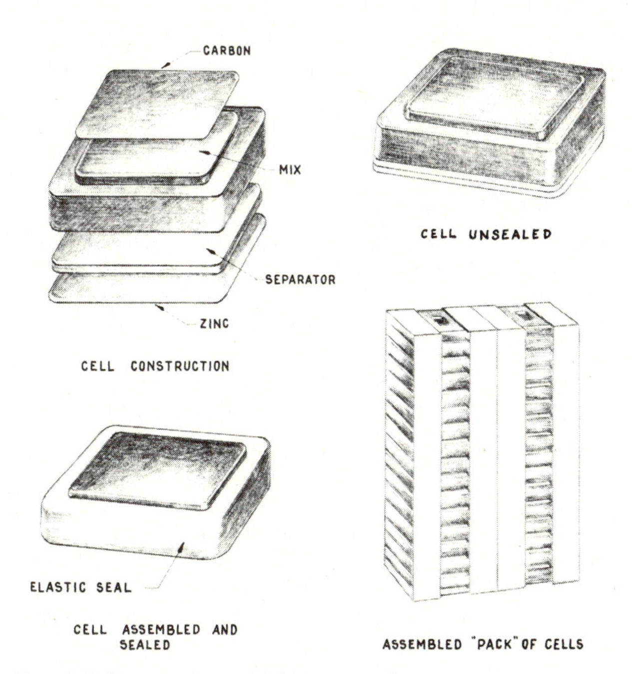

Fig. 6-4 Construction of a battery pack.

Another type of cell used for batteries which are light and compact is the wafer cell. This cell is rectangular in shape with the corners slightly rounded. It consists of a sandwich of artificial manganese dioxide mix between disks of flat zinc and carbon. The electrolyte is an ammonium chloride mixture. The sandwich is wrapped in a plastic film and sealed as shown in Fig. 6-5. A spot of silver wax on the positive and negative sides of the cells provides an electrical contact between cells as they are stacked together. The stack of cells is wrapped in plastic and packaged as a finished battery. The cells are made in a variety of sizes to meet the requirements of many different applications.

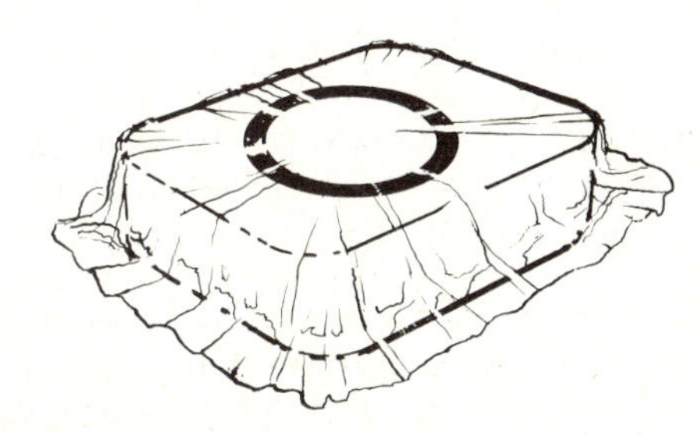

Fig. 6-5 Sandwich cell.

ALKALINE CELLS

Voltaic cells utilizing an alkaline electrolyte are usually termed **alkaline cells.** The electrolyte consists primarily of a potassium hydroxide solution. Potassium hydroxide (KOH) is a powerful caustic similar to household lye and can cause severe burns if it comes into contact with the skin. The electrodes of such cells can be of several different types of materials such as manganese dioxide and zinc, silver oxide and zinc, silver oxide and cadmium, mercuric oxide and zinc, or nickel and cadmium. Some of these cells are rechargeable and can be classed as **secondary cells.**

The capacity of any small cell is rated in milliamperehours since the normal load drain is considerably less than 1 A. A cell which can provide a load drain of 15 mA for 10 h is rated at 150 milliampere-hours (mAh). Large cells which can provide upward from several hundred milliamperes for a normal load are usually rated in ampere-hours. The open-circuit voltage for alkaline dry cells varies according to the materials from which they are made. Table 6-1 shows the characteristics of some types of alkaline cells.

Mercury cells are commonly used for many applications and are made in sizes to fit the device in which they are to operate. An illustration of the variety of such cells is shown in Fig. 6-6.

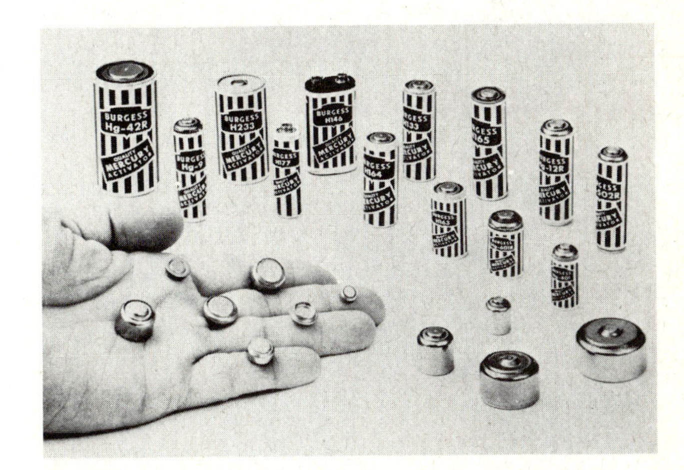

Fig. 6-6 A variety of mercury cells. (Burgess Battery Div., Clevite Corporation.)

A mercury cell consists of a positive electrode of mercuric oxide mixed with a conductive material and a negative electrode of finely divided zinc. The electrodes and the caustic electrolyte are assembled in sealed steel cans. Some electrodes are pressed into flat circular shapes and others are formed into hollow cylindrical shapes, depending upon the type of cell for which they are made. The electrolyte is immobilized in an absorbent material between the electrodes.

NICKEL-CADMIUM CELLS

Nickel-cadmium electric cells and batteries have been developed to a high degree of efficiency and dependability. They are used in small devices which formerly used carbon-zinc dry cells and in other devices where carbon-zinc cells cannot meet the load requirements. They are also being manufactured in large sizes for use in aircraft

TABLE 6-1 Comparative Characteristics of Alkaline Cells

	Open-circuit volts	Average operating volts	W/h		W/h		A/h	
			Per lb	Per km	Per in^3	Per cm^3	Per lb	Per km
Alkaline-manganese	1.46	1.15	50	110	23	377.2	43	94.6
Silver oxide-zinc	1.86	1.5	26–60	57.2–132	1.4–4.0	22.96–65.6	18–40	39.6–88
Silver oxide-cadmium	1.4	1.1	12–40	26.4–88	0.7–3.0	11.48–49.2	10–38	22–83.6
Air-depolarized	1.4	1.2	53	116.6	2.2	36.08	40–45	88–99

where they replace the lead-acid batteries which were formerly used. The service and maintenance of nickel-cadmium aircraft batteries is discussed later in this chapter.

As mentioned previously, a secondary cell is one which can be charged and discharged repeatedly without appreciable deterioration of the active elements. An advantage of the nickel-cadmium secondary cell is that it can stand in a discharged condition indefinitely at normal temperatures without deterioration. If a lead-acid battery is left in the discharged condition for a substantial period of time, sulfation of the plates occurs, and the cells lose much of their capacity.

Some typical nickel-cadmium dry cells and batteries are shown in Fig. 6-7. These cells may be made with various electrode designs, but the active elements are the same. The negative electrode consists of metallic cadmium, and the positive electrode is nickel oxyhydroxide (NiOOH). When the cell is discharged, the negative electrode becomes cadmium hydroxide and the positive electrode becomes nickel hydroxide [$Ni(OH)_2$].

Fig. 6-7 Nickel-cadmium dry cells and batteries.

The most common electrode designs for nickel-cadmium cells consist of perforated steel pockets to hold the active materials or perforated nickel plates or woven nickel screens into which the active materials are impregnated by sintering. **Sintering** is a process of heating finely divided metal particles in a mold to approximately melting temperature. The metal particles weld together where they are in contact with other particles, and this results in a porous material. In the case of nickel-cadmium electrodes, the sintered material is nickel or nickel carbonyl for the positive plates and cadmium for the negative plate. A nickel-cadmium cell which has been cut away to show construction is illustrated in Fig. 6-8.

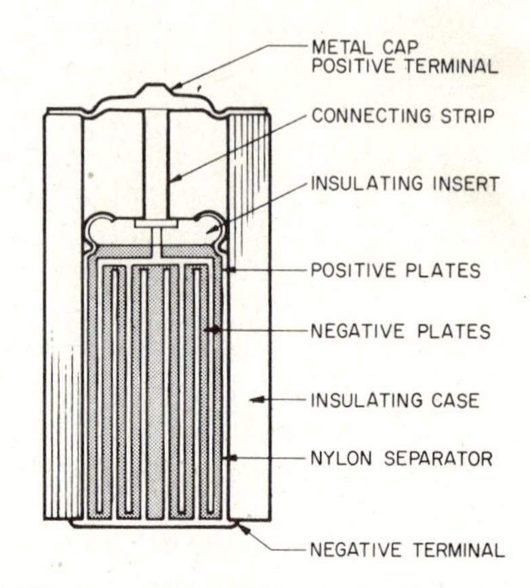

Fig. 6-8 Ni-Cd cutaway to show construction.

During the discharge of a nickel-cadmium cell, electrons are released in the negative material as chemical change takes place. These electrons flow through the outer electric circuit and return to the positive electrode. Positive ions in the electrolyte remove the electrons from the positive electrode. During charge, the reverse action takes place, and the negative electrode is restored to a metallic cadmium state.

Nickel-cadmium cells generate gas during the latter part of a charge cycle and during overcharge. Hydrogen is formed at the negative electrode and oxygen is formed at the positive electrode. In vented-type batteries, the hydrogen and oxygen generated during overcharge is released to the atmosphere together with some electrolyte fumes. In a sealed dry cell, it is necessary to provide a means for absorbing the gases. This is accomplished by designing the cadmium electrode with excess capacity. This makes it possible for the positive electrode to become fully charged before the negative electrode. When this occurs, oxygen is released at the positive electrode while hydrogen cannot yet be generated because the negative electrode is not fully charged. The cell is so designed that the oxygen can travel to the negative electrode where it reacts to form chemical equivalents of cadmium oxide. Thus, when a cell is subject to overcharge, the cadmium electrode is oxidized at a rate just sufficient to offset input energy, and the cell is kept at equilibrium at full charge.

If a cell is charged at the recommended rate, overcharging can occur for as long as 200 or 300 charge cycles without damage to the cell. If the charge rate is too high, the oxygen pressure in the cell may become so great that it will rupture the seal. For this reason, charge rates must be carefully controlled.

Recharging of a nickel-cadmium cell should be at a rate in milliamperes which is equal to approximately 10 percent of the nominal milliampere-hour capacity. For example, a 900 mAh cell or battery should be charged at 90 mA. Cells should be charged at this rate for 14 to 16 h.

STORAGE BATTERIES

The term *storage battery* has been used for many years as the name for a battery of secondary cells and particularly for lead-acid and nickel-cadmium batteries.

Lead-acid secondary cells consist of lead-compound plates immersed in a solution of sulfuric acid and water which is the **electrolyte.** Each cell has an open-circuit voltage of approximately 2.1 when fully charged. When connected to a substantial load, the voltage is approximately 2. Aircraft storage batteries of the lead-acid type are generally rated at 12 or 24 V; that is, they have either 6 or 12 cells connected in series (see Fig. 6-9).

The arrangement of a common 12-V automobile storage battery is shown in Fig. 6-10. Note that there are six cells connected in series to produce 12 V. Actually the emf of a 12-V battery is somewhat more than 12 V because each cell when fully charged produces about 2.1 V. The storage batteries shown in Figs. 6-9 and 6-10 show the cell connectors exposed. Batteries manufactured in recent years have the cell connectors sealed under an asphaltic or plastic material to reduce leakage and improve durability.

Schematic diagrams of cells connected in series and parallel are shown in Fig. 6-11. In the series diagram, four 2-V cells are connected in series to produce 8 V. If the same four cells are connected in parallel, as shown in Fig. 6-11*b*, the total voltage is the same as that of one cell; however, the capacity of a group, in amperes, is four times the capacity of a single cell.

To increase both the voltage and the amperage by combining single cells, the cells are connected in a series-parallel circuit like that shown in Fig. 6-12. When 16 cells are connected in this manner, the voltage is four times as great as that of a single cell, and the current capacity of the combined cell is four times as great as that of a single cell.

When batteries or cells are connected incorrectly, they may be damaged. For example, if a technician intends to connect three batteries in parallel and connects one of them incorrectly, as shown in Fig. 6-13, there will be a short circuit from the center battery to the two end batteries. This will either burn out the wiring or discharge the batteries and possibly damage them beyond repair. It is essential that the technician understand well the characteristics of battery circuits and the proper methods for connecting batteries and cells.

Storage batteries are convenient for aircraft use because their weight is not excessive for the power developed and because they can be kept in a nearly fully charged state by means of an engine-driven generator. It must be remembered that the aircraft storage battery is used only when other sources of electric power are not available. On small airplanes the battery is used directly for starting the engine, but on larger craft an auxiliary unit is usually plugged into the airplane. Jet engines of the larger type require special types of starters such as air-turbine, fuel-air combustion, solid-fuel combustion, and high-powered electric starters. In all cases except the fuel-air combustion and solid-fuel starters, an auxiliary power supply is needed to effect a start.

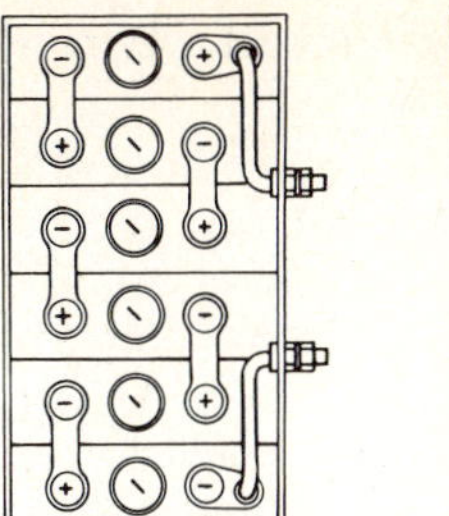
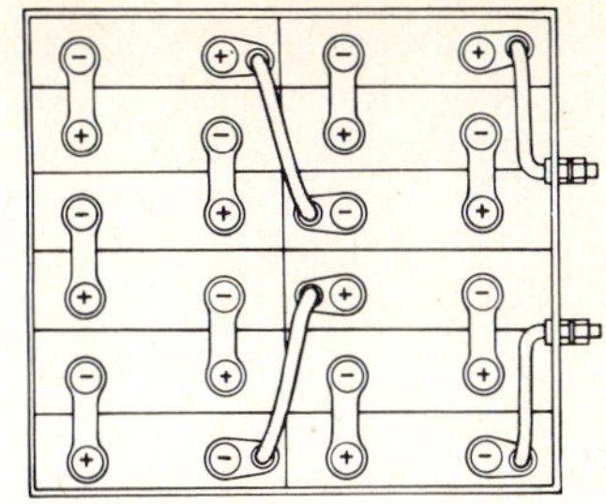

Fig. 6-9 Arrangement of cells in lead-acid storage batteries for aircraft.

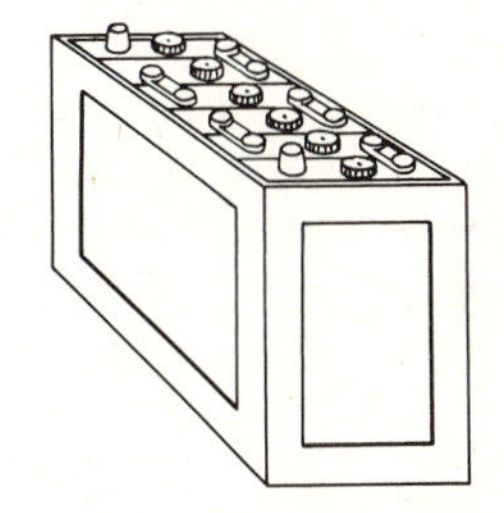

Fig. 6-10 Arrangement of a 12-volt automobile storage battery.

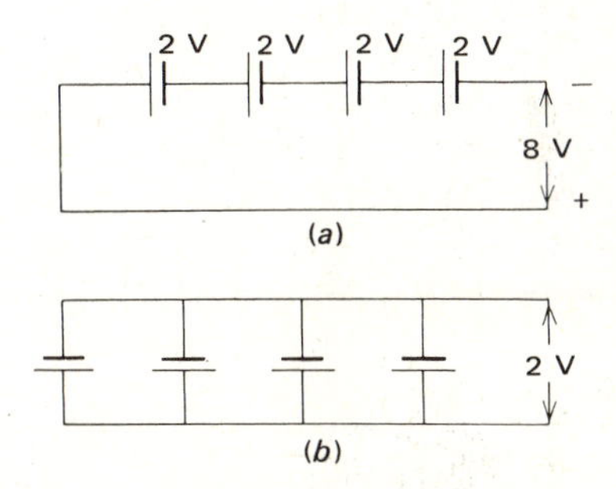

Fig. 6-11 Series and parallel cell connections.

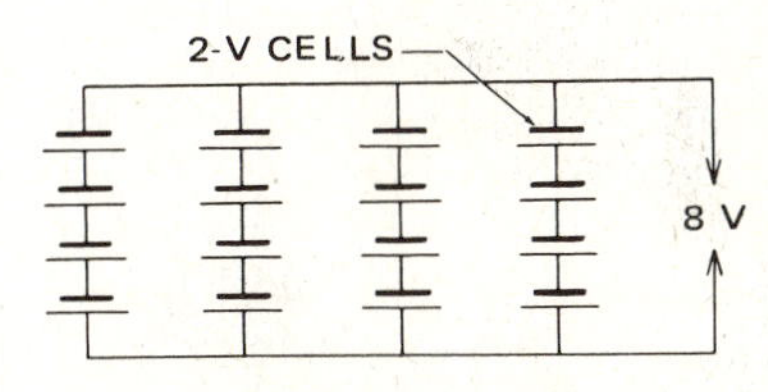

Fig. 6-12 Series-parallel cell connections.

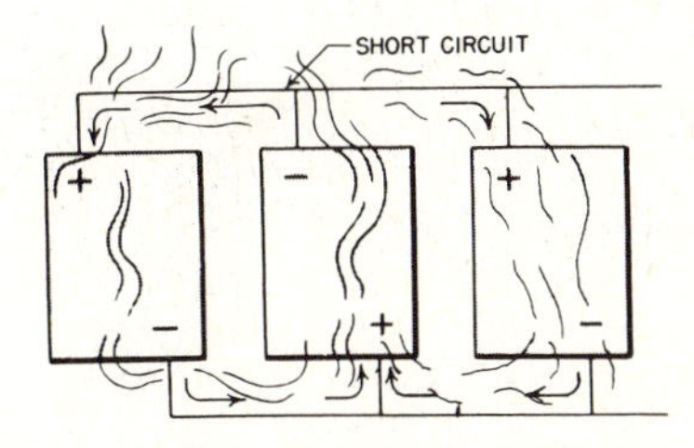

Fig. 6-13 Incorrect battery connections.

Aircraft storage batteries differ from automobile batteries in that they are much lighter in construction. Weight is an important factor in aircraft, missiles, and space vehicles, and hence it is of prime importance that electric power be supplied with the least possible sacrifice of useful load. The battery used in such cases must be as light as possible and still provide power sufficient to meet requirements. Aircraft storage batteries are usually of higher voltage and lower ampere-hour rating than automobile storage batteries; hence, it is necessary that the aircraft battery be kept in a high state of charge at all times. Because of its lighter construction, an aircraft battery must be handled with great care to avoid the possibility of damage.

Another requirement for airborne storage batteries is that they be so constructed that they may be inverted without spilling the electrolyte. Methods for accomplishing this will be discussed in the section describing battery construction. Batteries needed for low-power output for a relatively short period of time may be of the dry-cell type.

The aircraft storage battery can deliver a relatively large amount of power for a short time, but if it is left connected to a heavy load it will soon become discharged. The aircraft storage battery should be used only for limited purposes and for emergencies when other sources of electric energy are not available. Also, it should be kept fully charged to prevent sulfation of the plates and the resulting loss of capacity.

Small airplanes that use a battery only for lights and radio may or may not be equipped with a generator. For such installations, a relatively high-capacity battery should be used. A 34-Ah battery will supply power for lights and a radio receiver for a few hours, but if the airplane is not equipped with a generator the battery should be removed and recharged after each flight. For local daylight flights a dry-cell battery pack may be used to operate a radio.

In large airplanes there are many electric-motor-operated devices which operate for intervals of a few seconds. The initial torque, or starting current, required by some of these motors is several times the capacity of the engine-driven generators, and in instances such as these the extra power is supplied by the batteries. This only applies to dc power systems since a battery cannot supply ac power except through a comparatively low-power inverter system.

THEORY OF THE LEAD-ACID CELL

The lead-acid secondary cell used in a storage battery consists of positive plates filled with lead peroxide (PbO_2); negative plates filled with pure spongy lead (Pb); an electrolyte consisting of a mixture of sulfuric acid and water mixed in such quantities that the solution has a specific-gravity range of 1.275 to 1.300 for a fully charged battery. **The specific gravity of a substance is defined as the ratio of the weight of a given volume of the substance to the weight of an equal volume of pure water at +4° centigrade.**

A chemical action takes place when a battery is delivering current as shown in Fig. 6-14. The sulfuric acid in the electrolyte breaks up into hydrogen ions (H_2) carrying a positive charge and sulfate ions (SO_4) carrying a negative charge. An ion is an atom or molecule which is either

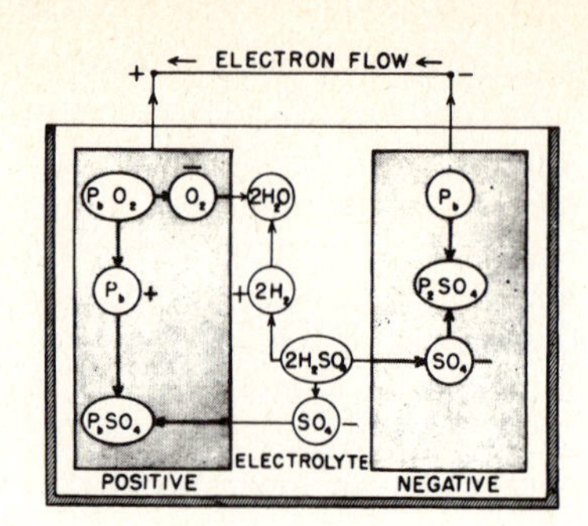

Fig. 6-14 Chemical action in a lead-acid secondary cell.

positively or negatively charged. A positively charged ion has a deficiency of electrons, and a negatively charged ion has an excess of electrons. The SO_4 ions combine with the lead plate and form lead sulfate ($PbSO_4$). At the same time, they give up their negative charge, thus creating an excess of electrons on the negative plate.

The H_2 ions go to the positive plate and combine with the oxygen of the lead peroxide (PbO_2), forming water (H_2O), and during the process they take electrons from the positive plate. The lead of the lead peroxide combines with some of the SO_4 ions to form lead sulfate on the positive plate. The result of this action is that the positive plate has a deficiency of electrons and the negative plate has an excess of electrons.

When the plates are connected together externally by a conductor, the electrons from the negative plate flow to the positive plate. This process will continue until both plates are coated with lead sulfate and no further chemical action is possible; the battery is then said to be discharged. The lead sulfate is highly resistant to the flow of current, and it is chiefly this formation of lead sulfate which gradually lowers the capacity of the battery until it is discharged.

During the charging process, current is passed through the storage battery in a reverse direction. A dc supply is applied to the battery with the positive pole connected to the positive plate of the battery and the negative pole connected to the negative plate. The emf of the source is greater than the emf of the battery. This causes the current to flow in a direction to charge the battery. The SO_4 ions are driven back into solution in the electrolyte, where they combine with the H_2 ions of the water, thus forming sulfuric acid. The plates then return to their original composition of lead peroxide and spongy lead. When this process is complete the battery is charged. Inasmuch as the sulfuric acid in the electrolyte is used up as the battery is discharged and returned to the electrolyte as it is charged, a test of the specific gravity of the electrolyte will give a good indication of the state of charge of the battery. This will be discussed further in the section on testing methods.

BATTERY CONSTRUCTION

A storage battery consists of a group of lead-acid cells connected in series and arranged somewhat as shown in Fig. 6-9. The closed-circuit voltage of the 6-cell battery is approximately 12 V, and that of the 12-cell battery is about 24 V. Closed-circuit voltage is the voltage of the battery when connected to a load.

Each cell of a storage battery has positive and negative plates arranged alternately and insulated from each other by separators. Each plate consists of a framework, called the **grid,** and the **active material** held in the grid. A standard formula for the grid material is 90 percent lead and 10 percent antimony. The purpose of the antimony is to harden the lead and make it less susceptible to chemical action. Other metals, such as silver, are also used in some grids to increase durability.

A typical grid is illustrated in Fig. 6-15. The heavy border adds strength to the plate, and the small horizontal and vertical bars form cavities to hold the active material. The structural bars also act as conductors for the current, which is distributed evenly throughout the plate. Each plate is provided with extensions, or feet, which rest upon ribs on the bottom of the cell container. These feet are arranged so the positive plates rest upon two of the ribs and the negative plates upon the two alternative ribs. The purpose of this arrangement is to avoid the short-circuiting which could occur as active material is shed from the plates and collects at the bottom of the cell.

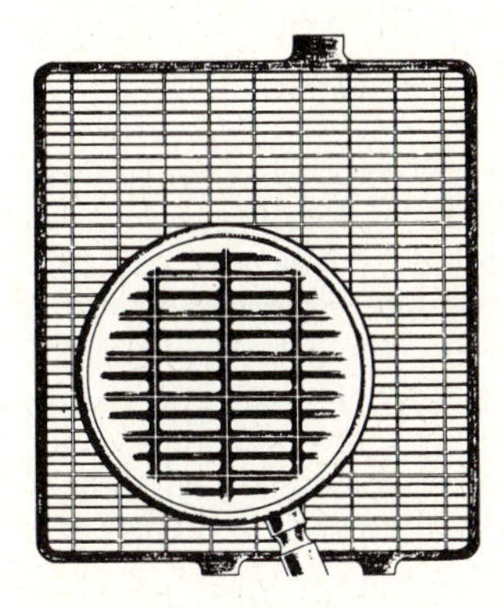

Fig. 6-15 Grid for a lead-acid cell plate.

The plates are made by applying a lead compound to the grid. The paste is mixed to the proper consistency with dilute sulfuric acid, magnesium sulfate, or ammonium sulfate, and is applied to the grid in much the same manner as plaster is applied to a lath wall. The paste for the positive plates is usually made of red lead (Pb_3O_4) and a small amount of litharge (PbO). In the case of the negative plates, the mixture is essentially litharge with a small percentage of red lead. The consistency and the manner of combining the various materials has considerable bearing on the capacity and life of the finished battery.

In compounding the negative-plate paste, a material called an *expander* is added. This material is relatively inert chemically and makes up less than 1 percent of the mixture. Its purpose is to prevent the loss of porosity of the negative material during the life of the battery. Without the use of an expander, the negative material contracts until it becomes quite dense, thus limiting the chemical action to the immediate surface. To obtain the maximum use of the plate material, the chemical action must take place throughout the plate from the surface to the center. Typical expanding materials are lampblack, barium sulfate, graphite, fine sawdust, and ground carbon. Other materials, known as hardness and porosity agents, are sometimes used to give the positive plates desired characteristics for certain applications. One or more manufacturers reinforce the active material of the battery plates with plastic fibers of 0.118 to 0.236 in [3 to 6 mm] in length. This adds substantially to the active life of the battery.

After the active material paste is applied to the grids, the plates are dried by a carefully controlled process until the paste is hardened. They are then given a forming treatment in which a large number of positive plates are connected to the positive terminal of a charging apparatus and a like number of negative plates, plus one, are connected to the negative terminal. They are placed in a solution of sulfuric acid and water (electrolyte) and charged slowly over a long period of time. A few cycles of charging and discharging converts the lead compounds in the plates to active material. The positive plates thus formed are chocolate brown in color and of a hard texture. The negative-plate material has been converted to spongy lead of a pearl-gray color. After forming, the plates are washed and dried. They are then ready to be assembled into **plate groups.**

The chemical control of all materials used in a storage battery is very important. Close temperature and humidity controls are used in the various phases of manufacture. All these factors are essential to producing a reliable storage battery. Because of the various processes by which plates are made, it is extremely important that the user follow the manufacturer's recommendations on placing the battery in service and on its subsequent maintenance. This information generally accompanies the packing case in either card or booklet form. If such information is not found with the battery, it may be obtained by communicating with a representative of the manufacturer.

Plate groups are made by joining a number of similar plates to a common terminal post (see Fig. 6-16). The number of plates in a group is determined by the capacity desired, inasmuch as capacity is determined by the amount (area) of active material exposed to the electrolyte. Each plate is made with a lug at the top to which the **plate strap** is fused. A positive-plate group consists of a number of positive plates connected to a plate strap, and a negative group is a number of negative plates connected in the same manner. The two groups meshed together with separators between the positive and negative plates constitute a **cell element** (see Fig. 6-17). It will be noted in the illustrations that there is one less positive plate than there are negative plates. This arrangement

Fig. 6-16 Plate groups.

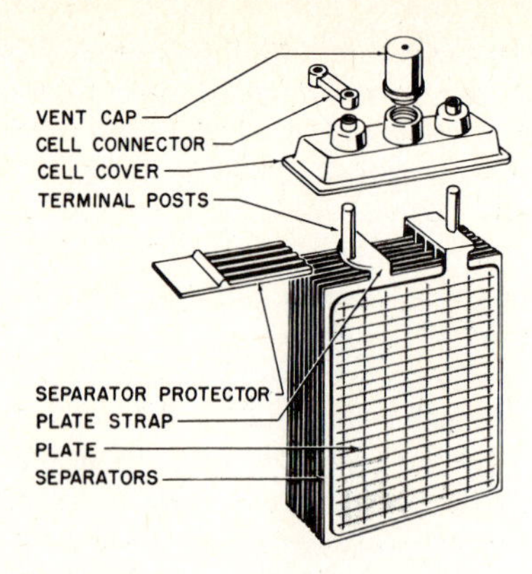

Fig. 6-17 Cell element for a lead-acid cell.

provides protection for the positive plates, inasmuch as they are more subject to warping and deterioration than the negative plates. By placing negative plates on each side of every positive plate, the chemical action is distributed evenly on both sides of the positive plate, and there is less tendency for the plate to warp.

The **separators** used in lead-acid storage batteries are made of wood, rubber, or other insulating materials. Their purpose is to keep the plates separated and thus prevent an internal short circuit. Without separators, even if the containers were slotted to keep the plates from touching, material might flake off the positive plates and fall against the negative plates. Negative material might expand sufficiently to come in contact with the positive plates, or the positive plates might buckle enough to touch the negative plates.

The material of the separators must be very porous so that it will offer a minimum of resistance to the current passing through. The separators are saturated with electrolyte during operation, and it is this electrolyte which conducts the electric current. It is obvious also that the separators must resist the chemical action of the electrolyte.

When wood separators are used, they are usually made of bass, poplar, some kinds of pine, Douglas fir, cedar, cypress, or redwood. Wood separators, after being cut to size, are chemically treated to remove objectionable chemicals such as wood acids. This treatment also serves to expand the wood pores. On the side of the separator adjacent to the positive plate, the separators are vertically grooved to permit normal shedding of active material.

Glass-wool separators are used by some manufacturers. Fine glass fibers are laid together at different angles and cemented on the surface with a soluble cement. The resulting sheet is used in conjunction with a wood or perforated-rubber separator. The glass wool is placed in the cell adjacent to the positive plate. Because of the compressibility of glass wool, it comes into very close contact with the positive plate and prevents the loosened active material from shedding. It is claimed that batteries with this type of separator have a longer life than those without it.

Another very effective method for providing plate separation is to enclose the positive plates in microporous, polyethylene pouches. This increases the efficiency of the battery, because the plates are much closer together—approximately 0.05 in [1.25 mm]—than they are with other types of separators. The pouches also prevent the shedding of active material from the positive plates.

When the cell elements are assembled, they are placed in the **cell container,** which is made of hard rubber or a plastic composition. Cell containers are usually made in a unit with as many compartments as there are cells in the battery. In the bottom of the container are four ribs. Two of these ribs support the positive plates, and the other two support the negative plates. This arrangement leaves a space underneath the plates for the accumulation of sediment, thus preventing the sediment from coming in contact with the plates and causing a short circuit. The construction of the cell bottom and of the plate-supporting ribs is shown in Fig. 6-18.

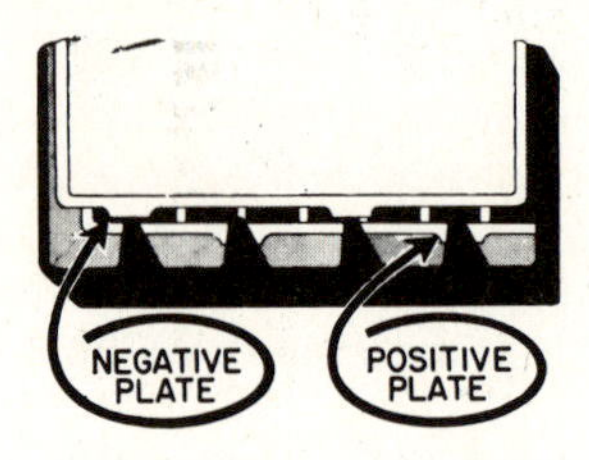

Fig. 6-18 Sediment space in a cell container.

The sediment space provided in storage batteries is of such capacity that it is not necessary to open the cells to clean out the sediment. When the sediment space is full to the point at which the spent material may come in contact with the plates, the cell is worn out.

The assembled cell of a storage battery has a cover made of material similar to that of the cell container. The cell cover is provided with two holes through which the terminal posts extend and a threaded hole into which is screwed the vented cell cap. When the cover is placed on the cell, it is sealed in with a special sealing compound. This is to prevent spillage and loss of electrolyte.

When a storage battery is on charge and approaching the full-charge point or is at the full-charge point, there is a liberal release of hydrogen and oxygen gases. It is necessary to provide a means whereby this gas may escape, and this is accomplished by placing a vent in the cell cap. This vent contains a lead valve which is so arranged that it will close the vent when the battery is inverted or in any position at which there is danger of spillage. A vent cap of this type is illustrated in Fig. 6-19. Another type of vent cap, also illustrated in Fig. 6-19, incorporates a tube which extends almost to the top of the plates. With this type of construction, the battery plates fill only slightly over one-half the cell container. The space in the top of the container is provided to hold the electrolyte when the battery is on its side or in an inverted position. A baffle plate is placed slightly above the plates to prevent splashing of the electrolyte. A hole in the baffle plate for the escape of gas and for access to the electrolyte is located to one side of the bottom of the gas-escape tube. If the electrolyte level is flush with the baffle plate, the end of the tube will always be above the electrolyte level, regardless of the position of the battery.

A development by the Exide Industrial Division of The Electric Storage Battery Company has provided an improved battery cap for light airplane batteries. On this battery, the vent plug contains a sintered alumina (alumi-

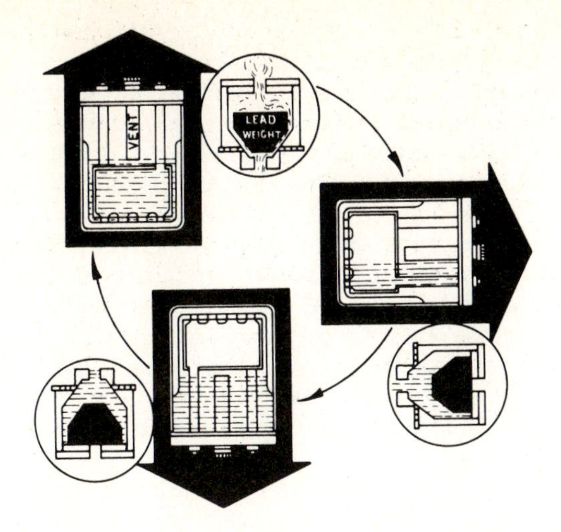

Fig. 6-19 Battery vent caps (plugs).

num oxide) plug instead of the heavy lead one used in many other types of batteries. This plug permits the diffusion of gases through it without letting fluids pass. It is much smaller and lighter than the lead-valve plug; hence, it saves both weight and space. The construction of this plug is illustrated in Fig. 6-20.

A battery vent cap that is particularly well adapted to acrobatic and military aircraft is shown in Fig. 6-21. As shown in the drawing, there is a valve in the bottom of the unit and this valve is opened and closed by the action of the conical weight in the upper part of the cap. When the battery is tilted approximately 45°, the weight drops against the side of the cap, pulling up on the valve stem and closing the valve. When the battery is brought back to a position approximately 32° from vertical, the weight centers itself again, allowing the valve stem to lower and open the vent valve.

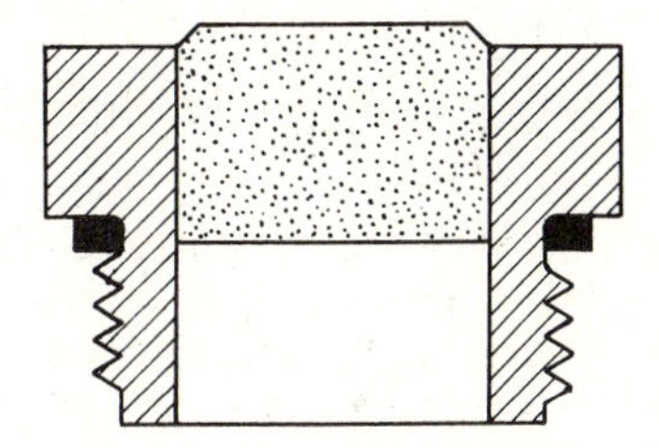

Fig. 6-20 Vent plug with a sintered alumina barrier.

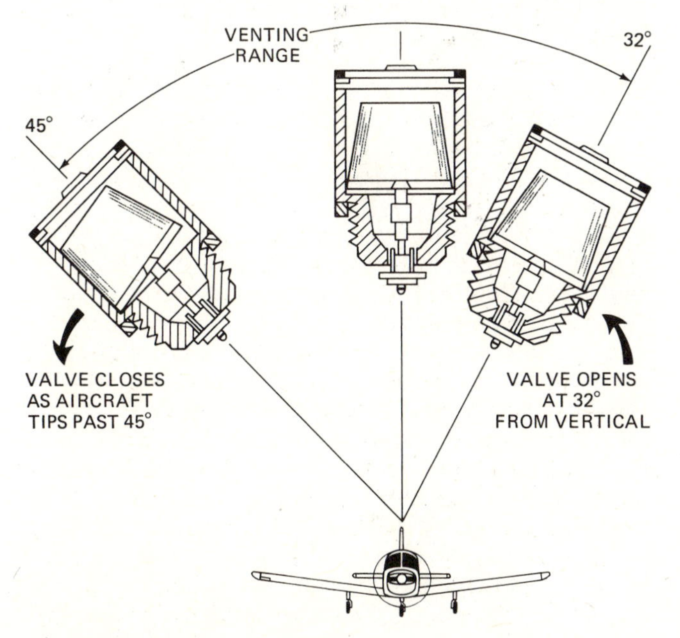

Fig. 6-21 Vent plug for acrobatic and military aircraft batteries. (Teledyne Battery Products)

BATTERY DESIGN FEATURES

Although the majority of lead-acid storage batteries are constructed with similar features, there are many differences in size and detail design, depending upon the use to which the battery is to be put. A completely assembled metal-encased battery for aircraft is shown in Fig. 6-22.

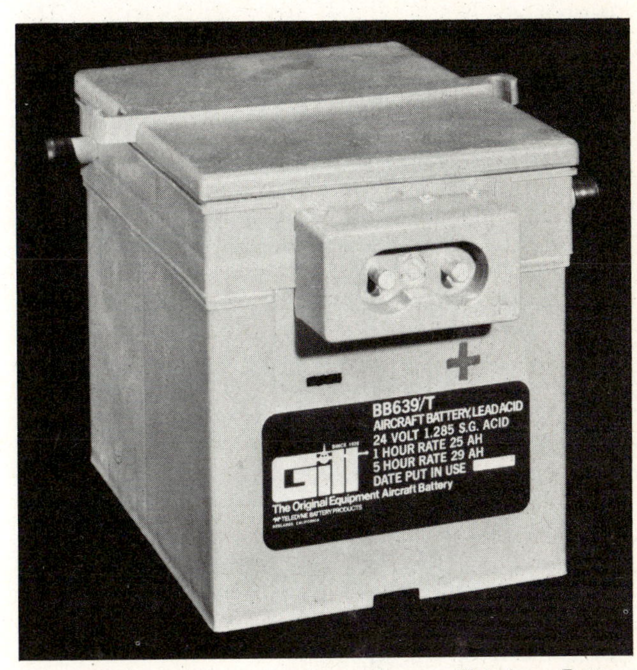

Fig. 6-22 Aircraft battery with metal case. (Teledyne Battery Products)

The cell containers are integral with a metal shielding box coated with acid-resistant paint. This box provides mechanical protection as well as electrical shielding and is fitted with a metal cover secured in place with hold-down rods. The design also provides an airtight space above the cells so that the gases being emitted will not escape into the aircraft in which the battery is installed. The vent space is provided with a connection for a tube installed to carry the battery gases overboard. This is a requirement for any battery which emits gases during operation.

The cell elements in a metal-encased battery are arranged in the case in a manner which makes it convenient to connect the terminal posts in series. The posts are connected together by means of lead straps called **cell connectors.**

The main negative and positive terminals of the battery are connected to external terminals in the side of the metal case. These terminals are adequately insulated from the case by washers and bushings.

A storage battery for light aircraft is shown in Fig. 6-23. This battery is made with a lightweight polystyrene case

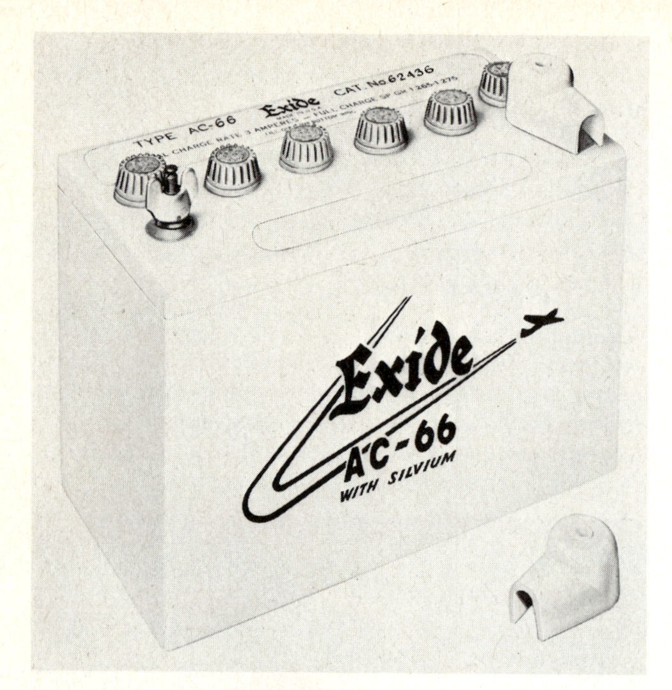

Fig. 6-23 Storage battery for light aircraft.

instead of the hard rubber formerly used; such casing provides an improvement in the power-weight factor. The grids in the battery are constructed of Silvium, a patented alloy developed by Exide, which provides greater grid life than the previously employed lead alloys. Another feature of this battery is the use of sintered alumina cell plugs, described in an earlier section.

A battery designed for use in an aircraft with an enclosed and ventilated battery compartment is shown in Fig. 6-24. The plates in this battery are reinforced with plastic fibers, and the positive plates are enclosed in microporous pouches to provide plate separation and protection. The intercell connectors are internal and permanently sealed with an epoxy resin. The construction prevents leakage between cells, decreases internal resistance, and ensures a cleaner battery.

Fig. 6-24 Battery for use in an enclosed and ventilated battery compartment. (Teledyne Battery Products)

The battery has a one-piece cover with no electric parts exposed except the positive and negative terminals. This design helps to prevent accidental short circuits and eliminates electric leakage between cells on top of the battery. The positive and negative terminals are locked in position with cast-on lugs that fit in notches in the cover. This secures the terminals and prevents the damage that might otherwise occur if excessive force is applied to the terminals when connecting and disconnecting battery cables. Positive and negative terminal identification is molded into the cover adjacent to each terminal.

A battery designed for use where there is no enclosed battery compartment is shown in Fig. 6-25. A gas-collecting manifold is provided by the sealed compartment at the top of the battery. This compartment is provided with

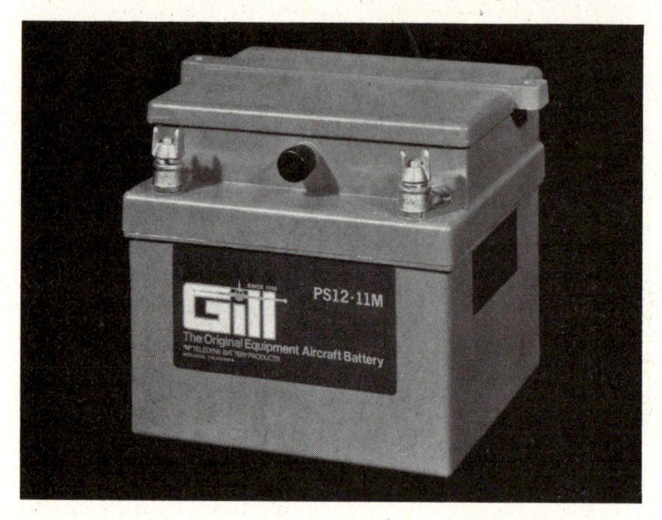

Fig. 6-25 Storage battery with sealed case for installation where there is no sealed battery compartment. (Teledyne Battery Products)

an air inlet and an air and gas outlet that are connected to an inlet air source and a discharge tube, respectively. During operation, airflow through the manifold carries gases and acid fumes into a neutralizing chamber and then overboard.

BATTERY RATINGS

VOLTAGE

Storage batteries of all types are rated according to voltage and ampere-hour capacity. It has been pointed out that the voltage of a fully charged lead-acid cell is approximately 2.1 V when the cell is not connected to a load. The silver-zinc cell has a rating of about 1.4 V, and the nickel-cadmium cell is rated at about 1.22 V.

Under a moderate load, the lead-acid cell will provide about 2 V. With an extremely heavy load, such as the operation of an engine starter, the voltage may drop to 1.6. A lead-acid cell that is partially discharged has a higher internal resistance than a fully charged cell; hence, it will have a higher voltage drop under the same load. This internal resistance is partially due to the accumulation of lead sulfate in the plates. The lead sulfate reduces the amount of active material exposed to the electrolyte; hence, it deters the chemical action and interferes with the current flow.

Figure 6-26 shows the discharge characteristics of a typical aircraft battery of the lead-acid type. The open-circuit voltage remains almost at 2.1 V until the battery is discharged. It then drops rapidly toward zero. The closed-circuit voltage gradually decreases from 2 to approximately 1.8 V as the cells discharge. Again, the voltage drops rapidly after 5 h of discharge.

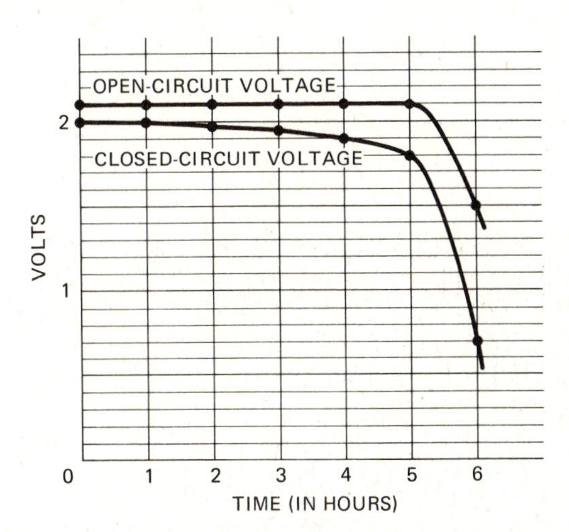

Fig. 6-26 Discharge characteristics of a lead-acid cell.

Even though battery cells vary considerably in voltage under various conditions, batteries are nominally rated as 6-V (3 cells), 12-V (6 cells), and 24-V (12 cells). In replacing a battery, the technician must ensure that the replacement battery is of the correct voltage rating.

CAPACITY RATING

The capacity of a storage battery is measured in **ampere-hours.** One ampere-hour is defined as a current flow of 1 A for a period of 1 h. Five ampere-hours means a current flow of 1 A for 5 h, 2.5 A for 2 h, or any combination of current flow and time that will give a product of 5. This relationship can be expressed as follows:

$$\text{Capacity (Ah)} = I \times T$$

where I = current, A
T = time, h

The capacity of a storage battery is based on a given discharge rate, since the capacity will vary according to the rate of discharge. The capacity of a lead-acid cell is generally based on a 5-h discharge rate. A 17-Ah battery will supply a current of approximately 3.4 A for a period of 5 h, because 3.4 ×5 =17. A 34-Ah battery will deliver twice that amount of current for the same period of time. When the capacity of a battery is based on a 5-h discharge rate, it does not mean that the battery will not become discharged in less than 5 h. If a very heavy load is applied to the battery, it may become discharged in a few minutes.

FIVE-MINUTE DISCHARGE RATE

Another rating applied to storage batteries is known as the 5-minute (min) discharge rate. This rating is based upon the maximum current a battery will deliver for a period of 5 min at a starting temperature of 80° F [26.7° C] and a final average voltage of 1.2 V per cell. This applies only to lead-acid batteries. The 5-min rating gives a good indication of the battery's performance for the normal starting of engines.

When a fully charged battery is connected to a very heavy load, it apparently becomes discharged in a short time. A good example of this is the starting of an automobile engine on a very cold morning. After turning the engine for a short time, the starter may refuse to operate. This failure occurs largely because the heavy flow of current has caused a rapid sulfation of the active material on the surface of the plates while the material inside the plates is still in a charged condition. The lead sulfate on the surface of the plates offers a high resistance to the flow of current; hence, the voltage drop within the battery becomes so large that there is not sufficient voltage to continue driving the heavy load. If the battery is allowed to remain idle for a time, it will again be able to deliver a substantial load current.

SIGNIFICANCE OF DISCHARGE RATE

The discharge rate of a battery has a very important effect on its capacity. For example, a certain silver-zinc alkaline cell which is rated at 7.5 Ah at a 1-h discharge rate, when discharged at 30 A will become discharged in 10 min. Since 10 min is $\frac{1}{6}$ h, the capacity at this discharge rate will be $\frac{1}{6}$ ×30 =5 Ah. When discharged at 60 A, the cell will have a rating of less than 3 Ah. Thus, it is found with all types of secondary cells, the Ah rating decreases as the rate of discharge increases.

TESTING LEAD-ACID BATTERIES

There are a number of different methods for testing secondary cells, the best method in each case depending upon the type and size of the cell. It is often desired to determine the state of charge, the capacity, and the condition so that it may be known whether the battery will continue to serve its function satisfactorily. Because of the differences existing in the characteristics of lead-acid, nickel-iron, nickel-cadmium, and silver-zinc cells, it is recommended that the manufacturer's instructions be followed in each particular case.

HYDROMETER TEST

For a lead-acid cell of the type used industrially in automobiles and in aircraft, a hydrometer is often the most convenient method for determining the state of charge. It has been previously explained that the specific gravity of the electrolyte in a lead-acid cell decreases as the charge in the cell decreases. This is because the acid in the electrolyte becomes chemically combined with the active material in the plates as the battery produces current; hence, less acid remains in the electrolyte. Since the specific gravity of the acid is considerably greater than that of water, the loss of acid causes the specific gravity of the electrolyte to drop.

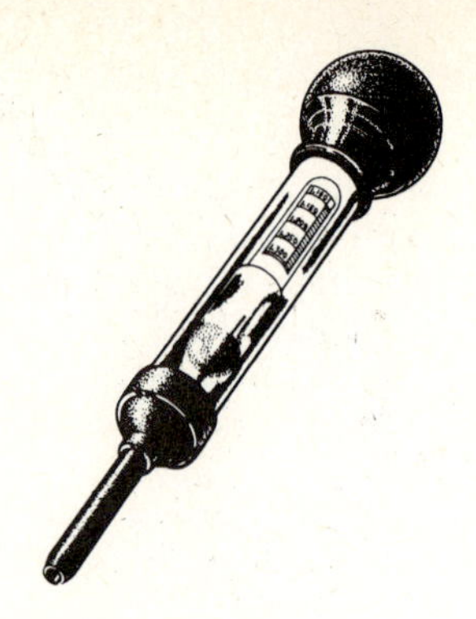

Fig. 6-27 Drawing of a hydrometer.

A hydrometer is used to determine the specific gravity of the electrolyte in a lead-acid cell. A typical hydrometer used for battery testing is shown in Fig. 6-27. It consists of a small sealed glass tube weighted at the end to make it float in an upright position. The amount of weight in the bottom of the tube is determined by the specific-gravity range of the fluid to be tested. In the case of a battery hydrometer, the specific-gravity range is from 1.100 to 1.300. The weight in the hydrometer must be such that when the specific gravity of the fluid is 1.100, only the tip of the stem will be above the fluid, and when the specific gravity is 1.300, almost the entire stem will be above the fluid level as shown in Fig. 6-28. A paper scale with readings of 1.100 to 1.300 is placed inside the glass stem, and the entire tube is placed inside a larger glass-tube syringe. With this arrangement, electrolyte may be drawn from a cell into the glass tube and the reading noted. The electrolyte is then returned to the cell from which it was taken.

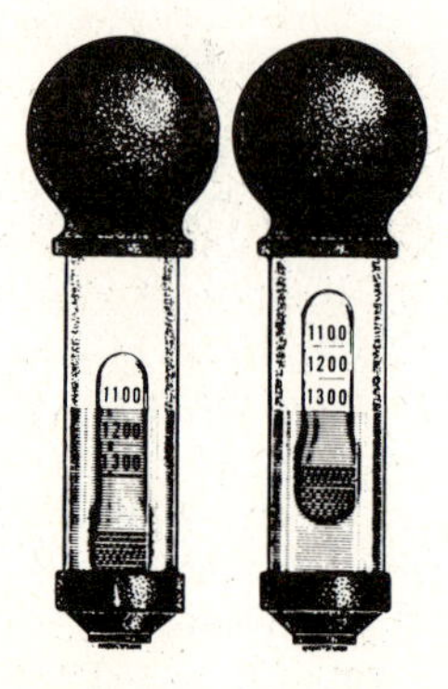

Fig. 6-28 Hydrometer readings.

The specific-gravity reading is taken at the fluid level on the stem of the hydrometer when it is floating freely in the electrolyte. It is important that the electrolyte be seen between the bottom of the float and the rubber plug in the bottom of the glass tube. Floats are often made with glass projections on the sides of the larger part so that the float will not tend to adhere to the side of the tube. The better hydrometers are so made that the top of the stem cannot touch the side of the glass tube regardless of the angle at which it is held. This feature eliminates the need for jarring the hydrometer to cause the float to assume a vertical position inside the tube.

It is a good idea to have more than one hydrometer available so that one can be checked against the other. If irregular readings are obtained, the float should be closely examined for hairline cracks that may have allowed electrolyte to seep inside and change the weight of the float. It is also necessary to determine whether the paper scale has slipped out of its proper place inside the stem of the hydrometer. It is good practice to disassemble the hydrometer from time to time, clean the inside of the tube, and inspect the float.

When using a hydrometer to test a lead-acid cell, a reading of 1.275 to 1.300 indicates a fully charged cell. If the reading is from 1.200 to 1.240, the charge is low. This does not mean that the cell is nearly discharged, but it does indicate that the battery may not be able to furnish power sufficient for such a heavy load as the starting of an engine. A reading of 1.240 in an automobile battery is sufficient for normal operation, even though the battery is not fully charged.

It must be emphasized that a lead-acid cell should be kept as near full charge as possible at all times to prolong its life. If such a cell is left partially charged, the lead sulfate forms hard crystals which cannot easily be changed back to active material.

The specific gravity of a lead-acid cell is not always an indication of the state of charge in the cell. If the electrolyte is removed from a discharged cell and replaced by an electrolyte of a high specific gravity, the cell will still be in a discharged condition even though the hydrometer shows it fully charged. Normally, electrolyte should not be added to or removed from a cell. The addition of water is necessary periodically to replace that which has decomposed into hydrogen and oxygen by electrolysis and that which has evaporated, but acid should never be added unless the electrolyte has been lost by spillage because acid does not evaporate. If it becomes necessary to add acid, the battery should be fully charged, on charge, and gassing freely. Then, by means of a syringe, the electrolyte should be drawn off to the level of the splash plate and replaced with electrolyte having a specific gravity of 1.320. The charge should be continued for 1 h at a low rate before another test is taken. If the specific gravity is still low, the operation should be repeated and the charge continued for another hour.

When processing new batteries for service, it is permissible to remove the electrolyte by turning the battery upside down. The battery may then be refilled with electrolyte of the correct specific gravity.

Electrolyte with a specific gravity of more than 1.325 should never be added to a battery. If a new battery that has a final specific gravity of 1.250 is emptied and refilled with 1.300 electrolyte, it will be found that after an hour's charge the specific gravity will be approximately 1.275 and not 1.300 as might be expected. This is due to the retention of considerable 1.250 electrolyte by the plates and separators when the battery is emptied. This electrolyte of low specific gravity dilutes the 1.300 electrolyte used for refilling the battery.

The reason for guarding against the removal of electrolyte from old batteries by turning them upside down is that there is a danger of depositing the sediment in the bottom of the cell containers on the lower part of the plates. This may result in short-circuiting the plates if the sediment adheres to them after the battery is returned to an upright position.

The only reason for the addition of electrolyte to an old battery is to replace electrolyte which has been lost by spillage. A battery fully charged at a specific gravity of 1.275 to 1.300 when new may have a specific gravity of 1.250 to 1.275 for a full charge near the end of its life. In this case, electrolyte should not be added. There is always the danger that the plates may be somewhat sulfated and that the addition of an electrolyte of higher specific gravity will aggravate this condition.

EFFECT OF TEMPERATURE

When a battery is tested with a hydrometer, the temperature of the electrolyte must be taken into consideration because the specific-gravity readings on the hydrometer will vary from the true specific gravity as the temperature goes above or below 80° F [26.7° C]. No correction is necessary when the temperature of the electrolyte is between 70 and 90° F [21.1 and 32.2° C] because the variation is not great enough to be considered. At higher or lower temperatures it is necessary to apply a correction according to Table 6-2.

TABLE 6-2

Electrolyte temperature		
°F	°C	Correction, points
120	48.9	Add 16
110	43.3	Add 12
100	37.8	Add 8
90	32.2	No correction
80	26.7	No correction
70	21.1	No correction
60	15.6	Subtract 8
50	10.0	Subtract 12
40	4.4	Subtract 16
30	−1.1	Subtract 20
20	−6.7	Subtract 24
10	−12.2	Subtract 28
0	−17.8	Subtract 32
−10	−23.3	Subtract 36
−20	−28.9	Subtract 40
−30	−34.4	Subtract 44

The corrections in Table 6-2 should be added to or subtracted from the reading on the hydrometer. For example, if the temperature of the electrolyte is 10°F [−12.2°C] and the hydrometer reading is 1.250, the corrected reading will be 1.250 −0.028, or 1.222. Notice that the correction points represent thousandths.

Some hydrometers are equipped with a correction scale inside the tube; the temperature correction then may be applied as the hydrometer reading is taken.

HIGH-RATE DISCHARGE TESTER

A high-rate discharge tester is used to determine the terminal voltage of battery cells when the battery is under load (see Fig. 6-29). The tester is constructed as a unit incorporating a zero-at-center voltmeter having a range of 2.5 V in either direction, a Nichrome shunt of low resistance, and two pointed prongs used to contact the cell terminals. The shunt is connected across the contact prongs to apply a heavy load to the cell being tested. The voltmeter is connected across the shunt and measures the

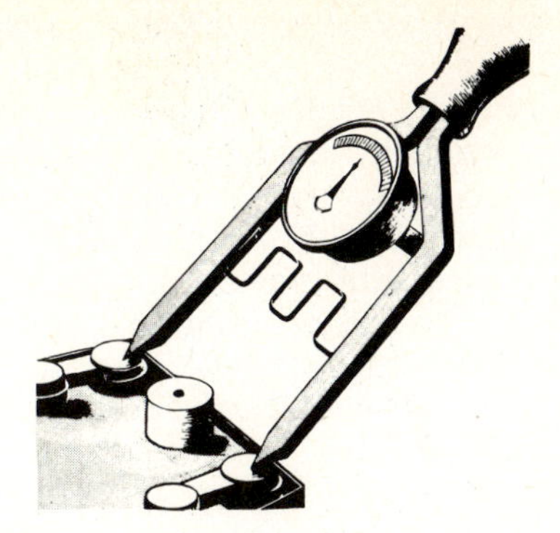

Fig. 6-29 A high-rate discharge tester.

voltage drop across the shunt resistance. Since the voltmeter is of the zero-at-center type, it is not necessary to observe the polarity of a cell when contact is made.

When a high-rate discharge tester is used, it is advisable to remove the vent plugs from the cells and blow any accumulated gas out of the cells. Following this, a heavy layer of cloth thoroughly saturated with water should be laid over the cover openings. This step is important because the space above the plates in the cells is often filled with explosive gases which, when ignited, will usually blow the cell container apart and scatter acid on nearby objects and persons. Since the high-rate discharge tester places a heavy load on the cells, an arc is formed when the contact is broken. This arc will ignite combustible gases in the vicinity.

To make a test of the cell condition, the battery should be fully charged. The contact prongs should be held firmly on the terminals of the cell for about 15 s. During this time the voltmeter should be observed to note any change in cell voltage. If the cell voltage begins to fall off or if any cell has a voltage 0.2 V lower than the average of the other cells, it is an indication that the cell has deteriorated internally. The history of the battery should be considered, and if it has been in service for a period near the normal life of this type of battery, it should be discarded. If the test shows a bad cell in a comparatively new battery, then the cell may be disassembled and repaired.

Another method for testing a battery for the terminal voltage of its cells is shown in Fig. 6-30. In this method, the battery is connected to a substantial load and the voltage of each cell tested with a voltmeter. Any cell that shows a voltage of 0.2 lower than any other cell should be repaired, or the battery should be discarded.

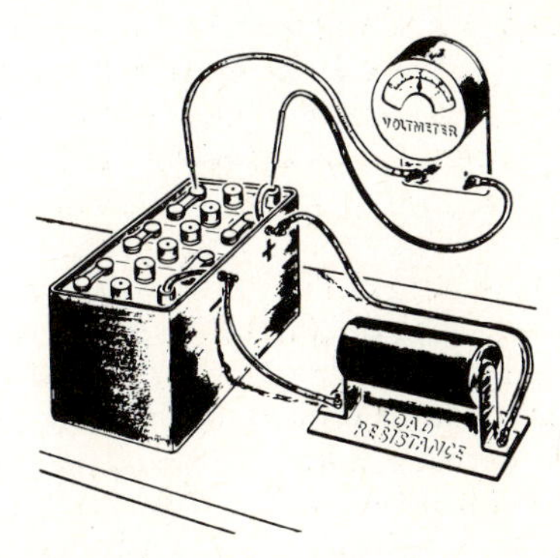

Fig. 6-30 Testing a cell with a voltmeter.

Since the internal resistance of a cell increases as the cell discharges, the foregoing tests give an indication of the state of charge in the cell. If a voltmeter is connected to a lead-acid cell when it is not under a load, a reading of 2.1 V can be obtained even though the cell may be nearly discharged.

Some technicians use a voltmeter to test a battery while it is on charge. The voltage of the cells increases as the battery is being charged and will reach approximately 2.6 V for a lead-acid type which fully charged and still on charge.

It must be noted that many aircraft batteries are constructed in such a manner that the individual cell terminals are not exposed and cannot be contacted with a voltmeter. In these cases, the testing methods described above cannot be employed. The manufacturer's service manual will give approved methods for testing particular batteries.

CAPACITY TESTER

Capacity-testing devices for storage batteries have been used for a number of years, and these instruments give a good indication of the condition of a battery. A capacity tester incorporates load resistances, a voltmeter, and a time clock, as shown in the circuit of Fig. 6-31. A fully charged battery is connected to a measured load until the voltage, as indicated on the voltmeter, drops to a predetermined level. At this time, the reading on the clock is noted, and from this the capacity of the battery is determined. After this test the battery should be recharged.

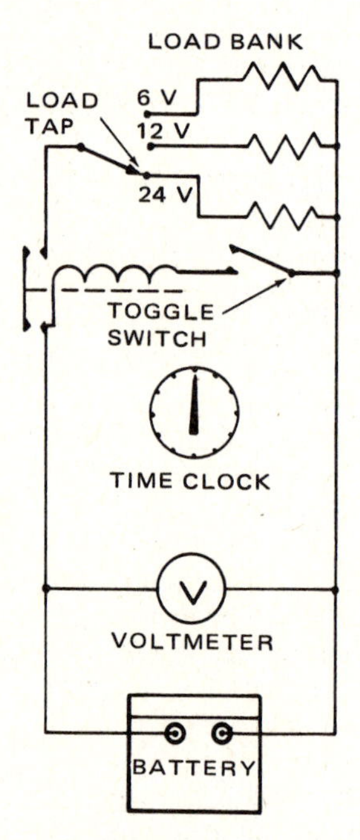

Fig. 6-31 Circuit for a capacity tester.

When connecting or disconnecting a battery in a circuit, it is best to see that the circuit is open when contact with the battery is made or broken. In this way, there will be no arc at the terminals of the battery, and the danger of an explosion will be reduced. The circuit can be controlled by means of a switch removed from the immediate area of the battery.

In a typical battery-capacity tester the operating-procedure instructions are included with the instrument. These are as follows:

1. Be sure the toggle switch is OFF.
2. Connect load tap to correct battery-size terminal.
3. Connect leads to terminals of battery.
4. Turn the time clock back to ZERO and throw the toggle switch to ON.
5. When voltage has dropped to 10.3 for 12-V batteries or 20.6 for 24-V batteries, throw toggle switch to OFF and read time clock.

CAUTION: *Keep the tester away from highly volatile and inflammable materials because of heat developed in the load compartment.*

CHARGING METHODS

Storage batteries are charged by passing a direct current through them in a direction opposite to that of the discharge current. That is, the supply must be connected to the battery, positive to positive and negative to negative. Various sources of direct current may be used, but the most commonly used devices are either rectifiers or dc generators. The manner in which batteries are connected to the power source varies; it is usually determined by the type and voltage of the batteries being charged. When batteries of different voltages must be charged from the same power supply, they are usually charged by the constant-current method. Another method used is the constant-potential (voltage) method. This system is universally used on aircraft, where an engine-driven generator is continuously charging the battery to its requirements.

Battery-charging methods may also be classified as manually cycled and system-governed methods. When batteries are charged in the hangar or shop, the manually cycled method is usually employed. This means simply that the voltage or current is controlled by an operator according to the requirements of the batteries being charged. In the system-governed method, the voltage of the power supply is automatically controlled by a carefully adjusted regulator.

CONSTANT-CURRENT CHARGING

Because batteries of several different voltages may thereby be charged at once, constant-current charging is the most convenient way to charge batteries not connected in an operating system. A constant-current system usually consists of a bulb-type rectifier, a dry-plate rectifier, or a diode rectifier which changes the normal alternating current into direct current. A transformer is used to reduce the available 115- or 220-V ac supply to the desired level before it is passed through the rectifier. As previously mentioned, rectifiers may be made with copper oxide, selenium, germanium, or silicon as the active element.

Figure 6-32 illustrates a typical constant-current system. The batteries are connected in series; hence, it is obvious that the current supply will be the same for each battery. The proper charging current for best results is usually marked or stamped on the battery case by the manufacturer. For example, a manufacturer may recommend a

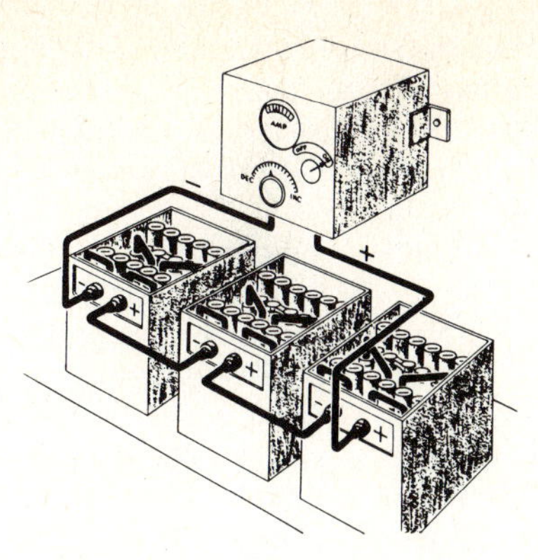

Fig. 6-32 Constant-current charging.

starting current of 6 A and a finishing current of 3 A. The battery should be placed on a 6-A charge until it starts gassing or until the temperature reaches 110°F [43.3°C], whichever comes first. It should then be put on a 3-A charge until it is fully charged. This applies to lead-acid cells. The charging rate for alkaline cells should be set according to the instructions for the particular battery.

The charging current may be varied to some extent if the temperature of a lead-acid cell is not allowed to exceed 110°F and if the battery is not allowed to bubble or gas excessively. If either of these conditions occurs, the charging current must be reduced. This type of charging cannot by highly efficient when batteries in various states of charge and capacity are being charged simultaneously, since the current must be regulated to avoid overheating any of the batteries and to prevent gassing. Under these conditions, batteries in a low state of charge do not receive sufficient current to charge them at the most efficient rate. When batteries of different capacities are being charged in the same system, it is frequently possible to connect some of them in parallel in order to obtain the proper charging current as shown in Fig. 6-33.

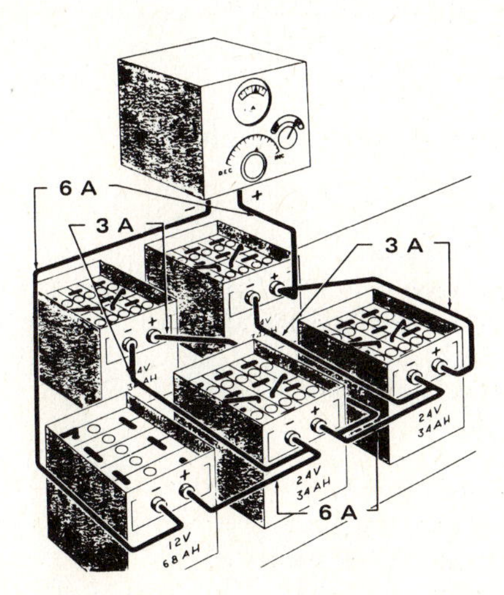

Fig. 6-33 Charging batteries of different capacities.

CONSTANT-POTENTIAL CHARGING

The battery-charging system in an airplane or long-range missile is of the constant-potential (voltage) type. An engine-driven or turbine-driven generator capable of supplying the required voltage is connected through the electric system directly to the battery. A battery switch is incorporated into the aircraft system so that the battery may be disconnected when the airplane is not in operation. The voltage of the generator is accurately controlled by means of a voltage regulator connected in the field circuit of the generator.

For a 12-V system, the voltage of the generator is adjusted to approximately 14.25 V. For a 24-V system, the generator voltage should be between 28 and 28.5 V. When these conditions exist, the starting current through the battery will be high, but as the state of charge increases, the battery voltage increases, thus causing a drop in the voltage difference between the generator and the battery. This produces a proportional decrease in the charging rate.

When the battery is fully charged, its voltage will be almost equal to the generator voltage; hence the charging current will drop to less than 1 A. When the charging current is low, the battery may remain on charge without any appreciable effect; however, the electrolyte level should be watched closely to see that it does not fall below the top of the plates. Figure 6-34 shows the circuit for a typical constant-potential system in an airplane. The power and battery-charging circuit for a system using an alternator is shown in Fig. 9-6.

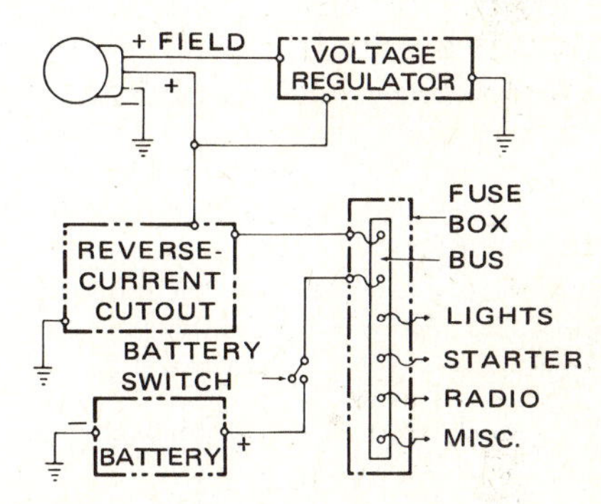

Fig. 6-34 Simplified aircraft charging system.

The figures given in the foregoing paragraphs for a lead-acid battery in an airplane represent average conditions for 12- and 24-V systems. When the airplane is to be used in normally hot or cold climates, however, the settings in Table 6-3 are recommended.

The reasons for these recommendations is that with a variation of battery temperature there is a consequent variation of the final or full-charge voltage on batteries charged at fixed-voltage settings. Table 6-4 is based on tests made at a setting of 28.5 V.

At extremely low battery temperatures, a setting of 28.5 V does not supply enough current to charge a bat-

TABLE 6-3 Voltages

Ambient temperature °F	Ambient temperature °C	12-V Battery	24-V Battery
65	18.3	14.1–14.9	29.2–29.8
80	26.7	13.9–14.7	27.8–29.4
105	40.6	13.7–14.5	27.4–29.0
125	51.7	13.5–14.3	27.0–28.6

TABLE 6-4

Battery temperature		
°F	°C	Charging rate, A
60	15.6	0.75
100	37.8	1.00
120	48.9	1.13

tery completely. At battery temperatures in excess of 90°F [32°C], the current input at 28.5 V tends to heat the battery.

When using a constant-potential system in a battery shop, the power source is usually a motor-driven generator. A voltage regulator which automatically maintains a constant voltage is incorporated into the system. A field rheostat may be used in place of a voltage regulator, in which case the voltage must be adjusted manually as required. The leads from the power supply are connected to two heavy bus bars across which the batteries are shunted as shown in Fig. 6-35. Any number of batteries may be charged at the same time with a system of this type, provided that the capacity of the charger is not exceeded and the batteries are of the same voltage. A battery of high capacity has lower resistance than a battery of low capacity. Hence, a high-capacity battery will draw a higher charging current than a low-capacity battery when both are in the same state of charge and when charging voltages are equal.

If there are no battery-charging facilities available, as is the case at some small airports, an aircraft battery may be taken to a garage or other shop where automobile batteries are charged. If a constant-current system is used, the aircraft battery may be connected in series with automobile batteries, provided that the charging current is kept at such a level that the aircraft battery does not overheat or gas excessively.

EFFECTS OF CHARGING ON BATTERY LIFE

The capacity of a storage battery is closely proportional to the area of its plates. This is particularly true at high rates of discharge. Thick plates have the advantage of providing a greater reserve of active material and are generally used when long life is desired and when space and weight permit the installation of such a battery. Operating a lead-acid battery on cycle service, that is, on a discharge-charge basis, will not provide so long a life as is obtainable from the same battery when it is operated by a properly adjusted voltage-regulated system. On the other hand, a nickel-cadmium battery or the other alkaline batteries will not suffer detrimental effects as a result of cycling.

The shedding of active material from the positive plates of a lead-acid battery is a gradual process and proportional to the number of cycles of charge and discharge. Batteries operated on this basis generally wear out from loss of material in the positive plates. In the case of batteries operated on voltage-regulated systems, the loss of positive-plate material is generally due to the slight washing action of the gas bubbles or to the increased volume of gas generated during bench charging. The amount of gassing of a lead-acid battery on a properly adjusted voltage-regulated system is very low, and as a rule, a worn-out battery on this type of charge deposits very little sediment.

When a comparatively new battery has worn-out positive plates, it is frequently concluded that overcharging has occurred, as a result of either too high a regulator setting or too much bench charging. However, if the history of the battery is known, the worn-out condition of the positive plates may be found to be the result of abuse to the negative plates early in the life of the battery. When its full-charge specific gravity is 1.275 to 1.300, allowing a battery to stand at a specific gravity below 1.250 for long periods of time results in the formation of excessive amounts of fixed sulfate in the negative plates. This sulfate will break down only after considerable overcharging, if it breaks down at all, and it is these necessary, prolonged charges that cause the positive plates to deteriorate. Even when this sulfate appears to be reduced, the efficiency of the battery has often been lowered, and it will lose its charge more rapidly than when it was new. Consequently, such a battery needs more frequent freshening charges, which naturally shorten its life.

The discharge voltage of a healthy battery does not decrease with age, although it will be found that on a bench charge, an old battery may not have as high a voltage when fully charged as it did when it was new.

PRECAUTIONS

There are several precautions which must be observed in the handling of storage batteries, especially when they are being charged.

When a storage battery is being charged, it generates a certain amount of hydrogen and oxygen. Since this pro-

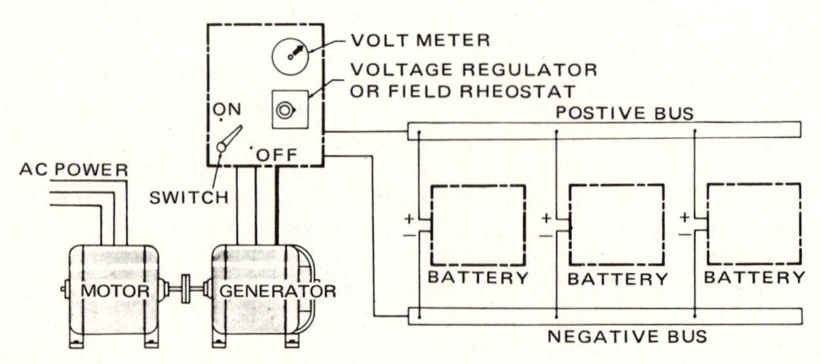

Fig. 6-35 Constant-voltage charging.

duces an explosive mixture, it is essential that precautions be taken to prevent ignition of the gas mixture. The vent caps of the battery should be loosened and left in place, and no open flames, sparks, or other means of ignition should be allowed in the vicinity. Power should always be turned off by means of a remote switch before a battery is connected or disconnected from the charger.

The electrolyte in a lead-acid cell contains sulfuric acid, and that of an alkaline cell contains potassium hydroxide. Both chemicals are highly corrosive and will burn the flesh as well as attack metals and other substances. If an electrolyte is spilled, it should be neutralized and washed away. Sulfuric acid can be neutralized with a solution of ammonia or bicarbonate of soda. Potassium hydroxide can be neutralized with boric acid, vinegar, or some other mildly acid solution. After an electrolyte has been neutralized, the affected area should be washed thoroughly with clean water.

The bench for storage batteries should be made of wood or metal and heavily coated with a paint which will resist the effects of the electrolyte. Adequate ventilation must be provided in the area used for charging so that an accumulation of explosive gases cannot occur.

INSTALLATION OF LEAD-ACID BATTERIES

BATTERY COMPARTMENT

The battery compartment in an airplane should be easily accessible so that the battery may be serviced and inspected regularly; it should also be isolated from fuel, oil, and ignition systems and from any other substance or condition which could be detrimental to its operation. Any compartment used for a storage battery which emits gases at any time during operation must be provided with a ventilation system. The inside of the compartment must be coated with a paint which will prevent corrosion caused by electrolyte.

The battery must be so installed that spilled electrolyte is drained or absorbed without coming into contact with the airplane structure. The shelf or base upon which the battery rests must be strong enough to support the battery under all flying and landing conditions. The battery must be held firmly in place with bolts secured to the aircraft structure. Metal-case batteries are held down by means of bolts which extend through ears on the battery cover. Nonmetallic batteries are held down by metal clamps which hook over the handles of the battery or over the edge of the battery case.

Batteries should not be located in engine compartments unless adequate measures are taken to guard against possible fire hazards and the injurious effects on a battery of excessively high temperatures. Battery manufacturers have determined that temperatures of 110 to 115°F [43 to 46°C] and higher are likely to cause rapid deterioration of the separators and plates. The critical temperature specified by the manufacturer should not be exceeded at any time. Forced ventilation of the battery compartment may be necessary to guard against excessive battery temperatures, and this may be provided by means of a tube leading from the slip stream into the container and a suitable vent tube leading out of it.

VENTILATING SYSTEMS

Ventilation of a battery system may be provided by placing the battery in a compartment through which air is circulated or by ventilating the battery case as shown in Fig. 6-36. Requirements for the adequate ventilation of batteries or battery compartments installed in civil aircraft are set forth in Federal Aviation Regulations of the Federal Aviation Administration. For military aircraft and missiles, the requirements are given in the specifications for the particular vehicle to be manufactured.

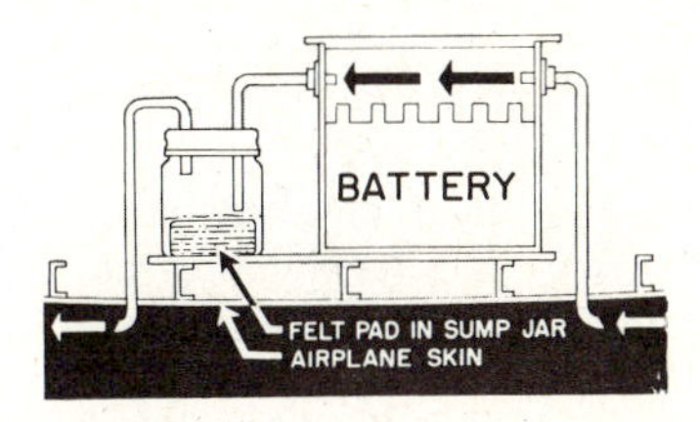

Fig. 6-36 Battery ventilating system.

The system illustrated in Fig. 6-36 is typical of those employed in many airplanes. Air is carried through a tube to the interior of the battery case from a scoop outside the airplane. After passing over the top of the battery, the air, battery gases, and acid fumes are carried through another tube to the battery sump.

The battery sump is a glass jar of at least 1-pint (pt) [0.47 liters (L)] capacity. In the jar is a felt pad about 1 in [2.54 cm] thick saturated with a 5 percent solution of bicarbonate of soda and water. The tube carrying fumes to the sump extends into the jar to within about $\frac{1}{4}$ in [0.635 cm] of the felt pad. An overboard-discharge tube leads from the top of the sump jar to a point outside the airplane. The outlet for this tube is so designed that there is a negative pressure on the tube whenever the airplane is in flight. This helps to ensure a continuous flow of air across the top of the battery, through the sump, and outside the airplane. The acid fumes going into the sump are neutralized by the action of the soda solution, thus preventing corrosion of the airplane's metal skin or damage to a fabric surface.

Certain types of nickel-cadmium batteries are designed so that the gas generated during operation is recombined chemically with the electrolyte. These batteries may be sealed to prevent leakage or the emission of gas. A battery of this type may be used in an airplane without a ventilation system. It is necessary, however, that such batteries be so installed that they will not become heated to a point beyond that specified by the manufacturer.

BATTERY CABLES

The electric leads to a battery in an airplane must be large enough to carry any load imposed on the battery at any time. They must be thoroughly insulated and protected from vibration or chafing and are usually attached to the airplane structure by means of rubber-lined or plastic-lined clamps or clips. Battery cables must be securely attached to the battery terminals; they are usually held in place as shown in Fig. 6-37. A heavy metal lug is soldered or swaged to the end of the cable and then attached to the terminal by means of a wing nut with a flat washer

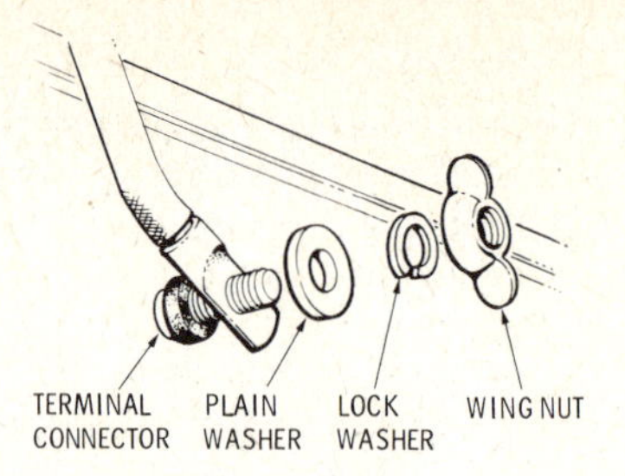

Fig. 6-37 Battery terminal connection.

and a lock washer. It must be noted that this is only one method for attaching battery terminals; others are also satisfactory.

Battery terminals must be protected from accidental shorting by means of a terminal cover. This may be a plastic or rubber shield over the terminal, or it may be a metal housing, as shown in Fig. 6-38. When a metal cover is used, the battery cables are passed through a rubber grommet in the end of the cover and connected to the battery terminals with a nut. Before being connected, the terminals are coated with a light film of terminal grease such as petrolatum to prevent corrosion. The terminal cover is then attached to the battery case with bolts or screws.

Fig. 6-38 Battery with terminal cover.

On certain types of aircraft batteries, when the connections are made to battery terminals supported by a rubber battery case, the leads must be made of flexible cables so that an undue strain will not be placed on the terminals. When stiff cables are used, any movement of the battery or leads will result in damage to the battery case.

QUICK-DISCONNECT PLUG

The quick-disconnect plug commonly used with standard AN batteries has proved to be very satisfactory for battery connection. It consists of an adapter secured to the battery case in place of the terminal cover and a plug to which are attached the battery leads. Two smooth contact prongs are screwed onto the battery terminals, and

Fig. 6-39 Battery quick-disconnect unit.

the plug is pulled into place on the battery by means of a large screw attached to a hand wheel. This screw also pulls the plug off the terminals to disconnect the battery. The cannon battery quick-disconnect plug is illustrated in Fig. 6-39.

Another very popular battery connector is the Elcon (see Fig. 6-40), manufactured by Icore International. This connector is more compact than the Cannon connector. The connector consists of two main assemblies: the terminal assembly attached to the battery to serve as a receptacle and the connector plug assembly to which the battery cables are connected. The plug assembly is inserted into the receptacle on the battery and is seated firmly by means of the center screw (worm) in the plug. The worm is a cam-locking device that seats firmly when the hand wheel or T handle is rotated fully to the right.

A principal feature of the Elcon connectors is a helical wire contact. The design of the contacts provides for many contact surfaces with the mating male pin, thus assuring a low-resistance contact. The helical contact is made of silver-plated, soft copper wire, and since it is flexible, it compensates automatically for any wear at the mating surfaces.

Fig. 6-40 Elcon battery connector. (Icore International)

Battery cables are attached to the Elcon plug by means of bolts extending through the ends of the connector strips that are metallically bonded to the helical contacts. The standard cable terminals fit over the bolts and are secured by means of a plain nut and lock washer. The plug assembly must be disassembled in order to attach or remove the battery cables. After the cables are attached and the nuts are correctly torqued, the two parts of the plug assembly are bolted together. It is not possible to assemble the plug incorrectly because of a lug on the outer part that mates with a hole in the inner part. Polarity signs are molded into the plug, both inside and outside, so there is no reason why the cables should be connected incorrectly.

The plug described in this section is utilized by many military and commercial aircraft. It conforms to specification MS25182-2.

PLUG-IN BATTERIES

In order to provide for rapid battery service on some older aircraft, plug-in batteries were employed. This type of battery is equipped with prongs that fit into receptacles mounted in the aircraft. With this battery, it is a comparatively simple task for the technician to disengage the battery supports and pull the battery out of the compartment. The replacement battery is then lifted into place and secured. Electric connections to the battery are automatically disconnected and connected by the bayonet-type connectors when the battery is removed and replaced. A battery with connectors of this type is shown in Fig. 6-41.

Fig. 6-41 A plug-in battery.

INSPECTION AND SERVICE

PERIODIC INSPECTIONS

Like other parts and systems in an airplane, batteries should be inspected periodically to make certain that they will continue to perform satisfactorily. In addition to inspections, storage batteries require servicing. Lead-acid batteries require the addition of distilled water at regular periods, depending upon the amount of activity they have had. Alkaline batteries require the addition of water at much greater intervals of time, even as long as a year apart. Sealed nickel-cadmium batteries, of course, do not require the addition of water.

The following inspections and services are recommended for lead-acid batteries about once a week for aircraft operated on a daily basis:

1. Inspect the mounting of the battery. Make sure that no part of the supporting structure is cracked or weakened in any way.

2. Remove the cover from the battery case, if it is the covered type, and inspect the interior. Look for evidence of leakage and corrosion. The top of the battery should be clean and dry. A small amount of corrosion around the terminals may be removed with a stiff brush and a mild soda solution. A wire brush should not be used because of the danger of short-circuiting the battery. When using a soda solution, take care to see that none of the solution enters the battery cells. If it does, the solution will neutralize the electrolyte, and the battery is likely to go "dead." After cleaning the battery with soda solution, rinse it with clean water and dry the top of the battery.

It is important to note that a lead-acid battery whose top is damp with electrolyte and dirt will discharge itself quite rapidly because of the conductance of the electrolyte; a steady current flows from the negative terminal of the battery to the positive terminal. Hence it is essential that the top of any storage battery be kept clean and dry.

Areas found to be corroded should be cleaned and then coated with a thin film of grease. If a large amount of corrosion is found in the battery case, the battery should be removed and cleaned thoroughly. If appreciable damage has been done, either to the battery case or to the battery mounting structure, the damaged parts should be repaired or replaced.

3. Check the amount of electrolyte in the battery. If the level is so low that a hydrometer reading cannot be taken, add distilled water until it is approximately $\frac{3}{8}$ in [0.95 cm] above the plates. If the battery has baffle plates above the cell plates, the electrolyte level should be brought up to the level of the hole in the baffle plate in each cell. The hole in the baffle plate is directly under the cell cap and is provided so that electrolyte may be drawn from the cell with a hydrometer.

4. Test the specific gravity of the battery cells with a hydrometer. If the reading, when corrected for temperature, is less than 1.220, the battery should be removed for charging. When using a hydrometer, be sure to return the electrolyte to the cell from which it was taken. If there is a difference of more than 30 points between the reading for any two cells, the battery should be removed for repairs. Do not take a hydrometer reading after adding water to a cell. This reading will be erroneous until the water and the electrolyte are thoroughly mixed.

5. Inspect the terminal connections. See that they are tight and free from corrosion. If a quick-disconnect plug is used on the battery, remove it and inspect the contacts. If they are dirty or corroded, clean them thoroughly and apply a small amount of terminal lubricant. Replace the plug, making sure that the hand wheel is tight.

6. Inspect the battery cables for condition of insulation, evidence of chafing, and security of connections.

7. Replace the cover on the battery case, making sure that the hold-down nuts are tightened sufficiently and safetied.

8. Inspect the ventilation tubes and sump jar. See that the felt pad in the sump is covered with soda solution; if necessary, add enough solution to cover it. See that the sump lid is tight, and check the vent tubes for cracks and for security of connections.

At intervals of 90 days, or after 200 h of service, batteries should be removed, cleaned, tested, and inspected thoroughly. Otherwise, the approved maintenance schedule established for the operating agency should be followed.

When removing a storage battery from an airplane, take care to prevent accidental short-circuiting of the battery terminals. Some operators recommend taping the terminals after the cables are mounted. Wrenches or pliers used to disconnect the terminals should have handles wrapped with insulating tape to prevent accidental shorting against the metal structure of the airplane. Short circuits of the battery terminals not only will place undue strain on the battery but also may cause structural damage to the airplane and physical injury to the mechanic. If gasoline fumes are present, there is the possibility of fire or explosion.

After removing the battery, wash it with clean water. If corrosion is present, it may be removed with a brush and a soda solution. When the battery has been thoroughly cleaned, it should be dried and inspected for condition. Small cracks or fissures in the sealing compound may be closed by passing the flame of a Prestolite torch over them carefully. Care must be taken that the flame is not allowed to burn the cell covers or other parts of the battery. Flame should *not* be used to seal cracks in plastic battery cell covers.

CAUTION: *Before a flame is used near a battery, the cell must be purged of flammable gases.*

Flammable gases may be eliminated from the cell by removing the cell caps and blowing a slow stream of air into the cell for about 1 min. The battery must not be on charge, nor must it have recently been on the charging line. There have been many cases of batteries exploding and causing physical injury to persons nearby. Some operators recommend the use of a soldering iron rather than a torch to repair cracks in the sealing compound of a battery, thus reducing the danger of an explosion.

After a battery has been thoroughly cleaned, inspected, repaired, and checked for electrolyte level, it should be placed on charge as previously described and checked occasionally during the charging period. After about 1 h of charging, the specific-gravity readings of all the cells should be nearly the same. The temperature of the battery should be noted during the charging period and should not be allowed to exceed 110°F [43.3°C]. If a thermometer is not available, the temperature may be checked by feel. Place the hand against the battery case, and if it is uncomfortably warm, the battery is too hot.

Allow the battery to charge until there is no longer a change in the specific-gravity readings. Then continue the charge for approximately 2 h. At this time, all the cells should be gassing freely, but the temperature should not be excessive. The fully charged battery can now be given a test for high-rate discharge or capacity. Either of the tests previously described will give a good indication of the condition of the battery.

ADJUSTING SPECIFIC GRAVITY

If the battery electrolyte is found to have a specific gravity above 1.300 after charge, it should be adjusted. This may be done by drawing off a part of the electrolyte with a hydrometer and replacing it with distilled water. The battery should then be placed on a slow charge for about 1 h and rechecked. If the specific gravity is still not correct, additional adjustments must be made.

If the specific gravity of the battery should read 1.260 or lower at the end of charge, electrolyte should not be added until after the battery is given additional charging. Several hours of slow charging (not over one-fourth the maximum rate for the battery) will not injure the battery, and it may result in converting some of the sulfated lead back to active material. If continued charging has not resulted in bringing the specific gravity up to 1.260 or more, it may be established at the 1.260 minimum by withdrawing the electrolyte with a syringe or hydrometer and adding 1.320 electrolyte.

MIXING THE ELECTROLYTE

Electrolyte of a given specific gravity can be purchased; it is sometimes more convenient and economical, however, to mix it at the shop. Table 6-5 gives the proper amount of water to be mixed with a given amount of sulfuric acid to obtain the desired specific gravity.

TABLE 6-5 Mixing Electrolytes

Specific gravity desired	Add 1 gal 1.400 acid to:	Add 1 gal 1.835 acid to:
1.240	$\frac{3}{4}$ gal water	$3\frac{1}{2}$ gal water
1.275	$\frac{1}{2}$ gal water	$2\frac{3}{4}$ gal water
1.300	$\frac{1}{3}$ gal water	$2\frac{1}{2}$ gal water
1.340	$\frac{1}{7}$ gal water	2 gal water
1.400	None	$1\frac{1}{2}$ gal water

The container in which the electrolyte is mixed should be made of glass, glazed earthenware, or other material which will not be attacked by the acid.

When mixing acid with water, **always pour the acid into the water. Never pour water into the acid.** The heat generated may cause the acid to be thrown out onto the operator, who may receive severe burns.

After the electrolyte is mixed, it should be tested for specific gravity. If the proper level has not been attained, it can easily be adjusted by the addition of acid or water as necessary. The electrolyte will be quite warm immediately after mixing, and so a temperature correction must be made when testing it for specific gravity.

Acid of a *commercial* grade should not be used in batteries. A grade known as *battery* grade should be used instead. It is free of the impurities which may contaminate a battery and is also cheaper than the chemically pure grade, commonly called the CP grade.

PLACING NEW BATTERIES IN SERVICE

The principal rule to observe when placing new batteries, of either the lead-acid or the alkaline type, in service is to follow the manufacturer's instructions. Because these instructions may vary considerably, care must be taken to follow them accurately.

New lead-acid batteries which are stored in warehouses for long periods of time or which are placed in storage pending sale are not filled with electrolyte. The plates are dry-charged before assembly, and no electrolyte is placed in the cells until the battery is put in service.

When new batteries are received in the dry state, they should be filled with electrolyte having the specific gravity recommended by the manufacturer. After one or more hours, the electrolyte level should be checked, and if it has fallen, more electrolyte should be added to bring it up to the recommended level. The battery may be placed in service after the electrolyte has been in the cells for at least 1 h; if time is available, however, it is better that the battery be charged slowly for approximately 18 h. The rate of charge will depend on the type and capacity of the battery; this information is usually included in the instructions supplied with the battery.

When the charging is finished, the specific gravity of the electrolyte should be checked. If it is below 1.275 or above 1.300, it should be adjusted by the addition of acid or water as required.

When a storage battery containing electrolyte is placed in storage, it should first be fully charged. All electrolyte spilled on the top of the battery should be removed with a solution of soda, and the battery should then be washed with clean water and thoroughly dried. The terminals of the battery should be coated with terminal grease or petroleum jelly. While the battery is in storage, it should be recharged every 30 days to compensate for the self-discharge which takes place when it is not in use.

COLD-WEATHER OPERATION

Temperature is a vital factor in the operation and life of a storage battery. Chemical action takes place more rapidly as temperature increases. For this reason, a battery will give much better performance in temperate or tropical climates than in cold climates. On the other hand, a battery will deteriorate faster in a warm climate. In some cases, a lower specific gravity is specified for warm-climate operation in order to add to the life of the battery. It is recommended that batteries being used in tropical countries be filled with electrolyte having a maximum specific gravity of 1.260 when fully charged.

In cold climates, the state of charge in a storage battery should be kept at a maximum. A fully charged battery will not freeze even under the most severe weather conditions, but a discharged battery will freeze very easily.

When water is added to a battery in extremely cold weather, the battery must be charged at once. If this is not done, the water will not mix with the acid and will freeze.

Table 6-6 gives the freezing point for various states of charge.

TABLE 6-6 Freezing Points for States of Charge in a Storage Battery

Specific gravity	Freezing point °F	°C
1.300	−95	−70.6
1.285	−85	−65
1.275	−80	−62.2
1.250	−62	−52.2
1.225	−35	−37.2
1.200	−16	−26.7
1.175	−4	−20.0
1.150	+5	−15.0
1.125	+13	−10.6
1.100	+19	−7.2

CAPACITY LOSS DUE TO LOW TEMPERATURES

Operating a storage battery in cold weather is equivalent to using a battery of lower capacity. For example, a fully charged battery at 80°F [26.6°C] may be capable of starting an engine twenty times. At 0°F [17.8°C] the same battery may start the engine only three times.

Low temperatures greatly increase the time necessary for charging a battery. A battery which could be recharged in 1 h at 80°F may require approximately 5 h of charging when the temperature is 0°F.

NICKEL-CADMIUM STORAGE BATTERIES

In the first part of this chapter, the construction and operation of nickel-cadmium cells which are constructed as dry cells were described. The term *dry cell* indicates only that the cell is so designed and constructed that there is no danger of spilling the electrolyte, because the electrolyte is immobilized and is not in a liquid form. The electro-

Troubleshooting Chart for Lead-Acid Batteries

Indication	Cause	Remedy
Battery will not hold its charge.	Battery worn out or deteriorated.	Remove the battery and test it for capacity and condition. Replace if necessary.
	Partial short circuit in the electric system.	Check electric system for shorts or load. See that the battery switch is turned off when the airplane is not in use. Remove and charge the battery.
	Battery switch left turned on.	
	Charging rate too low.	Adjust voltage regulator for proper charging rate.
	Electrical load too large for battery and generator.	Install a battery and generator with sufficient capacity for applied loads.
	Battery left standing too long.	Remove and recharge the battery. When the airplane is not in use for a long period of time, see that the battery is recharged every 30 days.
	Broken internal cell walls.	Replace the battery.
	Plates "shorted" internally.	Replace the battery.
Battery will not take a charge.	Battery is worn out.	Replace the battery.
	Battery plates sulfated.	Place the battery on charge until the specific gravity does not change for a period of 2 h. Then charge at 1 A for 60 h.
Battery life is too short.	Overcharging.	Adjust voltage regulator for correct charging rate.
Battery life is too short.	Electrolyte level below top of plates.	Check the electrolyte level frequently and add water to keep the level $\frac{3}{8}$ in [0.95 cm] above the plates.
	Battery discharged too frequently due to excessive use for starting engines.	Use an outside power supply for starting engines if possible.
	Battery plates sulfated.	Place the battery on protracted overcharge as above.
Electrolyte runs out of vent caps.	Electrolyte level too high.	See that the electrolyte is maintained at the correct level. Remove spilled electrolyte.
	Charging rate too high, causing gassing and overcharge.	Adjust voltage regulator for correct charging rate.
Excessive corrosion in battery case and compartment.	Overcharging.	Adjust voltage regulator for correct charging rate.
	Electrolyte level too high.	Correct the electrolyte level.
	Vent lines clogged.	Clear the vent lines and see that they are properly connected.
	Electrolyte leaking around cell covers. Battery case cracked.	Remove the battery and make necessary repairs. Replace if the case is cracked.
Battery case cracked or broken.	Improper installation. Hold-down bolts loose, or tightened too much.	Replace the battery. Tighten hold-down bolts correctly.
	Battery frozen.	Replace the battery. See that the battery is kept fully charged when operating in low temperatures. When water is added, charge the battery for 1 h to mix the water with the electrolyte.
Battery polarity is reversed.	Caused by connecting the battery incorrectly.	Completely discharge the battery and recharge it in the correct manner.

lyte does, however, have the characteristics of a liquid to the extent that it permits the movement of ions from the solution to the plates and vice versa.

In the nickel-cadmium storage batteries discussed in this section, the electrolyte is in liquid form and consists of a solution of potassium hydroxide (KOH). This is the same chemical utilized for the electrolyte in nickel-cadmium dry cells.

CELL AND BATTERY CONSTRUCTION

The nickel-cadmium cells and batteries described herein refer to vented-cell batteries. Vented-cell batteries have vent caps in the covers of the cells to permit the escape of hydrogen and oxygen, particularly when the battery is on overcharge.

Each cell of the battery consists of negative and positive plates, separators, electrolyte, cell container, cell cover, and vent cap. The plates are made from sintered metal plaques impregnated with the active materials for the negative and positive plates. The plaques are made of nickel carbonyl powder sintered at a high temperature to a perforated nickel-plated steel base or a woven nickel wire base. This results in a porous material that is 80 to 85 percent open volume and 15 to 20 percent solid material. The porous plaque is impregnated with nickel salts to make the positive plates and cadmium salts to make the negative plates. After the plaques have absorbed sufficient active material to provide the desired capacity, they are placed in an electrolyte and subjected to an electric current which converts the nickel and cadmium salts to the final form. The plaques are then washed and dried and cut into plates. A nickel tab is welded to a corner by which the plates are joined into plate groups.

The **separator** in a nickel-cadmium cell is a continuous thin porous multilaminate of woven nylon with a layer of cellophane. The separator serves to prevent contact between the negative and positive plates. The separator is continuous and is interposed between the plates as each successive plate is added to the plate pack or stackup. The cellophane portion of the separator acts as a barrier membrane inhibiting the oxygen which is formed at the positive plates during overcharge from reaching the negative plates. Oxygen at the negative plates would recombine with cadmium and create heat that might lead to **thermal runaway;** thus the cellophane serves to inhibit thermal runaway.

The electrolyte for a nickel-cadmium battery is a solution of 70 percent distilled water and 30 percent potassium hydroxide, which gives a specific gravity of 1.3. Specific gravities for nickel-cadmium batteries may range between 1.24 and 1.32 without appreciably affecting battery operation. In a nickel-cadmium cell, the specific gravity of the electrolyte gives no indication of the state of charge of the cell.

The cell container consists of a cell jar and a matching cover which are permanently joined at assembly. It is designed to provide a sealed enclosure for the cell, preventing electrolyte leakage or contamination. The vent cap is mounted in the cover of the cell and is constructed of plastic. It is fitted with an elastomer (flexible rubber or plastic) sleeve valve to permit release of gases as necessary, especially when the battery is on overcharge. The cap can be removed whenever necessary to adjust the electrolyte level. The vent valve automatically seals the cap to prevent leakage of electrolyte.

A cell core and the assembly of a complete cell for a nickel-cadmium battery are shown in the drawings of Fig. 6-42.

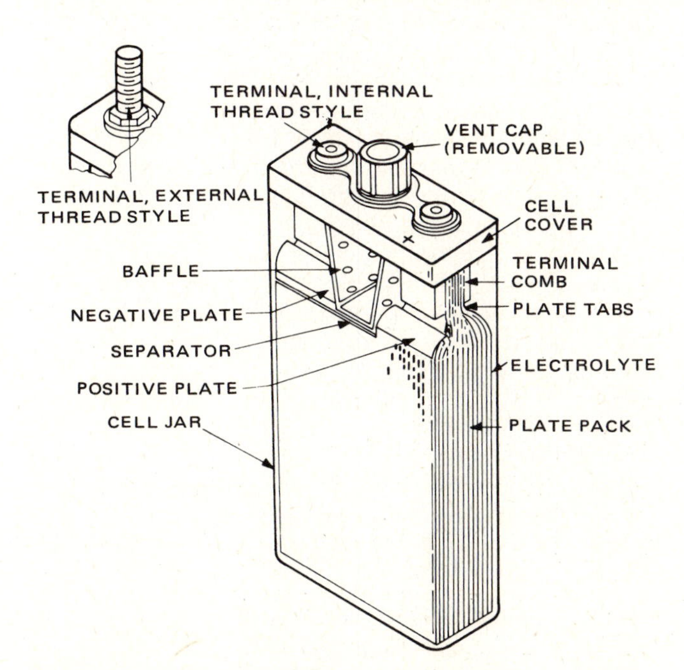

Fig. 6-42 Nickel-cadmium cell for an aircraft battery. (Battery Division, General Electric Company)

Another type of complete nickel-cadmium cell is illustrated in the drawing of Fig. 6-43. The cell is assembled by welding the tabs of the plates to their respective terminal posts. The terminal and plate-pack assembly is then inserted into the cell container, and the baffle, cover, and terminal seal are installed. The cover is permanently joined to the jar to produce a sealed assembly.

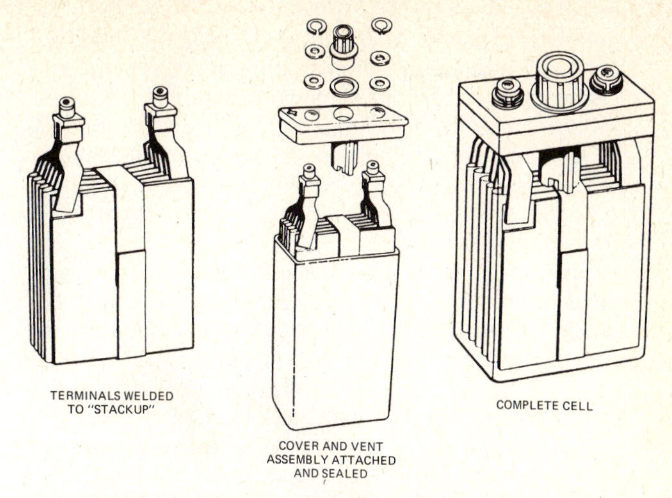

Fig. 6-43 Cell construction. (Marathon Battery Company)

Fig. 6-44 Nickel-cadmium aircraft battery. (Marathon Battery Co.)

Nickel-cadmium batteries are designed to allow the replacement of individual cells. The cells are placed in an insulated metal, plastic, or wood case in proper order and then connected in series with the cell connectors. The end cells may be connected to external posts or to a quick-disconnect unit. A complete battery is illustrated in Fig. 6-44.

PRINCIPLES OF OPERATION

The principles of operation of a nickel-cadmium cell have been described briefly in an earlier section; however, it is appropriate to review the principles and give additional information at this point. The exact chemical reactions which occur during charge and discharge of a nickel-cadmium cell are somewhat complex; however, they are similar to those described for carbon-zinc cells and lead-acid cells in that ions in the electrolyte remove electrons from the positive plate and deliver electrons to the negative plate during discharge. The reverse occurs during charge.

As previously explained, the active material of the negative plate of a charged nickel-cadmium cell consists of metallic cadmium (Cd), and the active material of the positive plate is nickel oxyhydroxide (N,OOH). As the battery discharges, hydroxide ions (OH) from the electrolyte combine with the cadmium in the negative plates and release electrons to the plate. The cadmium is converted to cadmium hydroxide [$Cd(OH)_2$] during the process. At the same time, hydroxide ions from the nickel oxyhydroxide positive plates go into the electrolyte carrying extra electrons with them. Thus electrons are removed from the positive plate and delivered to the negative plate during discharge. The composition of the electrolyte remains a solution of potassium hydroxide because hydroxide ions are added to the electrolyte as rapidly as they are removed. For this reason the specific gravity of the electrolyte remains essentially constant at any state of discharge. It is, therefore, impossible to use specific gravity as an indicator of the state of charge.

When a nickel-cadmium battery is being charged, the hydroxide ions are caused to leave the negative plate and enter the electrolyte. Thus the cadmium hydroxide of the negative plate is converted back to metallic cadmium. Hydroxide ions from the electrolyte recombine with the nickel hydroxide of the positive plates, and the active material is brought to a higher state of oxidization called nickel oxyhydroxide. This process continues until all the active material of the plates has been converted. If charging is continued, the battery will be in overcharge, and the water of the electrolyte will be decomposed by electrolysis. Hydrogen will be released at the negative plates, and oxygen will be released at the positive plates. This combination of gases is highly explosive, and care must be exercised to avoid any possibility of ignition of the gases.

Water is lost from the electrolyte during overcharge because of electrolysis. Some water is also lost by evaporation and entrainment of water particles during the venting of cell gases. By theory, 1 cubic centimeter (cm^3) of water will be lost by electrolysis for every 3 h of overcharge. In practice the loss is not this high because there is some recombination of hydrogen and oxygen within the cell.

The separator acts as an electrical insulator and a gas barrier between the negative and positive plates. The nylon fabric provides separation to prevent contact between plates of opposite polarity. The cellophane acts as a gas barrier to prevent oxygen from reaching the negative plates. Oxygen reaching the negative plates will cause the plates to heat with resulting plate damage, as explained earlier.

As mentioned previously, the state of charge of a nickel-cadmium cell cannot be determined by the specific gravity of the electrolyte. Furthermore, the voltage of a nickel-cadmium cell remains essentially constant, either open or under load, until the cell is completely discharged. Thus, the only way to determine the state of charge of a cell is by a complete and measured discharge or by timing the rate of charge from a completely discharged state.

PERFORMANCE OF NICKEL-CADMIUM BATTERIES

The performance of nickel-cadmium batteries is affected by a number of factors. Among these are temperature, duty cycle, quality of electric connections, contamination, and improper charging.

The batteries can be operated over a wide range of temperatures; however, at extremes, there will be a decrease in effective capacity. The best operating temperatures are in the range between 60 and 90°F [15.56 and 32.2°C]. The maximum range for discharge is from −65 to +160°F [−54 to 71°C]. For charge, the maximum range is from 20 to +120°F [−6.67 to 49°C].

Batteries may develop cell imbalance and loss of capacity when subjected to duty cycles consisting of shallow, rapid discharges and constant-potential (voltage) charging. Capacity reconditioning must be employed to correct the imbalance condition. This process will be described later in this section.

Poor electric connections interfere with the operation of a nickel-cadmium battery just as they do in any electric circuit. Loose, corroded, dirty, or burned connectors can cause complete stoppage of current flow, reduced operating voltage, or arcing. In any event, the installation of the battery should be accomplished with clean, tight connections.

Contamination of the electrolyte with tap water, acids, or other noncompatible substances will result in poor performance or complete failure of the battery. Battery operation with damaged, missing, or loose vent caps will cause the potassium carbonate level to increase and will reduce discharge voltages. Potassium carbonate is formed by the chemical reaction between carbon dioxide in the air and the potassium hydroxide in the electrolyte. Oxygen will enter the cell through a loose or open vent cap and discharge the negative plate through chemical action. It is therefore important that the vent caps of nickel-cadmium cells be kept firmly in place at all times except when water or electrolyte is being added.

The charging of nickel-cadmium cells should be done according to manufacturers' instructions. These instructions are prepared to give the user the most dependable service and longest life that the battery can provide. High overcharge currents cause the battery to overheat and can result in separator damage. Low charge current or incorrect charging methods can result in reduced battery output. Recommended charging methods for nickel-cadmium batteries are discussed in this chapter.

NICKEL-CADMIUM BATTERY INSTALLATION

The installation of nickel-cadmium batteries in an airplane requires the same care and attention to details that is necessary when installing lead-acid batteries. Condition of the battery compartment, cleanliness, secure mounting, and good electric connections are all of prime importance.

A new battery should be thoroughly inspected for shipping damage when it is first unpacked. Any evidence of damage, such as dented or broken cases, cracked or leaking cells, or evidence of careless handling should be reported to the shipping company immediately. The battery, if damaged, cannot be placed into service. Shorting devices and transportation plugs must be removed. These items are only employed for shipping and storage purposes. An instruction tag or sheet will usually accompany the battery. The instructions should be followed carefully to assure successful operation.

A new battery is normally shipped in the discharged condition. Unlike lead-acid batteries, nickel-cadmium batteries do not deteriorate or suffer damage in the discharged state. Short-circuiting clips are often connected across cell terminals to assure that the battery is in the discharged state. A discharged battery may appear to have insufficient electrolyte; however, electrolyte or water should not be added to the new battery unless specific instructions are given that this should be done.

The screws and connectors of a new battery should be checked for torque and security. Screws and nuts should be torqued according to instructions. Loose connectors will cause arcing and excessive heating, and this may deform the seal around the terminal. The arcing during charge can also cause an explosion because of the mixture of hydrogen and oxygen gases generated in the cells.

Since the electrolyte used in a nickel-cadmium battery is a strong caustic, care must be used to make sure that it does not come into contact with any part of the body. As explained earlier, potassium hydroxide is similar to household lye and will cause severe burns if allowed to contact the skin. Rubber gloves, a rubber apron, and protective goggles should be worn when handling the electrolyte. If the electrolyte should get on the skin, it should immediately be washed off with water, and the area should be treated with 3 percent acetic acid, vinegar, or lemon juice. This will neutralize the potassium hydroxide and stop its burning action. If the electrolyte should get into the eyes, they should be flushed at once with water and it is important that medical treatment be obtained immediately.

Before the new battery is placed in service, it should be charged according to the manufacturer's instructions. When connecting the battery to the charging unit, it must be with the positive terminal to the positive bus of the charger and the negative terminal to the negative bus. If the battery is connected with reverse polarity, permanent damage will result.

The battery compartment in an airplane or any other vehicle must be clean and dry. If the compartment has been previously used for lead-acid batteries, it should be washed out with an alkaline solution to neutralize any acid residue. The compartment should then be dried and painted with an alkaline-resistant paint.

Since hydrogen and oxygen are released from a battery during charge, ventilation of the battery compartment is necessary. Some battery cases contain vent nozzles which can be connected to tubes to remove gas. With others, provisions are made in the case or cover to permit escape of the gases into the battery compartment. In either case, the minimum air-flow requirement for ventilating the compartment is 5 cubic feet (ft^3) [0.15 m^3 or 150 L] per minute or enough to keep the hydrogen concentration below 4 percent.

After the battery is mounted in the aircraft, the quick-disconnect unit is connected. It is always wise to consult the aircraft manufacturer's service manual to assure that all points of battery installation are properly performed.

CHARGING METHODS FOR NICKEL-CADMIUM BATTERIES

As is true of lead-acid batteries, nickel-cadmium batteries can be charged either by the constant-current method or the constant-voltage (constant-potential) method. The constant-current method is slower, but it is the preferred method of charging when time permits and equipment is available. The constant-current method is more effective in maintaining cell balance and capacity, and it permits easy computation of the charge input in ampere-hours.

CONSTANT-VOLTAGE CHARGING

Nickel-cadmium batteries can be charged quickly by the constant-voltage method, but the charging time will depend upon the current delivery capability of the charging equipment. On an aircraft system, the battery will be charged by the constant-voltage method and will usually be held at nearly full charge. The aircraft charging system quickly restores a partially discharged battery to full charge without excessive water loss. Over a period of time, this type of charging leads to cell imbalance, and it becomes necessary to apply **capacity reconditioning** occasionally.

The time required to charge a nickel-cadmium battery by the constant-voltage method may vary; however, approximately 90 percent of the rated capacity can be restored in 1 h, provided that the charging equipment is capable of delivering current equal to ten times the ampere-hour rating of the battery. Current flow versus time is indicated in Fig. 6-45 for the constant-voltage charging method when starting the charge with a fully discharged battery. A very high inrush of current takes place at the beginning of the charge because the battery voltage is

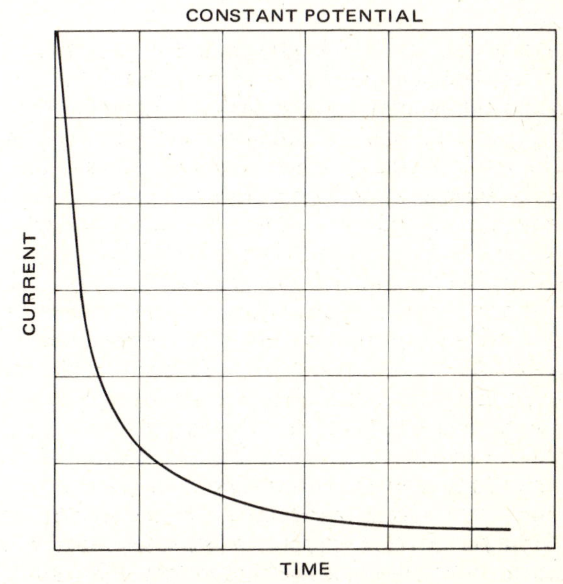

Fig. 6-45 Current flow versus time for a nickel-cadmium cell.

very low. The current is limited only by the capacity of the charging system and the internal resistance of the charging circuit, including the internal resistance of the battery. The charging current drops off quickly because the battery voltage soon rises to 1.2 V per cell or more, depending upon the temperature.

The charging voltage in an aircraft electric system should be checked frequently and adjusted as necessary to maintain the battery at optimum performance and to obtain a long battery life. If the charging voltage is too low at a given battery temperature, the battery will not receive a full charge. On the other hand, if the charging voltage is too high at a given battery temperature, the battery can be damaged by a charging current that remains high after the battery is fully charged. Too high a charging voltage may cause excessive water loss, which will result in an increase in battery temperature. This could eventually result in thermal runaway and destruction of the battery.

The voltage regulator in the aircraft system should be set at a level consistent with the ambient temperature band. Setting of the regulator should be done after a start and a few minutes into the charging period. The chart of Fig. 6-46 shows recommended charging voltages per cell for various battery temperatures. Table 6-7 gives charging voltage ranges for batteries when the ambient temperature is at or near 75°F [24°C].

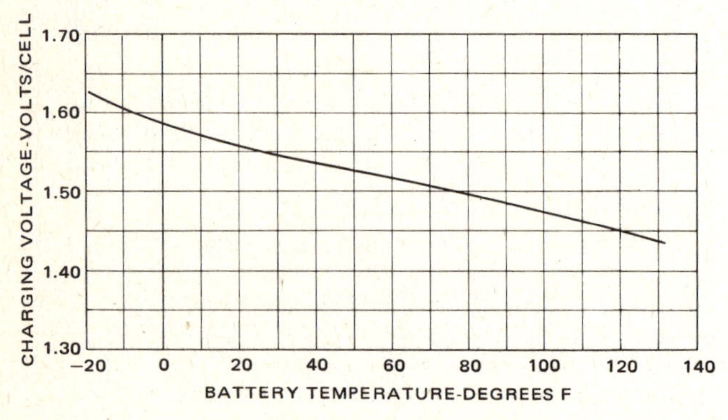

Fig. 6-46 Charging voltage per cell for various temperatures. (Saft America Inc.)

TABLE 6-7 Recommended Regulator Voltage Settings with Temperature at 75°F [24°C]

Nominal capacity, Ah	Number of cells	Nominal battery voltage	Time, h	Charging voltage
3 to 80	5	6	2–4	7.5–7.75
	10	12	2–4	15.0–15.5
	12	15	2–4	18.0–18.5
	19	24	2–4	28.0–29.0
	20	25	2–4	30.0–30.5
	22	27.5	2–4	33.0–33.5

CONSTANT-CURRENT CHARGING

The basic principle involved in the charging of a nickel-cadmium battery is to return to the battery the electrical energy necessary to replace the energy that was consumed during discharge. Because batteries are not 100 percent efficient, it is common practice to charge a battery with 140 percent of its nominal rated capacity. For example, if a battery is rated at 40 Ah, it is charged at 8 A for 7 h or at 4 A for 14 h. Thus, 56 Ah is returned to the battery in either case. Table 6-8 gives various charge rates with respect to battery capacity for constant-current charging.

TABLE 6-8 Charge Rates for Nickel-Cadmium Batteries with Constant-Current Charging

Cell capacity rating, Ah	5-h Rate-amp, charge time 7 h	10-h Rate-amp, charge time 14 h
13.5	2.7	1.4
15	3.0	1.5
25	5.0	2.5
30	6.0	3.0
40	8.0	4.0
45	9.0	4.5
85	—	8.5
15	—	11.5

The following notes should be remembered when charging nickel-cadmium batteries:

1. Charging is most efficient at ambient temperatures between 60 and 90°F [15.6 and 32.2°C].

2. Batteries should be thoroughly inspected for condition before charging.

3. Polarity of connections should be double-checked before turning on the charger.

4. Individual cells should not be charged unless the cell-container walls are supported to prevent swelling or distortion of the cell.

5. Two or more batteries should not be charged in series by the constant-voltage method unless the batteries are of the same type and capacity and in the same state of charge.

6. Two or more batteries can be charged in series by the constant-current method if they require the same charge rate.

7. Two or more batteries can be charged in parallel on the same bus using the constant-voltage method provided that they use the same charge voltage.

8. Moderate foaming in cells during charge does not necessarily indicate a defect, particularly after water has been added. Continued foaming, even after several cycles of operation, does indicate contamination of the electrolyte.

9. Improper adding of water or electrolyte to a nickel-cadmium cell or battery can be damaging, especially if the cell or battery is not fully charged. The manufacturer's instructions should be consulted to make sure that the electrolyte level is correct. Water should be added only as specified and to the level set forth in the instructions.

10. Battery temperature should be checked periodically to avoid overtemperature and possible **thermal runaway.**

SPECIAL CHARGING SYSTEMS FOR NICKEL-CADMIUM BATTERIES

In addition to the two basic charging systems previously described, there are chargers which use a combination of the two methods and chargers which use current pulses with feedback networks. Many of these chargers are designed for specific applications and are used only for charging nickel-cadmium batteries.

The **constant-potential current-limited** charger operates initially in a current-limiting region, then switches to the constant-voltage system to complete the charge. This charger requires more time for charging than the types previously described.

The **constant-potential constant-current** charger starts as a constant-voltage charger and switches to constant current near the end of the charge. The final rate is at the level of a trickle charge. The reduced charge rate at the end reduces overcharging and subsequent water loss.

The **constant-current temperature-compensated constant-potential** charger operates in a manner similar to the constant-potential current-limited charger except that the initial charging current is lower. The charger remains in the constant-current mode for almost the entire charge, then switches to a temperature-controlled constant-voltage mode.

The **pulse-type** charger delivers the charging current in pulses. A feedback network senses the open-circuit battery voltage between the pulses and regulates the charging current for the best balance between maximum charge efficiency and minimum water loss.

Since nickel-cadmium batteries will self-discharge at the rate of approximately 1.2 percent per day at normal temperatures, standby charging is required to maintain a full charge. This charging should be in the temperature range between 60 and 90°F [15.6 to 32.2°C] and at the rate of 3 mA/Ah rated capacity of the battery. This means that a 30-Ah battery would require a 90-mA charging current to maintain full-charge when not in use. Electrolyte level should be checked regularly for batteries on standby charge.

EFFECT OF TEMPERATURE IN CHARGING

Nickel-cadmium batteries are capable of sustaining very high current drains and good performance at temperature extremes. However, sustained high current and high temperature will have a degrading effect on performance. The effective capacity of a battery will decrease when subjected to high current demands and extreme temperatures. The available capacity of a nickel-cadmium battery with respect to temperature is indicated in Fig. 6-47. As explained previously, the optimum operating temperature for nickel-cadmium cells is in the range of 60 to 90°F [15.6 to 32.2°C].

At lower temperatures, the charge acceptance is good if proper corrections are made in the charging voltage. If the charging voltage is not corrected for low temperature, the battery will not deliver its rated capacity. Operating the battery at elevated temperatures subjects the battery to the danger of thermal runaway, particularly in overcharge at constant voltage. This condition is characterized by a continuously rising current and increasing battery temperatures during constant-voltage charge.

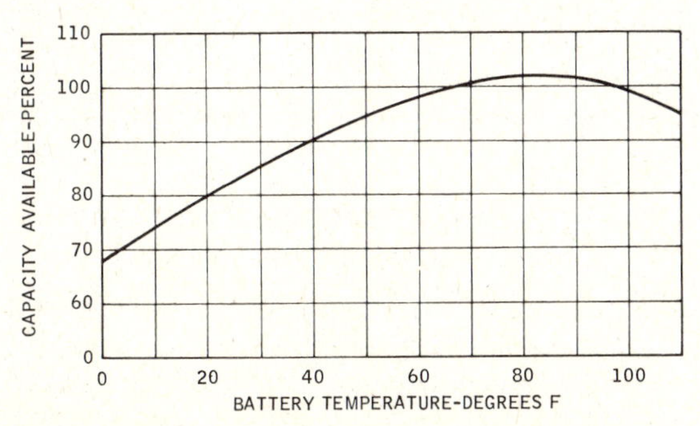

Fig. 6-47 Capacity of nickel-cadmium batteries with respect to temperature. (Battery Division, General Electric Co.)

The input current to a fully charged battery results in water loss and heat generation. At high ambient temperatures, the heat loss of the battery through radiation and conduction is lower than the heat generating rate. This net increase of battery heat raises the battery temperature and reduces the resistance of the battery to input current flow. Input current flow, therefore, increases if the battery is on constant-voltage charge. This higher current further increases the temperature, and the resistance continues to decline.

If allowed to continue, the preceding mode of operation will completely destroy the battery. The separator will be ruined by heat disintegration, and the nickel hydroxide of the positive plate will be dehydrated and converted to an electrochemically irreversible form of nickel oxide. This material is an insulator, and its formation permanently reduces the battery capacity.

If charging is required at high ambient temperatures, adjustments must be made in the charge voltage to prevent thermal runaway. The chart of Fig. 6-46 gives recommended voltages at various temperatures. Multiply the recommended voltages per cell by the number of cells to obtain the correct voltage setting for battery charging at constant voltage.

Some nickel-cadmium batteries are equipped with temperature sensors. These make it possible to continuously monitor battery temperature by means of instruments in the cockpit or the flight engineer's station.

CHARGING EFFICIENCY

Charging efficiency is the ratio of ampere-hours available on discharge to ampere-hours returned to the battery during charge. Efficiency is always less than 100 percent. Therefore, the ampere-hours returned to the battery must always be greater than the ampere-hours removed by discharging. The total charge necessary may be as low as 110 percent or much greater, depending upon the temperature and the method of charging. As stated previously, it is generally recommended that charging be continued long enough to return 140 percent of the ampere-hours delivered by the battery during discharge.

STORAGE OF NICKEL-CADMIUM BATTERIES

Nickel-cadmium batteries may be stored in either the charged or discharged condition without any appreciable deterioration. If stored in a charged condition, the battery will slowly discharge unless it is connected to a standby charger. If the battery is connected to a standby charger, it must be checked periodically, and distilled water will have to be added from time to time to prevent drying up of the electrolyte. Furthermore, the area where the battery is stored should be well ventilated to prevent the accumulation of explosive gases. Unless it is necessary to keep batteries immediately available for use, it is recommended that they be stored in the discharged condition.

Nickel-cadmium batteries can be stored at temperatures between −65 and +160°F [−53.9 and +71.1°C]; however, long-term storage above 120°F [48.9°C] is not recommended.

When nickel-cadmium batteries are returned to service after storage, the following steps should be taken to assure satisfactory performance:

1. Check the vent plugs to make certain that potassium carbonate deposits have not formed within the vents and made them inoperative. If deposits are found, the caps should be washed in hot water of 120 to 150°F [48.9 to 65.6°C] and then dried before replacing in the battery.
2. Allow the battery temperature to return to the correct operating range, if it has been stored at high or low temperatures, before placing on charge.
3. Charge the battery at the constant-current rate, level the electrolyte with distilled water, then discharge at the 2-h rate.
4. Recharge the battery and place it in service.

PREVENTIVE MAINTENANCE OF NICKEL-CADMIUM BATTERIES

The performance of any storage battery is dependent upon proper maintenance, and this is particularly true of nickel-cadmium batteries. The operator of aircraft equipped with nickel-cadmium batteries should schedule maintenance procedures to fit the type of aircraft operation. For an airplane which flies regularly, maintenance procedures can be scheduled according to increments of flight time. For example, capacity reconditioning can be scheduled for every 100 h of flight.

In service, powdery deposits may accumulate on the cell links and the upper surfaces of the cells in a battery. These deposits may be removed by wiping with a clean cloth or brushing with a plastic brush.

The outside of a battery can be cleaned by washing with soap and water. After washing, the battery should be rinsed and dried. For a thorough cleaning of the cell covers and connectors, the vents should be sealed and the top of the battery flushed with water. Excess water is then removed, and the battery is dried.

After cleaning, the battery is checked for mechanical defects. Damaged parts must be repaired or replaced. Nickel-cadmium batteries are generally constructed so they can be rebuilt by removing damaged cells and replacing them with new ones, thus making it possible to salvage the case and the cells which are still in good condition. The individual cells are easily removed from the case by disconnecting the links and pulling the cell slowly upward. A cell-puller attached to the terminal posts provides the best means for removing cells. In any case, the work must be performed with care to prevent damage to good cells and to prevent the spillage of the electrolyte.

PRECAUTIONS IN HANDLING NICKEL-CADMIUM BATTERIES

Certain safety practices should be observed in handling nickel-cadmium batteries to avoid personal injury and to prevent damage to the batteries. The following are particularly important:

1. Metal articles should be removed from the hands and wrists before starting work on a battery. Inadvertent contact of rings, watch bands, bracelets, or any other metal item with metallic parts or connectors of opposite polarity can result in fusing of the metal and severe burns to the wearer. It must be remembered that the connectors and terminals of the cells are bare metal and can be short-circuited easily by allowing any metal object to come into contact with them.
2. Tools used to service nickel-cadmium batteries should be of the insulated type. Metal tools dropped onto the battery are likely to short-circuit the connectors and cause arcing which could not only damage the battery but also injure the technician.
3. Care must be taken in handling the electrolyte. As explained previously, the electrolyte consists of potassium hydroxide and water and is highly caustic. Severe burns can occur if the electrolyte comes in contact with the skin. Rubber gloves, goggles, and other protective clothing should be worn when one is working with electrolyte. If electrolyte should contact the skin, the area should be flushed immediately with water and then neutralized with a 3 percent solution of acetic acid, vinegar, lemon juice, or a 10 percent solution of boric acid. Electrolyte in the eyes is most serious. Should this problem occur, the eyes should be flushed immediately with water and treatment by a physician should be obtained as quickly as possible.
4. During charging, a mixture of hydrogen and oxygen gases is released by the battery. The same care described for lead-acid batteries should be taken in working around nickel-cadmium batteries. It is particularly important to ensure that no arcing of connections or terminals takes place near the batteries when vent caps are removed from the cells. Connections to a charging system should be made only with power off. Intercell connectors and terminals of the battery should be clean and properly torqued.
5. Nickel-cadmium batteries should not be serviced in the same area as lead-acid batteries. The electrolyte for lead-acid batteries is acid, and the electrolyte for nickel-cadmium batteries is alkaline. Inadvertent mixing of the different electrolytes will cause a strong chemical reaction and render the electrolytes useless. Only pure electrolyte or distilled water should be used in servicing the batteries. Contamination by foreign substances must be prevented. Vent caps should be kept in the cells at all times except during adjustment of the electrolyte.

CAPACITY RECONDITIONING

It has been mentioned previously that nickel-cadmium batteries will develop cell imbalance after a certain period of operation, depending upon the type of service to which the battery has been subjected and the charging system employed. Cell imbalance simply means that some cells may be fully charged while others may be only partially charged. The capacity of the battery will then be limited by the cells which have a partial charge. As imbalance increases, the capacity of the battery will decrease. To eliminate cell imbalance, the battery must have all cells completely discharged and recharged. This process is called **capacity reconditioning.**

For aircraft service, it is recommended that capacity reconditioning be performed after 100 h of flight time, provided that the battery gives satisfactory service for this period of time. A 50-h inspection is recommended to determine if the battery is functioning in a normal manner and to see if the electrolyte level is correct.

The first step in capacity reconditioning is to discharge the battery completely at the rate specified in the service manual. The rate is usually such that the battery can be discharged in 1 or 2 h. If one or more of the cells drop to zero voltage or assume a reverse polarity before the average voltage drops to the specified level, it is recommended that a shorting clip be placed across the terminals of each such cell while the discharge continues. After the cells are discharged to approximately 0.5 V each, open-circuit voltage, all the cells are short-circuited by means of shorting clips to reduce the individual cell voltage to zero.

Starting with all cells at zero voltage, the battery is recharged at the 5-h rate for 7 h or at the 10-h rate for 14 h. After about 5 min of charging, the cells are tested for individual voltage readings. If any cell reads more than 1.6 V, distilled water is added to that cell. Any cell which reads less than 1.2 V is replaced. At the end of the charge, the cells are checked again for voltage. Any cell that measures less than 1.5 V or more than 1.75 V while still on charge is replaced or reconditioned as specified in the service manual.

After the battery is fully charged, it should be allowed to stand for 2 to 4 h. The electrolyte level is then checked, and if it is not correct, it is adjusted.

CAPACITY CHECK

When it is desired to determine the condition of a nickel-cadmium battery, it is necessary to perform a capacity check. This is accomplished by measuring the electric energy that can be obtained from the fully charged battery. If it is desired to determine whether an aircraft charging system is maintaining the proper charge in the battery or batteries, the capacity check is made by testing the battery in the state of charge it held upon removal from the aircraft.

The capacity check is made by discharging the battery at a measured rate and noting the time it takes to reduce the voltage to an average of 1 V per cell. A 19-cell battery should be discharged until the voltage is 19 V. For a capacity check, the discharge current load should be at the 1-h rate, that is, 24 A for a 24-Ah battery, 40 A for a 40-Ah battery, and so on. If the discharge capacity is at least 85 percent of rated capacity, the battery is considered satisfactory. If a battery is being discharged at the 1-h rate and it discharges to a level of 1 V per cell in 51 min, its discharge capacity is 85 percent of its rated capacity.

If, after a capacity check, the capacity of the battery is less than 85 percent of rated capacity, the battery should be deep-cycled (completely discharged by means of shorting clips) and recharged as for capacity reconditioning. Another capacity check should be made to assure that the reconditioning was successful. The battery should be fully charged before being replaced in an aircraft.

MAINTENANCE PRACTICES

Maintenance practices for nickel-cadmium batteries will vary to a limited extent for different makes and models of batteries; however, general practices are similar. Any differences will be specified in service manuals.

Inspections

When a battery is received in the shop for routine servicing, it should first be inspected visually for damage to the can or case, cover, and external battery connectors. Any defects are corrected by repair or replacement. The battery should be discharged before making repairs. The top of the battery should be inspected for cleanliness, loose or corroded connections, leaking cells, and damaged hardware. Vent plugs (caps) should be checked for tightness and damage. Loose plugs should be tightened and damaged plugs must be replaced.

Cleaning

A battery may be found to have a white, powdery deposit on the top. This is caused by spilled or leaking electrolyte. This and any other foreign material may be removed by washing the top of the battery with clean water. The battery should be tilted at an angle of about 45° so the water will flow off rapidly. The vent plugs must be tight to ensure that water does not enter the cells. Material that does not wash off easily may be removed with a nonconducting bristle brush. A metal brush must not be used because it will cause short circuits between cells. The tops of batteries must not be cleaned with solvents, chemical solutions or any other liquid except clean water. After a battery is cleaned and the water drained off, it may be dried with oil-free compressed air.

Disassembly

Nickel-cadmium batteries are designed and constructed so they may be completely or partially disassembled for the replacement of cells. Before disassembly, the battery should be cleaned and completely discharged as explained previously. Discharge must be accomplished using shorting clips on all cells at the end of the discharge. Disassembly steps are as follows:

1. Remove the shorting clips.
2. Remove all intercell connectors.
3. Loosen the vent plugs using the proper vent plug wrench.
4. Remove cells by attaching the correct cell puller to each cell in order and pulling the cell straight up.
5. After each cell is removed, tighten the vent plug.

Reassembly

The reassembly of a nickel-cadmium battery is essentially the reverse of disassembly; however, care must be taken to perform the operation properly. Before a battery is assembled, all parts must be inspected for condition. Any defects must be corrected. The cells must be in good condition. Replacement cells must be new or rebuilt by the manufacturer of the battery. The following are typical procedures for the reassembly of one type of battery:

1. Insert cells into the case or can, being sure that they are correctly positioned with respect to polarity for the particular model of battery being assembled. Do not force cells into place by hammering or other excessive pressure. Use a steady force on the terminals to press them

into place. The best procedure is for the cell at the middle of a row to be inserted last.
2. Lightly polish the cells' terminal surfaces with fine emery cloth and wipe clean.
3. Recheck cell polarity and then place intercell connectors in their correct positions.
4. Install nuts or cap screws, depending upon the model of the battery, finger-tight at all cell terminals.
5. Starting at the positive terminal of the battery, tighten each terminal nut or screw to the torque specified in the service manual. Care should be taken to ensure that the terminal screw is not binding, owing to thread damage or bottoming, but is actually tightening the connector terminal assembly. During tightening of terminal screws, do not skip around or leave the job partially completed and then come back to it. Finish the complete battery assembly once it is started. If one screw or nut is missed, permanent damage to a battery may result or arcing at the loose terminal can cause an explosion.

Final Inspection

After a battery has been reassembled, certain inspections should be made to ensure that the battery is ready for charging. These are as follows:

1. Check each cell with a voltmeter, following the battery circuit, to ensure that the cells are in proper order with respect to polarity. If the battery should be charged with a cell in reverse polarity, the cell would be permanently damaged.
2. Recheck the torque of all terminals.
3. Check vent caps for correct seating and assembly.

Troubleshooting Chart for Nickel-Cadmium Batteries

Indication	Cause	Remedy
Excessive electrolyte spewage.	Electrolyte level too high.	Clean and adjust electrolyte level.
	Vent caps loose or damaged.	Replace or tighten caps, recondition, and check electrolyte level and clean.
	Damaged cell case.	Short out all cells to 0 V, clean, replace defective cell, recondition, and check electrolyte level.
	Charging voltage too high.	Clean and recondition, check electrolyte level and correct charge voltage.
Excessive use of water by one or more cells.	Leaky or defective cells or vent caps.	Inspect for electrolyte leakage and remove and replace defective cells/vent caps.
	Cell imbalance.	Reconditioning cycle.
Foaming during charge after water addition.	Low concentration electrolyte.	Reconditioning cycle, replace defective cells.
Continued foaming during charging.	Oil and grease contamination in the electrolyte.	Replace defective cells.
No battery output.	Broken or disconnected links. Loose battery connector.	Replace or tighten links and connector.
	Cell open internally.	Replace defective cell.
Loss of capacity.	Electrolyte level too low.	Recondition, adjust electrolyte level, and capacity test.
	Charging rate too low in service.	Recondition and capacity test and adjust charging rate.
	Too little usage or shallow discharges.	Perform reconditioning cycle and capacity test.
Below normal output voltage.	Battery left on load.	Reconditioning and capacity test.
	Charger regulator set too low.	Recondition and capacity test correct regulator setting.
	Internal connecting links loose.	Torque to proper value, recondition, and capacity test.
	External connector burned or pitted.	Clean or replace, recondition and capacity test.
	Defective or reversed cell.	Replace defective cell, recondition and capacity test.
	Cell case current leakage.	Discharge, disassemble and clean and/or replace defective cells, recondition and capacity test.
Terminal links discolored or burned.	Loose terminal hardware.	Clean or replace links and torque all terminal hardware to specified values.
Burned or arced terminals.	Foreign metallic objects in the battery; tools dropped into the open battery.	Locate and remove foreign objects and replace defective parts.
Burned or pitted output receptacle terminals.	Improperly fitting or loose mating output connections.	Clean, repair or replace output receptacle and inspect connector plug, replace if loose.
Distortion of cell case and cover.	Cell with internal short.	Discharge, disassemble, and replace defective cells.
	Charger failure.	Discharge, disassemble, and replace defective cells.
	Plugged vent caps, minor explosion.	Discharge, disassemble, and replace defective cells.
	Overheated battery, improper cooling.	Discharge, disassemble, and replace defective cells.
Distortion of battery case or cover.	Major explosion: loose links, dry cells, high charge voltage, charger failure.	Discharge, disassemble, and replace defective cells, links, vent caps, hardware and case. Repair or replace battery case and cover.

4. Perform the capacity reconditioning procedure on the battery as described previously.
5. Check each cell for electrolyte level and adjust as necessary, using only distilled water to raise the level.

TROUBLESHOOTING

The chart on page 100 gives recommended troubleshooting procedures for one type of nickel-cadmium battery. Service manuals usually include troubleshooting charts for particular models of batteries, and these charts should be followed.

REVIEW QUESTIONS

1. Briefly describe a *voltaic cell.*
2. What is the difference between a *primary* cell and a *secondary* cell?
3. What voltage is developed by a carbon-zinc cell?
4. What is a *dry cell?*
5. Describe a *battery* of dry cells and explain how various voltages are obtained in different batteries.
6. What electrolyte material is used in an *alkaline* cell?
7. What is the composition of the electrodes in a *mercury* cell?
8. Give some advantages of the nickel-cadmium dry cell over the carbon-zinc dry cell.
9. What are the active materials in a nickel-cadmium cell?
10. What are the active materials in a lead-acid storage cell?
11. Describe the construction of a lead-acid storage cell.
12. Describe a *plate group.*
13. What electrolyte is used in a lead-acid storage cell?
14. Why does the number of negative plates in a lead-acid cell exceed the number of positive plates?
15. What materials are used for separators?
16. What advantages are provided by the employment of microporous polyethylene pouches to enclose the positive plates of a battery?
17. Describe *cell containers.*
18. What means is employed to prevent the sediment in the bottom of a cell container from short-circuiting the plates?
19. Explain the means used to prevent the spillage of electrolyte from an aircraft storage battery.
20. Describe three different types of vent caps.
21. Describe the various construction features of aircraft storage batteries.
22. How are aircraft storage batteries constructed to provide for elimination of explosive gases?
23. What determines the voltage of an aircraft storage battery?
24. What ratings are used to describe aircraft storage batteries?
25. What is the approximate open-circuit voltage of a fully charged lead-acid cell?
26. If a storage cell will deliver 20 A for 5 h, what is the ampere-hour rating?
27. Why is it that a lead-acid storage battery will appear to be discharged after the application of a heavy load for a short time, but will again deliver power after disconnecting the load for a few minutes?
28. What occurs with respect to ampere-hour rating when the discharge rate is increased above that used to establish the rating?
29. What is the most common method for determining the state of charge of a lead-acid battery?
30. Give the hydrometer readings for full charge, half charge, and discharged condition in a lead-acid cell.
31. Describe the process for testing a lead-acid cell with a hydrometer.
32. Under what condition may new electrolyte be added to a lead-acid cell?
33. How does temperature affect a hydrometer reading?
34. Describe the method for testing a lead-acid cell under load by means of a voltmeter.
35. Give the principal safety precaution which must be observed in working with lead-acid storage batteries.
36. Describe a capacity tester.
37. Explain the difference between constant-voltage charging and constant-current charging.
38. What type of charging is employed in an aircraft electric system?
39. What adjustments should be made in charging voltage when ambient temperatures are high?
40. Describe the effects of charging on battery life.
41. What hazards exist with respect to lead-acid batteries during charging?
42. Describe a battery compartment in an aircraft.
43. How are explosive gases from a battery eliminated from an aircraft?
44. Discuss the connection of battery cables to a battery.
45. Describe the Elcon battery connector plug.
46. List the inspections that should be made periodically for batteries and battery compartments.
47. Explain the detrimental effects of short circuits that may occur in careless handling of batteries.
48. How would you adjust the specific gravity for a lead-acid battery?
49. Explain how electrolyte should be mixed for a lead-acid storage battery.
50. Why should acid be poured into water rather than water be poured into acid when electrolyte is being mixed?
51. Describe the procedure for placing a new battery in service.
52. What precautions should be observed in operating batteries in very cold weather conditions?
53. What are likely causes for a battery's failure to hold its charge?
54. What electrolyte is used in nickel-cadmium batteries?
55. Describe the construction of a nickel-cadmium cell.
56. Describe the means used to separate the plates in a nickel-cadmium cell.
57. What is the function of the cellophane strip in the separator?
58. Explain the operation of a nickel-cadmium cell.
59. What are the factors affecting the performance of nickel-cadmium batteries?
60. Why should the vent caps be kept securely in place except during adjustment of the electrolyte in a nickel-cadmium battery?
61. Describe the procedures for placing a new nickel-cadmium battery in service.

62. What is the danger caused by loose cell connectors in a nickel-cadmium battery?
63. What are satisfactory charging methods for nickel-cadmium batteries?
64. Why can a nickel-cadmium battery be charged more rapidly by the constant-voltage method than by the constant-current method?
65. Explain the importance of keeping the charging voltage in an aircraft system at a particular level.
66. What is the correct voltage range for charging a 24-V nickel-cadmium battery with ambient temperatures of 70 to 80°F [21.1 to 26.7°C]?
67. When a completely discharged nickel-cadmium battery is being charged, how much electric energy must be returned to it in percent of its Ah rating for a full charge?
68. List factors that should be considered in charging nickel-cadmium batteries.
69. Describe the effect of temperature in charging nickel-cadmium batteries.
70. Explain a *thermal runaway*.
71. What condition of charge is most suitable for nickel-cadmium batteries that are to be stored?
72. What steps should be taken when a battery that has been stored is returned to service?
73. Discuss preventive maintenance with respect to nickel-cadmium batteries.
74. List precautions that should be observed in handling nickel-cadmium batteries.
75. What protective measures should be taken in handling electrolyte?
76. Why should service areas for lead-acid and nickel-cadmium batteries be separated?
77. What is *capacity reconditioning* and why is it performed?
78. How is a capacity check performed?
79. What conditions are observed in making an inspection on nickel-cadmium batteries?
80. How should a nickel-cadmium battery be cleaned?
81. Describe the procedure for disassembling a nickel-cadmium battery. Describe reassembly.
82. What practice should be observed in installing and tightening cell connectors?
83. What final inspections should be made when the assembly of a nickel-cadmium battery has been completed?

CHAPTER 7

Generator Theory

Electric generators have been used for many years to produce electric power for a multitude of purposes. From the original crude "dynamos" employed to produce power in the early days of our electrical age, the generator has been developed to a very high level of dependability and efficiency, especially for aircraft use.

Ever since the first aircraft to use any kind of electric equipment was launched, the electrical loads on airplanes and other flying devices have increased. Today, modern jet airliners are equipped with scores of different electric systems, each requiring a substantial amount of electric energy. To supply the power for their electric systems, these giant airplanes are equipped with generating equipment sufficient to furnish lights for a small town. The generators (alternators) for the Boeing 747 jumbo jets can produce a total of 240 000 kilovolt-amperes (kVA) or more. One generator is driven by each of the engines, and each one produces power far in excess of that produced by any commercial generator of similar weight.

An electric generator may be defined as a machine which changes mechanical energy to electric energy. The mechanical energy may be supplied by any one of many devices such as gasoline engines, steam engines, diesel engines, steam or water turbines, wind-driven propellers, gas-turbine engines, or atomic-powered engines. On airplanes, generators are usually driven by the airplane engine.

Generators are classified according to the type of current they supply, their capacity, or power output, type of windings, number of phases, internal connections, and frequency if the generator is an ac type. For example, one typical generator may be described as a 28-V dc 200-A generator, and another as a 117/208-V three-phase 400-Hz 20-kW generator. Dc generators are further described as shunt-wound or compound-wound. Shunt-wound and compound-wound generators are used on aircraft. Ac generators, also called **alternators,** are usually rated according to their voltage, number of phases, power output, and operating frequency.

When an airplane is in flight, its own generator or generators supply power for all its electrical operations. It is well to remember that the battery serves no particular purpose as long as the generator system is working properly. The generator system must be designed so that it will carry all operating loads during flight.

Large aircraft like jet airliners employ three-phase ac power systems rated at 117/208 V, with a frequency of 400 Hz. Compared with a 28-V dc system, an ac system will develop several times as much power for the same weight; hence it is a great advantage in large aircraft to use the ac systems.

Light aircraft manufactured during recent years are commonly equipped with 12- or 24-V alternators similar to those used for automobiles. The alternating current is rectified to produce the dc requirements for the aircraft electric systems.

PRINCIPLES OF ELECTRIC GENERATION

Electricity is produced in a generator by electromagnetic induction. As explained in an earlier section of this text, it is a fundamental principle that when there is a relative movement between a magnetic field and a conductor held perpendicular to the line of flux, an emf is produced in the conductor. If the ends of the conductor are connected together, the emf will cause a current to flow as shown in Fig. 7-1. The direction of current flow is determined by the direction of the magnetic flux and the direction in which the conductor is moved through the flux.

A simple way to determine the direction of current flow is to use **the left-hand rule for generators. Extend the thumb, index finger, and middle finger so they are at right angles to one another as illustrated in Fig. 7-2. Turn the hand so the thumb points in the direction of movement of the conductor and the index finger**

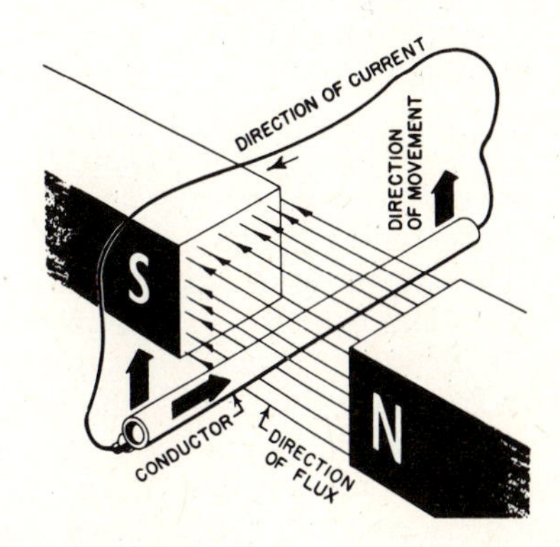

Fig. 7-1 Generator action.

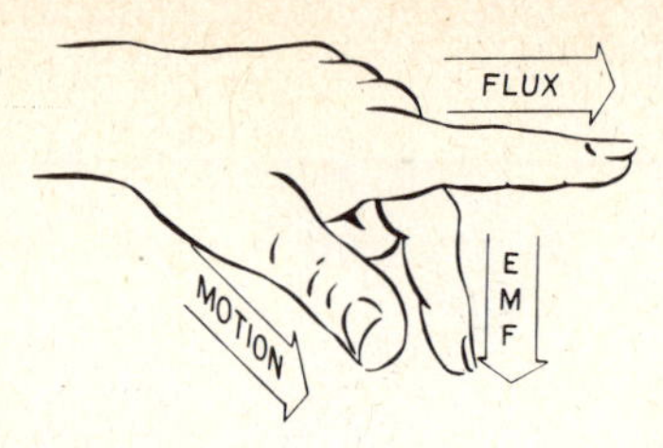

Fig. 7-2 Left-hand rule.

points in the direction of the magnetic flux. Then the middle finger will be pointing in the direction of the current flow. Remember, current flow is from negative to positive. Flux direction is considered to be from north to south.

SIMPLE AC GENERATOR

A simple ac generator can be constructed by placing a single loop of wire between the poles of a permanent magnet and arranging it so that it may be rotated as shown in Fig. 7-3. The current is taken from the wire loop by means of brushes which make continuous contact with the collector rings (slip rings). One collector ring is connected to each end of the wire loop. In Fig. 7-3, the sides of the loop are designated *AB* and *CD*. As the loop rotates in the direction indicated by the arrow, the side *AB* will be moving up through the magnetic field. If we apply the lefthand rule for generators, we find that a voltage is induced which will cause current to flow from *A* to *B* in one side of the loop and from *C* to *D* in the other side of the loop. This is because *AB* is moving **up** through the field and *CD* is moving **down** through the field.

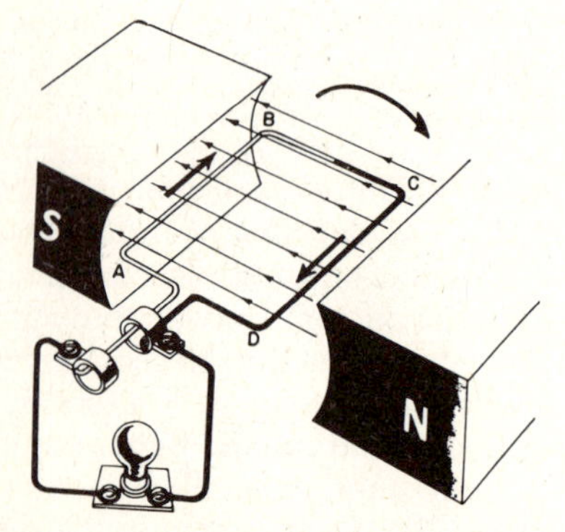

Fig. 7-3 Simple ac generator.

The voltage induced in the two sides of the loop add together and cause the current to flow in the direction *ABCD*, through the external circuit, and then back to the loop. As the loop continues to rotate toward a vertical position, the sides will be cutting fewer lines of flux, and when it reaches the vertical position, the sides of the loop will not be cutting any lines of flux but will be moving parallel to them. At this position, no voltage is induced in the loop because a conductor must cut across flux lines in order to induce a voltage. By rotating the loop through the vertical position and back to the horizontal, a voltage will be induced again, but it will be in the opposite direction in the loop because side *AB* will now be moving **down** through the field and side *CD* will be moving **up** through the field. Thus, a reversal of current takes place in the circuit outside the generator.

The values of the voltage induced in the loop may be shown by a sine curve as explained in Chapter 4. The voltage is at a zero value when the loop is in a vertical position, and then it climbs to a maximum value when the loop is in the horizontal position. This is indicated on the sine curve from 0 to 90°. As the loop continues to turn, we find that the voltage is maximum at 90°, zero at 180°, maximum at 270°, and zero again at 360°.

ESSENTIAL PARTS OF A GENERATOR

The essential parts of a simple ac generator are shown in Fig. 7-4. These are a **magnetic field** which may be produced by a permanent magnet or by field coils, a rotating loop or coil called the **armature,** or rotor, **collector rings,** and **brushes** by which the current may be taken from the armature. The poles of the magnet are called **field poles.** In most generators, these poles are wound with coils of wire called **field coils.** The path of the magnetic flux is called the magnetic circuit and includes the yoke connecting the field poles as well as the armature.

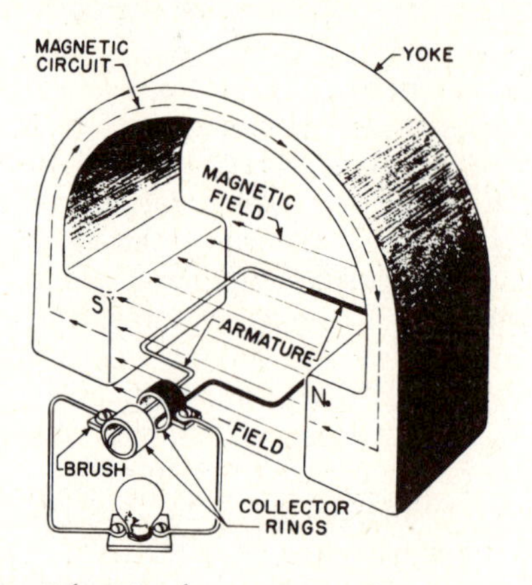

Fig. 7-4 Essential parts of an ac generator.

VALUE OF INDUCED VOLTAGE

The voltage induced in a conductor moving across a magnetic field depends upon two principal factors: the strength of the field (the number of lines of force per unit area) and the speed with which the conductor moves across the lines of force. In other words, the voltage depends upon the number of lines of force cut per second. For example, if a conductor cuts lines of force at the rate of 100 000 000 lines/s, an emf of 1 V will be established between the ends of the conductor.

SIMPLE DC GENERATOR

Dc generators are needed for many aircraft electric systems, for battery charging, and for various other applications. For this reason, an ac generator will not meet all power requirements unless a means of rectifying the alternating current is provided. Figure 7-5 shows an elementary type of dc generator quite similar to the simple ac generator explained previously.

A pulsating direct current may be obtained from the illustrated generator by using a commutator in place of the

Fig. 7-5 Simple dc generator.

collector rings on the ac generator. A **commutator** is a switching device which reverses the external connections to the armature at the same time that the current reverses in the armature. The commutator in Fig. 7-5 is a split ring which turns with the armature. One end of the rotating loop connects to one half of the ring, and the other end of the loop connects to the opposite half of the ring. The two sections of the commutator are insulated from each other. Two brushes are placed in a position relative to the commutator so that as the commutator turns, the brushes pass from one segment of the commutator to the other at the same time that the current is reversing; there is then practically no emf between the two segments. This system of changing the alternating current of the armature to direct current in the external circuit is called **commutation.**

Referring to Fig. 7-5, observe that the side of the loop moving up through the field will always be connected to the positive brush and that the side of the loop moving down through the field will always be connected to the negative brush. The current from the generator will then be traveling in one direction in the external circuit, but it will pulsate; that is, it will vary in intensity from zero to maximum and back to zero through each half turn of the armature. A current of this type is called a pulsating direct current and is not suitable for many uses.

ELIMINATION OF RIPPLE

Since the pulsating direct current of the simple generator is not satisfactory for all purposes, it is necessary to construct a generator which will produce an almost constant voltage. This is accomplished by increasing the number of coils in the armature and the number of field coils. Figure 7-6*a* illustrates the nature of the voltage from a single-coil generator, and Fig. 7-6*b* shows the curve for a generator with four armature coils. Notice the great difference in the nature of the voltage.

Armature coils are wound on a laminated soft-iron core. The iron core concentrates the field flux and greatly increases the voltage generated. The laminations reduce the effects of eddy currents induced in the core.

The current from any dc generator will have a slight pulsation known as **commutator ripple,** but this ripple does not interfere in ordinary electric circuits for purposes such as lighting and operating electric motors. For radio circuits, the commutator ripple must be eliminated because it causes a hum in the radio output. A capacitor of correct capacitance shunted across the dc power leads of a radio receiver will greatly reduce the amount of ripple. For a still more effective filter, an inductance or *choke* coil is connected in series with the dc line along with the capacitor in parallel with the dc line. Remember that an inductance opposes any change in the current flow. A capacitor, choke coil, or a combination of the two connected in a circuit to reduce ripple is called a **ripple filter.**

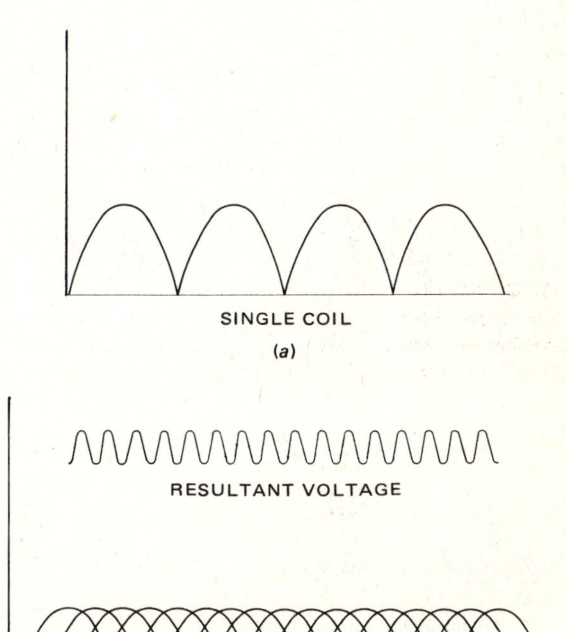

Fig. 7-6 Comparing voltages from single-coil and multiple-coil armatures.

RESIDUAL MAGNETISM

A permanent-magnet field is not satisfactory for practical generators because it is impossible to regulate the voltage of such a generator to compensate for changes in load and speed. For this reason, field coils are used to provide the magnetism required for the generation of current. In dc generators the field coils are usually energized by current from the generator. Fortunately, any substance which has been magnetized will retain a certain amount of magnetism. Materials such as soft iron give up most of their magnetism very quickly when removed from the magnetizing influence. However, they do retain a small amount which is known as **residual magnetism.**

It is this residual magnetism which makes it possible to start a generator without exciting the field from an outside source of magnetism. The residual magnetism in the field poles causes a weak voltage to be generated when a generator begins to rotate. This voltage sends a small amount of current through the winding of the shunt field of the generator, causing the field strength to increase. The increase in field strength causes a corresponding increase in generator voltage, and a mutual increase in field strength and voltage continues until the voltage reaches the proper value for the generator. If the residual magnetism should be lost because of excessive heat or shock, it can be restored to the field by passing a direct current through the field windings in the correct direction. This procedure is called *flashing the field* and is discussed in a later section of this text.

CHARACTERISTICS OF DC GENERATORS

Dc generators are classified as shunt-wound, series-wound, or compound-wound, according to the manner of connecting the field coils and the armature.

The internal connections for a shunt-wound generator are shown in Fig. 7-7, where it can be seen that the field coils are connected in parallel with the armature. In this type of generator, it is necessary to have a resistance or some other means of regulation in the field circuit to prevent the development of excessive voltage. If such a generator should be running without a load, the entire output would be going through the field coils, thus producing a very strong field. This field would, of course, increase the voltage of the generator, and the field strength would also increase. The result would be a continued increase of both the field strength and voltage until the generator burned out. Before practical voltage regulators were developed, the output of a generator was controlled by means of a movable field brush. The position of this brush on the commutator limited the amount of current through the field and thus controlled the voltage. Modern generator systems employ voltage regulators in the circuit to govern automatically the amount of current through the field windings. These regulators will be explained in detail in a later section of this text.

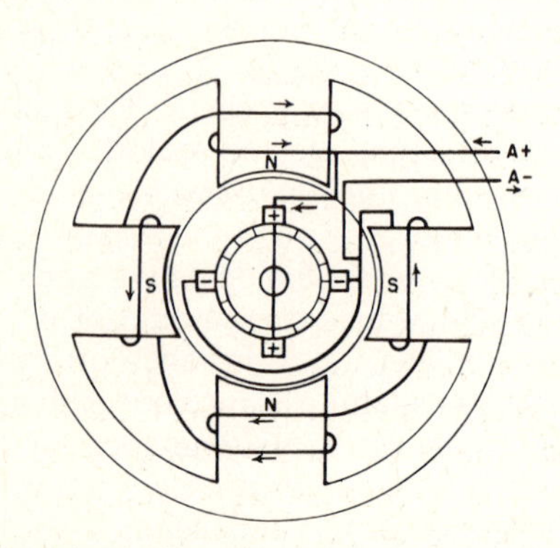

Fig. 7-7 Circuit for a shunt-wound generator.

Shunt generators without field-current regulation are satisfactory only for operation at a constant speed and with a constant load. In practice it is doubtful that such a generator can be used except experimentally. In such a generator there would be a change in voltage for every change in speed or load. As the load increases, the terminal voltage decreases. This is partly due to the internal resistance of the armature winding, which causes a voltage drop in the armature and thus reduces the terminal voltage. A decrease in terminal voltage causes a corresponding decrease in field current, which in turn lowers the field strength and produces a further reduction of terminal voltage. It is obvious that voltage stabilization could not be attained under such conditions.

Compound-wound generators combine the features of series and shunt generators. A compound generator with an external shunt field connection is shown in Fig. 7-8.

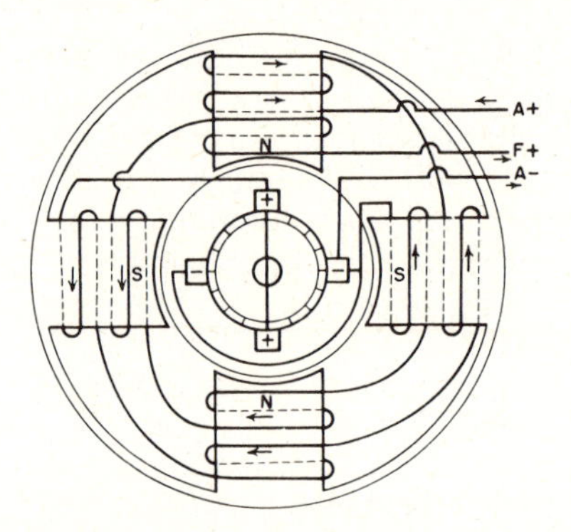

Fig. 7-8 Circuit for a compound-wound generator.

$A+$ is the main positive terminal, and $A-$ is the main negative terminal. $F+$ is the terminal for the shunt field. When the two fields are correctly balanced, the generator will maintain a constant voltage through the full range of normal loads and is said to be *flat-compounded.* When the series field is of more effect than the shunt field, the generator voltage will rise as the load is increased, and the generator is described as *overcompounded.* An *undercompounded* generator is one in which the shunt field has a greater effect than the series field, and the voltage will decrease as the load is increased.

The best type of generator for an aircraft electric system is either a shunt generator or one which has a shunt field which can become stronger than the series field. The voltage is controlled by a voltage regulator and is maintained at a constant level regardless of speed or load. Remember that the series field in a generator increases in strength as the load increases. This permits the generator to operate at a high power level without overloading the shunt field . As mentioned elsewhere in this chapter, the series-field winding also serves to compensate for armature reaction through the operation of interpoles in certain high-performance generators.

ANALYSIS OF AN ARMATURE CIRCUIT

In a previous paragraph it was explained that a practical generator has many coils of wire in the armature. These coils are connected to the commutator segments in such a manner that they are in series with one another. Figure 7-9*a* shows the connections for a typical commutator in a two-pole generator. Assume that the armature has eight coils of two turns each wound around the armature through oppositely positioned slots. If the magnetic flux is horizontal, no voltage will be induced in the vertical coils because the coil sides will be moving parallel to the lines of force and will not be cutting any of them. The coils in position *B* and *B′* will be cutting across a maximum number of flux lines and will therefore have a maximum emf induced in them.

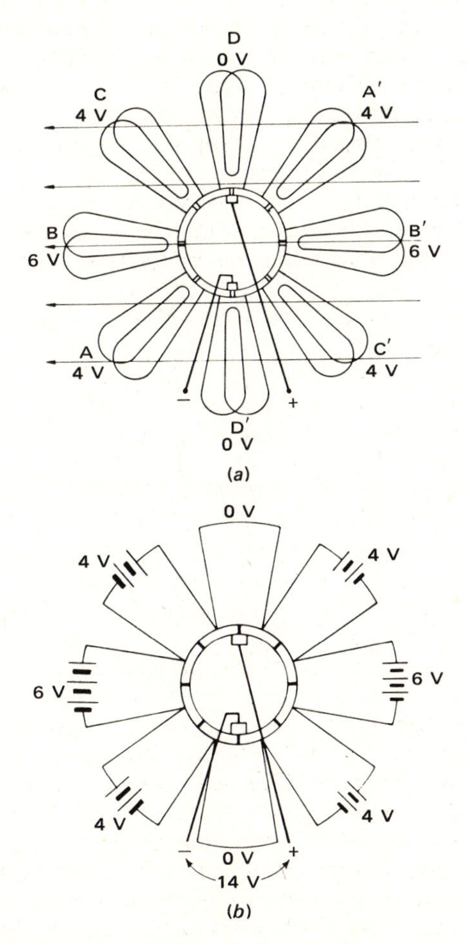

Fig. 7-9 Armature circuit with battery analogy.

For the purpose of illustration, we shall assume that this emf is 6 V. The coils at positions *A*, *A′*, *C*, and *C′* will then have an induced emf of approximately 4 V each. The result is that there are three voltage-producing coils connected in series in each half of the armature.

Figure 7-9*b* shows a battery analogy of the armature circuit. In each of the two series circuits in the armature, there are two 4-V coils and one 6-V coil. The total emf from each series is 14 V, and since the two circuits are connected in parallel, the amperage will be twice that of one series circuit.

The armature-winding arrangement illustrated in Fig. 7-9 is known as **progressive lap winding.** There are several different types of windings used for generators and motors, but the one shown here is adequate for the purpose of this discussion.

ARMATURE REACTION

Since an armature is wound with coils of wire, a magnetic field is set up in the armature whenever a current flows in the coils, as in Fig. 7-10*a*. This field is at right angles to the generator field shown in Fig. 7-10*b* and is called **cross magnetization** of the armature. The effect of the armature field is to distort the generator field and shift the neutral plane, as illustrated in Fig. 7-10*c*. This effect is known as **armature reaction** and is proportional to the current flowing in the armature coils.

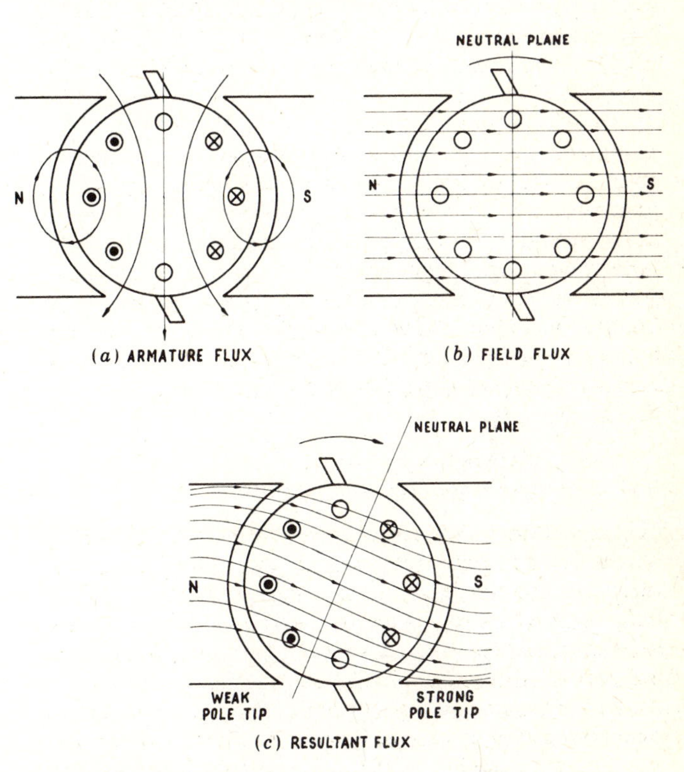

Fig. 7-10 Armature reaction.

The brushes of a generator must be set in a position known as the **neutral plane;** that is, they must contact segments of the commutator which are connected to armature coils having no induced emf. If the brushes were contacting commutator segments outside the neutral plane, they would short-circuit "live" coils and cause arcing and loss of power. Armature reaction causes the neutral plane to shift in the direction of rotation, and if the brushes are in the neutral plane at no load, that is, when no armature current is flowing, they will not be in the neutral plane when armature current is flowing. For this reason it is desirable to incorporate a corrective system into the generator design.

There are two principal methods by which the effect of armature reaction is overcome. The first method is to shift the position of the brushes so that they are in the neutral plane when the generator is producing its normal load current. In the other method, special field poles, called **interpoles,** are installed in the generator to counteract the effect of armature reaction.

The brush-setting method is satisfactory in installations in which the generator operates under a fairly constant load. If the load varies to a marked degree, the neutral plane will shift proportionately, and the brushes will not be in the correct position at all times.

INTERPOLES

The use of interpoles is the most satisfactory method for maintaining a constant neutral plane in a generator. The windings of the interpoles are in series with the load; hence, the interpole effect is proportional to the load. The polarity of the interpoles is such that their effect is opposite to that of the armature field; that is, each interpole is of the same polarity as the next field pole in the direction of rotation. With this polarity, the interpole may be said to pull the generator field into the correct position. A typical interpole system is shown in Fig. 7-11.

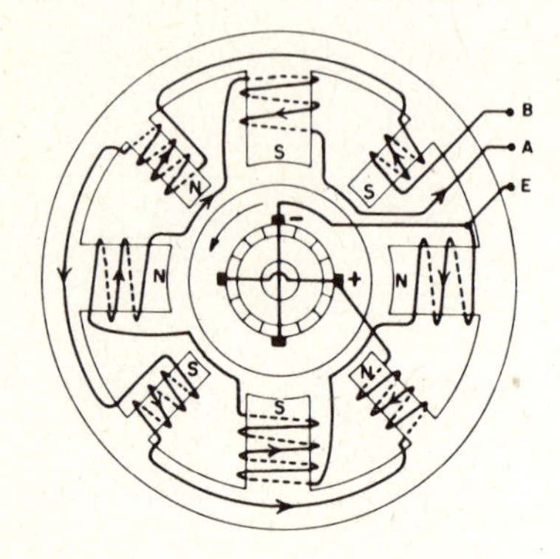

Fig. 7-11 Generator circuit with interpoles.

In many generators, a **compensating winding** is used to help overcome armature reaction. This winding consists of conductors embedded in the field-pole faces with one coil surrounding sections of two field poles of opposite polarity (see Fig. 7-12). The compensating winding is in series with the interpole windings; hence, it works with the interpoles and increases their effectiveness. The sparkless commutation obtained by the use of interpoles and compensating windings increases the life of the

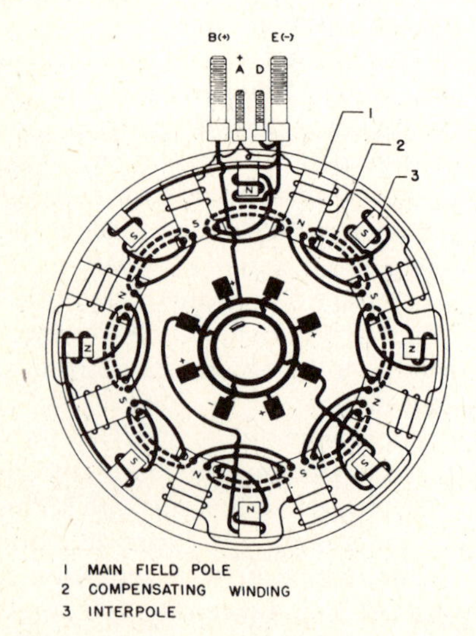

Fig. 7-12 Generator with interpoles and compensating winding.

brushes and commutator, reduces radio interference, and greatly improves the efficiency of the generator.

AC GENERATORS

AC generators, often called **alternators,** were not used extensively in aircraft until the late 1950s. Since that time, the ac generator has become the principal source of electric power in almost all types of aircraft. On jet airliners and military aircraft, the ac system supplies almost all the electric power required for the aircraft. Where dc is needed, rectifiers are used. Ac power systems provide a maximum amount of power for the weight of electric equipment in the aircraft. In large aircraft, this is particularly important. For light aircraft, ac generators are used, and rectifiers are used to provide dc for the operating systems. It is therefore essential that the technician be thoroughly familiar with ac theory and systems.

There are three principal advantages in the use of alternating current for electric-power systems. (1) The voltage of ac power may be changed at will by means of transformers. This makes it possible to transmit power at a high voltage with low current, thus reducing the size and weight of wire required. (2) Alternating current can be produced in a three-phase system, thus making it possible to use motors of less weight for the same amount of power developed. (3) Ac machinery, such as alternators and motors, do not require the use of commutators; hence, service and upkeep are greatly reduced.

A simple example will demonstrate the advantage of using high voltages for power transmission. We shall assume that we have a 1-hp [746-W] motor which must be driven at a distance of 100 ft [30.5 m] from the source of electric power. With a dc source of 10 V, the motor will require approximately 125 A, assuming that the motor is 60 percent efficient. Now, when we consider the current-carrying capacity of copper wire we find that a No. 1 cable is required to carry the current for the motor. One hundred feet of this wire weighs approximately 25 lb [11 kg]. If we substitute a 1-hp 200-V ac motor for the 10-V dc motor, the current required is only about 5 or 6 A, depending upon the efficiency of the motor. This will require a No. 18 wire which weighs about 1 lb [0.5 kg] for 100 ft. This comparison clearly demonstrates the advantage of higher voltages for power transmission.

A simple two-pole alternator is shown in Fig. 7-13. The stationary part of the alternator circuit is called the **stator,** and the rotating part is called the **rotor.** The stator is actually a stationary armature, and the rotor is a rotating

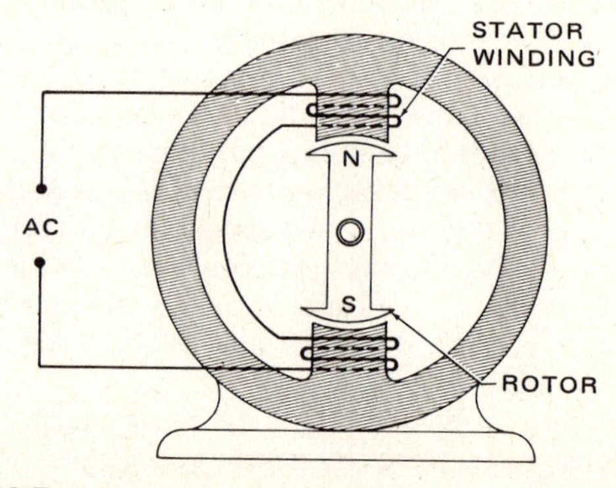

Fig. 7-13 Two-pole alternator.

field which may be either a permanent magnet or an electromagnet. As the rotor turns, the magnetic flux cuts across the stator poles and induces a voltage in the stator winding. The induced emf will reverse polarity every half revolution of the rotor because the flux will reverse in direction as the opposite poles of the rotor pass the stator poles. One complete revolution of the rotor in a two-pole alternator will produce 1 cycle of alternating current; that is, the emf in the stator coils will increase from zero to maximum once in each direction and then return to zero.

The number of cycles of alternating current per second is called the **frequency.** Since a two-pole alternator produces 1 cycle per revolution (cpr), it is apparent that an alternator produces 1 cycle of alternating current from each pair of poles in the rotor. If we wish to determine the frequency of any given alternator we proceed as follows: Divide the number of poles by 2 and multiply the result by the speed in rpm to obtain the number of cycles per minute. To find the cycles per second, divide the cycles per minute by 60.

Let us assume that we wish to determine the frequency of an alternator having 4 poles and turning at 1800 rpm. Dividing 4 by 2 gives 2 cpr, or 3600 cycles/min. Then, dividing 3600 by 60, we obtain 60 Hz.

If the alternator does not have a permanent magnet for the rotor, an electromagnet must be used and a direct current used to excite the rotor. An alternator with a four-pole electromagnetic rotor is illustrated in Fig. 7-14. A permanent magnet rotor is not satisfactory because the strength of the field flux always remains the same and voltage regulation cannot be accomplished. For this reason the separately excited type of alternator is almost universally used.

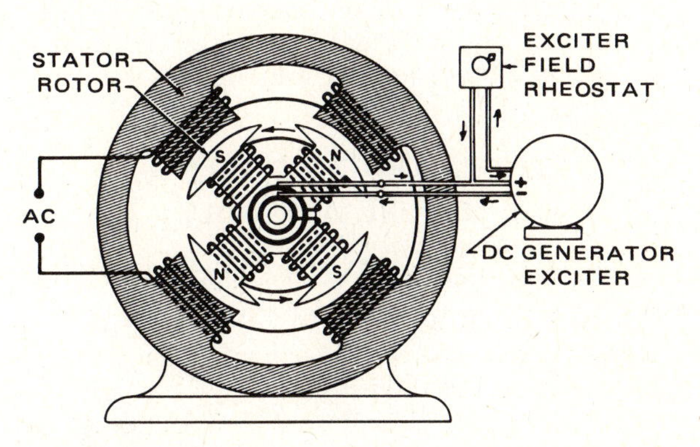

Fig. **7-14** Alternator with dc-excited field.

Alternators are classified according to voltage, amperage, phase, power output (watts or kilovoltamperes), and power factor. The phase classification of an alternator is the number of separate voltages which it will produce. Usually alternators are single-phase or three-phase, depending upon the number of separate sets of windings in the stator. Three-phase alternators are constructed with three separate windings spaced so that their voltages are 120° apart.

POWER FACTOR

The power factor of an alternator is the ratio of active power to apparent power produced by the alternator. If the output of the alternator is 110 V and 100 A, it would appear that the power output is 11 000 W. This is the apparent power. The active power is the power actually produced, and it will depend upon the phase relationship of the voltage and amperage. When voltage and amperage are in phase, the power factor is 1, or 100 percent.

The power factor of an alternator is equal to the cosine of the phase angle between the voltage and current when the characteristic curves of the voltage and current are true sine curves. The cosine of an angle may be obtained from a table of trigonometric functions. If the current from an alternator lags behind the voltage by 30°, the power factor will be 0.866 because cos 30° =0.866. To determine the active power, multiply the apparent power by the power factor.

The cosine of an angle is determined by using a right triangle as shown in Fig. 7-15. The ratio of the adjacent side (*AC*) to the hypotenuse (*AB*) is the cosine of the angle *A*. In Fig. 7-15, this angle is 45°. If the hypotenuse (*AB*) is given a value of 1, the value of the adjacent side (*AC*) will be 0.7071, or the cosine of 45°.

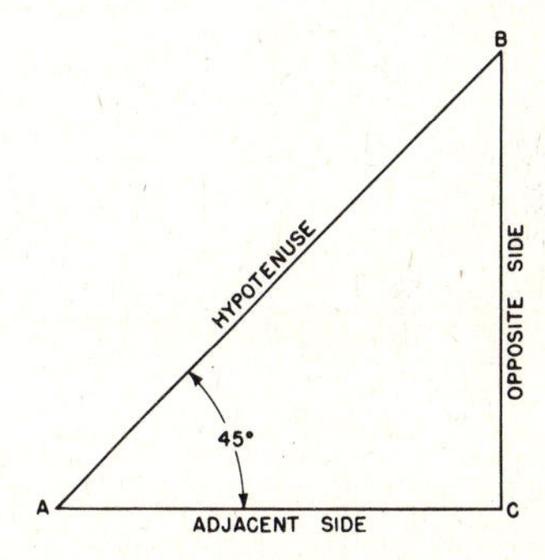

Fig. 7-15 Right triangle for determining the cosine of an angle.

REVIEW QUESTIONS

1. Give the basic definition of a *generator.*
2. How may generators be classified for output?
3. What is a commonly used term for an ac generator?
4. Explain the electrical principle by which electricity is produced in a generator.
5. How may the direction of current flow in an armature be determined?
6. Name the essential parts of a dc generator.
7. What determines the voltage value in a generator?
8. How may commutator ripple be reduced?
9. How is residual magnetism utilized in a generator?
10. Give two classifications for dc generators with respect to internal circuit connections.
11. Explain *armature reaction.*
12. How may armature reaction be reduced?
13. What are some of the advantages of ac generators?
14. Describe the *rotor* and *stator* for an ac generator.
15. What determines the frequency of the current produced by an alternator?
16. What is meant by *excitation* in an alternator?
17. How is excitation accomplished in an alternator?
18. How is the power factor for an ac generator determined?
19. If an alternator produces 1000 kVA, what is the actual power in wattage when the power factor is 0.866?

CHAPTER 8

DC Generators and Controls

Aircraft generators differ considerably from generators for other purposes, such as those built for automobiles or for stationary power plants. The main difference is that aircraft generators have a much higher power-weight ratio than the other types. For example, a certain 12-V 30-A automobile generator weighs at least half as much as a 30-V 300-A aircraft generator. The power output of the automobile generator is 360 W, but the aircraft generator will deliver 9000 W. The ac generators used on jet airliners will deliver from 30 to 60 kW or more, and yet their weight is only a little more than that of the lower-power dc generators.

Several factors contribute to the efficiency and light weight of the aircraft generator. The enamels, varnishes, and insulating materials are highly heat-resistant; hence, the generator can operate at a high temperature and carry a maximum load. High-output generators are also cooled by a forced air stream, and this further increases the load which the generator may carry. The air stream is produced by a fan built into the generator, by ram air from air ducts, or by bleed air from the compressors of jet engines. Passages through which the cooling air may flow are provided in the armature and between field coils. The magnetic materials used in the armature are of high permeability, and so they offer little opposition to the rapid reversals of magnetic flux. This contributes to high-speed operation and permits fewer turns of wire in the armature.

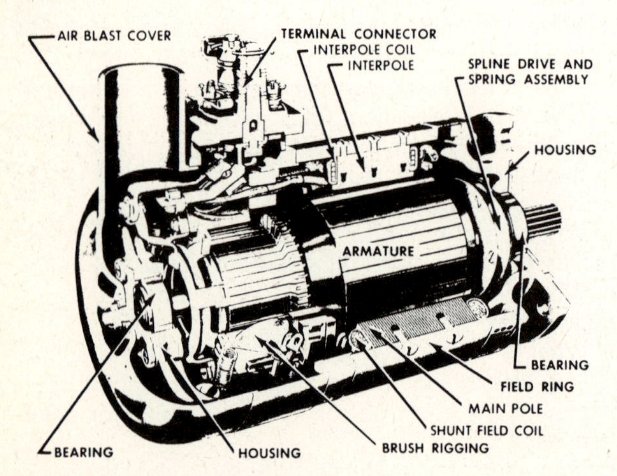

Fig. 8-1 High-output aircraft generator.

Figure 8-1 is a cutaway view of a Leece-Neville type P-2 generator typical of those used for dc electric systems on large aircraft during World War II and for several years afterward. This type of generator is still employed on some older aircraft. A study of the illustration reveals many of the features of a comparatively high-power dc aircraft generator.

Generators of simpler construction are manufactured for use in small aircraft, where the load is comparatively low. Figure 8-2 illustrates a typical dc generator for light aircraft. This generator has an output of 14 V at 30 to 40 A. Generators of this type do not usually employ interpoles or other special features typical of the high-amperage types.

Ac generators (alternators) are employed in aircraft built in recent years; however, many older aircraft are still in operation and these usually are equipped with dc generators. It is therefore important that the aviation maintenance technician be familiar with dc generators and controls.

GENERATOR COMPONENTS

ARMATURE ASSEMBLY

The armature assembly (Fig. 8-3) consists of a laminated soft-iron core mounted on a steel shaft, the com-

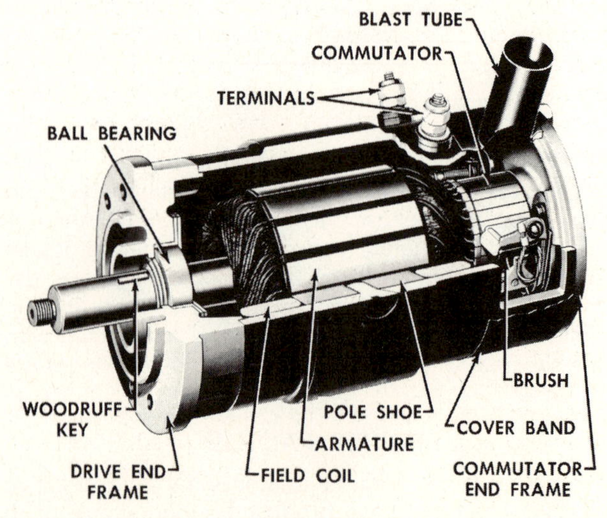

Fig. 8-2 Dc generator for light aircraft.

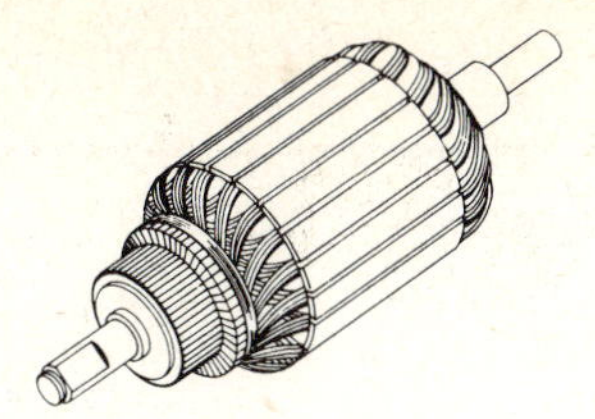

Fig. 8-3 Armature assembly.

mutator at one end of the assembly, and armature coils wound through the slots of the armature core. The core is made of many soft-iron laminations coated with an insulating varnish and then stacked together. The purpose of the laminations is to eliminate or reduce the eddy currents which would be induced in a solid core. The effect of these currents has been explained previously. The laminations for the armature core are stacked together in such a manner that the slots are lined up so the armature coils may be placed in them. Before the coil windings are installed, insulating paper or fabric is placed in the slots to protect the windings from wear and abrasion.

Insulated copper wire of a size large enough to carry the maximum armature currents is wound in coils through the slots of the armature. The number of armature segments between the sides of each coil depends upon the number of poles in the generator. For example, if the armature of a four-pole generator has 24 segments, each side of an armature coil should be separated by six segments. Aircraft generators employ what is called a **drum** armature winding. Drum windings are classified as lap or wave windings, depending upon the relative position of the coil terminals connecting to the commutator. It is not practical for the aircraft technician to rewind armatures, and for that reason armature-winding theory will not be covered in this text.

After an armature is wound, the coils are held in place by means of nonmetallic wedges placed in the slots. On some models, bands of steel are placed around the armature to prevent the windings from being thrown out by centrifugal force when the armature is driven at high speeds.

The commutator consists of a number of copper segments insulated from the armature structure, and from each other, with mica. The segments are constructed to be held in place by wedges located between the shaft and the segments. A cross section of a typical commutator is shown in Fig. 8-4. Each commutator segment has a riser to which are soldered the leads from the armature coils. The surface of the commutator is cut and ground to a very smooth cylindrical surface. The mica insulation between the segments is undercut approximately 0.020 in [0.051 cm] to make certain that it doe not interfere with the contact of the brushes with the commutator.

The armature core is attached to the shaft by various methods. Because of the severe twisting action that takes place during changes of engine speed or generator load, a shock-absorbing device is usually introduced between the armature body and the drive shaft. The **spline drive and spring assembly** shown in Fig. 8-1 forms a shock-absorbing coupling. Another device used to reduce the effect of twisting forces is a slip clutch. This clutch consists of surfaces held firmly together by springs whose

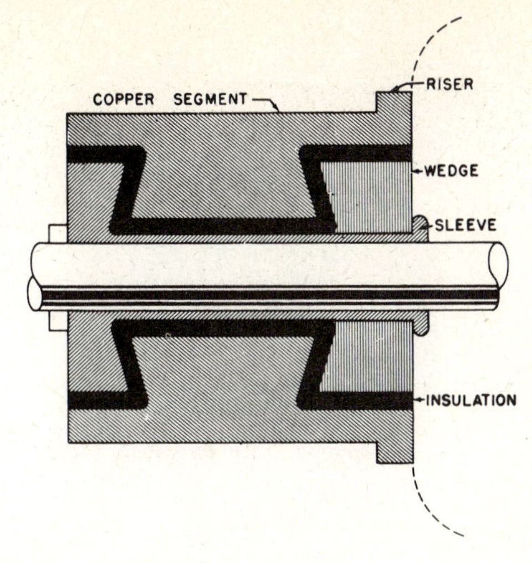

Fig. 8-4 Cross section of a commutator.

strength, however, is such that it will allow the two surfaces to slip when excessive twisting loads are applied. Some generators have a small intershaft running the full length of the main shaft. This is called a **quill** shaft and is connected solidly inside the main shaft at the end opposite the drive end. At the drive end it is free to turn inside the hollow main shaft. This allows it to take up the shocks and vibrations by twisting to a limited degree when sudden loads are encountered.

FIELD-FRAME ASSEMBLY

The heavy iron or steel housing which supports the field poles is called the **field frame, field ring,** or **field housing.** It not only supports the field poles but also forms a part of the magnetic circuit of the field. The pole shoes are held in place by large countersunk screws which pass through the housing and into the shoes. The screws are usually **staked** into place so that the field poles may not easily shift position.

Small generators usually have two to four poles mounted in the field-frame assembly, and large generators may have as many as eight main poles and eight interpoles. The pole pieces are rectangular and in most instances are laminated. The main shunt field windings consist of many turns of comparatively small insulated copper wire. Series windings, such as those on the interpoles, consist of a few turns of insulated copper wire large enough to carry the entire load current without overheating. A typical field-frame assembly is shown in Fig. 8-5.

BRUSH RIGGING

The brush-rigging assembly (Fig. 8-6) is located at the commutator end of the generator. The brushes are small blocks of a carbon and graphite compound soft enough to give minimum commutator wear but sufficiently hard to provide long service. Special brushes have been designed for generators used in extremely high-altitude operation. These are needed because arcing increases at high altitudes and will cause the rapid deterioration of ordinary brushes.

The brushes slide freely in metal holders and are held firmly against the commutator by means of springs. The tension of these springs should be sufficient to provide a brush pressure of approximately 6 psi [41 kPa] of contact

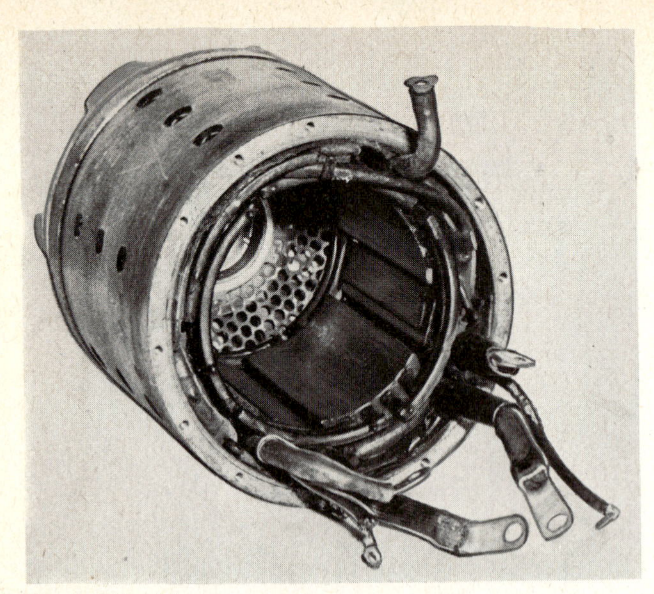

Fig. 8-5 Field-frame assembly.

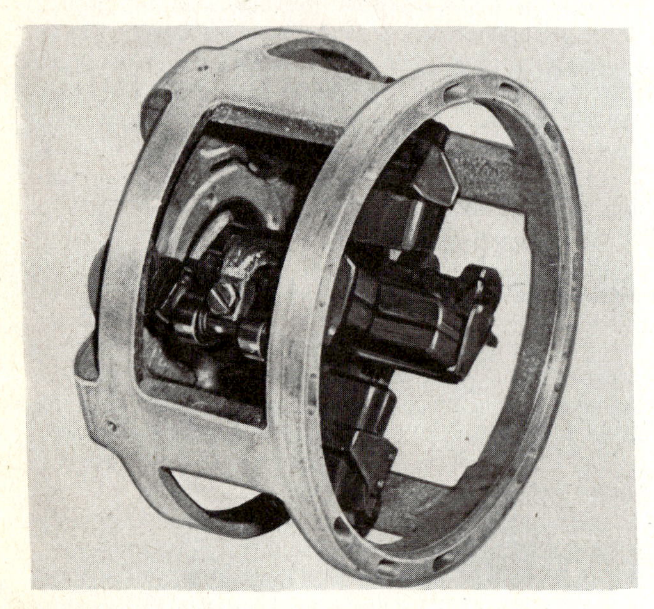

Fig. 8-6 Brush-rigging assembly.

surface. A flexible lead is connected from the brush to the brush frame to ensure a good electrical connection. Brushes of similar polarity are connected together electrically with a metal strip or wire.

END FRAMES

The generator end frames support the armature bearings and are mounted at each end of the field frame. The frame at the commutator end of the generator also supports the brush-rigging assembly. The frame at the drive end is flanged to provide a mounting structure. On some generators, the end frames are attached to the field-frame assembly by means of long bolts extending entirely through the field frame. On others, the end frames are attached by machine screws into the ends of the field frame.

Generator bearings are usually of the ball type, prelubricated and sealed by the manufacturer. Prelubricated bearings do not require any service except at overhaul or in case of damage. The bearings fit snugly into the recesses in the end frames and are held in place by retainers attached to the end frames with screws.

COOLING FEATURES

Since a generator operating at full capacity develops a large amount of heat, it is necessary to provide cooling. This is accomplished by means of passages leading through the generator housing between the field coils. In high-output generators there are also cooling-air passages through the armature. Cooling air is forced through the passages either by a fan mounted on the generator shaft, or by pressure from a ram-air duct leading into an air scoop mounted on the end of the generator, or by bleed air from the compressor of a jet engine. Openings are provided in the end frame opposite the fan or air fittings to allow the heated air to pass out of the generator housing.

NAMEPLATE DATA

Every generator is supplied with a nameplate securely attached to the field frame. The nameplate will usually give the generator's voltage, amperage capacity, make, model, and serial number. In some cases, the direction of rotation will also be given. Although dc electric systems are usually 12- or 24-V and the generators for these systems are marked accordingly, the actual voltage is somewhat higher; that is, a 24-V generator will usually be adjusted to produce about 28 V. These higher voltages are necessary to keep the batteries charged.

Direction of rotation should be shown either on the data plate or elsewhere. In any event, it is essential that the generator be driven in the correct direction; otherwise, little or no voltage will be produced.

GENERATOR CONTROL

PRINCIPLES OF VOLTAGE REGULATION

In the section of this text describing generator theory, it was explained that voltage produced by electromagnetic induction depends upon the number of lines of force being cut per second by a conductor. In a generator, the voltage produced depends upon three factors: (1) the speed at which the armature rotates; (2) the number of conductors in series in the armature; and (3) the strength of the magnetic field. In order to maintain a constant voltage from the generator under all conditions of speed and load, one of the foregoing conditions must be varied in accordance with operational requirements.

It is obvious that the speed of the generator cannot be varied according to requirements if the generator is being driven directly by the engine. Also, it is impossible to change the number of turns of wire in the armature during operation. Therefore, the only practical means of regulating the generator voltage is to control the **strength of the field.** This is easily accomplished, because the strength of the field is determined by the current flowing through the field coils and this current can be controlled by a variable resistor in the field circuit outside the generator.

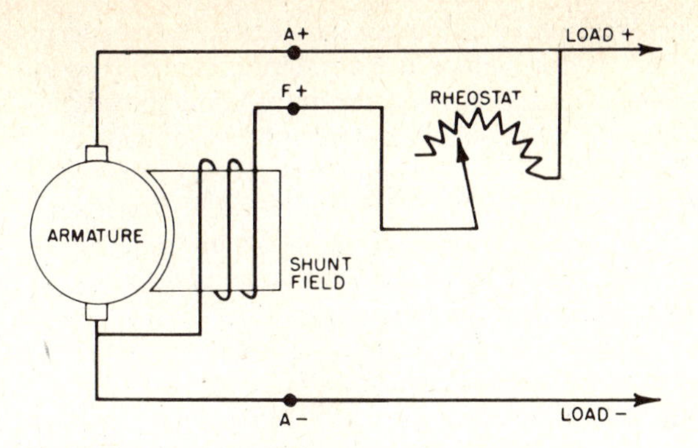

Fig. 8-7 Regulation of voltage in a dc generator.

The simplest type of voltage regulation is accomplished as shown in Fig. 8-7. In this arrangement, a rheostat (variable resistor) is placed in series with the shunt field circuit. If the voltage rises above the desired value, the operator may reduce the field current with the rheostat, thus weakening the field and lowering the generator voltage. An increase in voltage is obtained by reducing the field-circuit resistance with the rheostat. All methods of voltage regulation in aircraft electric systems employ the principle of a variable or intermittent field resistance. Modern voltage regulators have been developed to such a high degree of efficiency that the emf of a generator will vary only a small fraction of a volt throughout extreme ranges of load and speed.

Voltage regulators or controls for modern aircraft are usually of the solid-state type. That is, they employ transistors and diodes as controlling elements. Because there are still many older airplanes in use that employ vibrator-type and variable-resistance voltage regulators, we shall examine these in the next section.

VIBRATOR-TYPE VOLTAGE REGULATOR

A generator system using a vibrator-type voltage regulator is shown in Fig. 8-8. A resistance which is intermittently cut in and out of the field circuit by means of vibrating contact points is placed in series with the field circuit. The contact points are controlled by a voltage coil connected in parallel with the generator output. When the generator voltage rises to the desired value, the voltage coil produces a magnetic field strong enough to open the contact points. When the points are open, the field current must pass through the resistance. This causes a substantial reduction in field current, with the result that the magnetic field in the generator is weakened. The generator voltage then drops immediately, causing the voltage-coil electromagnet to lose strength so that a spring may close the contact points. This allows the generator voltage to rise, and the cycle is then repeated. The contact points open and close many times a second, but the actual time that they are open depends upon the load being carried by the generator. As the generator load is increased, the time that the contact points remain closed increases and the time that they are open decreases. Adjustment of the generator voltage is made by increasing or decreasing the tension of the spring which controls the contact points.

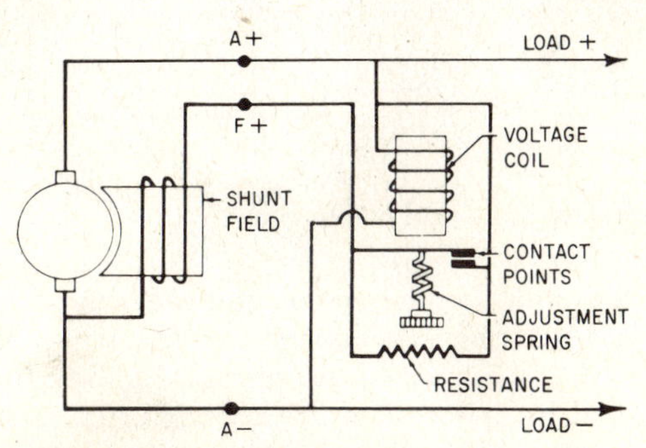

Fig. 8-8 Vibrator-type regulator circuit.

Because the contact points do not burn or pit appreciably, vibrator-type voltage regulators are satisfactory for generators which require a low field current. In a system in which the generator field requires a current as high as 8 A, the vibrating contact points would soon burn and probably fuse together. For this reason, a different type of regulator is required for heavy-duty generator systems.

If the regulating resistance becomes disconnected or burned out, the generator voltage will fluctuate and excessive arcing will occur at the contact points. When inspecting the vibrator-type voltage regulator, be sure to see that the connections to the resistance are secure and that the resistance is in good condition.

CARBON-PILE VOLTAGE REGULATOR

The carbon-pile voltage regulator derives its name from the fact that the regulating element (variable resistance) consists of a stack, or **pile,** of carbon disks (see Fig. 8-9).

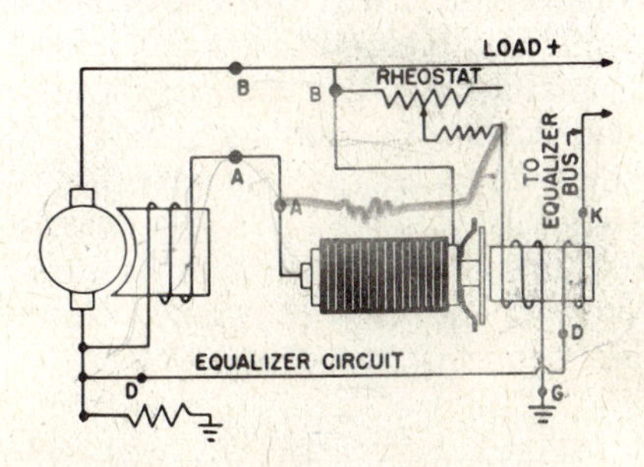

Fig. 8-9 Carbon-pile voltage regulator circuit.

Usually the carbon pile has alternate hard-carbon and soft-carbon (graphite) disks contained in a ceramic tube with a carbon or metal contact plug at each end. At one end of the pile, a number of radially arranged leaf springs exert pressure against the contact plug, thus holding the disks pressed firmly together. For as long as the disks are compressed, the resistance of the pile is very low. If the pressure on the carbon pile is reduced, the resistance increases. By placing an electromagnet in a position where it will release the spring pressure on the disks as the voltage rises above a predetermined value, a stable and efficient voltage regulator is obtained.

The carbon-pile voltage regulator is connected in a generator system in the same manner as any other regulator, that is, with a resistance in the field circuit and an electromagnet to control the resistance. The carbon pile is in series with the generator field, and the voltage coil is shunted across the generator output. A small manually operated rheostat is connected in series with the voltage coil to provide for a limited amount of adjustment, which is necessary when two or more generators are connected in parallel to the same electric system.

Adjustment of a carbon-pile regulator, other than that which can be obtained with the rheostat in the voltage-coil circuit, is very critical and requires considerable care and precision. The core of the electromagnet is threaded to provide for adjustment of the gap between the core and the armature. The width of this gap must be correct in order to balance the strength of the magnet against the leaf springs which apply pressure to the carbon pile. A screw adjustment at the opposite end of the pile makes it possible to vary the pressure which the leaf springs exert against the pile. Because the other adjustments should be made on a test bench with proper instruments, it is not ordinarily recommended that the technician attempt to adjust a carbon-pile regulator except by using the rheostat.

EQUALIZING CIRCUITS

When two or more generators are connected in parallel to a power system, generators should share the load equally. If the voltage of one generator is slightly higher than that of the other generators in parallel, that generator will assume the greater part of the load. For this reason, an equalizing circuit must be provided which will cause the load to be distributed evenly among the generators. An equalizing circuit includes an equalizing coil wound with the voltage coil in each of the voltage regulators, an equalizing bus to which all equalizing circuits are connected, and a low-resistance shunt in the ground lead of each generator (see Fig. 8-10). The equalizing coil will either strengthen or weaken the effect of the voltage coil, depending upon the direction of current flow through the equalizing circuit. The low-resistance shunt in the ground lead of each generator causes a difference of potential between the negative terminals of the generators proportional to the difference in load current. The shunt is of such a value that there will be a potential difference of 0.5 V across it at maximum generator load.

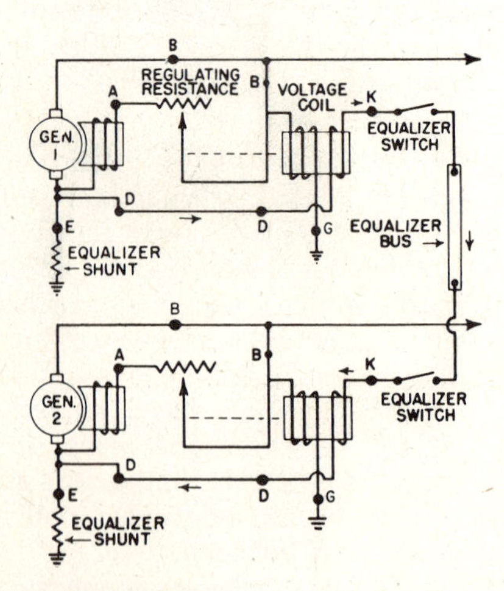

Fig. 8-10 Equalizing circuits.

Assume that generator 1 in Fig. 8-10 is delivering 200 A (full load) and that generator 2 is delivering 100 A (half load). Under these conditions there will be a potential difference of $\frac{1}{2}$ V across the shunt of generator 1 and $\frac{1}{4}$ V across the shunt of generator 2. This will make a net potential difference of $\frac{1}{4}$ V between the negative terminals of the generators. Since the equalizing circuit is connected between these points, a current will flow through the circuit. The current flowing through the equalizing coil of voltage regulator 1 will be in a direction to strengthen the effect of the voltage coil. This will cause more resistance to be placed in the field circuit of generator 1, thus weakening the field strength and causing the voltage to be reduced. The drop in voltage will result in the generator taking less load. The current flowing through the equalizing coil of voltage regulator 2 will be in a direction that will oppose the effect of the voltage coil, thus causing a decrease in the resistance in the field circuit of generator 2. The generator voltage will increase because of increased current in the field windings, and the generator will take more of the load. To summarize, the effect of an equalizing circuit is to lower the voltage of a generator which is taking too much of the load and to increase the voltage of the generator which is not taking its share of the load.

Equalizing circuits can correct for only small differences in generator voltage; hence, the generators should be adjusted to be as nearly equal in voltage as possible. If the generator voltages are adjusted so that there is a difference of less than $\frac{1}{2}$ V between any of them, the equalizing circuit will maintain a satisfactory load balance. A periodic inspection of the ammeters should be made during flight to see that the generator loads are remaining properly balanced.

REVERSE-CURRENT CUTOUT RELAY

In every generator system in which the generator is used to charge batteries as well as to supply operating power, an automatic means must be provided for disconnecting the generator from the battery when the generator voltage is lower than the battery voltage. If this is not done, the battery will discharge through the generator and may burn out the armature. Numerous devices have been manufactured for the purpose of automatically disconnecting the generator, the simplest being the **reverse-current cutout relay.** Fig. 8-11 is a schematic diagram illustrating the operation of such a relay.

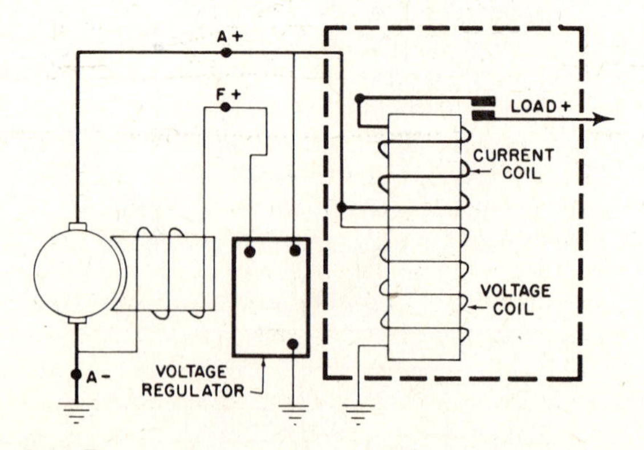

Fig. 8-11 Reverse-current cutout circuit.

A **voltage coil** and a **current coil** are wound on the same soft-iron core. The voltage coil has many turns of fine wire and is connected in parallel with the generator output; that is, one end of the voltage winding is connected to the positive side of the generator output, and the other end of the winding is connected to ground, which is the negative side of the generator output. This is clearly shown in the diagram. The current coil consists of a few turns of large wire connected in series with the generator output; hence, it must carry the entire load current of the generator. A pair of heavy contact points are placed where they will be controlled by the magnetic field of the soft-iron core. When the generator is not operating, these contact points are held in an open position by a spring.

When the generator voltage reaches a value slightly above that of the battery in the system, the voltage coil in the relay magnetizes the soft-iron core sufficiently to overcome the spring tension holding the contact points open. The magnetic field closes the contact points and thus connects the generator to the electric system of the airplane. As long as the generator voltage remains higher than the battery voltage, the current flow through the current coil will be in a direction which aids the voltage coil in keeping the points closed. This means that the field of the current coil will be in the same direction as the field of the voltage coil and that the two will strengthen each other.

When an airplane engine is slowed down or stopped, the generator voltage will decrease and fall below that of the battery. In this case the battery voltage will cause current to start flowing toward the generator through the relay current coil. When this happens, the current flow will be in a direction which creates a field opposing the field of the voltage winding. This results in a weakening of the total field of the relay, and the contact points are opened by the spring, thus disconnecting the generator from the battery. The contact points do not open in normal operation until the reverse current has reached a value of 5 to 10 A.

The tension of the spring controlling the contact point should be adjusted so that the points will close at approximately 13.5 V in a 12-V system and at 26.6 to 27 V in a 24-V system. To make this adjustment, a precision voltmeter should be used.

CURRENT LIMITER

In some generator systems a device is installed which will reduce the generator voltage whenever the maximum safe load is exceeded. This device is called a **current limiter** and is designed to protect the generator from loads which will cause it to overheat and eventually burn the insulation and windings. Current limiters are almost invariably used with automobile generator systems.

The current limiter operates on a principle similar to that of the vibrator-type voltage regulator. Instead of having a voltage coil to regulate the resistance in the field circuit of the generator, the current limiter has a current coil connected in series with the generator load circuit (see Fig. 8-12).

When the load current becomes excessive, the current coil magnetizes the iron core sufficiently to open the contact points and bring a resistance into the generator field circuit. This causes the generator voltage to decrease with a corresponding decrease in generator current. Since the magnetism produced by the current-limiter coil is proportional to the current flowing through it, the decrease in generator load current also weakens the magnetic field of the current coil and thus permits the contact points to

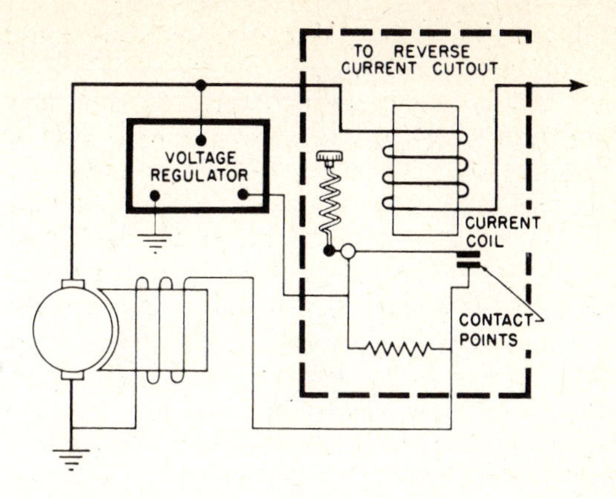

Fig. 8-12 Current-limiter circuit.

close. This removes the resistance from the generator field circuit and allows the voltage to rise again. If an excessive load remains connected to the generator, the contacts of the current limiter will continue to vibrate, thus holding the current output at or below the maximum safe limit. The contact points are usually set to open when the current flow is 10 percent above the rated capacity of the generator.

The current limiter described above should not be confused with the fuse-type current limiter. The fuse-type limiter is merely a high-capacity fuse which permits a short period of overload in a circuit before the fuse link melts and breaks the circuit.

To prevent damage to the power system, many electric circuits for aircraft employ an overload relay. Such a relay will automatically disconnect a generator from the system if its voltage or load becomes excessive.

TWO-UNIT CONTROL PANEL

Generator systems for light aircraft often have the generator control units mounted on a single panel. When the voltage regulator and reverse-current cutout relay are mounted on a single panel, the system is called a two-unit control panel or box (see Fig. 8-13). In the voltage regulator on this panel, an extra coil, called an accelerator winding, is wound with the voltage coil. This coil is connected in series with the field-regulating resistance and is wound in a direction opposite to that of the voltage winding. Its purpose is to reduce the magnetism of the core when the contact points open; this reduction causes the points to close more quickly than they would without

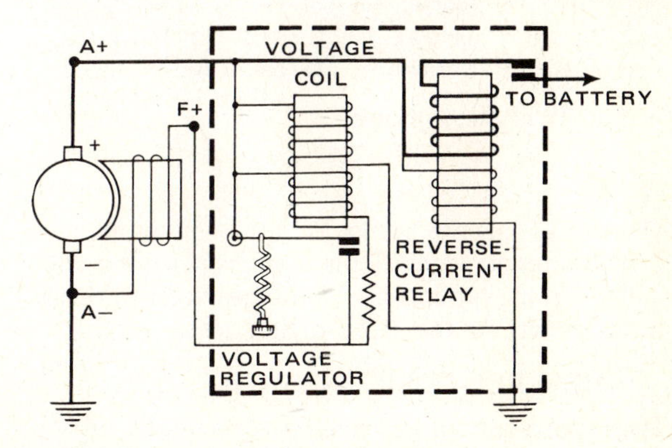

Fig. 8-13 Two-unit control panel circuit.

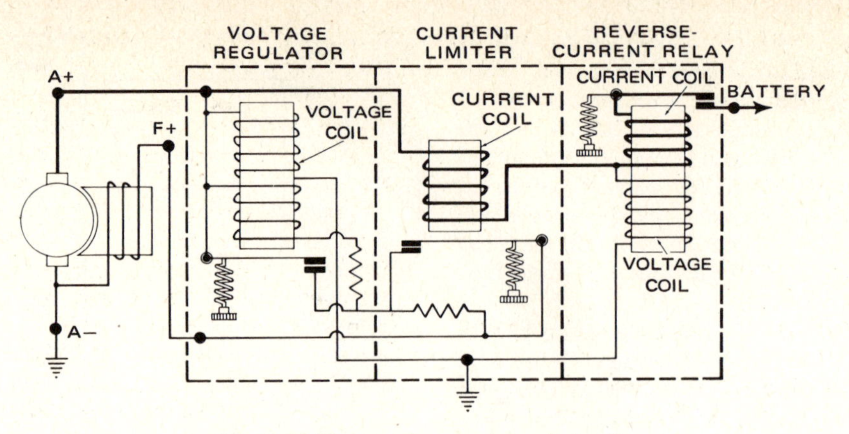

Fig. 8-14 Three-unit control panel circuit.

the neutralizing effect of a reverse coil. The result of this arrangement is that the contact points vibrate more rapidly and produce a steadier voltage from the generator.

THREE UNIT CONTROL PANEL

A three-unit control panel consists of a voltage regulator, a current limiter, and a reverse-current cutout relay mounted as a single unit (see Fig. 8-14). This combination will provide for both voltage regulation and protection from excessive loads. A photograph of such a unit with the cover off is shown in Fig. 8-15.

The three-unit control panel has proved very successful for the control of 6- and 12-V generator systems. Because of its dependability and low cost, it was used almost exclusively in light-aircraft generator systems before the development of transistor regulators.

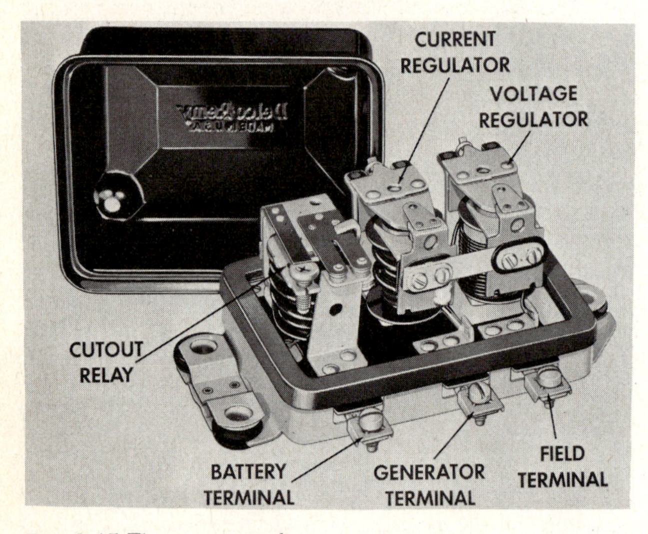

Fig. 8-15 Three-unit regulator.

The wiring diagram in Fig. 8-14 is only one of several possible arrangements. Some systems place the voltage regulator in the ground side of the generator field circuit, but the results are the same in either case.

REVERSE-CURRENT SWITCH RELAY

The reverse-current switch relay, often used in 24-V systems, serves two principal functions. It acts as a reverse-current cutout and also serves as a remote-control switch by means of which the generator may be connected to or disconnected from the electric system. The switch feature is essential in large aircraft because it eliminates the necessity for heavy cables and switches, thus providing for less weight and reducing the attendant hazards of high-current-carrying cables leading into the cockpit.

Figure 8-16 is a schematic diagram of a typical reverse-current switch relay. This unit includes two separate relays, one called the **main contactor** relay and the other called the **pilot** relay. The pilot relay controls the current through the coil of the main contactor relay; that is, when the pilot contact points are closed, the current may flow through the main contactor coil if the generator switch is on. The pilot-relay winding is arranged in two sections. One of these is a voltage coil consisting of many turns of fine wire, and the other is a current coil of only one or two turns of heavy copper wire or strip.

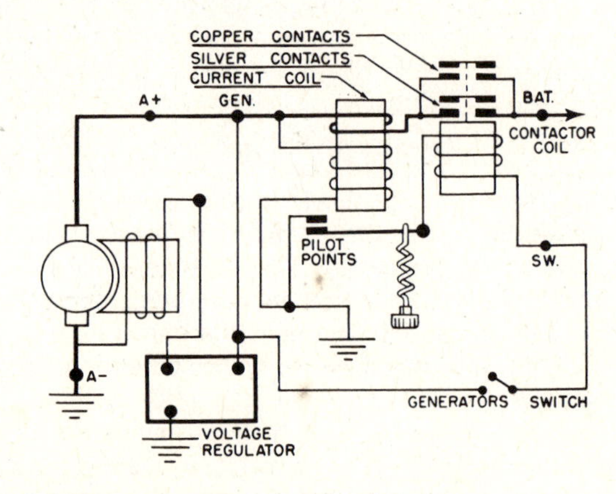

Fig. 8-16 Reverse-current switch relay circuit.

When the generator voltage rises to approximately 26.5 V, the current flow through the voltage coil will provide sufficient magnetic strength to close the pilot contact points. If the generator switch is closed, current will flow through the contactor coil, thus closing the main contact points and connecting the generator to the electric system.

The main contact points are specially constructed to resist the effects of arcing. Some relays have as many as four contacts, two of copper, which make the first contact and break contact last, and two of silver, which carry the heavy current during operation. The copper contact points take all the arcing when the circuit is closed or opened. It is important during operation to turn off all load in the aircraft before turning the generator switch off. This will eliminate the severe arcing which would occur if the circuit were broken while a heavy current was flowing through the relay.

A study of the diagram in Fig. 8-16 will show that the current for the main contactor relay must pass through the manual generator switch. With this arrangement a relatively small switch may be used to open and close the main contactor and control the main circuit. A small wire, AWG No. 18 or 20, is adequate for the switch circuit in the cockpit.

The reverse-current feature of the reverse-current switch relay is provided by the current coil on the pilot relay. When the generator voltage falls below the voltage of the battery or below the voltage of another generator in the system, a current will flow in reverse through the relay. This current flowing in the current coil neutralizes the effect of the voltage winding and causes the pilot contact points to open. This opens the circuit through the main contactor coil, the main contact points open, and the generator is thus disconnected from the system. The reverse current required to cause the relay to open is usually between 10 and 20 A.

DIFFERENTIAL REVERSE-CURRENT RELAY

The reverse-current relay in most prevalent use by airlines and military services for dc systems is the differential-voltage type. This relay is constructed in a manner similar to the reverse-current relay previously described; in operation, however, the differential relay is controlled by the differential voltage between the battery bus and the generator rather than by the voltage level of the generator.

Figure 8-17 is a schematic circuit showing the operating principles of the differential relay. The differential voltage coil is connected between the generator terminal of the relay and the battery terminal. Current can flow in this coil only when there is a voltage differential between the battery bus and the generator. The differential-voltage-coil circuit is broken by a voltage relay when the generator is not operating or when the voltage is low. Observe in the circuit diagram that the voltage coil receives its power

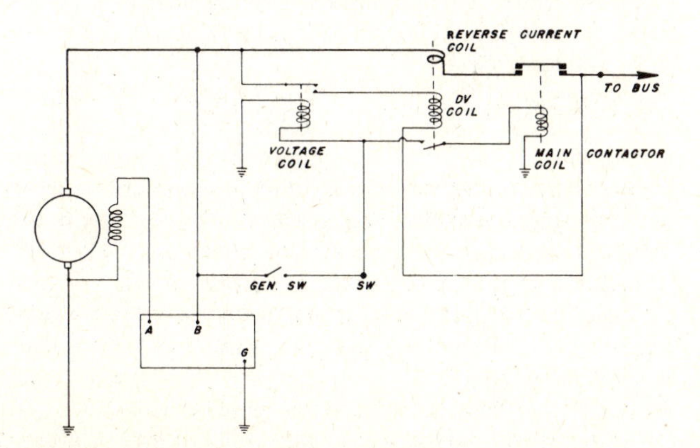

Fig. 8-17 Differential reverse-current relay circuit.

through the generator switch. If the switch is off, the voltage coil will not close the circuit through the differential coil. The purpose of the voltage relay is to prevent a feedback of current through the differential relay when the generator is not operating. In some types of relays a ballast resistance is used for this purpose.

During operation of the differential reverse-current relay, the main contacts of the relay are held in a closed position by the action of the main contactor coil, which receives current directly from the generator. When the generator voltage reaches a level of 0.35 to 0.65 V higher than that of the battery or the main power bus, the differential voltage coil closes the circuit through the main contactor coil. The main points close and the generator is connected to the power system. When the generator voltage drops and there is a substantial reverse current, the differential-voltage-relay contact points open, thus deenergizing the main contactor relay and causing the main contact points to open. This disconnects the generator from the electric system. The relay is usually adjusted to open when there is a reverse current of 10 to 20 A.

COMBINATION GENERATOR CONTROL PANELS

To provide a rapid method of field service for generator control systems, manufacturers have developed plug-in control panels. In case of the failure of any of the control units, the panel may be removed and replaced in a few seconds. The faulty control panel may then be repaired and adjusted in the electrical shop and returned to service later. These panels afford excellent automatic operation and protection for the generator system.

A combination control panel combines a standard carbon-pile voltage regulator with a differential reverse-current system. The main contactor of the reverse-current relay is not included in the panel, since it must be located at a point as near to the generator as design will permit; all the controlling units, however, are included in the panel.

Additional protection for the electric system is provided by means of an overvoltage relay in connection with the control panel. The overvoltage relay may be included in the panel or connected separately. With either method the operation is the same. The overvoltage relay coil is connected so that it will sense the voltage in the system as illustrated in Fig. 8-18. When the voltage reaches a level above the setting determined to be safe, the contact will close and the trip coil of the generator field relay will be energized. As shown in the diagram, the trip coil, when energized, will open the field circuit, the generator-switch circuit, the equalizing circuit, and the trip-coil circuit. It will also close the circuit to an indicator and to the reset coil. Thus the generator will be completely disconnected from the system, and since the field circuit will also be broken, it will stop producing voltage which might cause damage. The equalizer circuit is disconnected by the relay so that a faulty generator will not act through the system to affect the operation of the other generators.

After a field-control relay is tripped, it must be reset before the generator can be returned to normal operation. If the relay is provided with a reset coil, as in Fig. 8-18, it may be reset by placing the generator switch in the RESET position. If the relay immediately trips again, no further

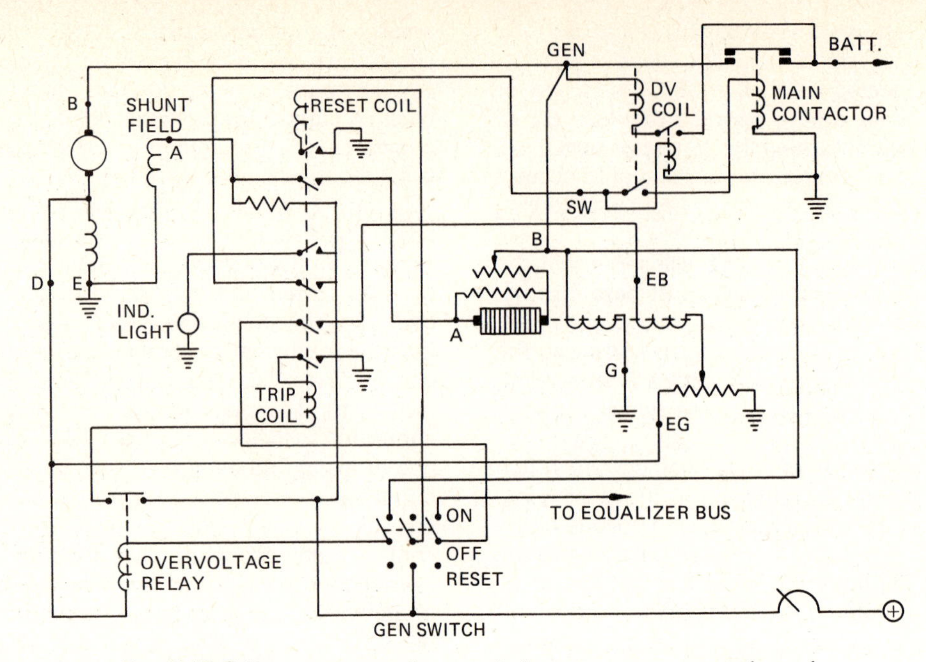

Fig. 8-18 Schematic circuit of a typical plug-in generator control panel.

attempt should be made to reset until the overvoltage condition has been corrected.

An important feature of the generator control system shown in Fig. 8-18 is the resistor connected between the generator field and the power supply. This resistor allows a very small current to flow in the generator-field windings even though the generator may not be operating. This current maintains correct polarity in the generator field under all conditions; hence, it is not necessary to flash the field in order to reestablish correct polarity. Sometimes a generator will lose its residual magnetism or become incorrectly polarized because of heat, shock, or a momentary current in the wrong direction. In this case a current must be passed through the field in the direction that will reestablish a field of correct polarity. This is called **flashing the field.**

REVERSE-CURRENT AIR CIRCUIT BREAKER

Further protection for generator systems has been provided on some large aircraft power systems by a reverse-current air circuit breaker. This unit is a double-pole relay which may be actuated by either one of two coils. One of the coils is in series with the generator power lead between the reverse-current relay and the main bus. Whenever a sustained surge of heavy reverse current takes place, the circuit breaker opens and disconnects the generator from the system. At the same time, another pair of contacts breaks the generator field circuit, thus causing the generator voltage to drop to a negligible level. The reverse-current air circuit breaker is also actuated whenever the overvoltage protector is tripped. Figure 8-19 illustrates a reverse-current air circuit breaker. Note that it is necessary to reset the circuit breaker manually when it has been tripped.

INSTALLATION OF CONTROL UNITS

The installation of the various units of a generator system is not difficult. It is of prime importance, however, that all such units be located where they are protected from fuel, oil, water, alcohol, hydraulic fluid, heat, and any other condition or substance which might prove detrimental. Voltage regulators and control panels should be attached to shockproof mounts to eliminate the damage which can result from vibration and landing shocks. On small airplanes the generator control panel is sometimes mounted on the forward side of the engine fire wall or in the forward part of the cockpit. In either case the panel must be protected from dust and other substances with a suitable cover.

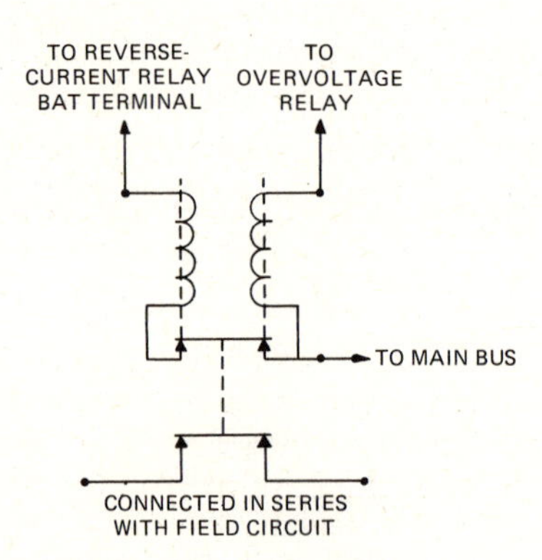

Fig. 8-19 Reverse-current air circuit breaker.

Voltage regulators for 24-V high-output generators must be mounted at a location where they will have adequate ventilation. These regulators often become very warm when the generator is under a heavy load, and if the heat becomes excessive, the regulator may fail. Carbon-pile regulators are supplied with cooling fins which radiate the heat generated in the carbon pile. These regulators must not be completely covered, because such covering would prevent the circulation of air necessary for cooling. Carbon-pile voltage regulators will give better service when they are mounted in a horizontal position.

When control panels or separate units of a control system are installed, it is of primary importance that all electric connections be clean and tight. Before either a plug-in control panel or a voltage regulator is installed, the contacts should be inspected to see that they are free of corrosion, oil, or dirt. The spring contacts of the base must bear firmly against the contact prongs when the unit is installed.

On large aircraft, reverse-current relays may be installed near the generators, or they may be in a power panel inside the airplane so that they are accessible during flight. In generator systems in which the reverse-current relay is the only means of disconnecting the generator from the battery bus, a reverse-current-relay failure becomes very serious. If the generator remains connected to the battery when the engine is slowed or stopped, the generator may be burned out. When the system is protected with a reverse-current circuit breaker or some other type of circuit protector, the danger of a generator's being burned out by a reverse current becomes remote.

INSPECTION AND SERVICE OF CONTROL EQUIPMENT

The time intervals for the inspection of various electrical units in an airplane are usually established by the manufacturer or by the operator, as in the case of scheduled-airline equipment. The inspection periods for equipment installed in light, privately owned aircraft may be given in an operator's handbook. Otherwise, the regulations of the Federal Aviation Administration give the rules for 100-h and other periodic inspections.

The following are typical inspections required for generator control equipment and associated circuits:

Inspect the area where the generator control panel is located for grease, oil, dust, or any other detrimental substance. Correct any unsatisfactory condition noted.

Inspect the wiring and terminal connections. See that the insulation of the wiring is in good condition and that all terminal connections are clean and tight.

Remove covers from units which are not sealed, and inspect the interior for dust, grease, and evidences of overheating. Inspect visible contact points for evidences of pitting.

Remove plug-in units from the mounting base and examine the electric contacts. These should be clean and should bear firmly against the contact point on the unit.

Inspect shock mountings. They should be firm and free of cracks, but flexible.

If a voltage regulator is the carbon-pile type, inspect the regulator housing and cooling fins for cracks or other damage.

Remove defective voltage regulators and reverse-current relays and send them to a properly equipped shop for repair or overhaul.

GENERATOR INSPECTION, SERVICE, AND REPAIR

BALANCING GENERATOR LOAD

When it is desired to balance the load among the generators in any system, the technician should always follow the procedure set forth by the manufacturer of the aircraft. This procedure will be found in the manufacturer's service or maintenance manual.

The balancing procedure is usually begun by checking all generators with a precision voltmeter. This is done after the generators and engines are warmed up to normal operating temperature, that is, while the aircraft is in flight. Under these conditions all generators are adjusted to exactly the same output voltage (28 to 29 V for a 24-V system). A substantial load is then turned on and the ammeters are examined. All generator loads should be within ± 10 percent of one another. If the generator loads are not within these limits, the generator with the greatest error should be adjusted first. A small movement of the paralleling knob on the voltage regulator should produce an instant change in the load current for the generator being adjusted. If the load on one generator is reduced, the other generators will pick up load. All ammeters should be watched while the adjustments are being made.

FLASHING THE FIELD

If a generator fails to show any voltage whatsoever when it is operating at the proper rpm, this condition is often due either to the loss or to the reversal of the polarity of the residual magnetism in the field. This can be corrected quickly by flashing the field.

The simplest method of flashing the field is to remove or disconnect the voltage regulator and connect a source of positive voltage momentarily to the field terminal at the regulator base. The voltage must be applied not to the regulator but only to the lead from the generator field terminal.

When it is necessary to flash the field of a generator on which the field terminal is negative, connecting a positive voltage to the field terminal would polarize the field in reverse or produce no results. To flash the field under these circumstances, it is necessary to isolate the field circuit of the generator. A simple method of doing this is to connect the negative terminal of a small battery to the generator field terminal and the positive terminal of the battery to the positive terminal of the generator. A current will then pass through the field coils in the correct direction and polarize the field for proper operation.

GENERATOR TROUBLESHOOTING

The first step in troubleshooting a generator circuit on an airplane is to determine what type of system is in use. If the system is on a small airplane, it is likely that the control unit is a three-element type, that is, that it contains a voltage regulator, a reverse-current cutout, and a current limiter. If it is a 24-V system, it is likely that the voltage regulator is a carbon-pile type and that the reverse-current relay is a differential type. The type of control system must be known before an intelligent approach can be made to the problem of locating trouble.

When a three-unit control is under inspection, it must be determined whether the voltage-regulator circuit is in the negative or the positive side of the field circuit. If the field terminal (*F*) of the generator circuit in Fig. 8-20*a* is connected to ground, the generator voltage will drop to only 2 or 3 V because the field strength will be at the residual-magnetism level. This is because no field current will be flowing. In the circuit of Fig. 8-20*b* where the voltage control is in the negative side of the field circuit, grounding the terminal *F* of the regulator will cause an immedi-

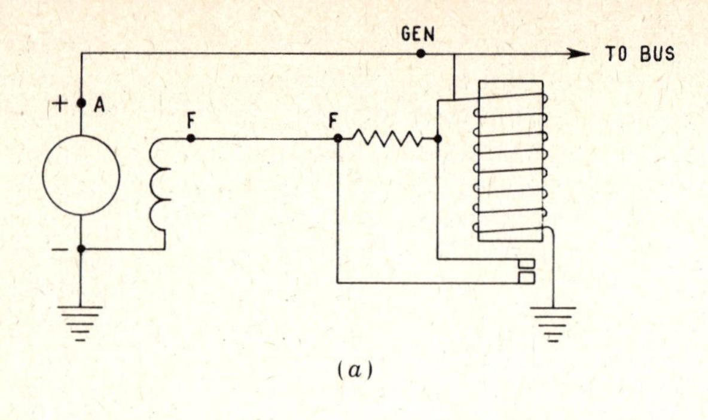

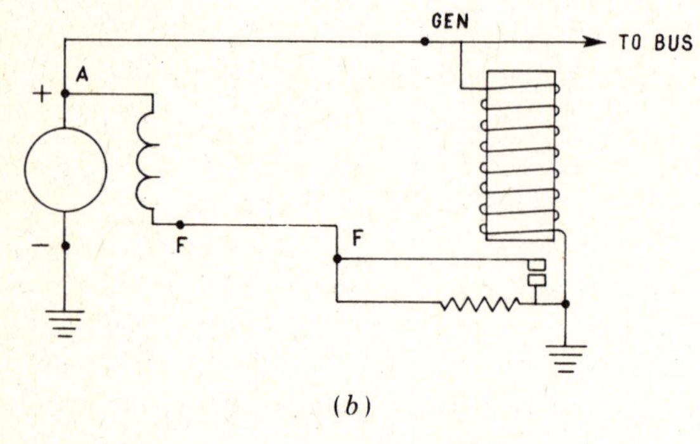

Fig. 8-20 Different arrangements for field control.

ate rise in generator voltage when the generator is running at normal speed. When a generator system of this type is being checked, the generator field terminal must not be grounded while the generator is connected to the airplane bus. Serious damage to aircraft electric equipment may occur if generator voltage is permitted to exceed its normal value when connected to the airplane system.

The two most likely indications of generator-system trouble are (1) no voltage and (2) residual voltage. Residual voltage is the result of residual magnetism in the field poles. When there is absolutely no voltage, the trouble may be a burned-out generator, bad brushes, an open circuit, or loss of residual magnetism. When residual voltage shows on the voltmeter, it indicates that the generator is operating but that there is no field current. This means that the field circuit is open because of a broken or loose wire or that the voltage regulator is defective. The first unit to examine in this case is the voltage regulator.

In an installation in which a plug-in regulator is employed, it is a simple matter to remove the regulator and install one known to be in good working condition. If the generator still does not produce voltage, it will then be necessary to check the wiring of the field circuit with a continuity tester or ohmmeter.

Since many light airplane systems use control units containing a voltage regulator, a current limiter, and a reverse-current cutout, it is often wise to make a preliminary test before removing the entire unit for replacement. To test such a system, a flow of current must be provided through the field circuit. This can be done by using a 10-Ω resistor with a capacity of at least 20 W to bypass the regulator. In a system in which the regulator is in the positive side of the field (see Fig. 8-20*a*), the resistor should be connected for a few seconds between the *GEN* terminal and the *F* terminal of the regulator, with the generator running at an adequate speed. As soon as the resistor is connected, a voltmeter connected across the generator output terminals should immediately show a rise in voltage.

To test a regulator in the ground side of the field circuit (see Fig. 8-20*b*), the resistor should be connected from the *F* terminal of the regulator to ground. In any case, if the generator voltage does not immediately increase from the residual level, the field circuit should be checked for continuity. If the generator voltage does increase, it is then apparent that the voltage regulator is at fault, and replacement of the voltage regulator should correct the trouble. It is important to make sure that the regulator being installed in any system is designed for operation with a generator of the type used in that system. This may be determined by referring to the manufacturer's service manual.

INSPECTION

Generators in service should be given a periodic inspection of external connections, wiring, brushes, commutator, mounting, and performance. These inspections should be carried out according to the manufacturer's instructions; however, certain general service and inspection operations may be carried out every 25 or 30 h even when the manufacturer's instructions are not available. The following inspections are considered essential:

1. Inspect the generator terminal connections to see that they are clean and tight.
2. Inspect the flange mounting for cracks or looseness of mounting bolts. See that there are no oil leaks around the mounting.
3. Remove the cap or band which covers the brushes and commutator. Blow out any accumulation of carbon dust with dry compressed air. Inspect the brushes for amount of wear, and see that they slide freely in the holders. If a brush is binding in the holder, it should be removed and cleaned with a clean cloth moistened with unleaded gasoline or a good petroleum solvent. If brushes are worn in excess of the tolerances specified by the manufacturer, they must be replaced. Inspect the tension of the brush springs by lifting them with the fingers. A weak spring should be adjusted or replaced. If a spring scale of the proper range is available, it may be used to measure the brush-spring tension; the spring tension should measure within the manufacturer's specifications.
4. Inspect the commutator for cleanliness, wear, and pitting. A commutator in good condition should be smooth and of a light chocolate color. If there is a slight amount of roughness on the commutator, it may be removed with No. 000 sandpaper or a special abrasive stick manufactured for cleaning commutators and seating brushes. If an abrasive stick is used, the proper application is to hold the end of the stick against the commutator while the generator is running. This is done until the commutator is smooth, clean, and bright. After smoothing, all sand and dust particles should be blown out with compressed air. Dirt may be removed from the commutator with a cloth moistened with a petroleum solvent. Oil on the brushes and commutator indicates a faulty oil seal in the engine. If this condition exists, the generator should be removed, disassembled, and thoroughly cleaned. If the end frame

of the generator has an oil-drain vent, it should be inspected to see that it is open. Before the generator is reinstalled, the engine oil seal at the generator drive must be replaced.

5. Inspect the area inside the commutator end of the generator case for lead particles. If particles of lead are visible, it is likely that the armature has been overheated. This may have been caused by overloading the generator for a sustained period, by short-circuited coils in the armature, by short-circuited segments of the commutator, or by the sticking of the reverse-current-relay points. The generator should be removed and a new or rebuilt armature installed. Before the generator is reinstalled, the cause of armature failure should be determined and corrected.

DISASSEMBLY

The disassembly procedure for specific generators cannot be discussed in detail in this text, inasmuch as it varies among different makes and models of generators. However, if it becomes necessary to disassemble a generator, the technician should refer to the instructions furnished by the manufacturer of that particular model. If these instructions are not available, the technician may proceed as follows with reasonably good results:

1. Remove the strap or cap which covers the brush and commutator.

2. Remove the brushes and disconnect the flexible leads from the brush holders. Mark the brushes for their proper position in the brush rigging.

3. Disconnect the field and terminal leads, and mark the connections so that they can be reconnected correctly.

4. Remove the screws or bolts which attach the end frames to the field frame. Some generators have a nut and washer which holds the armature shaft in the bearing, and in this case, the nut and washer must be removed before the end frame is taken off.

5. When both end frames are free, remove them and take out the armature. *Note:* Further disassembly may be accomplished as required, but for inspection and cleaning purposes, the removal of brushes, end frames, and armature is usually sufficient.

6. After disassembly blow the brush dust from the field assembly, using dry compressed air.

7. Use a cloth moistened with unleaded gasoline or a good petroleum solvent to clean the field-frame assembly, commutator, and brush rigging. Do not immerse any of these parts in gasoline or other cleaning solvent.

REPAIR OF THE COMMUTATOR

If the commutator is slightly rough or pitted, it may be smoothed with No. 000 sandpaper. After smoothing, blow out all sand and dust particles with dry compressed air. If the commutator is very rough or badly pitted, the armature should be placed on a metal-turning lathe and a light cut taken across the surface of the commutator. This is most easily accomplished with equipment especially designed for the purpose. The cut on the commutator should be only deep enough to remove the irregularities on the surface. This cut will also correct any eccentricity which has developed as a result of uneven wear.

After the commutator has been turned on a lathe, it is necessary to undercut the mica insulation between the segments to a depth of approximately 0.02 in [0.051 cm]. To assure a clean cut to the required depth, use a cutting tool slightly wider than the thickness of the mica. If a tool of the proper type is not available, a hacksaw blade may be used. A few careful strokes in each slot will accomplish the desired results. After undercutting the mica, smooth any burrs or sharp edges on the commutator segments with No. 000 sandpaper.

TESTING

For testing armatures, a device called a **growler** is used. This device consists of many turns of wire wound around a laminated core with two heavy pole shoes extended upward to form a V into which an armature may be placed. Actually, the growler is nothing more than a large specially designed electromagnet. Figure 8-21 shows an armature being tested on a growler. The power supply for the growler is standard 110-V alternating current. The current causes a noticeable hum when an armature is placed between the pole shoes; hence, the name *growler.*

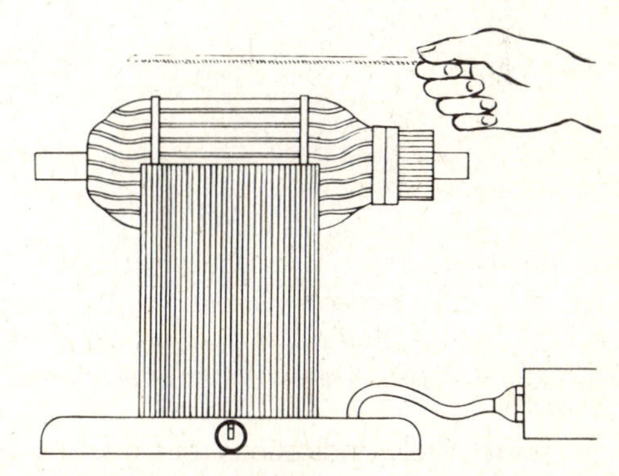

Fig. 8-21 Armature on a growler.

When placed on a growler, an armature forms the secondary of a transformer. The winding of the growler is the primary. The rapidly moving field produced by the winding of the growler induces an alternating current in the windings of the armature. By connecting a test lamp between segments of the commutator and the armature on the growler (see Fig. 8-22), it can be determined whether an open circuit exists in any of the coils. The lamp will light when connected across segments of a good coil. To test for a short circuit in the windings, a thin strip of steel is placed on the armature segments and the armature

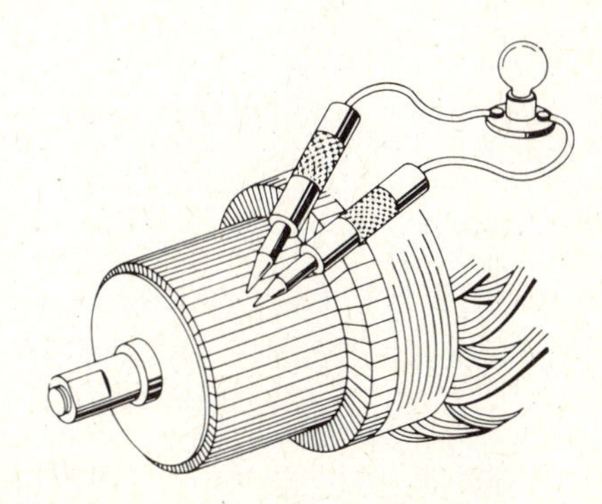

Fig. 8-22 Test for an open coil in an armature.

slowly rotated between the poles of the growler. If there are no shorts, a weak magnetic attraction will be noticed. One or more shorted coils will cause a strong vibration of the metal strip at certain points on the armature surface.

To test for a ground between the windings and the core of the armature, a 110-V power supply and a test lamp may be used. Some growlers are designed with this test light as a part of the equipment. Connect one of the test leads to the commutator and the other to the armature shaft or core as shown in Fig. 8-23. If there is a grounded coil, the lamp will light. An ohmmeter also may be used for this test. Connect one of the test prods of the ohmmeter to a segment of the commutator and the other prod to the shaft or core. With a good armature the ohmmeter should show infinite resistance, that is, an open circuit. A continuity tester consisting of a battery and a test lamp may also be used for the foregoing test by connecting it in the same manner as the ohmmeter.

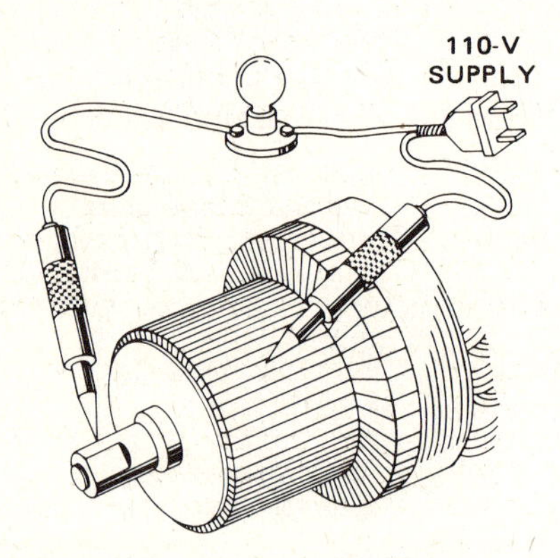

Fig. 8-23 Test for a ground in the armature winding.

To test a field coil for continuity, the prods of an ohmmeter are connected to the terminals of the coil. The shunt field coil should show low resistance, approximately 2 to 30 Ω, depending upon the type of generator in which the coil is used. A series field coil should show practically no resistance because it carries the entire load to the generator and the internal resistance of the generator must be as low as possible. A continuity tester also may be used to test a field coil for an open circuit. When an ohmmeter is used to test the field coils, it must be of a type designed to test for low resistance. A good combination meter usually has a low-resistance circuit in connection with the ohmmeter.

Field coils should also be tested for grounds between the coils and the field frame. To make this test, a 110-V power supply and a test lamp are connected in series with the field frame and one of the terminals of the field coil. If the lamp lights, there is a ground and the field will have to be discarded or rewound. An ohmmeter may be used for this test and should indicate infinite resistance between the windings and field frame.

SERVICE OF BEARINGS

As stated previously in this text, modern aircraft generators are usually equipped with prelubricated sealed bearings. During normal service inspections, it is not necessary to lubricate or otherwise service bearings of this type. If a bearing seizes or becomes rough, it should be replaced with a new one. Bearings may be checked by rotating the armature of the assembled generator by hand. The armature should turn freely and smoothly. If any roughness is noted, the bearings should be replaced.

The removal and installation of bearings requires great care. When a bearing is removed from its mounting, force should be applied only at the outer race. Every ball-bearing unit has an inner race and an outer race. The bearing races are curved grooves in which the ball bearings roll. The inside surface of the inner race fits on the armature shaft, and the outside surface of the outer race fits into the bearing mount. If force is applied to the inner race when the bearing is being removed from the mounting, the balls and the races are likely to be damaged. When a bearing is removed from the armature shaft, force should be applied only to the inner race. For this operation a bearing puller of the proper type may be used. In replacing bearings, it is important that they be squarely aligned with the mounting. Otherwise, the outer race may be broken, or the bearings will bind. A soft-metal or fiber tube which will fit the outer rim of the bearing should be used to press the bearing into place. When a bearing is placed on the armature shaft, a tube of suitable material and size should be used to exert force against its inner race.

If a sealed bearing has lost some of its lubrication, because of excessive heat over a long period of time, it may be restored sufficiently for continued use if it has not also been damaged. This is accomplished by cleaning the bearing thoroughly and then rotating it a few times while it is submerged in a bath of warm SAE No. 20 oil. A small amount of oil will seep into the bearing and provide protection against further wear.

SEATING OF NEW BRUSHES

If new brushes are to be installed in a generator, they must be seated so that the face of the brush will have maximum surface contact with the commutator. The brushes should be installed after the generator is assembled and then seated as follows: Place a strip of No. 000 sandpaper around the commutator with the sand surface against the brush face, and turn the armature in the normal direction of rotation. This causes the sandpaper to grind the face of the brush on a contour with the commutator. When the face of the brush is ground sufficiently to make maximum contact with the sandpaper of the commutator, remove the sandpaper and blow out all sand and brush particles with dry compressed air.

Another method of seating brushes recommended by some manufacturers is carried out as follows: Mount the generator on a test stand so it may be rotated at normal operating speeds. Install the brushes in their proper positions, and run the generator at approximately 1500 revolutions per minute (rpm) [157 radians per second (rad/s)]. Fold a strip of No. 000 sandpaper over the end of a rigid piece of insulating material with the sand surface outside. Hold the sand surface against the commutator while the generator is rotating. Fine sand particles will be carried across the face of the brushes and the brushes shaped into the contour of the commutator. After the brushes are seated, blow out all sand and dust with dry compressed air.

A third method of seating brushes, which has proved very satisfactory, is to use an abrasive stick specially de-

Troubleshooting Chart

Indication	Probable cause	Remedy
Generator produces voltage but ammeter reads zero when the load is turned on.	Generator switch is not turned on.	Turn on the switch.
	Current limiter or fuse in the generator power lead is burned out.	Replace fuse or current limiter.
	Defective or inoperative reverse-current cutout or relay.	Replace defective unit.
	Wiring or connections defective.	Inspect wiring. Make necessary repairs or corrections.
	Defective or improperly connected ammeter.	Check wiring to ammeter. Replace ammeter if necessary.
No voltage or amperage from generator.	Polarity of generator field reversed.	Flash the field.
	High resistance between the brushes and commutator.	Flash the field and clean the commutator.
	Faulty connections at voltage regulator. Plug-in type not making contact at the terminals.	Check connections and contacts at voltage regulator.
	Open circuit in voltage regulator.	Repair or replace voltage regulator.
	Open field circuit, or open field in generator.	Check continuity of field circuit including generator.
	Generator armature burned out.	Replace generator.
	Brushes excessively worn.	Install new brushes.
	Generator drive shaft broken.	Replace generator.
	Generator armature or field grounded or short-circuited.	Replace generator.
	Generator terminal connections faulty.	Make necessary repairs or adjustments of terminal connections.
	Generator brushes binding in holders.	Remove and clean the brushes.
	Brush spring tension too low.	Adjust or replace brush springs.
	Commutator dirty, rough, pitted, or eccentric.	Clean or resurface commutator as required.
Generator voltage too high.	Voltage regulator not properly adjusted.	Adjust voltage regulator.
	Contact points in voltage regulator stuck.	Repair or replace voltage regulator.
	Connections to regulating resistance are short-circuited.	Check wiring in voltage regulator and make necessary repairs.
	Short circuit between field terminal and positive generator terminal.	Repair generator internal wiring.
	Open circuit to voltage coil in voltage regulator.	Repair or replace voltage regulator.
	Negative or ground lead to regulator not properly connected.	Check and repair ground connection.
	Defective rheostat in carbon-pile regulator.	Repair rheostat or replace regulator.
	Poor contact at ground terminal of regulator base.	Clean contacts or bend base contacts so they will bear firmly against terminal pin.
	Faulty voltmeter.	Replace voltmeter.
	Open circuit or ground in series resistance of voltage coil.	Replace resistor.
Generator voltage too low.	Voltage regulator not properly adjusted.	Readjust voltage regulator.
	Poor connections in field circuit.	Check wiring connections and correct those that are faulty.
	Insufficient resistance in voltage coil circuit of regulator.	Adjust or replace fixed resistor in voltage coil circuit.
	Poor contact at terminal in base of voltage regulator.	Clean and adjust contacts.
	Faulty voltmeter.	Replace voltmeter.
	Defective regulating resistance in voltage regulator.	Repair or replace voltage regulator.
Generator voltage fluctuates.	Loose or dirty connections in generator field circuits.	Clean and adjust or tighten connections.
	Faulty connections in voltmeter circuit.	Correct faulty connections.
	Voltage regulator contact points dirty or pitted.	Repair or replace contact points.
	Faulty connections in voltage regulator.	Repair or replace voltage regulator.
	Regulator base contacts loose or dirty.	Clean and tighten contacts.
	Voltage regulator not properly adjusted.	Adjust regulator.
	Generator brushes worn or binding in holders.	Clean or replace brushes.
	Commutator dirty, rough, or eccentric.	Clean or resurface commutator.
Excessive arcing at generator brushes.	Worn or binding brushes.	Clean or replace brushes.
	Commutator dirty, rough, or eccentric.	Clean or resurface commutator.
	Brush spring tension too low.	Adjust or replace brush springs.
	Brushes not located in neutral plane.	Correct brush location.

Troubleshooting Chart (*Continued*)

Indication	Probable cause	Remedy
Generator burned out after operation.	Main contact points in reverse-current relay stuck closed.	Replace main contact points and replace generator.
Battery has low state of charge.	Capacity of generator too low for load.	Replace generator with one of sufficient capacity.
	Generator voltage too low.	Adjust generator voltage and check with precision voltmeter.
	Generator switch not turned on.	Have generator switch turned on during all flight operations.
Improper division of load in parallel system.	Voltage adjustments not properly balanced.	Adjust voltage regulators in accordance with instructions.
	Loose or defective connections in equalizer circuits.	Correct defective or loose connections.
	Equalizer connections reversed.	Correct the connections.
	Incorrect resistance in generator negative lead.	Install the correct resistance.
	Equalizer switch defective.	Replace the defective switch.
	Equalizing circuit contact points in the field relay dirty or defective.	Repair contact points or replace field relay.
Generator drops out of system when voltage is normal.	Overvoltage relay defective or out of adjustment.	Replace overvoltage relay.

signed for the purpose. This abrasive stick should be used in the same manner as the sandpaper described in the above paragraph. As the abrasive stick is held against the commutator, small particles of abrasive material are carried under the brushes and grind them to the contour of the commutator.

INSTALLATION

To install a generator on an engine, remove the mounting-pad cover and install the proper gasket over the studs. Fit the generator spline or gear into place, being careful not to damage them. Tighten the nuts on the hold-down studs, applying torque as recommended for the size of the nuts used. Connect the generator cables to the proper terminals, and see that all connections are clean and tight. If the generator employs an air duct for cooling, make sure that it is connected properly to avoid the possibility of its coming loose during operation.

It is recommended that a fuse or circuit breaker be installed in the circuit between the generator and the battery in all installations. This will provide protection for both the generator and the wiring. Wiring must be protected against wear due to chafing and against deterioration caused by oil or other substances. The proper installation of electric wiring is covered in this text in the chapter on aircraft electric systems.

The generator troubleshooting charts lists the most probable generator troubles, their causes, and their remedies.

REVIEW QUESTIONS

1. What features make it possible for an aircraft generator to carry greater loads than automobile generators of similar weight?
2. Describe the *armature assembly* for a typical aircraft generator.
3. What method is used to reduce the shock loading on the shaft of an aircraft generator?
4. Compare the *shunt field windings* in an aircraft generator with the *series windings.*
5. Give the approximate brush pressure required for aircraft generators.
6. What type of bearings are commonly used in aircraft generators?
7. What service is required for prelubricated sealed bearings?
8. Describe means for cooling aircraft generators.
9. For voltage regulation, which of the following are varied: rpm, field strength, number of armature windings?
10. Describe the action of a vibrator-type voltage regulator.
11. What is the function of the carbon pile in a carbon-pile voltage regulator?
12. Describe the operation of a carbon-pile voltage regulator.
13. Describe the operation of the equalizing circuit.
14. Why is a reverse-current relay required in a generator system?
15. Explain the operation of a differential reverse-current relay.
16. What voltage is required of a generator in a 24-V system?
17. Explain the operation of a current limiter used with a vibrator-type voltage regulator.
18. What is the purpose of an overvoltage relay in a generator system?
19. Describe the operation of a reverse-current air circuit breaker.
20. Give the requirements for the proper installation of a voltage regulator.
21. List inspection procedures required for generator control equipment.
22. Give the procedure for balancing generator load in an airplane.
23. What is meant by *flashing the field*?
24. If a generator produces only 2 or 3 V, what is likely to be the trouble?
25. What troubles may exist if there is no voltage from the generator?
26. If a generator shows full voltage but no amperage shows on the ammeter, what is the cause?

27. List typical generator inspections.
28. If you wish to overhaul a generator, what information should you have available?
29. How would you smooth the commutator of a generator?
30. What is the proper method for seating new generator brushes?
31. What is used to test the armature of a dc generator?
32. How would you test an armature for grounded windings?
33. Describe the use of an ohmmeter to test a field for grounded windings.
34. Describe the steps for the installation of a typical aircraft generator.

CHAPTER 9

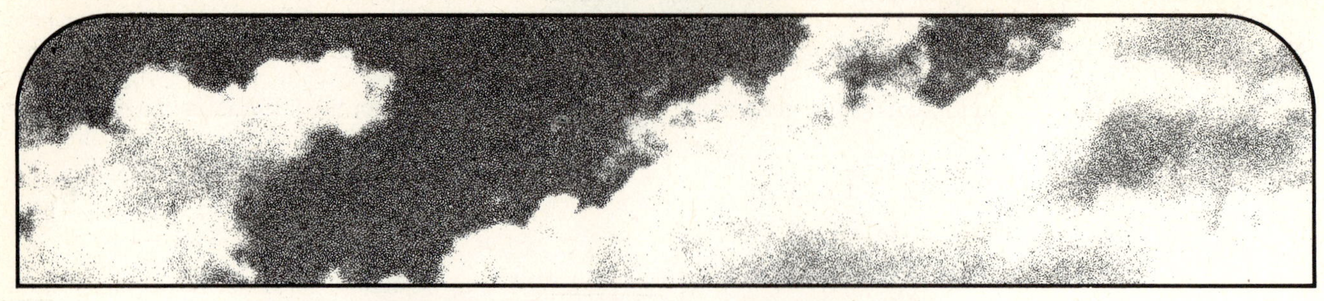

AC Generators, Controls, and Systems

The use of ac generators for the production of electric power in modern aircraft of all types is widespread. As explained previously in this text, the transmission of electric power can be accomplished must more efficiently and with less weight involved by utilizing elevated voltages and lower current in electric power systems. In addition, three-phase ac generators (alternators) and motors are more efficient and lighter in weight than comparable dc generators and motors. Ac power can easily be transformed to lower or increased voltages, and it can be rectified to produce necessary dc power.

In large aircraft, ac power is used directly to perform the majority of power functions for the operation of control systems and electric motors for a variety of purposes. In light aircraft, alternating current is changed to direct current by rectifiers in the generator and the direct current is supplied to the electric system.

AC GENERATION

PRINCIPLES OF AC GENERATION

The principle of **electromagnetic induction** has previously been explained as it relates to both dc and ac generators. To repeat briefly, when a conductor is cut by magnetic lines of force, a voltage will be induced in the conductor, and the direction of the induced voltage will depend upon the direction of the magnetic flux and the direction of movement across the flux. (Refer back to Fig. 7-2 to refresh your memory on the left-hand rule for generators.)

Consider the simple generator (alternator) illustrated in Fig. 9-1. A bar magnet is mounted to rotate between the faces of a soft-iron yoke on which is wound a coil of insulated wire. As the magnet rotates, a field will build first in one direction and then in the other. As this occurs, an alternating voltage will appear across the terminals of the coil. The waveshape of the ac voltage will roughly approximate a sine wave.

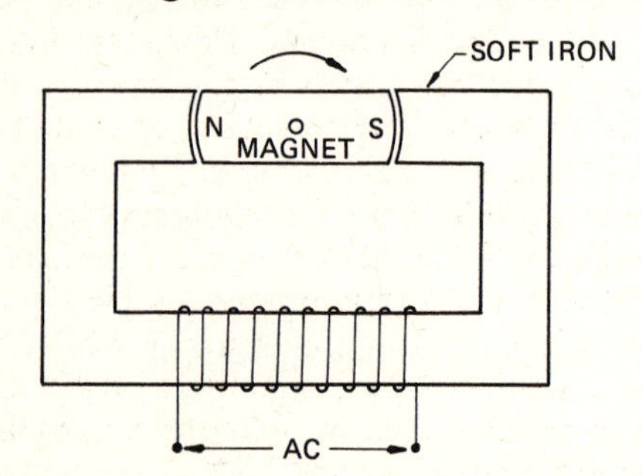

Fig. 9-1 Simple ac generator.

AIRCRAFT ALTERNATORS

Almost all alternators for aircraft power systems are constructed with a rotating field and a stationary armature. Some small alternators are made with permanent-magnet rotors, but when a steady voltage must be provided at a fixed frequency, the field strength of such an alternator must be varied according to load requirements. For this purpose an **exciter** is employed which can furnish a variable direct current to the rotor winding of the alternator, and a voltage-regulator system is used to change this current as required to maintain a constant voltage. This variable-exciter current must be supplied by a dc source.

AC POWER SYSTEMS

Ac generators or alternators are manufactured in many sizes to meet the requirements of electric systems for different types of aircraft. The alternators for light aircraft are quite similar to those used for automobiles, and the voltage controls or regulators are also similar. For large aircraft such as jet airliners, the alternators and control systems are much more complex.

In light-aircraft installations, the electric power is produced as alternating current and immediately changed to direct current by means of rectifiers to supply the system and to charge the batteries. In the electric power systems for large aircraft, the electric power is distributed as alternating current with a constant frequency in order to take advantage of the more efficient power transmission and reduced weight that is made possible with alternating current.

PRINCIPLES OF AIRCRAFT ALTERNATORS

The aircraft alternator is a three-phase unit rather than the single-phase type shown in Fig. 9-1. This means that the **stator** (stationary armature) has three separate windings, effectively 120° apart. The field rotates and is called the **rotor.** The schematic illustration of Fig. 9-2 will serve to

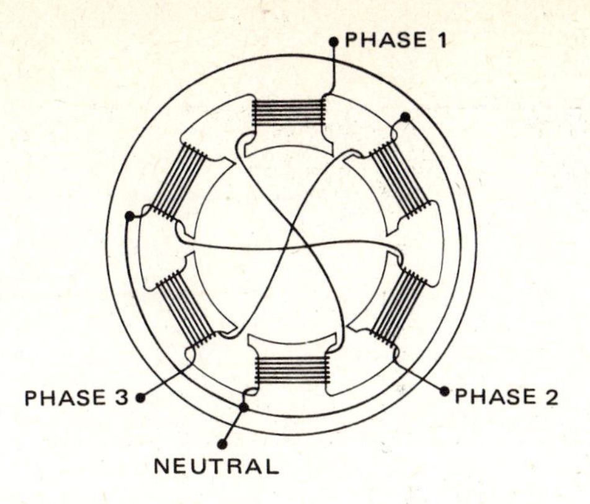

Fig. 9-2 A Y-connected stator.

indicate how the stator windings are arranged, although the windings in an actual stator will appear different. Also, it will be found that some stators will be wound in the Y configuration, and others will be wound in the delta configuration. Schematic diagrams of these arrangements are shown in Fig. 9-3.

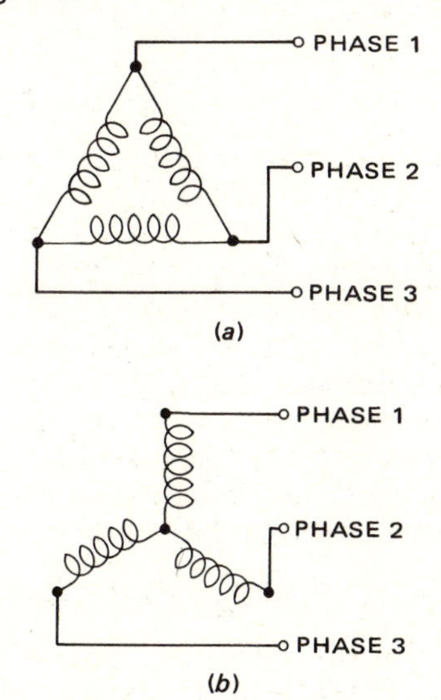

Fig. 9-3 Schematic diagrams of delta and Y-connected stators.

The output of a three-phase alternator is indicated in the drawing of Fig. 9-4. Note that there are three separate voltages 120° apart. That is, each voltage attains a maximum value in the same direction at points 120° apart. As the rotor of the alternator turns, each phase goes through a complete cycle in 360° of rotation. That is, each voltage reaches maximum in one direction, passes through zero, and reaches maximum in the opposite direction, then returns to the starting point in 360°.

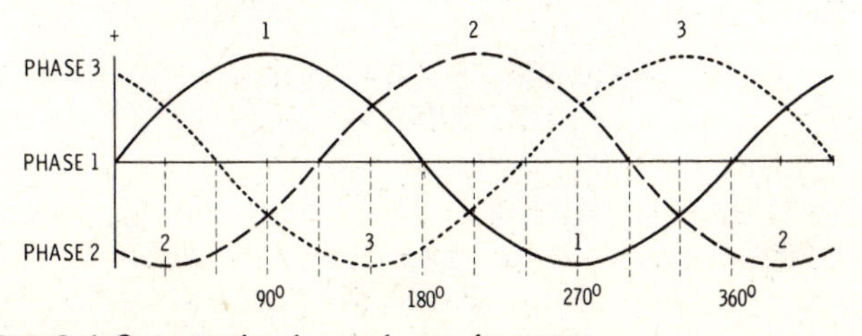

Fig. 9-4 Output of a three-phase alternator.

It may appear that the voltages in the diagram of Fig. 9-4 would be opposing one another, but this is not the case. When we observe the rotation of a wheel, we note that when one side of the wheel moves in one direction, the opposite side of the wheel is moving in the opposite direction, and both sides are contributing to the rotation of the wheel. The same principle may be applied to a three-phase alternating current.

ALTERNATOR SYSTEM FOR LIGHT AIRCRAFT

The alternator system for light aircraft is similar to a power system with a dc generator; however, the actual output of the alternator is alternating current. To use this current in a light-aircraft power system, it must first be converted to direct current. This is accomplished by means of a three-phase full-wave rectifier. A rectifier for three-phase alternating current consists of six silicon diodes if the rectifier is designed for full-wave rectification. A schematic diagram of a delta-wound stator with a three-phase, full-wave rectifier is shown in Fig. 9-5. The

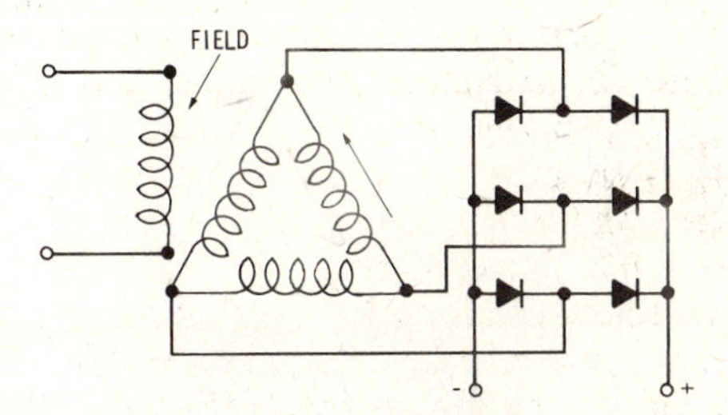

Fig. 9-5 Schematic diagram showing a full-wave, three-phase rectifier connected to the stator of an alternator.

arrowheads which represent the diodes point in a direction opposite the actual electron flow. Under the conventional system (current flow from positive to negative), the arrows would be in the direction of flow. In the diagram of Fig. 9-5, it can be seen that the current can be flowing in all sections of the stator at the same time. In the external circuit, the current is dc. In the aircraft system, the dc is applied to the battery, the power buses, and the voltage regulator. A typical electric power circuit is shown in Fig. 9-6. Since the rectifier is mounted in the end frame of the alternator, the alternator output terminals are marked for direct current.

LIGHT-PLANE ALTERNATOR

A typical alternator for light aircraft is shown in Fig. 9-7. Units similar to this are manufactured by such companies as the Ford Motor Company, Prestolite, the Chrysler Corporation, and the Delco-Remy Division of General Motors Corporation. The particular type of alternator to be used in an aircraft system may be determined from the aircraft manufacturer's parts catalog or from the catalog prepared by the manufacturer of the alternator.

The alternator is a comparatively simple device and is designed to give many hours of troublefree service. The principal components are the three-phase stator, the rotor (rotating field) with its windings for excitation, the slip rings and brushes through which current for excitation is transferred to the rotating unit, and the rectifier as-

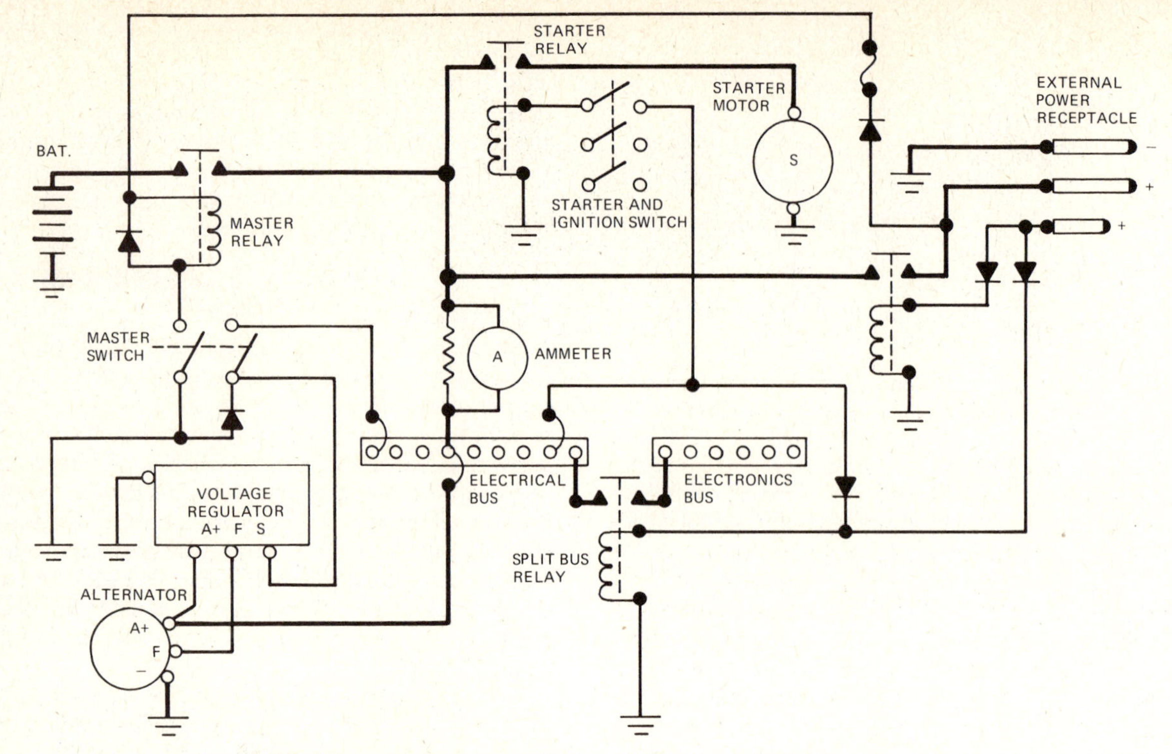

Fig. 9-6 Electrical power circuit for a light airplane equipped with an alternator.

Fig. 9-7 Alternator for a light airplane.

sembly which consists of six diodes connected to form the three-phase full-wave rectifier.

The operation of alternators was explained in Chapter 7. It will be remembered that many alternators (ac generators) utilize a rotating field (rotor) and a stationary armature (stator). This is the case with the alternators for aircraft. A typical alternator for light aircraft has a rotor with 8 or 12 poles alternately spaced with north and south polarity. This provides the rotating field within the stator which is generally Y-wound. The strength of the rotating field is controlled by the amount of current flowing in the rotor winding. This current is governed by the voltage regulator. The output of the stator is applied to a full-wave rectifier consisting of six diodes mounted within the alternator housing. The output of the alternator is, therefore, direct current as it is supplied to the aircraft electric power system.

Alternators for light aircraft may be driven by a belt and pulleys or be gear-driven and flange-mounted on the engine. In the latter case, the engine manufacturer must provide the correct mounting and gear drive for the alternator.

Figure 9-8 shows a gear-driven alternator. The internal construction of the alternators is the same and consists of the components shown in Fig. 9-9. The drive-end head contains a prelubricated bearing, an oil seal, a collar and shaft seal, and a blast tube connection for ventilation.

Fig. 9-8 A flangemounted gear-driven alternator. (Prestolite)

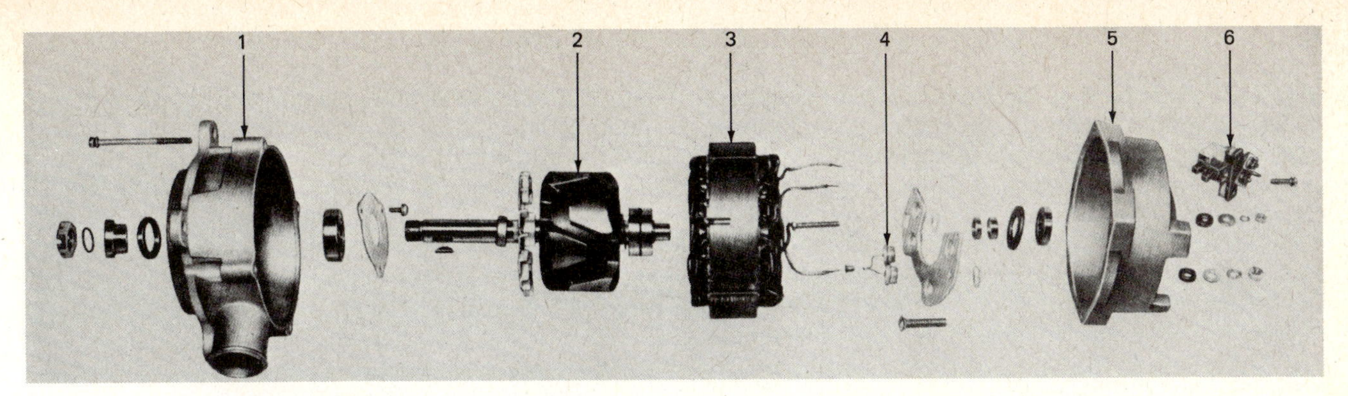

Fig. 9-9 Components of an alternator: 1. Drive and lead 2. Rotor 3. Stator 4. Rectifiers 5.Slip ring and head 6. Brush and holder assembly. (Prestolite)

The rotor has eight poles and is mounted on a shaft with a ventilating fan on the drive end. The slip rings, slip-ring end bearing inner race, and spacer are on the other end of the shaft. The rotor windings and winding leads are treated with high-temperature epoxy cement to provide vibration and temperature resistance. High-temperature solder is used to secure the winding leads to the slip rings.

The stator of the alternator has a special electric lead that is connected to the center of the three-phase windings. This lead is used to activate low-voltage warning systems or relays. The stator is treated with epoxy varnish for high-temperature resistance.

The rectifiers are crystal diodes connected to produce full-wave rectification of the three-phase alternating current that is produced by the alternator. These rectifiers are rated at 150 PIV (peak inverse voltage, or transient voltage) so they will not be damaged by any transient voltage that may occur. Three positive rectifiers (diodes) are mounted in the rectifier mounting plate, and three rectifiers are mounted in the slip-ring end head. Each pair of rectifiers is connected to a stator lead with high-temperature solder. The stator leads are anchored to the rectifier mounting plate with epoxy cement for vibration protection.

The slip-ring end head provides the mounting for the rectifiers and rectifier mounting plate, output and auxiliary terminal studs, and the brush and holder assembly. The slip-ring end head also contains the roller bearing and outer race assembly plus a grease seal.

The brush and holder assembly contains two brushes, two brush springs, a brush holder, and insulators. Each brush is connected to a separate terminal stud and is insulated from ground. The brush and holder assembly can be easily removed for brush inspection or replacement purposes.

MAINTENANCE OF ALTERNATORS

Maintenance of alternators follows the principles of good mechanical and electrical practice and should be accomplished according to the instructions given in the maintenance manual for the particular unit requiring service. In general, the disassembly procedure is similar to that of other generators. Care must be taken to assure that the parts are marked and identified in such a manner that they can be reassembled correctly.

The rotor winding can be tested with an ohmmeter or continuity tester. The reading is taken with the test prods of the instrument applied to the slip rings. The resistance of the rotor winding should be relatively low and within the limits specified by the manufacturer. Grounding of the rotor winding can be tested by connecting one test prod of an ohmmeter to the rotor shaft and the other to one of the slip rings. The reading should indicate infinite resistance. If current flow is indicated, the rotor must be replaced.

The stator winding can be tested by checking between the stator leads with the ohmmeter. The reading in each case should be within specifications. Normally the reading will show low resistance. If the resistance is above or below the limits specified by the manufacturer, the stator must be replaced. To test for grounded windings in the stator, the ohmmeter is connected between one stator lead and the stator frame. The ohmmeter should show infinite resistance.

To test for open windings in the stator, one test probe of the ohmmeter is connected to the auxiliary terminal or to the stator winding center connection. The other probe is connected to each of the three stator leads, one at a time. The ohmmeter should show continuity in each case, and the resistance should be in the range specified by the manufacturer.

Slip rings are inspected for wear or damage. Small nicks or scratches can be removed with fine sandpaper; however, if the slip rings cannot be easily smoothed, they should be turned on a lathe or the rotor should be replaced. Wear should not exceed the amount specified by the manufacturer. Brushes should usually exceed $\frac{1}{2}$ in. [1.27 cm] in length to be acceptable.

Alternator bearings should be inspected and serviced in the same manner as those for a generator or motor. Care must be taken to assure that the correct servicing procedures are followed for the particular type of bearing being serviced.

TRANSISTORIZED VOLTAGE REGULATORS

Although the voltage of an alternator can be controlled by the same types of voltage regulators as those described previously for dc generators, it is common practice to use **transistorized** or **transistor** voltage regulators for alternator systems. The transistorized voltage regulator utilizes a voltage-control coil and contact points to control a transistor which, in turn, adjusts the field current for the alternator. A transistor voltage regulator contains no coils, contact points, or other moving parts. In this type of regulator, the current for the alternator field is entirely controlled through the interaction of transistors, diodes, capacitors, and resistors. The transistor voltage regulator will be described later in this section.

The theory and construction of diodes, transistors, and other solid-state devices are given in Chapters 4 and 12 of this text; however, it is useful to give a brief explanation of transistor operation before showing how transistors function in voltage regulators. In the previous discussion of diodes, it was explained that semiconductor materials such as germanium and silicon can have their characteristics altered by the addition of small amounts of "impurities" such as indium or antimony, thus forming *p*-type (positive-type) or *n*-type (negative-type) materials.

A triode transistor is formed by joining two pieces of *n*-type or two pieces of *p*-type material with a thin layer of the opposite type between the two pieces of the same type. This arrangement is shown in the drawings of Fig. 9-10. The three sections of the transistor are called the **emitter,** the **base,** and the **collector.** The base is the very thin section of *n*-type material between the two sections of *p*-type material. When the emitter and collector only are connected to a voltage source, no current can flow through the transistor. Current **can** flow from the *n*-type base to a *p*-type section when the base is connected to a negative voltage source and the *p*-type section is connected to the positive terminal of the same voltage source. Hence, when the base of the *pnp* transistor is biased negative with respect to the emitter, current (electrons) can flow from the base to the emitter. Furthermore, when the collector is made negative with respect to the emitter, and the base is biased negatively, current will flow from the collector to the emitter. At this point we are speaking of current flowing in the true direction from negative to positive.

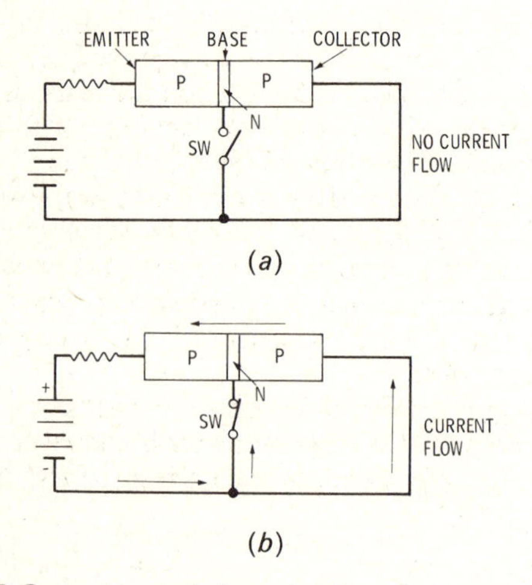

Fig. 9-10 Current-control characteristics of a transistor.

Because of the characteristics we have just described, the triode transistor can be used as an electric valve. A small amount of current flow through the base-emitter circuit can be used to control a large amount of current through the collector-emitter circuit. If the transistor is connected as shown in Fig. 9-10*a*, current cannot flow; however, when the base is connected to the negative source of voltage, the collector-emitter section of the transistor becomes a good conductor. This is shown in Fig. 9-10*b*. Since a small base current can control a large collector-emitter current, the transistor can be used to control the field current of an alternator, as will be shown.

Transistorized regulators are usually of a four-or three-terminal type. The four-terminal type consists of a field relay, a transistor, a voltage-regulator winding, a diode, and resistors. A voltage regulator of this type is shown in Fig. 9-11.

Fig. 9-11 A four-terminal transistorized voltage regulator.

The field relay is similar to relays previously discussed. Essentially the relay is an electromagnetically controlled switch used to connect the alternator output to one terminal of the alternator field. The other side of the field circuit is controlled by the transistor and the voltage-regulator coil.

The field-control circuit of the transistorized voltage regulator is shown in a simplified form in Fig. 9-12. In this circuit, the field relay is controlled by the ignition switch. The switch is not shown in the drawing of Fig. 9-12. When the ignition switch is closed, the field relay closes and connects the field to the alternator output. In a typical system this will allow about 4.5 A to flow through the field winding, provided that the transistor-base circuit is connected to ground through the voltage-regulator contacts. This would be the condition when the alternator

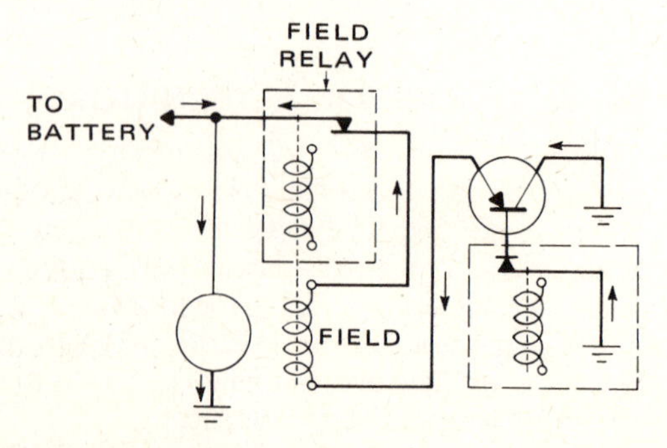

Fig. 9-12 Simplified field-control circuit for a transistorized voltage regulator.

voltage is below the level for which the regulator is adjusted.

In the previous discussion of transistors, it was explained that the transistor becomes a good conductor through the emitter-collector section when the base circuit has negative bias. In the case of the regulator under discussion, the base circuit carries approximately 0.35 A when the voltage-regulator contact points are closed.

When the alternator voltage attains the value for which the regulator is adjusted, the regulator contact points are opened by the magnetic force of the regulator coil, thus cutting off the base current to the transistor. The emitter-collector section then becomes nonconductive, and the field current is blocked. The alternator voltage then drops, and the regulator points close again to provide bias for the transistor-base circuit. Field current can then flow through the transistor and the voltage rises to the regulated value. This cycle continues with the regulator contact points vibrating rapidly (about 2000 times per second) to maintain the alternator voltage at the required value.

A more complete schematic diagram of the voltage regulator and alternator circuits is shown in Fig. 9-13. The diagram of the alternator shows the stator in which the alternating current is generated, the field coil and slip rings through which current flows to the field coil, and the six-diode rectifier. When the ignition switch is closed, the battery and alternator are connected to the field-relay coil to produce a magnetic field that closes the relay contacts. When this happens, current flows from ground, through the collector-emitter circuit of the control transistor and the F_1 terminal of the regulator, to the F_1 terminal of the alternator. After passing through the alternator field, the current enters the F_2 terminal of the regulator and passes through the closed field-relay contact points and out the *BAT* terminal of the regulator. From this point it flows to the positive (+) terminal of the alternator, through the rectifier network, and to ground.

When the field windings of the alternator are energized, a dc voltage will be delivered to the system by the alternator, provided, of course, that the alternator is running. As alternator voltage increases, current flow through the two windings of the voltage-regulator coil will increase. When the voltage reaches the value for which the regulator is adjusted, the voltage-regulator contacts open. This stops the emitter-base current in the transistor, and the transistor becomes nonconductive, as explained previously.

When no current is flowing in the alternator field windings, the alternator voltage immediately decreases, the spring at the end of the contact arm opens the points, and the cycle repeats as previously explained. The high rate of vibration of the contact points provides a steady voltage for all practical purposes. Since the voltage-regulator contact points are held closed by a spring when the voltage is below the desired value, and the points are opened as a result of alternator voltage reaching this value, an increase of spring tension will cause the alternator voltage to increase. Voltage adjustments are therefore made by turning the screw which controls the spring tension.

It is important to note that the voltage-regulator contact points carry only about 0.35 A when the alternator field current is over 4 A. On regulators without transistor control, full generator field current must pass through the regulator contact points. This amounts to approximately 1.5 A for 12-V systems and 1 A for 24-V systems. Since vibrating contact points are burned by higher amperages, the use of a transistor makes it possible to increase the life of the contact points because of the lower current through the points.

In the diagram of Fig. 9-13, observe that the **accelerator winding** on the voltage regulator is connected to the regulator F_2 terminal and through a resistor and the con-

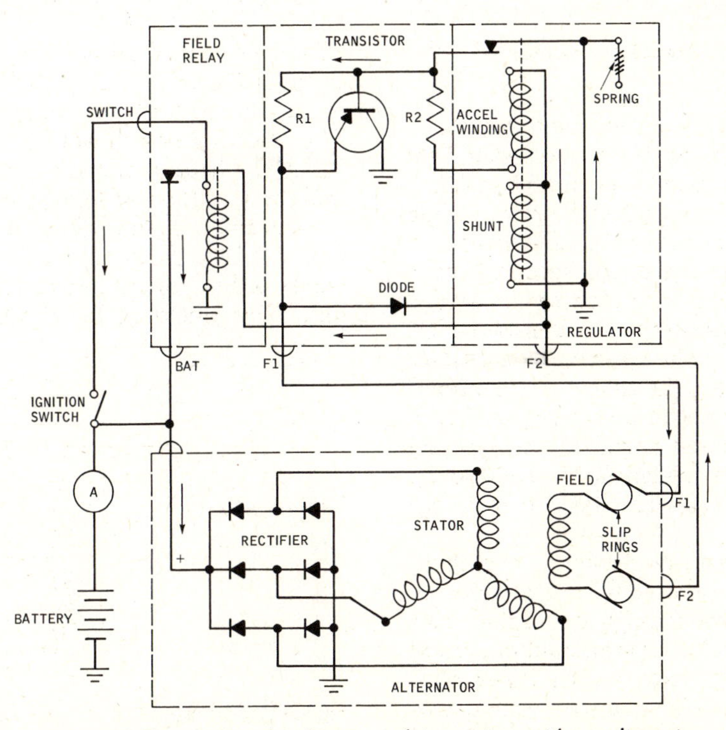

Fig. 9-13 Transistorized voltage-regulator system with an alternator.

tact points to ground. This winding will therefore carry no current when the contact points are open. The effect of this arrangement is to reduce the magnetic pull on the contacts as soon as the contact points open, thus making the spring more effective in closing them again. When the contact points are closed, the magnetic effect of the accelerator winding is added to the total magnetic force again, and the points reopen very quickly. The effect of the accelerator winding thus causes the contact points to vibrate (open and close) much more rapidly than they would with the shunt coil only. This is the reason for the term *accelerator winding.*

The resistor R_1 connected between the emitter and base of the transistor acts to prevent emitter-to-collector current leakage when the voltage-regulator contact points are open under high-temperature conditions. There is a tendency for some current to leak through the transistor when there are too many free electrons present at high temperatures, even though the regulator contacts are open. The resistor R_2 in series with the voltage-regulator accelerator winding and the contact points is a design feature to limit the current through the winding without the necessity of using exceptionally fine wire for the winding.

The diode in the regulator is connected directly across the field winding. If the voltage contacts open without a diode in this circuit, the sudden interruption of field current and the resulting high voltage induced in the field winding would damage or destroy the power transistor.

When the ignition switch is open (turned off), the field relay is deenergized, and the main contacts open. This disconnects the regulator from the battery and the alternator, and the alternator voltage drops to zero. There is no need for a reverse-current relay in an alternator circuit of this type because the diodes of the full-wave rectifier prevent battery current from flowing through the alternator stator.

In the use of transistors, it is important to note that high temperatures can cause improper functioning and permanent damage to the transistors. For this reason, the transistors used with voltage regulators or in any other circuit must be kept at safe operating temperatures. The transistors used with voltage regulators usually have heavy metal bases which act as heat sinks to carry the heat away from the active elements. In any installation, the maximum safe operating temperature for the transistors must be known, and provision must be made to assure that these temperatures cannot be exceeded.

Some alternator systems utilize an indicator light to show when the alternator is not charging the battery. In these cases, the three-terminal voltage regulator is used with an external field relay and an indicator light. The circuit for such a system is shown in Fig. 9-14. An examination of this circuit shows that the ignition switch connects the battery to the field-relay winding and causes the contact points to close. The battery is also connected to the indicator light. The indicator-light circuit is completed to ground through the contact points of the indicator-light relay. These points are closed when the relay is not energized; hence, the light will be on. Since the field relay closes when the ignition switch is turned on, the battery is connected to the field windings of the alternator. The alternator output voltage then opens the indicator-light relay, thus turning off the indicator light. The operation of the transistor- and voltage-regulator unit are the same as that previously described.

TRANSISTOR VOLTAGE REGULATORS

A typical transistor voltage regulator is shown in the photograph of Fig. 9-15. This regulator has no moving parts and consists primarily of a control diode, a control resistor, a power transistor, field diodes, capacitors, and resistors. The principles of operation for diodes and transistors have been described previously and should be reviewed to obtain a good understanding of the operation of the transistor voltage regulator.

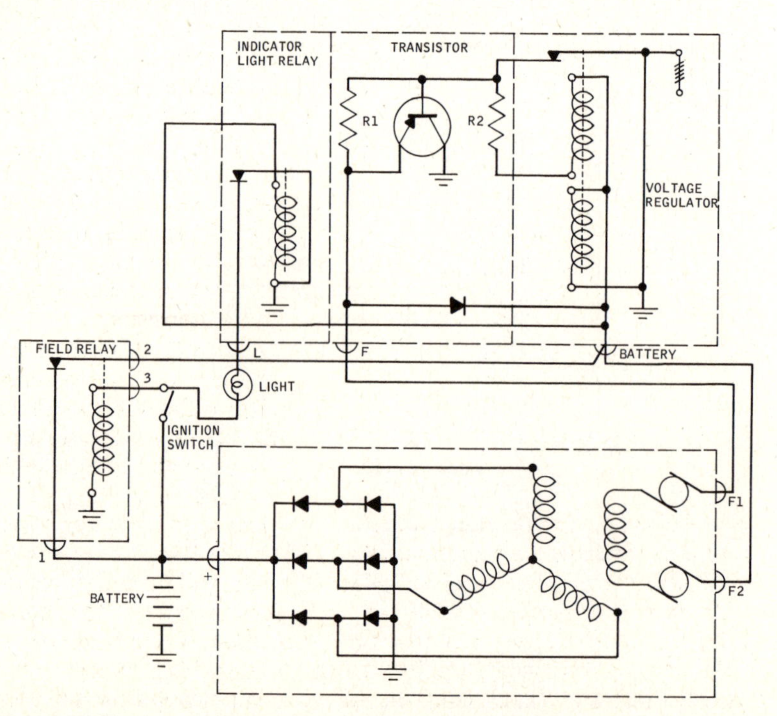

Fig. 9-14 Transistorized voltage regulator system with an indicator light circuit.

Fig. 9-15 A transistor voltage regulator.

A circuit diagram to illustrate the operation of a transistor voltage regulator is shown in Fig. 9-16. In this description of the operation, each item will be explained in terms of actual current flow from negative to positive. For example, when the battery is furnishing current to the circuit, current flows from the negative terminal to ground and current flows from the circuit into the positive terminal of the battery. When the battery is being charged, current is flowing into the negative terminal of the battery and out the positive terminal.

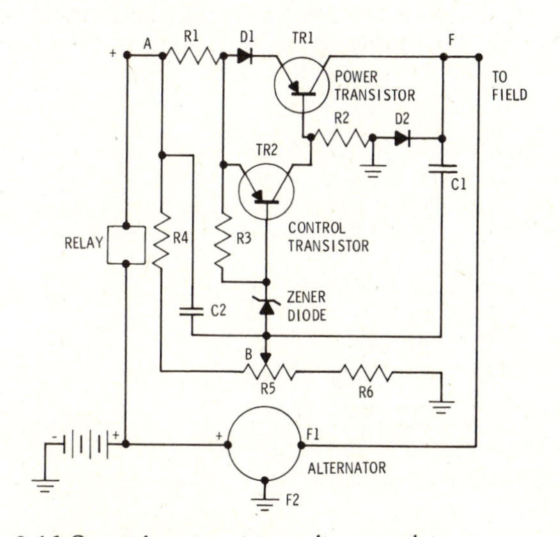

Fig. 9-16 Circuit for a transistor voltage regulator.

In the circuit of Fig. 9-16, when the ignition switch is closed, the battery and alternator are connected to the relay and through the relay to the positive terminal (A) of the regulator. There is then a complete circuit from ground through the resistor R_2, the base of the **power transistor** TR_1, the diode D_1, the resistor R_1, and back to the battery and the positive terminal of the alternator. If the output of the alternator is below the voltage for which the regulator is set, transistor TR_1 will have forward bias and current will flow from the F_1 terminal of the alternator to the F terminal of the regulator and through the collector-emitter circuit of TR_1. The circuit is completed through D_1, R_1, and the relay to the alternator. This current flow excites the field of the alternator, and the output of the alternator quickly rises to the desired level. In the circuit of Fig. 9-16, it can be seen that there is a circuit from ground through R_6, R_5, the **zener diode,** R_3, and R_1 to A. There is also a circuit from the zener diode through the base-emitter circuit of the **control transistor** TR_2 and through R_1 to terminal A of the regulator. The zener diode blocks the flow of current from R_5 until the voltage between ground and A reaches approximately 14.5 V. At this point the zener diode begins to conduct and applies a forward bias through the base-emitter circuit of TR_2, the **control resistor.** TR_2 then becomes conductive, and current flows through the collector-emitter section from ground. This current flow is from ground through R_2, TR_2, R_1, and out A. The effect of this is to short-circuit the base-emitter circuit of TR_1, and this causes TR_1 to stop conducting field current for the alternator. The alternator voltage immediately drops, and the zener diode stops conducting, thus removing the forward bias from TR_2, which also stops conducting. This returns the forward bias to TR_1, which starts conducting field current again, and the cycle repeats. This cycle repeats about 2000 times per second, thus producing a reasonably steady voltage of approximately 14.5 V from the alternator.

The two key points to understand with respect to the operation of the transistor voltage regulator are the zener-diode operation and the control of the power transistor by the control transistor. The zener diode may be compared to a relief valve which opens at a given pressure in a hydraulic system. When the zener diode conducts current, it causes the control transistor to shut off the power transistor. The reason that the control transistor can stop the flow of current through the emitter-base circuit of the power transistor is that there is a difference in the voltage drops across the emitter-base circuits of the two transistors when the control-transistor emitter-base circuit is conducting. The diode D_1 causes approximately 1 V drop in potential across the emitter-base circuit of the power transistor when the circuit is conducting. When the emitter-collector circuit of the control transistor begins to conduct, there is no appreciable voltage drop across the control transistor; hence, a 1 V reverse bias becomes effective across the emitter-base circuit of the power transistor. This, of course, stops the emitter-base current in the power transistor.

Adjustment of alternator voltage output is accomplished through the variable resistor R_5. A change in the resistance of this resistor will change the voltage level across the zener diode, thus raising or lowering the level of alternator output voltage required to cause the zener diode to conduct.

The resistor R_1 and capacitor C_1 act to reduce the time required for the field voltage to change between maximum and minimum values. This prevents overheating of the transistors. The capacitor C_2 reduces the voltage variations which appear across the resistors R_4 and R_5, thus making the regulator more accurate. Resistor R_3 prevents leakage current from the emitter to the collector in the control transistor. Resistor R_4 is a special temperature-sensitive type which acts to increase the alternator voltage slightly at lower temperatures. This aids in maintaining adequate charge current for low-temperature operation. Diode D_2 aids in controlling field-current flow as the power transistor rapidly turns the field current on and off.

Another transistor voltage regulator is illustrated in Fig. 9-17. This regulator is similar to the one described previously, but it utilizes three transistors rather than two. The internal circuitry of the regulator is shown in the circuit diagram of Fig. 9-18.

Fig. 9-17 Transistor voltage regulator with three transistors. (Prestolite)

The operating principles are the same as previously explained. The power transistor, T_3, is turned on when the control transistor, T_2, provides forward bias from ground through the emitter-base circuit of T_3, the diode D_2, the emitter-collector circuit of T_2, and the resistor R_6. Thus, whenever T_2 is conducting, T_3 will also be conducting.

When the ignition switch is turned on, current will flow from ground through a voltage-dividing network consisting of P_1, R_2, and R_1 to the *I* terminal of the regulator. This network determines the system operating voltage relative to the zener diode Z_1 reverse conducting voltage. When this voltage reaches a predetermined level, the zener diode will begin to conduct. Current then flows from ground through the emitter-base circuit of T_1 and through the zener diode and R_1 to the *I* terminal of the regulator. This causes T_1 to conduct through its collector-emitter circuit, which, in turn, removes the bias from T_2. T_2 stops conducting, thus removing the base forward bias from T_3 and causing it to stop conducting. This stops current flow through the alternator field, and the alternator voltage drops. The zener diode Z_1 then stops conducting, which turns T_1 off. This allows T_2 and T_3 to turn on, and T_3 then conducts field current to the alternator field and voltage increases until Z_1 starts to conduct again. This cycle repeats many times per second and provides for a reasonably steady voltage.

The diode D_1 connected across the field circuit protects the regulator components from the transient voltages produced by the field windings as field current is stopped and started. The capacitor C_1 is used to filter ripple and alternator diode switching spike when the alternator is operating without a battery. The lamp L_1 provides transient voltage protection by absorbing surge voltages. The control P_1 is used to provide a limited range of voltage adjustment.

TROUBLESHOOTING AN ALTERNATOR SYSTEM

To provide typical procedures for troubleshooting a light-aircraft alternator system, the troubleshooting chart for the alternator system in a Cessna 182 airplane is reproduced here. It must be remembered that there are variations with different systems; hence, it is important to consult the manufacturer's instructions for a particular system. A general rule to remember in all cases is that alternator or generator voltage is controlled by varying the amount of field excitation current.

AC SYSTEMS FOR LARGE AIRCRAFT

Although the alternating-current power systems for large aircraft are extremely complex when compared with power systems for light aircraft, it is important that the maintenance technician have some understanding of the

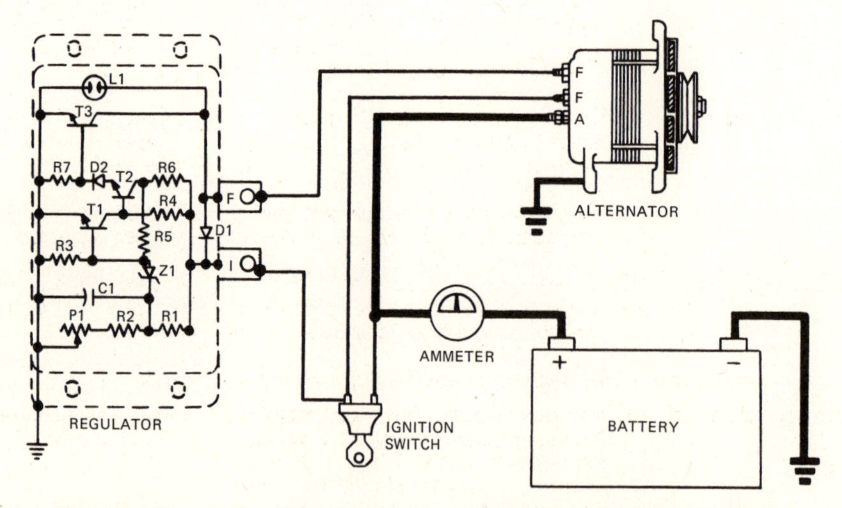

Fig. 9-18 Transistor voltage regulator in circuit with alternator and battery. (Prestolite)

Troubleshooting Chart for an Alternator System

Probable cause	Isolation procedure	Remedy
	Ammeter indicates heavy discharge with engine not running or alternator circuit breaker opens when master switch is turned on.	
Shorted field in alternator.	1. Remove plug from regulator with master switch on and observe if heavy drain persists.	If heavy drain is reduced, go to step 2. If heavy drain is not reduced, go to step 3.
	2. Check resistance from terminal *F* on alternator to the alternator case. Normal indication is 6 to 7 Ω.	If resistance is too low, repair or replace alternator.
Shorted radio-noise filter or shorted wire.	3. Remove cable from output terminal of alternator. Check resistance from end of cable to ground (MASTER SWITCH MUST BE OFF).	If resistance does not indicate a direct short, go to step 6. If resistance indicates a direct short, go to step 4.
	4. Remove cable connections from radio-noise filter. Check resistance from the filter input terminal to ground. Normal indication is infinite resistance.	If reading indicates a direct short, replace filter. If no short is evident, go to step 5.
	5. Check resistance from ground to the free ends of the wires which were connected to the radio-noise filter (or alternator if no noise filter is installed). Normal indication does not show a direct short.	If a short exists in wires, repair or replace wiring.
Shorted diodes in alternator.	6. Check resistance from output terminal of alternator to alternator case. Reverse leads and check again. Resistance reading may show continuity in one direction, but should show an infinite reading in the other direction.	If an infinite reading is not obtained in at least one direction, repair or replace the alternator.
	Alternator system will not keep battery charged.	
Regulator faulty or improperly adjusted.	1. Start engine and adjust for 1500 rpm [157 rad/s]. Ammeter should indicate a heavy charge rate with all electric equipment turned off. Rate should taper off in 1 to 3 min. A voltage check should indicate a reading consistent with the appropriate voltage-temperature chart.	If charge rate tapers off very quickly and voltage is normal, check battery for malfunction. If ammeter shows a low charge rate or any discharge rate, and voltage is low, proceed to step 2.
	2. Stop engine, remove cowl, and remove cover from voltage regulator. Turn master switch on and off several times and observe field relay in regulator. Relay should open and close with master switch, and a small arc should be seen as contacts open.	If relay is inoperative, proceed to step 3. If relay operates, proceed to step 4.
	3. Check voltage at *A*+ terminal of regulator with master switch closed. Meter should indicate bus voltage.	If voltage is present, replace the regulator. If voltage is not present, check wiring between regulator and bus.
	4. Remove plug from regulator and start engine. Momentarily connect the *A*+ and *F* terminals together on the plug. The ammeter should show a heavy rate of charge.	If a heavy rate of charge is observed, replace the regulator. If a heavy charge rate is not observed, proceed to step 5.
Faulty wiring between alternator and regulator, or faulty alternator.	5. Check resistance from *F* terminal of alternator to *F* terminal of regulator. Normal indication is very low resistance.	If reading indicates no or poor continuity, repair or replace wiring from regulator to alternator.
	6. Check resistance from *F* terminal of alternator to alternator case. Normal indication is 6 to 7 Ω.	If resistance is high or low, repair or replace alternator.
	7. Check resistance from case of alternator to airframe ground. Normal indication is very low resistance.	If reading indicates no or poor continuity, repair or replace alternator ground wiring.
	Alternator overcharges battery. Battery uses excessive water.	
Regulator faulty or improperly adjusted.	Check bus voltage with engine running. Normal indication agrees with voltage-temperature chart. Observe airplane ammeter. Ammeter should read near zero after a few minutes of engine operation.	Replace regulator if it does not check out satisfactorily.

systems and their functions. In this section we shall examine the general construction of large alternators and the functions of control units necessary for effective operation, particularly in systems where several alternators operate in a parallel system.

HIGH-OUTPUT ALTERNATORS (GENERATORS)

High-output brushless alternators were developed early in the 1950s for the purpose of eliminating some of the problems of alternators which employ slip rings and brushes to carry exciter current to the rotating field. Among the advantages of a brushless generator are the following:

1. Lower maintenance cost, since there is no brush or slip-ring wear.
2. High stability and consistency of output, because variations of resistance and conductivity at the brushes and slip rings are eliminated.
3. Better performance at high altitudes, because arcing at the brushes is eliminated.

A brushless alternator consists of rotor which includes the rotating armature of the exciter generator, a half-wave or full-wave, three-phase rectifier, and the rotating field for the main generator (alternator); the stationary field of the exciter generator and the stationary armature (stator) of the main alternator; and necessary framework and bearings to make the unit functional. A schematic diagram of the electric circuit in one type of brushless alternator is shown in Fig. 9-19.

The exciter generator for the alternator is a three-phase alternator, the output of which is converted to dc by means of the rectifier. The direct current from the rectifier is fed to the windings of the main alternator rotor. Voltage regulation is provided through the field of the exciter in much the same manner as that described previously. The Westinghouse 8QL40S alternator is shown in Fig. 9-20. Note that the main rotating field (rotor) and the exciter armature are mounted on the same hollow shaft. The exciter-armature output passes through conductors to the inside of the shaft and to the three-phase rectifier. The dc output from the rectifier is carried outside the shaft to the main rotor windings. Since the exciter armature, the recti-

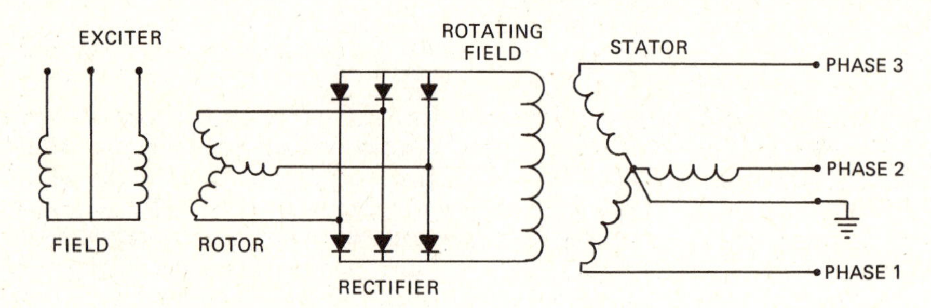

Fig. 9-19 Schematic circuit for a brushless alternator.

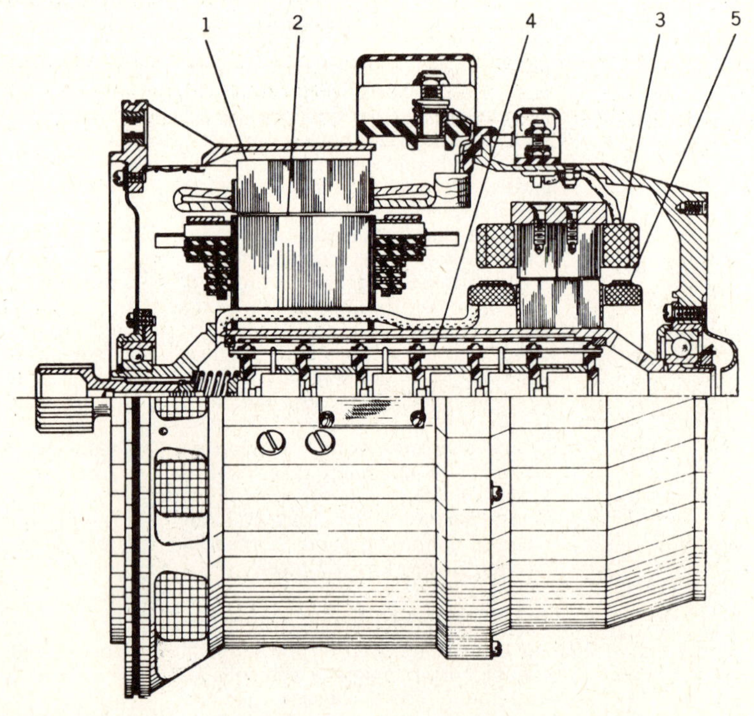

Fig. 9-20 Westinghouse brushless alternator. 1. Three-phase ac stator 2. Main rotating field 3. Exciter stator 4. Rotating rectifier assembly 5. Three-phase ac exciter rotor

fier, and the main rotor are all mounted on or in the same shaft, there is no need for slip rings, brushes, or a commutator.

The need for flashing the field of the exciter to restore residual magnetism has been eliminated by the installation of permanent magnets in the main field poles. Thus there is always a magnetic field to start the generation of current.

The design of the brushless ac generator is made possible by the development of the silicon recitifier. This three-phase rectifier consists of six single rectifiers connected together to form one unit which is provided with cooling fins. Even though the unit is comparatively small, its capacity is adequate to handle the maximum field load required by the alternator. To provide a full-wave rectifier for three-phase alternating current using six rectifiers, the individual rectifier diodes are joined in series pairs with three input terminals, one between the individual units of each pair. The positive sides of the pairs are connected together for the positive output, and the negative sides are connected together for the negative output.

The field for the alternator is provided with two windings, one to provide the field flux and the other, a damping winding, to improve stability and prevent overcorrection. The damping winding acts in opposition to the effects of the main winding when flux changes are too great, the effect being in proportion to the rate of flux change; that is, when the rate of flux change is high, the damping effect will be stronger than when the rate is low.

BENDIX 60-KVA ALTERNATOR

Many different high-output alternators have been built for large aircraft; however, the brushless alternator in various makes, designs, and sizes is generally accepted as the most dependable and troublefree. Typical of the larger alternators for jet aircraft is the unit illustrated in Fig. 9-21. This generator is designed and built by the Bendix Corporation for use in the Boeing 747 jet airliner. Four of these generators are used in the main electric power system and are driven by the airplane engines. These can be and usually are operated in parallel to provide all operating power for flight. Two of the same types of generators are used with the auxiliary power units (APUs) and cannot be paralleled with the main power generators.

The Bendix Model 28B263-13 generator under discussion is a three-phase, 8000 rpm, brushless, air-cooled machine rated at 120/208 V, 400 Hz, and is capable of producing more than 60 kVA over a power-factor range of 0.75 lagging to 0.95 leading. The generator weighs 71.75 lb [32.55 kg] including the quick attach-detach adapter (QAD).

The voltage rating 120/208 means that the generator is Y-connected and that the voltage across any single phase is 120 V, whereas the voltage across any two of the main output terminals is 208 V. This is illustrated in Fig. 9-22.

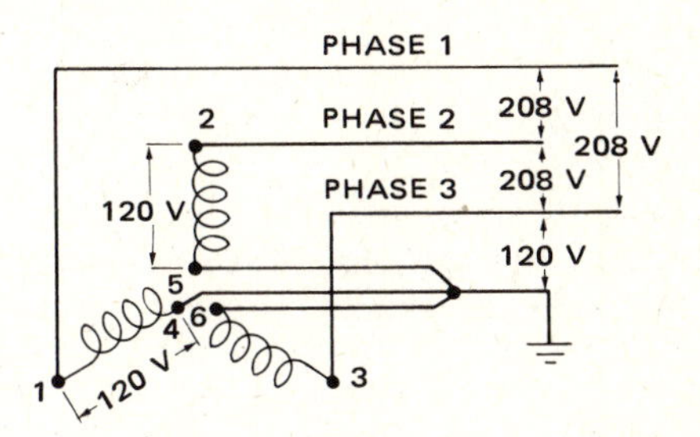

Fig. 9-22 Y-connected stator for an alternator.

One terminal of each separate stator winding is connected to ground, and the other terminal of the winding is the main output terminal. For aircraft circuits requiring 115/120 V, single phase, the circuit is connected between one main phase and ground. For power circuits such as those for motors, all three main phases are connected to the three-phase stator of the motor.

The exciter field of the generator includes permanent magnets to make the unit self-starting. The exciter armature is a three-phase unit and rotates within the stationary field. The output of the exciter is rectified by means of a three-phase half-wave rectifier to produce the dc which is fed directly to the rotating field of the main generator. The internal circuitry of the generator is shown in the schematic drawing of Fig. 9-23.

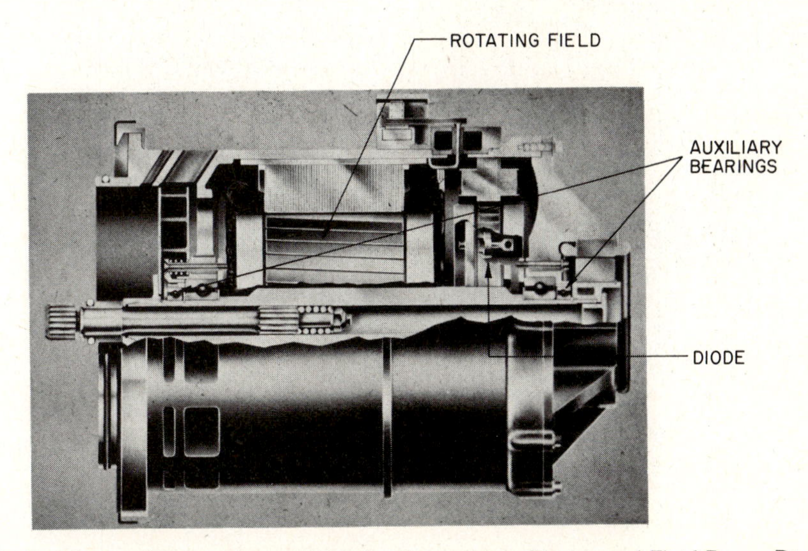

Fig. 9-21 Bendix brushless generator. (Bendix Corporation, Electric and Fluid Power Division)

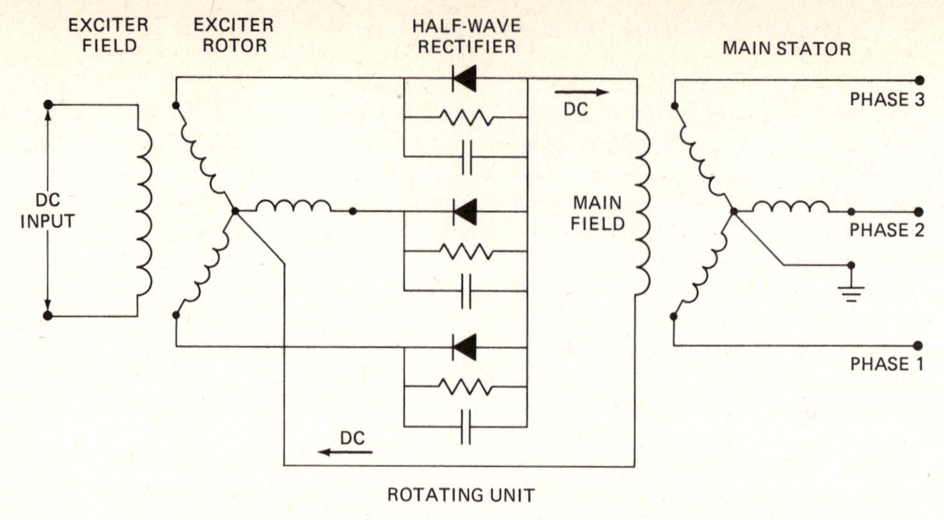

Fig. 9-23 Internal circuit for Bendix brushless alternator.

The housing and end bell of the generator are constructed of high-strength magnesium alloy to provide maximum strength with minimum weight. The input shaft spline is oil lubricated by oil from the constant-speed drive (CSD) unit. For the APU generators, this lubrication is provided by the APU lubrication system.

The rotor of the generator is supported by two double-width ball bearings lubricated with a special bearing grease. The generator shown in Fig. 9-21 also has an auxiliary bearing system. This system consists of small ball bearings mounted on the generator shaft adjacent to the main bearings together with insulated sensing circuits installed in the outer race housing. A radial clearance of approximately 0.002 in [0.0051 cm] is provided between the outer race of the auxiliary bearing and its housing, so that in normal operation, the auxiliary bearing turns as a unit with the shaft and does not contact the stationary housing. When the main bearings begin to wear or fail in such a way that the shaft becomes displaced from its normal rotational center by 0.002 in, the outer race of the auxiliary bearing will contact its stationary housing, thus bridging the gap between the insulated detector ring and adjacent steel surface. This grounds the detector circuit, thus energizing a light to alert the flight engineer that a generator failure is imminent. The auxiliary bearings are designed so they alone can support the generator rotor for several hours of operation, thus making it possible to avoid a generator failure in flight.

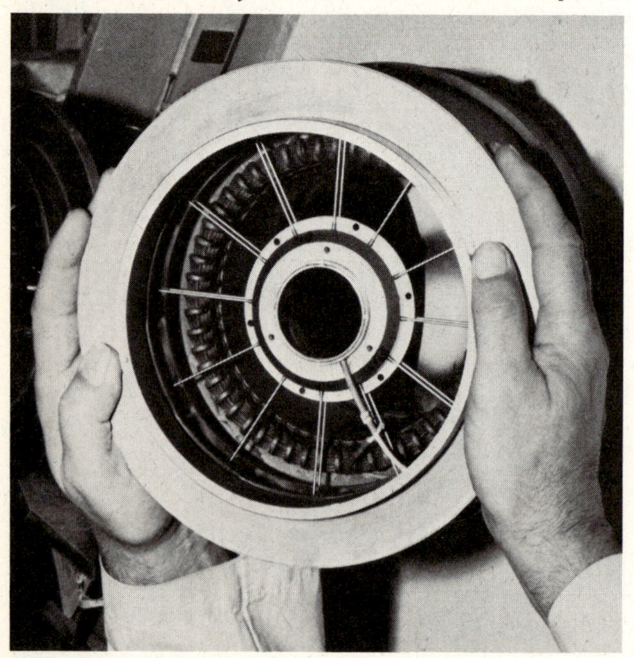

Fig. 9-24 Bearing support spokes. (Bendix Corp., Electric and Fluid Power Division)

An important feature of the generator design is a 24-spoke bearing support used at the drive end. This is shown in Fig. 9-24. The spokes are spaced 30° apart on both sides of the support, which holds an oversized bearing to accommodate abnormal loads. The spoke arrangement allows the generator rotor to move longitudinally with temperature changes.

Another feature of the generator is an oil-lubricated drive shaft that contains an oil reservoir. This reservoir furnishes lubrication for both the drive shaft and the rotor shaft splines. The design increases the life of the shaft.

VOLTAGE REGULATION AND GENERATOR CONTROL

The purpose of this section is to explain some of the factors involved in the paralleling of three-phase ac generators in a manner which will permit each generator to function most efficiently without appreciably affecting the operation of the others in the system. The paralleling of dc generators is a comparatively simple operation involving only the voltage control of the generator output. With ac systems, the voltage, frequency (controlled by generator rpm), load current in each phase and from each generator, reactive load division, and other elements must be controlled.

The voltage control for each generator is not difficult. As mentioned previously, the exciter-field current for the alternator exciter is controlled by the voltage-regulator section of the **generator-control unit (GCU)** and is increased or decreased according to the requirements of the alternator. As exciter-field current is increased, exciter voltage increases, and this strengthens the alternator field. When the field strength increases, the output voltage increases.

The frequency for the output of each generator could be controlled easily if all the generators were driven by the same engine. Since this is not true, and it is impossible to keep each engine at exactly the same rpm, it is most difficult to keep all generators at the same frequency and synchronized with one another at a frequency of 400 Hz. To do this, a constant-speed drive (CSD) is employed with each generator on each engine. The CSD has the ability to convert a variable engine speed to a constant rpm. The principles by which this is accomplished will be explained later in this chapter.

Alternating-current power systems for large aircraft require generator-control systems and units which are highly complex, solid-state devices that are beyond the scope of maintenance technicians' unless they are electronics specialists. We shall not, therefore, attempt to study in detail all the functions of generator-control system units. We shall, however, examine one of the earlier, less complex, systems to provide a general understanding of the many factors involved in controlling an ac power system for large jet aircraft.

EARLY-TYPE GENERATOR CONTROL

The voltage-regulator systems for the first large jet aircraft utilized **saturable reactors,** or **magnetic amplifiers,** commonly called **magamps,** as the controlling devices for exciter current to the exciter fields. Their disadvantage when compared with the solid-state type control units is in weight. The solid-state units weigh but a small fraction of the weight of the magamp regulators.

Even though the technician is not likely to encounter a generator control unit that utilizes magamps, such a system will be described here because it will give the student an understanding of the many variables involved in the control of power systems utilizing synchronized ac generators operating in parallel. The functions controlled by this system must also be controlled by a system utilizing a solid-state GCU.

SATURABLE REACTOR

A magamp, or magnetic amplifier, utilizes the principle of a **saturable reactor,** which is nothing more nor less than a special type of inductance coil with a magnetic core capable of being easily saturated with magnetism. It has been explained previously that inductive reactance is the result of a changing magnetic field in the core of an inductance coil. When an ac voltage is connected to the terminals of such a coil, a continuously changing magnetic field induces a reactive voltage which opposes the changes in the magnetic field. This opposition is called inductive reactance and effectively reduces the ac flow; hence, it is measured in ohms.

If we place two separate windings on a magnetic core as shown in Fig. 9-25 and pass sufficient direct current through one of the windings to saturate the core magnetically, we greatly reduce the inductive reactance of the other winding. This is because there is no change in the magnetic field as long as the core is saturated.

The magnetizing curve for an electromagnet is shown in Fig. 9-26. The curves combine to form a **hysteresis loop;** the horizontal scale *HH'* represents the flux density. When a current starts to flow in a coil, the core is not

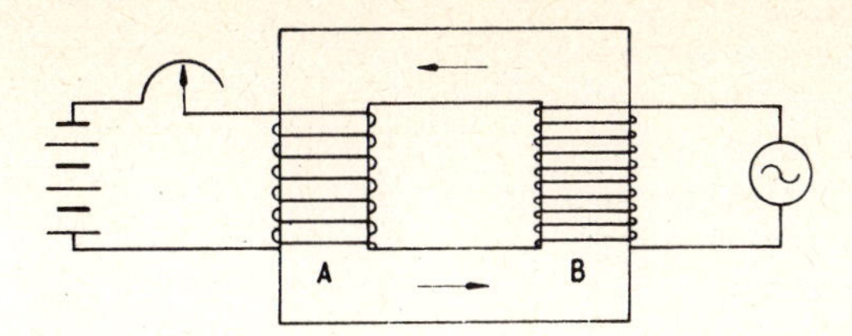

Fig. 9-25 A saturable reactor.

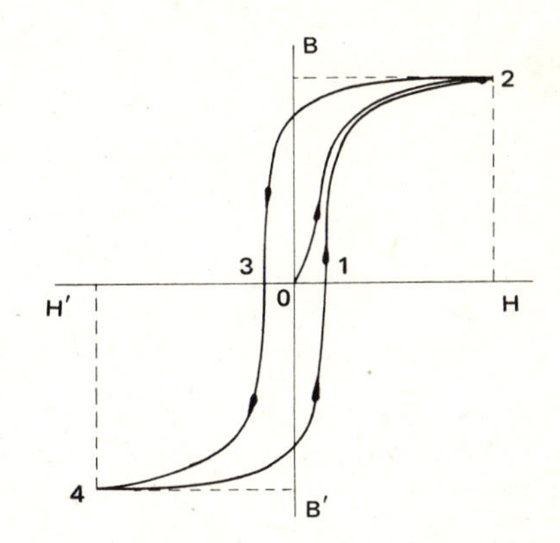

Fig. 9-26 Hysteresis loop.

magnetized instantly but follows a curve somewhat like 0—2 in Fig. 9-26. If the current is reversed, the magnetization follows the curve 2—3—4. If the current reverses again, the magnetization follows 4—1—2. The reason that the magnetization does not follow the same curve in both directions is that hysteresis causes magnetization to lag behind the magnetizing force.

The points 2 and 4 of the hysteresis curve represent a condition of saturation; flux density in the core cannot increase beyond this value.

If there is no direct current flowing in winding *A* of Fig. 9-25, the alternating current in winding *B* will be greatly reduced by inductive reactance. Now, if we apply direct current to winding *A*, a steady magnetic field will appear in the core, and this dc field will reduce the amount of change which can take place in the ac field and thus reduce the inductive reactance. This, of course, results in an increase of alternating current through winding *B*. If we increase the direct current to a point at which the core is saturated, there will be no appreciable inductive reactance and the maximum alternating current will flow in winding *B*.

The characteristics of a saturable reactor make it possible to control current flow in one circuit by changing the current flow in another circuit. For example, by varying the direct current in winding *A* of the reactor in Fig. 9-25, we can vary the alternating current through winding *B*.

CONTROLLING CIRCUITS FOR A MULTIPLE-ALTERNATOR SYSTEM

A voltage-regulating system and power-controlling system used with a multiple-alternator electric power system in a large aircraft must sense a variety of parameters (quantities or measurements) and adjust these parameters to keep the system in balance. In the system to be described here, the following are the circuits we must consider:

- Voltage reference
- Line-voltage sensing
- First-stage magamp
- Second-stage magamp
- Bias power supply
- Starter relay
- Feedback circuits
- Underfrequency compensation
- Reactive-load division

VOLTAGE REFERENCE

The voltage-regulating system continuously compares an exact reference voltage with the line voltage of the three-phase alternator to provide a signal for increase or decrease of line voltage. The voltage reference circuit utilizes a gas-filled voltage-regulator tube to establish the reference voltage. A simplified voltage reference circuit is shown in Fig. 9-27. Three-phase current at 225 V is

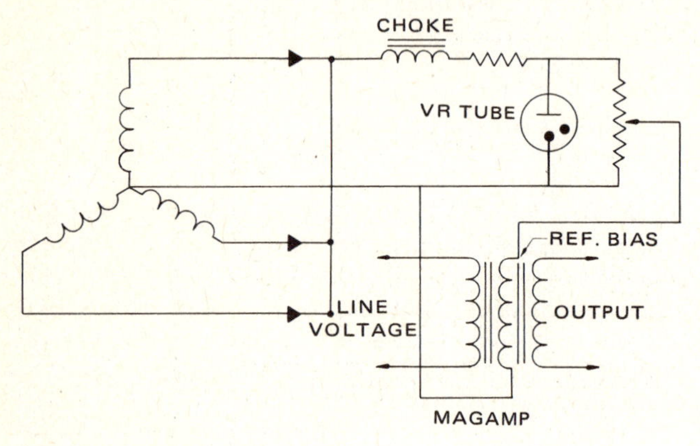

Fig. 9-27 Voltage reference circuit.

taken from the bias transformer and passed through three rectifiers to provide half-wave rectification. The resulting pulsating direct current is passed through a choke coil and resistor filter and connected across a gas-filled tube. When this tube fires, it maintains a constant 105 V across its terminals. This voltage is applied across a bleeder system (resistors connected as a voltage divider), and 65 V is taken from this system as a reference voltage for the first-stage magamp (magnetic amplifier). This provides the exact voltage reference for the regulator system.

LINE-VOLTAGE SENSING

The circuit which senses the main-line voltage is also called the **high-phase sensing** circuit because it senses the voltage of the highest-voltage phase of the main line. The sensing transformer is a three-phase type with a Y-connected primary (see Fig. 9-28). Three separate center-tapped secondary windings provide full-wave rectification of each phase. The three rectified voltages are passed through a choke coil and resistor filters to become steady direct current. These three dc voltages are connected to separate blocking diodes. Each diode will pass current only when its voltage is higher than the other two. The output terminals of the diodes are connected to a common conductor which leads to the first-stage magamp winding.

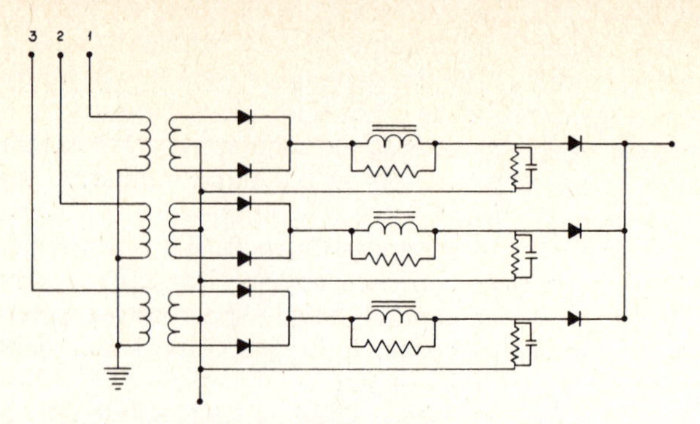

Fig. 9-28 Line-voltage sensing-transformer circuit.

FIRST-STAGE MAGAMP

In the first-stage magamp, also called the magnetic preamplifier, of the voltage-regulator system, the bias is supplied from the voltage reference circuit and applied through a series resistor to the bias winding. This establishes the correct level of saturation for the operation of the magamp.

The control winding of the magamp is fed by the difference voltage between the reference voltage coming from the voltage divider across the voltage-regulator tube and the voltage coming from the high-phase (line-voltage) sensing circuit. The reference voltage is 65 V, and the high-phase circuit produces approximately 97 V. The difference is then approximately 32 V.

During the operation of this system, a decrease in generator voltage will cause a decrease in the high-phase sensing-circuit output, but the 65-V reference voltage will remain constant. Thus there will be a decrease in the difference voltage which will produce a corresponding decrease in current flow through the control winding of the preamplifier. The effect of this decrease is to reduce the magnetic saturation of the core and thus decrease the output of the power circuit.

SECOND-STAGE MAGAMP

The output of the power circuit of the first-stage magamp is connected to the control winding of the second stage. The control winding in this stage opposes the field of the bias circuit. Hence, when current flow decreases in the control winding, the magnetic saturation is increased, and the power-circuit output to the generator exciter field also increases. This results in a build-up of exciter current through the ac generator rotor and a corresponding increase in generator voltage. Schematic circuits for the two magamps are shown in Fig. 9-29, which illustrates a complete voltage-regulator-system circuit.

BIAS POWER SUPPLY

The bias-power-supply section of the voltage-regulator system furnishes bias for the second-stage magamp and power for the **starter relay.** (*Note:* This starter relay has no connection with the engine starter.) The power is taken from one of three separate three-phase secondary windings of a three-phase transformer connected to the main line. The bias-power-supply section produces 29 V which is fed through a three-phase full-wave rectifier. The output of the rectifier is divided between the bias winding of the second-stage (power) magamp and the starter relay.

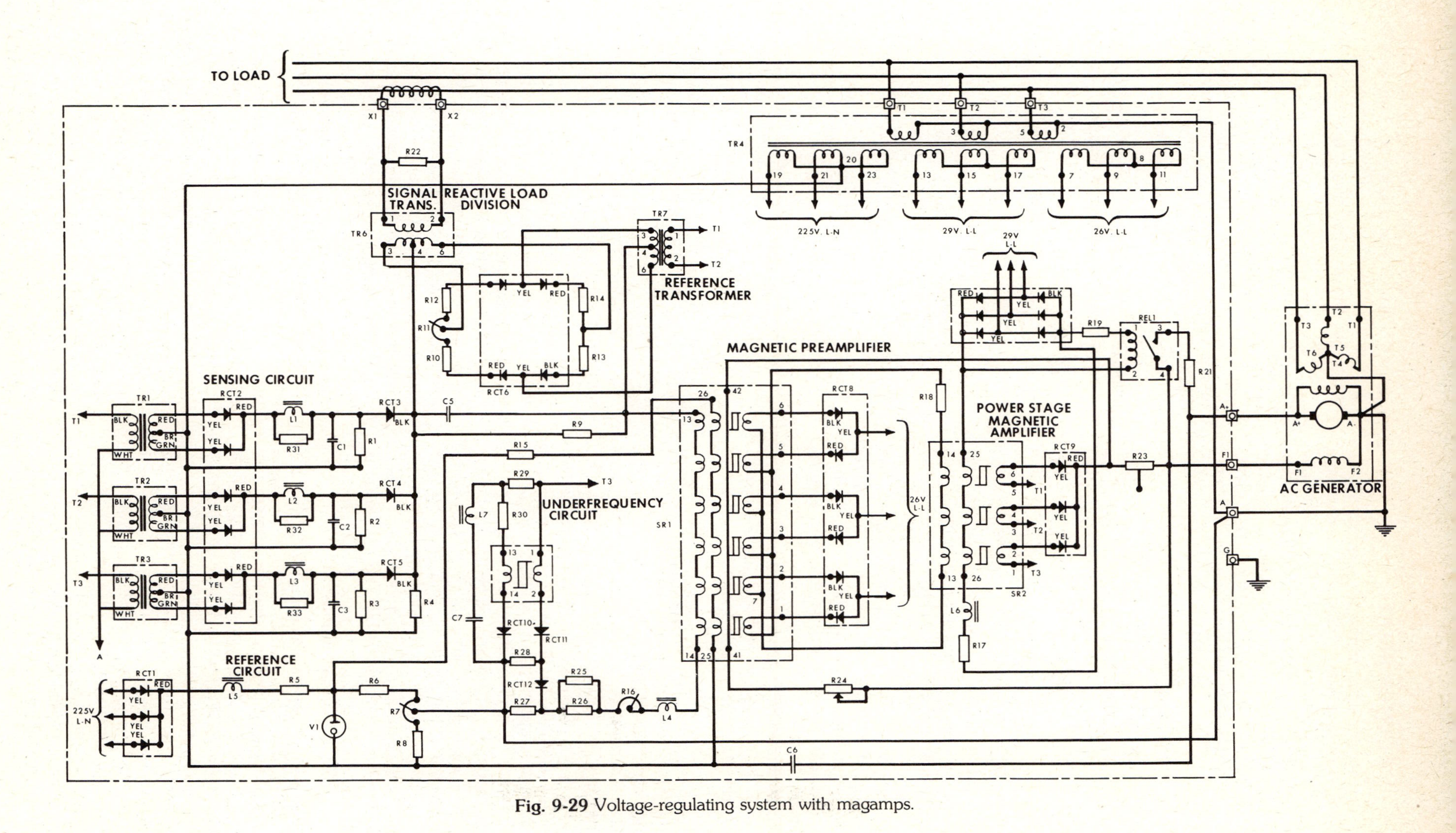

Fig. 9-29 Voltage-regulating system with magamps.

STARTER RELAY

When the ac generator is first started, the output voltage is too low to provide a rapid buildup. The starter relay (*REL* 1 in Fig. 9-29) is connected to allow the exciter to be self-energizing; that is, the exciter output is fed through the starter relay back to the exciter field. As soon as the ac output attains a satisfactory level, the starter relay opens, and the exciter field is then fed from the second-stage magamp.

FEEDBACK CIRCUITS

There are two feedback circuits in the voltage-regulator system. The **positive-feedback** circuit is designed to increase the sensitivity of the system, and the **negative-feedback** circuit serves to stabilize the operation of the regulator.

The positive-feedback system utilizes a resistor (R_{23}) in the exciter-field circuit. The voltage drop across this resistor increases in proportion to the exciter-field current and is used to assist the bias in the first-stage magamp. Thus, as load increases, the positive feedback increases and drives the magamp nearer to saturation. This builds up the output of the first-stage magamp and increases the output of the regulator to the exciter field. An adjustable series resistor is connected in the feedback circuit to limit current flow through the feedback winding and to provide for adjustment.

It will be noted that in Fig. 9-29 a capacitor, C_6 is connected between *A* of the exciter and one side of the voltage reference circuit. During steady operation, this circuit has no effect because the capacitor is charged to the steady values in the circuit and no current can flow in either direction. When a sudden load is applied to the system, a voltage drop occurs at both the ac generator and the dc exciter. This is reflected as a negative transient voltage at the capacitor, and because it is a changing voltage, it is "passed through" the capacitor to the voltage reference circuit. (Remember that current does not actually pass through a capacitor, but the effect of a change at the capacitor is passed along the circuit.)

The voltage drop in the main line will be sensed by the high-phase sensing circuit, which will then attempt to increase the regulator output to the exciter field. But the negative-feedback signal will oppose the signal from the high-phase sensing circuit and prevent too sudden a fluctuation in the voltage-regulator circuit. The final result is that sudden load changes do not cause violent fluctuations and oscillation in the regulator system.

UNDERFREQUENCY COMPENSATION

The purpose of the underfrequency-compensation circuit is to provide for a decrease in generator voltage when engine speed is reduced to a point at which the generator drive unit is turning the generator at below 350 Hz. If the voltage-regulator system were regulated to maintain full voltage at very low frequencies, damage would be caused to units of the system because they would be forced to operate beyond safe limits. At normal frequencies (around 400 Hz) and above 350 Hz, the underfrequency-compensation circuit has no effect.

The underfrequency-compensation circuit operates by means of a saturable reactor having a **reset winding** and a **gate winding,** both connected to the same phase-to-neutral voltage. The reactor is connected in the circuit with rectifiers arranged so that the reset winding will carry current in one direction during one half cycle and the gate winding will carry current in the opposite direction during the other half cycle (alternation).

It will be noted that in the underfrequency circuit in Fig. 9-29, both windings of the reactor are connected to phase 3 of the main line. During normal operation, current flow through the gate winding passes through a rectifier and the resistor R_{28} to ground. Current flow through the reset winding passes through two resistors and a rectifier. In parallel with the reset circuit are a choke coil and a capacitor in series. This circuit is designed to draw more current through the resistor R_{29} and reduce current flow through the reset winding, thus increasing the sensitivity of the circuit.

It must be noted that current flow in the reset winding establishes a field opposite to that of the gate winding; hence, a decrease in reset current will tend to increase current through the gate winding.

When the frequency of the generator decreases below 350 Hz, the gate-winding current increases rapidly, flowing through R_{28}. The voltage across R_{28} increases to a point at which the zener diode RCT_{12} begins to conduct in reverse and allows a current flow through R_{27}. (The zener diode is designed to act as any other rectifier until a certain inverse voltage level is reached. It will then begin to conduct in the reverse direction.) The current flow through R_{27} produces a voltage signal which is added algebraically to the control-winding reference voltage and the high-phase sensing voltage. The effect of the underfrequency signal will be to increase the output of the first-stage magamp; this increase will decrease the excitation output of the second-stage magamp, and the voltage of the ac generator will thus be reduced.

REACTIVE-LOAD DIVISION

It has been explained in a previous chapter that inductance in an ac circuit causes current to lag the voltage and capacitance causes current to lead the voltage. It was also explained that when the voltage and current are out of phase, negative power is developed. When two or more alternators are operated in parallel, a certain amount of **reactive power** exists because of inductance in the circuits. It is important that this reactive load be balanced among the generators because an unbalanced reactive load will increase the possibility of one generator's pulling out of synchronization with the others; it may therefore cause damage by overheating the generator in which too much reactive current is flowing. It should be understood that when an ac generator in a parallel system is underexcited, it will draw reactive current from the others in the system and will, in effect, be driven by them. When this condition is severe, the generator may become overheated. Ac power systems are designed so that an excessive unbalance of reactive power will cause the fault-protection circuits to disconnect the offending generator from the system.

Real power in an ac system is expressed in kilowatts (kW) and is governed by the driving torque applied to the generator. Reactive power is expressed in kilovolt-amperes reactive (kvar) or volt-amperes reactive (var) and is controlled by generator excitation. The degree of excitation is determined by the amount of direct current flowing in the rotor (rotating-field) windings of the alternator.

Real current is that portion of the current in an ac system which is in phase with the voltage, and reactive current is that which is out of phase with the voltage. Reactive power is determined by the amount of reactive current flowing in the system.

It is now apparent that the parallel operation of ac generators requires that both driving torque and excitation be accurately controlled to provide suitable division of real and reactive loads among the generators. In this section we are concerned with the method by which reactive load is sensed and with how the excitation is governed to balance the reactive load among the generators.

The circuit for reactive-load division is designed to compare the reactive load of each generator with the others and then provide signals to each voltage-regulator system for increasing or decreasing the load as needed. The first requirement for such a circuit is to make the comparison of reactive loads for each generator in the system. This is accomplished by means of the **reactive-power-sensing equalizing loop,** which is illustrated in Fig. 9-30. A current transformer (CT) is connected in phase C of each generator output, and these transformers are connected together in an equalizing loop as shown. This loop senses the reactive load of each generator and compares it with the average of all the loads. If all the loads are equal, no voltage will appear across any of the R_{22} load resistors, and hence no correction will be called for.

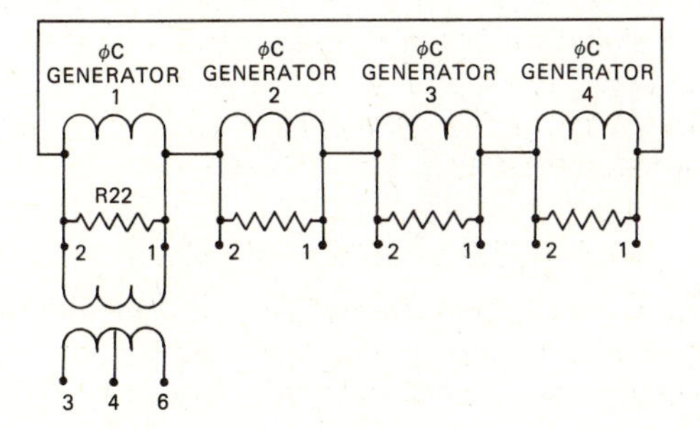

Fig. 9-30 Reactive-power sensing loop.

The section of the reactive-load circuit which makes the comparison of the reactive loads and the reference voltage is called the **ring demodulator** or discriminator. In Fig. 9-29, this section includes resistors R_{10}, R_{11}, R_{12}, R_{13}, and R_{14} connected in a series bridge-type circuit with four rectifiers. The reference voltage for the circuit is taken between phases A and B of the three-phase circuit through the primary of the reference transformer TR_7. The voltage across the secondary of this transformer will normally be 90° out of phase with the voltage across the reactive-load-signal transformer TR_6. The secondary windings of TR_6 and TR_7 are center-tapped and connected together through resistor R_9. When the reactive loads are in balance, there will be no current flow through R_9 and no voltage signal from R_9.

A mathematical analysis of the ring demodulator of this current is beyond the scope of this text. It may be stated, however, that the ring demodulator constantly compares the signal from phase C, if any, with the reference signal. If the two voltage signals are 90° out of phase, there will be no corrective action through the voltage-regulator system. If the phase angle is less than 90°, a positive voltage will be developed across R_9. If the angle is more than 90°, the signal across R_9 will be negative. The signal from R_9 is connected across the control winding of the first-stage magamp. This can be done because the internal resistances of both the voltage control circuit and the reactive-load circuit are much higher than the resistance of the control-circuit winding. The effect of the signal voltage from R_9 is to adjust the excitation of each alternator to maintain a balanced reactive-load division among all the alternators paralleled in the system.

SOLID-STATE GCU

A solid-state GCU for large aircraft is shown in the photograph of Fig. 9-31. This unit combines the functions of voltage regulation, power control, contactor and relay control, and a variety of protective functions. The GCU shown is completely solid-state (static) except for the relays, which are hermetically sealed. The circuitry is mounted in a standard case described as a $\frac{3}{8}$ short ATR

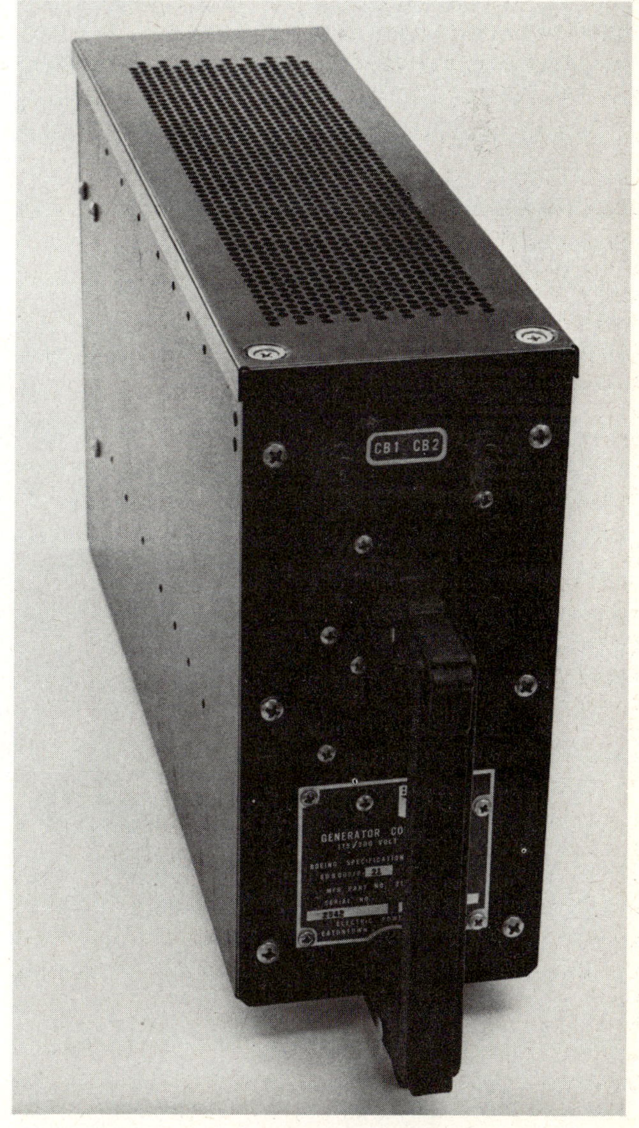

Fig. 9-31 A sold-state generator control unit. (Bendix Corp., Electric and Fluid Power Division)

case. It is easily fitted to or removed from its installation in a standard ARINC (Aeronautical Radio, Inc.) rack by use of the handle on the front of the case. A latch device on the handle holds the unit securely in place and also serves as a jack during the engaging or disengaging of the external connector.

The major circuit assemblies of the GCU are shown outside the case in Fig. 9-32. It will be noted that the individual circuits are mounted on printed circuit boards that can be easily plugged in or removed from the case. This type of construction greatly simplifies maintenance and troubleshooting. Circuit boards having defects can be quickly removed and replaced with functional circuits.

When the circuit boards (modules) are installed in the case, they are interconnected by means of circuitry in the bottom of the case. This circuitry is shown in Fig. 9-33.

VOLTAGE-REGULATOR SECTION

Voltage regulation in the GCU shown in Fig. 9-31 is accomplished primarily by means of a zener diode connected in a circuit with a voltage divider system and other controlling elements. In addition to maintaining a constant voltage of 115/200 V, three-phase average, the voltage regulator also provides for equal division of the reactive portion of the load when the generators are operating in parallel and limits the output current to 635 A maximum. It further limits the demand of the generator to 240 hp (179 kW).

Reactive load is sensed by means of current transformers in the A phase of each generator. If the reactive load is evenly divided, no correction signal will be developed. When reactive load is not evenly divided, a signal will be

Fig. 9-32 Printed-circuit boards employed in solid-state GCU. (Bendix Corp., Electric and Fluid Power Div.)

Fig. 9-33 Interconnecting circuitry in bottom of GCU. (Bendix Corp., Electric and Fluid Power Div.)

produced that causes the regulator to reduce the excitation of any generator carrying a high portion of the load and increase the excitation of generators carrying a low portion of the reactive load.

CONTROL SECTION

The control section of the GCU controls various contactors and relays so they are closed only when satisfactory conditions exist, and it opens them automatically when designated conditions exist. The following control functions are handled by the control section:

- Generator field control and indication
- Generator circuit-breaker (GCB) control
- Bus-tie circuit-breaker (BTB) control
- Essential-power relay control
- Automatic paralleling
- Anticycling
- Bus-tie circuit-breaker automatic reclose

PROTECTIVE FUNCTIONS

Although the control functions mentioned in the foregoing paragraph may be considered protective in a number of respects, the GCU exercises other functions for protection when system faults occur. These protective functions prevent damage to generators and system components that would occur without the actions of the protective system. The following protective functions are provided:

- Underspeed of any generator
- Overvoltage
- Undervoltage
- Differential fault
- Overexcitation
- Underexcitation
- Difference current
- Overfrequency
- Open phase

The protective system automatically disconnects and deactivates any generator in the system when any of the listed faults exist beyond a predetermined length of time.

FAULT ANNUNCIATION

The GCU system described here is designed to provide command signals to an external maintenance annunciator to identify certain system components that have developed faults. Fault annunciation is provided for the following line-replaceable system components:

- Generator
- GCU
- CSD
- Load controller
- Generator feeders

ELECTRIC POWER SYSTEM FOR A LARGE AIRLINER

A complete description of the electric power system for the Boeing 747 or any other airliner is much too extensive to include in a general text on aircraft electric systems; however, a general description of the system and its components can be provided to give the student an understanding of the complexity of such a system and the functions of the various units. The student should learn the names of the components of the systems and the purpose of each unit.

SYSTEM DIAGRAM

The drawing of Fig. 9-34 is a single-line diagram of the power system for the Boeing 747 aircraft. It is not a circuit diagram, but it does show the relationships existing among the principal components of the system. The abbreviations used in the drawing may be defined as follows:

APB	Auxiliary power-unit circuit breaker
APU	Auxiliary power unit
BPCU	Bus power-control unit
BTB	Bus-tie circuit breaker
CSD	Constant-speed drive
CT	Current transformer
DP	Differential power
ESS PWR	Essential power
GCB	Generator circuit breaker
GCU	Generator-control unit
GEN	Generator
SSB	Split-system circuit breaker
XPC	External-power contactor

The foregoing abbreviations are not all that technicians may encounter; however, it is well that all those listed be remembered. In technical manuals, the abbreviations are often used without explanation, and it is necessary for technicians to know the terms and abbreviations if they are to understand what they are reading.

In the diagram of Fig. 9-34, it will be noted that the four main generators and the two APU generators may be

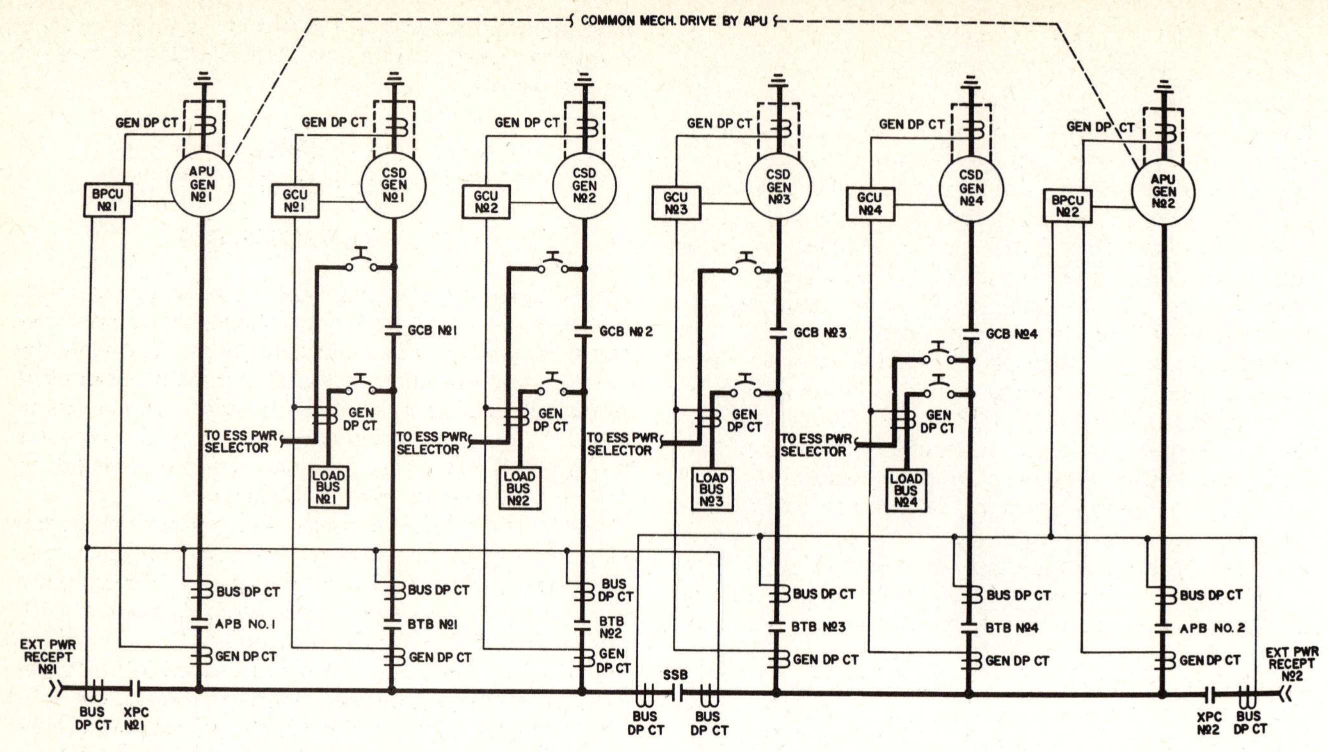

Fig. 9-34 Single-line diagram of the Boeing 747 electric power system. (Boeing Co.)

connected to the main bus. It is not possible, however, for the APU generators to be connected to the bus at the same time that the main generators are connected. The bus-tie breakers and the auxiliary-power breakers (APB No. 1 and APB No. 2) will not allow simultaneous operation.

PRIMARY GENERATING SYSTEM

As previously stated, the main generators for the 747 power system are three-phase, 60 kVA, 400 Hz, brushless alternators. These are Bendix 28B263-13 units in current systems. It must be emphasized that modified versions of these units or newer designs may be installed at any time to increase efficiency or dependability or to reduce weight. For example, liquid-cooled generators have been manufactured, and these units can be installed to reduce weight.

Each of the main generators is connected to its generator bus through the GCB. Each main generator bus may also be connected to the other power sources through its BTB.

The generator buses are tied together through a symmetrical, two-section synchronizing bus. The sections are connected to each other through the SSB to provide a four-channel parallel system, or the SSB may be left open to allow operation of two separate systems, each consisting of two generators operating in parallel.

The two APU generators are driven by a single gas-turbine engine. These generators are rated at 90 kVA because extra air cooling is provided, thus permitting them to operate at higher loads without overheating. As mentioned, the APU generators cannot be operated in parallel with the main generators or with each other.

The ac power system of the Boeing 747 aircraft can also be supplied by ground units through two 250-A external power receptacles located near the nose gear. An XPC is associated with each external power receptacle to connect it to the synchronizing bus. Ground power sources cannot be paralleled with each other or with the main or APU generators.

MAIN AC LOAD BUSES

Each of the four main ac load buses is connected directly to its associated bus by a thermal circuit breaker. Individual loads are likewise connected to the buses through thermal circuit breakers. Physically, the main load buses consist of the common terminals of the respective circuit breakers and the wiring interconnecting these terminals.

All 115-V ac loads, except essential, standby, ground service, ground handling, and galley loads, are connected to the main ac load buses. These buses are divided into several sections so that some loads are supplied from portions of buses located in the below-deck main power center and other loads are supplied from portions of buses located in the cockpit.

ESSENTIAL BUS

Essential power is that which is necessary for the safe operation of the aircraft although many circuits may be inoperative. The essential-power buses can be connected on the generator side of GCB's 1, 2, or 3 or to the main No. 4 ac bus by means of the essential-power selector relays. Normally the essential bus is supplied from the main ac bus No. 4.

The **standby ac bus** supplies 115-V equipment needed for safe flight and landing. During normal operation, it is supplied from the essential ac bus. In the event of loss of essential ac power, the standby ac can be supplied by a battery-powered inverter.

DC SYSTEM

A 28-V dc system is provided to supply those loads requiring dc power. This power is supplied by four 75-A transformer-rectifier (TR) units which are energized from the Nos. 1, 2, 3, and essential 115-V ac buses, respectively. Each TR unit powers an associated 28-V dc load bus. The TR's are normally operated in parallel; however, a blocking diode prevents the essential TR from feeding other than the essential dc bus. A 20-A TR unit energized by the ac ground-handling bus supplies dc loads required during ground operation.

In addition to the TR units, a 24-V battery provides a standby source. The battery is of the nickel-cadmium type described previously. The battery permits starting the electric system without an external source, supplies back up electric-system power, and can supply power to minimum navigation, instrument, and communication systems, should all other sources fail. A battery charger energized by the ground service bus maintains full capacity in the battery. The hot battery bus, for those loads requiring uninterrupted power, is always connected to the battery. The battery bus is normally energized by the essential dc bus. In the event of loss of essential dc bus power, however, it is automatically transferred to the hot battery bus. Similarly, the standby dc bus is normally energized by the essential bus but can be transferred to the battery bus.

THE GCU

The GCU for the Boeing 747 main generators is the Bendix 21B73-1-AC, described earlier in this chapter. Four such units, each weighing 7.9 lb [3.58 kg], are used, one for each generator in the system.

BUS POWER-CONTROL UNIT

The BPCU provides voltage regulation, control, and protection of the APU generators and also provides the control and protection of an external power channel and an associated half of the synchronizing bus. The BPCU is similar to the GCU, but has additional functions of external power control and split-system breaker control. The most important function is to control the many contactors and power relays associated with the APUs and external power systems.

AC CONSTANT-SPEED GENERATOR-DRIVE SYSTEM

GENERAL PRINCIPLES

In an ac power system it is usually necessary to maintain a fairly constant speed in the ac generator. This is because the frequency of the ac generator is determined by the speed with which it is driven. It is especially important to maintain constant generator speed in installations in which the generators operate in parallel. In this case it is absolutely essential that generator speed be maintained constant within extremely close limits.

In order to provide constant-speed generator operation in modern ac electric systems, it is common practice to use a constant-speed generator drive such as that manufactured by the Sundstrand Aviation Electric Power Division of the Sundstrand Corporation.

Sundstrand CSD units are manufactured in many designs to fit a variety of applications. For the Lockheed L-1011 airplane, each CSD unit is integral with an ac generator. The complete unit is called an **integrated-drive generator (IDG).** The principle of operation for all CSDs is essentially the same.

The complete CSD system consists of an axial-gear differential (AGD) whose output speed relative to input speed is controlled by a flyweight-type governor that controls a variable-delivery hydraulic pump. The pump supplies hydraulic pressure to a hydraulic motor which varies the ratio of input rpm to output rpm for the AGD in order to maintain a constant output rpm to drive the generator and maintain an ac frequency of 400 Hz.

The external appearance of an integrated-drive generator is shown in the drawing of Fig. 9-35. In this view, the CSD is on the left end of the unit and the generator is at the right end. The generator is cooled by oil spray delivered by the CSD section.

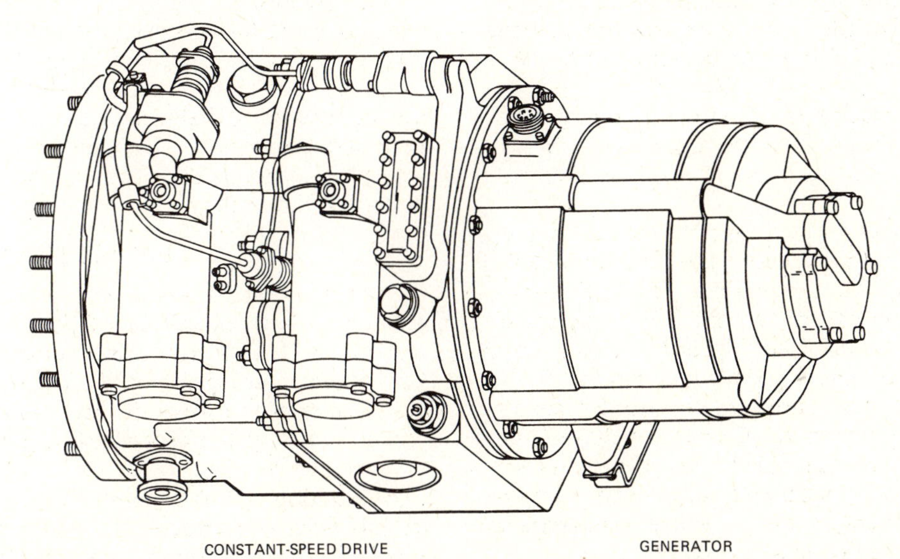

Fig. 9-35 An integrated-drive generator. (Sundstrand Aviation Electric Power, Div. of Sundstrand Corp.)

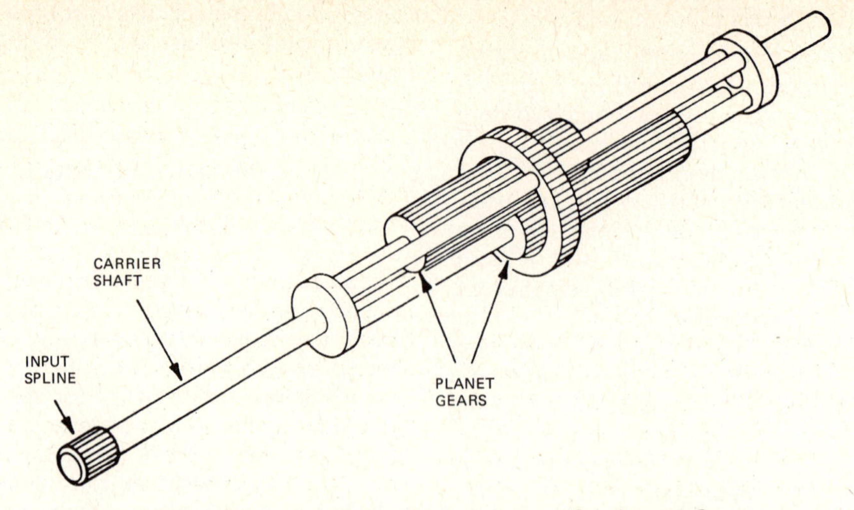

Fig. 9-36 Carrier shaft with planet gears. (Sundstrand Corp.)

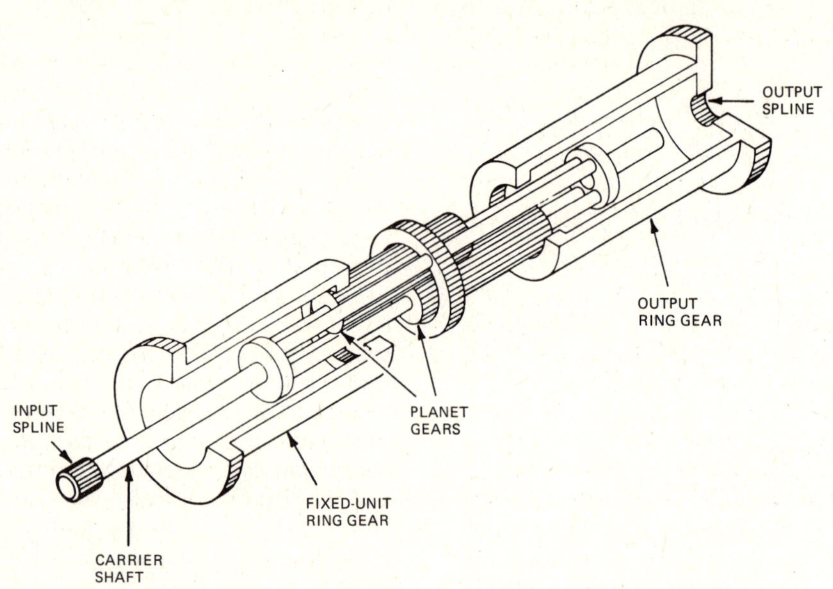

Fig. 9-37 Carrier shaft with fixed unit ring gear and output ring gear.

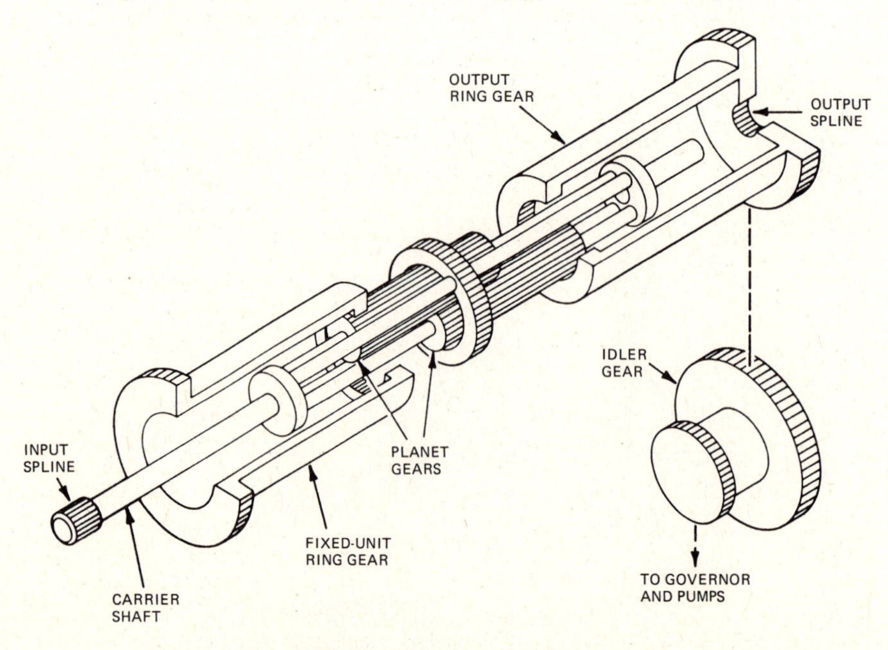

Fig. 9-38 Assembly showing idler gear.

The operation of the CSD may be understood by tracing the mechanical actions that take place from the engine drive through the AGD as it is controlled by the rotation of the hydraulic motor. A drawing of a portion of the AGD is shown in Fig. 9-36. This is the carrier shaft with its associated planet gears. The input spline couples the carrier shaft directly to the engine gearbox so its speed is always proportional to engine speed and it always turns in the same direction.

There are two planet gears on the carrier shaft which mesh with each other. These gears are mounted on their own individual journal shafts independently of the carrier shaft as they orbit about the centerline of the carrier shaft. A fixed ring gear and an output ring gear are mounted to mesh with the two planet gears as shown in Fig. 9-37. The fixed-unit ring gear meshes with the first planet gear, while the output ring gear meshes with the second planet gear. When the fixed-unit ring gear is at zero rpm, the rotation of the carrier shaft causes the first planet gear to rotate while orbiting around the inside of the fixed-unit gear ring. Then the second planet gear, which is constantly in mesh with the first planet gear, transfers its rotation to the output ring gear, causing it to rotate at twice the speed of the carrier shaft. An idler gear, shown in Fig. 9-38, transfers rotation from the output ring gear to the governor, the generator scavenge pump, and the charge and scavenge pump.

Because of the 2:1 ratio between the output ring gear and the carrier shaft, when the fixed-unit ring gear is held stationary and the carrier shaft is turned at 6000 rpm, the output speed will be 12 000 rpm [1256 rad/s]. If input speed is less than 6000 rpm, speed must be added to the input to maintain a constant output speed of 12 000 rpm. To do this, the fixed-unit ring gear is rotated in the opposite direction of the carrier shaft, thereby applying increased rotational speed to the first planet gear in mesh with the fixed-unit ring gear. As the first planet gear turns faster, the second planet gear also turns faster, thereby causing the output ring gear and the output speed to increase. This is known as an **overdrive** condition.

When the input speed to the carrier shaft is above 6000 rpm [628 rad/s], it must be reduced in order to maintain a constant output speed of 12 000 rpm. To do this, the fixed-unit ring gear is rotated in the same direction as the carrier shaft, thus reducing the rotational speed on the first planet in mesh with the fixed-unit ring gear. Because the second planet gear is in mesh with the fixed-unit ring gear, its speed is also reduced, thereby reducing the speed of the output ring gear. This is known as the **underdrive** condition. It can be understood, therefore, that the output speed can be controlled by the rotational direction of the fixed-unit ring gear. Directional control of the fixed-unit ring gear rotation is accomplished by using a variable displacement and a fixed displacement hydraulic unit. This application is shown in Fig. 9-39.

There are two fixed-displacement hydraulic units (motors) in the CSD system and two variable-displacement hydraulic pumps. Each fixed hydraulic unit consists of a cylinder block containing nine equally spaced pistons. The cylinder block is splined to a shaft and gear that is in mesh with the fixed-unit ring gear. The pistons in the cylinder block reciprocate (move in and out) by moving on a fixed wobbler plate that forms an inclined plane. Since the unit is driven by hydraulic pressure, the pistons always move from the top of the wobbler plate toward the bottom. When pressure is applied to the pistons, they will reciprocate, causing the cylinder block to rotate. This is the same principle involved in the operation of many other hydraulic motors. The direction of rotation that the fixed unit assumes is determined by the pressurization of the fixed pistons on a particular side of the fixed wobbler

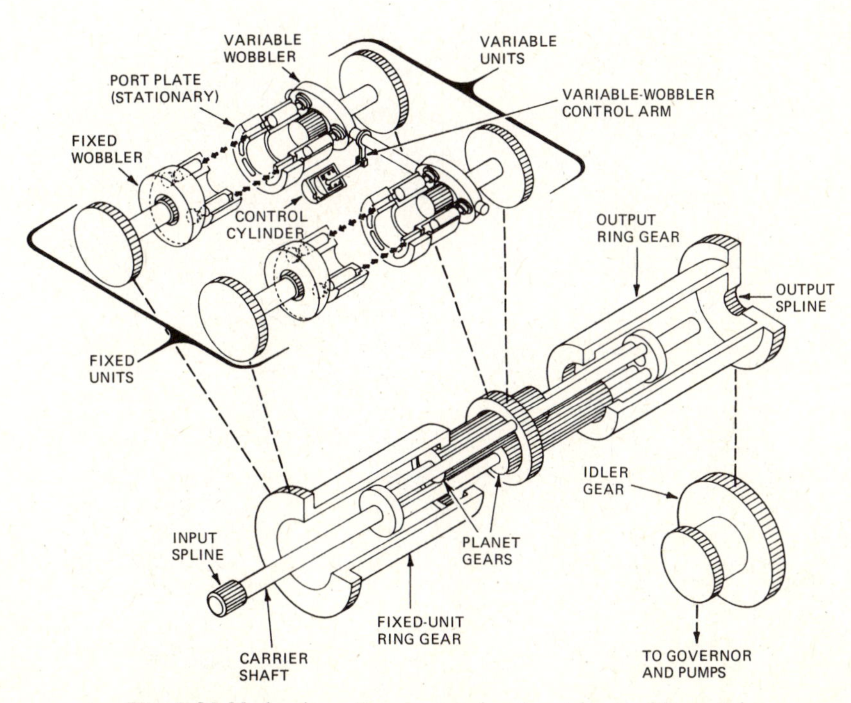

Fig. 9-39 Hydraulic units associated with axial gear differential system. (Sundstrand Corp.)

plate. The pressure being applied to the pistons is supplied by the variable-displacement hydraulic units (pumps).

As explained previously, there are two variable-displacement hydraulic units in the system. Each unit consists of a cylinder block with nine equally spaced pistons, which is the same arrangement as for the fixed units described previously. The difference between the fixed units and the variable displacement units is that the variable displacements have wobbler plates whose angles can be changed, thus varying the hydraulic fluid output. The cylinder blocks of the variable displacement units are splined to shafts that are driven by the carrier shaft. Because the carrier shaft is coupled to the input shaft, the variable hydraulic always rotates in one direction and at a speed proportional to the speed of the engine. The pistons in the variable hydraulic unit also work against the face of the variable wobbler plate. Because the wobbler plate is movable, it can be positioned to an angle opposite or parallel to the angle of the fixed wobbler in the fixed hydraulic unit (motor).

The angle of the variable wobbler is determined by the governor which directs oil pressure to the control cylinder. As the angle of the wobbler plate is increased, the pistons in the cylinder block take longer strokes and pump a greater volume of oil. This increases the rotational speed of the motors that rotate the fixed-unit gear.

When the input drive speed is too low to provide the desired output speed to drive the generator, oil is ported from the governor to the control cylinder. The pressure acting against the control piston in the control cylinder changes the angle of the variable wobbler plate. Hydraulic pressure is then ported to the fixed-unit piston. The fixed unit rotates in a direction opposite that of the variable unit. The fixed-unit ring gear, being in mesh with the fixed unit, rotates in a direction opposite that of the carrier shaft, thereby adding to the output speed. Thus the system is in the overdrive condition. This condition is illustrated in Fig. 9-40.

The underdrive phase of the system is essentially the opposite of the overdrive phase. In this situation, the fixed hydraulic units turn in the same direction as the variable hydraulic units. In the **straight-through** condition, there is no rotation of the fixed-unit ring gear. This is the situation where the input rpm is 6000.

The governor for the CSD system is a spring-biased flyweight-operated hydraulic control valve. Its function is to control porting of transmission charge oil to the control cylinder that changes the angle of the variable wobble plates. The rotating sleeve in the governor is driven by the output ring gear and an idler gear through the governor drive gear. In an overdrive or underdrive condition, the flyweights of the governor move to direct oil to or from the control cylinder to produce the required correction in output drive speed. The system for the operation and control of the CSD is shown in the schematic drawing of Fig. 9-41.

LOAD CONTROLLER

When ac generators are operating in parallel, a means must be devised to assure that each generator is carrying its share of the real load. As explained previously, the GCU controls reactive-load division through excitation of the generator fields; however, real load must be controlled through the power delivered to the generator through the generator drive or CSD.

The load that each generator in a parallel system is carrying is sensed by CTs, one for each generator. The CT senses the current flow in one phase, perhaps phase *A*, for each generator and delivers the information to the load controller. The load controllers are interconnected in a manner to permit them to make comparisons of the load of each generator. If a generator is carrying more

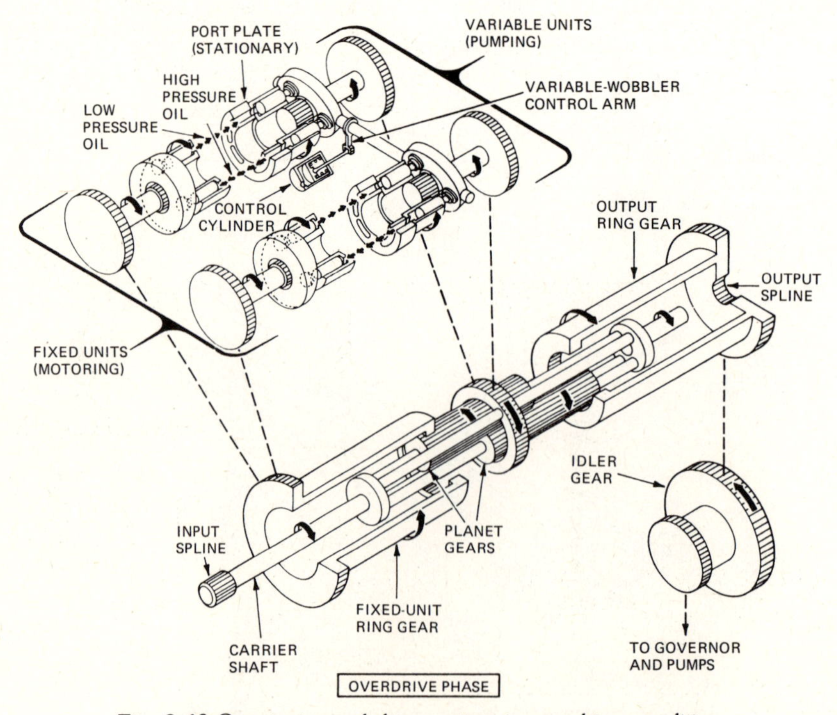

Fig. 9-40 Constant-speed drive system in overdrive condition.

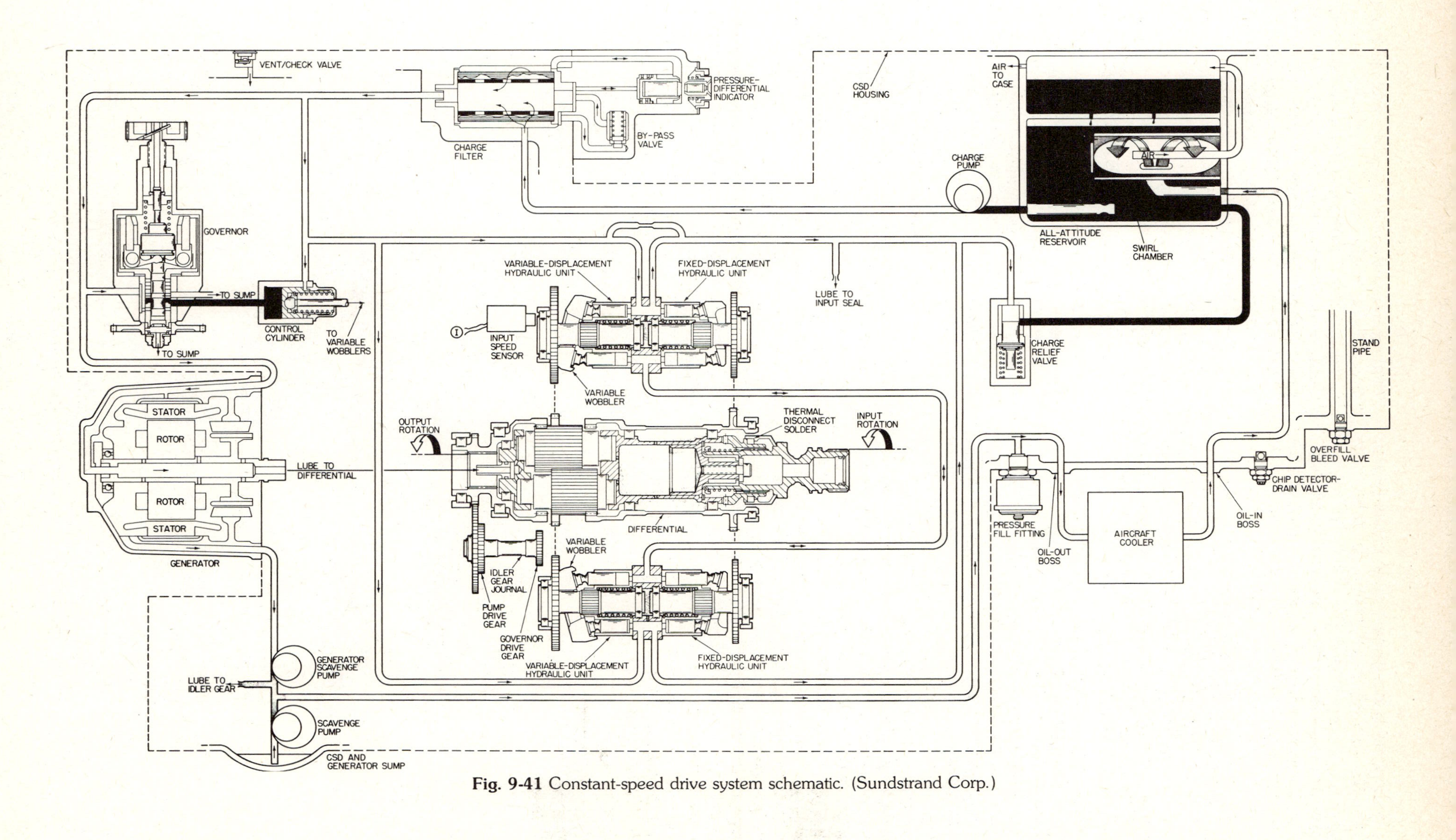

Fig. 9-41 Constant-speed drive system schematic. (Sundstrand Corp.)

than its share of the real load, the load controller for that generator will send a signal to the trim head of the CSD for that generator requiring a slight reduction in the torque driving the generator. The adjustments are made automatically and continuously to keep the real loads for the generators closely balanced.

VARIABLE-SPEED CONSTANT-FREQUENCY POWER SYSTEMS

In an effort to simplify and improve the production of ac power for aircraft and to get away from the need for hydromechanical constant-speed drives, a number of systems have been devised for producing 400 Hz three-phase electric power through electronic circuitry. This has been made by the great advances in solid-state technology developed in recent years.

Variable-speed constant-frequency systems are referred to as VSCF, VASCOF, CFG (constant-frequency generator), and ECEPS (electronic convertor, electric power supply). Basically, the systems employ a generator whose variable speed and variable-frequency power would not be suitable for power needs in aircraft system; however, the variable-frequency power is converted to constant-frequency power by means of solid-state circuitry, and this makes the power suitable for aircraft use.

The drawing of Fig. 9-42 is a block diagram showing the principal elements of an ECEPS system. The brushless ac generator is similar to those described previously; however, since it is driven directly by the engine, its speed and output frequency will vary as engine speed varies. The variable three-phase power is fed to the full-wave crystal-diode rectifier, where it is converted to direct current and filtered. This direct current is fed to the conversion circuitry, where it is chopped into square-wave outputs that are separated and summed up to produce three-phase 400-Hz alternating current.

Variable-speed constant-frequency systems can be designed with separate components or as an integrated unit as shown in Fig. 9-43. In this drawing it will be seen that the generator and static convertor are mounted as a unit on the engine. The GCU which contains the voltage regulator is mounted in the aircraft.

INVERTERS

An **inverter** is a device for converting direct current to alternating current at the frequency and voltage required for particular purposes. Certain systems and equipment in aircraft electric or electronic systems require 26-V 400-Hz ac power, and others require 115-V 400-Hz power. To provide this power, it is often necessary to employ an inverter.

ROTARY INVERTERS

For many years, inverters were simply special types of motor-generators. That is, a constant-speed motor was employed to drive an alternator that was designed to produce the particular type of power required.

A typical rotary inverter is shown in Fig. 9-44. The rotary section of this unit consists of a dc motor driving an ac generator. The rotors of the motor and the alternator are dynamically balanced and are mounted on the same shaft. Fans are also mounted on the shaft to provide for air-cooling.

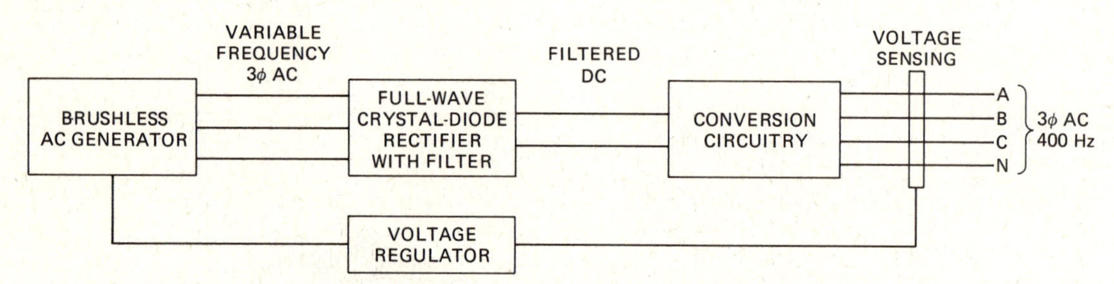

Fig. 9-42 Basic elements of a variable-speed, constant-frequency electric power system.

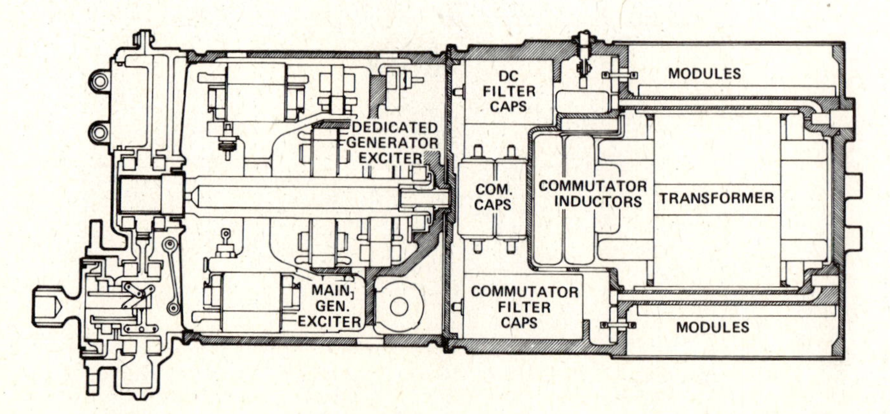

Fig. 9-43 Drawing of an ECEPS generator and static converter designed as an integrated unit. (Bendix Corp.)

Fig. 9-44 A rotary inverter. (Bendix Corp., Electric and Fluid Power Division)

A four-pole, compound, compensated field winding and a wave-wound armature are utilized in the motor. A damper winding in the salient poles of the alternator aids in maintaining output waveshape under single-phase operating conditions.

This particular inverter utilizes in input voltage of 26 to 29 V direct current. At 26 V the amperage is 165, at 27.5 V the amperage is 155, and at 29 V the amperage is 155. The output is 115 V, single phase; 115 V, three-phase; and 200 V, three-phase. Frequency is 400 Hz for all phases.

Maintenance of rotary inverters is similar to that for motors and generators. Maintenance practices are set forth in the manufacturer's maintenance or service manual.

The control box on top of the inverter should not be disassembled in the field. If it appears to be defective, the entire inverter should be sent to a repair shop that is equipped to perform the electrical and electronic work and tests that may be required.

STATIC INVERTERS

A static or solid-state inverter serves the same functions as other inverters. However, it has no moving parts and is therefore less subject to maintenance problems than the rotary inverters. A typical unit is shown in Fig. 9-43.

The internal circuitry of a static inverter contains standard electric and electronic components such as crystal diodes, transistors, capacitors, transformers, etc. By means of an oscillator circuit, it develops the 400-Hz frequency for which it is designed. This current is passed through a transformer and filtered to produce the proper waveshape and voltage. The unit shown in Fig. 9-45 utilizes an input of 18 to 30 V direct current and produces an output of 115 V single-phase alternating current with a frequency of 400 Hz. The weight of the unit is 18.5 lb [8.4 kg].

Static inverters are easily removed for testing. If they require repair, they should be sent to an approved facility that is equipped to perform the work required.

REVIEW QUESTIONS

1. Describe the construction of a typical aircraft alternator.

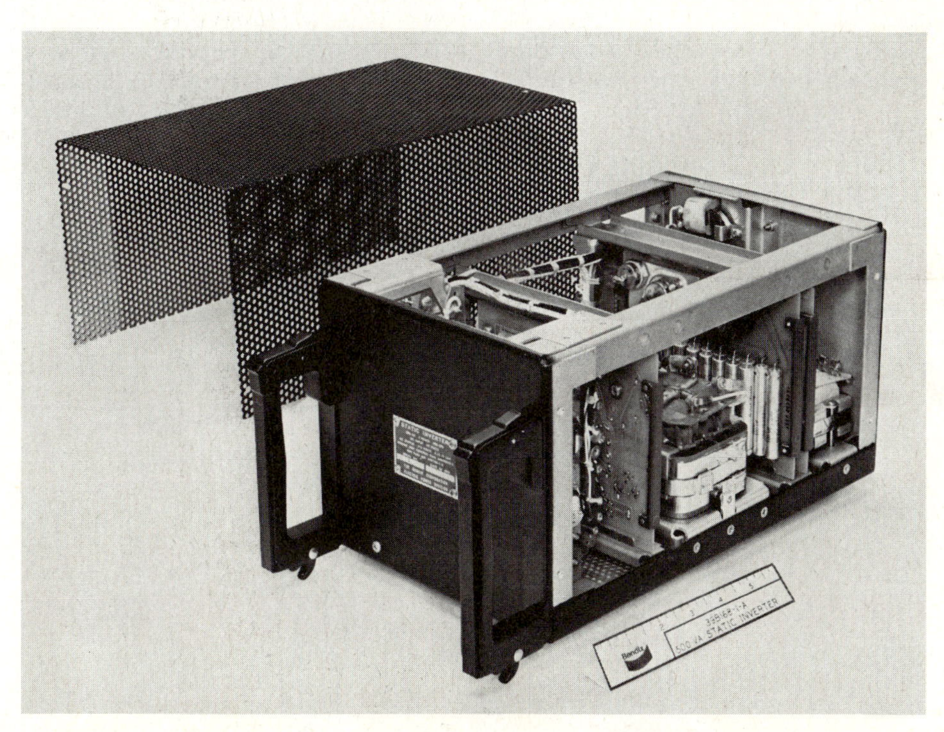

Fig. 9-45 A static inverter. (Bendix Corp., Electric and Fluid Power Division)

2. What are the means by which the rotor of an alternator is excited?
3. How is the production of a three-phase alternating current accomplished in an aircraft alternator?
4. In a light-aircraft electric system, how is the direct current needed for the system obtained when the principal power source is an alternator?
5. What is meant by a *three-phase* stator?
6. Compare a *delta* winding with a Y winding.
7. In a three-phase system, how many degrees separate the phases?
8. What is the composition of a three-phase full-wave rectifier?
9. Describe a three-phase alternator for light aircraft.
10. By what means are alternators for light aircraft driven?
11. Describe the tests that may be made on a small alternator in order to determine that it will operate satisfactorily.
12. What is the advantage of using a *transistorized* voltage regulator?
13. Describe the operation of a transistorized voltage regulator.
14. How is voltage adjusted when using a transistorized voltage regulator?
15. What is the function of the *accelerator winding* on the voltage-regulator coil in a transistorized voltage regulator?
16. Why is it not necessary to include a reverse-current relay in a system using an alternator?
17. Briefly explain the difference between a *transitorized* voltage regulator and a *transistor* voltage regulator.
18. In a transistor voltage regulator, which transistor carries the excitation current for the alternator?
19. What is the function of the *zener diode* in a transistor voltage regulator?
20. What unit is used in a transistor voltage regulator to regulate alternator voltage?
21. If the airplane ammeter shows a heavy discharge when the engine is not running, what may be the causes?
22. What troubles may exist if the alternator does not keep the battery charged?
23. Describe the construction of a brushless alternator for a large aircraft.
24. Explain how field excitation is accomplished in a brushless generator.
25. What design feature makes the brushless alternator self-starting?
26. What factors must be controlled when ac generators are operated in parallel?
27. Describe the operation of a magnetic amplifier. Explain the necessity for controlling the balance of *reactive power.*
28. What is the meaning of *kVA*? *kvar*?
29. What are the advantages of the solid-state GCU?
30. Describe the voltage-regulator function of a solid-state GCU.
31. What are the functions of the control section of the solid-state GCU?
32. List the protective functions of the solid-state GCU.
33. What is the meaning of the following abbreviations: APB, APU, BPCU, BTB, CSD, CT, GCU, SSB?
34. How are the APU generators driven?
35. What unit controls the APUs and external power?
36. What is meant by *essential power*?
37. What is the function of the *standby ac bus*?
38. How are dc circuits supplied with power?
39. How is it possible to keep an alternator at a constant speed when the engine by which it is driven changes rpm?
40. Give a brief explanation of the operation of a CSD.
41. By what means does a CSD sense overspeed or underspeed?
42. By what means does the governor of the CSD sense an overload or underload condition in an alternator?
43. Explain the basic principles of the variable-speed constant-frequency electric power system.
44. What is the advantage of a VSCF system?
45. Explain the purpose of an inverter.
46. What is the advantage of a static inverter?
47. Describe the principle of a static inverter.

CHAPTER 10

Electric Motors

An electric motor is a device which changes electric power to mechanical energy; that is, its function is opposite to that of a generator. Electric motors and generators are very much alike in construction, and some generators are actually used as motors under some conditions. This is true of engine-driven generators which are used as motors to start the engine. Certain turbojet engines employ starter-generators which are constructed with special series field windings for use in starting. These field windings are only energized during starting, and their purpose is to provide a very high starting torque. Usually it is not good practice to use a generator as a motor, because certain features which make a generator more efficient will have the opposite effect when the unit is used as a motor.

Electric motors may be classified in many ways; the number of different types of motors is so great, however, that it would be impossible to describe them with simple classifications. There are a few basic features which are common to all dc motors, and these will help to indicate the type of motor to be used for a specific purpose. Dc motors are described in part by the type of internal winding they have. There are **series-wound, shunt-wound,** and **compound-wound** motors, named according to the relationship between the field coil connections and the armature winding. Motors of all types are usually rated according to horsepower. Usually the data plate will also show the voltage and amperage. Additional information on dc motors includes rpm, type of duty, and some other points descriptive of the motor design.

Ac motors are classified according to horsepower, phase, operating frequency, and type of construction. Usually the power factor is also stated. In any event, all the characteristics of the motor must be considered in making a selection for a particular duty.

Electric motors are used in aircraft, missiles, and spacecraft for many purposes. Among the many units and systems requiring electric motors are engine starters, cowl flaps, intercooler or heat-exchanger shutter or control valves, heat-control valves, landing gear, flaps, trim tabs, flight controls, fuel pumps, hydraulic pumps, vacuum pumps, controllable propellers, gyrostabilizing units, navigation devices, and tracking devices. It is not intended that this text cover the details of all electric motors, but a thorough exposition of motor theory and functions will be given. This should enable the student to work out an understanding of any installation encountered.

MOTOR THEORY

MAGNETIC ATTRACTION AND REPULSION

Electric motors utilize the principles of magnetism and electromagnetic induction. The repulsion of like magnetic poles and the attraction of unlike poles work together to produce the torque which causes a motor to rotate. In the section of this text describing generator theory it was shown that if a conductor is moved across a magnetic field, a voltage will be induced in the conductor. In that case, the movement of the conductor in the field caused a current to flow. Conversely, if a current from an external source is passed through a conductor while it is in a magnetic field, the conductor will tend to move across the field; hence, the flow of current causes a movement of the conductor (see Fig. 10-1).

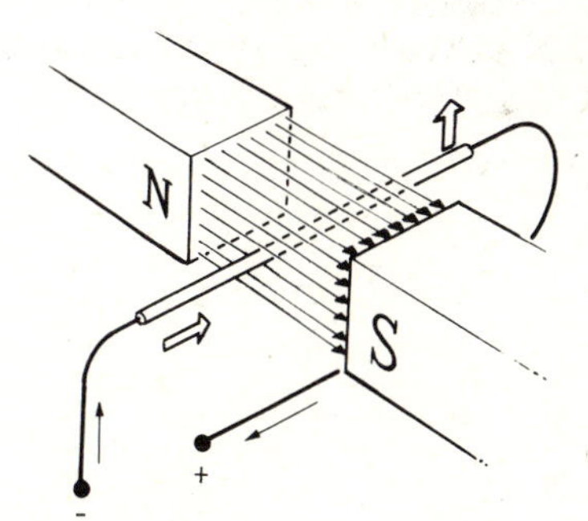

Fig. 10-1 Current-carrying conductor in a magnetic field.

The direction in which a current-carrying conductor in a magnetic field tends to move may be determined by the use of the **right-hand motor rule.** This rule is applied as follows: **Extend the thumb, index finger, and middle finger of the right hand so that they are at right angles to one another as shown in Fig. 10-2. Turn the hand so that the index finger points in the direction of the magnetic flux and the middle finger points in the direction of the current flowing in the conductor. The thumb will then point in the direction of the conductor movement.**

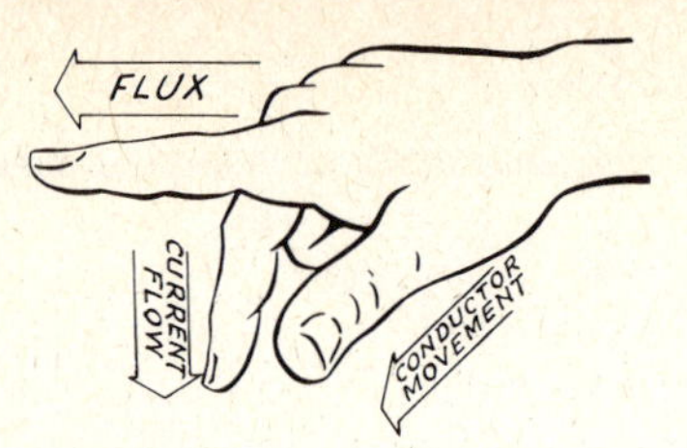

Fig. 10-2 Right-hand rule for motors.

The movement of a conductor in a magnetic field, as described above, is caused by the flux which the current produces around the current-carrying conductor. This flux reacts with the flux of the magnets, thus imposing on the conductor a force which causes it to move. A study of the fundamental principles of electricity will show that a coil of wire will have magnetic polarity when a current is passed through it, and if a soft-iron core is placed in the coil, the result is an electromagnet. If this electromagnet is placed between the poles of a field magnet and is free to rotate, the flux of the electromagnet will react with the flux of the field magnets and produce torque which will cause the electromagnet to turn (see Fig. 10-3*a*). The north pole of the electromagnet is attracted by the south pole of the field magnet and repelled by the north pole. The electromagnet, called the armature, will continue to rotate until it is lined up with the field. At this point it would normally stop because the conditions of repulsion and attraction would be satisfied. In an electric motor, however, the armature polarity is reversed at this point through the action of the commutator which reverses the connections to the input current (see Fig. 10-3*b*). It will be noted that the flux reversal takes place just before the armature becomes aligned with the field, thus causing the armature to continue to rotate as it attempts to line up with the new conditions (Fig. 10-3*c*). The flux reversal in the armature takes place each time the armature becomes nearly aligned; hence, it continues to rotate for as long as electric energy is applied.

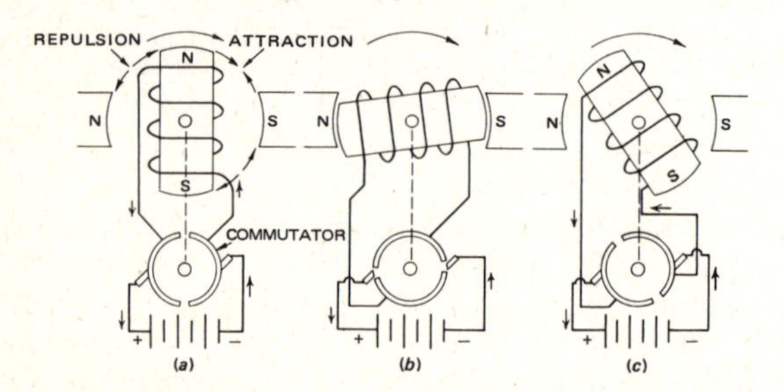

Fig. 10-3 Principle of the dc motor.

A simple motor of the type described above does not deliver a smooth flow of power because the torque is high when the armature is at right angles to the field poles and there is no torque at the moment the armature is in line with the field poles. In order to deliver smooth power, the armature is provided with additional coils so that there will always be a high torque. Figure 10-4 shows a motor with four armature poles. With this arrangement the torque on one set of poles will increase as the torque on the other set decreases, and the motor will deliver a reasonably smooth power output. The addition of more coils

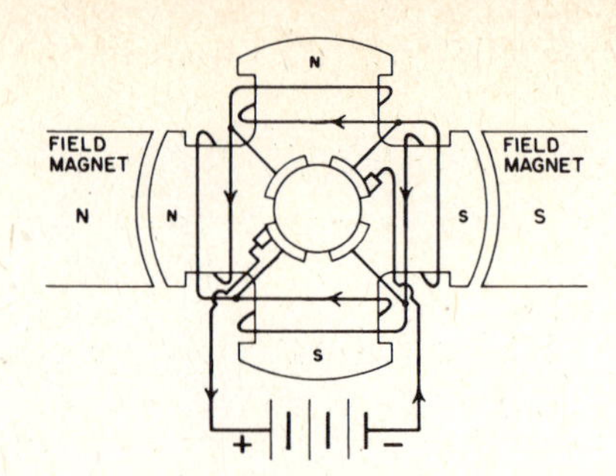

Fig. 10-4 Dc motor with a four-pole armature.

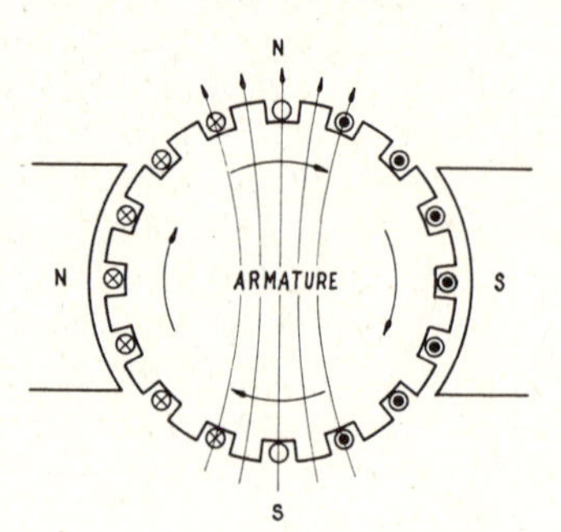

Fig. 10-5 Cross-section diagram of an armature.

will provide still smoother power. This is normally accomplished by winding the coils around the armature through oppositely positioned slots (see Fig. 10-5). If the motor had four field poles, the sides of the armature coils would be spaced a distance of one-fourth the circumference of the armature.

The action of a drum-wound armature in a magnetic field is illustrated in Fig. 10-5. This diagram represents a cross section of an armature with the conductors in the armature slots shown as small circles. The cross in the small circle indicates current flowing away from the observer, and the dot indicates current flowing toward the observer. By applying the right-hand motor rule, the direction of the torque on the armature may be determined. For example, the current on the left of the armature is flowing into the page, and the field flux is from left to right. The right-hand rule then indicates that the conductor would move up through the field. On the opposite side of the armature, the conductor would move down through the field, and the armature would turn in a clockwise direction.

Another method for finding the direction of torque is to determine the direction of the armature flux. This may be done by using the left-hand rule for electromagnets. If the left hand is held with the fingers pointing in the direction of the current around the armature, the thumb will be pointing upward. This, of course, indicates that the north pole is at the top of the armature. Since the north field pole is at the left, the top of the armature will be repelled and will move to the right. Since the south field pole is at the right, it will attract the north pole of the armature and add to the force which moves the top of the armature to the right. In a similar manner the bottom of the armature is moved to the left.

The action of the commutator continually switches the input current to new sections of the armature winding so that the top of the armature is always a north pole; hence, the armature continues to rotate in an effort to align itself with the field poles.

COUNTER EMF AND NET EMF

It has been pointed out in previous discussions that a conductor moving across a magnetic field will have an emf induced within itself. Since the conductors in the armature of a motor are cutting across a magnetic field as the armature rotates, an emf is produced in the conductors and this emf opposes the current being applied to the armature from the outside source. This induced voltage is called **counter emf,** and it acts to reduce the amount of current flowing in the armature. The **net emf** is the difference between the applied emf and the counter emf.

An engine-driven generator unit, such as an auxiliary power plant, gives an excellent example of the action of counter emf in a generator. When battery voltage is applied to the generator, it acts as a motor to start the engine. When the engine starts and begins to run at normal speed, the counter emf produced in the armature becomes greater than the applied battery voltage. The current then flows in the opposite direction and charges the battery.

Counter emf plays a large part in the design of a motor. Motors must be designed to operate efficiently on the net emf, which is only a fraction of the applied emf; hence, the resistance of the armature coils must be relatively low. Before a motor gains speed, the current through the armature is determined by the applied emf and the armature resistance. Since the armature resistance is low, the current is very high. As the speed of the motor increases, the counter emf builds up and opposes the applied emf, thus reducing the current flow through the armature. This explains the facts that there is a large surge of current when a motor is first started and that the current then rapidly falls off to a fraction of its initial value.

With some electric-motor installations the starting current is so high that it would overheat and damage the wiring or the armature, and so resistance must be inserted into the circuit until the motor has gained speed. The resistance may be automatically cut out as the speed of the motor increases, or it may be controlled manually.

DC MOTORS

ESSENTIAL PARTS

A dc motor has the same essential parts as a dc generator, that is, an armature, field, commutator, and brushes (see Fig. 10-6). It has already been explained that, except for minor features of design, a dc motor and a dc generator are identical. As in a generator, the field in a motor may be a permanent magnet, or it may be a core wound with coils to form an electromagnet. The electromagnetic field is almost always employed because such a field lends itself to a wide range of operation.

The armature, commutator, and brushes of a dc motor are nearly identical with those of a generator. A further description of these parts will be given in a later section of this chapter.

TYPES AND CHARACTERISTICS OF DC MOTORS

DC motors are **series-wound, shunt-wound,** or **compound-wound,** depending upon the arrangement of the

Fig. 10-6 Essential parts of a dc motor.

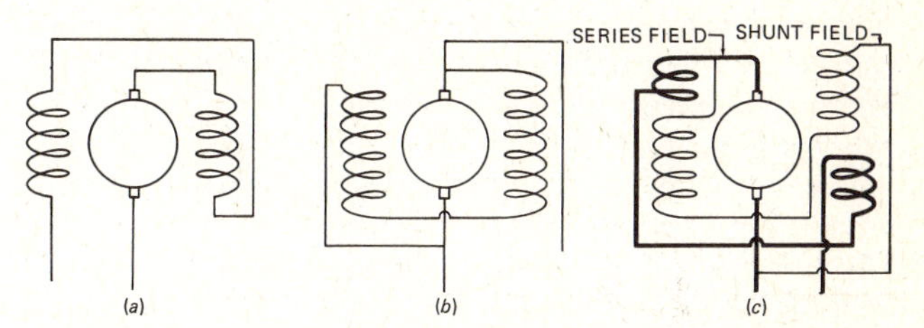

Fig. 10-7 Schematic diagrams of different types of motor circuits.

field windings with respect to the armature circuit (see Fig. 10-7).

In a series motor the field coils are connected in series with the armature as shown in Fig. 10-7*a*. Since all the current used by the motor must flow through both the field and the armature, it is apparent that the flux of both the armature and the field will be strong. The greatest flow of current through the motor will take place when the motor is being started; hence, the starting torque will be high. A motor of this type is very useful in installations in which the load is continually applied to the motor and in which the load is heavy when the motor starts. In aircraft, series motors are used to operate engine starters, landing gear, cowl flaps, and similar equipment. In each case the motor must start with a fairly heavy load; the high starting torque of the series motor is particularly well suited to this condition.

If a series motor is not connected mechanically to a load, the speed of the motor will continue to increase for as long as the counter emf is substantially below the applied emf. The speed may increase far above the normal operating speed of the motor, and this may result in the armature flying apart because of the centrifugal force developed by the rapid rotation. A series motor should always be connected mechanically to a load to prevent it from "running away."

The reason for the increase in speed when a series motor is not driving a load may be understood if the behavior of the field in such a motor is considered. As the speed of the motor increases, the counter emf increases. As the counter emf increases, however, the field current decreases. Remember that the field is in series with the armature and that since the counter emf causes the armature current to decrease, it must necessarily cause a decrease in the field current. This weakens the field so that the counter emf cannot build up sufficiently to oppose the applied voltage. A current continues to flow through both the armature and the field, and the resulting torque increases the armature speed still further. This increase of speed will continue until the centrifugal force tears the armature apart, or, as is the case with very small motors, the friction and other losses in the motor balance the armature torque.

In a shunt motor the field coils are connected in parallel with the armature (see Fig. 10-7*b*). The shunt field must have sufficient resistance to limit the field current to that required for normal operation because the counter emf of the armature will not act to reduce the field current. Since the voltage applied to the field at operating speed will be practically the same as the voltage applied to the motor as a whole, regardless of counter emf, the resistance of the field must be many times the resistance of the armature. This is usually accomplished by winding the field coils with many turns of fine wire. The result of this arrangement is that the motor will have a low starting torque because of a weak field. The reason for the weak field is that the armature, owing to its low resistance, draws most of the current when the motor is first starting.

As the armature of a shunt motor gains in speed, the armature current will decrease because of counter emf, and the field current will increase. This will cause a corresponding increase in torque until the counter emf is almost equal to the applied emf, at which time the motor is operating at its normal speed. This speed is almost constant for all reasonable loads.

When a load is applied to a shunt motor, there is a slight reduction in speed which causes the counter emf to decrease and the net emf across the armature to increase. Since the resistance of the armature is low, a slight rise in net emf will cause a comparatively large increase in armature current, which in turn increases the torque. This prevents a further decrease in speed and actually holds the speed to a point only slightly less than the no-load speed. The current flow increases to a level sufficient to hold the speed against the increased load. Because of the ability of the shunt motor to maintain an almost constant speed under a variety of loads, it is often called a **constant-speed motor.**

Shunt motors are used when the load is small at the start and increases as the motor speed increases. Typical of such loads are electric fans, centrifugal pumps, and motor-generator units.

When a motor has both a series field and a shunt field (Fig. 10-7*c*), it is called a compound motor. This type of motor combines the features of series and shunt motors; that is, it has a strong starting torque like the series motor but will not overspeed when the load is light. This is because the shunt winding maintains a field which allows the counter emf to increase sufficiently to balance the applied emf. When the load on a compound motor is increased, the speed of the motor will decrease more than it does in a shunt motor, but it provides speed sufficiently constant for many practical applications.

Compound motors are used to operate machines subject to a wide variety of loads. In aircraft they are used to drive hydraulic pumps which may operate from a no-load condition to a maximum-load condition. Neither a shunt motor nor a series motor would satisfactorily fulfill these requirements.

AC MOTORS

THEORY OF OPERATION

The basic principles of magnetism and electromagnetic induction are the same for ac and dc motors, but the application of the principles is different because of the rapid reversals of direction and changes in magnitude characteristic of alternating current. Certain characteristics of ac motors make most types more efficient than dc motors; hence, such motors are used commercially whenever possible. During recent years ac power systems have been developed for large aircraft with the result that a much larger amount of electric power is available on aircraft than would be available with dc systems of the same weight. Thus, one of the main advantages of the ac power system is that it provides more power for less weight.

There are three principal types of ac motors. These are the **universal** motor, the **induction** motor, and the **synchronous** motor. There are many variations of these types including combinations of features to meet different requirements. Among such motors are repulsion motors, split-phase motors, capacitor motors, and synchronous motors that utilize induction principles for starting torque.

A universal motor is identical with a dc motor and may be operated on either alternating or direct current. Since the direction of current flow in the field and the armature changes simultaneously when alternating current is applied to a universal motor, the torque continues in the same direction at all times. For this reason the motor will turn steadily in one direction regardless of the type of current applied. Typical of universal motors are those used in vacuum cleaners, small electric appliances, and electric drill motors. Universal motors are not used in aircraft electric systems because the alternating current has a frequency of 400 Hz, and at this frequency very substantial losses occur in a universal motor. These losses are the result of hysteresis and eddy currents which occur whenever there is a rapid change of magnetic fields.

The induction motor has a wide variety of applications because of its operating characteristics. It does not require special starting devices or excitation from an auxiliary source and will handle a wide range of loads. It is adaptable to almost all loads when an exact and constant speed is not required.

The essential parts of an induction motor are the **rotor** and the **stator.** The stator is in the form of a shell with longitudinal slots on the inner surface (see Fig. 10-8). The stator windings are placed in these slots in a manner similar to the winding of a dc armature.

If a source of direct current is connected to two terminals of a stator winding, it will be found that sections of the interior surface of the stator have a definite polarity. If the dc connections are reversed, the polarity of the stator will also reverse. When an alternating current is applied to the connections of the stator, the polarity of the stator will reverse twice each cycle. Ordinarily the stator of an ac motor has two or three separate windings, depending upon whether the motor is designed for two-phase or three-phase current. When multiphase currents are applied to the windings of a stator, a rotating magnetic field is established within the stator (see Fig. 10-9). As the current in each phase changes direction and magnitude, the

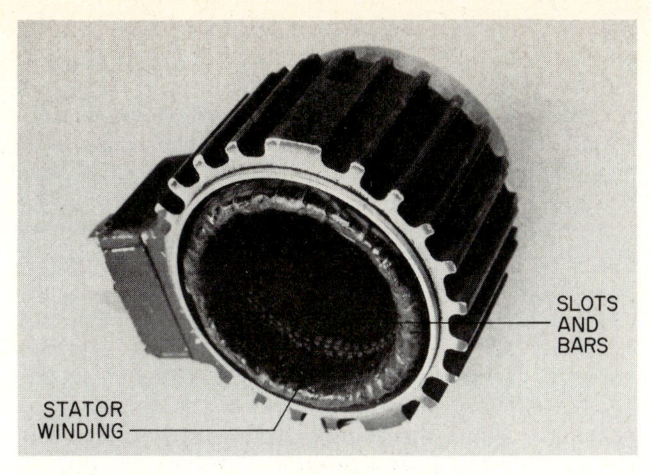

Fig. 10-8 Stator of an ac motor.

combined field of the stator will rotate at the frequency of the alternating current.

If we study carefully the diagram and the graph for position 1 in Fig. 10-9, we will find that phase 1 is positive with maximum current and that the stator field is vertical. The current is negative in both phase 2 and phase 3 with all the current flowing through the phase 1 winding in both the stator and the generator. The generator, or alternator, is represented by the inverted Y coils at the bottom of each diagram. In position 1 we see that approximately one half the current flows through phase 2 and the other half through phase 3. This results in the vertical field shown in the diagram.

When the current has changed through an angle of 30° to position 2, the current in phase 1 is still positive but has decreased, the current in phase 2 is at zero, and the current in phase 3 has increased in the negative direction. This results in a field produced entirely by the poles of phases 1 and 3 in the stator, and the position of the field is 30° clockwise from vertical. If we study the diagrams for positions 3 and 4 and determine the current flow through each phase winding, we will find that the stator field turns 30° further for each position. If the current values are plotted for a complete cycle, we find that the field rotates through 360° for each cycle.

The rotor in an induction motor consists of a laminated iron core in which are placed longitudinal conductors (see Fig. 10-10). In a **squirrel-cage** rotor these conductors are

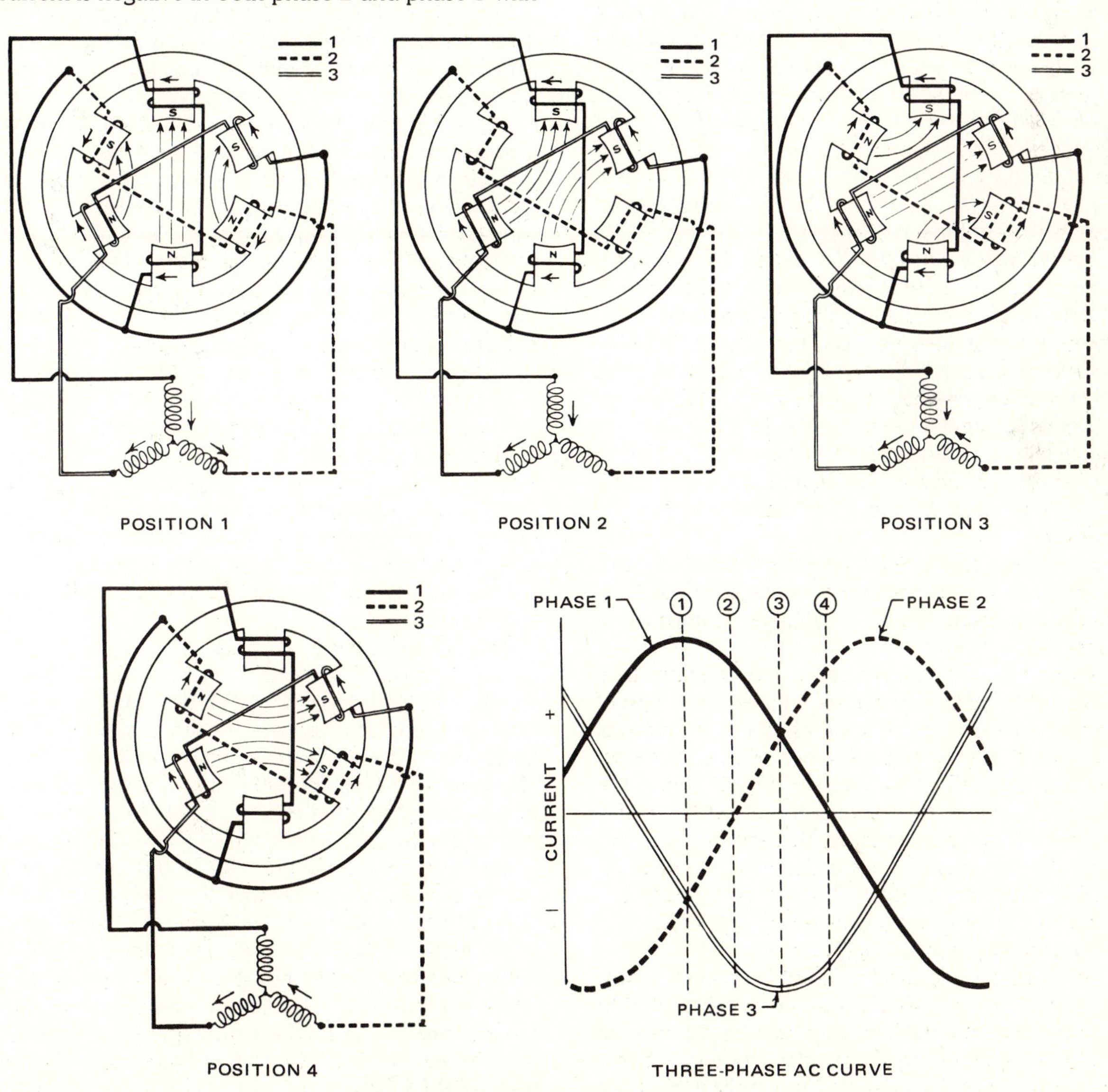

Fig. 10-9 Rotating field of an ac motor.

Fig. 10-10 Squirrel-cage rotor.

usually copper bars connected together at the ends by rings. When this assembly is placed in the rotating field produced by the stator, a current is induced in the conductors. Since the conductors are short-circuited, there is a flow of current from those on one side of the rotor, through the rings at the ends of the rotor, to the conductors on the other side. This current produces a magnetic field which is at an angle to the field of the stator. If the rotor field should come into line with the stator field, there would be no torque; hence, the rotor field must always be a few degrees behind the stator field. The percentage of difference in the speeds of the stator and rotor fields is called the **slip.** It must be emphasized that this slip is absolutely necessary. The only field provided initially by the input of current to the motor is the field produced by the stator. The rotor has no electrical connection with the external power, and the only way it can produce a field is by having current induced within itself as the flux of the rotating stator field cuts across it. The interaction of the rotor field with the stator field then produces the torque which causes the rotor to turn.

When the motor is mechanically connected to a load, the load tends to slow the rotation of the rotor. This causes the slip to increase, and the rotor conductors cut a greater number of lines of force per time interval, thus increasing the rotor current and the rotor field. This stronger field produces an increased torque which enables the motor to carry the increased load.

There is another effect which must be considered when a load is applied to an induction motor. This is a lowering of the power factor caused by the inductive reactance of the rotor. When the rotor is turning at almost synchronous speed, that is, the speed of the stator field, the frequency of the rotor current and the inductive reactance of the rotor are low. As the load is applied to the motor, the slip increases, and there is a corresponding increase in the frequency of the rotor current. This increases the inductive reactance of the rotor, and the power factor of the motor consequently decreases. It will be remembered that the power factor is equal to the cosine of the phase angle between the voltage and the current and that inductive reactance increases this phase angle.

When the load on an induction motor becomes so great that the torque of the rotor cannot carry it, the motor will stop. This is called the **pull-out point.**

IMPROVING STARTING QUALITIES

An induction motor will start satisfactorily under no load without any special starting devices. However, when such a motor is connected directly to a substantial load which must be moved when the motor starts, it is usually necessary to add resistance in the rotor circuits. There are several methods for accomplishing this, but the explanation of one method is sufficient for this text.

From the study of alternating current it will be remembered that the power factor for alternating current flowing in a purely resistive circuit is 100 percent. On the other hand, alternating current flowing in a purely inductive circuit would have a power factor of 0 percent if such a circuit were possible. Therefore, the addition of resistance to an inductive circuit will have the effect of improving the power factor. To add the necessary resistance to the rotor circuit for starting purposes, two squirrel-cage windings are used. One of these windings is of copper and has low resistance, and the other is of german silver and has high resistance. When the starting current is applied to the motor, the high-resistance winding produces the starting torque because of its high power factor. As the rotor gains speed, the effect of the high-resistance winding decreases, and the effect of the low-resistance winding increases. When the motor is operating at normal speed it has the advantage of a low-resistance rotor winding.

SPLIT-PHASE MOTORS

Single-phase induction motors have no torque when the rotor is at rest; hence, it is necessary to incorporate into them devices for providing a starting torque. This may be accomplished by providing the motor with two or three separate windings and using combinations of inductance, capacitance, and resistance to change the phase of the voltages applied to the different windings. This is known as **phase splitting.** A motor having devices for this purpose is called a split-phase motor. Figure 10-11 shows a motor circuit in which a capacitor is used to cause the current in one winding to lead the current in the other winding. This, in effect, causes the motor to act as a two-phase motor during starting.

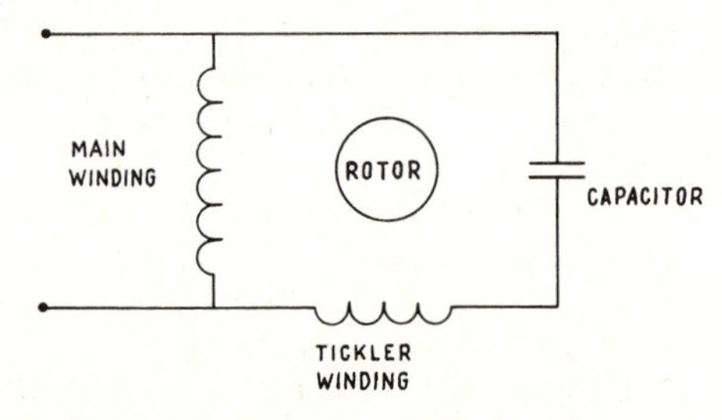

Fig. 10-11 Circuit for a capacitor motor.

Split-phase motors of the capacitor type are used extensively in industry for low-power applications such as drill presses, grinders, small lathes, and small saws. In private homes the split-phase motor is most commonly used for washing machines. In large aircraft the split-phase motor is used as an actuator for various types of comparatively small loads.

As shown in Fig. 10-11, a capacitor is often used in order to provide starting torque. When the motor has attained a certain rpm, a centrifugal switch opens and cuts out the capacitor circuit. Motors that employ capacitors for starting or for continuous operation are often called **capacitor motors.**

REPULSION MOTORS

A repulsion motor utilizes the repulsion of like poles to produce the torque for operation. The rotor is wound like an armature and employs a commutator and brushes. The brushes are short-circuited across the commutator at an angle that causes the induced current in the windings to produce a polarity in the rotor that will be in opposition to the polarity of the stator. That is, a north pole produced in the rotor will be near a north pole in the stator. The rotor is therefore caused to rotate because of the repulsion between the like poles. As the rotor turns, the brushes on the commutator remain in the same position, so the polarity of the rotor remains in the same position even though the rotor turns. The repulsion principle is used in some motors to provide starting torque, after which the motor operates as an induction motor.

SYNCHRONOUS MOTORS

Synchronous motors, as the name implies, rotate at a speed which is synchronized with the applied alternating current. These motors have some features in common with induction motors and a construction similar to that of alternators. The stator consists of a laminated soft-iron shell with coils wound through slots on the inner surface. A three-phase synchronous motor has three separate windings in the stator and produces a rotating field like the stator of an induction motor. The rotor may be a permanent magnet in a very small synchronous motor, but in larger motors the rotor is an electromagnet excited by an external source of direct current.

The theory of operation of a synchronous motor is very simple. If a magnet is free to turn and is placed in a rotating field, it will align itself with the field and rotate at the same speed. If no load is placed on such a motor, the center of the rotor poles will be exactly in line with the center of the stator field poles. In practice this does not occur because of friction. Friction and load cause the center of the rotor poles to lag behind the center of the field poles formed by the stator. The angle between the rotor field and stator field is called the lag, and it increases as the load on the motor is increased. If the load becomes so great that it overcomes the magnetic reaction, the pull-out point is reached and the motor will stop. At this time the incoming current will increase to a short-circuit value, and the torque will become negligible.

When operating within its load limits, a synchronous motor will rotate at the same speed as the alternator supplying the current, provided that the alternator has the same number of poles as the motor. Since the speed of a synchronous motor depends entirely on the frequency of the current supply, such motors are useful when constant speeds and frequencies are desired. One of the common uses of synchronous motors is to change the frequency of alternating current. Since the motor will turn at a precisely constant speed, it can be used to drive an alternator through a differential gear system to provide an exact frequency of any desired value.

Synchronous motors are commonly used on airplanes in the electric tachometer. A three-phase alternator is connected to a drive on the engine, and the alternator output is connected to a synchronous motor in the tachometer indicator. The frequency of the current is directly proportional to the engine speed; hence, the synchronous motor in the indicator will rotate at a speed proportional to engine speed. The indicating needle is coupled to the synchronous motor through a permanent magnet and drag cup. The distance that the needle moves along the rpm scale is proportional to the speed of the motor.

A synchronous motor differs from an alternator in that it has a high-resistance squirrel-cage winding placed in the rotor to give a good starting torque. This winding causes the motor to start as an induction motor and run as a synchronous motor. When the motor has reached synchronous speed, it is turning with the magnetic field, and the conductors of the squirrel-cage winding are not cutting lines of force. If the rotor tends to hunt or oscillate, however, the squirrel-cage winding will have an induced current which tends to dampen out the oscillations and prevent hunting.

MOTOR LOSSES

The efficiency of electric motors of any type is largely determined by the losses of power resulting from friction, resistance, eddy currents, and hysteresis. The power used to overcome the friction of bearings is called the **friction loss.** This loss may also include the loss due to wind friction, which is sometimes called windage loss and is comparatively high when a motor is equipped with a fan to provide cooling by forced ventilation. The power used to overcome the resistance of the windings is called **resistance,** or **copper, loss.** Copper losses are dissipated in the form of heat.

The current induced in the armature core and the field poles are called **eddy currents** and are responsible for considerable loss in the form of heat. These losses are reduced by constructing the armature and field cores of laminated soft iron, the laminations being insulated from one another.

Hysteresis losses occur when a material is magnetized first in one direction and then in the other in rapid succession. The effect of hysteresis is to cause the change in strength of the magnetic flux to lag behind the magnetizing force and is presumably due to the friction between the molecules of the material as they are shifted in direction by the magnetizing force. Hysteresis losses are noticeable because of their heating effect. Any condition which produces heat in a motor causes a loss of power, or energy, because heat is one of the principal forms of energy and requires power to produce it.

Motor and generator losses are not particularly important to technicians, who are not concerned with the design of the equipment with which they work. Nevertheless, it is well that technicians understand these losses because there may be cases in which the knowledge will be helpful. For example, if an armature has become overheated to the extent that the insulating coating between the laminations of the core is burned out, the technician should know that the armature will not perform efficiently unless the laminations are separated, cleaned, and reinsulated. Usually the construction of an armature makes it impractical to rebuild the core; hence, one that is badly burned should be discarded.

The construction of electric motors with laminated armatures and field-pole cores helps to solve cooling problems because much of the heat encountered during operation is the result of the losses described above. This type of construction is particularly important for high-speed actuator motors. Actuator motors must have a high power-weight ratio, and to attain this it is necessary to operate them at relatively high speeds. For this reason all losses must be reduced to a minimum.

MOTOR CONSTRUCTION

CHARACTERISTICS OF AIRCRAFT ELECTRIC MOTORS

As in aircraft generators, the power-weight ratio of electric motors in aircraft and missiles must be high; that is, a small motor must deliver a maximum amount of power for a minimum of weight. A commercial motor may weigh as much as 100 lb/hp [60.8 kg/kW], but for aircraft purposes there are motors which weigh less than 5 lb/hp [3 kg/kW]. Reduced weight is attained by operating the motors at high speeds, high frequencies, and with relatively high currents. This necessitates the use of heat-resistant insulation and enamels in the armature and field windings.

Some fractional-horsepower motors used in aircraft rotate at over 40 000 rpm [4138 rad/s] with no load and at about 20 000 rpm [2069 rad/s] with a normal load. Since horsepower means the rate of doing work, it is apparent that a motor turning at 20 000 rpm develops twice the power of a similar motor turning at 10 000 rpm [1035 rad/s]. To reduce the effect of centrifugal force on the armature of the motor rotating at a very high speed, the armature diameter is made to be relatively small compared with its length.

CONTINUOUS- AND INTERMITTENT-DUTY MOTORS

Many electric motors used in aircraft are not required to operate continuously. Because the heat developed in a short time is not sufficient to cause any damage, a motor in this type of service is designed to deliver more power for its weight than a motor used for continuous service. If such a motor were used continuously it would overheat and burn the insulation and thus become useless. Motors designed for short periods of operation are called intermittent-duty motors, and those which operate continuously are called continuous-duty motors. The type of duty for which a motor is designed is sometimes stated on the nameplate and, if not on the nameplate, may be found in the manufacturer's specifications.

STARTER MOTOR

A typical direct-cranking starter motor for small aircraft engines is illustrated in Fig. 10-12. The armature winding is of heavy copper wire capable of withstanding very high amperage. The windings are insulated with a special heat-resistant enamel, and after being placed in the armature, the entire assembly is doubly impregnated with a special insulating varnish. The leads from the armature coils are staked in place in the commutator bars and then soldered with high-melting-point solder. An armature constructed in this manner will withstand the severe loads imposed for brief intervals while the engine is being started.

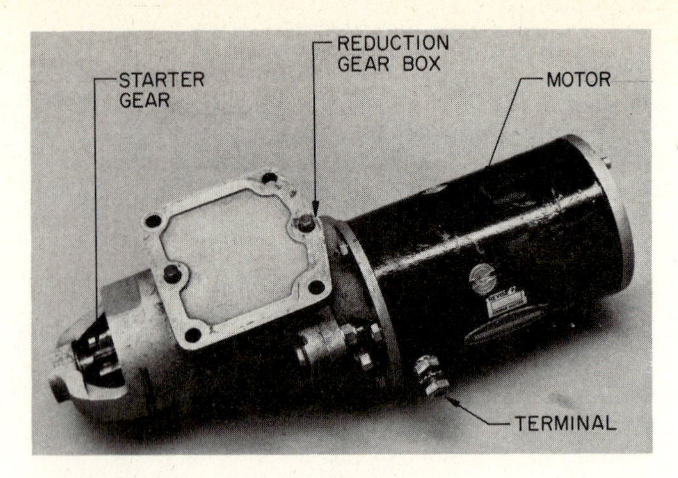

Fig. 10-12 Starter motor for a light aircraft.

The field-frame assembly is of cast-steel construction with the four field poles held in place by countersunk screws threaded into the pole pieces. The pole pieces are closely fitted to the inside contour of the field frame to provide the best possible magnetic circuit, because the field frame carries the magnetic flux from one field to the others. In other words, the field frame acts as a conductor for the magnetic lines of force; hence, it is a part of the magnetic circuit for the field poles. Since a motor of this type is series-wound, the field windings must be of heavy copper wire of a size sufficient to carry the high starting current.

An exploded view of a starter motor and drive is shown in Fig. 10-13. This complete assembly consists of six major components. These are the **commutator end head assembly,** the **armature,** the **frame-and-field assembly,** the **gear housing,** the **Bendix-drive assembly,** and the **pinion housing assembly.**

The gear cut on the drive end of the armature shaft extends through the gear housing, where it is supported by a roller bearing. The gear mates with the teeth of the reduction gear that drives the Bendix shaft. The shaft is keyed to the reduction gear, and the Bendix drive is held in position on the shaft by a roll pin. The shaft is supported in the gear housing by a closed-end roller bearing and in the pinion housing by a graphitized bronze bearing.

When the armature turns the reduction gear, the Bendix-drive pinion meshes with the flywheel ring gear on the engine by inertia and the action of the screw threads within the Bendix sleeve. A detent pin engages in a notch in the screw threads, which prevents demeshing if the engine fails to start and the starting circuit is deenergized. When the engine starts and reaches a predetermined speed, centrifugal force moves the detent pin out of the notch in the screw shaft and allows the pinion to demesh from the flywheel gear.

REVERSIBLE DC MOTORS

Motors used for the operation of landing gear, flaps, cowl flaps, and certain other types of apparatus must be designed to operate in either direction and are therefore called reversible motors.

The direction of rotation of a dc motor may be changed by reversing the electrical connections to the field or to

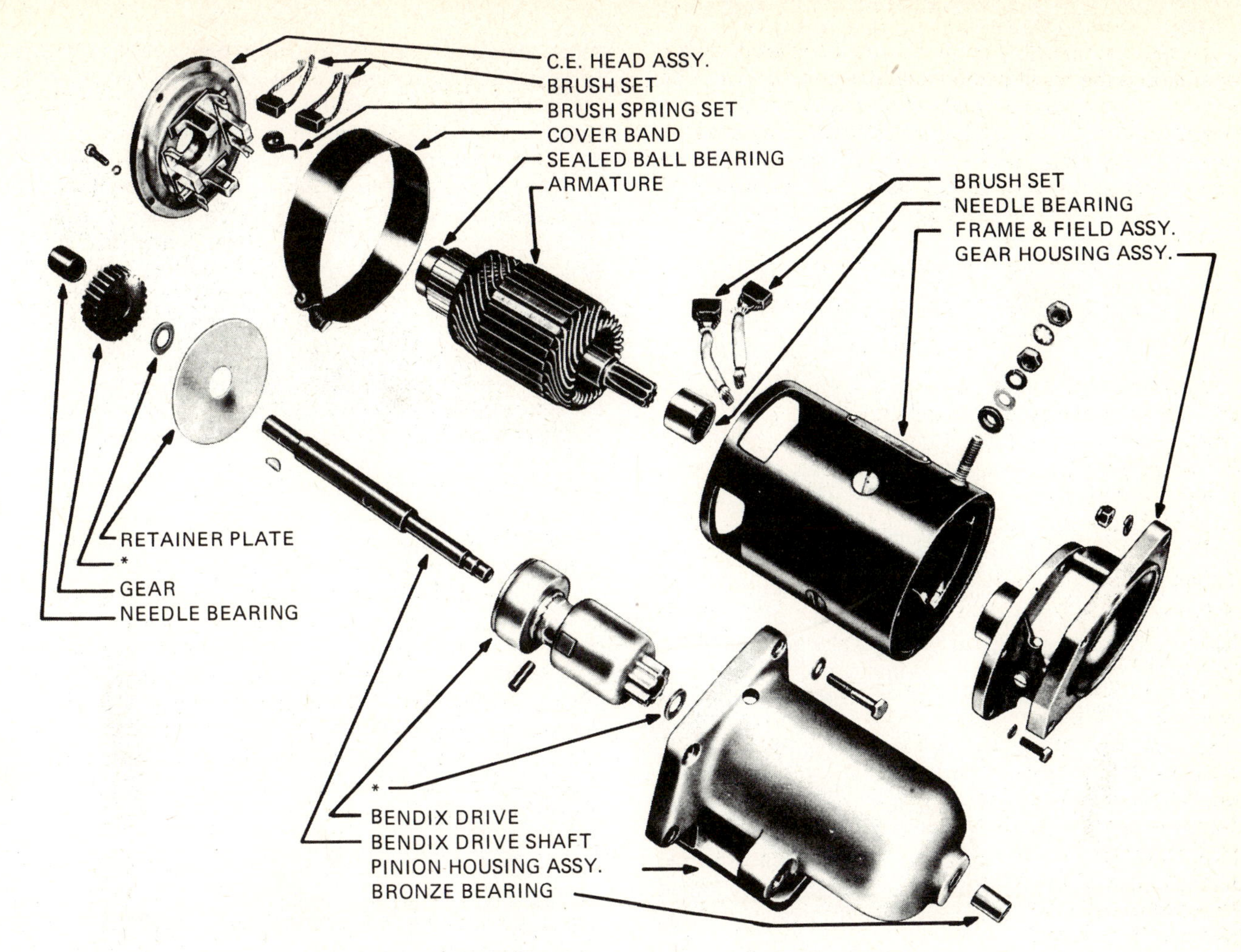

Fig. 10-13 Exploded view of starter motor and drive. (Prestolite)

the armature. Because reversing a motor by this method would require a complex external circuit, a simpler method is normally employed that provides a double field winding known as a **split field.** A schematic diagram of the circuit for a split-field motor is shown in Fig. 10-14. Note that a separate circuit is provided for each field winding. This makes it possible to change the direction of the motor at will by placing the switch in the desired position. The motor is reversed by changing field polarity in relation to the armature polarity when the different field windings are energized.

Reversible dc motors are controlled directly by double-throw single-pole switches or indirectly by relays controlled by similar switches. The use of relays is dictated by the amount of current which the motor draws while in operation. Any motor requiring more than 20 to 30 A will operate more satisfactorily with a relay-controlled circuit.

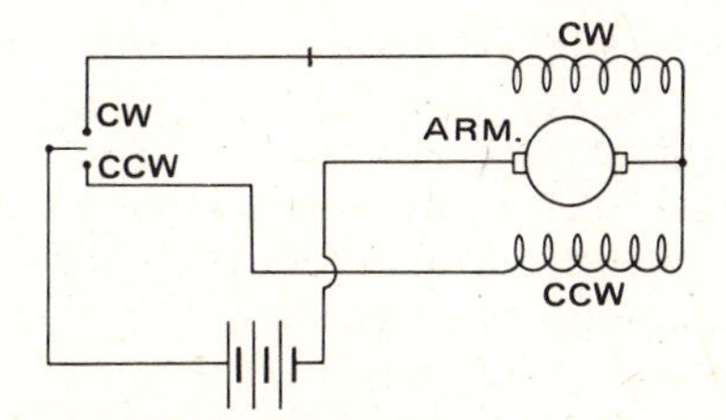

Fig. 10-14 Reversible motor circuit.

The separate field coils of a reversible motor are usually wound either in opposite directions on the same poles or on alternate poles. Since the field coils are in series with the armature, they must be wound with wire of a size large enough to carry the entire motor current. Remember that the entire load current passes through both the field and the armature.

The brushes in a reversible motor are usually held in box-type holders in line with the center of the motor shaft. With this arrangement the brushes are perpendicular to a plane tangent to the commutator at the point of brush contact, and the brushes will wear evenly regardless of the direction of motor rotation. On small motors the field and brush housing is sometimes made in one piece. The brush holders are inserted through openings at the end of the housing and are insulated from the housing by composition bushings. Each brush assembly consists of the brush, a helical spring, a flexible connector inside the spring, and a metal contact. When a brush is installed in the motor, it is held in place by a screw plug (see Fig. 10-15).

BRAKES AND CLUTCHES

Many motor-driven devices used in aircraft or missiles must be designed so that the operated mechanism will stop at a precise point. For example, when landing gear is being retracted or extended, it must stop instantly when

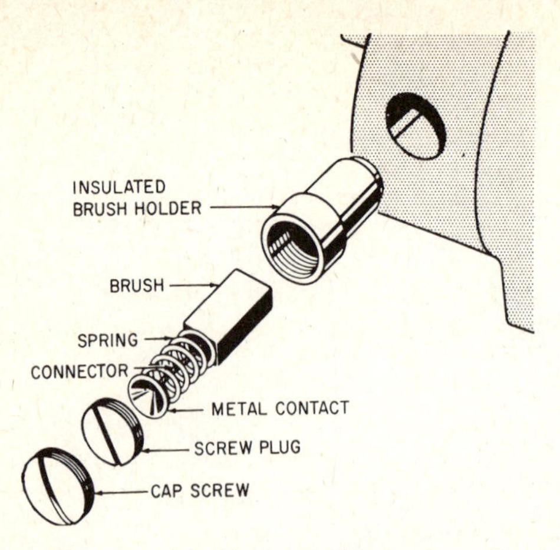

Fig. 10-15 Brush and holder assembly for a small motor.

the operation is complete. If the driving motor is connected directly to the operating mechanism, a great amount of strain will be imposed upon the motor when it is forced to stop because of the momentum of the armature and other moving parts. In installations requiring an instantaneous stop, a clutch and brake mechanism is employed to prevent damage when the mechanism is stopped.

One type of brake mechanism for actuator motors is illustrated in Fig. 10-16. This brake consists of a drum mounted on the armature shaft and internal brake shoes controlled by a magnetizing coil. The coil is placed inside the brake shoes, and when the motor current is turned off, the coil is deenergized and the brake shoes are forced against the drum by spring pressure. Conversely, when the power is turned on, the coil pulls the brake shoes away from the drum.

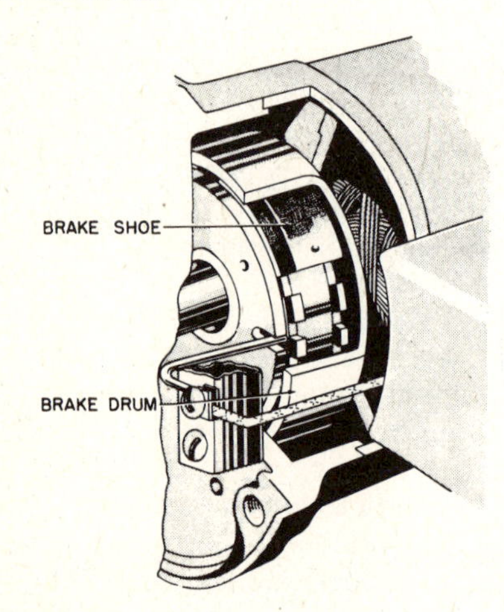

Fig. 10-16 Drum-type brake.

A disk-type brake, commonly used in actuator motors, consists of a rotating disk mounted on the armature shaft and a cork-lined braking surface on the stationary structure of the motor. A magnetizing coil is used to release the brake when the motor is energized, and a spring engages the brake when the current to the motor is turned off. A small amount of end play is allowed in the armature assembly mounting to provide clearance when the brake is released. When the brake coil is energized, the entire armature assembly moves slightly in a direction which will move the brake disk away from the braking surface. When the current is turned off, a spring moves the assembly in the opposite direction, and the friction produced between the brake disk and the cork-lined brake surface causes the armature to stop very quickly.

Clutches of several types have been designed for the purpose of disengaging the motor from the load when the power is cut off. All such clutches are engaged by magnetic attraction when the power is turned on and disengaged by spring action. A typical magnetic clutch is shown in Fig. 10-17. Two clutch faces are located within the clutch coil. One of the faces is mounted solidly on the armature shaft, and the other is connected through a diaphragm spring to the drive mechanism. When the clutch coil is energized, the two faces are magnetized with opposite polarities; hence, they are drawn together firmly. The friction thus produced causes the driven mechanism to turn with the motor. When the power is cut off, the diaphragm spring separates the faces, thus disengaging the motor.

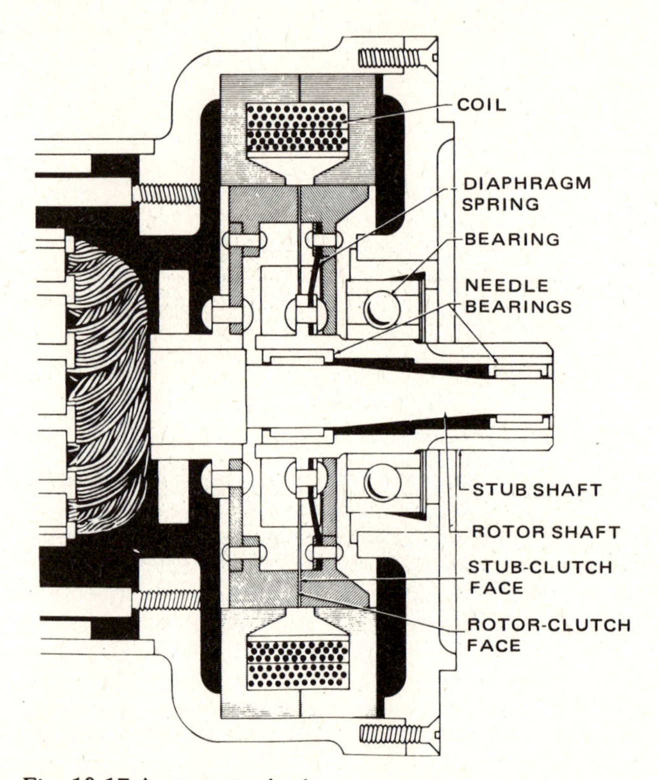

Fig. 10-17 A magnetic clutch.

Some actuating motors are provided with a combination brake and clutch. A magnetizing coil is located in the end of the motor housing as shown in Fig. 10-18. This coil, when energized, magnetizes a driving disk attached to the armature shaft. A driven disk is keyed to the output shaft, and when power is turned on, this disk moves against spring pressure until it engages with the driving disk. When the current is cut off, the driven disk is pulled away from the driving disk by the spring and is pressed against the brake plate at the opposite face, thus causing the driven mechanism to stop immediately.

Motors subject to sudden heavy loads are usually equipped with overload release clutches. A clutch of this type is called a slip clutch, and its function is to disconnect

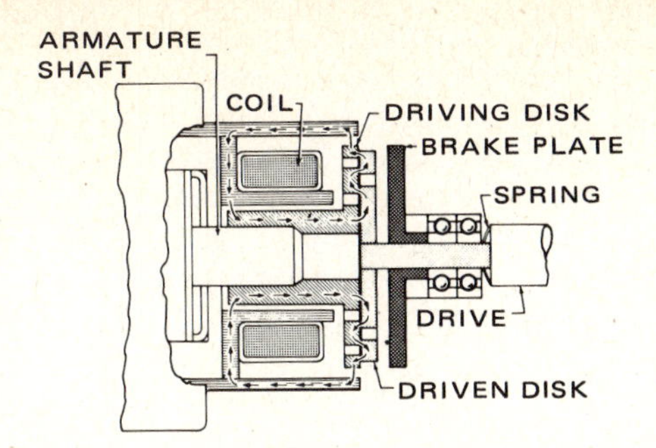

Fig. 10-18 Clutch and brake assembly.

the motor from the driven mechanism when the load is great enough to cause damage. This clutch consists of two groups of disks, alternately arranged, with one group splined to the motor drive and the other group splined to the motor-driven mechanism. These disks are pressed together by one or more springs designed to create sufficient pressure to cause the disks to rotate as one unit when the load is normal. When the load is excessive, the disks slip, thus preventing damage due to excessive torque.

LIMIT SWITCHES AND PROTECTIVE DEVICES

Because of the limited distance of travel permitted in the driven mechanism, reversible actuating motors are usually limited in their amount of rotation in each direction. It is essential, therefore, that the motor circuits be provided with switches which will cut off the power when the driven mechanism has reached the limit of its travel. Switches of this type are called **limit switches** and are actuated by cams or levers linked or geared to the driven mechanism. The adjustment of these switches is critical because severe damage may result if the motor continues to run after the limit of operation is reached. Stripped gears and broken shafts are often the result of improperly adjusted limit switches. If the driven mechanism is strong enough to withstand the torque imposed by the motor, the fuse or circuit breaker in the motor circuit will usually cut off the current to the motor.

Adjustment of the limit switches is accomplished by running the motor to the limit of travel and then adjusting the switch-actuating mechanism so that it has just opened the switch. The switches should be adjusted to open slightly before the extreme limit is reached.

Some actuating motors are provided with a thermal circuit breaker, or *thermal protector,* to protect the motor from overload and excessive heat. This device is mounted on the motor frame, and when heat reaches a predetermined limit, the circuit breaker will open and cut off the current to the motor. After the motor has cooled sufficiently, the circuit breaker will automatically close, thus permitting normal operation.

Figure 10-19 is a schematic diagram of a reversible-motor circuit with a thermal protective device and a coil for operating the clutch and brake. A circuit of the type shown would be used for operating cowl flaps, oil cooler shutters, air valves, and a variety of other devices. Both the limit switches shown in Fig. 10-18 are normally closed. Since they open only when the motor has reached the limit of travel in one direction or the other, it is readily apparent that there will never be a time when both switches are open. Notice that the thermal circuit breaker and the clutch coil are both in the ground circuit and will therefore be in operation for either direction of travel.

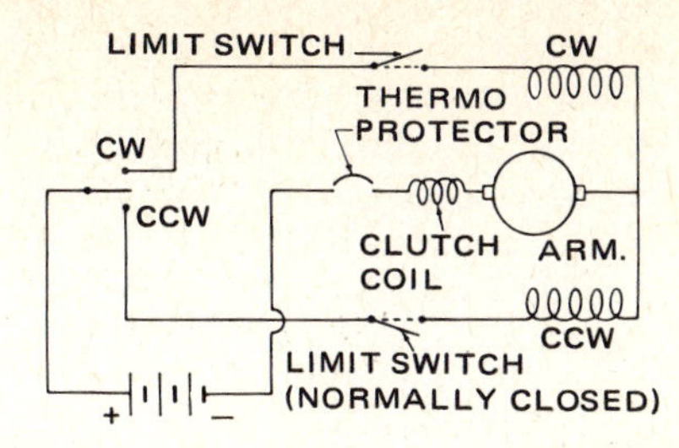

Fig. 10-19 Schematic circuit for a reversible actuator motor.

DC MOTOR CONSTRUCTION

An exploded view of a typical dc actuator motor is shown in Fig. 10-20. The principal sections of the motor assembly are the armature, the field coils and field frame, the brake assembly, and the thermal-protector assembly. The armature is a standard drum type wound on a laminated soft-iron core. Also mounted on the armature shaft is the commutator at one end and the brake-lining disk at the other.

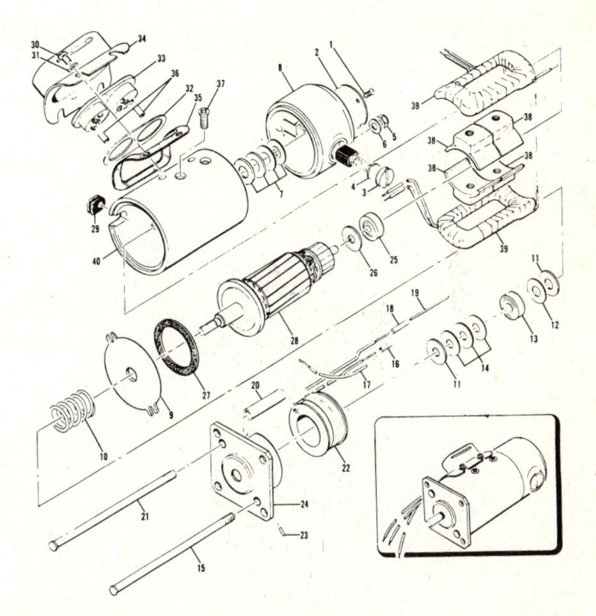

Fig. 10-20 Exploded view of a dc actuator motor. (1) Screw; (2) name plate; (3) brush retainer; (4) brush assembly; (5) nut; (6) washer; (7) shim washers; (8) motor cap; (9) brake armature; (10) brake armature spring; (11) spacer washer; (12) shim washer; (13) ball bearing; (14) shim washers; (15) motor assembly stud; (16) insulating sleeve; (17) wire; (18) insulating sleeving; (19) brush connector; (20) insulating sleeving; (21) motor assembly stud; (22) brake-coil assembly; (23) base-registering pin; (24) motor-base assembly; (25) ball bearing; (26) shim washer; (27) brake lining; (28) armature assembly; (29) motor lead grommet; (30) thermal-protector-case screw; (31) washer; (32) thermal-protector retainer; (33) thermal protector; (34) thermal-protector case; (35) thermal-protector gasket; (36) insulating sleeving; (37) field-pole screw; (38) field pole; (39) field winding; (40) motor housing.

The field for the motor is provided by two poles formed to fit around the armature with a clearance of about 0.01 in [0.025 cm]. The field coils are double-wound to provide for the reversal of field polarity necessary to reverse the motor rotation. Thermal protectors are connected in the circuit for each field (see Fig. 10-21).

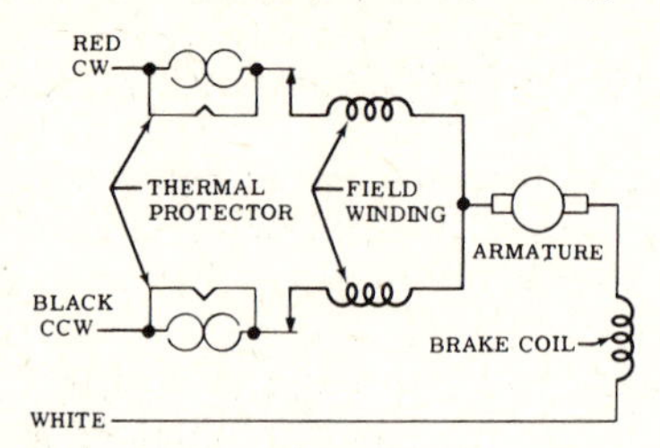

Fig. 10-21 Circuit for an actuator motor.

The brake assembly consists of a coil, a brake armature, and a brake lining mounted on the lining disk on the motor armature. The brake armature is a disk held in place by the motor studs, which pass through slots on the outer periphery of the armature. When the motor is not energized, the brake armature is held against the brake lining of the motor armature by a coil spring. This prevents the motor from turning. When the motor is energized, the magnetic brake coil draws the brake armature away from the brake lining, and the motor is free to turn.

Both ac and dc actuator motors have been manufactured in very small sizes for use in aircraft and missiles. Figure 10-22 is a photograph of an actuator assembly which may be operated by either a dc or an ac motor. Even though the actuator and motor assembly is very small, it can exert tremendous force.

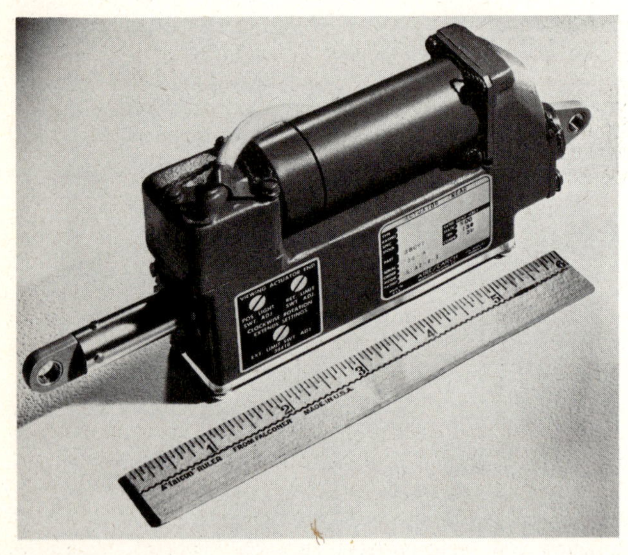

Fig. 10-22 Photograph of an electrically-driven linear actuator. (AiResearch Div., Garrett Corp.)

SINGLE-PHASE AC ACTUATOR MOTORS

Ac actuator motors are usually manufactured in the single-phase split-phase type when it is necessary that they be reversible; however, it is possible to reverse the direction of rotation in a three-phase motor by reversing connections to two of the input phases. This can be done with a double-throw double-pole switch in the external circuit. The construction of a single-phase reversible actuator motor consists principally of a squirrel-cage rotor, a double-wound stator, and a brake assembly (see Fig. 10-23). The core of the rotor is constructed of laminated soft iron. Slots are provided in the surface of the core for the copper bars which form the squirrel cage. At each end the copper bars are soldered or welded to copper rings.

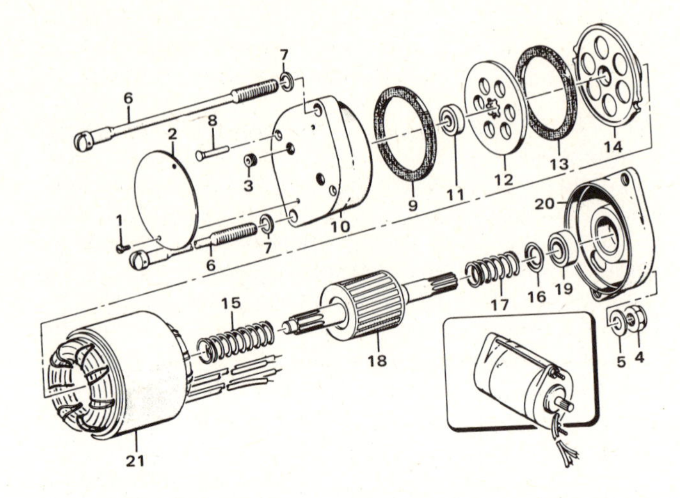

Fig. 10-23 Exploded view of an ac actuator motor. (1) identification-plate screw; (2) identification plate; (3) adjusting screw; (4) main assembly nut; (5) washer; (6) main assembly screw; (7) washer; (8) aligning pin; (9) brake lining; (10) end bell; (11) ball bearing; (12) break disk; (13) brake lining; (14) brake armature; (15) brake spring; (16) spring retainer washer; (17) compression spring; (18) rotor assembly; (19) ball bearing; (20) motor base;)21) stator assembly.

The double-wound stator provides the split field which is necessary to establish torque for starting under load. The stator leads are brought to the outside of the motor, where they are connected to a capacitor as shown in Fig. 10-24. Note that when the control switch is placed in the

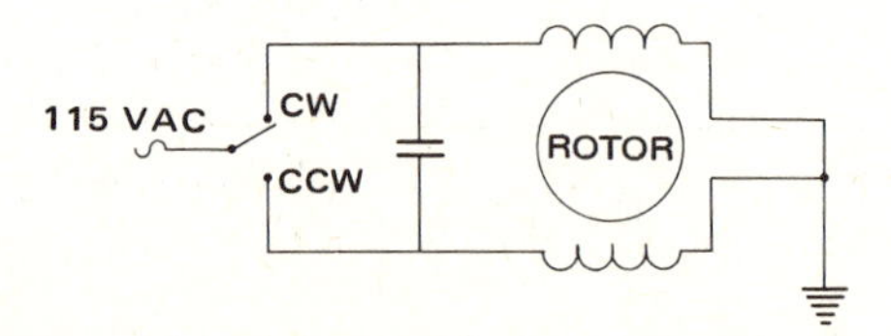

Fig. 10-24 Circuit for a split-phase reversible motor.

clockwise position, current flow will be through the clockwise coil directly and through the capacitor to the counterclockwise coil. This causes the current in the counterclockwise coil to lead the current in the clockwise coil, thus creating a clockwise-rotating field. Conversely, when the switch is placed in the counterclockwise position, the rotating field is in a counterclockwise direction, and the motor turns accordingly.

THREE-PHASE AC MOTORS

Three-phase ac motors for aircraft are quite similar to three-phase induction motors of the commercial or industrial type. The principal difference is that the aircraft motor operates at a frequency of 400 Hz; thus making it possible to employ a motor of lighter weight for the same power output. For example, a certain Freon compressor motor manufactured by the AiResearch Division of the Garrett Corporation and weighing less than 20 lb [9.1 kg] will deliver approximately 25 hp [18.6 kW]. One of the features that makes this performance possible is cooling by means of the liquid Freon which is pumped through the motor. Liquid cooling is also employed for high-performance motors which drive hydraulic pumps and pumps for other nonconductive and noncorrosive liquids.

The three-phase induction motor consists essentially of a three-phase Y-wound stator and a conventional squirrel-cage rotor. The three-phase stator produces a rotating field as explained previously, and this field induces a current in the rotor. The rotor current creates a field which opposes the stator field, with the result that the rotor attempts to turn at a speed which will keep it ahead of the stator field.

One type of three-phase actuator motor for aircraft is internally wired as shown in Fig. 10-25. It will be noted that the motor has a Y-wound stator; however, the neutral connections from each phase winding are individually connected to three separate legs of a full-wave rectifier. The rectifier output is directed through a clutch coil that is split to accommodate the alternating current in the neutral line.

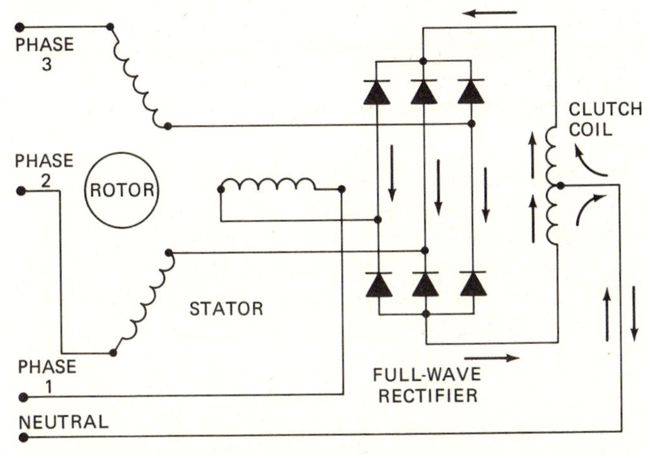

Fig. 10-25 Circuit for a three-phase ac motor with a full-wave rectifier to provide dc for the clutch coil.

The effect of this type of internal circuit is to allow for a high surge of current when the motor is started and a comparatively low current as soon as the motor gains full operating speed. Since the rotor windings are in series with the clutch coil, the clutch coil will receive the benefit of the high starting current to engage the clutch. When the clutch is engaged, it requires a relatively low current to hold it in that position. This low current is the result of the inductive reactance developed by the stator windings and the clutch coil.

Since the variety of electric motors is extensive and many look exactly alike even though they have different characteristics and specifications, the technician must be certain that a replacement motor in any system be identified by the correct part number. A motor installed in a system where the electrical characteristics do not match those of the motor will very likely be damaged and may cause damage to other elements of the circuit.

INSPECTION AND MAINTENANCE OF MOTORS

GENERAL INSPECTION PROCEDURES

In the section discussing generators, instructions were given for the inspection and maintenance of generators. Many of the instructions apply to motors as well because of the similarities between the two.

Preflight inspections of motors are usually operational. The switches for the various motor-driven units are turned on, and if operation is satisfactory, no further inspection is made. It is obvious that landing-gear actuators cannot be tested in this manner, but if the latest pilot's report is satisfactory, only a visual inspection need be made.

Depending upon the amount of operation to which a motor has been subjected, inspections should be carried out on it at intervals set forth in the manufacturer's operation manual. This type of inspection will probably include a check of the mounting, electric connections, wiring, brushes, brush springs, and commutator. For ac motors it is not usually necessary to consider brushes and commutator because there are none, except in univeral motors.

The construction of many small dc actuating motors makes it difficult to inspect the commutator without removal and disassembly. But because these motors are usually of the intermittent type, the wear on the commutator is negligible. A periodic inspection and replacement of the brushes, if necessary, will assure satisfactory operation until the time for overhaul. New brushes should be seated as outlined in the section covering generator maintenance. If the construction of a motor makes it impossible to seat the brushes after they are installed in the motor, they may be ground to the correct curvature with No. 000 sandpaper wrapped around a piece of wood or metal which has been turned to the diameter of the commutator. Usually brushes for small motors can be ordered specifically for a particular model; the brush face will be already ground to the correct curvature. The seating of the brushes may be checked by removing the brushes from the motor after a few minutes of operation and examining the area which has been polished by the commutator.

When the brush assemblies are covered with an end cap or cover band, they may be inspected or removed in the same manner used for those of a generator. Removal of the cap or cover band will give access to the brushes and holders. If a brush is held on a pivoted arm, it may be removed by lifting the arm and removing the brush screw. For a brush in a box-type holder, merely lift the brush spring and slide the brush out of the holder.

REMOVAL AND INSTALLATION

Because of the many different types of electric motors, specific instructions cannot be given here for their removal and installation. For any particular motor on an aircraft, the technician should consult the maintenance or overhaul manual supplied by the manufacturer.

Motor Troubleshooting Chart

Indication	Probable cause	Remedy
Motor fails to operate. No voltage.	Defective wiring, loose connections, or defective switch.	Test motor circuit and switch with continuity tester or ohmmeter.
	Electric power switch not turned on.	Turn on the power switch.
	Brushes and commutator dirty.	Clean the brushes and commutator.
	Loose or broken connections inside the motor.	Tighten or repair the connections.
	Open field or armature winding.	Remove and overhaul the motor.
	High mica insulation between commutator bars.	Undercut the mica. If necessary, turn the commutator on a lathe.
	Brushes worn out or sticking in holders.	Replace the brushes. See that they move freely in holder.
	Brush springs weak or broken.	Replace the brush springs.
Motor fails to operate. Draws high current when switch is turned on.	Short circuit in motor circuit.	Locate the short circuit and make repair.
	Short circuit inside motor.	Remove and repair the motor.
	Open field in shunt motor.	Remove and repair the motor.
	Motor overloaded.	Correct the amount of load imposed on the motor or install a motor of higher power.
	Mechanical stoppage such as seized bearings or binding parts in driven mechanism.	Locate and repair the defective parts.
Motor speed too slow.	Motor-driven parts need lubrication.	Lubricate parts as necessary.
	Low voltage.	Compare voltage supply with motor voltage as indicated on the name plate.
	Defective wiring or poor connections.	Check the motor circuit for defective wiring and for loose or dirty connections.
	Motor overloaded.	Correct the amount of load or install a motor of higher power.
	Short circuit in armature winding.	Remove and repair the motor.
	Dirty commutator or brushes.	Clean the commutator and brushes.
	Brushes sticking in holder.	Clean the brushes and see that they move freely in the holders.
	Brushes worn out.	Replace the brushes.
	Brush springs weak.	Replace the springs.
Motor runs too fast.	Voltage too high.	Check supply voltage against the motor voltage as indicated on the name plate.
	Short circuit in the field windings.	Remove and repair the motor.
Motor overheats.	Short circuit in the field windings.	Remove and repair the motor.
	Bearings of motor or drive mechanism require lubrication.	Lubricate parts as required.
	Armature dragging on field poles.	Remove and repair the motor.
	Voltage too high.	Check circuit for correct voltage.
	Arcing at brushes.	Refer to section of the chart on excessive arcing.
Motor vibrating and noisy.	Worn bearings or bent shaft.	Remove and overhaul the motor.
	Mounting loose or broken.	Repair motor mounting.
	Driven mechanism damaged or broken.	Repair driven mechanism as required.
Excessive arcing at brushes.	Commutator dirty, pitted, or worn.	Clean or repair the commutator.
	Brushes worn out or sticking.	Replace or clean the brushes.
	Brush springs weak.	Replace the springs.
	Poor commutation. Brushes not correctly located.	Adjust the position of the brushes.
	Open circuit in one or more armature coils.	Remove and repair the motor.
Direction of rotation incorrect.	Internal connections wrong.	Check the internal connections against a circuit diagram and make corrections.
	Two leads to a three-phase a-c motor reversed.	Reverse connections of two incoming power leads.

When the removal of a motor is necessary, the technician must give due consideration to the driven mechanism. In many cases a gear-train assembly must be removed with the motor. In any event, a brief visual inspection will usually enable the technician to determine the procedure to be followed.

Care must be taken to make sure that electric wiring and connector plugs are not damaged when a motor is disconnected. It is best to tape or otherwise insulate disconnected terminals which might accidentally become short-circuited if the battery switch were to be inadvertantly turned on.

If the removal of a motor leaves an opening through which dirt or other foreign matter may gain access to vital parts of a mechanism, the opening should be covered with a cloth or a plate. This is particularly important when removing a starter motor from an engine. If a nut, bolt, or other object should fall inside the engine, great damage may be caused, and the engine may require a complete overhaul.

The important points to be considered in the installation of a motor are as follows:

1. See that the mounting area is clean and properly prepared. Install the correct type of gasket if a gasket is required.
2. Be very careful not to cause damage when moving the motor into place. A nick or scratch in the mounting could develop into a crack and eventually cause failure.
3. Tighten screws or hold-down bolts evenly and with the correct torque. Make sure that nuts, bolts, or studs are properly safetied.
4. See that electric connections are clean; then tighten and safety them as required.

DISASSEMBLY AND TESTING

The disassembly, inspection, overhaul, and assembly of an aircraft electric motor should be performed in accordance with manufacturer's instructions. If manufacturer's instructions are not available, the general repair procedures may be carried out as for a generator. The following general rules apply:

1. Use the proper tools for each operation.
2. Mark and lay out parts in an order which will aid in assembly.
3. Do not use excessive force in any operation. If parts are stuck, determine the cause. If necessary, use a soft mallet to disengage parts. Sometimes parts are joined by means of metal pins or keys which may be overlooked by the technician; be sure that such devices are removed before attempting to separate parts which have been joined in this manner.
4. When bearings are pressed on a shaft, or when they are stuck because of corrosion, use a bearing puller for removal.
5. The use of an arbor press is recommended for the removal and installation of bearings and bushings which are press-fit. If an arbor press is not available, a fiber tube which fits the inner or outer race of the bearing may be used.
6. Keep all parts of an assembly clean. The workbench should be free from dirt and grease. When greasy parts are removed, they should be cleaned in a nonrust solvent.

CAUTION: *Never use carbon tetrachloride for cleaning parts. The fumes from this solvent are very poisonous.*

The testing of the parts of an electric motor is carried out in the same manner as tests for generator parts. A growler is used to test armatures for shorted or open coils. An ohmmeter or continuity tester is used to test for a ground between the armature windings and the core. Field coils may be tested with an ohmmeter or continuity tester for open circuits, short circuits, and grounds.

After a motor has been assembled, it should be given an operational test before it is installed in an airplane or missile. First the armature should be turned by hand to see that it rotates freely; there must be no roughness or unusual noise when this is done. The motor should then be operated with a low load for about 10 min to seat the brushes. The value of the voltage applied should be according to overhaul specifications. During the time that the motor is being tested, it should be observed closely for excessive heating and vibration. Directions for testing specific motors are usually included with the manufacturer's maintenance and overhaul instructions. When these instructions are available, they should be followed carefully.

REVIEW QUESTIONS

1. Define an *electric motor.*
2. Describe *series-wound, shunt-wound,* and *compound-wound* motors.
3. What is the principal characteristic of a series-wound motor?
4. For what type of load would a series-wound motor be most suitable?
5. What is the principal characteristic of a shunt-wound motor?
6. Explain why a typical dc motor rotates when connected to a proper power source.
7. Describe the method for determining the direction of rotation of a dc motor.
8. Why does the current being drawn by a shunt-wound motor decrease as the motor rpm increases?
9. What may happen to a series motor if it is connected to power without having a load?
10. What is the resistance requirement for the field winding of a shunt-wound motor?
11. What is the nature of the magnetic field in an induction motor?
12. How many poles are found in the stator of a three-phase motor?
13. Why does the rotor of an induction motor react with the field of the stator?
14. Describe a *squirrel-cage rotor.*
15. What is meant by the *pull-out point* of an ac motor?
16. Explain how the starting torque of an induction motor may be improved.
17. Explain the operation of a split-phase motor.
18. What is the principal characteristic of a synchronous motor?
19. Name some of the internal motor losses which occur in the operation of an ac motor.
20. How is reduced weight attained in the design of motors for use in aircraft?
21. Why are heavy windings used in the armature of a dc starter motor?
22. Describe the construction of a reversible motor.
23. Draw a circuit for the operation of a typical reversible motor.
24. Explain the operation of magnetic clutch and brake assemblies.
25. Why is it necessary to disengage an actuator motor from its drive when it is turned off?
26. Explain the adjustment of the limit switches in an actuator circuit.

27. Draw the circuit for a reversible single-phase ac motor.
28. How may the direction of rotation of a three-phase motor be reversed?
29. When high-performance motors are used for driving pumps, what design feature is employed to greatly increase the output of the motor with respect to its weight?
30. List some of the typical precautions that must be observed in the removal and installation of electric motors.
31. List general rules for the disassembly of an electric motor.
32. What may be the cause if an electric motor runs too slowly?
33. What may cause excessive arcing of the brushes in an electric motor?
34. If a newly installed three-phase motor turns in the wrong direction, what is the cause?

CHAPTER 11

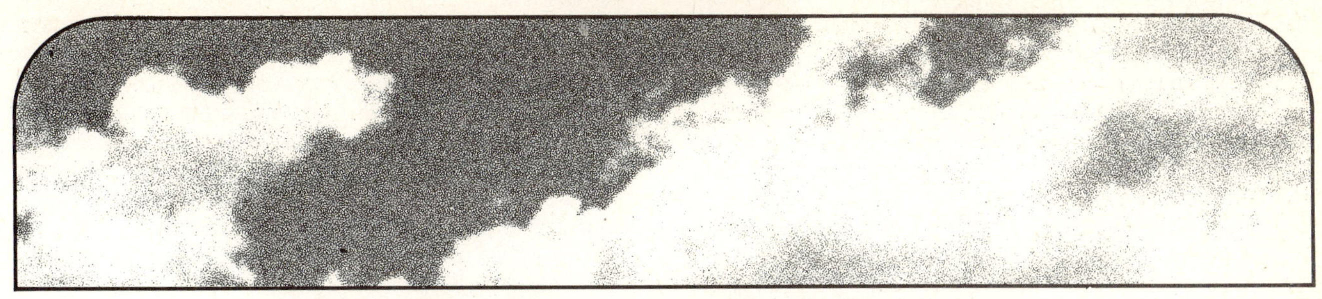

Installation and Maintenance of Electrical Systems

Modern aircraft are dependent upon the proper operation of their electric systems to obtain safe and satisfactory operation. Electric systems are required for powerplant control, systems control, navigation, communications, flight control, lights, galley operation, and other functions. With many aircraft, flight operations cannot be conducted safely without certain **essential** electric systems. It is therefore apparent that the proper maintenance of aircraft requires that the electric systems be kept in the best possible condition through inspections, testing, and the exercise of approved maintenance procedures.

To attain reliability in electric systems, it is essential that great care be exercised in the selection of components and materials and that each part be installed in such a manner that it will not be subjected to damaging conditions of any kind. For commercial and other civil aircraft, the requirements for the installation and approval of electric components and materials are established by the Federal Aviation Administration (FAA) and published in Federal Aviation Regulations (FAR). The regulations and directives of the FAA should always be observed in the maintenance of civil aircraft. For specific types of aircraft and equipment, the manufacturer's overhaul and maintenance manuals should be followed. During the design and manufacture of aircraft, the manufacturer makes certain that the requirements of the FAA are met to assure safe aircraft which can be certificated for civil use. For military aircraft, missiles, and spacecraft, specifications are established by the appropriate agency in cooperation with the manufacturers. The purpose of this chapter is to point out requirements established by all agencies for the correct installation of electric systems and components and to describe the types of wiring and other components which make up a complete electric system.

REQUIREMENTS FOR ELECTRIC SYSTEMS

GENERAL REQUIREMENTS

In general, requirements for aircraft electric systems are established to assure that the systems will perform their functions reliably and effectively. The requirements for normal, utility, and acrobatic aircraft are set forth in FAR Part 23. FAR Part 25 gives the requirements for transport category aircraft. Various changes are made in these requirements from time to time, and it is the responsibility of the FAA, manufacturers, and maintenance personnel to ensure that required changes are incorporated in certificated aircraft.

In this section it is not possible to list all current requirements in detail; however, we shall consider the principal factors that assure safe and effective electric systems. For the current requirements on an aircraft that a technician may be inspecting and maintaining, the appropriate manufacturer's bulletins and FAR should be consulted.

Electric systems for all aircraft must be adequate for the intended use. Electric power sources, their transmission cables, and associated control and protective devices must be able to furnish the required power at the proper voltage to each load circuit essential for the safe operation of the aircraft. Compliance with the foregoing requirement must be substantiated by an electrical load analysis, or by electrical measurements, that account for the electrical loads applied to the electric system in probable combinations and for probable durations.

Electric systems, when installed, must be free from hazards in themselves, in their methods of operation and in their effects on other parts of the aircraft. They must be protected from fuel, oil, water, and other detrimental substances and from mechanical damage such as abrasion or physically applied force. The systems must be designed so that the risk of electric shock to the crew, passengers, and ground personnel is reduced to a minimum.

Electric power sources must function properly when connected in combination or independently except that alternators may depend on a battery for initial excitation or for stabilization. No failure or malfunction of any electric power source may impair the ability of any remaining source to supply load circuits essential for safe operation of the aircraft except that an alternator that depends on a battery for initial excitation or for stabilization may be stopped by failure of that battery.

Each electric power source control must allow the independent operation of each source except that controls associated with alternators that depend on a battery for initial excitation or for stabilization need not break the connection between the alternator and its battery. A design of this type makes it possible to disconnect an alternator in a parallel system without affecting the operation of other alternators or generators in the system other than to increase the load on the active generators.

There must be at least one generator in an electric system if the system supplies power to circuits that are essential for safe operation of the aircraft. Each generator must be able to deliver its continuous rated power. If the design of the generator and its associated circuit is such that a reverse current could flow from the battery to the generator, a reverse-current relay or cutout must be provided in the circuit to disconnect the generator from the other generators and the battery when enough reverse current exists to damage the generator. Alternator systems do not require a reverse-current cutout because the diode rectifiers in the alternators prevent a reverse current flow.

Generator-voltage-control equipment must be able to regulate the generator output within rated limits on a continuous and dependable bases. Each generator or alternator must have an overvoltage control designed and installed to prevent damage to the electric system, or equipment supplied by the system, that could result if the generator were to develop an overvoltage condition. There must be a means to warn the flight crew immediately in case any generator in the system should fail.

There must be a means to indicate to appropriate flight crew members the electrical quantities in the system essential for safe operation of the aircraft. Generally, one or more ammeters are required. For dc systems, an ammeter that can be switched into each generator feeder may be used, and if there is only one generator, the ammeter may be in the battery feeder.

Electric equipment in a system must be so designed that in the event of a fire in the engine compartment, during which the surface of the firewall adjacent to the fire is heated to 2000°F [1093°C] for 5 min or to a lesser temperature substantiated by the applicant, the equipment essential to continued safe operation of the aircraft and located behind the firewall will function satisfactorily and will not create an additional fire hazard.

If provisions are made for connecting an external power source to the aircraft and that external power can be electrically connected to equipment other than that used for engine starting, means must be provided to ensure that no external power source having a reverse polarity or a reverse phase sequence can supply power to the aircraft's electric power system. This is usually accomplished by the use of a plug with different sized prongs. This makes it impossible to insert the plug in the receptacle incorrectly.

REQUIREMENTS FOR TRANSPORT AIRCRAFT

All systems and equipment installed in transport category aircraft must meet certain basic safety requirements, and these are set forth in FAR Part 25. All systems must be designed so they will perform their intended functions under foreseeable operating conditions. The electric system and associated components, when considered separately and in relation to other systems, must be designed so that the occurrence of any failure condition which would prevent the continued safe flight and landing of the aircraft is extremely improbable. Any failure that would reduce the capability of the aircraft or the ability of the crew to cope with adverse operating conditions must be improbable.

Warning information must be provided to alert the crew to unsafe operating conditions, thus enabling them to take appropriate corrective action. Systems, controls, and associated monitoring and warning equipment must be designed to minimize crew errors that could cause additional hazards. Compliance with requirements must be shown by analysis and, where necessary, by appropriate ground, flight, or simulator tests. The analysis must consider possible modes of failure, including malfunctions and damage from external sources. It must deal with the probability of multiple failures and undetected failures and the resulting effects on the aircraft and occupants, considering the stage of flight and operating conditions. The analysis must also consider the crew warning cues, the corrective action required, and the capability of detecting faults.

The power sources and the electric system must be able to supply the following power loads in probable operating combinations and for probable durations:

1. Loads connected to the system with the system functioning normally
2. Essential loads, after failure of any one prime mover, power converter, or energy storage device (battery)
3. Essential loads after the failure of any one engine on two-engine aircraft
4. Essential loads after the failure of any two engines on three-or-more-engine aircraft
5. Essential loads for which an alternate source of power is required, after any failure or malfunction in any one power supply system, distribution system, or other utilization system

Further requirements for electric systems in transport category aircraft specify that the generating capacity for the system and the number and kinds of power sources must be determined by a **load analysis.** The generating system includes electric power sources, main power buses, transmission cables, and associated control, regulation, and protective devices. The system must be designed so that power sources function properly when independent and when connected in combination with other sources. No failure or malfunction of any power source can create a hazard or impair the ability of remaining sources to supply essential loads. The design of the system must be such that the system voltage and frequency, as applicable, at the terminals of all essential load equipment can be maintained within the limits for which the equipment is designed, during any probable operating condition. System transients (variations in voltage and frequency) due to switching, fault clearing, or other causes must not make essential loads inoperative and must not cause a smoke or fire hazard.

There must be means accessible in flight to appropriate crew members for the individual and collective disconnection of the electric power sources from the system. The system must include instruments such as voltmeters and ammeters to indicate to appropriate crew members that the generating system is providing the electric quantities essential for the safe operation of the system.

It must be shown by analysis, tests, or both, that the aircraft can be operated safely in VFR (visual flight rules) conditions for a period of not less than 5 min with the normal electric power sources, excluding the battery, inoperative; with critical type fuel, from the standpoint of flameout and restart capability; and with the airplane initially at the maximum certificated altitude. Parts of the electric system may remain on if a single malfunction, including a wire-bundle or junction-box fire, cannot result in loss of the part turned off and the part turned on; the parts turned on are electrically and mechanically isolated from the parts turned off; and the electric wire and cable insulation, and other materials, of the parts turned on are self-extinguishing as proved by required tests.

INSTALLATIONS

The electric equipment, controls, and wiring for an aircraft must be installed so that operation of any one unit or system of units will not adversely affect the simultaneous operation of any other electric unit or system essential to the safe operation of the aircraft. Cables and wires must be grouped, routed, and spaced so that damage to essential circuits will be minimized if there are faults in heavy-current-carrying cables. This means that cables that might be subject to burning in case of a short circuit should not be grouped with essential circuit cables, because the burning of the shorted cable could also damage the essential circuit to the extent that it would not be operable.

The installations designed for an aircraft by the manufacturer are usually acceptable; however, changes are sometimes required, and these are called to the attention of the aircraft owner or operator by means of manufacturer's bulletins or Airworthiness Directives issued by the FAA.

NEED FOR PROTECTIVE DEVICES

Short circuits in electric systems constitute a serious fire hazard and also may cause the destruction of electric wiring and damage to units of electric equipment. For these reasons adequate protective devices and systems must be provided. Such devices include fuses, circuit breakers, and cutout relays.

In the generating system the protective devices must be of a type which will deenergize and disconnect faulty power sources and power-transmission equipment from their associated buses with sufficient rapidity to provide protection against hazardous overvoltage and other malfunctioning.

All resettable circuit protective devices should be so designed that when an overload or circuit fault exists, they will open the circuit irrespective of the position of the operating control. This means, of course, that a circuit protective device must not be of a type which can be overridden manually. Protective devices must, however, be clearly identified and accessible for resetting in flight if they are in an essential circuit. Resetting may be done only after the fault is corrected.

When fuses are used in an aircraft electric system, spare fuses must be provided for use in flight in a quantity equal to at least 50 percent of the number of fuses of each rating required for complete circuit protection. If oly one fuse of one particular rating is used in the aircraft system, then one spare should be carried for that rating.

Protective devices are not required in the main circuits of starter motors or in circuits where no hazard is presented by their omission. Each circuit for essential loads must have individual circuit protection; however, individual protection for each circuit in an essential load system is not required.

All fuses, circuit breakers, switches, and other electric controls in an airplane must be clearly identified so that the pilot or other member of the crew may quickly and easily perform in flight any necessary service to the unit. A **master switch** must be provided which will make it possible to disconnect all power sources from the distribution system. By means of relays, the actual disconnect should be made as near to the power source as possible.

ELECTRICAL LOAD

The electrical load of an aircraft is determined by the load requirements of the electric units or systems which may be operated simultaneously. It is essential that the electrical load of any aircraft be known by the owner or operator, or at least by the person responsible for maintenance of the aircraft. No electric equipment can be added to an aircraft's electric system until or unless the total load is computed, and it is found that the electric power source for the aircraft has sufficient capacity to operate the additional equipment.

To determine the electrical load of an aircraft, an **electrical-load analysis** is made. This is done by adding all the possible loads which can be operating at any one time. Loads may be **continuous** or **intermittent,** depending upon the nature of the operation. Examples of continuous loads are navigation lights, rotating beacon, radio receiver, radio navigation equipment, electric instruments, electric fuel pumps, electric vacuum pumps, air-conditioning system, and all other units or systems which may be operated continuously during flight.

Intermittent loads are those which are operated for only a few seconds or minutes and are then turned off. Examples of intermittent loads are landing gear, flaps, emergency hydraulic pumps, trim motors, landing lights, and circuits for other electrically operated devices which are normally operated for a very short period of time.

In computing the electrical load for an aircraft, all circuits which can or may be operated at any one time must be considered. The total **probable continuous load** is the basis for selecting the capacity of the power source. It is recommended that the probable continuous load be not more than 80 percent of the generator capacity on aircraft where special placards or monitoring devices are not installed. This permits the generator or alternator to supply the load and also keep the battery charged.

During periods when a heavy intermittent load such as landing gear is operated, an overload will probably exist, and the overload will be met for a short time by the battery and generator together. The operator of the aircraft should understand that prolonged operation under overload conditions will cause the battery to discharge to the extent that it cannot provide emergency electric power.

On twin-engine aircraft where two generators are used to supply the electric power, the capacity of the two generators operating together is used when computing power requirements. The probable continuous load is not excessive if the two generators can supply the power. When the total continuous load is greater than the capability of one generator to supply, it is necessary to provide for load reduction if one of the generators or one engine fails. The load should be reduced as soon as possible to a level which can be supplied by the operating generator.

The load condition during operation can be determined by observing the ammeter and voltmeter. When the ammeter is connected between the battery and the battery bus so that it will indicate CHARGE or DISCHARGE, it will be known that the system is not overloaded as long as the ammeter shows a charge condition. In this case, a voltmeter connected to the main power bus will show that the system is operating at rated system voltage. If there is an overload, the ammeter will show a discharge, and the voltmeter will give a low reading, the value of which is determined by the amount of the overload.

When the ammeter is connected in the generator lead, and the system is not current limited, an overload will be indicated when the ammeter reading is above the 100 percent mark. The ammeter should be "red-lined" so the pilot can determine easily when an overload exists. The pilot can then shut off some equipment and reduce the load to a suitable level.

The principal concern of the aviation maintenance technician with respect to electrical load in an aircraft is a situation where it is desired to add electric equipment. If the addition of such equipment has been tested and approved by the FAA for a particular installation, instructions will be available from the manufacturer of the equipment or the aircraft setting forth all requirements of the installation. These instructions should be followed carefully.

ELECTRIC WIRING

CHARACTERISTICS OF AIRCRAFT ELECTRIC WIRE

Because of the severe conditions that may be imposed upon the electric wiring or cable in aircraft service, the wire must meet certain standards. Wire or cable approved for aircraft use must be of the twisted or stranded type to provide flexibility. Depending upon the flexibility desired, the wire may have very fine strands or relatively coarse strands. For example, an American Wire Gage (AWG) 20 wire may have seven strands of AWG28 wire or 19 strands of AWG32 wire. A typical AWG22 wire has 19 strands of AWG34 wire.

The insulation for aircraft wire may be polyvinyl chloride (PVC), nylon, polyvinylidene fluoride, polyethylene, fluorinated ethylene propylene (FEP), polytetrafluoroethylene (TFE or Teflon), polymide, or some other material. In any case, the type of insulation is governed by operating temperature, insulation resistance, abrasion resistance, chemical resistance, and strength. Design engineers determine the requirements and specify the type of wire that will meet the requirements for each circuit.

In addition to the basic insulation of electric wire, an outer sheath is often applied to protect the wire from abrasion. This sheath may be PVC, TFE, nylon or other material and may be reinforced with glass fiber.

Typical specification numbers for approved aircraft wire are MIL-W-5086, MIL-W-16878, MIL-W-22759, and MIL-W-81381. Approved wire is also covered by MS standard numbers MS25190, 25191, 25471, 27110, 17331, 17332, and others. They often carry both the MS and the MIL specification numbers.

The specification number for aluminum aircraft wire is MIL-W-7072. When copper wire is replaced with aluminum wire, the aluminum wire should be two sizes larger than the copper wire because of the greater resistance of aluminum wire.

Characteristics of some typical aircraft wires are as follows:

MIL-W-5086/1 (MS25190)
Tin-coated copper conductor, PVC insulator, nylon outer sheath, 600 V, 221°F [105°C]

MIL-W-5086/2
Tin-coated copper conductor, PVC insulator, PVC-glass-nylon outer sheath, 600 V, 105°C

MIL-W-16878/1
Tin-coated copper conductor, PVC insulator, 600 V, 105°C

MIL-W-22759/16
Tin-coated copper conductor, ETFE insulation, 600 V, 302°F [150°C]

MIL-W-22759/19
Silver-plated high-strength copper alloy conductor; ETFE insulation; 600 V, 150°C

MIL-W-22759 (MS17411)
Silver-coated copper conductor, mineral-reinforced TFE insulation, 1000 V, 392°F [200°C]

The wires and cables described above are a few that are approved for aircraft use. Other types are also approved and are selected by engineers to meet certain specifications as required by the circuit design. Typical wires and cables are illustrated in Fig. 11-1.

Some circuits in an airplane require the use of shielded wiring to eliminate radio interference or to prevent undesirable voltages from being induced in the circuit. In this case the wire is manufactured with an outer metal sheath (shielding) of woven wire over the insulation. The center wire is the hot wire and the metal shielding is the ground. Shielded wires and cables are made with the shielding exposed and also with additional insulation over the shielding. In the latter case the cable may be called **coaxial** cable. Cable of this type is used for microphone leads, antenna transmission lines, and certain other special applications.

Electric wiring or cable used in areas where high temperatures exist must have heat resistance insulation. Fiberglas, asbestos, teflon, silicone products and similar materials are used for insulation. In specific installations, the manufacturer's specification for the particular wire or cable should be used. All standard wiring or cable should be protected from heat to prevent deterioration of the insulation.

Electric wire for aircraft circuits may be white or colored. If colored wire is used, it is generally employed to aid in identifying certain circuits and is indicated by color on the circuit diagram. Colored wire may be ordered by specifying a particular code number or letter provided with the specification number in the wire catalog.

It should be noted that the terms *wire* and *cable* are often used interchangeably. The term *wire* is used to indicate single strand and small stranded conductors. *Cable*

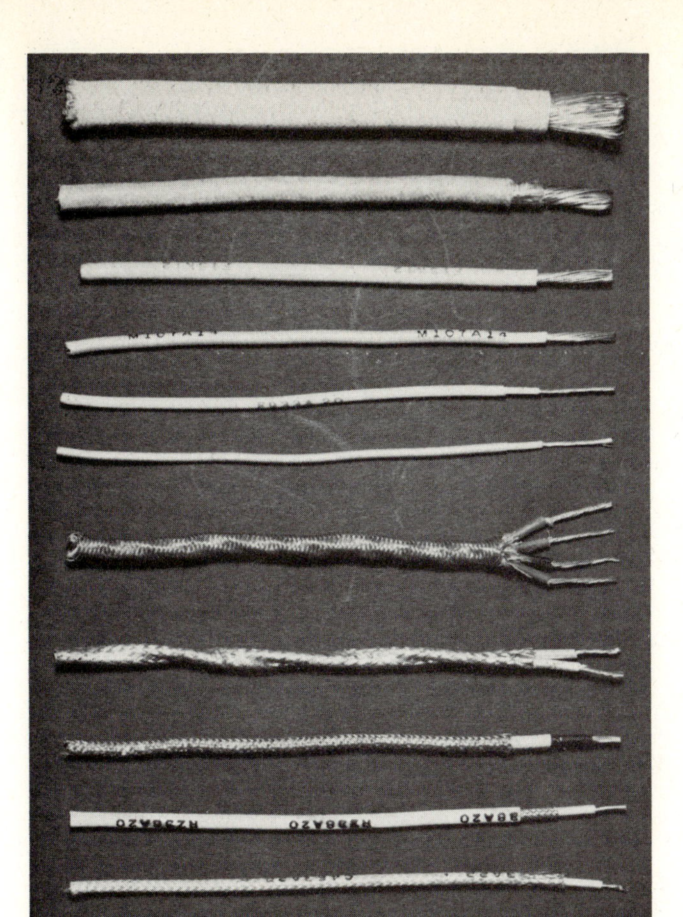

Fig. 11-1 Typical electrical wire and cable. (Prestolite)

generally refers to larger stranded conductors and multiple conductors in one outer sheath.

WIRE SIZE

The size of wire is specified according to the American Wire Gauge with the cross-sectional area indicated in circular mils. A **circular mil** is the area of a circle having a diameter of 0.001 in [0.0254 mm]. One circular mil, therefore, is a circular area of 0.000007854 in^2 [0.0005067 mm^2]. An AWG20 wire has a cross-sectional area equal to approximately 1119 cmil, which is 0.00084529 in^2 [0.567 mm^2]. An AWG10 size wire is 10443 cmil in cross-sectional area, which is 0.0082 in^2 [5.29 mm^2].

There are two principal requirements for any wire carrying current in an aircraft electric system: The wire must be able to carry the required current without overheating and burning, and it must carry the required current without producing a voltage drop greater than that which is permissible for aircraft circuits. For the guidance of technicians engaged in the replacement or installation of electrical wiring in civil aircraft, the FAA has prepared charts and tables setting forth the wire sizes needed to meet various conditions of installation and load. Table 11-1 gives the allowable voltage drop for each of the standard systems according to the nominal voltage of the system.

Table 11-1 establishes the maximum voltage drop which may occur between the power bus and any unit of electric equipment. The voltage drop between the generator and the bus or between the battery and the bus shall not exceed 2 percent of the regulated voltage when the generator is carrying its rated load or when the battery is discharging at the 5-min rate.

TABLE 11-1

Nominal system voltage	Allowable voltage drop: Continuous operation	Allowable voltage drop: Intermittent operation
14	0.5	1.0
28	1.0	2.0
115	4.0	8.0
200	7.0	14.0

A chart for determining the correct wire size for various loads and operating conditions is given in Fig. 11-2. This chart also gives the necessary size and length of cable to avoid a voltage drop of more than specified for any given circuit voltage. Assume that we wish to install a cable 25 ft [7.62 m] in length to carry a load of 15 A in a 28-V circuit. First we locate the figure 25 at the left of the chart in the 28-V column. We then follow the line horizontally to the right from 25 until it intersects the diagonal line for 15 A. This intersection falls between the vertical lines numbered 12 and 14, and we select No. 12 cable for the load because No. 14 would be too small to meet the requirements. Following the diagonal line for 15 A upward to the right until it intersects the No. 12 vertical line, we note that the intersection is above curve 1. This indicates that the cable can be used continuously for a 15-A load even though it is installed in a conduit or bundle.

Referring to the chart of Fig. 11-2, the columns headed 115, 200, 14, and 28 allow for voltage drops of 4 V, 7 V, 0.5 V and 1 V, respectively. This is in accordance with the information given in Table 11-1. The figures in the columns at the left of the chart of Fig. 11-2 are needed for determining wire length for voltage drop only. They are not needed for determining current-carrying capacity with respect to overheating.

If we wish to determine how much current a conductor can carry safely in a conduit or bundle on a continuous basis, we utilize curve 1 in the chart. For example, if a load in a particular circuit is known to be 20 A, we follow the diagonal 20-A line downward to the left and note that it intersects curve 1 between the No. 12 and No. 14 wire size lines. A No. 12 wire would be selected to carry the 20-A load. If the wire were single and in free air, a No. 16 could be used, because the 20-A line and curve 2 intersect between the lines for wire sizes No. 16 and No. 18. Note that the larger size wire is always selected when the indicated wire size falls between two size lines. This assures that the wire can carry the load without overheating.

When we use the chart of Fig. 11-2 to determine how much current a particular wire can carry, we start with the wire size line. If we want to know how much current a No. 12 wire can carry safely in a conduit or bundle, we follow the No. 12 size line upward on the chart to a point above curve 1. It will be noted that the intersection of the line and curve 1 falls between the 20-A and 30-A lines. By measuring, it can be seen that the intersection point is at approximately 23 A, so this current could be carried safely by the conductor.

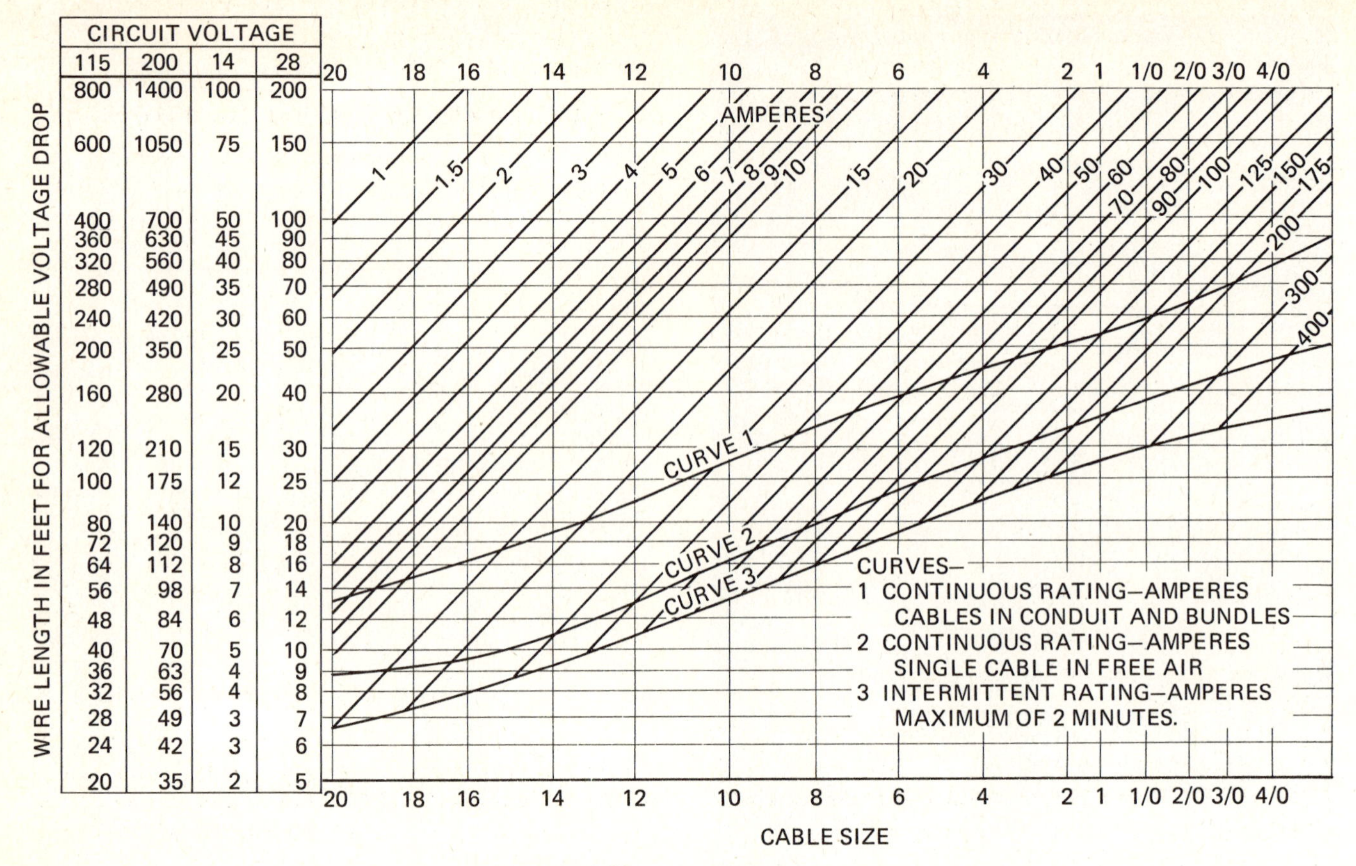

Fig. 11-2 Wire size and load chart.

To determine what length of a certain size of cable will carry a given load without more than a 1-V voltage drop, we must first make sure that the cable is large enough to carry the load. If we are using No. 8 cable to carry a continuous load of 40 A, we follow the No. 8 line vertically to its intersection with the 40-A diagonal line. From this intersection we project horizontally and find that the cable length should be about 35 to 36 ft [10.7 to 11 m]. The intersection of the amperage line and the cable-size line is above curve 1; hence, the circuit will be suitable for continuous service in a conduit or bundle.

For circuits in which more or less than a 1-V drop is allowed, the columns of figures to the left of the 28-V column in Fig. 11-2 are employed. If a certain circuit in a large airplane requires 140 ft [42.7 m] of wire and the voltage of the circuit is 200, the wire length of 140 ft is first located in the 200-V column. If the circuit carries 10 A, follow the horizontal line to the right from the 140 ft location to the diagonal 10-A line. The intersection of the 140-ft line with the 10-A line is just to the left of the No. 16 wire size line. A No. 16 wire would therefore be used to assure that the voltage drop would be no greater than 7 V, which is allowed in a 200-V circuit. This can be proved by the use of Ohm's law. Number 16 stranded copper wire has a resistance of 4.76 Ω per 1000 ft [304.8 m]. A length of 140 ft would then have 0.14 ×4.76 or 0.666 Ω. Since the voltage drop is equal to $I \times R$, then the voltage drop in this case would be 10 ×0.666, or 6.66 V. This is below the 7-V requirement, so the wire is satisfactory.

To determine whether the No. 16 wire in the foregoing circuit will carry the 10-A load without overheating in a bundle or conduit, observe the position of the intersection of the wire size line and the 10-A line. It will be noted that this intersection is well above curve 1; hence, the wire will not overheat with a 10-A load.

CURRENT-CARRYING CAPACITY OF WIRE

It is often desirable to obtain more information about wire capacity and characteristics than is provided in Fig. 11-2. For this purpose Table 11-2 is useful.

TABLE 11-2 Capacity, Weight, and Resistance for Stranded Copper Electric Wire

	Maximum amperes				
Wire size, AWG	Free air	Conduit or bundled	Resistance Ω/1000 ft (20°C)	Area, cmil	Weight, lb/1000 ft
20	11	7.5	10.25	1 119	5.6
18	16	10	6.44	1 779	8.4
16	22	13	4.76	2 409	10.8
14	32	17	2.99	3 830	17.1
12	41	23	1.88	6 088	25.0
10	55	33	1.10	10 443	42.7
8	73	46	0.70	16 864	69.2
6	101	60	0.436	26 813	102.7
4	135	80	0.274	42 613	162.5
2	181	100	0.179	66 832	247.6
1	211	125	0.146	81 807	288.0
0	245	150	0.114	104 118	382
00	283	175	0.090	133 665	482
000	328	200	0.072	167 332	620
0000	380	225	0.057	211 954	770

It will be noted that the indicated current-carrying capacity for copper electric wire is approximately the same in Fig. 11-2 and Table 11-2, although small differences show up in certain instances. Note that No. 6 wire is rated to carry 60 A safely in a conduit or bundle in both the chart and the table. In free air the No. 6 wire is rated for about 96 A in the chart and for 101 A in the table. These apparent discrepancies are not great enough to make much difference; however, for civil aircraft the chart ratings of Fig. 11-2 should be used.

From Table 11-2 it is possible to compute the voltage drop for any length of copper wire with any given load. For example, if it is desired to know the voltage drop in 100 ft [30.5 m] of No. 18 wire carrying 10 A, we use Ohm's law, but we must first determine the resistance from the figures given in the table. Note that the resistance of 1000 ft [304.8 m] of No. 18 wire is 6.44 Ω. Then, for 100 ft of the same wire the resistance would be 0.644 Ω. Then, by Ohm's law,

$$E = 10 \times 0.644 = 6.44 \text{ V}$$

Thus, we see that 100 ft of No. 18 wire will produce a voltage drop of 6.44 V when carrying a current of 10 A. To find the length of this wire which will produce a voltage drop of 1 V with a 10-A load, we merely divide 100 by 6.44. This produces a result of approximately 15.5 ft [4.57 m], which is the same length of cable for a 1-V drop as was indicated by the chart in Fig. 11-2 for No. 18 wire.

Although it is permissible to use aluminum wire in aircraft installations, the size of the wire must be larger than that of a copper wire for the same load. In general, an aluminum wire two sizes larger than the copper wire will be adaptable. Table 11-3 gives capacity, resistance, size, and weight for MIL-W-7072 aluminum wire approved for aircraft use.

Table 11-3 lists aluminum cable only of sizes 6 through 0000 because smaller aluminum cables are not recommended for aircraft use. It is interesting to note that aluminum cables of the larger sizes can be used advantageously to save weight, even though the aluminum cable is larger in diameter. Note that No. 00 aluminum cable has almost as much capacity as No. 0 copper cable but that in lengths of 1000 ft [304.8 m] the aluminum cable weighs only 204 lb [92.5 kg] against 382 lb [173.3 kg] for the copper cable. This is a saving of 178 lb [80.7 kg] for 1000 ft of cable. When substituting No. 0000 aluminum cable for No. 000 copper cable, it is found that the weight of the aluminum cable is less than half the weight of the copper cable.

TABLE 11-3 Capacity, Weight, and Resistance for Aluminum Wire

Wire or cable size, AWG	Maximum amperes: Free air	Maximum amperes: Conduit or bundled	Resistance, Ω/1000 ft (20°)	Area, cmil	Weight. lb/1000 ft
AL-6	83	50	0.641	28 280	
AL-4	108	66	0.427	42 420	
AL-2	152	90	0.268	67 872	
AL-0	202	123	0.169	107 464	166
AL-00	235	145	0.133	138 168	204
AL-000	266	162	0.109	168 872	250
AL-0000	303	190	0.085	214 928	303

REQUIREMENTS FOR OPEN WIRING

Open wiring is more vulnerable to wear, abrasion, and damage from liquids than wiring installed in conduits, hence, care must be taken to see that it is installed where it is not exposed to these hazards and in a manner to prevent damage. The number of wires grouped in a bundle should be limited in order to reduce the problems of maintenance and to limit damage in case a short circuit should occur and burn one of the wires in the bundle. Shielded cable, ignition cable, and cable which is not protected by a circuit breaker or fuse should be routed separately. The bending radius of a wire bundle should not be less than 10 times the outer diameter of the bundle. This is required to avoid excessive stresses on the insulation.

Since it provides only a delaying action, soft insulating tubing, sometimes called *spaghetti,* cannot be considered good mechanical protection against the external abrasion of electric wire. When such protection is required, conduit or ducting should be used.

CABLE LACING

The lacing of wire bundles should be performed according to accepted specifications. Approved lacing cord complying with specification MIL-C-5649 or twine specification JAN-T-713 may be used for cable lacing. If cable bundles will not be exposed to temperatures greater than 248°F [120°C] cable straps complying with specification MS3367, MS17821, or MS17822 can be used. Straps should not be used where strap failure would create an unsafe condition. Typical tie straps are shown in Fig. 11-3.

Single-cord lacing is used for cable bundles 1 in [2.5 cm] in diameter or less. For larger bundles, double-cord lacing should be employed. Cable bundles inside a junction box should be laced securely at frequent inter-

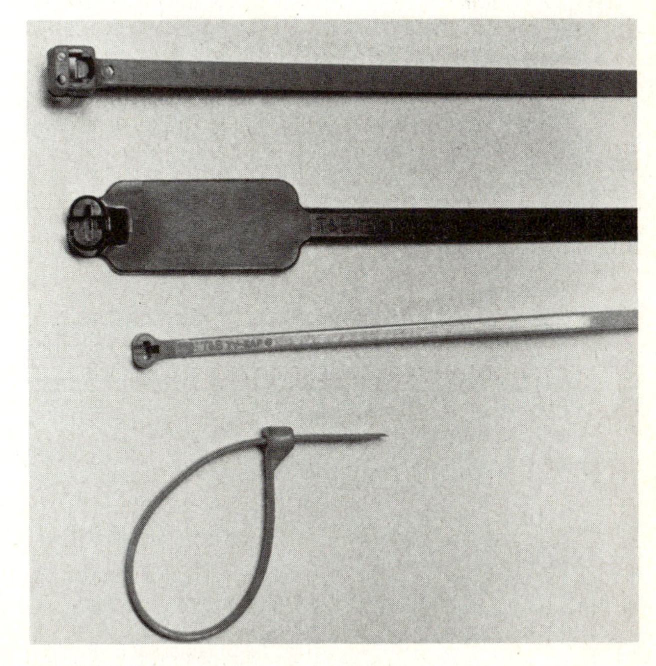

Fig. 11-3 Strap ties for wire bundles.

vals to assure that a minimum of movement can take place. In open areas, the bundles should be laced or tied if supports for the cable are more than 12 in [30.5 cm] apart.

Wire bundles may be laced with a continuous series of loops around the bundle as shown in Fig. 11-4 or with single ties as in Fig. 11-5. When the continuous lacing is applied, the first loop is a clove hitch locked with a double overhand knot as shown in Fig. 11-4*a*. The knot is pulled tight as shown in Fig. 11-4*b*, and the continuing end is then looped around the wire bundle with the cord brought over and under the cord from the previous loop to form the type of loop shown in Fig. 11-4*b*. These loops are continued at suitable intervals, and the series is then terminated with another clove hitch. The free end is wrapped twice around the cord from the previous loop and is then pulled tight to lock the loop. The terminating ends of the cord are trimmed to provide a minimum length of $\frac{3}{8}$ in [0.95 cm]. The method for making the terminal loop is illustrated in Fig. 11-4*c*.

When it is desired to use single ties to secure a wire bundle, the locked clove hitch is used. The clove hitch is formed as shown, and it is then locked with a square knot. Single ties are sometimes used to separate a group of wires from a bundle for identification purposes as shown in Fig. 11-6. This is to aid maintenance technicians in locating particular circuit wiring.

When double-cord lacing is required for large cable bundles, the first loop is made with a special type of slip knot similar to the "bowline-on-a-bight." This is shown in Fig. 11-7. The double cord is then used to make additional loops as required in the same manner as the single cord is used. The terminal lock knot is made by forming two single loops around the bundle and then tying the two ends with a square knot. Additional approved lacing methods are shown in Fig. 11-8.

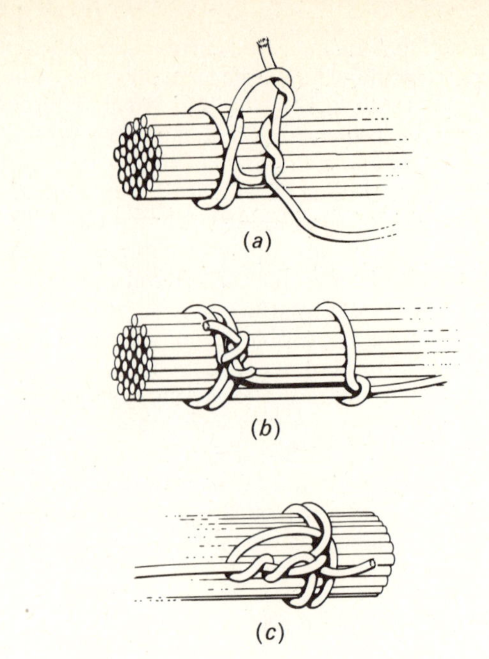

Fig. 11-4 Lacing for harnesses or wire bundles.

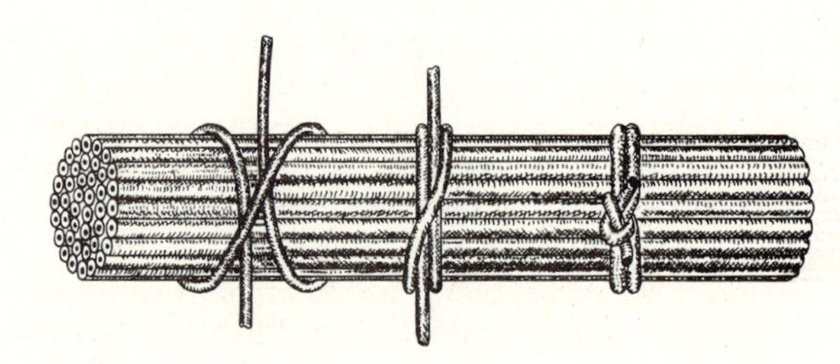

Fig. 11-5 Single-tie lacing.

WIRE AND CABLE CLAMPING

Electric cables or wire bundles are secured to the aircraft structure by means of metal clamps lined with synthetic rubber or a similar material. Specification MS-21919 cable clamp meets the requirement for civil-aircraft use. Such a clamp is illustrated in Fig. 11-9.

In the installation of cable clamps, care must be taken to assure that the stress applied by the cable to the clamp is not in a direction which will tend to bend the clamp. When a clamp is mounted on a vertical member, the loop

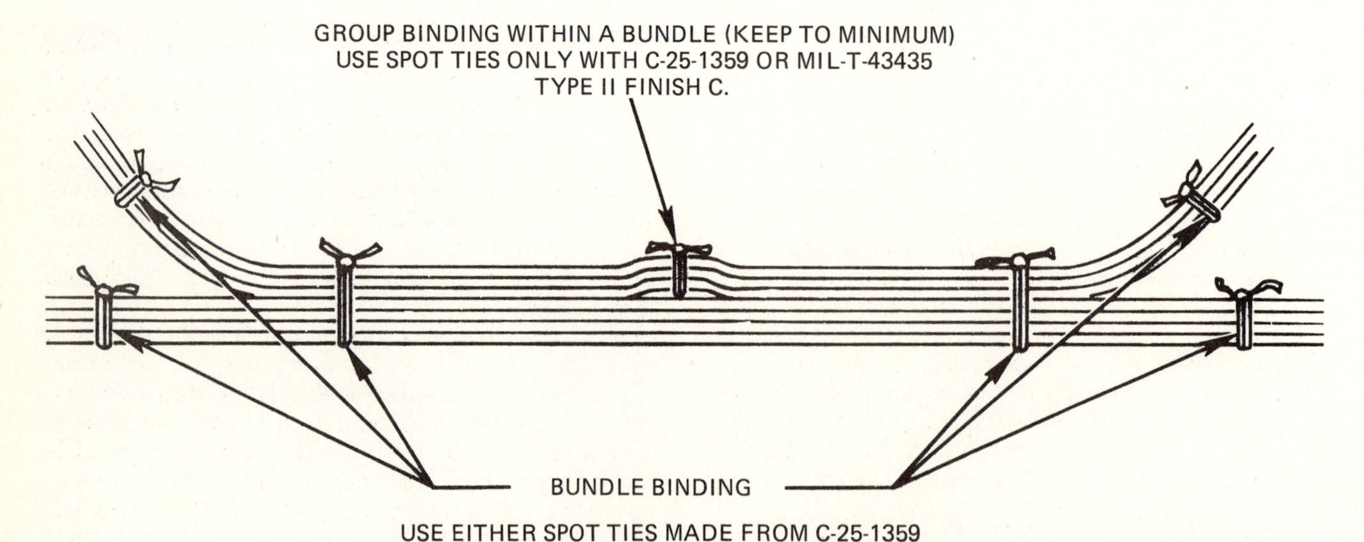

Fig. 11-6 Separation of wire groups for identification. (Lockheed)

Fig. 11-7 Beginning loop for double-cord lacing.

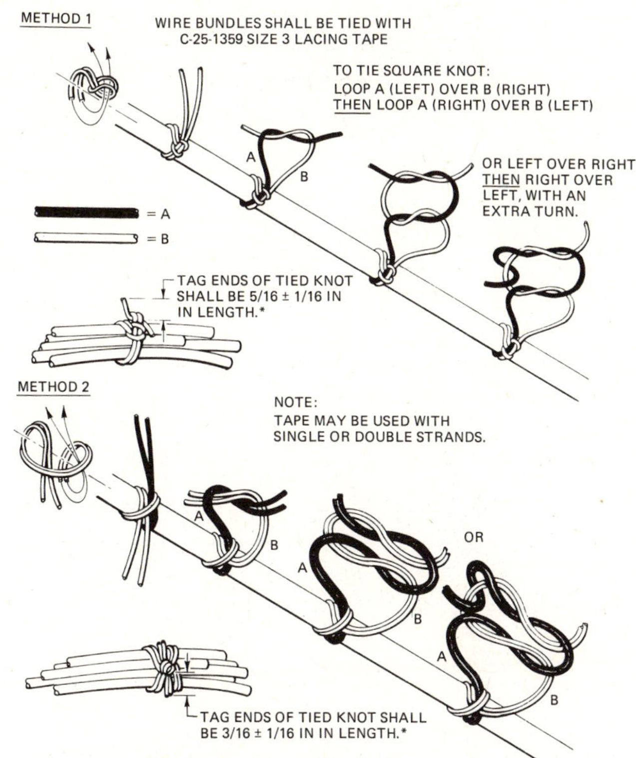

Fig. 11-8 Additional lacing methods. (Lockheed)

Fig. 11-9 Clamp for electrical cable.

of the clamp should always be at the bottom. Correct methods for installing clamps are shown in Fig. 11-10.

When a wire bundle is routed through a clamp, the bundle must be held within the rubber lining of the clamp, and no wires must be pinched between the flanges of the clamp. Pinching of the wire could cause the insulation to be damaged and a short circuit could result.

In installing electric wiring in a particular make and model of aircraft, it is the best practice to make the installation in accordance with the manufacturer's original design unless a specific change has been ordered. The clamps, wiring, and connectors should be of the same types specified and used by the manufacturer.

ROUTING OF ELECTRIC WIRE BUNDLES

The routing of electric wire should be done in a manner that will provide the protections previously mentioned,

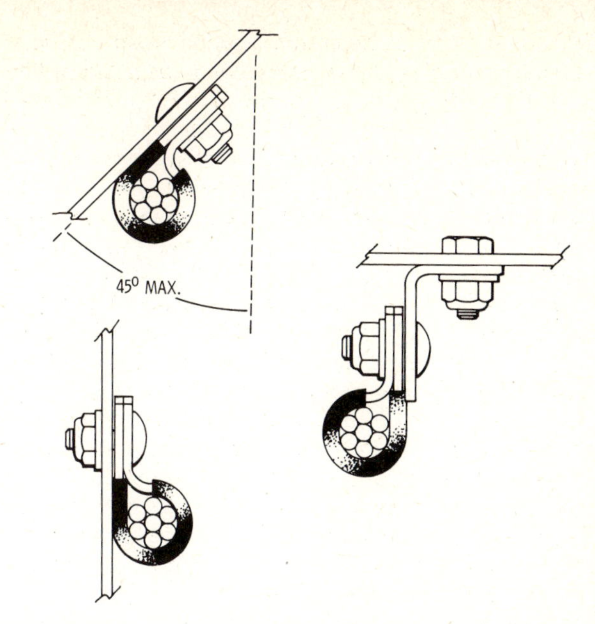

Fig. 11-10 Correct methods for installing clamps.

namely: heat, liquid, abrasion, and wear. Clamps should be installed in such a manner that the wires do not come in contact with other parts of the aircraft when subjected to vibration. Sufficient slack should be left between the last clamp and the electric equipment to prevent strain at the wire terminals and to minimize adverse effects on shock-mounted equipment. Where wire bundles pass through bulkheads or other structural members, a grommet or suitable clamping device should be provided to prevent abrasion as shown in Fig. 11-11. If a wire bundle is held by a clamp in the center of a hole through the bulkhead, and the clearance between the edge of the hole and the bundle is more than $\frac{1}{4}$ in [0.64 cm], a grommet is not required.

At points in an installation where electric cable may be exposed to oil, hydraulic fluid, battery acid, or other liquid, the cable should be enclosed in a plastic sleeve. At the lowest point in the sleeve, a hole $\frac{1}{8}$ in [0.32 cm] in diameter should be cut to provide for drainage. The sleeve can be held in place by clamps or by lacing.

If a hot cable terminal should come into contact with a metal line carrying a flammable fluid, the line might be punctured and the fluid ignited. This, of course, would result in a serious fire and probable loss of the airplane. Consequently, every effort should be made to avoid this hazard by physical separation of the cables from lines carrying oil, fuel, hydraulic fluid, or alcohol. When separation is impractical, the electric cable should be placed above the flammable fluid line and securely clamped to the structure.

Particular care must be used in installing electric wire on and in the vicinity of landing gear, flaps, and other moving structures. Slack must be allowed for required movement, but the wire must not be too loose. Routing of the wire must be such that it is not rubbed or pinched by moving parts during operation of the mechanism. An examination of the wire during a ground check of the operation of the mechanism will usually reveal any hazards.

Electric wiring must be protected from excessive heat. As noted previously, electric wiring is insulated and pro-

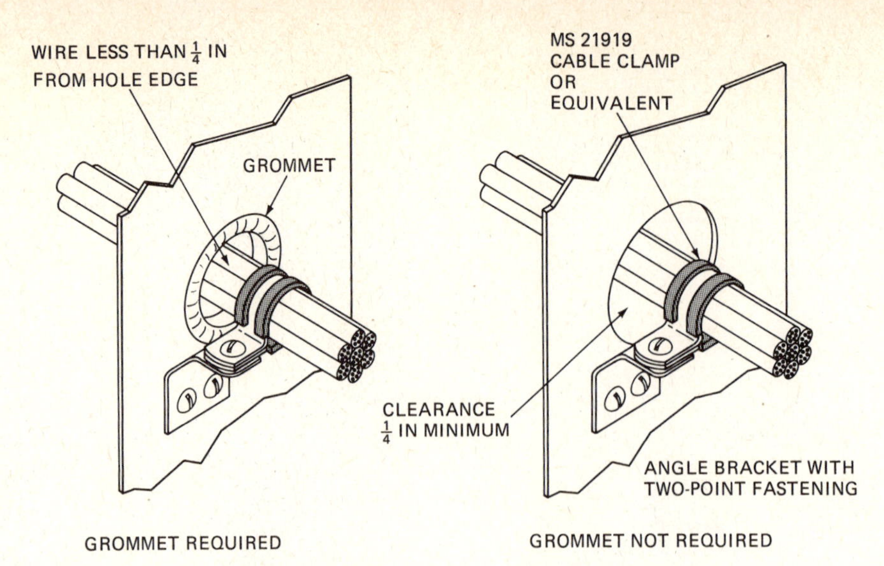

Fig. 11-11 Wire bundle passing through bulkhead.

tected with various types of materials, some of which can withstand temperatures as high as 392°F [200°C]. In areas where a wire must be subjected to high temperatures, it is necessary to use wiring with insulation made of asbestos or some other heat-resistant material. Wires should not be routed near exhaust pipes, resistors, or other devices that produce high temperatures except as required for special purposes and then only if the wire is protected with adequate heat-resistant insulation.

ELECTRICAL CONDUIT

Electrical conduit consists of thin-walled aluminum tubing, braided metal tubing called **flexible conduit,** and nonmetallic tubing. The purpose of conduit is to provide mechanical protection, and metal conduit is often used as a means of shielding electric wiring to prevent radio interference.

Approved flexible conduit is covered by specification MIL-C-6136 for aluminum and specification MIL-C-7931 for brass. The aluminum conduit is made in two types. Type I is bare and type II is rubber-covered.

The size of conduit should be such that the inside diameter is about 25 percent larger than the largest diameter of the cable bundle. This must be taken into consideration, as well as the fact that the specified diameter of conduit gives the outside diameter. The inside diameter is therefore the specified diameter less twice the wall thickness.

The inside of the conduit should be clean and free from burrs, sharp edges, or obstructions. When conduit is being cut and prepared, all edges and holes should be deburred to assure a smooth surface which will not damage the cable. The conduit should be inspected carefully after installing the end fittings to assure that the interior is clean and smooth. If a fitting is not installed on the end of a conduit section, the end should be flared to prevent the edge of the tubing from rubbing and wearing the insulation of the cable.

Installation of conduit should be such that it is protected from damage of all types. It should be securely attached to the structure with metal clamps so there can be no movement or vibration. A clean metal-to-metal contact will assure good bonding to aid in shielding. The installed conduit should not be under appreciable stress and should not be located where it may be stepped upon or used as a hand support by a member of the crew. Drain holes must be provided at the lowest point in any conduit run.

Rigid conduit which is cut or has appreciable dents should be replaced to prevent damage to the electric cable. Bends in the conduit must not be wrinkled and must not be flattened to the extent that the minor diameter is less than 75 percent of the nominal tubing diameter. Table 11-4 shows the minimum tubing-bend radii for conduit.

TABLE 11-4 Minimum Bend Radii for Rigid Conduit

Nominal tube outside diameter, in	Nominal tube outside diameter, cm	Minimum bend radii, in	Minimum bend radii, cm
$\frac{1}{8}$	0.32	$\frac{3}{8}$	0.96
$\frac{3}{16}$	0.48	$\frac{7}{16}$	1.11
$\frac{1}{4}$	0.64	$\frac{9}{16}$	1.43
$\frac{3}{8}$	0.96	$\frac{15}{16}$	2.38
$\frac{1}{2}$	1.27	$1\frac{1}{4}$	3.18
$\frac{5}{8}$	1.60	$1\frac{1}{2}$	3.81
$\frac{3}{4}$	1.92	$1\frac{3}{4}$	4.46
1	2.54	3	7.62
$1\frac{1}{4}$	3.18	$3\frac{3}{4}$	9.53
$1\frac{1}{2}$	3.81	5	12.7
$1\frac{3}{4}$	4.46	7	17.8
2	5.08	8	20.3

Flexible conduit cannot be bent as sharply as rigid conduit. This is indicated by Table 11-5, which gives the minimum bending radii for flexible aluminum or brass conduit.

When sections of flexible conduit are being replaced and it is necessary to cut the conduit, the operation can be greatly improved by wrapping the area of the cut with transparent adhesive tape. Fraying of the end will be greatly reduced because the tape will hold the fine wires in place as the cut is made with a hacksaw. Before a wire or cable bundle is placed in a conduit, the bundle should be liberally sprinkled with talc.

TABLE 11-5 Minimum Bend Radii for Flexible Conduit

Nominal internal diameter of conduit, in	cm	Minimum bending radius inside, in	cm
$\frac{3}{16}$	0.48	$2\frac{1}{4}$	5.72
$\frac{1}{4}$	0.64	$2\frac{3}{4}$	6.99
$\frac{3}{8}$	0.96	$3\frac{3}{4}$	9.53
$\frac{1}{2}$	1.28	$3\frac{3}{4}$	9.53
$\frac{5}{8}$	1.60	$3\frac{3}{4}$	9.53
$\frac{3}{4}$	1.92	$4\frac{1}{4}$	10.80
1	1.54	$5\frac{3}{4}$	14.61
$1\frac{1}{4}$	3.18	8	20.32
$1\frac{1}{2}$	3.82	$8\frac{1}{4}$	20.96
$1\frac{3}{4}$	4.46	9	20.86
2	5.08	$9\frac{3}{4}$	24.77
$2\frac{1}{2}$	6.35	10	25.40

CONNECTING DEVICES

WIRE AND CABLE TERMINALS

Since aircraft electric wires are seldom solid but are usually strands of small-gage soft-drawn tinned copper or bare aluminum twisted together to provide flexibility, the separate strands must be held together and fastened to connectors. These connectors are commonly called **terminals** or terminal lugs and are required to connect the wires to terminal posts on electric equipment or on terminal strips.

Approved terminals of the swaged or crimped type are available from several manufacturers. They are designed according to wire size and the size of the terminal stud to which the terminal is to be connected. One size of terminal will usually fit two or three different sizes of wire; for example, one size terminal will fit wire from Nos. 18 to 22. The terminals are attached to the wire by means of a special crimping tool. First, the insulation is removed with a wire stripper as shown in Fig. 11-12. Care must be taken that the stripper is of the correct size so that none of the wire strands are cut. For each type of terminal the length of insulation to be removed from the wire is specified. The bare wire is then inserted in the end of the terminal, and the terminal is crimped with the proper tool. The result must be such that the terminal attachment has a tensile strength at least equivalent to the tensile strength of the wire.

In the stripping of electric wire with a tool designed for wire stripping, the technician should see that the tool is sharp and is correctly adjusted. If the tool is not used correctly, it is likely that some of the wire strands will be cut or nicked. For wire sizes of No. 12 and below, it is not permissible to have any strands cut nicked. Number 10 copper wire may have 2 strands cut or nicked, and wire sizes from Nos. 8 to 4, inclusive, may have as many as 14 strands cut or nicked. Numbers 2 and 0 cable may have as many as 12 strands cut or nicked. Aluminum cable of any size is not permitted to have cut or nicked strands.

Figure 11-13 illustrates a group of AMP terminals properly attached to electric wire. Note that the terminal sleeves are crimped on both bottom and top. At the bottom of the illustration is a picture of an installing tool for AMP terminals. It can be seen that the tool is color-coded to match the color of the terminal sleeve. This coding as-

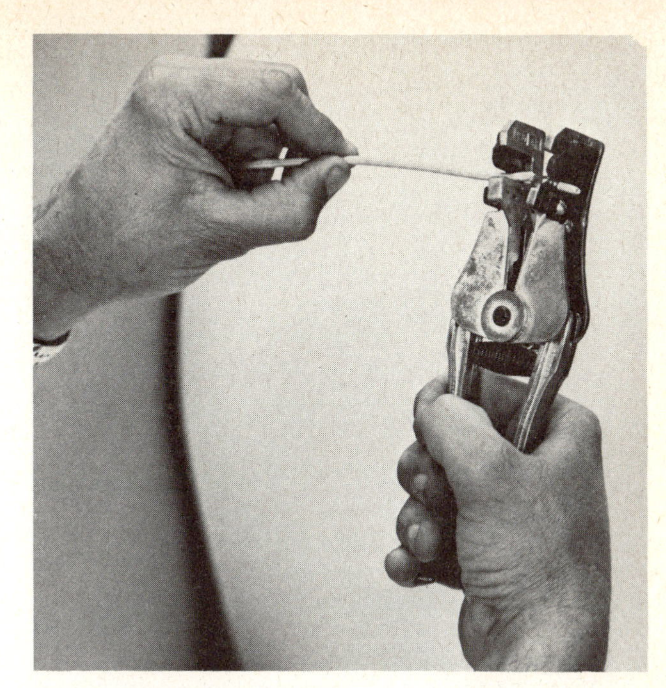

Fig. 11-12 Use of a wire stripper.

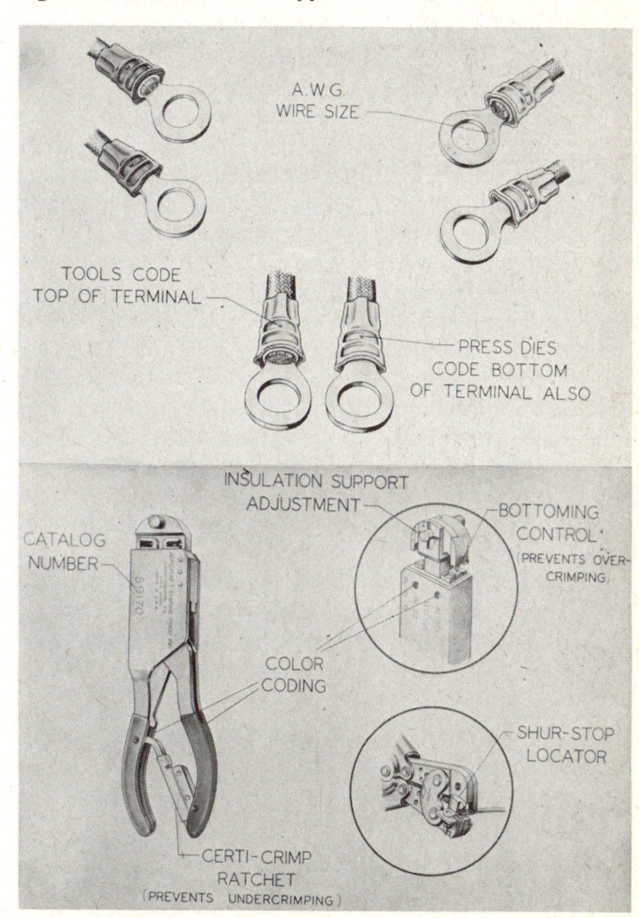

Fig. 11-13 Terminals and installing tool. (AMP Specialties)

sures that the proper size of tool will be used for each terminal. Figure 11-14 shows the construction of an AMP terminal. The terminal is equipped with a plastic insulating sleeve which makes it unnecessary to install insulation after the terminal is attached to the wire. It is important to note that after the terminal is installed, the strands of the

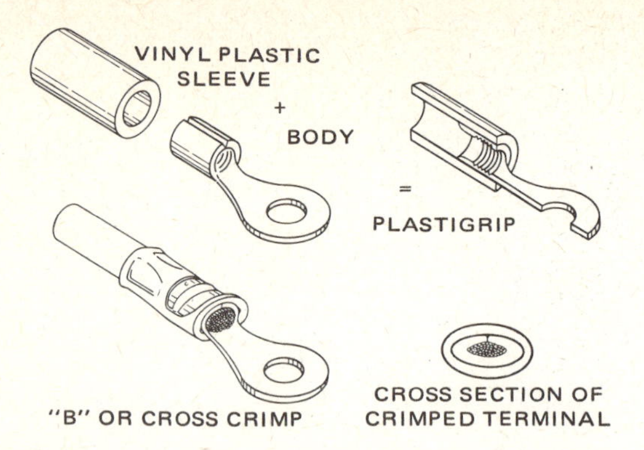

Fig. 11-14 Construction of a terminal. (AMP Specialties)

wire extend approximately $\frac{1}{32}$ in [0.079 cm] beyond the terminal sleeve. This condition is required to make sure that the terminal has sufficient grip on the wire.

The crimping tool used with AMP terminals is so constructed that it will not release until the terminal has been sufficiently crimped. This feature is provided by a ratchet installed between the handles of the tool. The use of this tool or any other special crimping tool should be attempted only by someone who is well informed about its proper operation. Manufacturers of terminals and installing tools supply instructions and specifications which give all the necessary information and data for proper installation.

In addition to the manual crimping tools, manufacturers provide power crimpers that are driven either by hydraulic or pneumatic power. When large numbers of crimps are to be made, the power tools save time and effort.

As mentioned previously, approved terminals for aircraft wire are produced by a number of different manufacturers. It is therefore important that the technician installing terminals identify the make and type of terminal and use the proper installing tools. If the wrong crimping tool is used on a terminal, it is likely that the crimping will be faulty and the wire and terminal may fail in service.

In replacing wires and terminals, the technician should use the same type of terminal that was used in the original installation if possible; however, the terminal need not be of the same make.

Wire terminal lugs are made in many styles to meet the requirements of different installations. A few of these styles are illustrated in the drawings of Fig. 11-15. As explained above, it is best practice to replace terminals with similar designs.

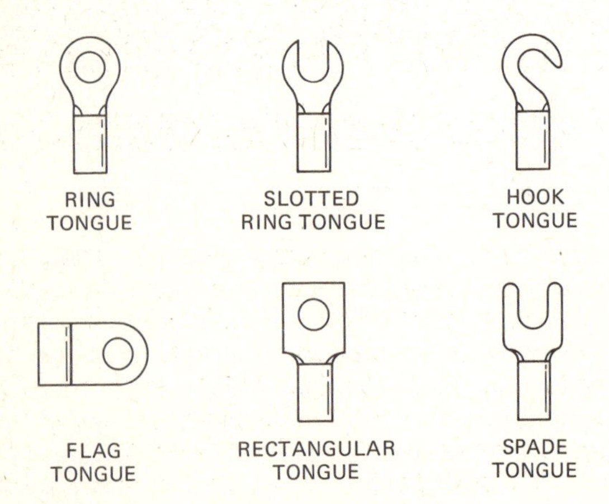

Fig. 11-15 Different types of terminal design.

Soldered terminals are considered unsatisfactory for general electrical use in aircraft electric systems, even though soldering is considered a good practice in electronic units such as radio receivers, radar equipment, and autopilot controlling circuits. Electric wires in the main electric system of an aircraft are of the flexible type. However, when a terminal is soldered to such a wire, the solder tends to penetrate the wire and make it rigid in the vicinity of the terminal. This makes the wire and terminal less resistant to vibration, with the result that it may become crystallized by fatigue and break off at the terminal. Another disadvantage of the soldered terminal is that the flux used for soldering may be of a corrosive type, thus bringing about failure through corrosion.

Because of the necessary unsoldering and resoldering, the maintenance of soldered terminal systems is more difficult than that of systems with swaged terminals. A technician well skilled in soldering techniques is required, because a poorly soldered terminal is a hazard in itself. An unskilled operator may burn the insulation, may fail to make sure that the solder is thoroughly sweated into the terminal, or may use a corrosive flux.

In the event that a joint must be soldered in an aircraft electric system, there are certain conditions which have to be observed. The flux should be of a noncorrosive type such as rosin. Rosin-core wire solder is most commonly used because the flux is automatically applied as the solder is melted on the joint. The two metal parts being joined by the solder must be brought up to the melting temperature of the solder so that the solder will flow smoothly into the joint and form a solid bond with the metal. Care must be exercised that adjacent insulation or electric units are not damaged by the heat.

SPLICING

The splicing of electric wires may be done if approved for a particular installation. In any case, the splice should be made with an approved splicing sleeve. The splicing sleeve is a metal tube with a plastic insulator on the outside or a plain metal tube that is covered with a plastic tube after the splice is made. The stripped wire is inserted into the end of the tube in the same manner in which wire is inserted into a terminal sleeve. The sleeve is then crimped with a terminal crimping tool. A typical crimp splice is shown in Fig. 11-16.

When splices are made in wires that are in a wire bundle, the spliced wires are placed on the outside of the bundle. The bundle ties or straps are located where there are no splices.

Fig. 11-16 A crimp type splice.

ELECTRIC TERMINAL STRIPS

The joining of separate sections of electric wire is usually accomplished by means of terminal strips like those illus-

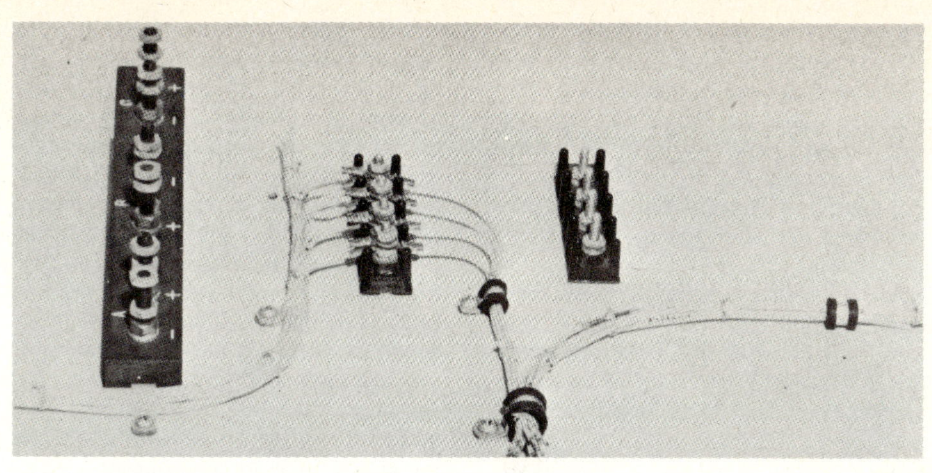

Fig. 11-17 Terminal strips.

trated in Fig. 11-17. A terminal strip is made of a strong insulating material with metal studs molded into the material or inserted through it. The studs are anchored so that they cannot turn and are of sufficient length to accommodate four terminals. Between each pair of studs are barriers to prevent wire terminals attached to different studs from coming into contact with each other. When it is necessary to join more than four terminals at a terminal strip, two or more of the studs are connected with a metal bus, and the terminals are then connected to the studs with no more than four terminals on any one stud.

The stud sizes in terminal strips must be adequate to withstand the stresses imposed during installation and tightening of the nut. For this reason it is common practice to use No. 10, or $\frac{3}{16}$ in [0.48 cm], studs for aircraft electric systems. This specification does not, of course, apply to electronic equipment and miniaturized installations, nor does it apply to equipment specifically designed and approved for a smaller terminal stud.

A stud in a terminal strip to which wire terminals are to be connected is usually mounted in the insulating strip with two flat washers, two lock washers, and two nuts as shown in Fig. 11-18. The stud is secured in the strip with a flat washer, a lock washer, and a plain nut. The terminal to be connected to the stud is placed directly upon the bottom nut, and a flat washer is placed over the terminal. The lock washer is placed upon the flat washer, and a plain nut is tightened against it. The nut should be tightened with a Spintite wrench or a socket on an extension. The torque applied to the nut must not exceed the safe limit, because excess stress can easily crack the material of the terminal strip. In all cases it is good practice to use the same combination of nuts and washers found on other terminals in the same system, provided the circuit types are similar. It is important to note that lock washers, either star type or split type, are never placed directly against a terminal or against the aircraft structure. A plain washer is always used under the lock washer to prevent damage to the terminal or structure.

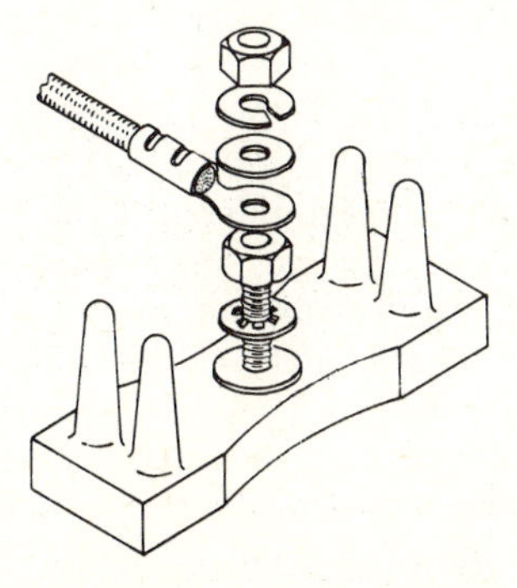

Fig. 11-18 Attachment of a terminal to a stud.

Terminal strips must be mounted in a manner and position such that loose objects cannot fall on the terminals. This may be accomplished by installing the strips on vertical bulkheads or overhead and providing them with suitable covers. A most effective method for protecting the terminal strips is to install them in a box made of metal or some other strong material. A junction box is illustrated in Fig. 11-19.

Fig. 11-19 A junction box.

Junction boxes may be constructed of any strong nonabsorbent fire-resistant material, such as aluminum alloy, Fiberglas, Formica, or stainless steel. Stainless steel is recommended when the installation must be fireproof. The box should be sufficiently rigid that the sides or top will not be subject to "oil-canning" (snapping in and out like the bottom of a hand oil can). Drain holes should be provided in the lower portion of the junction box to permit drainage of any condensed moisture accumulated.

The interior of the junction box should be designed to permit easy access to all installed items of equipment, terminals, and wire. Where marginal clearances are unavoidable, an insulating material should be placed between current-carrying parts and any grounded surface. Items of equipment should not be installed on the cover or door of a junction box, because it is not possible to inspect the clearances inside when the cover is on or the door closed.

Junction boxes should be installed so that the open face is downward to permit any loose items to fall out of the box rather than into it. Such items as nuts, bolts, washers, and other small metal units might be overlooked, become wedged between terminals during flight, and cause a short circuit.

Electric wiring inside a junction box should be laced or clamped in such a manner that terminals are not hidden, operation of equipment is not hampered, and motion between wires and equipment is prevented. The openings through which the wire bundles enter the box must be provided with clamps or grommets so that the insulation on the wires cannot become worn or otherwise damaged.

ALUMINUM WIRES AND TERMINALS

The installation of aluminum wiring and terminals requires exceptional care to ensure satisfactory operation. Aluminum wire hardens as a result of vibration more quickly than copper wire. For this reason aluminum wire should not be used where there is appreciable movement of the wire during operation of the aircraft.

Terminals, nuts, bolts, and washers used with aluminum wiring must be compatible with aluminum to avoid the electrolytic corrosion that takes place between dissimilar metals when they are in contact. Such hardware should be made of aluminum or an aluminum alloy or should be plated with cadmium or other compatible metal.

Since aluminum always has an oxide coating on the surface, special procedures must be used to eliminate the coating when joining aluminum parts for electrical contact. In installing terminal lugs on aluminum wire, means must be provided to destroy the oxide coating on the wire and inside the terminal sleeve. One method for doing this is to procure aluminum lugs with a petrolatum and zinc powder compound inside the sleeve. The terminal is wrapped with a plastic or foil cover to keep the compound inside. When the terminal is to be installed on the wire, the protective cover is removed and the wire is inserted in the sleeve. An inspection hole is provided to ensure that the wire is inserted to the correct distance. A finger is held over the inspection hole during the insertion of the wire to prevent the petrolatum compound from being squeezed out of the terminal. When the terminal is crimped, the abrasive action of the zinc dust in the compound removes the oxide film from the aluminum wire and terminal.

Some manufacturers have designed special terminal lugs for aluminum wire for "dry" installation. The material and construction are such that the crimping process automatically destroys the oxide film and produces a good contact.

CONNECTORS

Electric connectors are designed in many sizes and shapes to facilitate the installation and maintenance of electric circuits and equipment in all types of flying vehicles. For example, it may be necessary to replace a damaged section of electric harness in an aircraft. If the section of harness is connected to other sections by means of connectors, it is a comparatively simple matter to unplug the section at both ends and remove the damaged section. A completely new section may then be quickly installed. If the damaged section were connected to other sections by terminal strips, it would be necessary to disconnect each wire from the terminal studs separately; this operation would consume considerable time, especially if the harness contained many wires. Connectors are also used to connect electric and electronic assemblies such as voltage regulators, inverters, and amplifiers. When it is necessary to replace such an assembly, the connector makes it possible to disconnect the unit quickly and to reconnect the new unit with no danger of connecting any of the leads incorrectly.

A connector assembly actually consists of two principal parts. These parts are often called the **plug** and the **receptacle.** The plug section generally contains the pin sockets, and the receptacle contains the pins. The reader may encounter confusion regarding the terms *plug* and *receptacle* because some authorities call the section containing the sockets the *receptacle* and the section containing the pins the *plug.* To avoid such confusion, it is well to call the parts the *pin section* and the *socket section.*

When connector assemblies are installed, the "hot" or "live" side of the circuits should be connected to the socket section, and the ground side of the circuits should be connected to the pin section. This arrangement will reduce the possibility of shorting the circuits when the connector is separated.

Problems experienced with connectors are usually due to corrosion caused by moisture condensation inside the shell of the connector. If a connector is to be installed where corrosion is a problem, the assembly should be coated with a chemically inert, waterproof jelly, or a special waterproof connector should be installed. In all cases, any unused contact hole should be filled with a wire or plug to prevent the entrance of moisture or other foreign matter. The free end of a stub wire should be covered with potting compound or some other material to prevent electric contact.

In working on large aircraft electric systems, a technician will encounter many different types and makes of connector plugs. Among the brand names that may be encountered are Cannon, Bendix, AMP, Deutsch, Amphenol, and others. The earlier connector plugs were designed for the wires to be soldered to the pin and socket contacts; however, most connector plug assemblies are now designed with crimp-type pins and sockets. The pins and sockets are first crimped to the wires and then are installed in the plugs by means of special tools.

A typical connector assembly of the AN type is shown in Fig. 11-20. The *AN* indicates that the connector is an Air Force–Navy or Army–Navy type: that is, the specifications for the unit are established by the military services. The assembly in Fig. 11-20 is of the AN3106B design, and the parts of the socket section are shown in the drawing at the bottom of the illustration. Note that the assembly consists of a group of small tubular sockets inserted in a Bakelite insulator consisting of a front and rear section. The socket and insulator assembly is contained in a metal

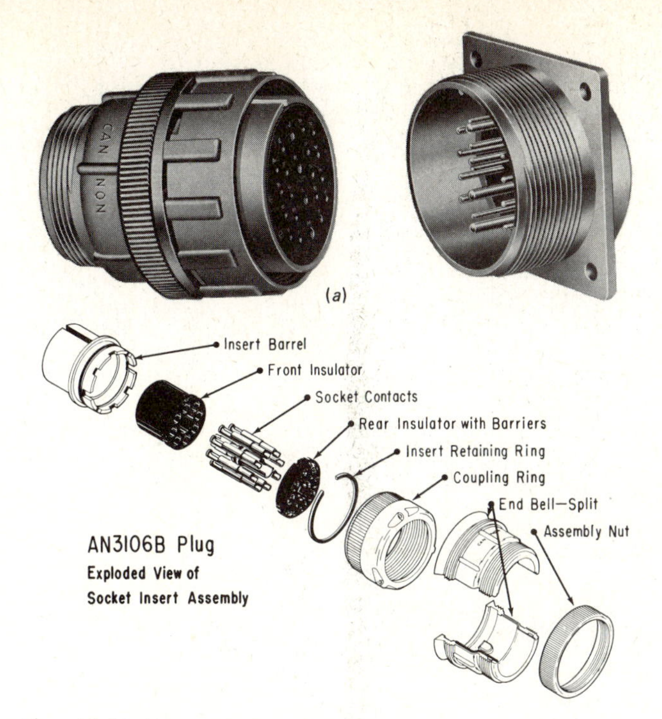

Fig. 11-20 AN-type connector.

shell which is held together with an assembly nut. Outside the entire assembly is a coupling ring by which the plug (socket section) is secured in the receptacle (pin section).

The AN number for a connector plug, together with dash numbers and letters, provides a description of the connector. For example, an AN3106-28-5S plug is defined as follows: *AN* means Army-Navy or Air Force–Navy, 3106 is the specification number, 28 is the size of the shell, and 5S identifies the socket arrangement. The receptacle for this plug is AN3102-28-5P. The meanings are the same as those for the numbers designating the plug. Note, however, that the letter *P* is found with the last dash number. This *P* indicates pin connectors; the *S* in the previous number indicates sockets.

The construction of the pins and sockets in an AN or other type of plug connector may be designed for solder connections to the electric wires, or the connector may be designed with crimp pins and sockets. At the end of the pin or socket in a solder-type connector is a small solder pocket. A short section of insulation is removed from the wire, and the bare stranded wire is then inserted in the pocket. Enough insulation should be removed from the wire so that none extends into the solder pocket. With the wire in the pocket, solder is applied with a small-pointed soldering iron or soldering gun. The solder should be of the rosin-core wire type and should be applied to the pocket as it is heated with the soldering iron. As soon as the solder flows smoothly into the pocket and penetrates the wire, the soldering iron should be removed to avoid the possibility of burning the insulation of either the wire being soldered or the adjacent wires. Only enough solder should be applied to fill the pocket, and all small drops of solder should be removed from between the pins. After each pin is soldered, a plastic sleeve insulator should be pushed down over the soldered joint and metal pin to prevent the possibility of short-circuiting. The insulating sleeves should be tied or clamped to prevent them from

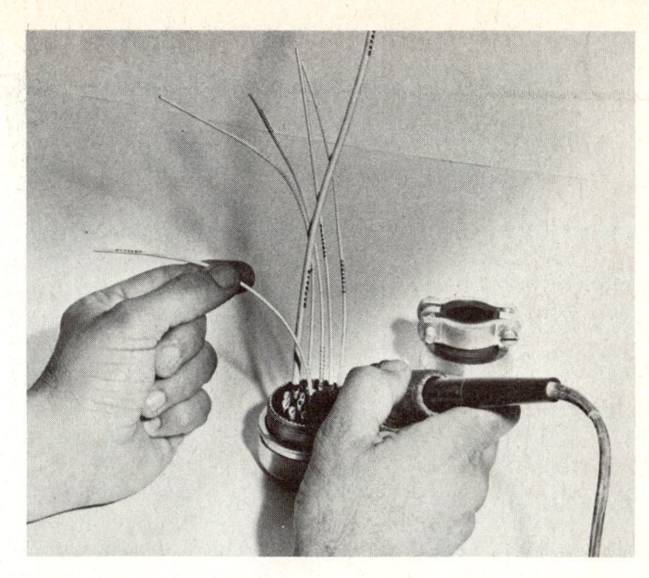

Fig. 11-21 Soldering wire to a connector.

slipping off the pins. Figure 11-21 shows a method for soldering cable to a connector plug.

Because of the almost infinite variety of possible electric circuits and installations, it is readily understandable that there must also be a wide variety of connector plugs and other connecting devices. Connectors may have from one contact pin and socket to more than one hundred. Several different types of connectors are shown in Fig. 11-22. For the installation of any particular plug assembly, the specification of the manufacturer or the appropriate governing agency must be followed.

Connectors currently being manufactured for aircraft use are often required to meet military specifications and are called MS electric connectors. Military specifications (MIL specs) are revised from time to time to incorporate performance requirements as dictated by design ad-

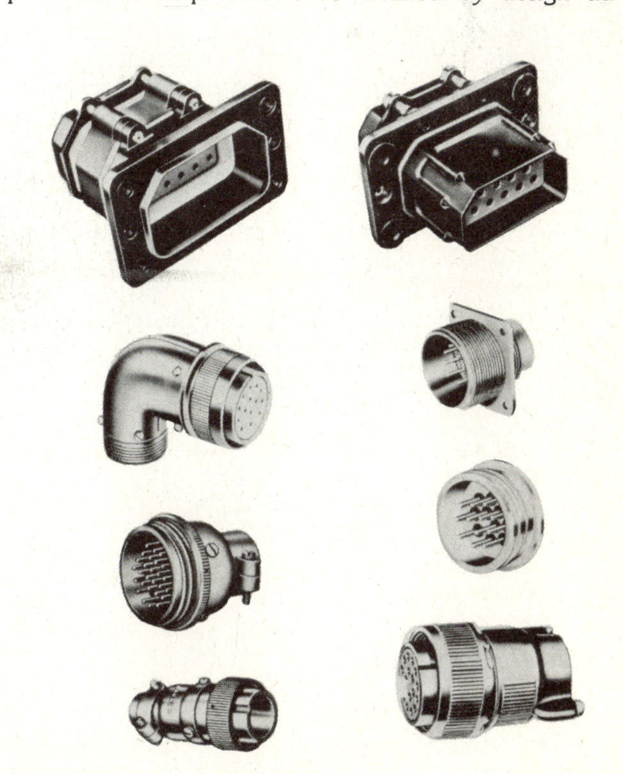

Fig. 11-22 Different types of connectors.

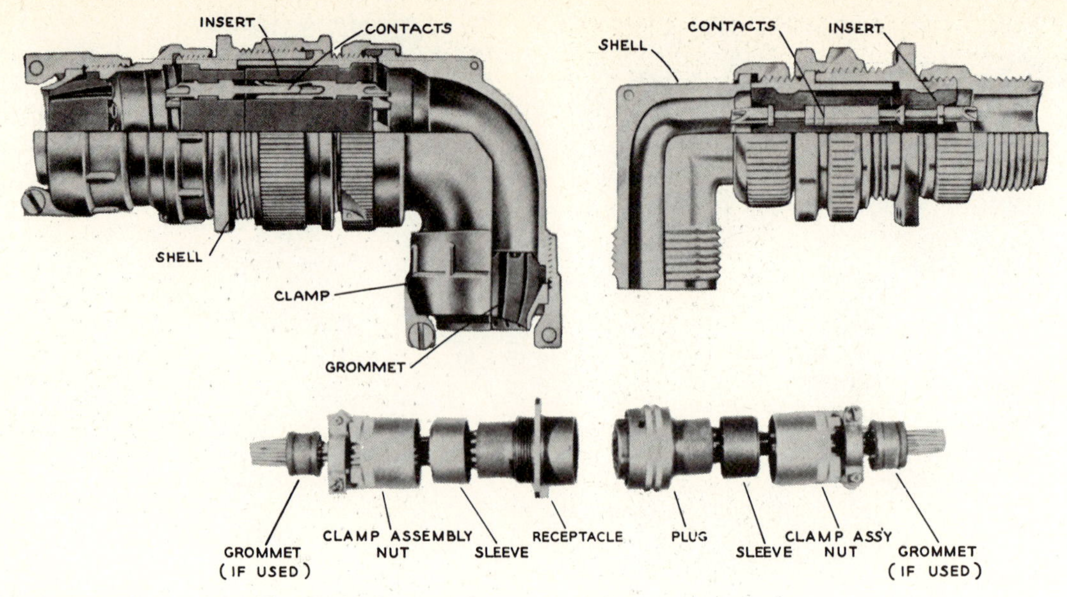

Fig. 11-23 Bendix MS-type connectors. (Bendix Corp.)

vances and more stringent operating requirements of allied equipment. The AN specifications are therefore superseded by MIL specs.

The general specification MIL-C-5015 provides for several designations of connectors to meet different requirements. These connectors carry MS numbers such as MS3100A-20-27S. In this designation, the number 3100 indicates a wall-mounted unit, the letter *A* indicates general utility usage, the number 20 indicates shell size, and 27*S* shows the socket arrangement. It will be noted that the numbering system is similar to that used for AN connectors. Typical Bendix MS connectors are shown in Fig. 11-23.

Connector assemblies are manufactured in many shapes and sizes to meet the requirements of modern electric and electronic equipment. The round connector is popular because it lends itself to easy joining and securing by means of a threaded collar. Many connectors are made in a rectangular shape, however, and these are often used when a harness is connected to an electronic unit.

A convenient method for connecting wires to connector plugs and other devices developed by manufacturers of terminals involves the use of **taper pins** and sockets in place of the standard solder technique. The taper-pin design was developed primarily to be used in miniaturized components to simplify assembly and eliminate the problems of soldered connections. Three different types of taper pins designed by AMP Incorporated are shown in Fig. 11-24. These pins are crimped on the end of the wire in the same manner in which other types of terminals are attached.

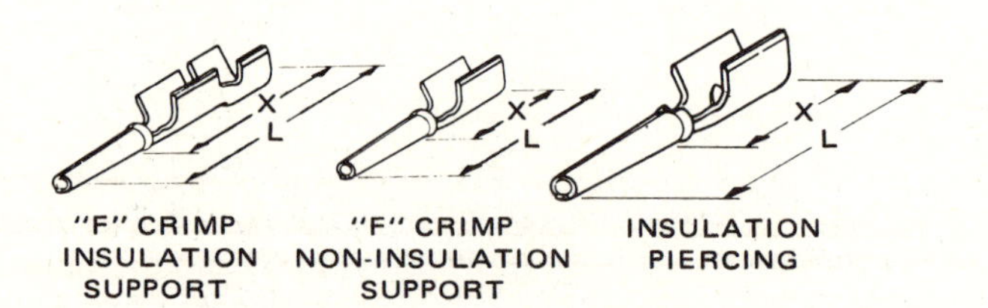

Fig. 11-24 Taper pins for crimp installation.

The insertion of a taper pin into a receptacle is accomplished with a special tool (see Fig. 11-25). In Fig. 11-25*a* the pin is shown before insertion into the tool. The wire-crimp portion of the pin is fitted into the slot in the end of the tool. In Fig. 11-25*b* the pin is shown in proper position in the tool. In Fig. 11-26 an operator is inserting a taper pin into a receptacle in the base of a connector plug.

Pins and sockets designed for crimped wire attachment are shown in Fig. 11-27. The pins and sockets are installed on the electric wires by crimping with a special tool which is color-coded for size. First, the insulation is stripped from the end of the wire, the length removed being dependent upon the size of the wire or contact. For example, $\frac{1}{4}$ in [0.64 cm] of insulation is stripped from a

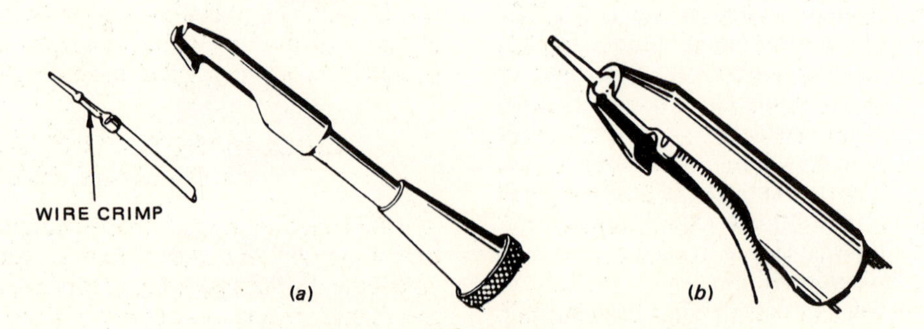

Fig. 11-25 Taper pin tool. (AMP Specialties)

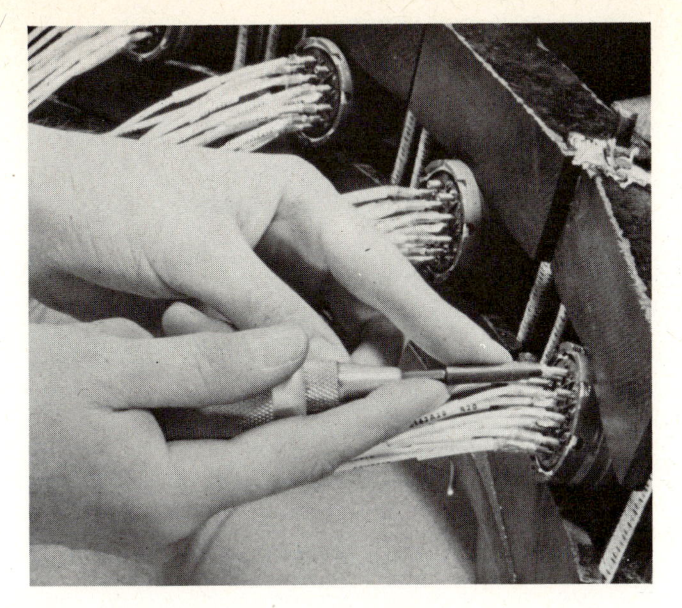

Fig. 11-26 Inserting a pin into a connector.

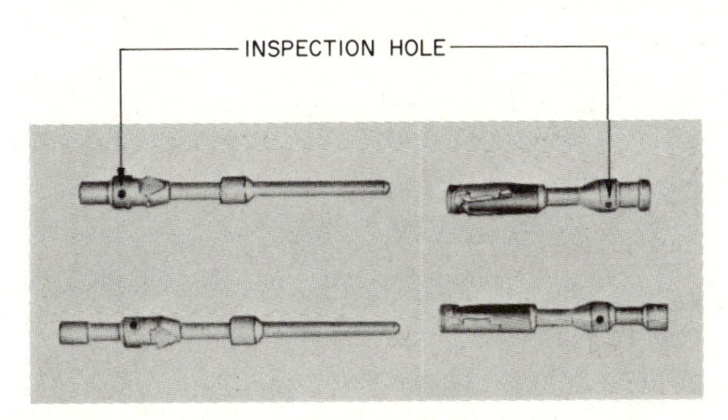

Fig. 11-27 Pins and sockets designed for crimp installation. (Bendix Corp.)

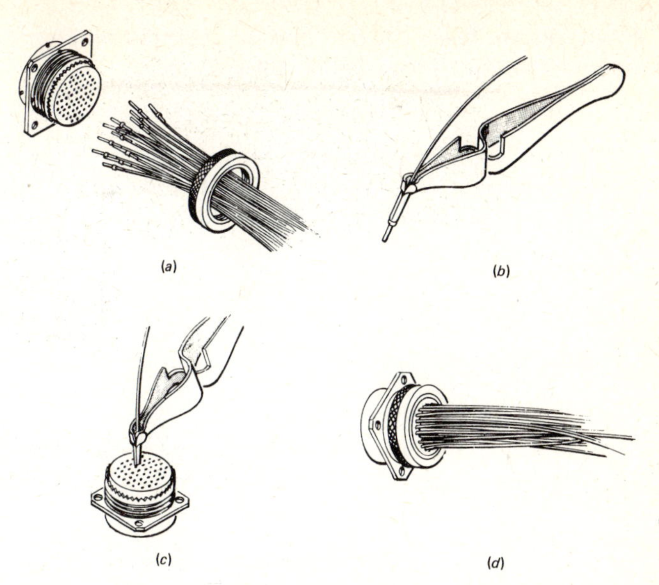

Fig. 11-28 Inserting pins into a connector. (Lockheed)

wire to be used in a No. 16 contact and $\frac{3}{16}$ in [0.48 cm] of insulation is stripped for the No. 20 contact. The wire is inserted in the wire well and must be visible in the inspection hole. The wire and contact are then inserted in the crimping tool after the color-coded positioner is installed in the tool. Closing the handles of the crimping tool to their full extent of travel will properly crimp the contact to the wire. The contacts are then inserted in the appropriate connector part by means of a special insertion tool.

Typical instructions for the insertion of pins in an MS27473 connector are as follows:

1. Remove the threaded backshell from the connector and slip it over all wiring to be terminated as shown in Fig. 11-28*a*.
2. Using the specified insertion tool, shown in Fig. 11-28*b*, position the wire in the tip of the tool so the tool tip butts up against the contact shoulder.
3. With slow, even pressure, insert the wired contact and tool tip straight into the contact cavity as shown in Fig. 11-28*c*. A slight click may be heard and/or felt as the metal retaining clip tines snap into place behind the contact shoulder. Do not rotate or spread the insertion tool tips while they are within the cavity (grommet).
4. Remove the tool from the grommet.
5. Insert unwired contacts and sealing plugs in all unused cavities.
6. Replace the threaded backshell on the connector and tighten by hand using the applicable holding fixture.

Assembled connector should appear as in Fig. 11-28*d*. During the assembly of connectors, the technician should carefully note the make, type, and specification numbers of all parts being used. Manufacturers' instructions specify the correct tools, parts, and procedures for each type of connector.

POTTING

The process of encapsulating electric wires and components in a plastic material is called **potting.** The process is sometimes recommended for certain components and should be accomplished as directed by the appropriate instructions. If potting is to be performed, the first requisite is that the item or items to be potted are perfectly clean. Petroleum solvent, methylene chloride, or other specified cleaner may be employed.

Potting is accomplished by forming a cavity with nylon or other plastic membrane around the item to be encapsulated. Potting compound meeting specification MIL-S-8516 or an equivalent is then mixed and poured into the cavity of the potting boot. The potting compound should cure for 24 h at 70 to 75°F [21.1 to 23.9°C] or may be placed in an oven at 100°F [37.8°C] for 4 h.

Potting compound in the unmixed state consists of the base compound and an accelerator or catalyst. After the two materials are mixed, they will harden beyond use within about 90 min. It is therefore important that work be planned so the material will be used within the working time of the compound. Manufacturer's instructions should be followed for the material being used.

CIRCUIT-PROTECTING DEVICES AND SWITCHES

As explained in the first chapter of this text, circuit protectors are devices designed to interrupt an electric circuit whenever the voltage or the current reaches a level which may cause damage to the wiring or to the electric units connected in the circuit. Among the principal types of protectors are **fuses, current limiters, circuit breakers,** and **overvoltage** or **overload relays.**

FUSES

A fuse is a strip of metal having a very low melting point. It is placed in a circuit in series with the load so that all load current must flow through it. The metal strip is made of lead, lead and tin, and tin and bismuth, or some other low-melting-temperature alloy.

When the current flowing through a fuse exceeds the capacity of the fuse, the metal strip melts and breaks the circuit. The strip must have low resistance, and yet it must melt at a comparatively low temperature. When the strip melts, it should not give off a vapor or gas which will serve as a good conductor, because this would create an arc between the melted ends of the strip. The metal or alloy used must be of a type which reduces the tendency toward arcing.

Fuses are generally enclosed in glass or some other heat-resistant insulating material to prevent an arc from causing damage to the electric equipment or other parts of the airplane. Fuses used in aircraft are classified mechanically as *cartridge type, plug-in type,* or *clip type,* although others are manufactured. All these types are easily inspected, removed, and replaced. Typical fuses are shown in Fig. 11-29.

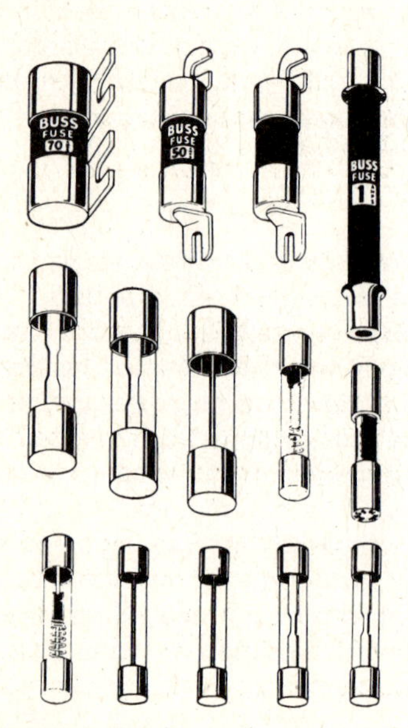

Fig. 11-29 Typical fuses.

The current limiter described here is essentially a *slow-blow* fuse. That is, when the circuit becomes overloaded, there is a short delay before the metal link melts and disconnects the circuit. This is because the link is made of copper, which has a higher melting point than the alloys used in other types of fuses. The current limiter will carry more than its rated capacity and will also carry a heavy overload for a short time. It is designed to be used in heavy power circuits where loads may occur of such short duration that they will not damage the circuit or equipment. The capacity of a current limiter for any circuit is so selected that it will always interrupt the circuit before an overload has had time to cause damage. Current limiters are shown in Fig. 11-30.

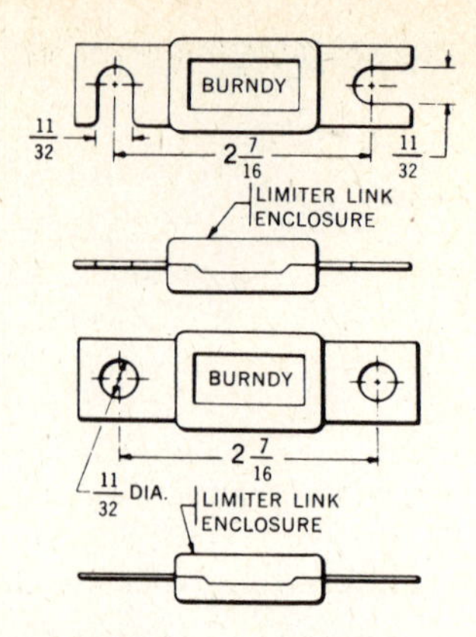

Fig. 11-30 Current limiters.

CIRCUIT BREAKERS

A circuit breaker serves a purpose similar to that of a fuse; however, the circuit breaker can usually be reset after a circuit fault is removed. The typical aircraft circuit breaker can be described as a manually operated switch which has an automatic tripping device. This tripping device breaks the circuit when the current reaches any predetermined value. The switch-type circuit breaker shown in Fig. 11-31 serves as both a fuse and a switch, and a single circuit breaker may be used to control several cir-

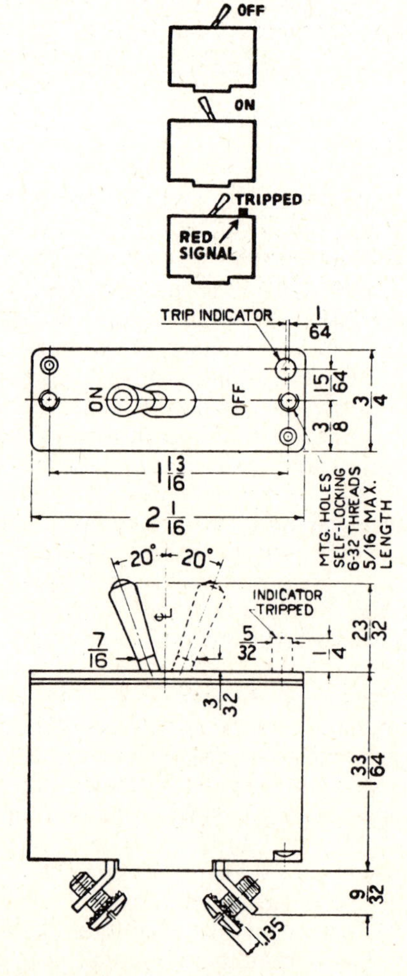

Fig. 11-31 Switch-type circuit breaker.

cuits. For example, it may receive current from a bus and pass it on to several parallel circuits.

One type of circuit breaker is known as the magnetic or electromagnetic type. The latter term is technically more correct, since the device operates through an electromagnet which pulls on a small armature and trips the breaker when energized with an overload current.

When a circuit breaker has opened a circuit, it must be reset after the circuit fault has been removed. To do this the switch lever is moved to the FULL OFF position and then returned to the ON position. If there is still too much current flowing in the circuit, that is, if the overload still exists, the circuit breaker will trip again without damaging the circuit. It is common practice to attempt to reset a circuit breaker immediately after it has "kicked out" because the reason for its having been tripped may have been a transient overload. In this case the circuit breaker will reset easily and the circuit again be operative. If there is a fault in the circuit, however, it must be removed before the breaker will reset.

Thermal circuit breakers, that is, those tripped by excessive temperature acting on a bimetallic strip, cannot be reset until the temperature has returned to normal. Since many circuit breakers are of this type, it is often necessary to wait for a short time before attempting to reset.

REQUIREMENTS FOR CIRCUIT-PROTECTION DEVICES

Circuit breakers and fuses should in all cases protect the wire in the circuit from overload and should be located as close as possible to the source bus. Remember that a bus is a metal strip to which a power supply is connected and from which other circuits receive power for operation. A bus is fitted with terminal posts to which the wire terminals are secured.

A circuit breaker or fuse should open the circuit before the wire becomes heated sufficiently to emit smoke. The time-current characteristic of the protective device should therefore be below that of the associated wire, with the result, of course, that the circuit protector will open the circuit before the wire is damaged. The term **time-current** refers to the product of multiplying the amount of current by the time during which it flows. In order to obtain maximum utilization of the connected equipment, the characteristics of the circuit protector should match as closely as possible those of the connected wire.

Table 11-6 is a guide to the selection of circuit-breaker and fuse ratings to protect copper cable. The conditions for the figures given in the table are as follows:

1. Wire bundles in 135°F [57.2°C] ambient temperature and altitudes up to 30 000 ft [9144 m].
2. Wire bundles of 15 or more wires (cables), with wires carrying no more than 20 percent of the total current-carrying capacity of the bundle as given in specification MIL-W-5088 [Aeronautical Standards Group (ASG)].
3. Protectors in 75 to 85°F [23.9 to 29.4°C] ambient temperatures.
4. Copper wire specification MIL-W-5086 (ASG) or equivalent.
5. Circuit breakers to specification MIL-C-5809 or equivalent.
6. Fuses to specification MIL-F-15160 or equivalent.

TABLE 11-6 Wire and Cable Protection Chart

Wire AN gage: copper	Circuit breaker, A	Fuse, A
22	5	5
20	7.5	5
18	10	10
16	15	20
14	20	15
12	25(30)*	20
10	35(40)	30
8	50	50
6	80	70
4	100	70
2	125	100
1		150
0		150

If the actual conditions of an installation deviate materially from those stated for Table 11-6, ratings above or below the values recommended may be justified. For example, a wire run individually in the open air may possibly be protected by a circuit breaker of the next higher rating to that shown on the chart. In general, the chart is conservative for all ordinary aircraft electric installations.

All resettable circuit breakers should be designed to open the circuit regardless of the position of the operating control when an overload or circuit fault exists. Such circuit breakers are described as **tripfree.** Automatic-reset circuit breakers should not be used as circuit protectors. These are the type which reset themselves periodically.

SWITCHES

A **switch** may be defined as a device for closing or opening (making or breaking) an electric circuit. It usually consists of one or more pairs of contacts, made of metal or a metal alloy, through which an electric current can flow when the contacts are closed. Switches of many types have been designed for a wide variety of applications. The switches can be manually operated, electrically operated, or electronically operated. The manual switch is usually operated by either a lever or by a push button. Electrically operated switches are generally called **relays,** as stated in previous sections of this text. An electronically operated switch utilizes an electron tube or a transistor to control the current in a circuit, and the "switch" is turned on or off by means of an electric signal applied as a voltage to the tube or transistor.

To be suitable for continued use, a switch must have contacts which are capable of withstanding thousands of cycles of operation without appreciable deterioration due to arcing or wear. The contacts are usually made of special alloys which are resistant to burning or corrosion. The operating mechanism of a switch must be ruggedly constructed so it will not fail due to wear or load stresses. For aircraft use, a switch must be of a type and design approved by appropriate governmental agencies and by the manufacturer of the aircraft.

The type of electrical load which a switch is required to control will determine, to some extent, the type and capacity of switch to be employed in a circuit. Some electric circuits will have a high surge of current when first connected, and then the current flow will decrease to the normal operating level. This is typical of circuits for incandescent lamps or electric motors. An incandescent lamp will draw a high current while the filament of the lamp is cold. The resistance of the filament increases severalfold as the temperature reaches maximum; hence, the current is reduced at this time. The switch for an incandescent-lamp circuit must be able to carry the high starting current without damage. An electric motor draws a high current when starting because no backvoltage is developed in the motor until it is rotating. When the motor reaches its normal operating speed, the backvoltage developed in the armature will substantially reduce the current flow. Inductive circuits, those which include electromagnetic coils of various types, have a momentary high voltage at the time the circuit is broken. This high voltage causes a strong arc to occur at the switch contacts. It will be remembered that a coil will develop an induced voltage due to the collapse of the magnetic field when the current flow in the coil is stopped. This is in accordance with Lenz's law, as previously explained.

It is apparent from the foregoing discussion that a switch must be able to carry a greater load than the nominal running load of the circuit in which it is installed. Accordingly, **derating factors** are applied in determining the capacity of a switch for a particular installation. The derating factor is a multiplier which is used to establish the capacity a switch should have in order to control a particular type of circuit without damage. For example, if an incandescent-lamp circuit operates continuously at 5 A in a 24-V system, the capacity of the switch should be 40 A because the derating factor is 8. That is, the surge current for the lamp circuit can be almost eight times the steady operating current. Table 11-7 gives the derating factors for aircraft switches in various types of dc circuits.

TABLE 11-7

Nominal system voltage	Type of load	Derating factor
24	Lamp	8
24	Inductive	4
24	Resistive	2
24	Motor	3
12	Lamp	5
12	Inductive	2
12	Resistive	1
12	Motor	2

The installation of switches should be in accordance with a standard practice so the operator will always tend to move the switch lever in the correct direction for any particular operation. Switches should always be installed in panels so the lever will be moved *up* or *forward* to turn the circuit *on*. For switches which operate movable parts of the aircraft, the switch should be installed so the switch lever is moved in the same direction that the aircraft part will be moved. The landing-gear switch should be installed so the switch lever will be moved *down* to lower the landing gear and *up* to raise the gear. The same principle should apply for wing-flap operation.

Switches are designed with varying numbers of contacts to make them suitable for controlling one or more electric circuits. The switch used to open and close a single circuit is called a single-pole single-throw (SPST) switch. A switch designed to turn two circuits on and off with a single lever is called a two-pole, or double-pole, single-throw (DPST) switch. A switch designed to route current to either of two separate circuits is called a double-throw switch. Schematic diagrams of several different types of switches are shown in Fig. 11-32.

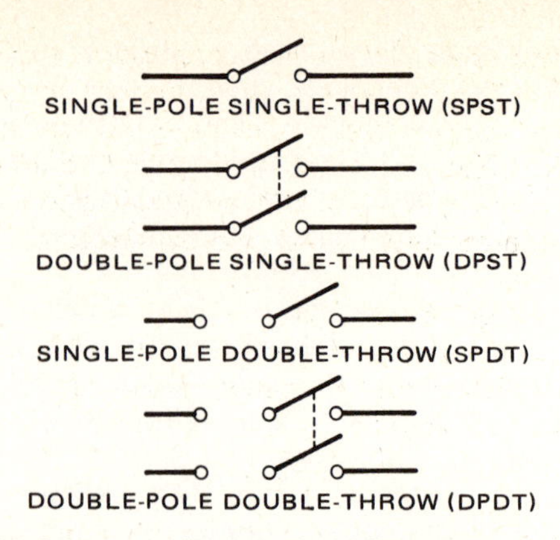

Fig. 11-32 Schematic diagrams for different types of switches.

BONDING AND SHIELDING

BONDING

Bonding is the process of electrically connecting the various metallic parts of an aircraft or other flight vehicle so that they will collectively form an integral electric unit. That is, there will be a very low-resistance path for current from any one part of the structure to any other part. The specific purposes of bonding are to provide a low-resistance path for electric equipment, thereby eliminating ground wires; to reduce radio interference; to decrease the probability of lightning damage to such aircraft elements as control hinges; and to prevent the buildup of static charges between parts of the structure, thus reducing the fire hazard which could result from spark discharges between these parts.

A bonding jumper is a short length of metal braid or metal strip with a terminal at each end for attaching to the structure. These jumpers should be as short as practicable and be installed in such a manner that the resistance of each connection does not exceed 0.003 Ω. They should also be installed in locations which provide reasonably easy access for inspection and maintenance. Care must be taken that bonding jumpers do not interfere with the operation of any movable part of the aircraft, nor should the normal movement of such parts result in damage to the bonding jumper.

When bonding jumpers are installed, it is important that all insulating coatings, such as anodizing, paint, oxides, and grease, be removed so that clean, bare metal surfaces come into contact. After the bonding is secured, it is good practice to coat the junctions with a sealing coating to prevent the entrance of moisture which could produce corrosion. Electrolytic corrosion may occur quickly at a bonding connection if adequate precautions are not taken. Aluminum-alloy jumpers are recommended in most cases, but copper jumpers are used to bond together parts made of stainless steel, cadmium-plated steel, copper, brass, and bronze. Where contact between dissimilar metals cannot be avoided, the choice of jumper and hardware should be such that corrosion is minimized and the part likely to corrode is the jumper or hardware.

A guide to the selection of metals which can be joined without the danger of corrosion is given in the following grouping of metals. Metals in any one group may be joined with a minimum likelihood of corrosion.

Group 1. Magnesium alloys
Group 2. Zinc, cadmium, lead, tin, steel
Group 3. Copper and its alloys, nickel and its alloys, chromium, stainless steel
Group 4. All aluminum alloys

The screws, washers, nuts, bolts, or other fasteners used for securing bonding jumpers must be of a material which is compatible with the metals being joined. For example, where aluminum jumpers are attached to aluminum-alloy structures, the fasteners should be made of aluminum.

The use of solder to attach bonding jumpers should be avoided for the same reasons that it is not recommended for electric-wire connections. Tubular members should be bonded by means of clamps or clamp blocks as illustrated in Fig. 11-33. In this installation a thin aluminum strip lines both inner surfaces of the clamp block, and the ends of the metal strips are carried around the ends of the clamp block so that they make contact with the aircraft structure. Bonding braid may be connected between electrically separate parts of a structure as shown in Fig. 11-34. Each terminal of the bonding braid is securely attached to the sheet-metal structure by means of machine screws and nuts. A typical attachment of a bonding jumper to an aluminum alloy structure is shown in Fig. 11-35.

When a bonding jumper is installed in a location where it will be required to carry a ground load for a unit of electric equipment, care must be taken to assure that the

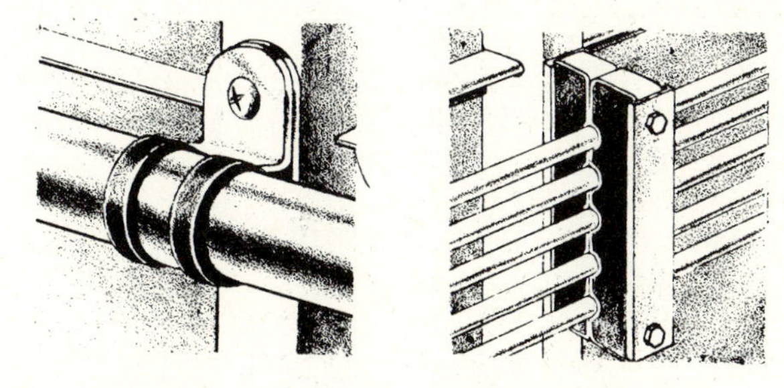

Fig. 11-33 Bonding for tubular members.

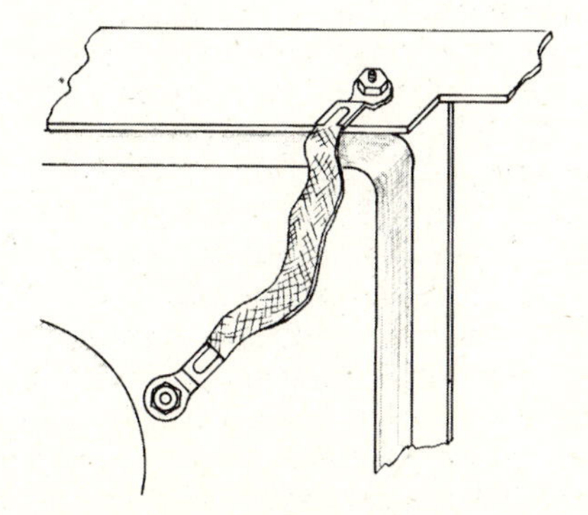

Fig. 11-34 Installation of a bonding braid.

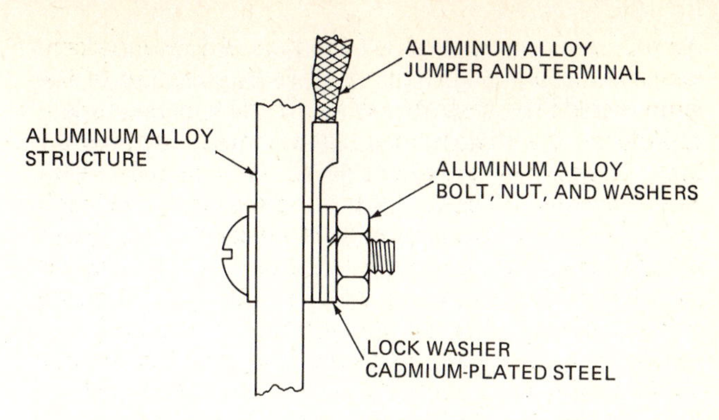

Fig. 11-35 Attachment of a bonding jumper.

jumper has sufficient capacity to carry the load. This could occur when an inverter or unit of radio equipment is mounted on shock mountings. If the equipment is grounded through the case, the ground current must be carried through the mounting structure and then through a bonding braid to the main structure. A bonding braid too small to carry the load could overload and melt, thus creating a fire hazard and also causing the equipment to fail.

In some cases, a unit of electric equipment is connected into a heavy-current circuit, perhaps as a control device such as a relay. This type of equipment should be insulated from the mounting structure since grounding the frame of the equipment may result in a serious ground fault in the event of the internal failure of the equipment. If a ground connection is required for a control coil, a separate small-gage wire may be used.

SHIELDING

Shielding is a metallic covering applied to wiring and equipment to eliminate interference with radio reception. The shielding is bonded to the main structure at frequent intervals. The word shielding is also applied to the process of enclosing wires or electric units with metal. The purpose of shielding is to cause HF voltages to be induced in the shield metal rather than in parts, units, or wires where it would cause interference. Shielding is used when one unit is to be protected from the effects of an HF current in an adjacent unit, or when a cable is to be protected from radio-frequency (RF) noise.

As mentioned previously in this section, shielding is accomplished by installing wires or cables in rigid or flexible conduit or by providing the wire with a braided metal outer sheath. The conduit must be bonded to the aircraft structure to be effective. Both mechanical protection and protection from electrical interference are accomplished in this manner.

WIRE IDENTIFICATION

To facilitate installation and maintenance, all wiring should be indelibly marked for identification. Any consistent numbering system is considered adequate if the numbers are placed at each end of each section of wire and also at intervals along the wire. To accomplish this marking, the wire is usually run through a numbering machine which stamps the numbers along the wire at specified intervals. The identification numbers and letters should clearly show the circuit in which the wire is installed, the particular wire in the circuit, the wire gage size, and other pertinent information. Care must be taken when marking coaxial cable with a machine. If too much pressure is applied to the cable, it may be flattened and this will change the electrical characteristics of the cable.

Electric wires or cables may be identified by both number and letters, especially on large aircraft. For example, on a typical airliner the following letter system is used to identify specific circuits:

AC power	*X*
Deicing and anti-icing	*D*
Engine control	*K*
Engine instrument	*E*
Flight control	*C*
Flight instrument	*F*
Fuel and oil	*Q*
Ground network	*N*
Heating and ventilating	*H*
Ignition	*J*
Inverter control	*V*
Lighting	*L*
Miscellaneous	*M*
Power	*P*
Radio navigation and communication	*R*
Warning devices	*W*

Numbers used with the letters for identification also have a specific purpose. In the identification number 2P281C-20, there are two letters and three separate numbers. The number 2 indicates that the wire is associated with the No. 2 engine, and the letter *P* means that the wire is a part of the electric power system. The number 281 is the basic wire number and remains unchanged between the electric units of any particular system regardless of the number of junctions it may have. The letter *C* identifies the particular section of wire in the circuit, and the number 20 indicates the gage of the wire. In this identification system, the numbers 5 and 6 are used as prefixes for circuits not associated with engines. The number 5 is used for left installations and the number 6 for right installations. Some wires are neither right nor left, and these do not have a number prefix. Sometimes two identification letters prefix the basic wire number; for example, the letter prefix *NC* would indicate the ground wire for a flight-control unit.

The foregoing system is used only for certain aircraft; many different systems have been devised by other manufacturers. In one of these systems the number 16-C40 indicates a specific section (40) of No. 16 cable in a flight-control circuit. No other section of cable in the airplane will carry the same number.

Identification markings are usually stamped directly on the insulation of the wire or cable. In an aircraft factory, the markings are stamped on the wires by means of a marking machine. The markings are placed at each end of a wire or cable section and at intervals of 12 to 15 in [30.5 to 38.1 cm] along the wire. Short sections of wire less than 3 in [7.6 cm] in length need not be marked. Wires of 3 to 7 in [17.8 cm] in length should be marked midway between the ends.

If the outer coating or surface of a wire sheath is such that it cannot be easily marked, sleeving or tubing can be marked and placed over the wire. High-temperature wires, shielded wiring, multiconductor cable, and thermocouple wires usually require special sleeves to carry identification marks. Metallic sleeves or bands must not be used on electric wires.

Wire harnesses are often identified by number to indicate the particular section installed in a system. These are identified by means of a marked sleeve or pressure-sensitive tape. Methods for marking wires and harnesses are shown in Fig. 11-36.

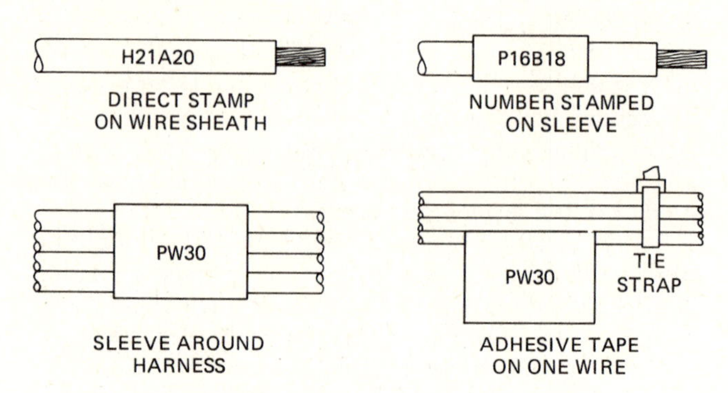

Fig. 11-36 Methods for marking wire and harnesses.

TYPICAL SYSTEMS

A SIMPLE ELECTRIC SYSTEM

A simple electric system for a light aircraft consists of a battery circuit, a generator circuit with associated controls, an engine-starter circuit, a bus bar with circuit breakers, control switches, an ammeter, lighting circuits, and radio circuits. A schematic diagram of the basic power circuits is shown in Fig. 11-37. The high-current-carrying cables in this system are connected from the bat-

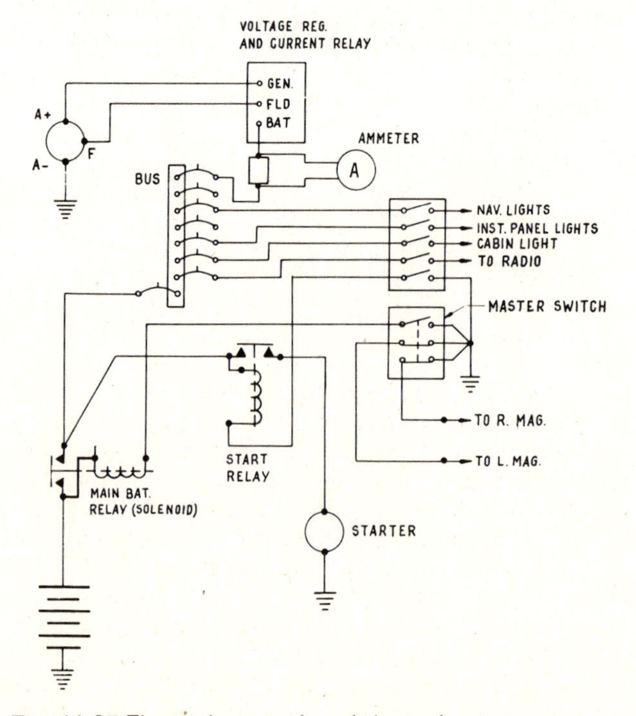

Fig. 11-37 Electrical system for a light airplane.

tery to the main battery relay, from the battery relay to the starter relay, and from the starter relay to the starter. The ground leads for the starter and the battery are also of heavy cable.

The main generator power cables are also considerably larger than the normal circuit wiring; however, they are usually smaller than the cables required to carry full battery current. This is because the battery is used for starting the engine and the starting current is very heavy. During operation of the aircraft, the battery is connected to the system but is not supplying power. Instead, it is taking power from the generator in order to maintain a charge. All the normal load currents are supplied by the generator during flight.

TYPICAL GENERATOR CIRCUITS

The generator circuits for a Cessna Model 310 airplane are shown in Fig. 11-38. The Cessna 310 is a twin-engine aircraft designed for use as a luxury private airplane or as an executive airplane for business travel. The generator system utilizes one generator on each engine paralleled through the bus (51). This bus is connected to the main bus (5), which supplies most of the aircraft circuits and to which the battery is connected. Each generator circuit contains a voltage regulator (48), a reverse-current cutout (52), paralleling resistors (44), and combination generator and equalizer switches. Note that when the generator switch is turned off, the field circuit of the generator is broken. When this occurs, the generator voltage drops almost to zero, and the reverse-current cutout opens to disconnect the generator from the bus.

Notice that each wire or cable in Fig. 11-38 is numbered for identification. In this particular circuit the principal number, used throughout the generator systems, is 479. The number following the principal number on each wire identifies the separate sections of wire in the circuit.

A schematic diagram of the alternator and power-distribution circuits for the Cessna 421 airplane is shown in Fig. 11-39. The Cessna 421 is a twin-engine high-performance aircraft for private or business use. From the circuit of Fig. 11-39, it will be noted that the alternators have both an ac and a dc output. The ac is used for the windshield heating system.

The circuit of Fig. 11-40 shows the use of crystal diodes connected across the windings of relays. As explained in a previous chapter of this text, when a current starts to flow in a coil, or when the current flow is stopped, the inductance of the coil creates a voltage opposing the change in current flow. Thus, when a switch is closed or opened in a relay coil circuit, an opposing voltage and current are produced. This can cause arcing at the switch contacts and also produce high transients in the circuit that cause radio interference and that may be damaging to some of the elements in the circuits affected. The diodes short-circuit the coils in the reverse direction, thus eliminating the effect of inductive voltages.

Another generator power circuit for a twin-engine aircraft is shown in Fig. 11-41. This circuit is installed in the Beechcraft Baron airplane and illustrates a current method of preparing electric circuit drawings. In the circuit shown, current flow is from the power bus (not shown) to the 50-A current limiters, XF2 and XF3. These are slow-blow, fuse-type current limiters. Current then

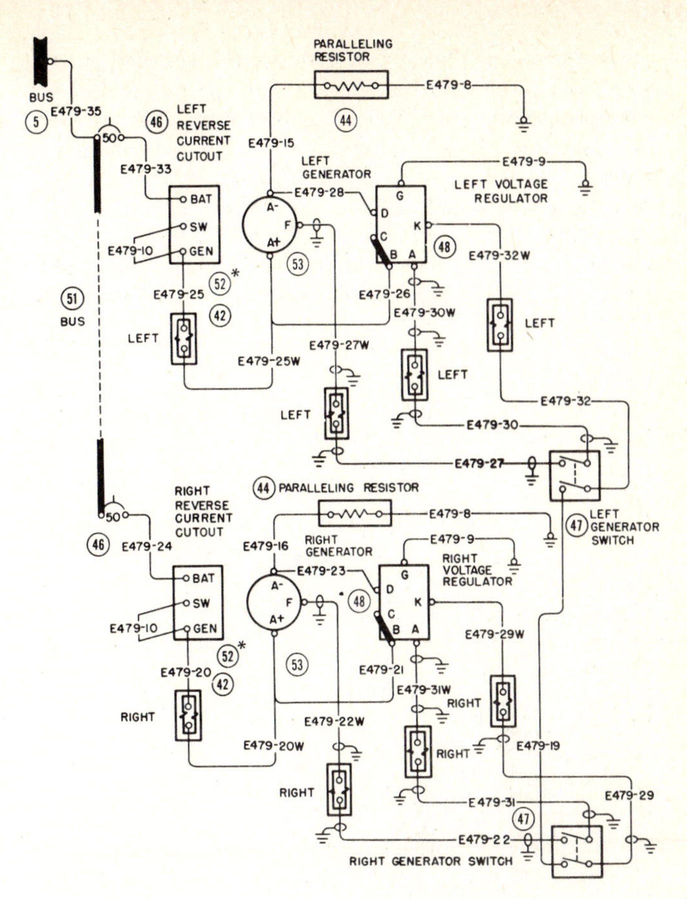

Fig. 11-38 Generator system for a light twin airplane. (Cessna)

flows through the ammeter shunts and to the positive terminal of each generator (alternator). The voltage across the shunts provides the signals to operate the loadmeters (ammeters). The loadmeter circuits are protected by 3-A fuses.

The circuit of Fig. 11-41 incorporates an **alternator-out** system in each alternator circuit. The sensor is connected to the AUX terminal of the generator in each case and causes a warning light to come on in the cockpit when either alternator fails.

BATTERY AND STARTER CIRCUITS

The battery and starter circuit for a Cessna 310 is shown in Fig. 11-42. Note that power from the battery is connected through the battery solenoid to a bus which supplies power for the two starter solenoids. When the battery switch is turned on, the battery solenoid is energized and closes the circuit to the bus. In this particular circuit the starter switches receive control power from the right fuel-boost switch or the right auxiliary-pump switch. When one of these switches is turned on, it is then possible to start the engines.

LANDING-GEAR CIRCUITS

Circuits involved in the operation of electrically powered landing gear are shown in Figs. 11-43 and 11-44. The circuit of Fig. 11-43 shows the circuitry associated with the reversible electric motor that raises and lowers the landing gear. The control circuit is protected by a 5-A circuit breaker in the circuit-breaker panel assembly. This circuit

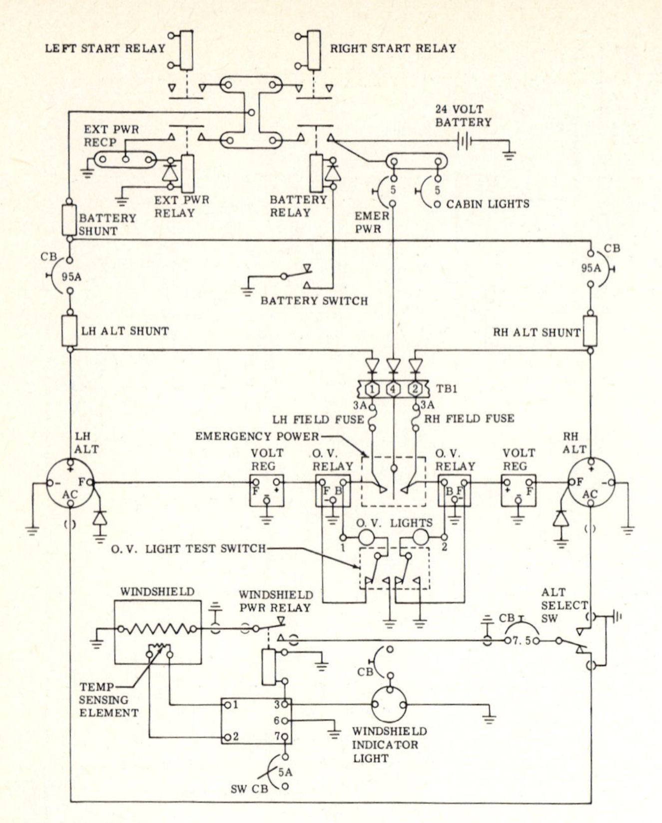

Fig. 11-39 Power distribution system. (Cessna Aircraft Co.)

incorporates the two landing-gear safety switches ("squat" switches) that prevent the operation of the landing gear as long as the airplane is on the ground. The landing-gear safety switches are identified as S36 and S37 in the circuit. The power circuit is connected to a 30-A circuit breaker in the circuit-breaker panel. This circuit provides power to the UP and DOWN power relays that are controlled through the control circuit. When the landing-gear switch, S38, is placed in the UP position with the airplane in flight, electric power flows from ground, through the UP relay, through the landing-gear UP limit switch, S39, through both safety switches, and to the circuit breaker CB18. This causes the relay to close and direct power to the landing-gear motor. When the landing gear reaches the UP position, the UP limit switch opens and cuts off power to the UP relay, thus stopping the motor. The reverse action takes place when the landing gear is lowered.

Figure 11-44 is a circuit diagram of a landing-gear-position-indicating system. This circuit operates in conjunction with the landing-gear control circuit of Fig. 11-43. The switches shown in the circuit represent the condition when the landing gear is in the down and locked position and the aircraft is on the ground. At this time, if electric power is turned on, the three green lights will be on to indicate that all three units of the landing gear are down and locked. The red gear-in-transit light will be out.

Another landing-gear circuit is shown in Fig. 11-45. The switches in the circuit are shown in the position for the gear down and the weight of the airplane resting on the landing gear. A careful study of this circuit reveals a number of safety features. For example, it is not possible to

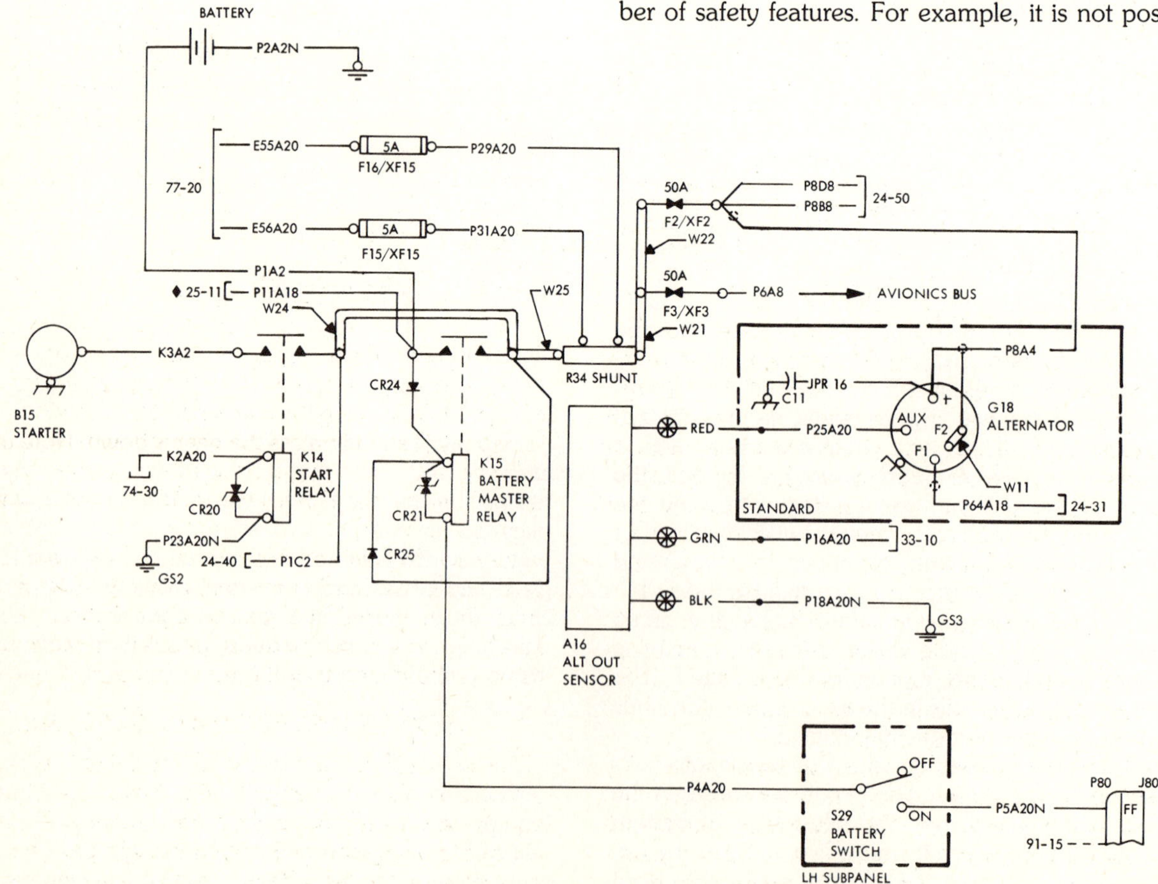

Fig. 11-40 Alternator circuit showing use of diodes. (Beech)

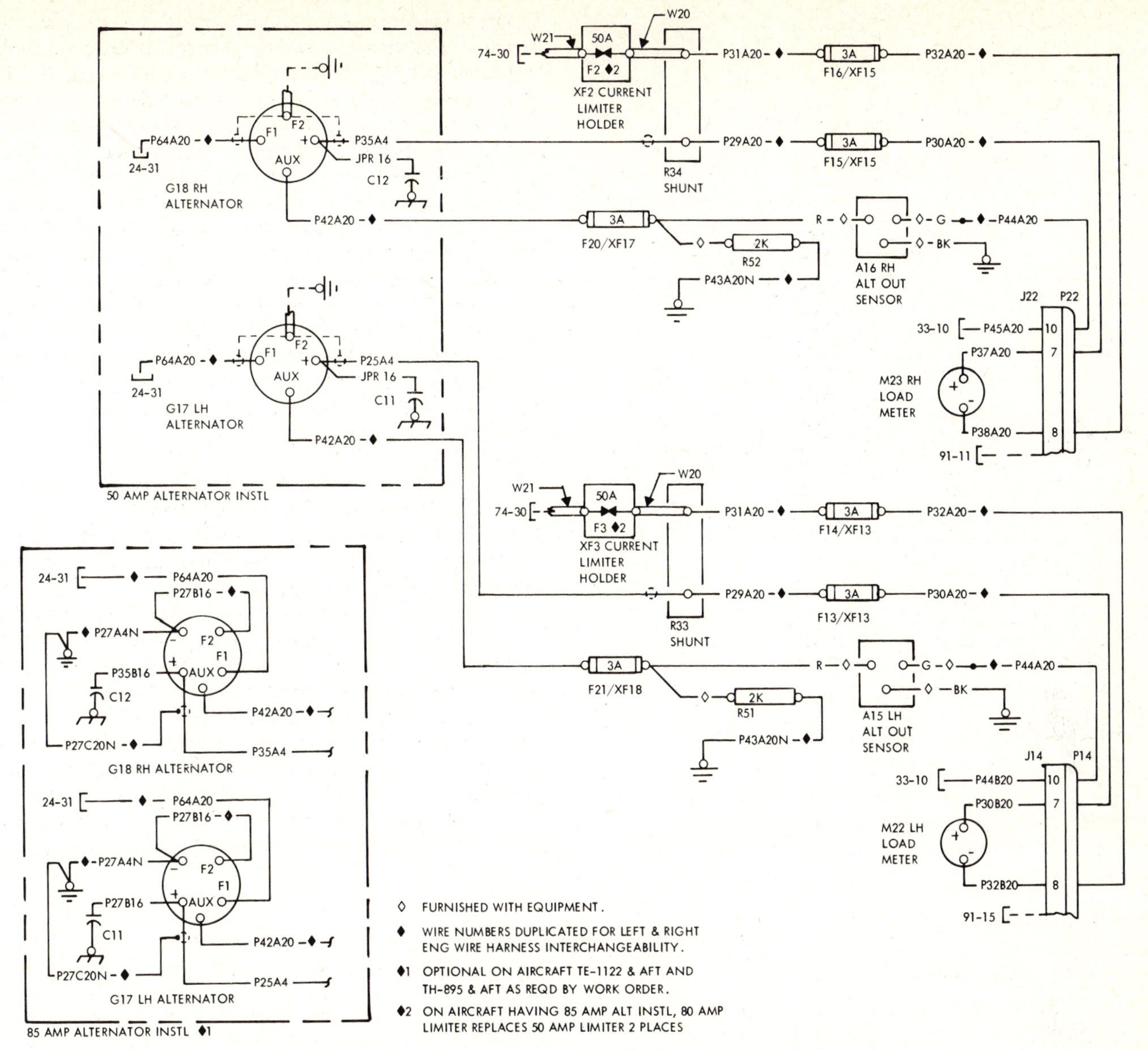

Fig. 11-41 Power circuit for a twin-engine airplane. (Beech Aircraft Corp.)

raise the gear when the airplane is on the ground, even though the landing-gear switch is placed in the UP position. Notice that the gear-relay control coil is fed through the left-gear safety (squat) switch. When the airplane is on the ground, this switch is open, and hence no current can be supplied to the relay coil. Furthermore, if the gear switch is in the UP position when the airplane is on the ground, a warning horn will sound.

When the airplane is in flight and the gear switch is placed in the UP position, the gear will rise, and when it is completely up, the up-limit switch will open and break the circuit to the control coil of the gear relay (183). When the limit switch opens, it also closes the circuit through the gear-up indicator light.

When the landing gear is up, the down-limit switch will be closed, thus making it possible to direct current to the down side of the gear motor. The gear lowers, but before the gear-down indicator light can come on, three microswitches must be closed. These are the right gear-down switch, the left gear-down switch, and the nose gear-down switch. These switches are connected in series; hence, no current can flow in the circuit unless all are closed. In flight, if the throttle is partially closed, the warning horn will sound unless the gear is down. Note that the warning horn must obtain power through one side of the down-limit switch. This switch is closed except when the gear is down.

The circuits shown in this section and the two previous sections are included in this text to show how typical circuits are presented in a maintenance manual. They are taken from service manuals and are presented here through the courtesy of the manufacturers.

POWER-DISTRIBUTION SYSTEMS

The power-distribution system for a light twin-engine airplane is shown in Fig. 11-46. Since the airplane is equipped with all the avionic equipment necessary for electronic navigation and optimum flight performance, it is necessary for the alternators to have comparatively high capacity.

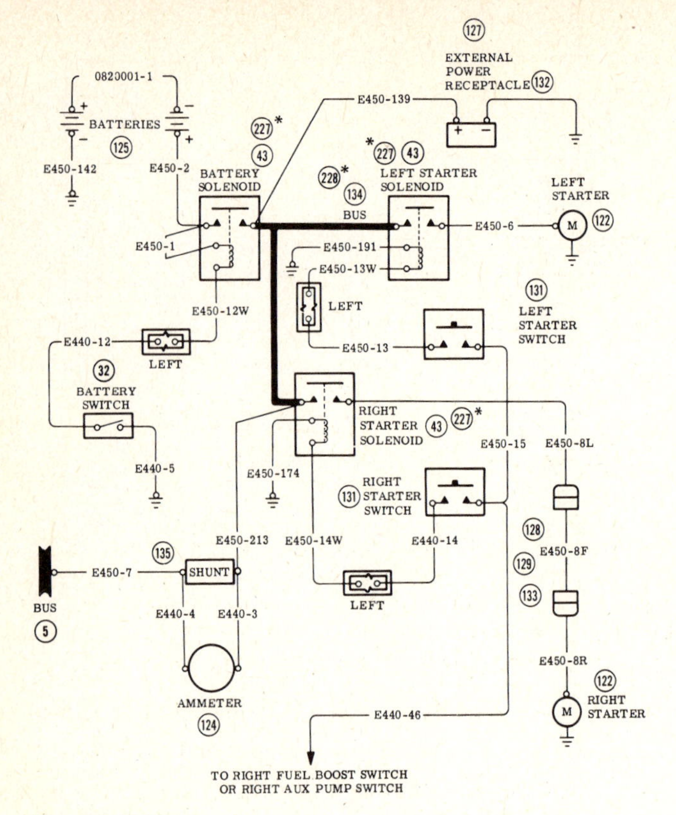

Fig. 11-42 Battery and starter circuits. (Cessna Aircraft Co.)

The electric system shown in Fig. 11-46 includes a 24-V, 17-AH battery enclosed in a sealed, stainless-steel battery box. Two 24-V, 70 A alternators driven by the engines supply all the normal power requirements of the aircraft and its equipment. The battery supplies power for starting the engines and for emergency peak loads.

The alternators are paralled by using one voltage regulator to control the field current for both alternators. The circuit diagram of Fig. 11-46 shows how this is accomplished.

An overvoltage relay in the system serves as a safety valve in case either one or both of the alternators should produce dangerously excessive voltage. This condition would exist in case of failure of the voltage regulator. In the event that the main voltage regulator fails and the overvoltage relay disconnects the alternator fields from the system, an auxiliary voltage regulator is available. Failure of the alternators may be detected by a discharge indication for the battery and a zero output on both alternator test positions.

When it is necessary to switch the system to the auxiliary voltage regulator, the manufacturer recommends the following procedure:

1. Reduce electric load to a minimum. This would usually mean that all avionic equipment is turned off.
2. Switch the **Voltage Regulator Selector** to the AUXILIARY position.
3. Reset any tripped circuit breakers, but *do not* reset the main voltage-regulator breaker.
4. Return the system to a normal required load.
5. If the overvoltage condition persists after switching to the auxiliary position, repeat the foregoing procedure with one alternator at a time. This can be done by tripping the alternator circuit breaker to cut one alternator out of the system.
6. When the faulty alternator is located, leave it off the line and continue flight with a reduced load.
7. In the event that both alternators are causing an overvoltage condition, remove both from the line, reduce electric loads to the barest minimum, and land as soon as possible.

The output of each alternator is checked by pressing a PRESS-TO-TEST switch and observing the ammeter in the overhead switch panel. The test switches are shown as LEFT ALTERNATOR SWITCH and RIGHT ALTERNATOR SWITCH in the circuit diagram of Fig. 11-46.

In the diagram it will be noted that the alternators are connected directly to the main bus at all times. This is possible because the rectifiers in the output circuits of the alternators prevent any current feedback through the alternator, as would be the case with a dc generator.

Electric switches for the various systems, including the MASTER SWITCH, are located on the circuit-breaker panel. The circuit breakers, located below the switches, automatically open their respective circuits in case of overload. The circuit breakers can be reset merely by pressing the RESET button. If a circuit breaker continues to disconnect, the trouble should be located and repaired before an attempt is made to operate the circuit.

The power distribution system for a gas-turbine-powered airplane with two engines, the Beechcraft Super King Air 200, is shown in Fig. 11-47. This schematic diagram is presented to show the complexity of a modern aircraft electric system and the many functions that require electric power.

It will be noted that the electric power for the system is supplied by two starter-generators. These serve as starters until the engines are started and then become generators as their speed of rotation increases. The starter field is a series winding separate from the generator field.

AIRCRAFT LIGHTS

All aircraft approved for flying at night must be equipped with various types of lights. Among these are position or navigation lights, anticollision lights, landing lights, instrument lights, warning lights, and cabin lights. In addition, other lights may be needed or required. Among these are taxi lights, ice-detection lights, cargo compartment lights, and all the special-purpose lights required in large passenger aircraft. All lighting equipment and installations must be approved by the FAA.

POSITION OR NAVIGATION LIGHTS

Each aircraft must have three position lights: two forward and one aft. The forward position lights are usually mounted on the tips of the wings because they are required to be as far outward as possible. The right position light is green, and the left is red. The forward position lights must show light through a 110° angle from directly forward to the right and the left as shown in Fig. 11-48.

The aft light is white and mounted as far to the rear as possible. It is common practice to mount the aft position

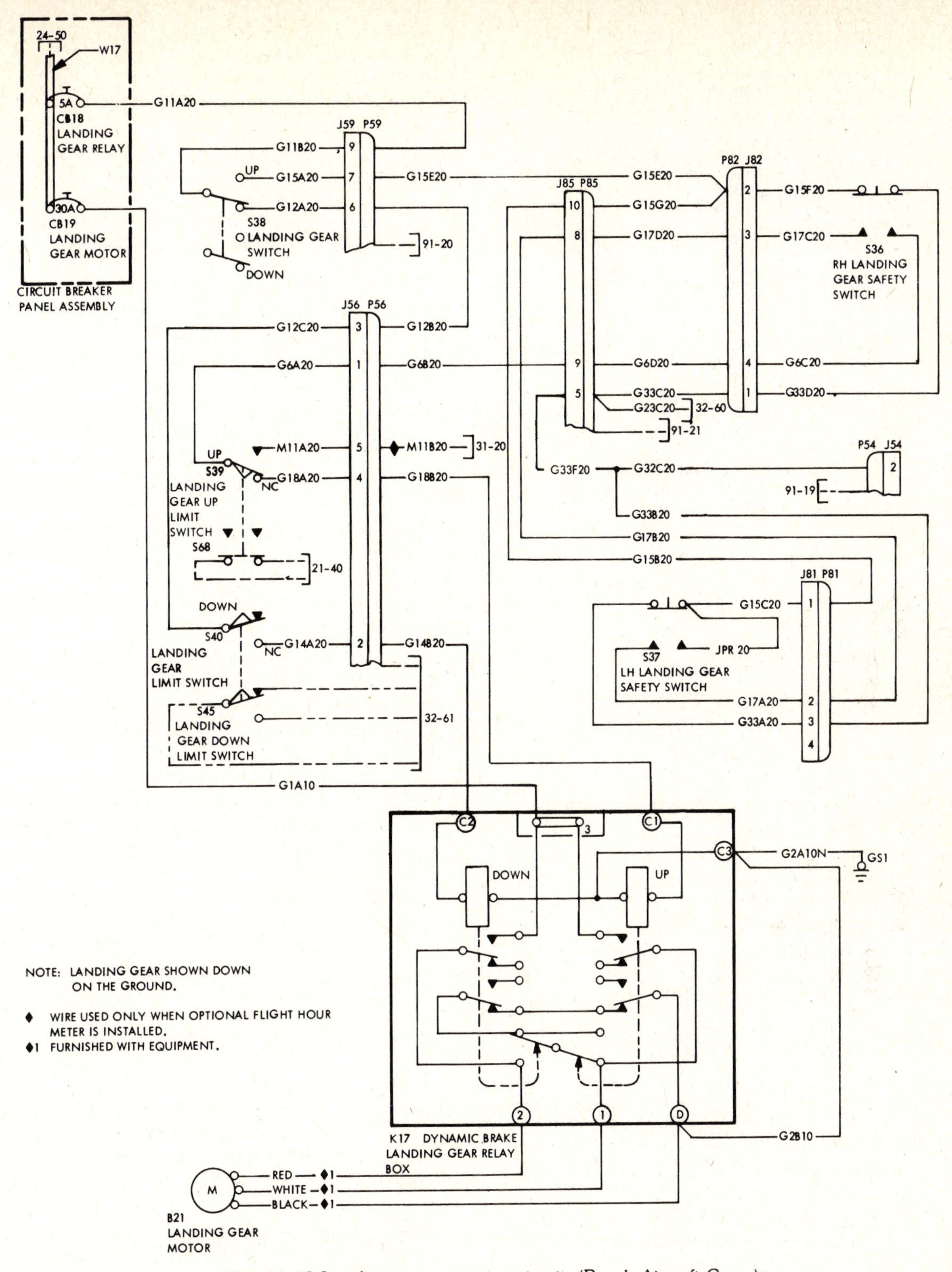

Fig. 11-43 Landing-gear actuating circuit. (Beech Aircraft Corp.)

light on the top of the vertical stabilizer (fin). The aft position light must show through an angle of 70° on each side of the centerline of the aircraft and to the rear.

The covers or color filters used on position lights must be of a material that is heat-resistant and will not shrink, fade, or become clouded or opaque.

All position lights must be in a single circuit and controlled by one switch. The power source is through one fuse or circuit breaker.

ANTICOLLISION LIGHTS

An anticollision light is designed to make the presence of an aircraft visible to pilots and crew members of other aircraft in the vicinity, particularly in areas of high-density aviation activity, at night and in conditions of reduced visibility. The anticollision light is of high intensity and flashes on and off not less than 40 or more than 100 cycles/min. The anticollision light system must consist of

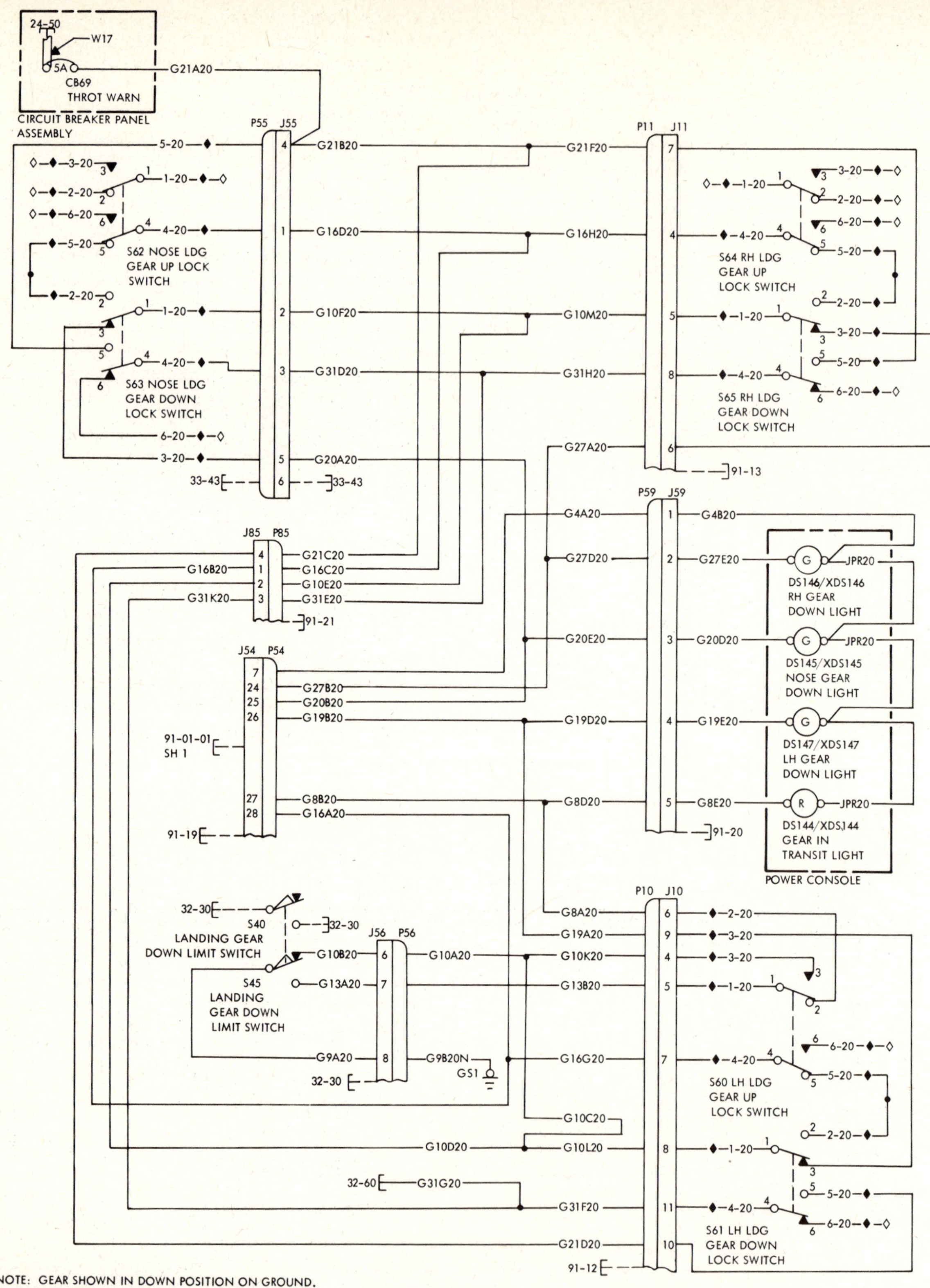

NOTE: GEAR SHOWN IN DOWN POSITION ON GROUND.
♦ STAMP REF DES OF APPLICABLE SWITCH ON HEAT SHRINKABLE TUBING & INSTALL ON WIRE FURNISHED WITH EQUIPMENT.
◊ TIE UNUSED WIRES INTO HARNESS. CUT OFF AT SPLICE LOCATION.

Fig. 11-44 Landing-gear position-indicating system. (Beech Aircraft Corp.)

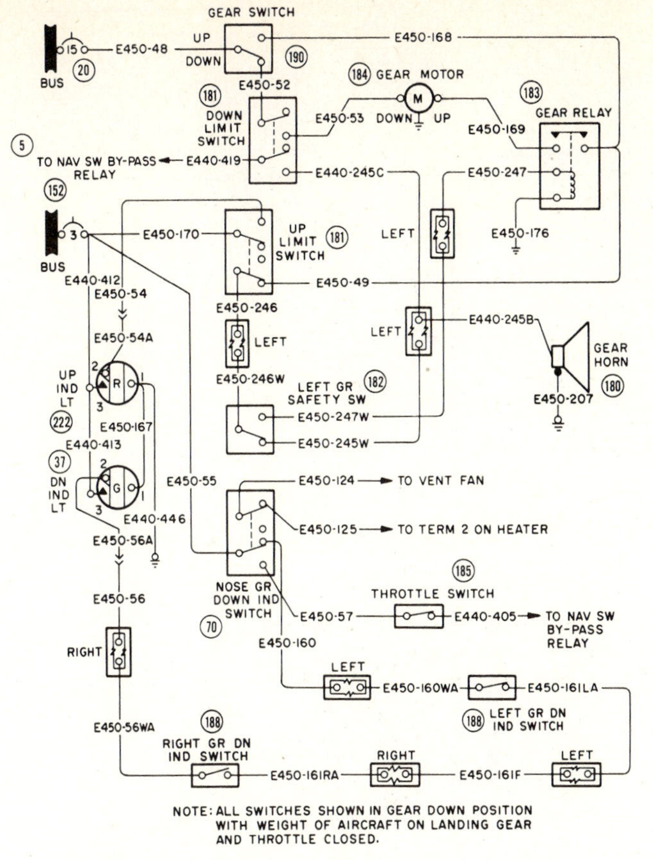

Fig. 11-45 Landing-gear circuit. (Cessna Aircraft Co.)

enough lights to illuminate the vital areas around the airplane, the precise number depending on the physical characteristics and configuration of the particular airplane.

Anticollision lights may be mounted on the top of the fuselage, the top of the vertical stabilizer, and on the bottom of the fuselage, depending upon the size of the aircraft. Small airplanes usually have one such light mounted on the top of the vertical stabilizer.

Flashing of anticollision lights is accomplished by one of two methods. One method is to mount two mirrors above a high-intensity light in such a manner that they reflect the light in two beams opposite each other. The mirrors are rotated by a small electric motor to provide the flashing effect.

The other type of anticollision light utilizes the principle of a stroboscope and is called a strobe light. In this system, a glass or quartz tube filled at low pressure with xenon gas is caused to flash by applying a high voltage to two electrodes in the tube and triggering the tube with an additional circuit. The current used to fire the tube is stored in a capacitor by means of a charging circuit. This circuit converts the low-voltage from the aircraft electric system to a high voltage (300 to 500 V) to charge the storage capacitor. A trigger circuit then applies the trigger signal to the trigger terminal of the tube and causes it to fire. The duration of the flash caused by the capacitor discharge may be a little more or less than 0.001 s, but the intensity of the light is very high and can be seen for many miles. The strobe principle is the same as that of a photographer's electronic flashgun.

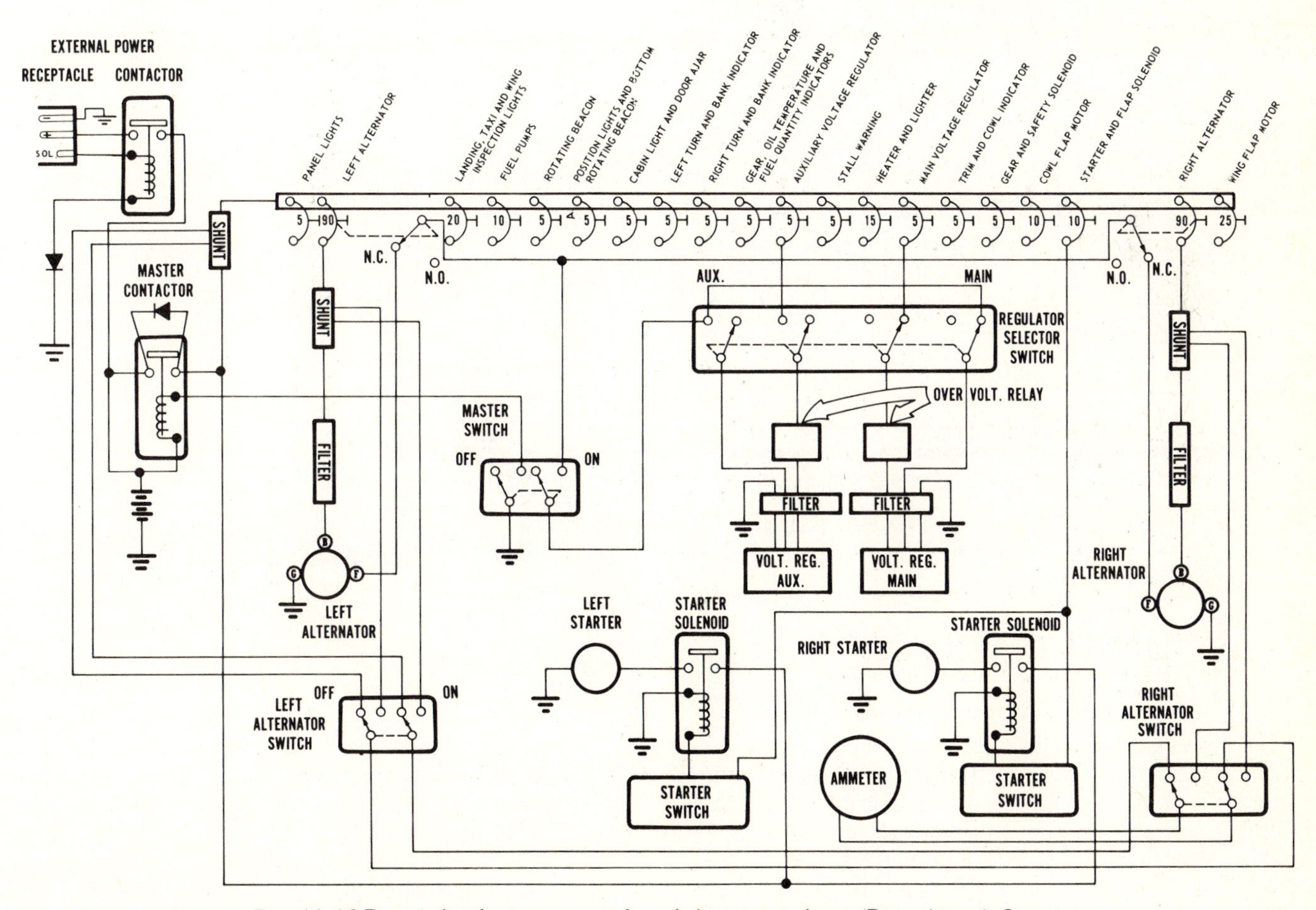

Fig. 11-46 Power distribution system for a light-twin airplane. (Piper Aircraft Corp.)

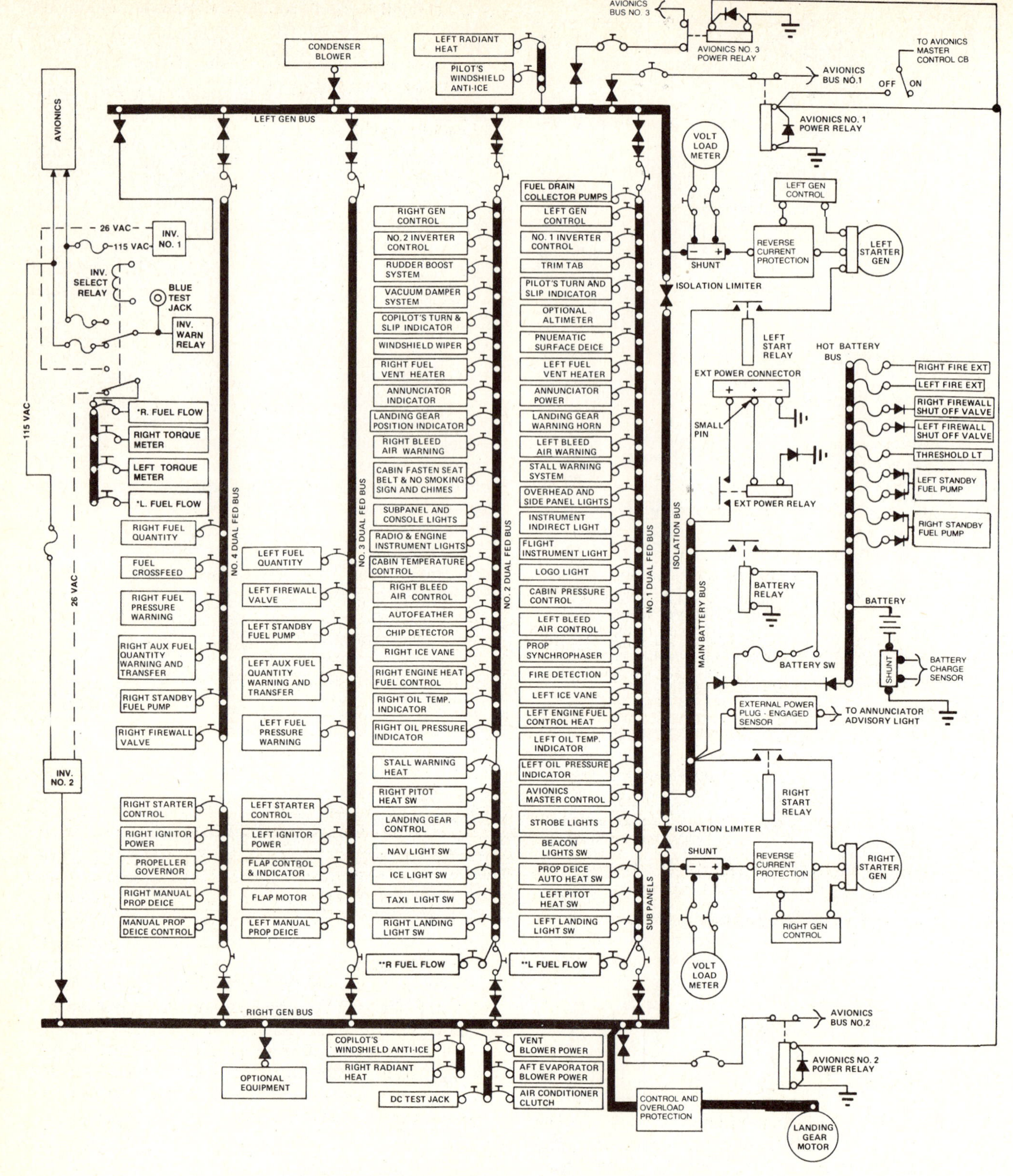

Fig. 11-47 Power distribution system for the Beechcraft Super King Air 200. (Beech Aircraft Corp.)

The installation of anticollision lights, if not done at the factory, must be carefully planned. The light must be located in a position where reflected light from the flash will not interfere with the pilot's vision. Methods for installing anticollision lights are set forth in the FAA publication AC 43.13-2A. Regulations concerning such lights are provided in FARs Part 23 and Part 25.

LANDING LIGHTS

Landing lights for aircraft are required to provide adequate light to illuminate the runway when the aircraft is making a landing. Landing lights are similar to the headlights of automobiles but are of a higher intensity. A parabolic reflector is utilized to concentrate the light into a beam of the desired width.

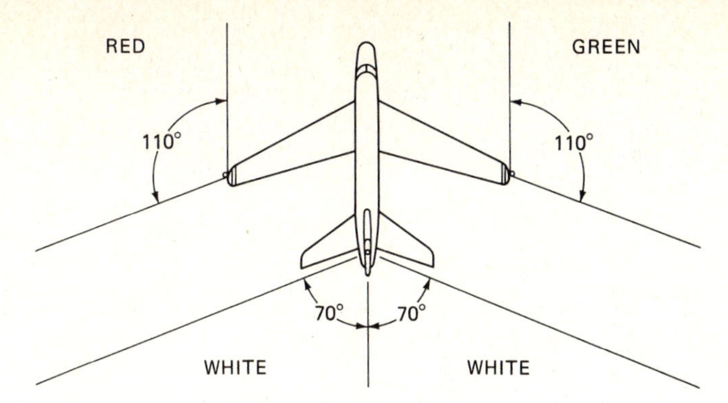

Fig. 11-48 Arrangement for position lights.

Landing lights may be attached to the stationary part of the nose gear, installed in the leading edges of wings, or, in the case of the retractable types, installed in the lower surface of the wings. Some large aircraft have landing lights in both the leading edges of the wings and retractable lights in the lower surfaces of the wings. The leading-edge landing lights may be turned on several miles away from the landing site, and the retractable lights are turned on shortly before landing.

Retractable landing lights are extended by means of a small but powerful motor that is able to move the light outward and forward against the force of the airstream. The extending motor and mechanism must be inspected and tested frequently to assure satisfactory operation.

INSTRUMENT LIGHTS

Instrument lights are installed behind the face panel of the instrument panel. The lights illuminate the instruments but do not shine directly toward the pilot or copilot. All instrument lights must be shielded in this manner. Instrument lights are provided with a dimming arrangement so the intensity can be adjusted to suit the needs of the pilot.

WARNING LIGHTS

Warning lights are provided to alert the pilot and crew to operating conditions within the aircraft systems. Red lights are used to indicate danger, amber lights to indicate caution, and green lights to indicate safe conditions. Indicator lights that are intended only for the purpose of providing information may be white.

ELECTRICAL TROUBLESHOOTING

Troubleshooting of an electric system follows the same general principles as the troubleshooting of other systems. When a fault exists in a system, it is necessary to determine the cause of the fault in the most effective manner and with the least expenditure of time and money. A thorough understanding of a system is essential if a fault in a system is to be eliminated quickly and effectively.

Faults in electric systems may be caused by defective electric components such as switches, fuses, circuit breakers, connectors, wiring, terminals, etc. In troubleshooting, it is the purpose of the technician to isolate the particular component that is causing the trouble.

TROUBLESHOOTING EQUIPMENT

The equipment normally used for troubleshooting of electric systems includes continuity testers, test lamps, voltmeters, ohmmeters, voltohmmeters, and multimeters. A multimeter is essentially a voltohmmeter with additional capabilities.

A continuity tester is a device for testing the continuity of an electric circuit or segment of a circuit. The segment of a circuit to be tested is connected between the leads of the continuity tester, and if the circuit is complete, the tester will respond. A continuity tester may be constructed by connecting a battery in series with a light bulb and two test leads as shown in Fig. 11-49. In this drawing, two flashlight cells are connected in series and to a 3-V flashlight bulb. The test leads are flexible, insulated wire with test prods at the ends of the leads. When the test prods are touched together, the light will light up, signifying continuity. If a section of unbroken wire is connected between the test prods, the light will glow to indicate that the wire is a complete conductor.

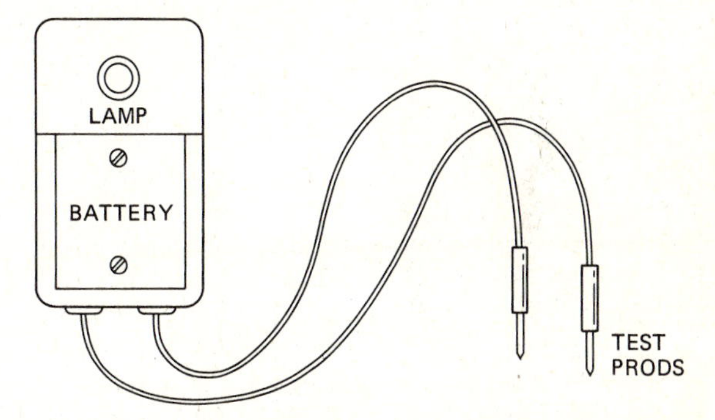

Fig. 11-49 A continuity tester.

A test light is used for testing "hot" circuits, that is, to determine whether electric power exists at the point being checked. A test light is simply a light bulb of the proper voltage for the circuit being tested with test leads connected to the socket terminals. Such an arrangement is shown in Fig. 11-50. If the test prods of the light are connected so that one is touching the terminal or point to be tested and the other to ground, and if the terminal being tested is "hot," the light will glow. Power must be turned on in the aircraft when one is testing with a test lamp.

Voltmeters, ohmmeters, and multimeters were described in Chapter 5 of this text. These instruments may be used instead of the continuity tester and test light. The ohmmeter is used as a continuity tester, and the voltmeter is used to test hot circuits.

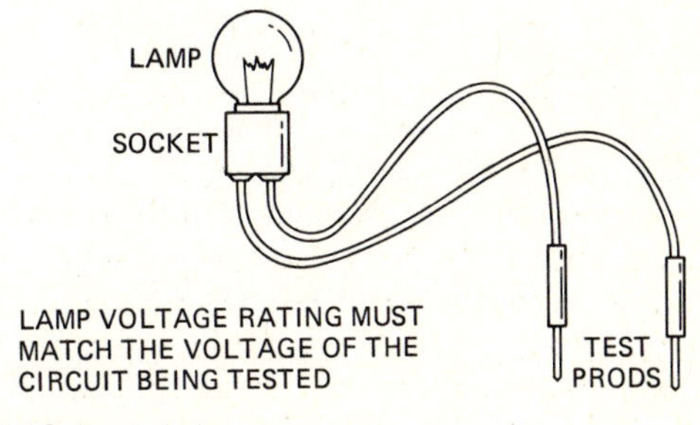

Fig. 11-50 A test light.

PROCEDURES

The procedures used in troubleshooting depend upon the circuit being tested, the type of electric power carried by the circuit, and the other circuits that may be interconnected with the circuit being checked. Proper procedures will often be described in the appropriate service manual.

An example of a troubleshooting procedure applied to a failed position light illustrates a suitable method for such a case. First, the technician must know that if one of the three position lights does not glow and the other two operate satisfactorily, the trouble is immediately isolated to the circuit for the one failed light. This eliminates any trouble with the switch, circuit breaker, or fuse. The first item to check is the light bulb. The bulb is removed and examined visually. If the filament can be seen, it can be checked for discontinuity. If it cannot be seen, the bulb can be checked with an ohmmeter. A good bulb should show very little resistance. If the ohmmeter indicates infinite resistance—that is, if the ohmmeter needle does not move—the bulb is burned out or otherwise defective and it is necessary merely to install a new bulb to correct the problem. If it is found that the bulb is in good condition, the next step is to examine the connections to the bulb socket in the aircraft. If these are good, a voltmeter is used to determine if power is delivered to the socket when the light switch is turned on. The voltmeter must be of a range that will handle the voltage of the system. If the voltmeter indicates that no voltage is present at the light socket, then the power lead should be checked for continuity. It may be necessary at this point to check the lighting circuit in the service manual to determine how the power lead for the light is connected. At the point where the power lead receives power from the light switch, voltage will be present because the other position lights operate satisfactorily. The continuity tester or ohmmeter should then be connected between the point where power exists to the end of the conductor at the light socket. This is to verify that a discontinuity exists. The continuity tester or ohmmeter is then connected between the ends of various segments of the power lead until the broken lead or defective terminal is located. Power in the airplane should be turned off during these tests.

If an electric motor will not operate, the circuit should be checked to determine whether the motor is switch-controlled or relay-controlled. Circuit breakers or fuses carrying power to the motor and relay are checked for condition. If these are satisfactory, the leads to the motor and relay are checked for tightness at terminals and for complete ground connections. If these appear to be satisfactory, the circuitry is checked with a voltmeter to see if power is reaching the motor. If power is reaching the motor and the motor will not run, it must be replaced.

TROUBLESHOOTING CHARTS

For many of the electric circuits and components, the manufacturers provide troubleshooting charts. By following the instructions given in these charts, the technician can usually isolate and correct any problem in the system.

REVIEW QUESTIONS

1. Why are dependable electric systems essential in modern aircraft?
2. How is reliability in aircraft electric systems attained?
3. List general requirements for aircraft electric systems.
4. What electric indicating instruments must be included in the electric power system of an aircraft?
5. List the requirements for an external power source?
6. Discuss the need for protective devices in aircraft electric systems.
7. Where should protective devices be located?
8. What is the purpose of a *master switch?*
9. What is an *electrical load analysis?*
10. What types of electrical loads may be considered as *intermittent loads?*
11. What must be done before adding electric equipment in an aircraft system?
12. Describe approved aircraft electric wire or cable.
13. Why is aircraft wire stranded?
14. What types of materials are used for wire insulation?
15. Describe *shielded* wire or cable.
16. What is a *coaxial* cable?
17. How is the cross-section area of wire indicated?
18. What voltage drop is allowed for a 115-V system in a continuously operating circuit?
19. What size copper wire should be used in a 28-V circuit for a 10-A continuous load when the wire is in a conduit and the distance from the bus to the load is 20 ft [6.1 m]?
20. What continuous load can be carried by a No. 14 copper wire with a length of 20 ft [6.1 m] and mounted in the open air?
21. What size copper cable should be installed for a starter that draws 200 A?
22. What size aluminum cable should be used for a 200-A continuous load?
23. What is the minimum bend radius for a wire bundle?
24. What can be done to protect wire and wire bundles from abrasion?
25. Describe a satisfactory method for wire-bundle lacing.
26. What method is employed to prevent cable or wire bundles from vibrating, swinging, or otherwise moving?
27. Describe an approved clamp for wire bundles.
28. Describe how wire bundles should be routed in an aircraft.
29. Describe electrical *conduit.*
30. When it is impractical to separate an electric wire from a line carrying flammable fluids, where would the wire be located with respect to the fluid line?
31. Describe typical terminals used with aircraft electric wire or cable.
32. Describe the use of wire strippers.
33. Why is it important to use specific types of crimping tools when installing wire terminals?
34. Why are soldered terminals not considered satisfactory for aircraft electric systems?
35. Describe the proper method of soldering if a soldered terminal is used.
36. How may aircraft wires be spliced?
37. Describe a *terminal strip.*
38. What is the maximum number of terminals that should be attached to a single stud?
39. What is the position requirement for mounting a terminal strip?

40. Describe the use of washers and lock washers in the installation of studs and bolts with terminals.
41. Describe the installation of a junction box.
42. Why is it not correct to install items on the inside of a junction-box cover?
43. When electric wires pass through holes in bulkheads or the side of a junction box, what protection must be provided?
44. Discuss the precautions necessary in the installation of aluminum cable in an aircraft system.
45. What is the advantage provided by the use of connector plugs in electric systems?
46. Describe the procedure for soldering wires to a connector plug.
47. What should be done regarding unused contact holes when a contactor is being wired?
48. Describe the installation of pins and sockets in crimp-type connectors.
49. Describe the process of *potting* the wire connections to an electric connector.
50. Describe a *fuse.*
51. What must be done before resetting a circuit breaker that is tripped?
52. Describe a *thermal circuit breaker.*
53. What should be the capacity of a circuit breaker used with No. 16 wire?
54. What is a *tripfree* circuit breaker?
55. Describe the principles of switch installation.
56. Why should switches be *derated* for some circuits?
57. Describe a DPST switch.
58. Explain the purposes of *bonding* in an aircraft.
59. What is the maximum resistance permitted for a bonding connection?
60. Describe the procedure for installing a bonding jumper.
61. What is the requirement for a bonding jumper that must carry a ground load for a unit of electric equipment?
62. Explain the importance of noting the types of metals joined by bonding.
63. What is the purpose of *shielding?*
64. How is shielding accomplished?
65. Explain how electric wires are identified in a system.
66. What precaution must be taken when marking coaxial cable with a machine stamp?
67. What methods are used for attaching identification numbers and letters to wires and harnesses?
68. Explain the value of electric circuit drawings.
69. What types of lights are required for aircraft that are certificated for night flight?
70. Give the requirements for position or navigation lights.
71. Describe two types of anticollision lights.
72. What are some requirements for instrument lights?
73. What equipment is used for electrical troubleshooting?
74. Give procedures for electrical troubleshooting.

CHAPTER 12

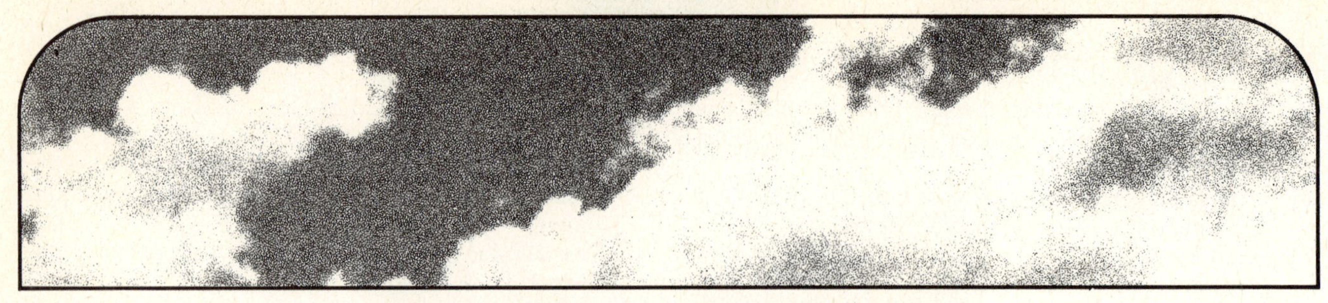

Principles of Electronics

The science, technology, and practices employed in the field of electronics have advanced greatly in recent years, largely because of the requirements of space flight where reliability and minimum weight are of prime importance. The computer industry has required and developed devices which can perform electronically at extremely high speeds, thus making it possible to perform calculations in seconds, whereas it would have taken days or even years to perform these same calculations manually or with comptometers. Electronic circuitry has been miniaturized and microminiaturized by means of transistors, integrated circuits, and special packaging techniques to the extent that some electronic circuits occupy only a tiny fraction of 1 percent of space once required for similar circuits.

The advances in electronic technology have been a boon to the aircraft industry because aircraft can now carry extremely complex systems for navigation, automatic flight, communications, and other functions with a very small weight penalty. To service the electronic equipment on aircraft, **avionics** technicians are employed. *Avionics* is the term for electronics as it is applied to aircraft, and highly trained technicians are needed to service and test the many components required in the industry. Airframe and powerplant maintenance technicians who find electronics interesting often take the additional training necessary to become avionics technicians.

It is quite obvious that the extensive field of electronics and avionics cannot be covered in one chapter or in one textbook. This chapter, however, will describe the basic principles of electronics so that students will have an understanding of how electronic systems function and thus be able to communicate intelligently with avionics technicians with whom they may work. The information gained from this chapter will also enable them to continue on to more advanced studies. In this chapter we shall consider such circuit elements as inductors, capacitors, resistors, electron tubes, transistors, integrated circuits, and other items used in the construction of electronic circuits. We shall examine the functions of these units and also the methods by which they are connected together to produce certain effects. In other chapters we shall see how the principles discussed in this chapter are used in radio receivers, radio transmitters, electronic control systems, radar, and other electronic devices.

ELECTRONIC CIRCUIT ELEMENTS

RESISTORS

A resistor is a circuit element designed to insert resistance in the circuit. A resistor may be of low value or of extremely high value. The nature of resistance and its effect in electric circuits have been discussed in a previous section; the construction of resistors and their use in circuits, however, will be described briefly here.

Resistors in electronic circuits are made in a variety of sizes and shapes. They are generally classed as fixed, adjustable, or variable, depending upon their construction and use. Some typical fixed resistors are illustrated in Fig. 12-1. These resistors are constructed of a small rod of

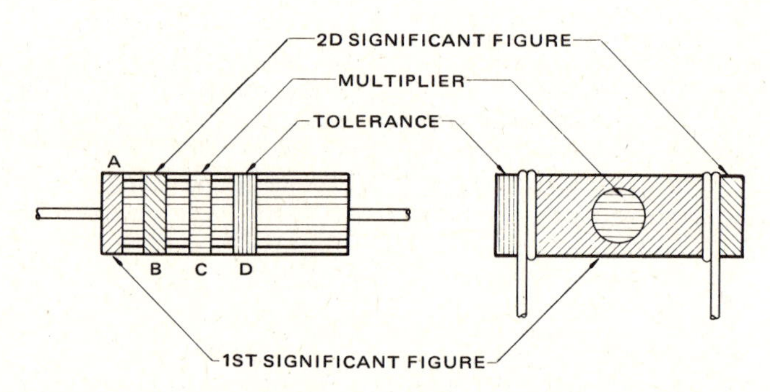

Fig. 12-1 Fixed, carbon-type resistors.

a carbon compound. The value of the resistance for each resistor is determined by the makeup and size of the carbon compound and may vary from only a few ohms up to several million ohms. The two important values associated with resistors are the value in **ohms** of resistance and the value in **watts** which represents the capacity of the resistor to dissipate power. The two types of fixed resistors shown in Fig. 12-1 are called axial-lead resistors and radial-lead resistors because of the way the electric leads are attached to the resistor.

The resistance value of small fixed resistors is indicated by a code color. The numerical values of the the colors used in this coding are as follows:

Black	0	Green	5
Brown	1	Blue	6
Red	2	Violet	7
Orange	3	Gray	8
Yellow	4	White	9

On the axial-lead resistor there are four color bands. The band at the end of the resistor is called band *A* and represents the first digit of the resistance value. The next, band *B*, represents the second digit of the resistance value; the third, band *C*, represents the number of zeros to be placed after the first two digits; and the fourth, band *D*, indicates the degree of accuracy or tolerance of the resistor.

If a resistor's four bands are colored with green (band *A*), blue (band *B*), orange (band *C*), and silver (band *D*), then the value of the resistor is determined as follows: the green band indicates the figure 5, the blue band the figure 6, and the orange band shows that three zeros are to follow the 5 and 6. Therefore, the resistor has a value of 56 000 Ω. The silver, or *D*, band represents a tolerance of 10 percent. If this band were gold, the tolerance would be 5 percent. When we wish to determine the value of a radial-lead resistor, such as that shown in Fig. 12-1, the body color represents the first digit, the colored tip represents the second digit, and the dot represents the number of zeros to follow the first two digits. The opposite end of the resistor will be colored silver, gold, or no color to indicate the tolerance. This color coding was established in 1938 by the Radio Manufacturers' Association, now known as the Electronic Industries Association (EIA).

Resistors required to carry a comparatively high current and dissipate high power are usually of the wire-wound ceramic type (see Fig. 12-2). A wire-wound resistor consists of a ceramic tube wound with fine resistance wire, which is then covered with a ceramic coating or glaze. This coating is often called vitreous enamel and prevents the wire from becoming damaged through use or mishandling. The terminals for the resistance wire are brought out at each end of the resistor as shown. The value of the wire-wound resistor is usually printed on the ceramic, or vitreous, coating.

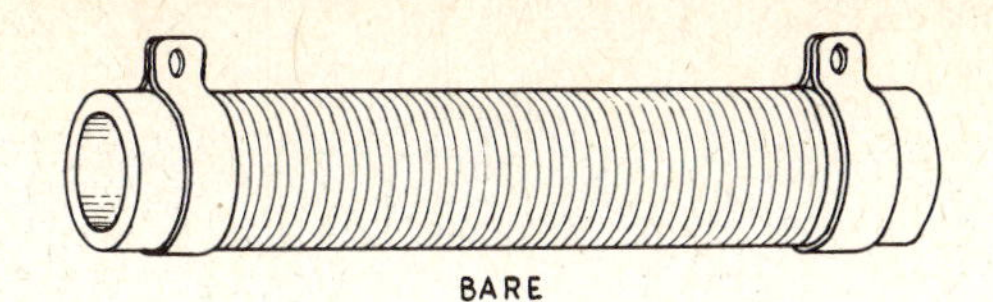

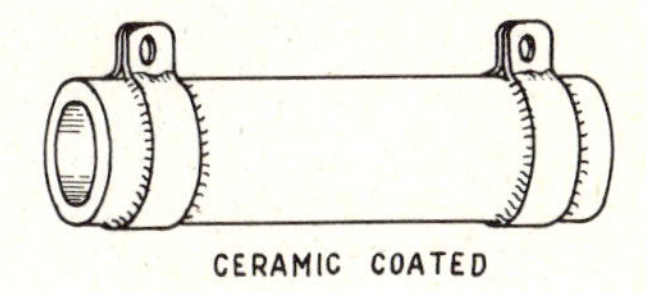

Fig. 12-2 Wire-wound resistors.

ADJUSTABLE AND VARIABLE RESISTORS

An adjustable resistor shown in Fig. 12-3, is usually of the wire-wound type with a metal collar which may be moved along the resistance wire to vary the value of the resistance placed in the circuit. In order to change the resistance, the contact band must be loosened and moved to the desired position and then tightened so that it will not slip. In this way the resistor becomes, for all practical purposes, a fixed resistor during operation.

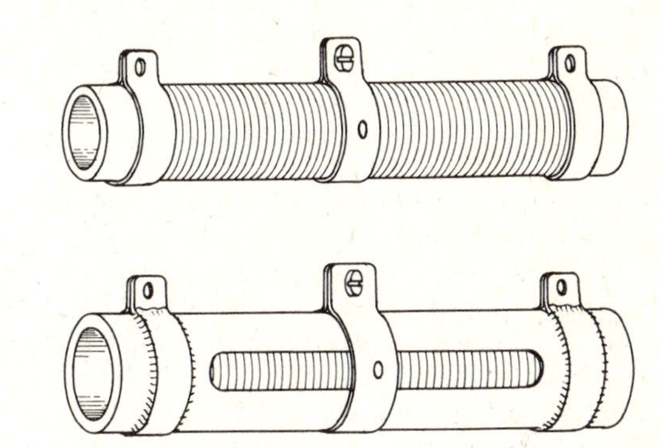

Fig. 12-3 Adjustable resistors.

A variable resistor is arranged so that it may be changed in value at any time by the operator of the electronic circuit. This change is usually accompanied by rotating a small adjustment knob or by turning a screw adjustment. Variable resistors are commonly known as **rheostats** or **potentiometers.** A rheostat is shown in Fig. 12-4*a* and a potentiometer is shown in Fig. 12-4*b*. A rheostat is normally connected in a circuit merely to change the current flow and has a comparatively low resistance value. Its circuit connections are as shown in Fig. 12-5. Note that the rheostat has two terminals, one connected to the wire-wound resistor and the other connected to the sliding arm or contact which moves along the resistor. This diagram

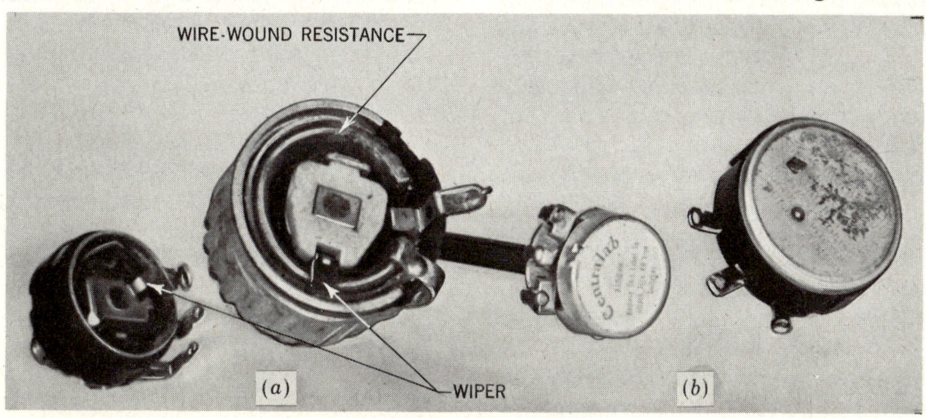

Fig. 12-4 Variable resistors.

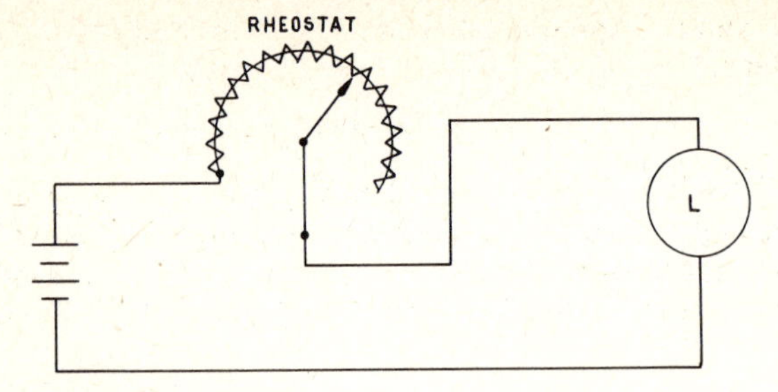

Fig. 12-5 Rheostat circuit.

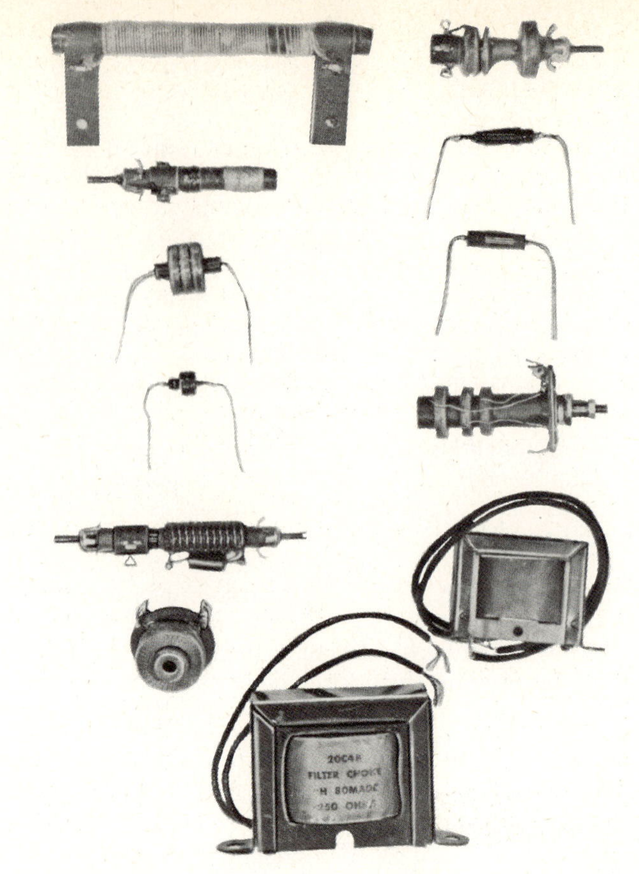

Fig. 12-7 Inductors.

for the rheostat is not the standard symbol but is used here to show the arrangement of a typical rheostat.

The potentiometer normally is connected with three terminals. One terminal is connected at each end of the resistor, and the third terminal is connected to the sliding contact arm. The value of the resistance of a potentiometer is comparatively high, and the resistance is normally made of a material such as a carbon or graphite compound. The purpose of a potentiometer is to vary the value of the voltage in a circuit. A diagram for a potentiometer in an experimental triode-tube circuit is shown in Fig. 12-6.

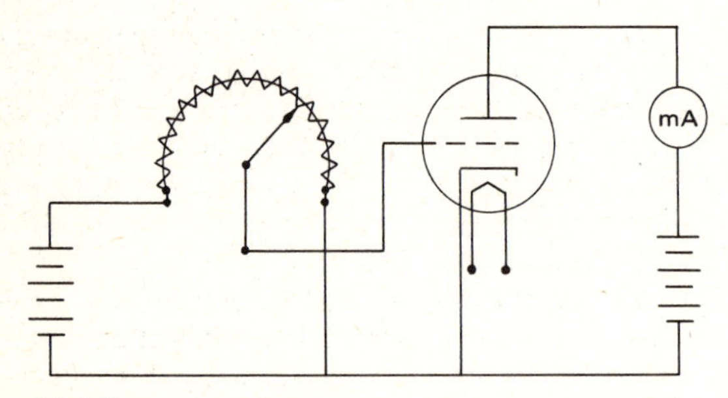

Fig. 12-6 Potentiometer circuit.

It must be pointed out that the use of a resistor of any type must be very carefully considered. The capacity of a fixed resistor, rheostat, or potentiometer must be such that it can handle the current through the circuit without damage. It is always necessary to compute the current through the resistor by means of Ohm's law before placing the circuit in operation.

INDUCTORS

The purpose of an inductor, or inductance coil, is to insert inductance into a circuit. As explained in a previous section, the effect of an inductance is to oppose any change in the existing current flow in a circuit. The opposition to current flow in an ac circuit by an inductor is called **inductive reactance** and is measured in ohms.

Inductors are made in many shapes and designs. An inductor used in extremely high-frequency circuits may consist of only one turn or even less than one turn of wire. On the other hand, an inductor used as a choke coil in a low-frequency circuit or in a filter circuit may contain many turns of wire and also be wound on an iron core to increase the inductance. Some typical inductors are shown in Fig. 12-7.

Inductors are often used in radio work in connection with capacitors to provide tuned circuits. These tuned circuits are most valuable in radio and television for filtering out unwanted frequencies and passing the desired frequencies.

In many electronic circuits it is desirable to use inductors which are variable in inductance. This means that devices must be provided which enable the operator to change the inductance of the inductance coil. A common method for changing the inductance is to use a powdered-iron core in the inductor and provide a means whereby this core may be moved in and out of the coil. An inductance coil which contains a movable core for tuning purposes is often called a *slug-tuned* inductor. Inductors of this type are often found in small radio receivers.

Inductance coils are rated as to value in **henrys.** In electronic work, 1 H is a comparatively large inductance. Therefore, many of the inductors used in electronic circuits are rated in **millihenrys.** One millihenry is one-thousandth of a henry. **One henry is the inductance of a coil which will produce a back voltage of 1 volt when the current change is at the rate of 1 ampere per second.**

All conducting elements of an electronic cirucit contain or possess inductance. The effect of this inductance is not great at low frequencies, but in electronic circuits, where frequencies are in the range of 500 to 1000 or 10 000 MHz, all the inductances of the circuit elements must be considered in the design and operation of the circuit. Some of these effects will be discussed later.

CAPACITORS

A capacitor may be defined as a device consisting of two or more conductor plates separated from one another by a dielectric (insulating material) and used for receiving and storing an electric charge. The effect of a capacitor in an electric circuit is to oppose any change in the existing voltage.

Capacitors are commonly used in dc circuits to reduce the effects of transient voltages and currents. **Electrical transients** are high voltages developed from time to time when the circuit is broken or reconnected, as when a switch is turned on or off. These transient voltages are usually caused by the inductance of a circuit. In an ac circuit the capacitor is often used to block direct current but permits the flow of the alternating current. In effect, the alternating current appears to flow through the capacitor but is actually being stored first on one plate of the capacitor and then on the other, as described in the section on alternating current.

Like many other electronic units, capacitors are manufactured in a wide variety of sizes and styles. Some very low-capacity capacitors are merely tiny wafers of metal separated by an insulator; large capacitors may weigh several pounds. Fixed capacitors are of two general types: one is the dry capacitor, which consists of metal plates separated by a dry dielectric such as mica or waxed paper, and the other the electrolytic capacitor, whose dielectric is a chemical paste or electrolyte. The electrolytic capacitor is effective in only one direction. This means that it must be connected in such a manner that the positive and negative polarities are correct. If it is connected in reverse, current will flow through the capacitor and destroy it. Fixed capacitors of both the dry and the electrolytic type are manufactured in a wide variety of shapes as shown in Fig. 12-8. The electrolytic capacitors are marked to indicate the correct method of connection into a circuit.

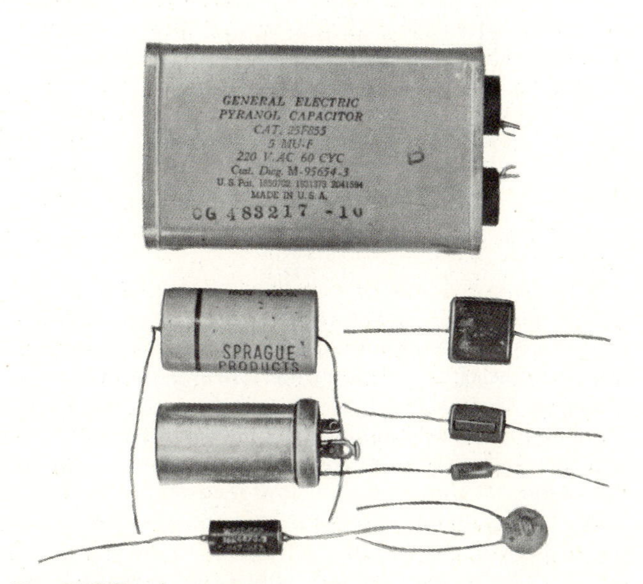

Fig. 12-8 Fixed capacitors.

As previously explained, the unit of capacitance is a **farad. A capacitor which will store 1 coulomb of electricity under an emf of 1 volt has a capacitance of 1 farad.** The farad is an extremely high value of capacitance; therefore capacitors used in standard electronic circuits are rated in **microfarads** (1 μF = one-millionth of a farad) or **picofarad** (1 pF = one millionth of a microfarad).

VARIABLE CAPACITORS

Variable capacitors are used in electronic circuits whenever it is necessary or desirable to change the capacitance in order to tune the circuit or to change the circuit values. Capacitors of this type are shown in Fig. 12-9. As is the case with any other type of capacitor, the capacitance of a variable capacitor is determined by the number and area of its plates. As shown in Fig. 12-9, the variable capacitor consists of a set of rotor plates and a set of stator plates. The rotor plates are mounted on a shaft in such a way that they may be rotated and caused to mesh with the stator plates. When the rotor plates are meshed with the stator plates, a thin layer of air exists between the two sets of plates, and there is no electric connection between them. The capacitance is greatest when the plates are completely meshed as shown in Fig. 12-10.

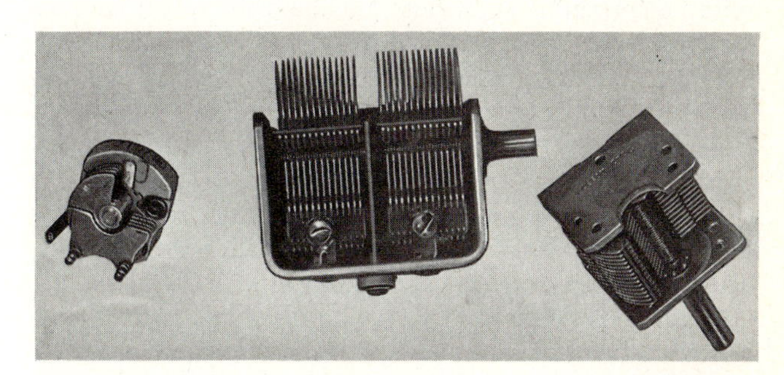

Fig. 12-9 Variable capacitors.

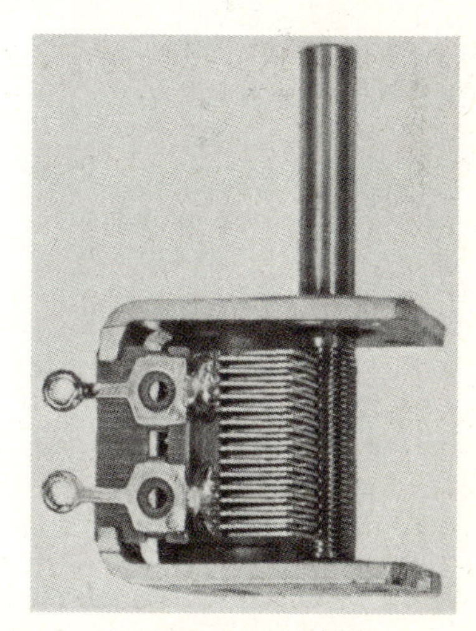

Fig. 12-10 Maximum capacitance position for a variable capacitor.

The effect of capacitance in an ac circuit is to give the circuit **capacitive reactance.** Capacitive reactance is measured in ohms and has an effect opposite that of inductive reactance. This fact makes it possible to use inductive reactance and capacitive reactance together to produce tuned circuits.

PRINCIPLES OF TUNING

RESONANT CIRCUITS

In the design and operation of electronic systems, resonant circuits provide the key to frequency control. When a certain frequency is to be produced, it is necessary to establish a circuit which is resonant at that frequency. Also, when a certain frequency is to be passed through a circuit and others eliminated, it is necessary to have a circuit which is resonant at the frequency to be passed. When a certain frequency is to be blocked, it is then necessary to place in the circuit a resonant tank circuit, which will block the frequency for which it is resonant. As can easily be seen, resonant circuits are most essential in radio and television receivers and transmitters.

From the study of alternating current we know that a resonant circuit is one in which the capacitive reactance (X_C) is equal to the inductive reactance (X_L). For any particular combination of capacitance and inductance we know that the resonant frequency is fixed; that is, the combination can have only one resonant frequency. This frequency may be determined by the formula

$$f = \frac{1}{2\pi\sqrt{LC}}$$

As an example, let us consider the circuit in Fig. 12-11.

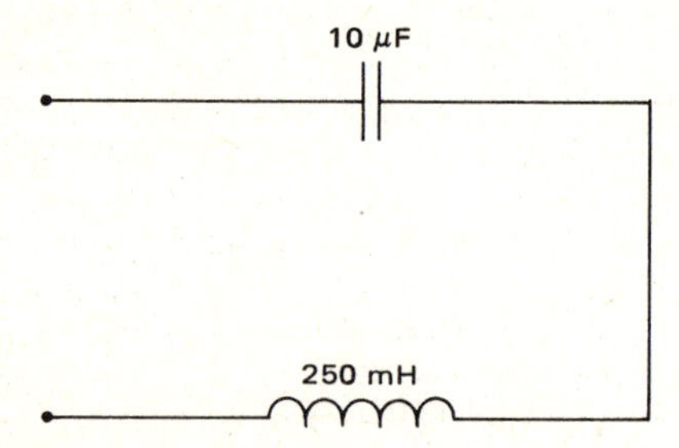

Fig. 12-11 Series LC circuit.

This is a series LC (inductance-capacitance) circuit containing a capacitance of 10 μF and an inductance of 250 mH. We shall determine the resonant frequency of this circuit as follows, remembering the foregoing formula:

$$f = \frac{1}{6.28\sqrt{10^{-5} \times 0.25}}$$

This may also be expressed

$$f = \frac{1}{6.28\sqrt{2.5 \times 10^{-6}}}$$

$$f = \frac{1}{6.28 \times 1.581 \times 0.001}$$

Then

$$f = \frac{1{,}000}{6.28 \times 1.581}$$

or

$$f = 100.7 \text{ Hz}$$

This is the resonant frequency of the circuit.

In a series LC circuit, such as that just described, the impedance at resonance is equal to the resistance of the circuit, inasmuch as the capacitive reactance and the inductive reactance cancel each other. This condition is true only in a series LC circuit. If we consider a parallel LC circuit, such as that shown in Fig. 12-12, we find that

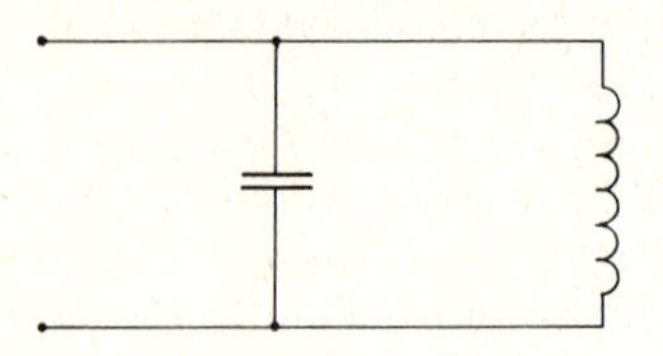

Fig. 12-12 Parallel LC circuit.

at resonance the impedance across the parallel circuit is almost infinite, and if it were not for resistance in the circuit, the impedance actually would be infinite. Figure 12-13 shows a curve which indicates the effect of resonance in a parallel LC circuit. Notice that at zero frequency the impedance is very low and that it then rises as the frequency approaches the resonant value. At resonance the impedance is at a maximum and then falls off as the frequency increases above resonance.

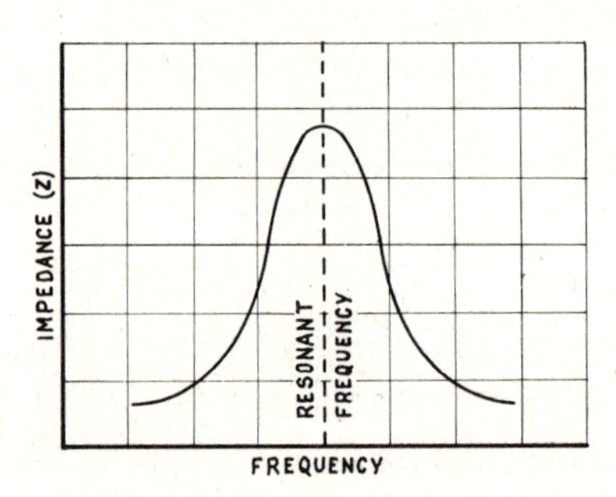

Fig. 12-13 Effect of resonance on impedance in a parallel LC circuit.

FILTERS

The characteristics of resonant circuits, as just described, make them very useful for filtering various frequencies in an electronic circuit. Among the types of filters used in electronic circuits are high-pass filters, low-pass filters, and bandpass filters. A high-pass filter tends to pass frequencies in the higher ranges and to attenuate or reduce the current at frequencies in low ranges. The low-pass filter will pass frequencies in the lower ranges and attenuate or reduce the current at frequencies of the higher ranges. A bandpass filter will allow a certain band of frequencies to pass and will reduce the current at frequencies below or above the band range.

A circuit for a **high-pass filter** is shown in Fig. 12-14. Notice that the capacitance is in series with the circuit and that the inductance is in parallel with the circuit. Since capacitive reactance decreases as frequency increases, at high frequency levels current will appear to flow through the capacitor; and since the inductive reactance increases

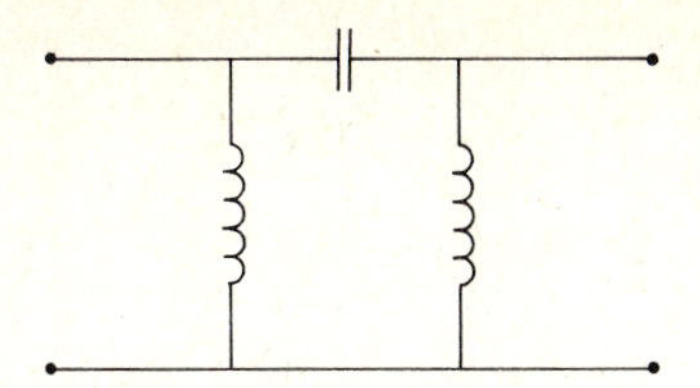

Fig. 12-14 High-pass filter.

as frequency increases, current will decrease through the inductance coil as the frequency increases. This circuit will therefore tend to pass high frequencies and eliminate or reduce low frequencies.

Figure 12-15 shows the circuit for a **low-pass filter.** In this circuit the inductance coil is in series, and the capacitance is in parallel. Low frequencies will pass easily through the inductance coil and will be reduced through the capacitors. As frequencies increase, they will be blocked by the inductance coil and passed by the capacitors.

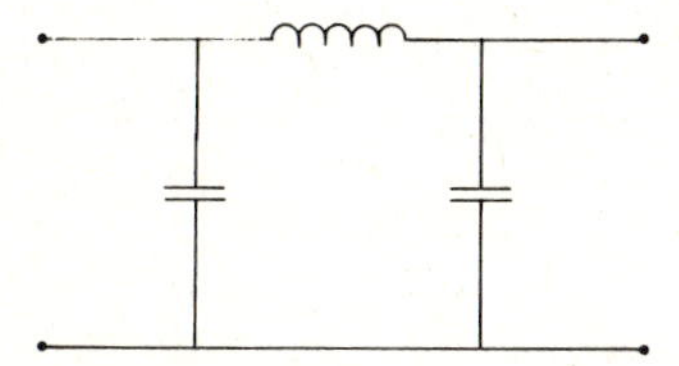

Fig. 12-15 Low-pass filter.

A **bandpass filter** is shown in Fig. 12-16. The impedance of the series *LC* circuit is high except at or near resonant frequency. Therefore, at resonant frequency the current flow will be comparatively high. At resonant frequency the impedance across the parallel portion of the circuit will also be high, thus preventing the current from being bypassed. The bandwidth of a bandpass filter is determined by the number of circuit elements and by the resistance of the circuit: the greater the resistance, the wider the band.

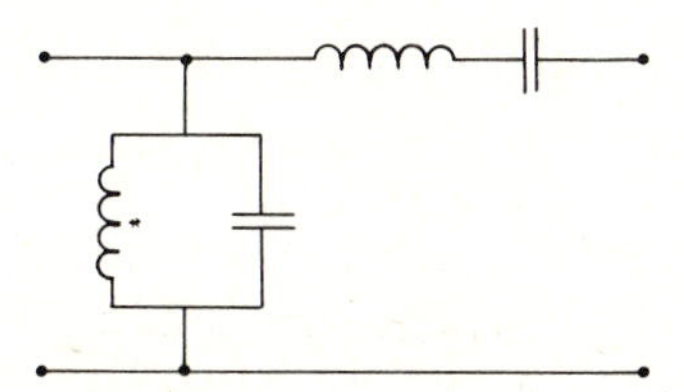

Fig. 12-16 Bandpass filter.

A **band-reject filter** is shown in Fig. 12-17. In this circuit the parallel *LC* circuit is in the line in series with the load, and the series portion or series *LC* circuit is in parallel. With this arrangement the resonant frequencies will be bypassed and blocked from the main line. All other frequencies will be passed along the main line.

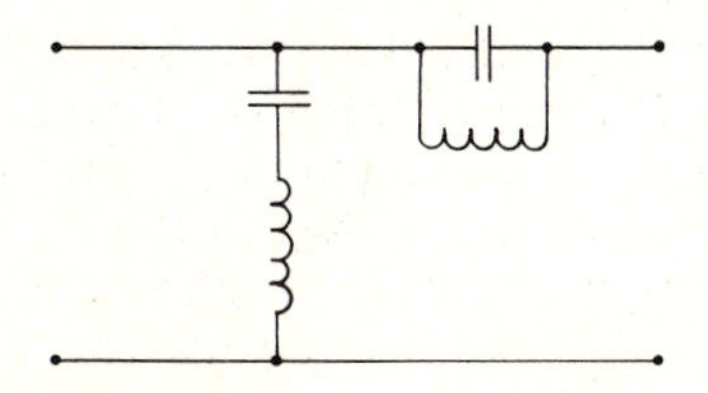

Fig. 12-17 Band-reject filter.

A filter may be made a **tuning circuit** by making either the inductance or the capacitance variable. A typical tuning circuit consists of a variable capacitor used with a fixed resistor. In some cases, however, the capacitor is fixed, and the inductance is tuned by means of a *slug*, or movable core. Tuning circuits are usually designed to have fairly high selectivity; that is, they allow only a very narrow band of frequencies to pass and reject all others.

THE ELECTRON TUBE

It may be stated without question that the modern electronic industry was born with the invention of the **electron tube.** The first discoveries in electron-tube phenomena were made by Thomas Edison in 1883 during his experiments with the incandescent lamp. Edison discovered that the heated filament of an incandescent lamp will give off electrons which pass to another electrode in the bulb and thus create an actual current flow from the filament to the other electrode, or plate. Edison did not follow through with any useful application of this discovery, but later, in 1889, J. J. Thomson formulated the theory that the heated filament gave off electrons. Since that time it has been well established that the heating of any material will cause electrons to separate from the substance and form a cloud around it. When the electrons leave the surface of the substance, the atoms in the substance retain a positive charge and attract the electrons back to the material. To cause the electrons to leave the material completely, it is necessary to provide a positive electrode in the vicinity of the material so that the electrons will be attracted to that positive charge. It is also necessary to provide a negative charge for the heated material to replace the electrons being removed. By varying the charges on the heated filament (cathode) and the plate (anode) of an electron tube, the current flow through the tube can be controlled. The tube therefore acts as an electric valve.

Since the advent of the transistor and other solid-state devices, the use of electron tubes in electronic circuits has greatly decreased. Many electronic technicians never encounter electronic tubes in their work other than cathode-ray tubes (CRT) used in oscilloscopes and other electronic equipment. Even though the aviation technician may not ever be involved with electron tubes, the theory and construction of such tubes will be described here because an understanding of such tubes helps in understanding other elements of electronic circuits, particularly transistors.

THE DIODE TUBE

An electron tube, also called a **vacuum tube,** consists of a glass or metal enclosure in which electrodes are placed and sealed in either a gaseous or an evacuated atmosphere. The simplest of electron tubes is the **diode,** which has two operating electrodes. One of these is the heated **cathode,** which emits the electrons, and the other is the **plate** or **anode.** The cathode may be directly heated or indirectly heated as shown in Fig. 12-18. The tube with the directly heated cathode utilizes the heated filament for the cathode. In this case the filament is coated with a special material which greatly increases the number of electrons emitted. If the tube has an indirectly heated cathode, the cathode consists of a metal tube in the

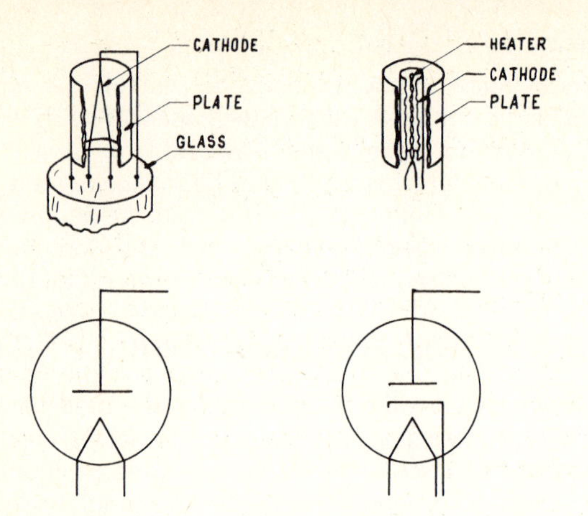

Fig. 12-18 Diode electron tubes.

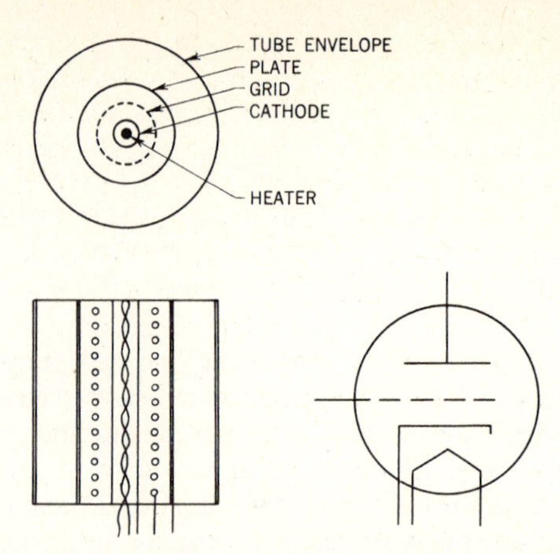

Fig. 12-20 Arrangement of triode tube elements and symbol for a triode tube.

center of which is a filament or heater. The heater is insulated from the metal tube. The outside of the cathode tube is covered with an electron-emitting material such as barium oxide, strontium oxide, or thorium oxide.

The principal advantage of the diode tube is that it will permit the flow of current in one direction only, that is, from the heated cathode to the anode. In order to accomplish this, the cathode must carry a negative charge with respect to the anode. If an alternating current is applied to the cathode, the tube will conduct only during one-half of each cycle, that is, while the cathode is negative and the anode or plate is positive. For this reason diode tubes are often used as rectifiers to change alternating current to direct current. Diode tubes are used almost universally in the power-supply circuits of such electronic devices as radio and television, which obtain their primary power from ac sources. Another important use of the diode tube is as a detector in a radio receiving circuit. In this application the tube changes the HF ac carrier wave into a direct current which displays the modulation of the AF signal. This, in effect, separates the audio portion of a radio signal from the RF portion, which is the carrier wave. Figure 12-19 is a circuit for a diode tube illustrating how the diode tube acts in an ac circuit. It will be noted that the output of the tube is a pulsating direct current.

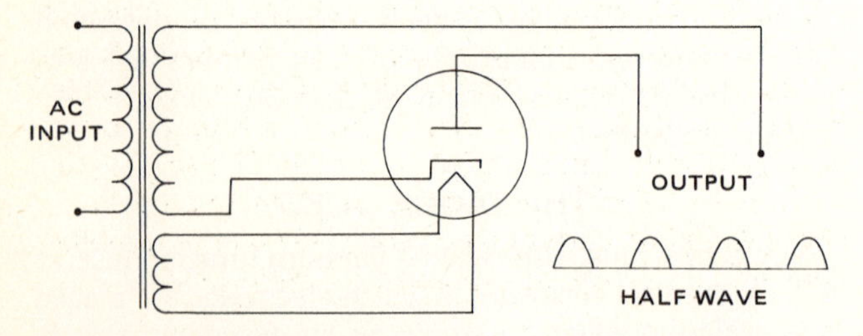

Fig. 12-19 Circuit for a diode tube.

THE TRIODE TUBE

The triode electron tube is similar to the diode tube except that it has one additional electrode. This is the **grid,** which is placed between the cathode and the plate. Figure 12-20 is a diagram showing the construction of a triode tube; it also shows a symbol for a triode tube as used in a standard electronic circuit.

The triode tube was discovered by Dr. Lee De Forest. De Forest found that by adding a third element to the diode tube, the electron flow from the cathode to the plate could be effectively controlled by changing the electric charge on the grid placed between them. Since like charges repel and unlike charges attract, it was found that a negative charge on the grid would tend to repel the electrons and prevent the flow from the cathode to the plate. On the other hand, a positive charge on the grid would accelerate the electrons and cause them to flow more rapidly and intensely from the cathode to the plate. Since the grid consisted of fine wire wound in such a manner that it surrounded the cathode and since the spacing between the grid wires was wide, it was easy for the electrons to travel through the grid to the plate.

The effect of the grid in a triode is to make it possible for the tube to act as an amplifier; that is, small changes in voltage on the grid will cause very substantial changes in the current flow from the cathode to the plate. Figure 12-21 illustrates the effects of small changes in grid voltage with positive-plate voltage constant.

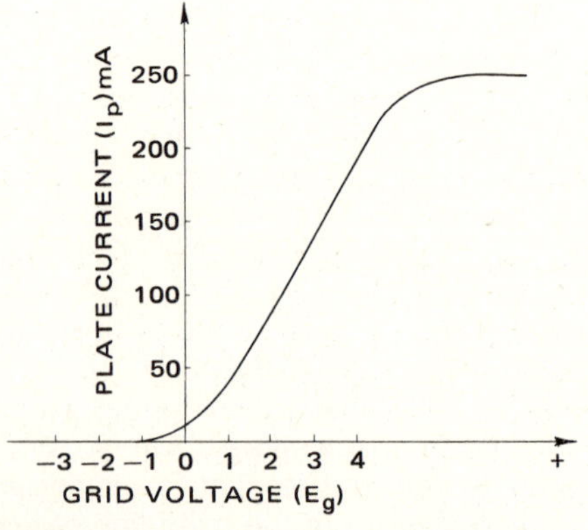

Fig. 12-21 Effect of grid voltage in a triode tube.

TETRODES AND PENTODES

The construction of a triode electron tube is such that capacitive effects become intolerable when the tube is used in RF circuits. It has been explained that a capacitor consists of conducting plates separated by an insulating material called the dielectric. In the triode tube the grid acts as one plate of a capacitor, and the tube plate acts as the other plate of a capacitor. The space between the grid and the plate acts as the dielectric. It has also been explained that the reactance of a capacitor decreases as the frequency increases. Thus, when frequencies become sufficiently high in a triode tube, a capacitive feedback occurs between the plate and the grid because of plate-to-grid capacitance (C_{gp}). This feedback causes oscillation and stops the normal operation of the tube. To overcome the effects of **interelectrode capacitance,** it is necessary to employ neutralizing circuits outside the tube or to use a tube designed to reduce interelectrode capacitance. Such a tube is the tetrode, whose name is derived from the Greek word *tetra,* meaning *four,* and the word *electrode.* Thus **tetrode** means four electrodes.

The tetrode tube employs a second grid called the **screen grid** located between the control grid and the plate (see Fig. 12-22*a*). The effect of this grid is to reduce greatly the capacitance between the control grid and the plate. When in operation, the screen grid usually carries a substantial positive charge with respect to the cathode but is also much less positive than the plate. A screen bypass capacitor is connected between the screen-grid circuit and ground to bypass alternating current, thus greatly reducing electrostatic effects between the control grid and the plate.

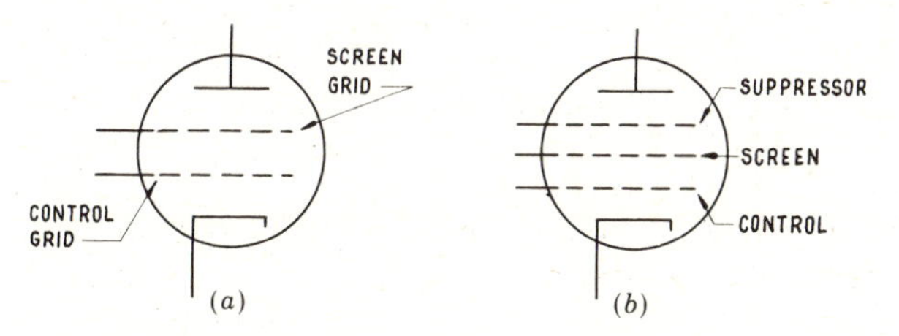

Fig. 12-22 Tetrode and pentode tube symbols.

After the development of the tetrode or screen-grid tube, it was found that such tubes could not be used for power amplification. The tetrode is so designed that electrons flowing from the cathode to the plate attain such very high velocities in power-amplification circuits that when they strike the plate they cause a secondary emission of electrons. Many of the secondary electrons are drawn away by the screen grid, and an appreciable amount of power is lost. To overcome the effect of secondary emission in the tetrode tube, an additional grid is installed between the screen grid and the plate. This grid is connected to the cathode or to another part of the circuit which provides a strong negative charge with respect to the screen grid. The negative charge on this grid drives the secondary electrons back to the plate so that they are not lost to the screen grid. This third grid is called a **suppressor grid** because it suppresses the effects of secondary emission. An electron tube with three grids is called a **pentode** because it contains five elements. The grids are designated *G*1 for the control grid, *G*2 for the screen grid, and *G*3 for the suppressor grid. A pentode tube symbol is shown in Fig. 12-22*b*.

THE PENTAGRID TUBE

The pentagrid tube is of particular interest because it is used widely in radio and television receivers as a converter and mixer to develop the modulated intermediate frequency which is sent on from the RF section to the intermediate-frequency (IF) amplifiers. The tube is called a pentagrid because it contains five grids; its operation will be explained in a later section.

THE TUBE MANUAL

Electron tubes are manufactured for a multitude of purposes and are made in many different sizes, shapes, and designs. Furthermore, their characteristics are as varied as their other features. It is quite apparent that a book of this type cannot adequately describe even a small number of tube types; however, the tube manuals published by manufacturers provide all necessary information for almost any type of tube.

TUBE CHARACTERISTICS

One of the important characteristics of an electron tube is the **amplification factor.** The symbol for the amplification factor is the Greek letter μ, spelled *mu*. The amplification factor is defined as **the ratio of change in plate voltage to the change in grid voltage required to give the same change in plate current.** For example, let us suppose that the plate current in a particular circuit is 20 mA. The plate voltage is 150 V and the grid voltage is 1 V. When we increase the plate voltage to 200 V, the plate current increases to 70 mA. Now we return the plate voltage to its original value and increase the grid voltage to 6 V. The plate current has again increased to 70 mA. Thus we have shown that a change of 50 V in plate voltage has caused an increase of 50 mA in plate current and that a change of 5 V in grid voltage will produce the same change of 50 mA in plate current. Therefore we see that a change of 5 V in grid voltage will cause the same change in plate current as is caused by a 50-V plate voltage change. The ratio is 5 to 50, or 1 to 10. The amplification factor is therefore 10.

The formula for amplification factor is the following:

$$\mu = \frac{\Delta E_p}{\Delta E_g} \qquad (I_p \text{ constant})$$

where ΔE_p = change in plate voltage
ΔE_g = change in grid voltage

Another important characteristic of electron tubes is **plate resistance.** One type of plate resistance is the dc resistance, which may be determined by Ohm's law, that is, $R = E/I$; in this case, the plate resistance is equal to the plate voltage divided by the plate current. The ac plate resistance is the **ratio of change in plate voltage to a change in plate current when the grid voltage remains constant.** This may be expressed in the formula

$$R_p = \frac{\Delta E_p}{\Delta I_p} \qquad (E_g \text{ constant})$$

Thus, the plate resistance in ohms is equal to a change in plate voltage divided by a change in plate current in amperes.

A third characteristic of electron tubes is called **transconductance.** This characteristic is the **ratio of a small change in plate current to the change in grid voltage which produces it.** Transconductance, being the opposite of resistance, is measured in **mhos; mho** is the word *ohm* spelled backward. The letter symbol for transconductance is G_m, and the formula may be stated

$$G_m = \frac{\Delta I_p}{\Delta E_g} \quad \text{(other voltages constant)}$$

Because the transconductance of an electron tube is generally only a small fraction of 1 mho, it is usually indicated in millionths of mhos, or **micromhos** (μmhos).

The three characteristics of electron tubes described above are interrelated as follows:

Amplification factor = plate resistance × transconductance

or

$$\mu = R_p \times G_m$$

CHARACTERISTIC CURVES

A graphical representation of the performance of an electron tube under varying conditions of voltage to the plate and grids is called a characteristic curve. In Fig. 12-23 are shown three characteristic curves illustrating how some of the characteristics of electron tubes may be determined. In Fig. 12-23*a* the curve shows how the plate current in a diode increases as plate voltage increases. Because changes in the temperature of the cathode would affect the emission of electrons, it is assumed that the filament, or heater, current is constant.

The curves in Fig. 12-23*b* show how the plate current in a triode increases as grid voltage changes to a less negative or more positive condition. For example, if the plate voltage is at +100 and the grid voltage is changed from −2 to +2, the plate current increases from 3 to 22 mA. We may also determine from the curves that it would take more than a 100-V change in plate voltage to produce the same change in plate current, with the grid voltage constant at 0.

The set, or "family," of curves in Fig. 12-23*c* shows the changes in plate current and changes in plate voltage for various values of grid voltage in a triode tube. This set of curves shows the effect of negative grid voltages only with respect to positive plate voltage. It will be noted that a very small change in grid voltage has a much greater effect than a large change in plate voltage. For example, if the grid voltage is −10 and the plate voltage is +200, the plate current will be about 1 mA. If the plate voltage is increased to +300, the plate current will rise to +10 mA. Now, if the plate voltage is at +200 and the grid voltage is changed from −10 to −2, the plate current will increase from 1 mA to 20 mA. It can be seen that a change of 8 V in grid voltage will produce a change of 19 mA in plate current and a change of 100 V in plate voltage will cause a change of 9 mA in plate current.

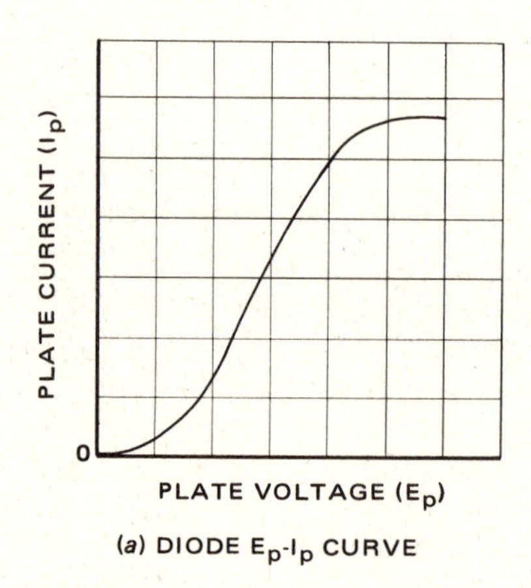

(*a*) DIODE E_p-I_p CURVE

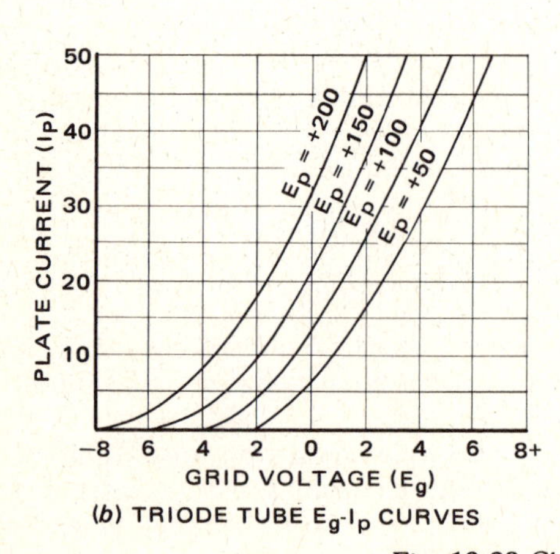

(*b*) TRIODE TUBE E_g-I_p CURVES

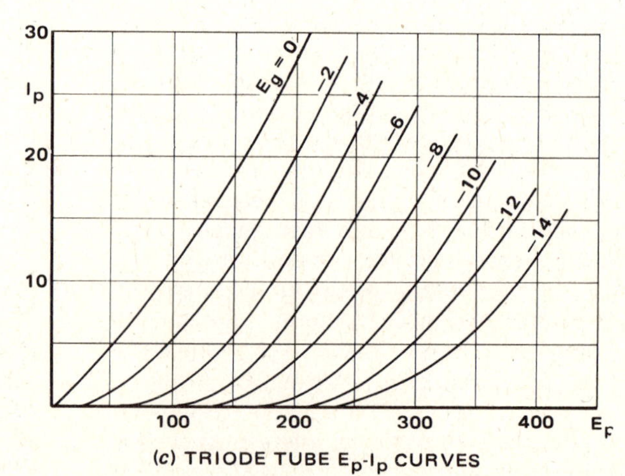

(*c*) TRIODE TUBE E_p-I_p CURVES

Fig. 12-23 Characteristic curves for electron tubes.

SOLID-STATE DEVICES

The basic principles of solid-state devices were explained in Chapter 4 of this text. In this section additional information will be provided to show some of the ways that various solid-state devices are used.

DIODES

A diode may be an electron tube or a solid-state device that is often used as a rectifier. In this case, crystal diodes are discussed.

A crystal diode is usually constructed with silicon as the active element. Pure silicon will not conduct electricity; however, by the addition of a small amount of an impurity such as aluminum, the silicon will become conductive. Aluminum added to silicon produces a *p*-type semiconductor, whereas antimony added to silicon will produce an *n*-type semiconductor. When *n*-type silicon and *p*-type silicon are joined, current can flow from the *n*-type material to the *p*-type material but current cannot flow in the opposite direction. This is illustrated in Fig. 12-24. In Fig. 12-24*a* a battery is connected to the silicon

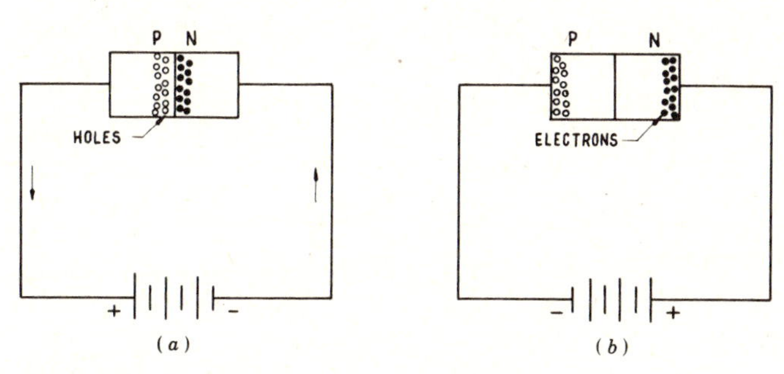

Fig. 12-24 Rectifying effect in a crystal diode.

crystal with the negative terminal of the battery connected to the *n*-type silicon and the positive terminal of the battery connected to the *p*-type silicon. The negative charge causes the electrons in the *n*-type material to move toward the junction and the positive charge causes the holes in the *p*-type material to move toward the junction. The electrons in the *n*-type material cross the junction and fill the holes on the *p* side. Electrons from the battery continue to replace the electrons that move across the junction and a current flows as the electrons are attracted to the positive terminal of the battery. If the battery is connected as shown in Fig. 12-24*b*, the electrons will be attracted by the positive charge and the holes will be attracted by the negative charge so there can be no movement of electrons across the junction and current cannot flow.

The single, unilateral-conduction diode represented by the drawings of Fig. 12-24 may utilize silicon, germanium or selenium as the semiconductor material. Impurities are added to produce the *n*-type or *p*-type material as required. As explained previously, the adding of impurities is called **doping.** The amount of doping is carefully controlled to produce the qualities desired in the diode.

Diodes are manufactured with a wide range of capacities. Some are designed to carry a small fraction of an ampere while others can carry a current as large as 3000 A. The aviation technician will most likely encounter diodes in alternators, certain actuator motors, full-wave rectifiers, and electronic circuits.

A principal consideration in the installation of power diodes is to ensure that the diode is firmly attached to the mounting which serves as a **heat sink.** Diodes that carry substantial current will become overheated and damaged or destroyed unless the heat developed is conducted away by the mounting structure. Many power diodes are provided with cooling fins by which heat is dissipated.

Large diodes are constructed with heavy metal bases to be mounted securely to a metal structure heavy enough to act as a heat sink. A mounting stud is provided on the base and is integral with it. Before a diode is mounted, the base of the diode should be inspected for cleanliness and smoothness and the mount to which it is to be attached should be similarly inspected. This is to assure that there will be maximum metal-to-metal contact between the base of the diode and the mounting. In some cases a heat-conducting jelly is placed between the diode base and the mounting to fill any gaps caused by irregularities in the surfaces to be joined. This assures maximum heat conductance from the diode to the mounting.

A diode designed for stud mounting is shown in Fig. 12-25. A hockey-puck diode is shown in Fig. 12-26. This diode must be mounted between metal plates, each of which must be perfectly smooth to make maximum metal-to-metal contact. The pressure that should be applied to the diode through the mounting surfaces is specified by the manufacturer. The heat-conducting jelly mentioned previously is often used in mounting hockey-puck diodes.

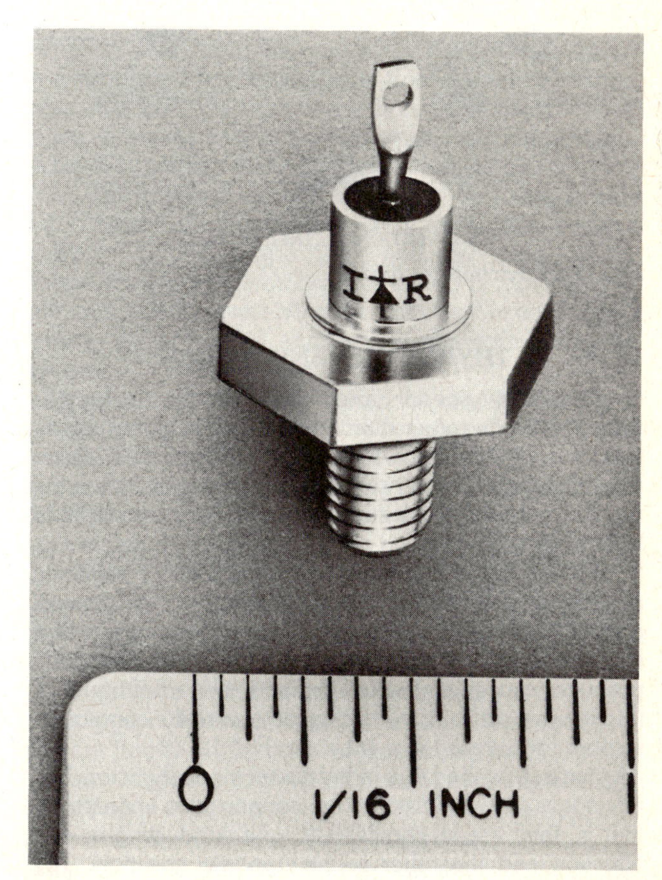

Fig. 12-25 A stud-mounted diode. (International Rectifier Co.)

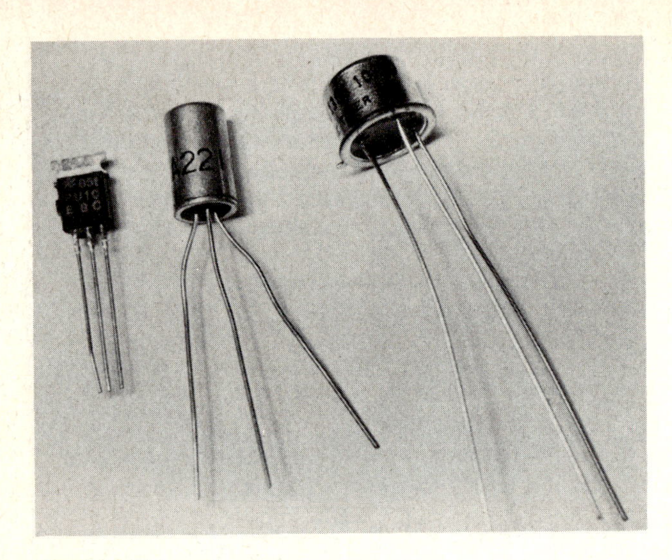

Fig. 12-35 Typical transistors.

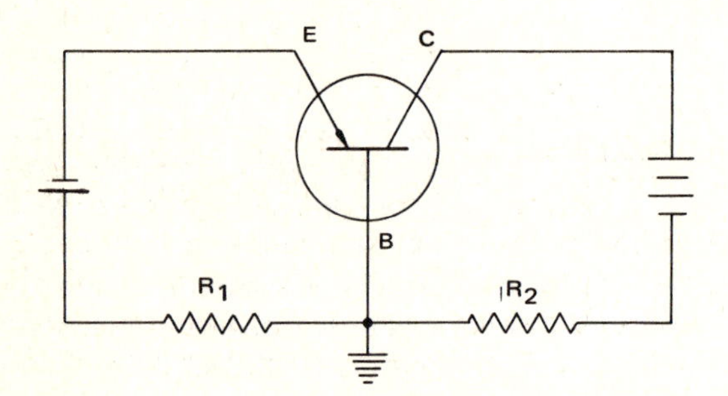

Fig. 12-36 Transistor circuit and symbol.

made in the oxide by a photo process at the point or points where it is desired to add new material by **diffusion.** Diffusion, as applied to the manufacture of a transistor or integrated circuit, is the process of heating the semiconductor material in a controlled atmosphere of the element to be added to the semiconductor. At the proper temperature, atoms of the gaseous atmosphere will penetrate the surface of the semiconductor. This will form the required type of material, either *n* type or *p* type, to the depth desired. The depth of penetration will be affected by both temperature and time of exposure. The transistor base material is formed to the desired depth, and this is then coated with an oxide. A window is opened in the oxide, and the emitter material is formed in the base material. The surfaces where connections are to be made are metallized, and the wafer is assembled in the package (case) and sealed. A cross section of the wafer formed into a transistor is shown in Fig. 12-37.

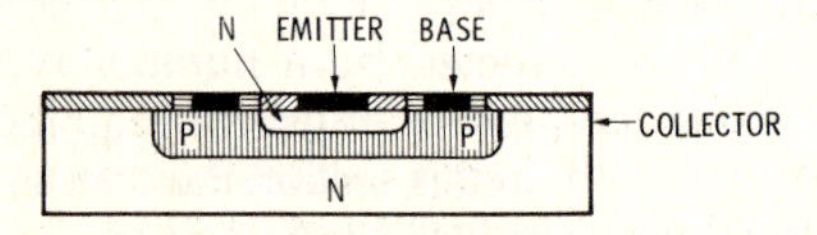

Fig. 12-37 Arrangement of a planar diffused transistor.

When it is realized that a transistor or integrated circuit is approximately 50 mils [1.27 mm] across and 10 mils [0.25 mm] in thickness, the question arises, why do transistors have to be so small? One reason for the size is that a transistor must often operate at very high frequencies (up to 1000 MHz or more), and small size reduces the effect of capacitive reactance and transit time that would otherwise render the transistor useless. It will be remembered that capacitive reactance decreases as frequency increases; hence, there is a limit to the frequency as long as capacitance exists in a circuit because capacitance will allow current to flow in the circuit, and the effect of the transistor will be reduced. Transit time is a factor because an electron must move a certain distance if it is to perform a function. If the transistor is comparatively large, the electrons must move a greater distance to accomplish their purpose. In transistors, the distance is often reduced to millionths of an inch and this makes it possible for the unit to operate at extremely high frequencies.

The packaging of a transistor is of prime importance. How the transistor is mounted and the type of case and contacts provided depend upon the use for which the transistor is designed. A power transistor which may be subject to considerable heating must have a heavy metal base and is often equipped with a heavy stud by which it may be mounted on a metal heat sink. The heat sink can then carry heat away rapidly enough to prevent damage to the transistor.

CHARACTERISTICS OF TRANSISTORS

The similarity of transistors to electron tubes has been mentioned. In an electron tube, grid bias and plate voltage both have an important effect on the tube operation. In like manner, the base-to-emitter (bias) voltage and the collector-to-emitter voltage determine the amount of current flow through the transistor. Figure 12-38 shows characteristic curves for collector current at different values of collector-to-emitter voltage and various values of base current. It will be noted that small changes in base current result in relatively large changes in collector current.

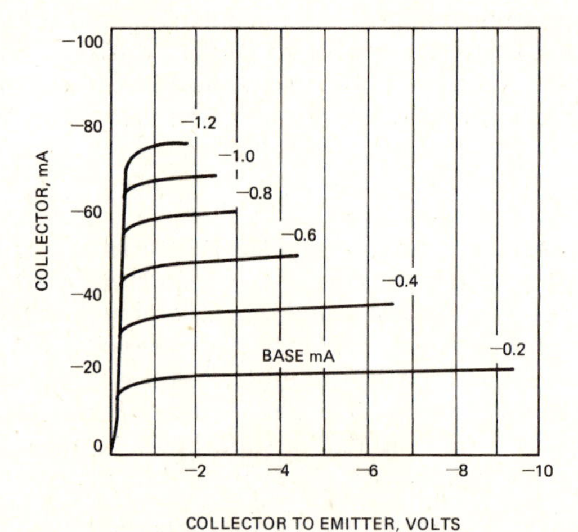

Fig. 12-38 Characteristic curves for a transistor.

We have considered only a few basic elements of transistors; however, the electronics technician may encounter hundreds of different types and sizes. To obtain information on most of the types, the technician should obtain a copy of the manufacturer's transistor manual. This book serves the same purpose as the tube manual in describing performance and characteristics of electron tubes.

FIELD-EFFECT TRANSISTORS

The transistors discussed thus far are controlled by a bias *current* applied to the base of the transistor. In an *npn*-type transistor, the electrons flowing from the base circuit into the *p*-type base make the base conductive, so electrons can flow from the emitter to the collector and out to the external circuit. In a field-effect transistor (FET) the current flow is controlled by a bias *voltage* that produces an electric field between the *gate* and the current path. The strength of this field determines the amount of current that can flow from the *source* to the *drain* in the transistor. In a field-effect transistor the gate, source, and drain are equivalent to the base, collector, and emitter in an ordinary transistor.

There are two general types of FETs. One type has a junction-type gate and is referred to as a JUGFET, and the other has an insulated gate and is called an IGFET. Both types are illustrated in the drawings of Fig. 12-39.

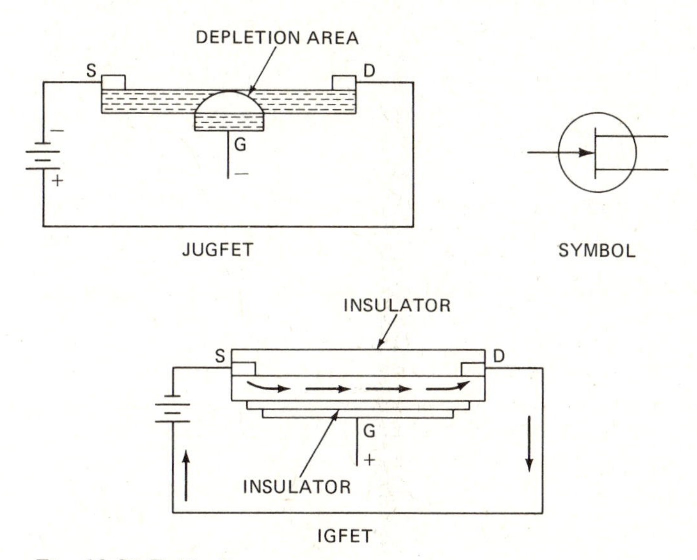

Fig. 12-39 Field-effect transistors.

The operation of the JUGFET shown in the drawing can be understood by noting that the gate is biased with negative voltage, electrons are repelled from the conducting channel, and this causes a reduction in current flow. If the bias voltage is increased sufficiently, all flow will be cut off. This is called the *pinch-off,* or *cutoff voltage*.

In the IGFET shown in Fig. 12-39, there is a silicon dioxide insulator between the metal gate and the *p*-type semiconductor material. This effectively prevents any current flow between the gate and the conducting material. In this particular FET, a positive bias voltage at the gate is required before current can flow from the source to the drain. The positive voltage attracts electrons to the conducting area, and these electrons provide the current flow together with electrons entering from the source. In effect, either type of FET becomes a voltage-controlled resistor and operates in much the same manner as an electron tube.

The principal value of an FET is that the control circuit draws no current. It can therefore function effectively with extremely low-power signals without disturbing the nature of the signals.

INTEGRATED CIRCUITS

The science of solid-state physics that led to the development of the transistor ultimately made it possible to combine transistors, capacitors, and resistors in one tiny chip of silicon crystal to produce a completely functional electronic circuit. The term *integrated circuit* indicates that all the components for a circuit are *integrated* into one unit. The silicon chips on which integrated circuits (ICs) are formed are often as small as 0.050 in [1.27 mm] square, and the chips for large-scale integrated circuits (LSIs) and very large integrated circuits (VLSIs) are usually less than $\frac{1}{4}$ in [6.35 mm] in length and width.

A typical integrated circuit contains transistors, diodes, capacitors, and resistors in quantities required for the needs of the circuit. The more complex circuits contain hundreds or thousands of these components all formed into one silicon chip. The photograph of Fig. 12-40*a* shows an LSI as seen under a microscope to reveal the arrangement of the circuit components. The greatest dimension of this chip is less than $\frac{1}{4}$ in. The photograph in Fig. 12-40*b* is a greatly enlarged view of a VLSI as integrated on a tiny silicon chip. Figure 12-40*c* shows an integrated circuit package ready for installation in an electronic system.

When integrated circuits are designed, the first step is to develop a circuit or circuit component that will meet the requirements for which the circuit is to be employed. Large drawings are then made of the circuitry after it has been tested for effectiveness. The drawings are made to show the layout of the various components as they are to be placed in the actual circuit when it is formed in the silicon wafer. These drawings are as large as necessary to include all the circuit components. From the main drawing, other drawings are made to form masks for the several steps needed in the diffusion processes. The drawings are reduced photographically to produce the actual size needed for manufacturing the circuit.

Connections from the circuit wafer to the outside connecting pins or wires are made with gold wire smaller than a human hair. The outside contact pins for the circuit package (case) are gold-plated to prevent corrosion.

The foregoing sequence described for the development and construction of an integrated circuit is only one of many that may be employed. It is presented here to give the technician a basic understanding of what is involved in microelectronic technology.

Integrated circuits of many types are used widely in the avionic systems of aircraft. Their use, together with printed circuitry and the use of modular components, has greatly reduced the weight of avionic systems and has simplified the maintenance and troubleshooting of such systems.

OSCILLATORS

USE OF OSCILLATORS

Among the various individual circuits, which may also be called subcircuits, in a complete electronic system are the oscillators. An oscillator is a circuit designed to generate an alternating current which may be of a comparatively low frequency or of a very high frequency, depending upon the design of the oscillator. Oscillators are used in radio and television transmitters to generate

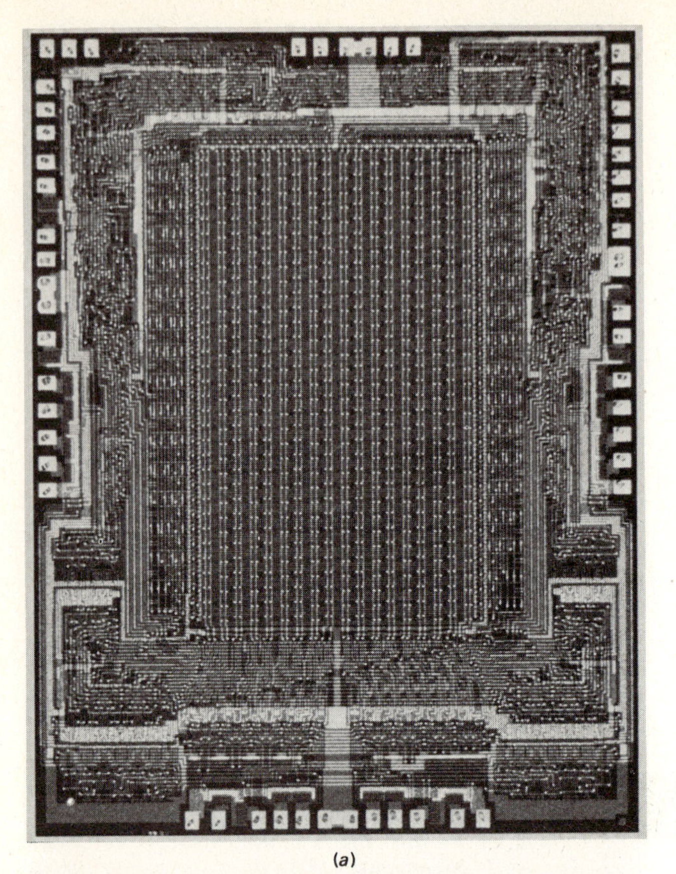

(a)

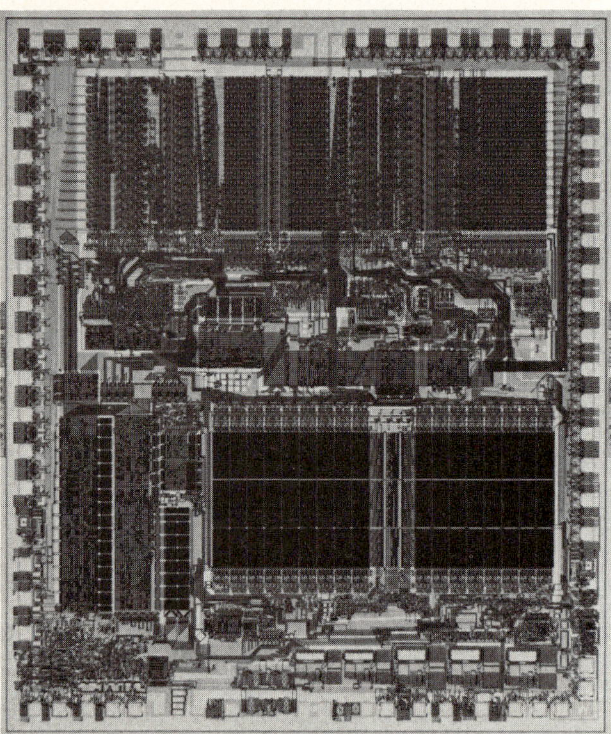

(b)

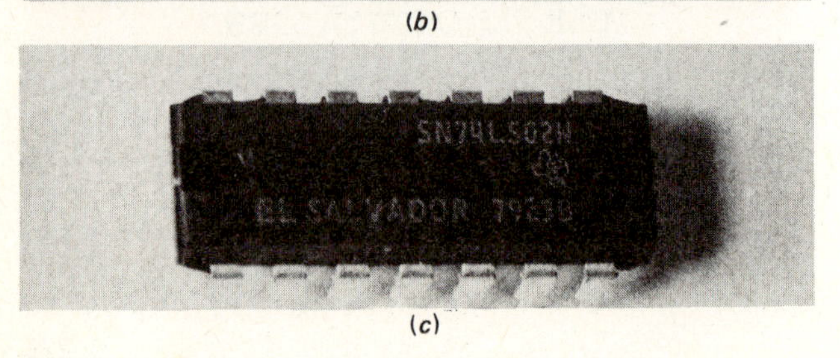

(c)

Fig. 12-40 Integrated Circuits. (Motorola Inc.)

the RF carrier waves, in receivers to produce the intermediate frequency, and in other circuits and systems in which it is necessary to develop an alternating current with a particular frequency.

OSCILLATOR THEORY

Fundamentally an oscillator consists of an *LC* tank circuit, an electron tube or transistor, and a means of feedback to supply power to replace tank losses. Although the action of a tank circuit was described in the section on alternating current, a brief review of its principles is desirable at this point.

Examine the circuit in Fig. 12-41. A battery is connected with a double-throw switch to a capacitor *C*. When the switch is thrown to position 1, the capacitor will charge to the voltage of the battery. If the switch is then placed in position 2, the battery will be disconnected from the capacitor, and the capacitor will retain its charge. Now, when the switch is moved to position 3, the capacitor will be connected to the inductance *L* and will discharge through the inductance. When the capacitor first starts to discharge, the current through *L* will build a magnetic field which induces an opposing voltage in *L*. This slows the discharge of *C*. When *C* becomes almost discharged, the field around *L* will begin to collapse, and this will induce a voltage which keeps the current flowing. This induced voltage charges the capacitor in a direction opposite to that in which the battery originally charged it.

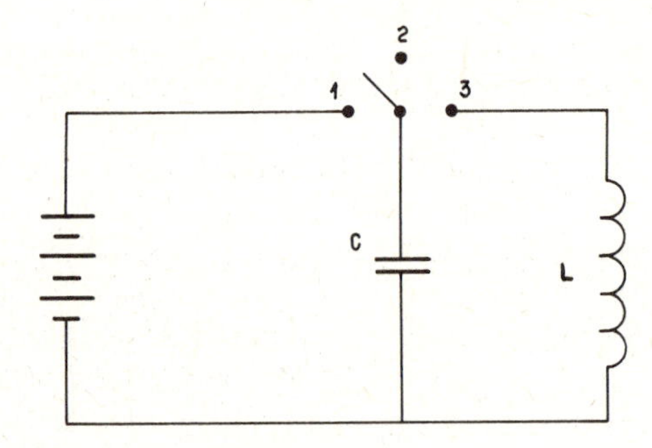

Fig. 12-41 A simple tank circuit.

When the charge of the capacitor reaches a voltage equal to the induced voltage in *L*, the current flow will stop, and the capacitor will start to discharge back through *L*. This action will be continuously repeated, with the energy stored first in the capacitor and then in the field of the inductance coil. An alternating current results which will degenerate to zero because of losses sustained in the circuit. If we could replace the small amount of energy lost during each cycle, we could prolong the generation of alternating current indefinitely. An oscillator circuit actually does this.

ARMSTRONG OSCILLATOR

If we connect a triode electron tube with a feedback coil to a tank circuit as shown in Fig. 12-42, we produce a simple Armstrong oscillator circuit. When the switch *S* is closed, a current will begin to flow from the cathode to the plate, through L_2, and back to the battery. The current through L_2 will produce a field which also induces a voltage in L_1 in such a way that current will flow through L_1 to the grid side of capacitor C_1. At this time the grid of

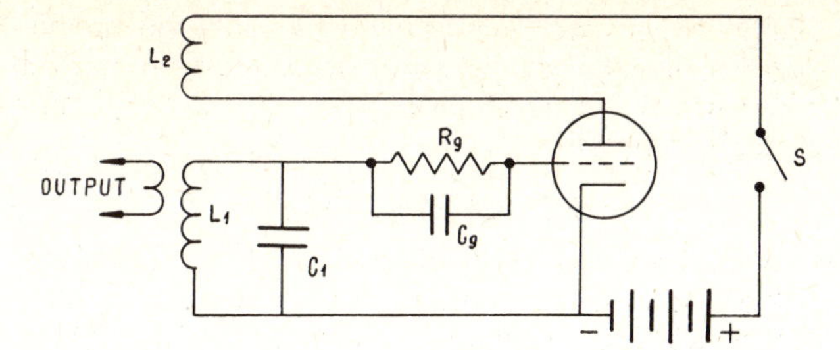

Fig. 12-42 Armstrong oscillator circuit.

the tube is positive, and the current will flow from the cathode to both the plate and the grid. The plate circuit quickly reaches a saturation point, and the expanding field of L_2 then becomes stationary. The voltage in L_1 therefore drops, and the field of L_1 collapses. This induces a voltage which tends to continue the current flow and charge the grid side of C negative. The grid now becomes negative, and the current flow from the cathode to the plate is cut off because of the negative grid bias. The action of the grid resistor and capacitor maintains the negative bias and keeps the tube at cutoff until the current flow is ready to reverse in the tank circuit. The cycle continues, and the feedback coil L_2 adds energy to the tank circuit once each cycle at the time that the tube starts to conduct.

The frequency of the ac output of an oscillator depends upon the values of the inductance and capacitance in the tank circuit. In the circuit diagram of Fig. 12-42 the tank circuit is produced by the coil L_1 and the capacitor C_1.

There are many types of oscillator circuits that operate on a principle similar to that of the Armstrong oscillator. Among the oscillator circuits in common use are the Hartley oscillator and the Colpitts oscillator shown in Fig. 12-43.

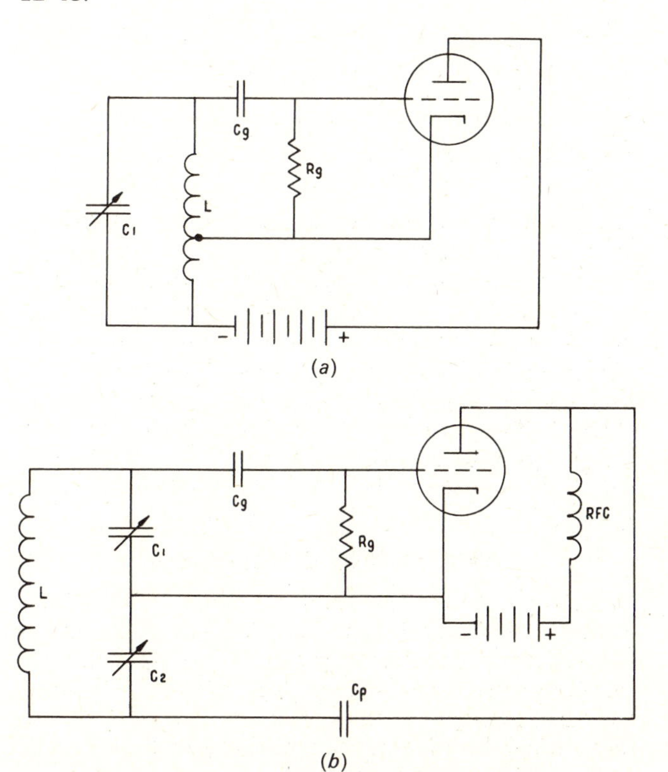

Fig. 12-43 (a) Hartley oscillator; (b) Colpitts oscillator.

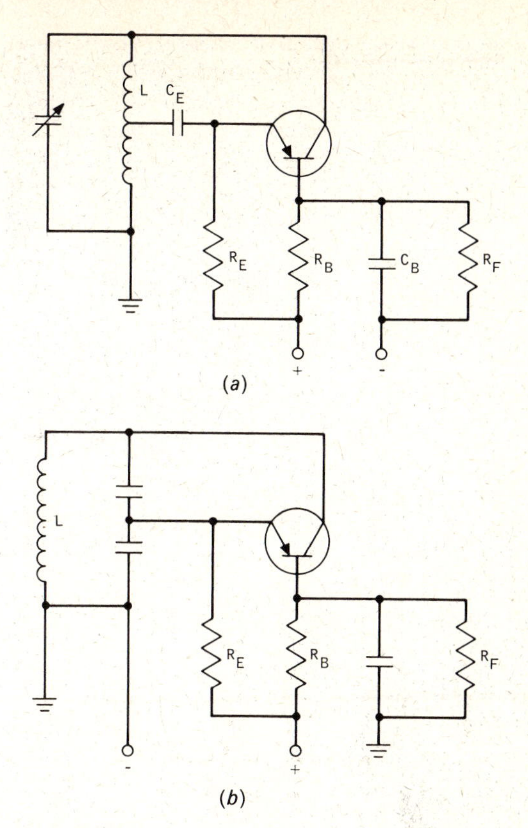

Fig. 12-44 Oscillators with transistor control.

Oscillators utilizing transistors instead of electron tubes are shown in Fig. 12-44. The circuit A is a Hartley oscillator and the circuit B is a Colpitts oscillator. The use of transistors in oscillators makes it possible to develop a much wider variety of performance than is possible with electron tubes.

SPECIAL-PURPOSE CIRCUITS AND COMPONENTS

THE *RC* CIRCUIT

The resistance-capacitance (RC) circuit shown in Fig. 12-45 is useful for timing in electronic circuits. It is quite obvious that the value of the resistance R will affect the time required to charge the capacitor C. Likewise it is apparent that the value of the capacitor C will affect the time required for charging. If the capacitance and resistance are both high in value, the time required for charging will be comparatively long.

The time for charging a capacitor to 63.2 percent of the applied voltage in a series RC circuit is called the **time constant** for the combination. The time constant in sec-

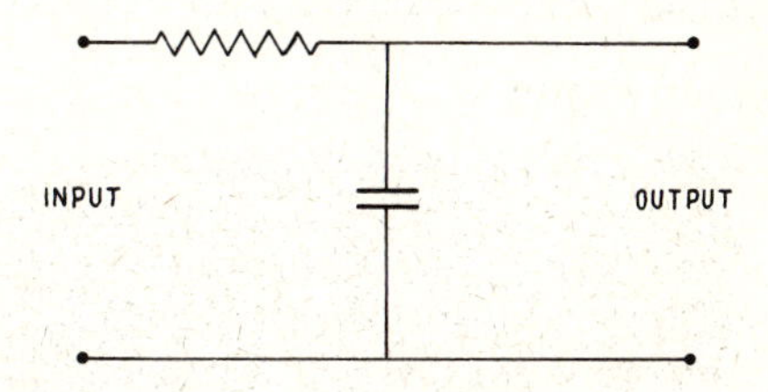

Fig. 12-45 A resistance-capacitance circuit.

onds is equal to the product of the resistance in megohms and the capacitance in microfarads. For example, if a series RC circuit has a resistance of 0.05 MΩ and a capacitance of 0.02 μF, the time constant will be 0.001 s.

The charging rate for a capacitor in an RC circuit is indicated by the curves in Fig. 12-46. Note that the voltage of the capacitor is at 63.2 percent at the first time constant. At the second time constant the voltage is up to 88.35 percent. During each time interval the capacitor charges 63.2 percent of the difference between the capacitor charge and the applied voltage.

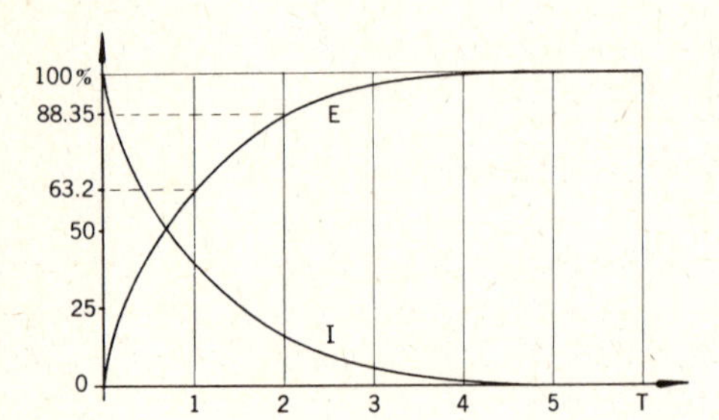

Fig. 12-46 Curves to show capacitor charge and discharge.

GAS-FILLED TUBES

A previous discussion explained briefly the operation of electron tubes of the vacuum type used in electronic circuits. There are special types of electron tubes, however, into which certain gases, such as neon or argon, are introduced to provide entirely different characteristics from those possessed by ordinary vacuum tubes. Among these gas-filled tubes are the neon glow lamp, the gas-filled phototube, the thermionic diode, and the thyratron which contains one or more control grids. Symbols for these tubes are shown in Fig. 12-47.

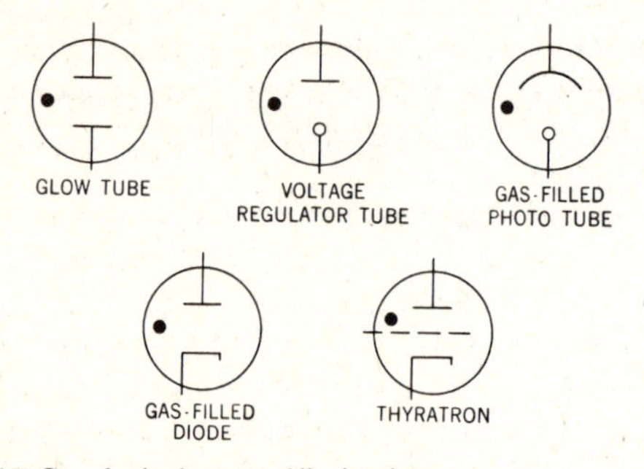

Fig. 12-47 Symbols for gas-filled tubes.

One of the principal characteristics of a gas-filled tube is the great change in its resistance when the gas becomes ionized. Ionization takes place in a gas-filled tube when a free electron, moving from the cathode to the anode, strikes or collides with an atom and causes other electrons to leave the atom. The electrons leaving the atom will move toward the anode (positively charged electrode), and the positive ion will move toward the cathode. Remember that an ion is an atom which possesses an electric charge because it has too many or too few electrons. When a positive ion is created in a gas-filled tube, it means that an atom has lost one or more of its normal number of electrons.

Ionization can occur in a gas-filled tube only when there is a difference of potential between the cathode and the anode great enough to cause an electron to strike an atom with sufficient force to free other electrons. This potential is called the **ionization voltage** (V_i). Furthermore, ionization can occur only when there are free electrons available from some source to start the process. In a gas-filled phototube the initial electrons are freed from the photosensitive cathode when light strikes it. These free initial electrons can also be produced by nuclear radiation, cosmic rays, or x-rays.

In the operation of a typical neon glow lamp, there is no appreciable current flow through the lamp (tube) until ionization potential has been reached. When the voltage across the lamp is increased, ionization of the gas will begin to occur, and a small current will start flowing before the lamp begins to glow. This is called **dark current.** A small increase in voltage above the point at which dark current flows will be the firing potential, also called starting potential or breakdown voltage. At this point the resistance of the lamp drops to a very low level, the current flow increases greatly, and the voltage across the lamp stabilizes or drops slightly. A photograph of a small neon glow tube, approximately $\frac{7}{8}$ in [2.22 cm] in length, is shown in Fig. 12-48.

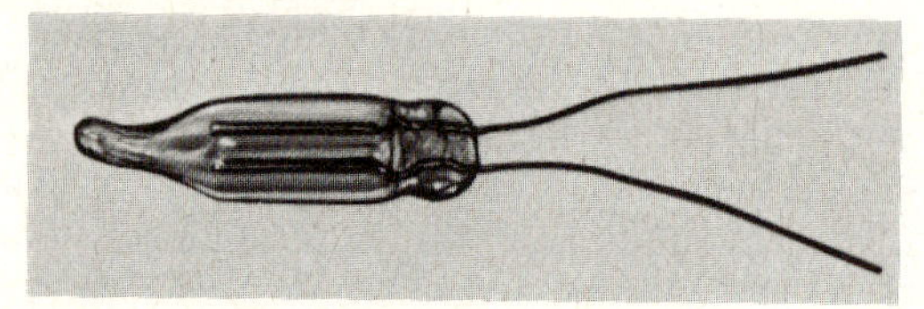

Fig. 12-48 A neon glow lamp.

The glow tube is commonly used to provide voltage regulation in electronic circuits. A simplified circuit of a glow tube used for voltage regulation is shown in Fig. 12-49. In this circuit the current through the voltage-dropping resistor R_1 is the sum of the currents through the flow tube and the load resistor R_2. If any voltage increase occurs across the tube, current through the tube immediately increases. This, of course, will cause an increase of current through R_1, with the result that the voltage drop across R_1 must increase and thus reduce the voltage across the tube to its rated level. In very simple terms we may say that all excess voltage is "bled off" across the glow tube; hence, the voltage across the tube and the load will remain very nearly constant.

When a thermionic (heated-cathode) diode tube is manufactured with a small amount of gas inside the tube envelope, the tube will function as a vacuum diode until a voltage above the ionization potential is applied. At this point the current flow will increase sharply, and the volt-

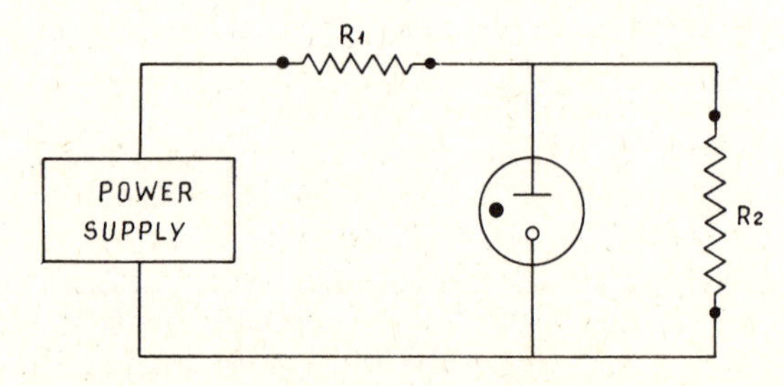

Fig. 12-49 Glow tube as a voltage regulator.

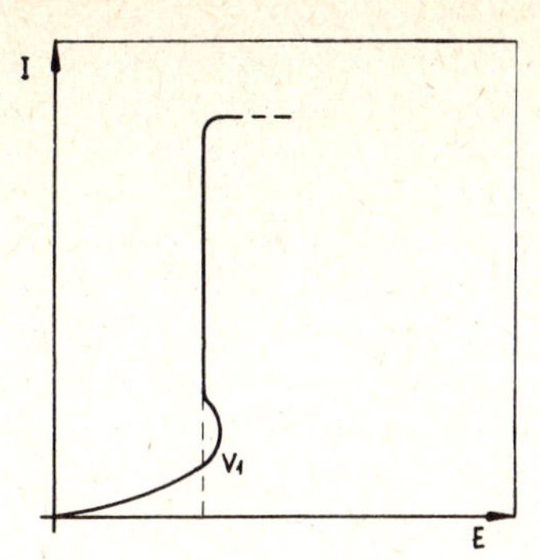

Fig. 12-50 Operation curve for a gas-filled tube.

age will stabilize or drop off slightly as shown in Fig. 12-50. When a gas-filled hot-cathode tube is provided with a control grid, it is called a **thyratron.** In this tube the control grid governs the firing point of the tube, and after the tube has fired, the grid has no effect. If a thyratron is connected as shown in Fig. 12-51, the firing point can be controlled by the potentiometer R_1. For as long as the

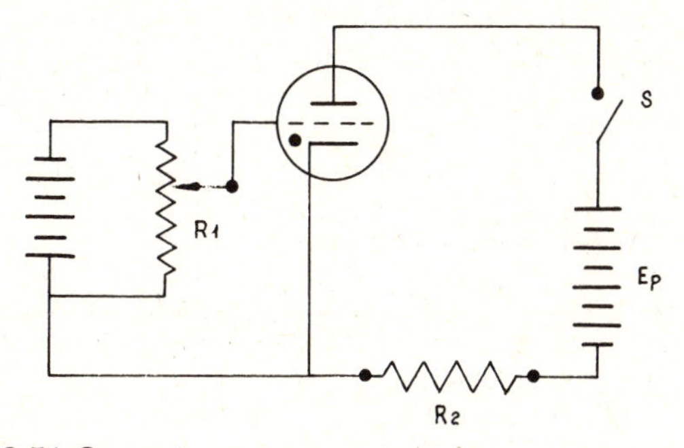

Fig. 12-51 Circuit for operation of a thyratron tube.

grid has a negative charge great enough to prevent electrons from passing through, the tube will not conduct even though the plate voltage is far above ionization potential. If the switch S is closed and E_p is above ionization potential, then the tube can be made to fire by adjusting R_1 for a smaller negative grid potential. When the tube fires, plate current will flow according to the limits of the ballast resistor R_2, and grid voltage will no longer have an effect. The ballast resistor R_2 is necessary to protect the tube from excessive current. The only means of stopping the current flow through the tube is to open the switch S. Thyratron tubes have many applications in electronic circuits, especially when it is desired to produce a switching action controlled by voltage.

Because of the proliferation of solid-state devices, gas-filled tubes are not used in many of their previous functions. For example, in voltage control, zener diodes are more effective and dependable than gas-filled tubes.

SWEEP CIRCUITS

A sweep circuit is designed to provide a reasonably linear rise in voltage to a required level and then a sudden decrease or *flyback* to a predetermined level. This circuit is used with a **cathode-ray tube** to provide the horizontal sweep of the electron beam. The electron beam must move at a comparatively slow rate across the fluorescent face of the tube as it traces a visible line, and then it must return to the opposite side of the tube face almost instantaneously. To accomplish this result, a *sawtooth* signal is used; such a signal is illustrated by the curve of Fig. 12-52.

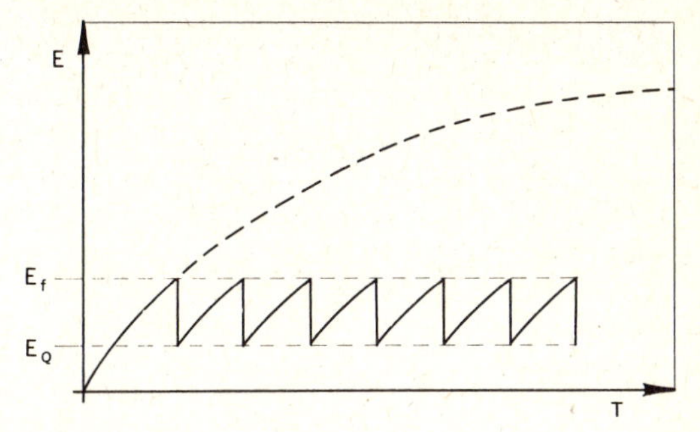

Fig. 12-52 A sawtooth signal.

The simplest type of circuit to produce a sawtooth wave pattern is a relaxation oscillator which uses a neon glow tube or a solid-state device. The circuit for this oscillator is shown in Fig. 12-53.

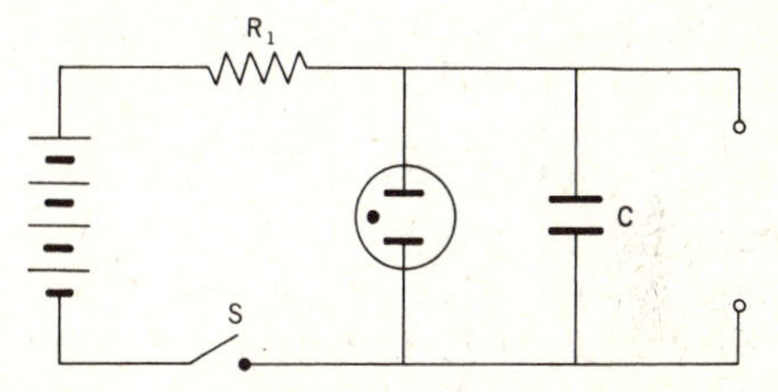

Fig. 12-53 Relaxation oscillator circuit.

When the switch S is closed, current will flow through the resistor R to charge the capacitor C. When the voltage of C has risen to the firing potential of the tube, the tube will fire and the capacitor will discharge through the tube. The firing potential is shown by E_f in Fig. 12-52. When the voltage of C drops below the quench voltage of the tube, the tube will stop conducting. This is shown by E_q in Fig. 12-52. At this point the capacitor will start to charge again and the cycle will repeat.

It will be noted that the curve of the rising voltage in Fig. 12-52 is not quite linear inasmuch as it follows the normal charging curve of a capacitor. By utilizing a small portion of the charging curve, it is possible to obtain a nearly linear signal. In actual practice more complete circuits are used to provide the sawtooth signal. These circuits make it possible to obtain an almost perfectly linear sweep voltage.

With modern electronic circuits it is possible to produce electric waves in almost any shape. The normal shape of a pure ac wave is a sine wave; this shape may be modified, however, so that square waves, sharply peaked waves, or individual pulses are obtained. These waveshapes may be produced at very low frequencies or at many millions of cycles per second. Such waveshapes are used for a variety of purposes, including radar, computers, and control circuits.

In testing and troubleshooting electronic circuits, there are many tests for which it is not possible to obtain correct voltage readings by the use of an ordinary electric voltmeter, even though the sensitivity of the meter may be very high. This is easily understood when we consider a

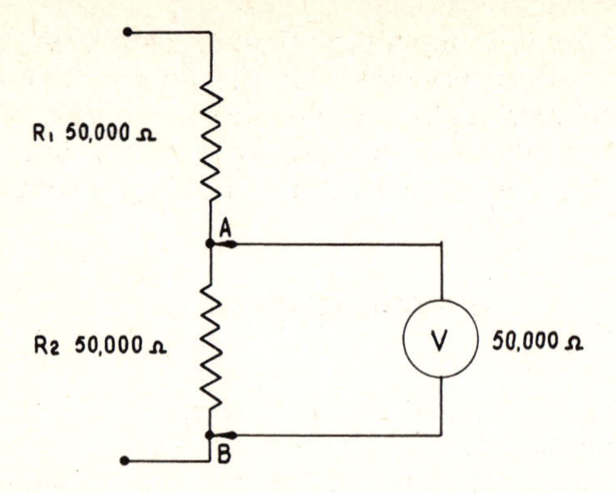

Fig. 12-54 Effect of a voltmeter on an electronic circuit.

typical electronic circuit. Assume that two resistors are connected in series, and each resistor has a value of 50 000 Ω, as shown in Fig. 12-54. If the voltage across the resistors is 200 V, the current flow through the circuit will normally be 2 mA, and the voltage across each resistor 100 V. We wish to check the voltage across the resistor R_2 with a voltmeter having a sensitivity of 5000 Ω/V and a range of 100 V. With these values, the resistance of the voltmeter will be 50 000 Ω, and in testing the voltage across R_2, the voltmeter will be in parallel with the resistor. When this is done, the resistance between *A* and *B* becomes 25 000 Ω, and resistance of the complete circuit becomes 75 000 Ω. The current flow through the circuit then is 200/75 000 or 2.67 mA. By computation, we then find that the voltage from *A* to *B* becomes 66.67 V.

It is obvious from the foregoing example that if a testing instrument draws power from the circuit being tested, the values of voltage and current in the circuit will be disturbed. This is particularly true of circuits having high resistance values and low current flow. To test the values in these circuits, it is necessary to use an instrument which will draw such a small current from the circuit that it will have no appreciable effect on the normal values of the circuit. To meet this requirement, a vacuum-tube voltmeter or a solid-state voltmeter is employed. An instrument of this type is described in Chapter 5.

Figure 12-55 is a simplified circuit of a dc VTVM. The series resistances form a voltage divider which makes it possible to change the range of the instrument as required. The total of these resistances may be as high as 20 MΩ; hence, the effect on the circuit being tested will be very small. A study of the circuit will reveal that the voltage across the meter *M* will vary as plate current

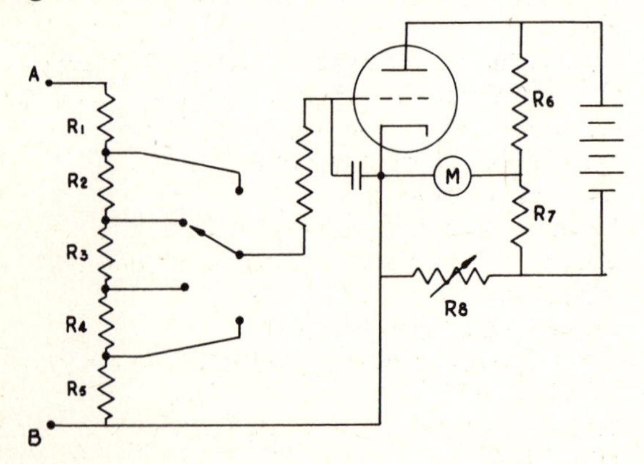

Fig. 12-55 A simplified vtvm circuit.

varies. Since plate current is determined by the charge on the grid, it is apparent that the voltages applied between *A* and *B* will determine the degree of needle movement in the meter.

CATHODE-RAY TUBE

For many years the cathode-ray tube (**CRT**) has become familiar to millions of persons as the picture tube in a television set. The CRT in television is designed to reproduce an undistorted picture on a screen. The picture is developed from a series of pulses and varying voltages applied to the elements of the tube.

Fundamentally, the CRT consists of an electron "gun," a phosphorescent screen, and deflecting devices to control the movement of the electron beam "shot" from the gun. A drawing to illustrate the CRT is provided in Fig. 12-56. As in any thermo-emitting tube, the heated

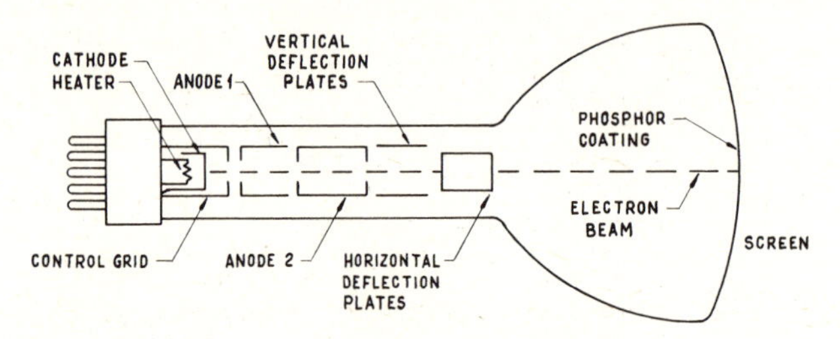

Fig. 12-56 Drawing to illustrate a cathode-ray tube.

cathode supplies the electron emission, and these electrons are accelerated toward the screen by the positive charges on the anodes. The intensity of the electron beam is regulated by means of the control grid charge. After the electron beam is accelerated and focused by the anodes, it is controlled in direction by the **deflection plates.** When the electrons strike the phosphor-coated screen, they cause a bright spot to appear. If an alternating voltage is applied to the vertical deflection plates, the spot will move up and down and form a straight line as shown in Fig. 12-57*a*. In like manner, if an alternating voltage is applied to the horizontal deflection plates, a horizontal straight line will appear on the screen as in Fig. 12-57*b*.

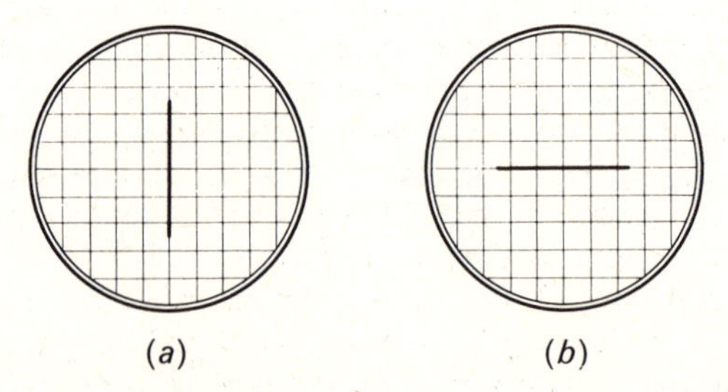

Fig. 12-57 Effects when ac charges are applied to deflection plates.

In practice the horizontal deflection of the electron beam is used to provide a time base. The output of a sawtooth oscillator is applied to the horizontal deflection plates so that the electron beam will sweep at a steady rate from left to right and at the end of the sweep will return instantly to the left side and start another sweep. As the voltage rises, the electron beam moves to the right; but when the voltage drops, the beam returns immediately to the left side of the screen.

If we set the horizontal timing for the CRT to 60 Hz and apply a 60-Hz alternating voltage to the vertical deflection plates, a stationary sine wave will appear on the screen.

THE OSCILLOSCOPE

The oscilloscope is an electronic instrument which employs a CRT and the necessary circuitry for the purpose of providing a visual analysis of recurring electrical values. The circuitry includes amplifiers for vertical and horizontal signals, a sawtooth signal generator—which may be a relaxation oscillator or a multivibrator—a synchronizing circuit, a power supply, a CRT, and all the necessary switches and adjustments to provide for a wide variety of applications. A block diagram of a simple oscilloscope is shown in Fig. 12-58.

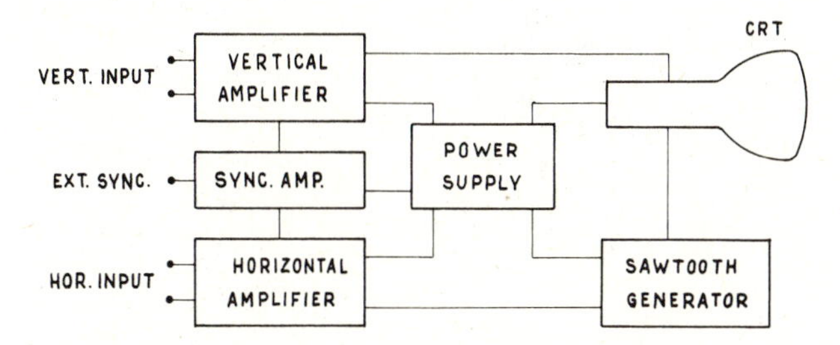

Fig. 12-58 Block diagram of an oscilloscope.

As oscilloscope may be used to analyze any type of recurring electrical value and is most valuable in determining where trouble exists in a complex electronic circuit. For example, if it is desired to examine the performance of an amplifier, the oscilloscope may be used to compare the input signal (waveshape) with the output signal. To do this, a square-wave generator is used to produce a square wave signal which is applied to the oscilloscope through an electronic switch directly to the input of the amplifier being tested. The output of the amplifier is also applied through the electronic switch to the oscilloscope. The electronic switch alternately applies the two signals to the oscilloscope but at such a rate that both signals appear on the oscilloscope screen. If the output wave does not have the same shape as the input wave, it is immediately apparent that distortion exists.

In tracing the trouble in an electronic cirucit, the oscilloscope may be used to study the waveshapes at various points throughout the circuit. When a waveshape is found that does not match what is normal at that point, then the trouble may be located.

Electronic instruments are essential in the troubleshooting and testing of electronic circuitry; if they are not used properly, however, great damage may be done to them, to the circuitry, or to both. Operators of these instruments must make sure that they know how the instruments are to be connected, and they must know that the voltages to be tested are not above the range of the instrument being used. They must also be aware of the effects of the testing instruments on the circuits being tested. It has been pointed out that a common electric voltmeter will often produce effects which make the readings erroneous.

Many electronic testing instruments have been developed for special purposes. Operators of any of these instruments must make sure that they thoroughly understand its use before attempting to operate it. In case of any doubt, the manufacturer's instructions should be studied.

REVIEW QUESTIONS

1. Define *resistor.*
2. Name the general types of resistors.
3. Give the color code for resistor values.
4. If the first color band of a fixed resistor is orange, the second is green, and the third is blue, what is its resistance value?
5. What is the principal difference between adjustable and variable resistors?
6. By what other names is a variable resistor commonly called?
7. What is the principal difference between a potentiometer and a rheostat?
8. What is the function of an inductor in an electronic circuit?
9. Explain *inductive reactance.*
10. What is an electrical *transient?*
11. Describe a variable capacitor.
12. Compare *inductance* with *capacitance.*
13. What precaution must be taken in connecting an electrolytic capacitor in a circuit?
14. What will happen if the voltage value of a capacitor is below that required for the circuit?
15. What is meant by the *wattage* of a resistor?
16. Define a *capacitance* of 1 F.
17. What units are usually used to indicate capacitance in an electronic circuit?
18. Compare the effects of inductive reactance and capacitive reactance.
19. What conditions exist in a resonant circuit?
20. Explain how a circuit may be tuned to a particular frequency.
21. What is the resonant frequency of a circuit having an inductance of 2 H and a capacitance of 0.5 μF?
22. Compare the impedance across a series resonant circuit with that across a parallel resonant circuit.
23. Describe the operation of a high-pass filter; a low-pass filter.
24. Explain a bandpass filter.
25. What is the principal function of filters in a radio circuit?
26. Explain the operation of a diode electron tube.
27. What materials are used to increase the electron emission from the cathode of an electron tube?
28. What is the principal use of a diode tube?
29. Describe a *triode tube.*
30. Explain how a tiode tube is used as an amplifier.
31. Why was it necessary to develop tetrode and pentode electron tubes?
32. Explain *capacitive feedback.*
33. What type of charge is carried on the screen grid of a tetrode tube?
34. What is the function of the suppressor grid in a pentode tube?
35. Briefly explain *grid bias.*

36. Define *amplification factor, plate resistance,* and *transconductance.*
37. Show how the characteristics mentioned in the previous question are interrelated.
38. What is indicated by the unit *mho?*
39. Of what value are characteristic curves?
40. Describe a crystal diode.
41. Why does electric current flow in one direction only in a unidirectional diode?
42. What considerations must be observed in the installation of power diodes?
43. Describe the purpose of a *heat sink.*
44. Describe the characteristics of a *zener* diode.
45. Describe a *thyristor* or *SRC.*
46. Compare a *triac* with a thyristor.
47. What is the difference between an *npn* and a *pnp* transistor?
48. What are some of the advantages of transistors in electronic circuits?
49. Name the three working parts of a transistor and give the function of each.
50. What section or element of a transistor serves a function similar to the grid in an electron tube?
51. Discuss the importance of temperature with respect to the operation of a transistor.
52. Explain the operation of two types of field-effect transistors.
53. Give a brief description of an integrated circuit.
54. What are the principal advantages of using integrated circuits in an electronic system?
55. What electronic units can be built into an integrated circuit?
56. What is the purpose of an oscillator?
57. Describe how an oscillator produces an alternating current.
58. What determines the frequency of the oscillator output?
59. What is the value of an *RC* circuit?
60. Define *time constant.*
61. In a gas-filled tube, what is the effect when the gas in the tube is ionized?
62. Explain *ionization voltage.*
63. Explain how a glow tube may be used for voltage regulation.
64. Describe the operation of a thyratron tube.
65. Explain the *sawtooth signal* used for the sweep circuit in a CRT.
66. What is the advantage of an SSVM or VTVM?
67. Why is it necessary to use an SSVM or VTVM for testing the operation of many electronic circuits?
68. Describe the construction and operation of a CRT.
69. What is the purpose of a CRT oscilloscope?
70. How may an oscilloscope be used to measure distortion in an amplifier?
71. What precautions must be observed in the use of electronic testing instruments?

CHAPTER 13

Radio Transmitters and Receivers

RADIO THEORY

The transmission and reception of radio signals involve the use of electronic equipment to develop electromagnetic and electric fields that are modulated to carry the type of intelligence desired, project these fields into the atmosphere, and then intercept these fields and convert them to usable information or data. In this section, the principles of radio transmission and reception are discussed.

Radio for aircraft includes communication equipment, navigation equipment, radar, and other electronic systems. In each of these areas there are many different types of electronic circuitry and devices designed to assure the safe and efficient operation of modern aircraft under all types of conditions of weather and traffic.

Radio transmitters and receivers are particularly important in the vicinity of large commercial airports. It is necessary for aircraft pilots to be able to contact the control tower of the airport from which they are to take off and the control tower of the airport where they plan to land. Special transmitters and receivers are used for navigation and flight control. Each airport control tower has assigned frequencies, and the radio-communications equipment on the airplane must be capable of operating at any control-tower frequency. Information regarding the control-tower frequencies is available in the Airmen's Guide and other publications used by certificated pilots.

RADIO WAVES

Radio signals emanate from the antenna of a transmitter partly in the form of electromagnetic waves. Such waves are radiated from any current-carrying conductor when the current periodically changes in magnitude and direction. The radiation of an electromagnetic wave from an antenna may be compared in some respects to a sound wave sent out by the string of a banjo when it is picked. The sound wave from the banjo string is a mechanical compression and rarification of the air caused by the vibration of the string (see Fig. 13-1).

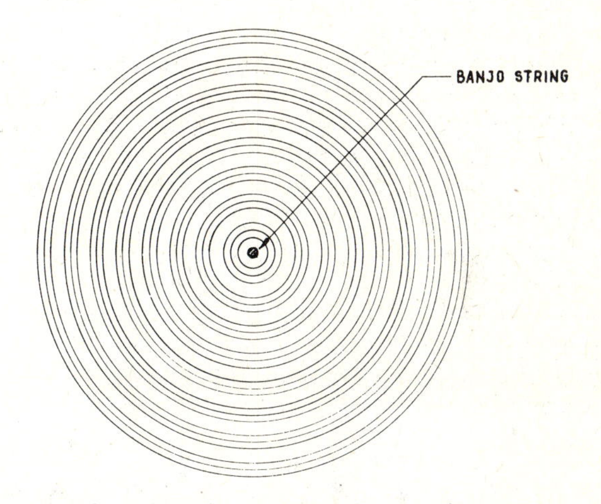

Fig. 13-1 Sound waves emanating from a vibrating string.

It has been previously explained that a magnetic field surrounds a current-carrying conductor. If the current flow in the conductor changes, the magnetic field will also change. The resulting movement of the field will cause the induction of a voltage in any conductor cut by the moving field.

An electric field is generated by the antenna in addition to the electromagnetic field produced. The two fields radiate from the antenna at the speed of light, which is approximately 186 300 mps [300 000 000 m/s].

Since a radio wave travels with the speed of light, it can readily be understood that when a transmitter starts operation, the signal from that transmitter may be detected instantly hundreds or thousands of miles from the transmitter, depending upon the power of the transmitter and the nature of the wave being transmitted.

The electromagnetic and electric fields produced by a radio transmitter antenna are at right angles to each other as shown in Fig. 13-2. The polarization of the fields with respect to vertical or horizontal positioning depends upon the design and position of the antenna from which they are being emitted. The polarity of the fields reverses rapidly, the rate of reversal being established by the frequency of the wave.

WAVELENGTH AND FREQUENCY

The length of a radio wave depends on its frequency. As in an ac sine wave, the wave emanating from an antenna increases to a maximum in one direction, drops to zero, and then increases to a maximum in the opposite direc-

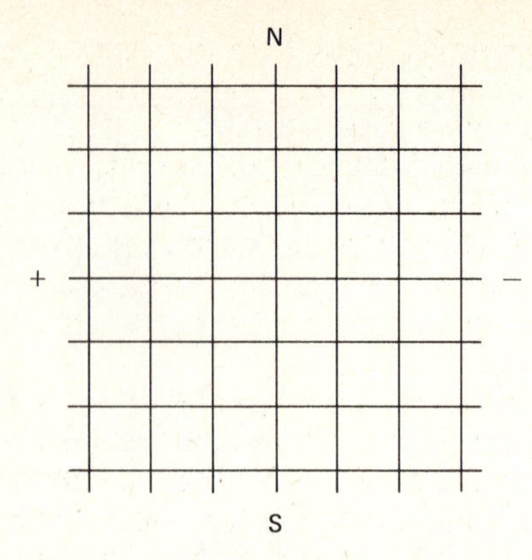

Fig. 13-2 Electric and electromagnetic fields in a radio wave.

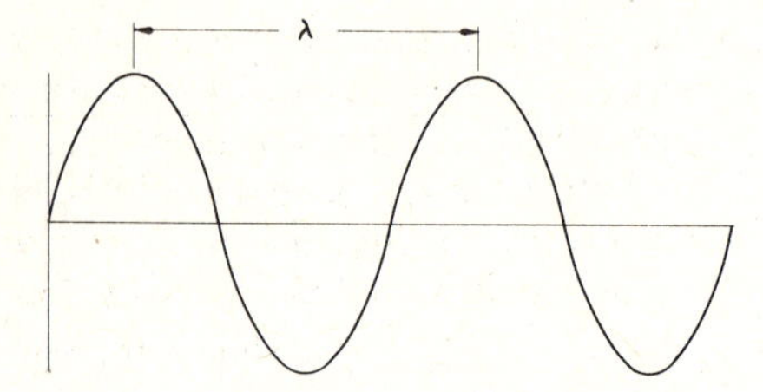

Fig. 13-3 Wavelength and wave motion.

tion as indicated by the curve of Fig. 13-3. The wavelength, indicated by the Greek letter lambda (λ), is the distance from the crest of one wave to the crest of the next. Since the wave travels at the rate of 300 000 000 m/s, the wavelength in meters is equal to 300 000 000 divided by the number of cycles per second (Hz), or f. If a wave is produced at the rate of 1 Hz, the length of the wave will be 300 000 000 m. If 300 cycles are produced per second, the wavelength will be 1 000 000 m [328 000 000 ft]. The formula for wavelength is

$$\lambda = \frac{300\,000\,000}{f}$$

If the wavelength is known, the frequency may be found by the formula

$$f = \frac{300\,000\,000}{\lambda}$$

For example, if an amateur radio station is operating on the 40-m band, the frequency will be approximately 7 500 000 Hz. This may be stated as 7500 kHz or 7.5 MHz.

FREQUENCY BANDS

Frequencies utilized in various types of radio systems range from 3 kHz to as high as 30 gigahertz (GHz). The frequencies are divided into seven bands, and these bands are assigned to certain types of operations. Table 13-1 shows the utilization of the various bands.

THE CARRIER WAVE

The field of electric and electromagnetic energy which carries the intelligence of a radio signal is called a carrier wave. The frequency of this carrier wave may be only a few hundred kilohertz or several thousand megahertz. It is usually in the **radio frequency** range, which is in excess of 20 000 Hz. Frequencies below 20 000 Hz, which are audible, are in the **audio-frequency** (AF) range.

In order to carry intelligence, an RF carrier wave must be modulated. This means that its form and characteristics are changed by means of some type of signal impressed upon it. Figure 13-4 illustrates an unmodulated carrier wave and a wave which has been modulated in amplitude by an AF signal. An RF carrier wave which has been modulated in amplitude is called an *amplitude-modulated* (AM) signal. If a voice signal is impressed upon a carrier, the modulation curve will follow the pattern of the voice frequencies.

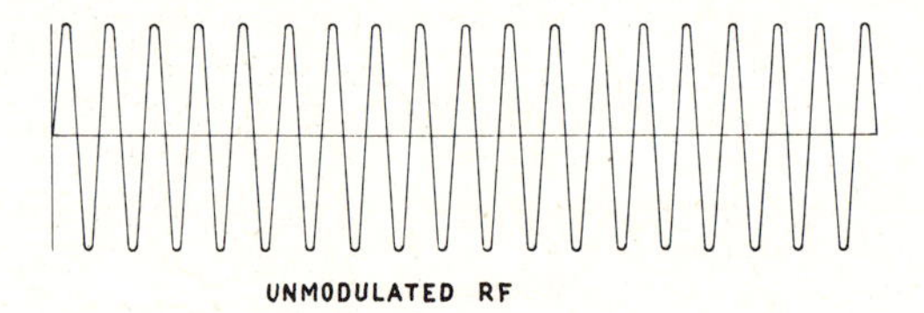

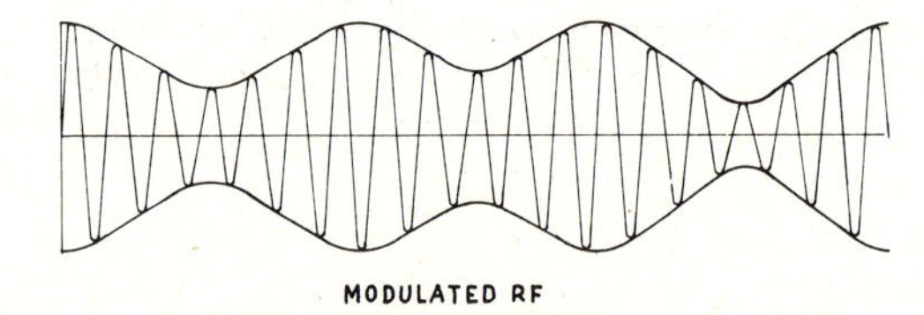

Fig. 13-4 Carrier waves unmodulated and modulated.

TABLE 13-1 Frequency Ranges and Utilization for Radio Waves

Designation	Frequency range	Wavelength	Utilization
Very low frequency (VLF)	3–30 kHZ	100 000–10 000 m	Navigation, time signals
Low frequency (LF)	30–300 kHz	10 000–1000 m	Navigation, broadcasting, maritime mobile, fixed
Medium frequency (MF)	300–3000 kHz	1000–100 m	Broadcasting, maritime mobile
High frequency (HF)	3–30 MHz	100–10 m	Broadcasting, amateur, maritime and aeronautical mobile, citizens' band (CB)
Very high frequency (VHF)	30–300 MHz	10–1 m	FM and TV broadcasting, aeronautical navigation and communication, amateur, maritime mobile
Ultrahigh frequency (UHF)	300–3000 MHz	1 m–10 cm	TV broadcasting, radar, aeronautical and maritime mobile, navigation, radio location, space communication, meteorological
Superhigh frequency (SHF)	3–30 GHz	10 cm–1 cm	Space and satellite communication, radio location and navigation, radar

Frequency modulation can be used in the VHF range and above. This type of modulation, commonly called FM, provides a signal that is much less affected by interference than amplitude modulation. As indicated by the name, frequency modulation is accomplished by varying the frequency of the carrier wave in accordance with the audio signal desired. Figure 13-5 shows how frequency modulation affects a carrier wave.

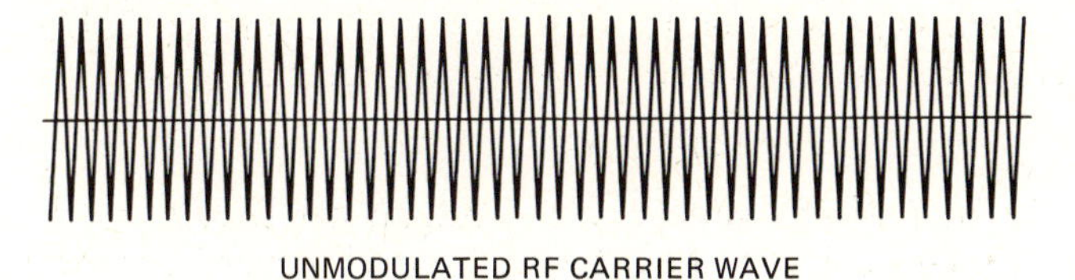

Fig. 13-5 Frequency modulation.

ANTENNAS

An antenna is a specially designed conductor that accepts energy from a transmitter and radiates it into the atmosphere. In receivers, it is the device that intercepts the energy from radio waves and conducts the energy to the receiver by means of a lead-in. Where transmitters and receivers are built into one unit, often called a **transceiver,** the same antenna serves for both transmitting and receiving.

The size and design of antennas vary in accordance with the frequency or frequencies of signals being handled. As frequencies increase, the length of the antenna must decrease. This is because wavelengths decrease as frequencies increase, and the length of antenna must be matched as closely as possible to the wavelengths of the carrier waves. Typical antenna lengths are full-wave (the same length as the carrier wave), half-wave, quarter-wave, or some other fraction of the wavelength.

The carrier wave emitted by the antenna of a radio transmitter travels both along the ground and through the air. The ground wave usually must travel in a direct line to a receiver; the sky wave is reflected from the layer of ionized gases which surrounds the earth at an altitude of about 60 to 250 miles (mi) [96.6 to 402.6 km], depending on the time of day, the season, and the location. This layer of ionized gases is called the **Heaviside layer** or **ionosphere.**

The ground wave from a transmitter is attenuated (reduced in strength) by irregularities such as hills, valleys, buildings, trees, etc., all of which block, reflect, or absorb the wave to some extent. For this reason a ground wave cannot be received at a great distance from a transmitter.

The sky wave, being reflected from the Heaviside layer, may be received thousands of miles from the transmitter. Because of the changes which take place in the ionosphere, the sky wave of a transmitter may vary considerably in strength and may even fade out completely. Moderate variations in wave strength are compensated for by means of an automatic volume control (AVC) in the receiver. The reflection of a sky wave and the attenuation of a ground wave are illustrated in Fig. 13-6. The ground wave disappears at *A*, but the reflected sky waves continue for thousands of miles.

When radio waves are emitted at very high frequencies, they are not usually reflected back to earth but continue through the ionosphere and out into space. Signals from very-high-frequency (VHF) or ultrahigh-frequency (UHF) transmitters must normally be received in a direct line from the transmitter. This means that there should be no intervening mountains, buildings, or other objects between the transmitter and the receiver if good reception is to be expected.

As explained previously, antenna length and arrangement are of primary importance in the transmission and reception of radio waves. The simplest form of receiving antenna is merely a length of wire insulated from the ground and connected to the antenna coil of the receiver. The wire is cut by radio waves, and these waves induce very small voltages of many frequencies in the wire. The signal for which the receiver is tuned will pass through the receiver and to the loudspeaker in a form suitable for sound reproduction.

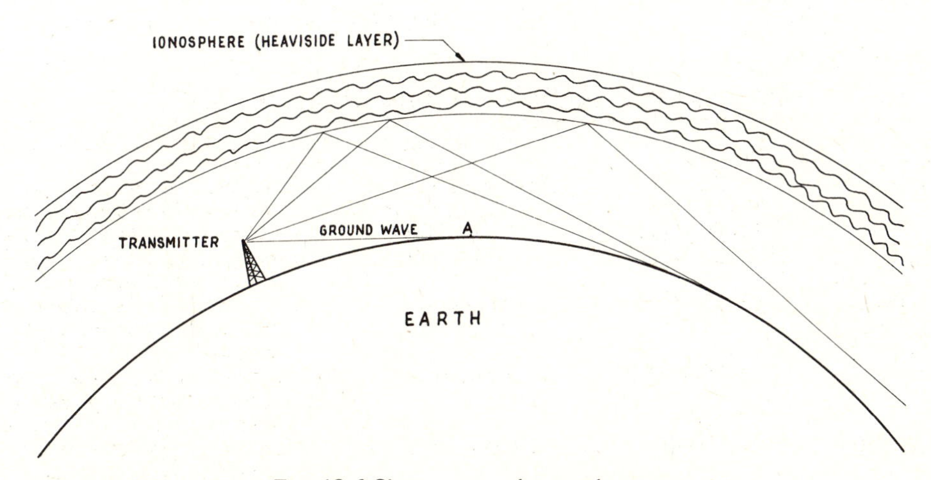

Fig. 13-6 Sky waves and ground waves.

The correct antenna length (half-wave) for either a transmitter or a receiver is determined by using the following formula:

$$l = \frac{468}{f}$$

where l = length, ft
f = frequency, MHz

This formula gives the length for a half-wave antenna. The figure 468 in the formula is a factor derived by converting meters per second to millions of feet per second, dividing by two, and multiplying by 0.95, the correction constant for antennas.

The formula for the length of a half-wave antenna is also expressed as

$$l = \frac{300\ 000\ 000 \times 3.28}{2 \times f}$$

In this formula the correction factor 0.95 is ignored. The principal objective in constructing an antenna and antenna-coupling system is to match the output impedance of the transmitter with the input impedance of the antenna system. The simplest types are the Hertz antenna and the Marconi or vertical antenna. The Hertz antenna consists of two lengths of wire extended in opposite directions as shown in Fig. 13-7. Each length of wire is $\frac{1}{4}$ wavelength ($\lambda/4$) long. (A study of a sine wave shows that $\frac{1}{4}$ wavelength permits voltage or current to increase from zero to maximum in one direction.) The two lengths of wire are fed by the transmitter at the center; hence, one length will become negative as the other becomes positive.

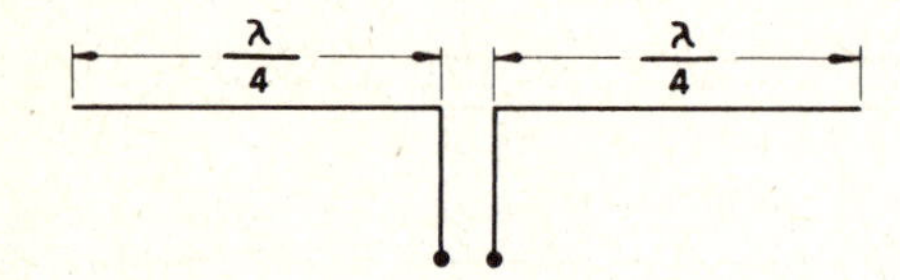

Fig. 13-7 The Hertz dipole antenna.

Coupling from a transmitter to the antenna is normally accomplished by means of *LC* circuits. A typical coupling circuit for a coaxial transmission line and a dipole antenna is shown in Fig. 13-8, but this is only one of the many possible arrangements.

The proper coupling of the antenna to a transmitter is essential for the maximum radiation of energy. The input impedance of the antenna must be as closely matched to the internal impedance of the transmitter as possible and the effective length of the antenna must be adjusted to the wavelength of the signal being broadcast. The coupling circuit in the transmitter accomplishes these purposes. When a transmitter is being prepared for operation, it is necessary to determine that the maximum signal strength is being radiated from the antenna. The antenna output is checked by means of a standing-wave ratio (SWR) meter. If the meter indicates poor signal radiation, the coupling circuit can be adjusted. This process is called "loading the antenna."

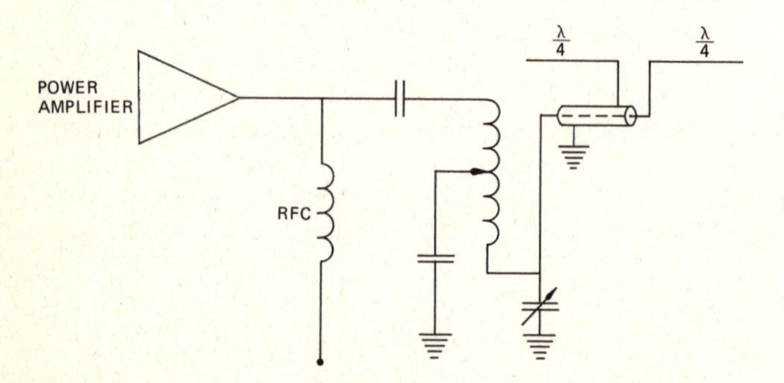

Fig. 13-8 Antenna coupling circuit for a dipole antenna.

FUNCTIONS OF A TRANSMITTER

A radio transmitter has several functions and one final objective. Briefly, the functions are: (1) to generate an HF carrier wave; (2) to amplify the carrier wave; (3) to modulate the carrier wave with sound, key, or other device to impress intelligence on it; (4) to amplify the modulated signal; (5) to couple the modulated signal to an antenna; and (6) to radiate the signal into the atmosphere. All these functions except the last are performed by circuits within the transmitter system so that the final objective is accomplished. The radiation of the signal is accomplished by the antenna.

A block diagram for a typical radio transmitter is shown in Fig. 13-9. This is only one of many possible arrangements.

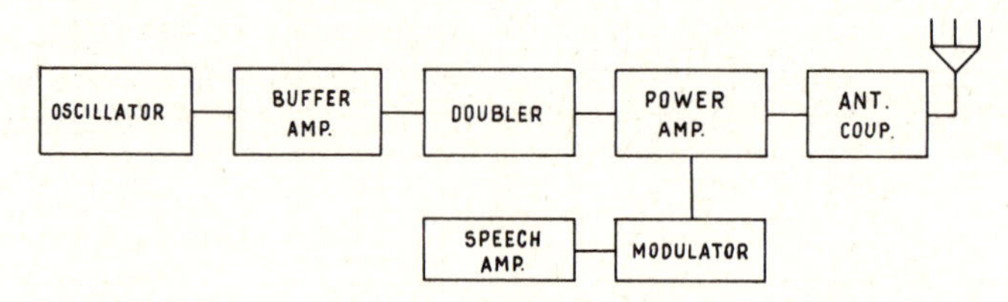

Fig. 13-9 Block diagram of a radio transmitter.

THE OSCILLATOR

It will be recalled that several types of oscillators have been discussed in previous sections. Among those which may be used to produce the RF carrier wave for a transmitter, the Armstrong, Hartley, Colpitts, and crystal-controlled oscillators are the four principal types. The oscillator most commonly used in transmitters is crystal-controlled in order to maintain an exact transmitting frequency. A mounted crystal serves the same function as a resonant circuit, with the thickness of the crystal determining the frequency.

A basic crystal-controlled oscillator circuit is shown in Fig. 13-10. In this circuit the crystal takes the place of the tank circuit employed in other oscillators, and feedback is provided by means of capacitive coupling. Maximum amplitude of the RF signal is obtained when the output circuit is tuned to the frequency of the crystal.

THE BUFFER AMPLIFIER

The purpose of a buffer amplifier is to prevent the loading of the oscillator circuit and a resulting change in the oscillator frequency. This means, of course, that the circuits following the oscillator must not draw power from the oscillator. Because the buffer amplifier must be of a type which will not draw power, it is usually an amplifier

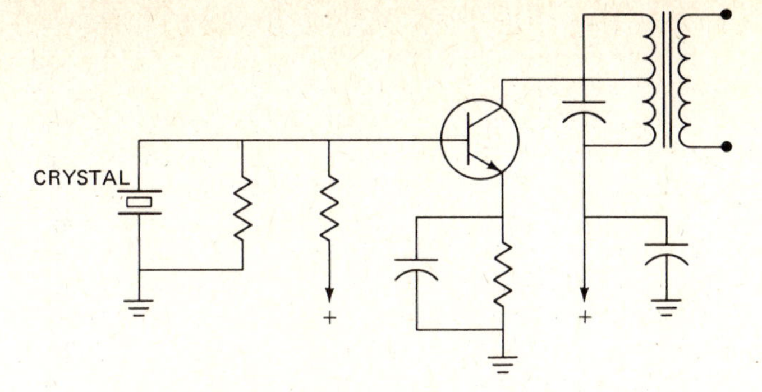

Fig. 13-10 Crystal-controlled oscillator circuit.

operated as a class A type, which has no appreciable base current flow. A field-effect transistor (FET) is ideal for this application, because emitter-collector current is controlled by the strength of an electric field rather than current flow through the base circuit. The RF current is coupled to the next amplifier stage through the coupling capacitor and a tank circuit. If it is desired, the tank may be used as a frequency multiplier by having it tuned to a higher multiple of the oscillator frequency. If the tank is tuned to twice the oscillator frequency, it is called a **doubler;** if it is tuned to three times the frequency of the oscillator, it is called a **tripler.**

FREQUENCY MULTIPLIERS

It is common practice to use a crystal in the oscillator circuit of a transmitter in order to hold the frequency at a fixed value. Since the frequency of a crystal is determined by its thickness, it will be apparent that there is a limit to the frequency which can be obtained with a crystal. This is because a crystal cannot be cut sufficiently thin for high frequencies without becoming so delicate that it could not be handled safely.

To overcome the frequency limit, frequency-multiplying circuits are employed. These may be doublers or triplers, depending upon the frequency of the tank circuit into which the output of the oscillator tube is fed.

The principal disadvantage of a frequency-multiplying circuit is that the output power is considerably less than it is from an amplifier operating *straight through,* that is, one in which no change in frequency takes place.

The principle of frequency multiplication may be understood by considering the action of a tank circuit. Remember that any tank circuit has a resonant frequency determined by the values of its capacitance and inductance according to the formula

$$f = \frac{1}{2\pi\sqrt{LC}}$$

If the capacitor in Fig. 13-11 is charged by means of the battery through the switch S in position 1 and the switch is then shifted to position 2, the electrons stored on one plate of the capacitor will start to flow through the inductance coil toward the opposite side of the capacitor. Current flow through the inductance will create a magnetic field in which is stored electric energy. When the current flow begins to decrease, the inductance tends to keep it flowing, and so the capacitor becomes charged in the opposite direction. Thus, the cycle continues back and forth

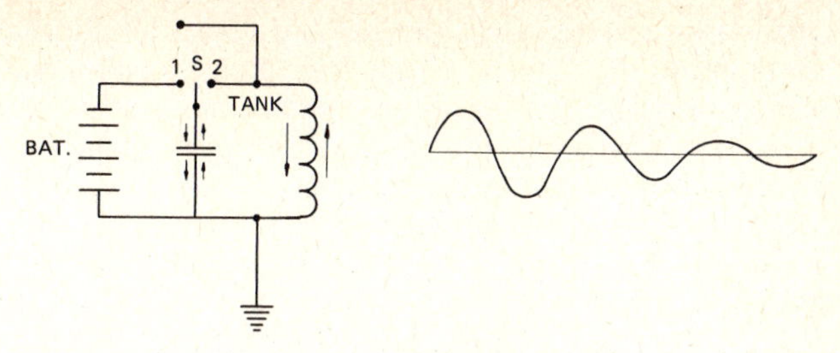

Fig. 13-11 Operation of a tank circuit.

with the energy alternately being stored in the electrostatic field of the capacitor and in the electromagnetic field of the inductor. Because of the resistance in the circuit, the alternating current will degenerate and disappear unless additional energy is supplied to keep it up.

In the frequency multiplier, the energy to maintain current flow is supplied by the tube or transistor output. If the tube or transistor is connected as a class C amplifier, the output will be in the form of widely separated pulses, as shown in Fig. 13-12. In this illustration the amplifier output is shown as separate pulses with a frequency of 1000 kHz; the tank current is sustained at 2000 kHz.

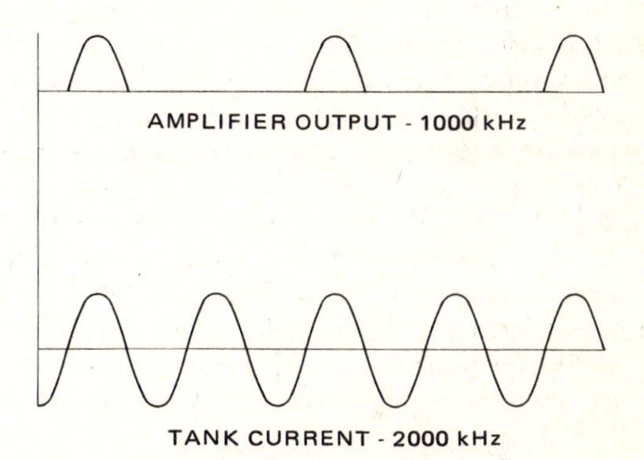

Fig. 13-12 Curves to illustrate the operation of a frequency doubler.

The action of a frequency multiplier may be compared to the action of a child in a swing. The swing may be kept going easily by applying a short push at every second or third swing. In Fig. 13-12, the short push is the amplifier pulse and the swing in motion is the tank current.

THE MODULATOR

The function of the modulator, or modulation circuit, in a transmitter is to impress a signal on the RF carrier wave. This signal usually consists of an AF sound wave; however, the carrier may also be modulated by means of a key to produce code signals.

The modulation of a carrier wave is illustrated in Fig. 13-13. It will be noted that the modulating wave modulates the **amplitude** of the carrier. As explained

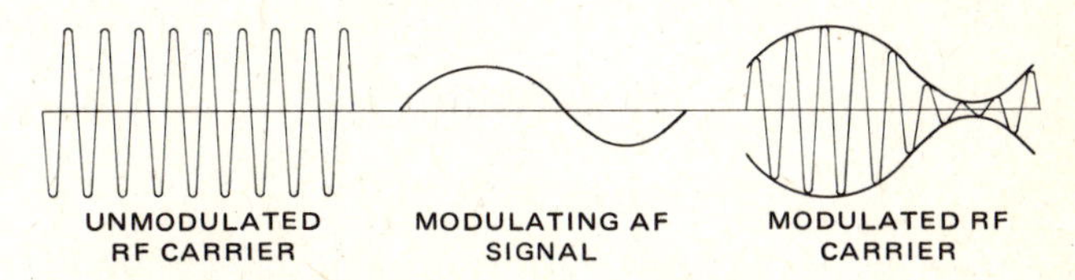

Fig. 13-13 Modulation of an r-f carrier wave.

previously, this type of modulation is called **amplitude modulation** (AM). To obtain the greatest efficiency from a modulator, it is necessary that the maximum modulation be of an amplitude which will increase the unmodulated RF carrier to twice its unmodulated amplitude. Likewise, the negative peaks of modulation power should be of a value which will reduce the RF carrier to zero amplitude. When these conditions exist, the modulation is 100 percent. Figure 13-14 illustrates 100 percent modulation. If a smaller degree of modulation takes place, the full potential of the carrier is not utilized. On the other hand, if the modulating wave has too great an amplitude, overmodulation will occur and the signal will be distorted.

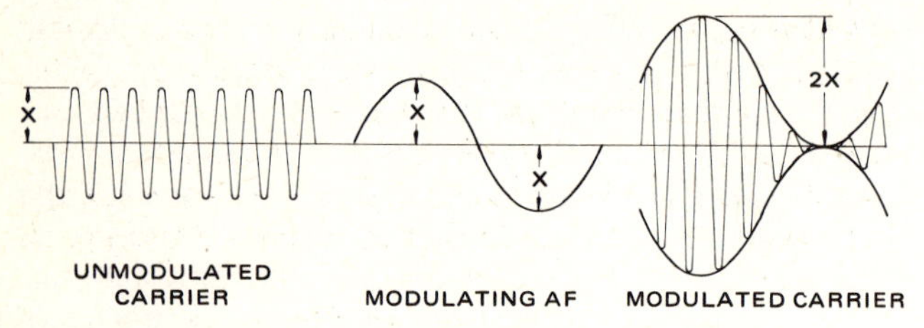

Fig. 13-14 Hundred percent modulation.

The actual practice of modulation in a modulator of a transmitter is the development of the modulating signal and the application of this signal to the carrier wave. As mentioned previously, the audio signal may modulate either the amplitude or the frequency of the carrier, depending upon how it is applied. In aircraft radio, amplitude modulation is employed for all voice transmissions.

In the modulator, the audio signal is applied to the base circuit of a transistor, thus modifying the emitter-collector current. In tube-type transmitters, the audio signal is applied to one of the grids or other elements of the tube.

POWER AMPLIFIER

The function of the power amplifier of a transmitter is to increase the power level of the modulated signal to the point that it meets the requirements of the transmitting system. Radio transmitters employ power transistors that are designed to carry the current required. The output of the power amplifier is coupled to the antenna by means of an adjustable *LC* circuit or in some other manner. One method of coupling is illustrated in Fig. 13-15.

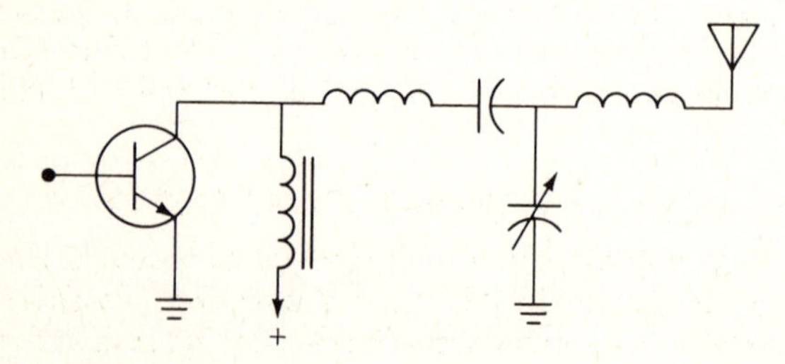

Fig. 13-15 Antenna coupling circuit for a UHF transmitter.

RECEIVERS

RADIO TUNING

When we think of the many radio and television transmitting stations located in almost every country of the earth, it immediately becomes apparent that there are thousands of radio signals in the air at all times. It then seems miraculous that a radio or television receiver is able to select one signal from all the others and reproduce the intelligence (picture, sound, or both) satisfactorily. This is the function of the tuning system in the receiver.

The heart of a tuner is a resonant circuit which consists of a capacitor and an inductor connected either in parallel or in series. It will be recalled that the series resonant circuit provides minimum impedance and that the parallel resonant circuit provides maximum impedance.

Figure 13-16 shows a simple circuit for a typical tuning unit. The radio signals cutting across the antenna induce signals of various frequencies which flow through the primary winding of the antenna coil to ground. These currents produce electromagnetic waves which induce voltages in the secondary winding of the antenna coil.

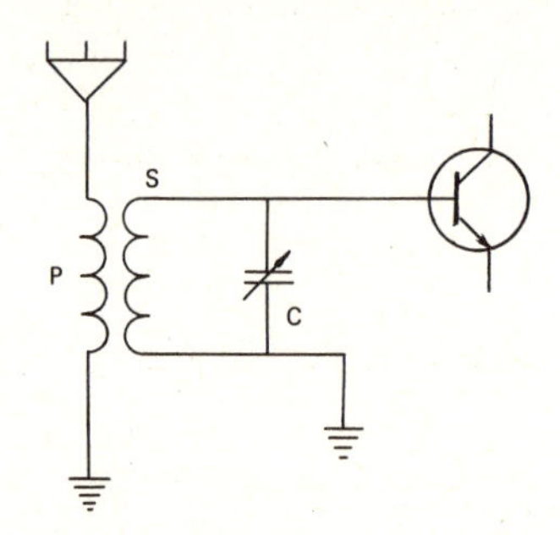

Fig. 13-16 A simple tuning circuit.

Since a variable capacitor *C* is connected across the secondary coil, a maximum of current will flow only at the resonant frequency of the coil and the capacitor. Hence, at resonant frequency a maximum voltage will be developed across the capacitor, and this same voltage will be apparent as the difference in potential between the control grid and the cathode of the electron tube. This voltage will be the input signal for the tube.

In some cases a series resonant circuit is provided in the primary system of the antenna coil as shown in Fig. 13-17. In this case, maximum current will flow in the primary only at resonant frequency. This provides for increased selectivity because unwanted frequencies are largely prevented from being induced in the secondary winding of the antenna coil.

The two circuits described are by no means the only methods of tuning, but they represent the basic principles used in all tuning circuits in the low and medium frequencies.

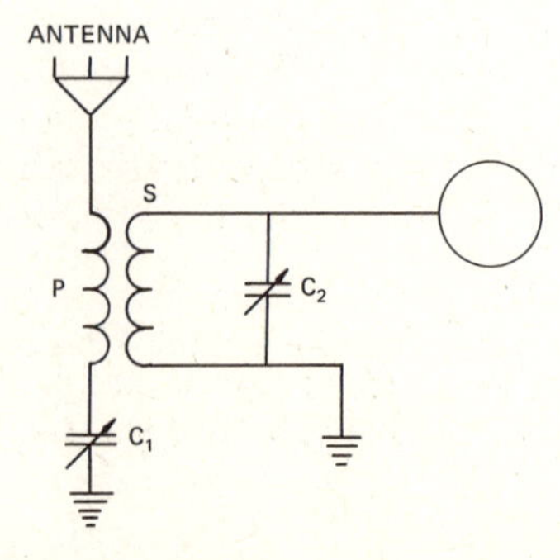

Fig. 13-17 Tuning circuit with series tuning in primary.

DETECTION

Detection of a radio signal is the process of separating the audio frequency from the RF carrier wave. This is accomplished by rectifying and filtering the modulated RF carrier so that only the AF signal passes to the speaker. Detection may be accomplished by means of a crystal detector, a crystal diode, a diode tube, a transistor, or a triode tube because any one of these devices may be connected as a rectifier.

Originally a crystal detector consisted of a small piece of galena or iron pyrite secured in a holder and contacted with a small wire point called a *cat's whisker.* The cat's whisker was moved about until the best point of conduction was located on the crystal. A crystal detector of the cat's whisker type is shown in Fig. 13-18. The crystal diode, also used as a detector, is usually a small manufactured crystal of a semiconductor enclosed in a small cartridge. As explained previously, a semiconductor conducts current freely in one direction and offers a high resistance in the opposite direction. When a transistor is used for detection, the unidirectional conducting feature is employed for rectification.

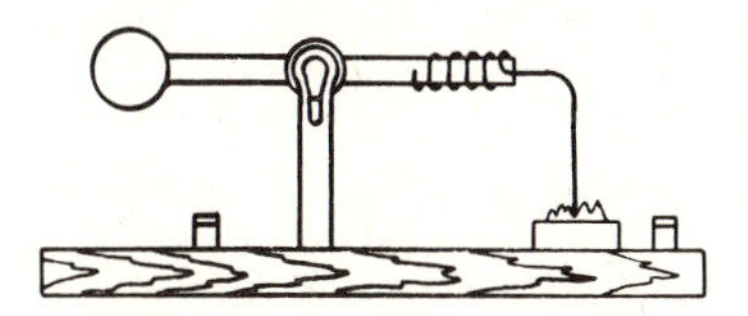

Fig. 13-18 Cat's-whisker crystal detector.

Since current will flow through the plate circuit of an electron tube in only one direction, such a tube will serve well as a detector. In many early radio receivers the first tube, which was usually a triode, acted as a detector and as an audio amplifier. In later receivers a diode section in the first audio amplifier serves as a detector.

The principle of detection is illustrated in Fig. 13-19. Figure 13-19*a* shows the modulated carrier wave, Fig. 13-19*b* shows the wave after detection, and Fig. 13-19*c* shows the audio signal after filtering. This audio signal is usually amplified before being reproduced through a speaker.

SOUND REPRODUCTION

The reproduction of sound in a radio receiver is a conversion of electric waves to sound waves. This is accomplished by means of earphones or a loudspeaker, both operating on the same general principle. Figure 13-20 is a simplified drawing of the method of sound reproduction in both the earphone and the speaker. A coil of many turns of very fine wire is wound on a permanent-magnet core. At a few thousandths of an inch from the magnet poles is mounted a magnetic diaphragm which will vibrate in response to changes in the magnetic field caused by varying current flow in the coil. This vibration causes the waves in the air which are recognized as sound.

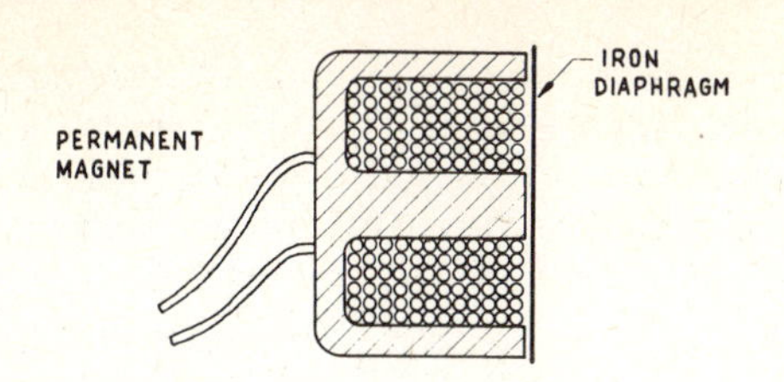

Fig. 13-20 Sound-reproducing mechanism.

THE CRYSTAL RECEIVER

The crystal receiver is the simplest of all types. Two circuits are shown in Fig. 13-21. In circuit *a* the tuning circuit is of the series type with a crystal or crystal diode in series with the tuning circuit and headphones. At resonant frequency a maximum of current will flow through the entire circuit. The headphones are connected with a capacitor in parallel to bypass the RF portion of the rectified signal.

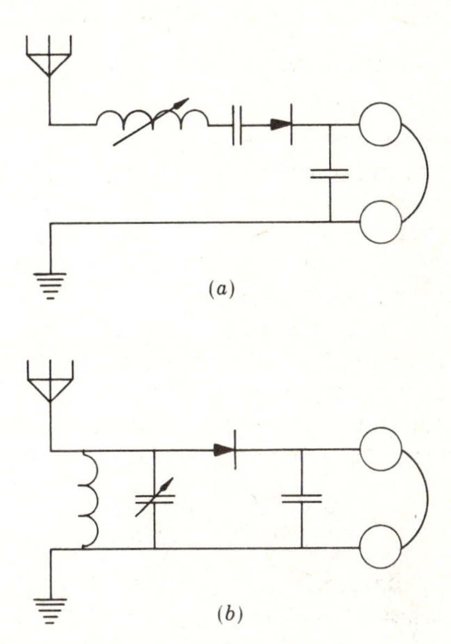

Fig. 13-21 Crystal or diode receiver circuits.

In circuit *b* of Fig. 13-21, tuning is accomplished with a parallel resonant circuit. In this case a maximum signal voltage will appear across the tank circuit at resonant frequency. This voltage will produce a current through the crystal detector and the headphones. The headphones are connected with a bypass capacitor as in the other circuit.

THE ELECTRON-TUBE RECEIVER

A very simple electron-tube receiver may have a circuit as shown in Fig. 13-22; the tuning system is similar to those used in other radio receivers. The grid leak and grid

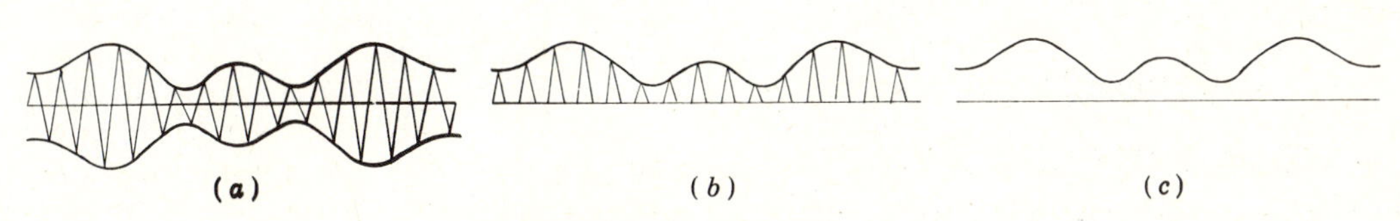

Fig. 13-19 Principle of detection.

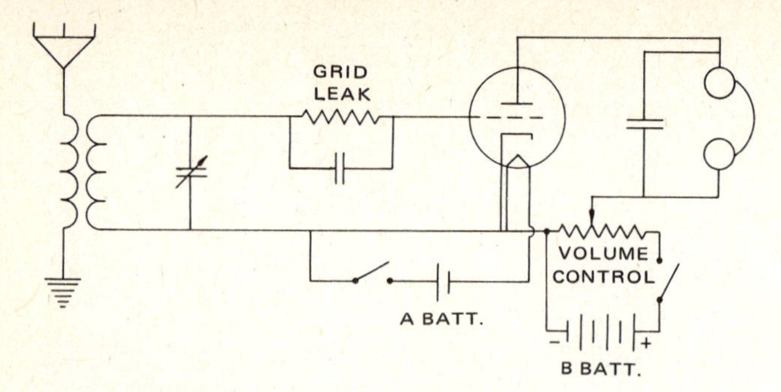

Fig. 13-22 Simplified single-tube receiver circuit.

capacitor maintain a negative bias; that is, they maintain a grid voltage which keeps the grid negative with respect to the cathode and prevents the flow of grid current which would otherwise cause distortion and loss of power.

Note that a B battery is connected in series with the plate circuit. The B battery provides a current flow through the plate circuit, and this current varies in accordance with the signal voltage impressed on the control grid of the tube. The B battery is connected with a high-resistance potentiometer in order to provide a volume control.

The basic receiver circuit shown in Fig. 13-22 can be designed to operate with a transistor. The transistor could act as the detector and as an amplifier to provide the audio signal for the headphones.

REGENERATION

Regeneration is the process of feeding back a portion of a receiver output to the input side of an electron tube in order to increase the gain.

The circuit shown in Fig. 13-23 is the same as that in Fig. 13-22 except that a "tickler" coil L_3 has been added in the plate circuit to feed back a portion of the output signal to coil L_2. The signal thus returned to the input increases the strength of the signal to the control grid, and this increased strength in turn provides a greater variation of plate current. In modern receivers regeneration is not commonly used as a means of increasing signal strength.

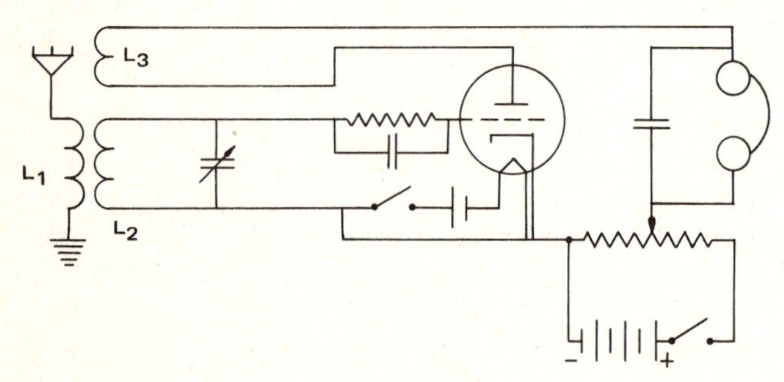

Fig. 13-23 Simple receiver with regeneration.

AMPLIFIERS

DEFINITION

An amplifier is a circuit which receives a signal of a certain amplitude and puts out a signal of greater amplitude. The amplification may affect voltage or power, or both, but its principal purpose is to increase the value of a weak signal so that it may be used to operate a speaker or some other electronic device.

CLASSIFICATION OF AMPLIFIERS

Amplifiers are classified according to function, operating level, or circuit design. The function may be to amplify power or voltage, and in this case the amplifier is described as a **power amplifier** or a **voltage amplifier.**

When the amplifier is classified according to operating level, the classification refers to the point on the characteristic curve through which the tube operates as established by the grid bias or the base-to-emitter bias for a transistor. A class A amplifier operates at a level such that plate current or emitter-collector current flows at all times because the voltage never reaches a sufficiently negative value to cut off the electron flow. The operation of this type of amplifier is shown by the curves in Fig. 13-24. It will be seen that the tube or transistor is biased near the center of the linear portion of the operating curve. The class A amplifier provides for a minimum of distortion of the signal; hence, it is used where maximum fidelity is desired.

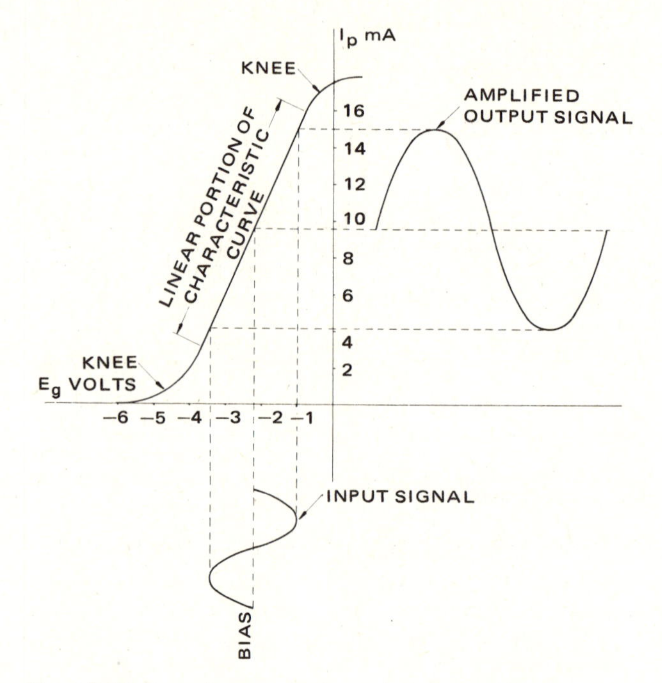

Fig. 13-24 Operation curves for a class A amplifier.

A class B amplifier is biased at approximately the cutoff point. With this arrangement only one half of the signal will be amplified, but the amplification can be carried to a much higher level than it can be by a class A amplifier because a much greater bias range is possible. Class B amplification is often used in **push-pull** amplifiers in which two tubes or transistors are employed, one amplifying one half of the signal and the other amplifying the other half. The two amplified halves of the signal are recombined in the output circuit to produce a signal of low distortion and high power. The curves in Fig. 13-25 illustrate the operating level of class B amplification, and the curves in Fig. 13-26 show how class B amplification performs in a push-pull circuit. Note that one half of the signal is amplified by tube 1 and the other half by tube 2. The circuit for a push-pull amplifier is shown in Fig. 13-27.

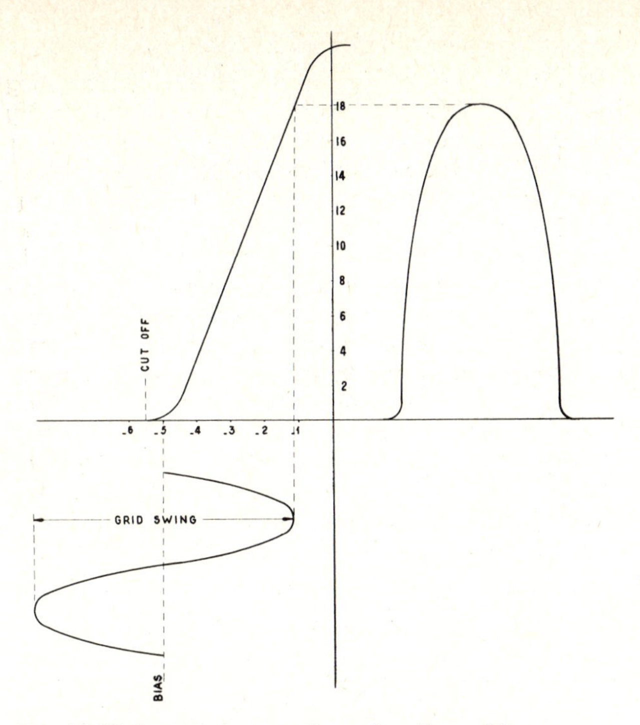

Fig. 13-25 Operation curves for a class B amplifier.

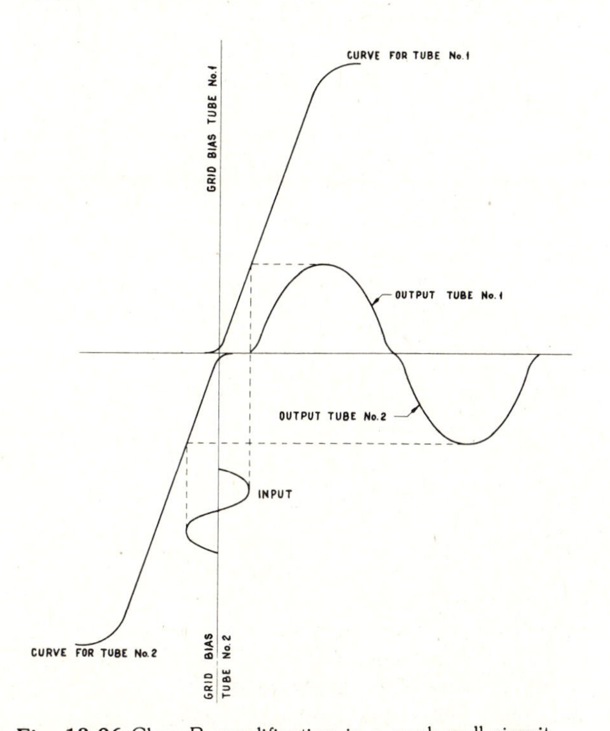

Fig. 13-26 Class B amplification in a push-pull circuit.

In class C amplification the grid of the amplifier tube or the base-emitter circuit of the transistor is biased well beyond cutoff so that only a small portion of the positive peaks of the signal are amplified. Current flows only during approximately 120° of the cycle. The use of class C amplifiers is limited to RF circuits because only a part of the signal curve is reproduced. In an RF circuit, when the output of the class C amplifier is fed into an *LC* system, the flywheel effect of the tank supplies the missing parts of the signal curves. The E_g-I_p curves for a class C amplifier are shown in Fig. 13-28.

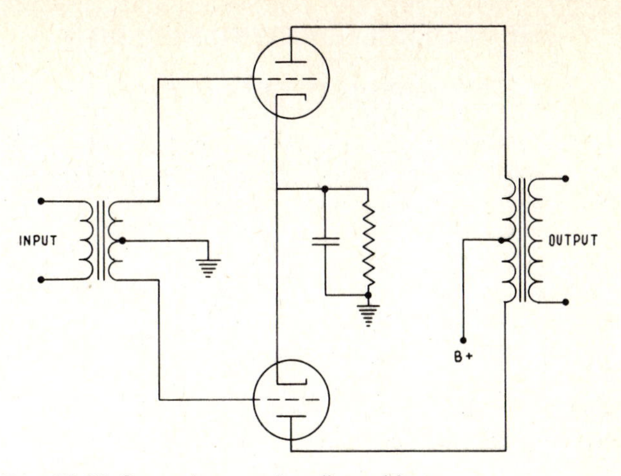

Fig. 13-27 Circuit for a push-pull amplifier.

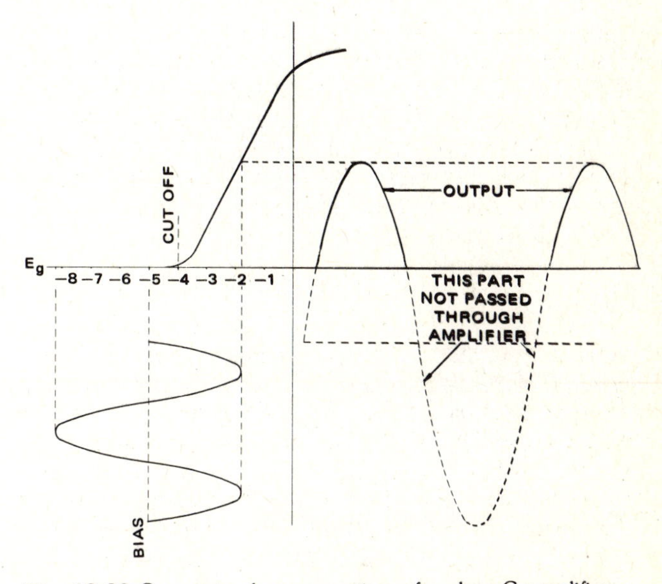

Fig. 13-28 Curves to show operation of a class C amplifier.

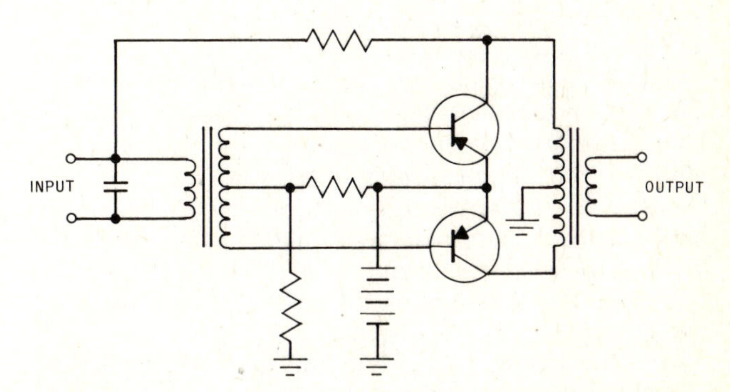

Fig. 13-29 Circuit for a push-pull amplifier in a transistor receiver.

With certain modifications of circuitry to accommodate different current and voltage values, transistors are commonly utilized in radio receivers. A class B or class AB amplifier circuit can be connected as shown in Fig. 13-29 to provide push-pull amplification with transistors.

THE SUPERHETERODYNE RECEIVER

INTERMEDIATE FREQUENCY

The superheterodyne receiver derives its name from the fact that a new signal frequency is generated in the receiver by means of a local oscillator called a beat-frequency oscillator (BFO). The BFO signal is fed into the converter or mixer system. The word *hetero* is a Greek term for *other,* and the word *dyne* means *power;* thus the term *heterodyne* literally means *other power.* It refers to the intermediate frequency developed in the mixer circuit.

A block diagram for a superheterodyne receiver is shown in Fig. 13-30. During operation the signal received by the antenna enters the RF amplifier, and the strengthened signal is then passed on to the mixer or converter stage. Here the incoming RF carrier and the oscillator (BFO) signal are combined to produce an intermediate signal of 455 kHz. Actually, when two frequencies are combined, four frequencies result. These are the two original frequencies and the sum and difference frequencies. If an RF frequency of 1000 kHz is combined with an oscillator frequency of 1455 kHz, the two new frequencies will be 2455 and 455 kHz. For the IF signal in the superheterodyne receiver, only the 455 kHz frequency is used.

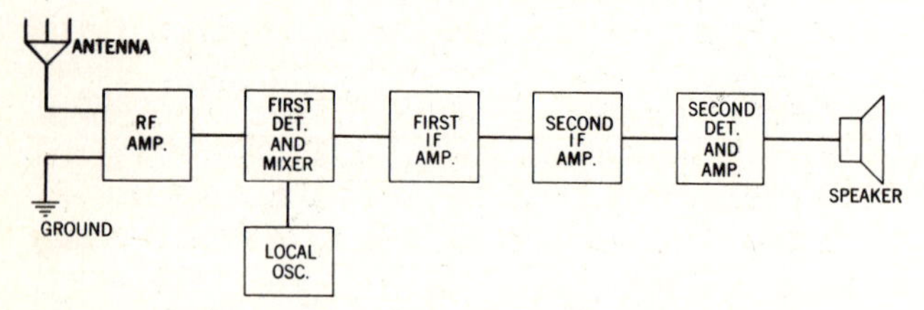

Fig. 13-30 Block diagram for a superheterodyne receiver.

The antenna circuit and the local oscillator circuit are tuned together to produce frequencies which have a difference of 455 kHz. This IF signal from the mixer or converter carries the audio signal which was originally on the RF signal. The IF signal is usually passed through two IF amplifier stages, each consisting of an IF transformer and an electron tube or a transistor with necessary resistors and capacitors. The two IF circuits are accurately tuned to 455 kHz with a bandwidth of approximately 10 kHz. This means that signals of 450 to 460 kHz will be passed through the IF amplifier stages and that all other frequencies will be attenuated. The 10 kHz bandwidth is necessary to accommodate the audio modulation which may be carried by the IF signal.

After the intermediate frequency has been amplified through the two IF amplifiers, it is passed through the second detector and first audio-amplifier stage. From this point, the signal may be directed to the speaker or to an additional stage of audio amplification.

Superheterodyne receivers are used in many electronic systems other than radio. They have a number of advantages, among which are simplified tuning, improved amplification, high selectivity, and good fidelity. In addition to their application in radio receivers, they are used in television and in other special devices. When they are used for frequencies other than those of the standard broadcast band, it is necessary to employ a different intermediate frequency from the 455 kHz one normally used for that band.

REVIEW QUESTIONS

1. Describe a *radio wave.*
2. What fields are found in a radio wave?
3. Explain *wavelength.*
4. Compare wavelength and frequency for a radio wave.
5. Give the formula for wavelength if the frequency is known.
6. What frequency bands are assigned to radio?
7. What is the difference between radio frequency and audio frequency?
8. Explain the functions of antennas.
9. How does the frequency of the radio waves transmitted or received relate to the length of the antenna?
10. Describe the *ionosphere.*
11. Of what value is the ionosphere in radio communications?
12. Give the formula for antenna length.
13. What is meant by *coupling* an antenna to a transmitter?
14. What are the functions of a radio transmitter?
15. What type of oscillator is commonly used in a radio transmitter?
16. What is the function of the *buffer amplifier?*
17. Describe the operation of a *frequency multiplier.*
18. What is meant by *100 percent modulation?*
19. How is modulation accomplished?
20. What is the purpose of the *power amplifier?*
21. How is it possible for a radio receiver to select one signal from the many thousands that exist in the atmosphere?
22. Draw a simple tuning circuit.
23. What is meant by signal *detection?*
24. By what means is an audio frequency signal converted to sound?
25. Describe the operation of a crystal receiver?
26. What is the purpose of the bypass capacitor across the headphones in a receiver?
27. What is the purpose of a *B battery* in a battery-operated radio receiver?
28. What is meant by *regeneration* in a radio receiver?
29. Describe an *amplifier.*
30. Name three classes of amplification.
31. What class of amplification produces the least distortion of the signal?
32. Explain class B amplification and how it is most commonly used in a receiver.
33. Explain the purpose of the *beat-frequency oscillator* in a superheterodyne receiver.
34. What intermediate frequency is most commonly used in a superheterodyne receiver?
35. What is the function of the *converter* in a superheterodyne receiver?
36. In what part of the superheterodyne receiver does detection take place?
37. Give the principal advantages of a superheterodyne receiver.

CHAPTER 14

Electric Instruments

In Chapter 5 of this text, electric measuring instruments were described. The basic elements of measuring instruments are characteristic also of many of the indicating instruments discussed in this chapter. These will be reviewed as the various types of indicating instruments are described.

Electric indicating instruments of various types have been described in other texts of this series; however, it is deemed appropriate to examine them further here because of the electrical factors involved. This chapter, therefore, focuses on electric indicating instruments for powerplants, aircraft systems, and controls.

The operation of modern aircraft, either of the small, private category or the transport type, could not exist without the use of instruments. On large aircraft, particularly, instruments are operated electrically or electronically. Instruments are needed to measure pressures, temperatures, attitudes, velocity, rate of flow, and numerous other conditions or parameters affecting the flight and operation of aircraft. Human beings are unable to react rapidly and accurately to the many and variable conditions which affect the flight of an airplane unless they have accurate and reliable instruments available for the quick and simple measurement of all needed information.

The pilot and flight engineer of an aircraft know that the engines, whether they be reciprocating or turbine types, will operate at their best only if the various operating temperatures are within the correct range and that other conditions are within definite operating limits. All operating conditions must be measured, and they are usually measured best by electricity. There is no quicker, easier, and more reliable method of conveying information from every section of the airplane to the flight crew than by means of electric or electronic circuits and devices. To the beginner, it may seem indirect and complicated to transform temperatures, speeds, and other nonelectrical quantities into electric currents or voltages, but no other system succeeds so well over the long distances required for the transmission of measurements in a large airplane. And no other method is so light, sensitive, reliable, and easily installed, inspected, and maintained. For these reasons, many modern airplanes have more miles of wiring than an average large city office building. Electric wires in all parts of the airplane carry messages from dozens, and sometimes hundreds, of sensitive feelers, or transmitters, to the instrument panel in the cockpit, just as the nervous system of the body carries messages to the brain.

The inspection and maintenance of instruments of any kind, including electric instruments, require the skill, patience, and accuracy of a watchmaker, as well as an all-around knowledge of the systems in which the instruments are to operate. Furthermore, instrument technicians should have a knowledge of the environmental conditions under which the instruments they service must function. The accuracy of an electric measuring instrument may be determined by the circuit and the mechanical arrangement in which it is used. For these and many other reasons, the servicing of electric instruments is a specialty which requires careful training and considerable experience before a technician can claim to be an expert.

There is a bright side to the picture, however. When a regular overhaul period for an airplane arrives, it is usually necessary only to inspect and test the instruments. This work may be accomplished by either a skilled airframe and powerplant technician or an instrument technician.

The purpose of this chapter is to present the basic principles of electric instruments not heretofore covered in this text. Knowledge of these principles is essential for students of aircraft and powerplant maintenance, for pilots, airframe and powerplant technicians, flight engineers, missile mechanics, test technicians, and others. There will be nothing mysterious or difficult about the subject if the reader will take one step at a time.

RPM MEASURING INSTRUMENTS

An rpm indicating instrument, or **tachometer,** is a most essential instrument in all aircraft. This instrument is used to show the rpm of reciprocating engines, the percent of power for turbine engines, the rpm of helicopter rotors, and the rotational speed of any other device where this information is critical. A typical tachometer for a reciprocating engine is shown in Fig. 14-1.

Since the tachometer registers engine rpm, it is a primary indicator of engine performance. The electric tachometer gives the same indications that are given by a mechanical tachometer, but the method for actuating the indicating unit is entirely different.

In multiengine airplanes, tachometers are especially important not only because of the value of engine speed as an indication of performance, but also because of the necessity for synchronizing the speeds of the engines. Both

Fig. 14-1 Tachometer for a reciprocating engine.

turbine engines and reciprocating engines operate most efficiently over a very narrow range of speed; hence, the operator must know the exact speed of each engine.

In the pioneer days of aviation, the pilot's seat was so close to the engine that the engine speed could be measured by mechanical means with equipment similar to that used for speedometers in automobiles. This required a flexible shaft connection from the tachometer drive of the airplane engine to the panel-mounted indicator in front of the pilot. In large modern airplanes, it is not practical to use a mechanical drive for the tachometer because of the great distance between the engine and the indicator and also because of the inherent weakness of long flexible shafts. Therefore, the long mechanical shafts are replaced by the electric wires of an electric tachometer.

The change from mechanical to electric connections was also due to the fact that more than one engine is usually necessary on a large airplane. Each engine of a multiengine airplane requires its own rpm counting device, and a tachometer for each engine may be installed at more than one station. For example, there is a tachometer for each engine mounted on the instrument panel where the pilot and copilot may read it, and there may also be one for each engine at the flight engineer's station. On this basis, a four-engine airplane would have at least eight tachometers. Obviously, the many connections needed for a multiple installation such as this are more practical in the form of electric wires than they would be in the form of mechanical shafts.

As mentioned previously, turbine engines employ **percent-of-power gages** in place of the tachometer used with reciprocating engines; however, the operation of these instruments is the same as that of the ac tachometer which will be described in this chapter. A percent-of-power indicator is shown in Fig. 14-2.

DC TACHOMETER

The crankshaft speed of the engine can be measured by generating voltage directly at the engine and then measuring this voltage by means of a voltmeter calibrated in terms of engine speed. Remember that when the field strength of a generator remains constant, the voltage output is proportional to the speed of rotation. The dc tachometer generator is a simple generator having a permanent-magnet field; hence, the voltage output of the generator is proportional to engine speed, and the rpm can be read on a voltmeter which has the dial marked in rpm instead of volts. Even though this system is obsolete, it is useful to illustrate electrical measurement of rpm.

Fig. 14-2 A percent-of-power indicator.

AC TACHOMETER

The dc tachometer requires a commutator and brushes, and these are subject to wear. The development of the ac tachometer eliminated this maintenance problem. The ac tachometer system in most common use consists of a three-phase ac generator (alternator) and the tachometer indicator which is driven by the alternator. The indicating mechanism consists of a mechanical tachometer driven by a three-phase synchronous motor as shown in Fig. 14-3.

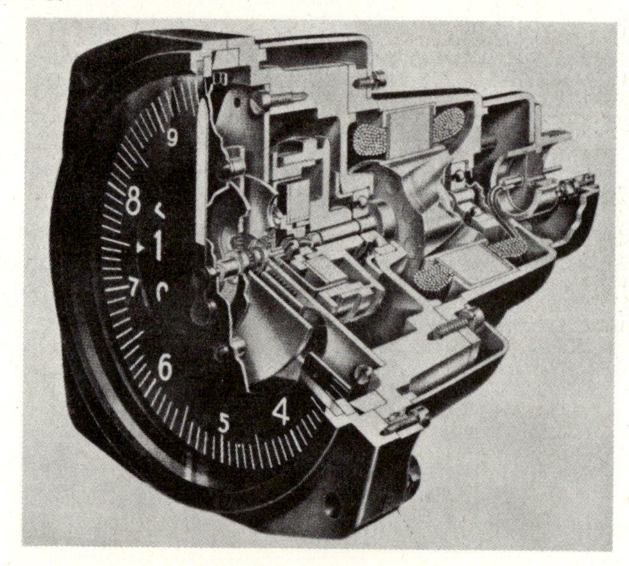

Fig. 14-3 An ac tachometer indicator.

An explanation of the alternator principle is given on pages 108 and 109 of this text. Briefly, the alternator, or transmitter, consists of a four-pole permanent magnet which rotates inside a three-phase stator; an external view of this unit is shown in Fig. 14-4. The stator of the alternator is connected by three wires to a similar stator in the synchronous motor, which operates the indicator.

A careful study of the diagram on page 159 and the explanation on pages 158 and 159 of this text will make clear the principle of operation of a synchronous motor. As the alternator is driven by the engine, the rotation of the permanent magnet induces a three-phase current in the stator. This current flows through the stator of the synchronous motor in the indicator and produces a rotat-

Fig. 14-4 Three-phase tachometer alternator.

ing field which turns at the same rate as the alternator rotor. The permanent-magnet rotor of the indicator keeps itself aligned with the rotating field and hence must also turn at the rate of the alternator rotor.

The synchronous motor in the indicator is directly coupled to a cylindrical permanent magnet which rotates inside a drag cup as shown in Fig. 14-5. As this magnet turns, it causes magnetic lines of force to drag through the metal cup and induce eddy currents in the metal. These eddy currents produce magnetic fields which oppose the field of the rotating magnet. The result is that as the speed of the rotating magnet increases, the drag or torque on the drag cup increases. The torque on the drag cup causes it to rotate against the force of the balancing hairspring and turn the pointer on the indicating dial. The distance through which the drag cup rotates is proportional to the speed of the synchronous motor; hence, it is also proportional to the engine speed.

TEMPERATURE INDICATORS

THERMOCOUPLE TEMPERATURE INDICATORS

Thermocouple temperature indicators are used most frequently when it is necessary to measure comparatively high temperatures. They are used in aircraft engines to measure cylinder-head temperature on reciprocating engines and tail-pipe or exhaust temperature on jet engines. The readings of these instruments provide information to the pilot or flight engineer, who can then operate the engine at its most efficient temperature and prevent damage to it from overheating. As previously explained in this text, thermocouple temperature indicators operate on the thermocouple principle. A simple circuit for such a system is shown in Fig. 14-6. The hot end consists of a junction of two dissimilar metals connected by dissimilar metal leads to the indicating instrument, which provides the cold end. The difference in temperature between the hot end and the cold end produces a current which is proportional to the difference.

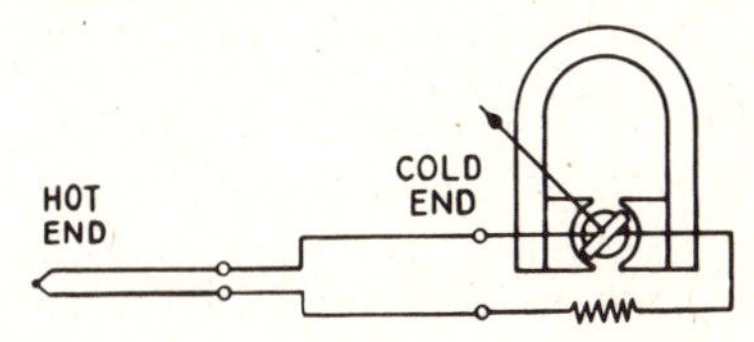

Fig. 14-6 Thermocouple system.

In order to obtain a temperature reading of the cylinder head or the tail pipe of a jet engine, the instrument must give the temperature reading of the cold end plus the difference in temperature between the cold end and the hot end. To accomplish this, the instrument (cold end or cold junction) must be continuously sensitive to its surrounding temperature and must add to the difference temperature the temperature at the instrument. This is accomplished by means of a bimetallic coil spring installed as shown in Fig. 14-7.

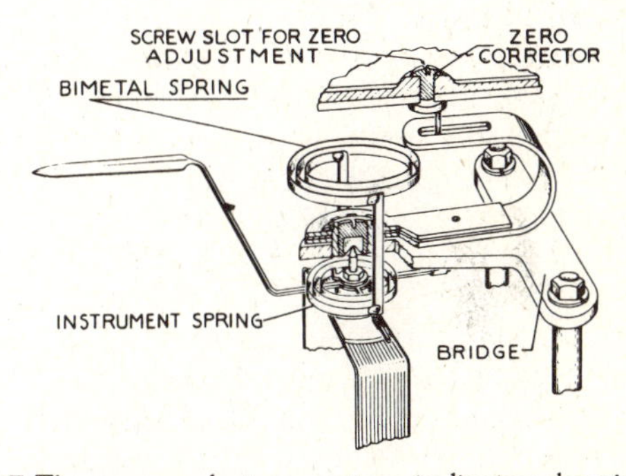

Fig. 14-7 Thermocouple temperature indicator showing bimetallic spring for temperature compensation.

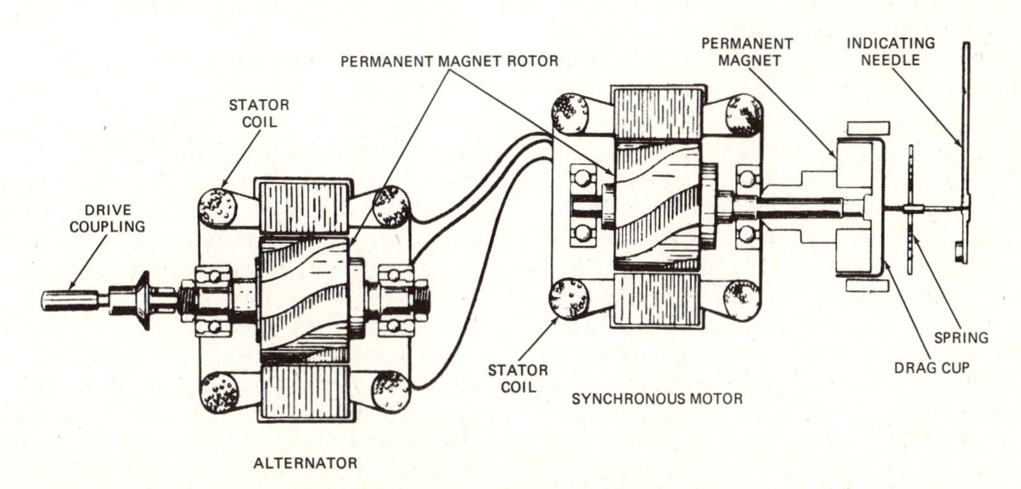

Fig. 14-5 Three-phase ac tachometer system.

It may be assumed that the scale for the indicator is uniformly divided and that a change in temperature of 100°F [55.6°C] is represented by a pointer deflection of 15 angular degrees on the scale. If the indicator, which is the cold junction, has a temperature of 0°F [−17.8°C] and the hot junction has a temperature of 600°F [315.6°C] the thermoelectric current will deflect the instrument pointer 6 × 15, or 90°.

On a hot day the instrument may have a temperature of 100°F. If the cylinder-head temperature at the hot junction is still 600°F, the difference in temperature between the two junctions is 500°F [260°C]. The pointer of the instrument now moves only 75 angular degrees, and the reading would be 500°F if some form of compensation were not provided as it is by the bimetallic spring. This spring causes the instrument to read its own temperature when it is not connected to the thermocouple leads. Hence, in the foregoing example, when the instrument temperature is 100°F, this 100°F will be registered on the instrument and the 500°F temperature difference between the junctions will be added to it to provide a correct reading of 600°F for the cylinder head.

Since the thermocouple instrument reads its own temperature when disconnected from the thermocouple leads, it may be checked for accuracy by comparing its reading with an accurate mercury thermometer placed adjacent to the instrument long enough to be sure that the instrument and the mercury thermometer are at the same temperature. The thermocouple leads **must be** disconnected to make this test. If the thermocouple instrument does not agree with the test instrument, it may be adjusted by means of the small adjusting screw on the face of the instrument. A typical cylinder-head temperature gage is shown in Fig. 14-8. An exhaust-temperature indicator for a jet engine is shown in Fig. 14-9.

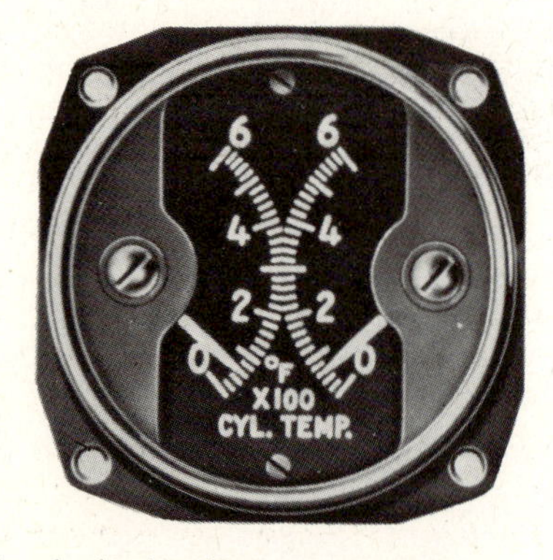

Fig. 14-8 A cylinder-head temperature gage.

THERMOCOUPLE LEADS

In accordance with Ohm's law, the emf generated in the thermocouple circuit produces a current inversely proportional to the resistance of the circuit at any given temperature difference between the hot junction and the cold junction. That is, if the resistance increases, the current will decrease; and if the resistance decreases, the current will increase. It is obvious that to obtain a constant current for a given temperature difference, the resistance of the circuit must also remain constant regardless of the temperature of the leads.

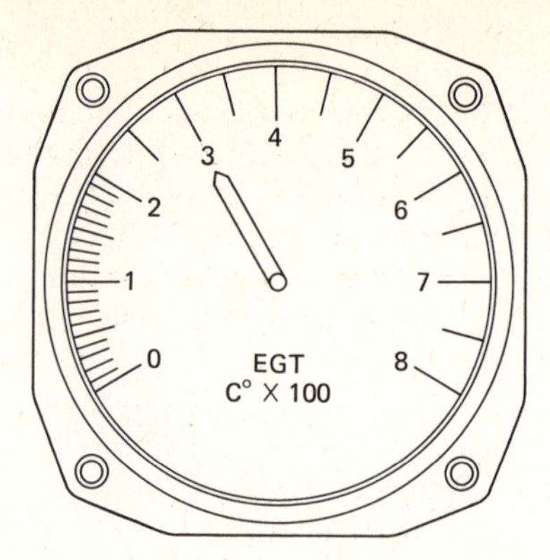

Fig. 14-9 An exhaust-gas temperature gage.

A combination of platinum and copper could be used in the leads, but this combination is not only too expensive but also subject to substantial changes in resistance. Instead, the thermocouple leads are usually made of either constantan and iron or constantan and copper. Constantan is a copper-nickel alloy which shows practically no change in resistance with changes in temperature. When constantan is used with either copper or iron, the thermoelectric effect is sufficient to operate the instrument even when the leads are quite long.

Thermocouple instruments used to measure jet tail-pipe temperature or exhaust gas temperature (EGT) utilize a thermocouple composed of Chromel and Alumel, which are special alloys manufactured by the Hoskins Manufacturing Company. These alloys are designed to operate at the high temperatures encountered in these applications. The thermocouple leads are made of the same alloys used in the thermocouple.

In order to compensate for the resistance changes which may occur as a result of temperature changes in the thermocouple leads, it is sometimes necessary to employ a **neutralizer** in the circuit. A neutralizer is a resistor unit made of a material which loses part of its resistance when temperature increases. It is designed with dimensions such that it loses as much resistance as the remainder of the circuit gains under any given temperature conditions; the total resistance of the circuit therefore remains constant.

When a thermocouple instrument is installed, it is essential that the correct leads be used. Standard thermocouple leads have resistances of 2 and 8 Ω, and the instrument must be provided with the type for which it is designed. Because of the very small amount of electric energy produced by a thermocouple, the electric connections must be clean and tight. Furthermore, it is absolutely essential that the leads are not crossed during installation. Iron must be connected to iron, constantan to constantan, copper to copper, Chromel to Chromel, Alumel to Alumel, etc. Usually, thermocouple leads are provided with connectors which make it impossible to connect them in reverse; however, it is always wise to examine the leads closely to make sure that the connections have been correctly made.

Since thermocouple leads are made with a specific resistance, they must never be cut or spliced. If there is extra length in the leads, they should be coiled up to take up the slack and secured.

Thermocouples are made in a variety of shapes to adapt to the specific installations for which they are designed. For example, the thermocouple for cylinder-head temperature is usually a washer which takes the place of the regular spark-plug washer. For the exhaust-gas temperature of a jet engine, the thermocouple is in the form of a probe installed to extend into the hot exhaust gases in the tail pipe. An EGT thermocouple is shown in Fig. 14-10.

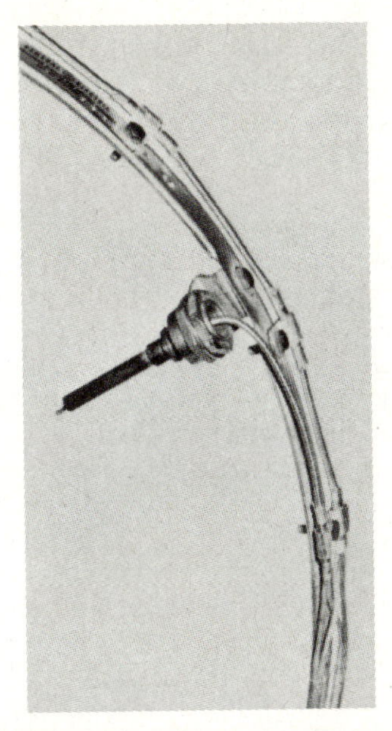

Fig. 14-10 An EGT thermocouple unit for a turbine engine.

RESISTANCE-TYPE TEMPERATURE INDICATORS

WHEATSTONE BRIDGE

When temperatures below 300°F [148.9°C] are to be measured, it is necessary to use an instrument different from the thermocouple type to obtain accurate indications. One of the common types of instrument used for these lower temperature indications is the **Wheatstone bridge** instrument. Among the temperatures it measures are those of **free air, carburetor air, cabin air,** and **coolant.**

It is a fundamental fact that the resistance of metals changes with changes in temperature. Also, since most electric instruments are basically devices for measuring current, changes in resistance may be converted into changes of current to obtain a pointer deflection across the scale of a direct-reading instrument. Resistance can be translated into current through the use of Ohm's law because $I = E/R$. Use of this law, however, assumes that the voltage is constant, a condition not always true when the voltage is supplied by a battery and generator system. Therefore, in order to translate resistance into current, the problem of the variable voltage must in some manner be solved. One solution is to use the Wheatstone bridge system.

The principle of the Wheatstone bridge can be explained by comparing it with a divided stream of water. Let us assume that a stream comes down from the mountains, meets a four-sided island, separates, and then unites on the other side of the island, as shown in Fig. 14-11. A canal is dug across the island, connecting a point in the middle of one branch of the stream with a similar point in the middle of the other. A paddle wheel is then mounted in the middle of this canal and a dam placed in each of the branches upstream from the canal and in each of the branches downstream from the canal. Each of these dams is provided with a floodgate to regulate the flow of water. By opening or closing these gates in the various dams, the water levels in the branches can be controlled.

Fig. 14-11 Water analogy for a Wheatstone bridge.

When the water level is the same at both ends of the canal across the island, the paddle wheel does not move because there is no water flow through the canal. If three gates are adjusted exactly alike, raising and lowering the fourth gate affects both the direction in which the water will flow through the canal and the speed of its flow, as indicated by the rotation of the paddle wheel. No matter which gate is being operated, however, the settings of the others must not be disturbed, or the paddle wheel will not give a reliable indication of the direction and velocity of the water flow.

The Wheatstone bridge circuit resembles the divided stream of the system just described. At some point in the electric circuit the current divides, flows through two branches, and then unites, just as the imaginary stream did. The dams with their floodgates in the branches of the stream are resistors in the electric circuit, and the canal is a conducting branch which includes a sensitive indicator (paddle wheel). We can regard the four resistors in the circuit as arms. Since three of them are to remain constant, they can be made of manganin wire, which is similar in properties to constantan, since it shows almost no change in resistance with changes in temperature. Any metal will be satisfactory for the fourth, since its resistance is to be measured and must therefore change with changes in temperature.

Before the instrument is connected in the bridge circuit, the pointer is set so that it rests at the **balance point.** This is the point on the scale at which the pointer rests when the entire system is connected to a source of current and the resistances are balanced so that no current is flowing through the instrument. Such conditions are described by saying that the bridge is in balance.

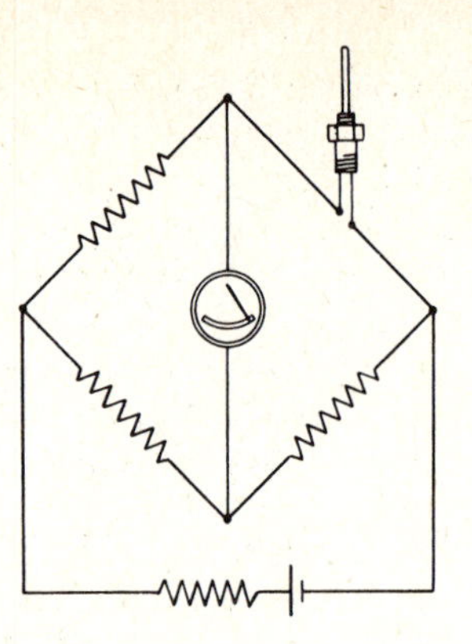

Fig. 14-12 A typical bridge circuit.

A typical bridge circuit is illustrated in Fig. 14-12. When the bridge is in balance, the resistances are equal to 100 Ω each. Three of the resistances remain at 100 Ω even though the temperature changes, but the fourth resistance (the temperature bulb) changes in value with the temperature. Let us assume that the balance point for this circuit is 82.4°F [28°C] and that the variable resistance has a value of 100 Ω at this temperature. Now, if the temperature of the variable resistance increases to 140°F [60°C], the resistance value will change to 112 Ω. Note that in Fig. 14-13, which is the same circuit as shown in

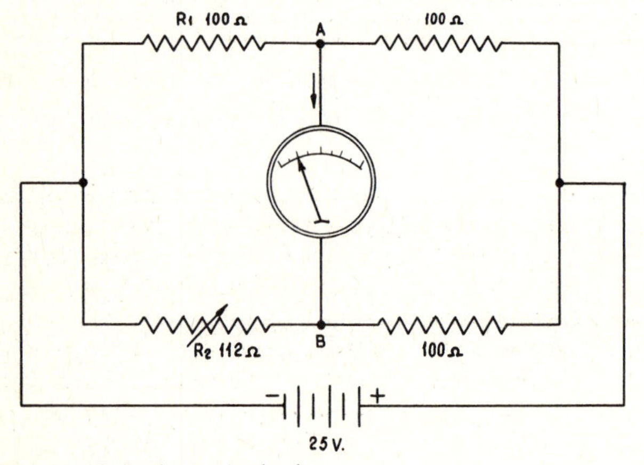

Fig. 14-13 Analysis of a bridge circuit.

Fig. 14-12, one side of the bridge will have a resistance of 200 Ω and the other side will have a resistance of 212 Ω. By Ohm's law we determine that when 25 V is connected across the bridge, the current through one side will be 0.125 A and the current through the other side about 0.118 A. Using these current values, we find that the voltage drop across the variable resistor will be about 13.1 V and that the voltage drop across the opposite resistor R_1 will be 12.5 V. It is apparent, then, that there will be a voltage difference of 0.6 V across the bridge between points *A* and *B*. Point *A* will be negative with respect to *B*; hence, current flow will be from *A* to *B* through the instrument.

In the foregoing example we have ignored the small effect of the instrument across *A* and *B*. Although this would cause some variation in the figures quoted, the results would be the same. Any unbalance in the circuit caused by a variation in the resistance value of the variable resistor will cause current to flow through the meter movement and register a temperature reading on the meter.

If the resistance of the indicating instrument in a Wheatstone bridge circuit is known, the formula for the solution of a resistance bridge may be used as explained in Chapter 1.

In practice, the fixed arms of the Wheatstone bridge are mounted in the case, as shown in Fig. 14-14. They may be wound on three spools, as shown, or on a single spool. The fourth or variable resistor is external because it is mounted at the location at which temperature is to be measured. Technically, the fourth resistor is called a **resistance bulb** (see Fig. 14-15).

Fig. 14-14 Interior of a Wheatstone-bridge instrument.

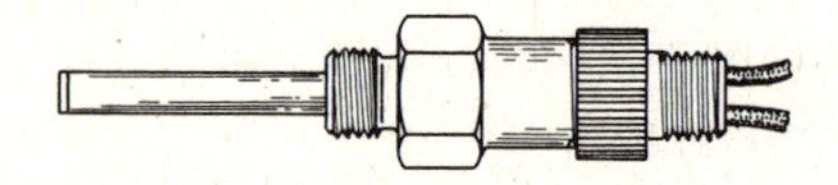

Fig. 14-15 Resistance bulb for temperature sensing.

The resistance wire, which is the essential feature of the resistance bulb, rests in the spiral grooves of an insulating material and is covered with a metal shield which conducts heat to and from it very quickly. This metal shield must be able to withstand the corroding influence of engine oils at high temperatures, the high flash temperatures in the carburetor of a backfiring engine, and the deteriorating influence of the atmosphere. Even though the resistance bulb is covered with a metal shell and substantial insulation, it responds to changes in temperature very rapidly. This sensitivity is important because the members of the flight crew are not interested in past temperatures; they want to know the situation at the exact second that the instrument is read.

The action of a resistance bulb may be understood by studying the graph in Fig. 14-16. It will be noted that the increase in resistance of a temperature bulb is almost linear with respect to temperature changes.

RATIOMETER TEMPERATURE GAGES

Since Wheatstone bridge instruments are sensitive to changes in supply voltage, it is apparent that the accuracy of such instruments is somewhat impaired in circuits in which there is any voltage variation. For this reason, instrument designers have developed a system which measures temperature by sensing the ratio between two currents. This is known as the **ratiometer principle** and provides reasonably accurate readings over a wide range of supply voltages.

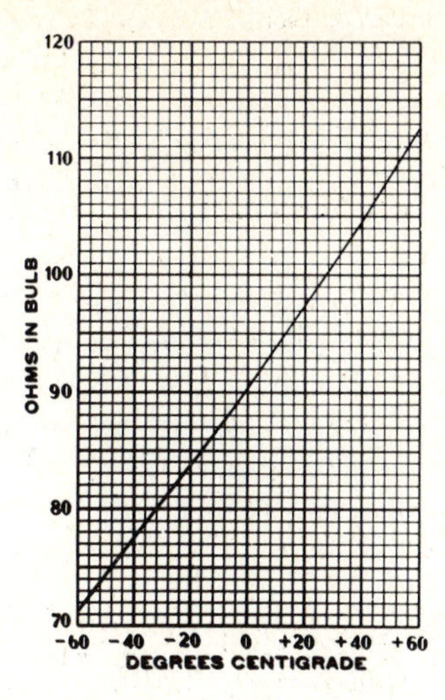

Fig. 14-16 Temperature-resistance curve for a typical resistance bulb.

Remember that a loop of current-carrying wire suspended in a magnetic field is acted upon by a force which attempts to rotate it; that is, the force produces torque on the loop. As the current increases in the loop, the torque increases. If the loop has two turns of wire instead of one, and the current and strength of the magnetic field remain the same, the torque will be twice as great as it is with only one turn. The torque is proportional to the current and also to the number of turns of wire through which the current flows. In other words, the torque is proportional to the product of the current and the number of turns of wire.

If the moving coil is between the poles of an unmagnetized electromagnet, there is no torque, regardless of the amount of current flowing through the winding and the number of turns of wire. If there is more magnetic flux in one place than in another, there is said to be more flux density where there are more magnetic lines of force. Torque is proportional not only to the current and the number of turns of wire through which the current flows, but also to the flux density.

Because magnetic flux can choose its path through the magnet and through the air gap between the poles and the core, it attempts to flow where there is as much iron and as little air as possible. If the air space between the core and the poles of a magnet is of equal width at all points, the magnetic flux is distributed uniformly, because all flux paths are of equal reluctance. This condition exists in many electric indicators, since they are designed so that the winding always intersects magnetic flux of the same density, regardless of the position assumed by the moving coil.

An entirely different situation prevails when the iron walls of the air gap are closer together in some places than they are in others. The magnetic flux then takes a short cut, and the lines of magnetic force crowd together at the narrow places. This crowding results in a nonuniform flux distribution, with the greatest flux density at the point where the air gap is the shortest, the density gradually decreasing as the distance from this point increases.

When a moving coil moves in a nonuniform field of flux density, the torque not only is a function of current and the number of turns of wire for a given constant magnetic field, but it also varies with the flux density. In this instance, the magnetic flux density varies from one position of the moving coil to another.

Fundamentally, the ratiometer consists of two moving coils mounted on a common shaft and rotating through nonuniform fields as shown in Fig. 14-17. Although the two coils are mounted rigidly on the same shaft, they are angularly displaced. If one is rotated, the other follows. Each coil revolves in a separate magnetic field so that as the field of one coil increases in density, the field of the other decreases. The coils are connected to two current sources of different polarity so that any torque produced in one coil opposes the torque produced in the other coil. Thus it is that the two coils will assume a position in which the torque of one balances the torque of the other. Very thin flexible conductors are used for the connections to the coils so that only magnetic forces will act upon them.

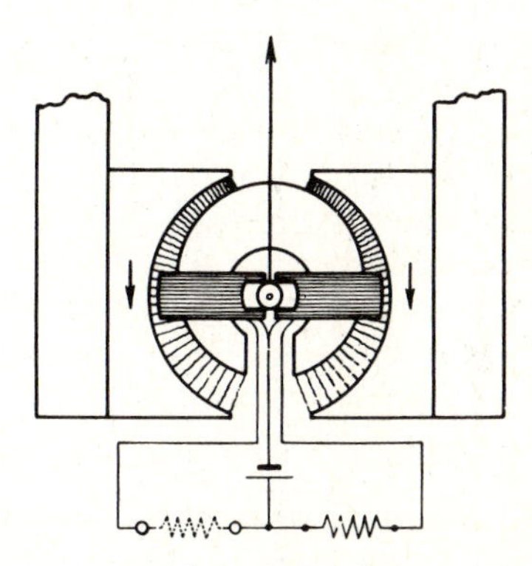

Fig. 14-7 Arrangement of moving coils in a ratiometer.

As mentioned previously, the two coils of the ratiometer come to rest in a position in which the product of ampere-turns and flux density is the same for both coils; that is, their torque values are balanced. The same situation can be presented in another manner by saying that for a given number of turns of wire in each coil and given dimensions of all parts of the system, the torque in each coil is a function of the current in its windings multiplied by the flux density in its part of the magnetic field.

If the ratio of the currents in the two coils is changed, the system will turn until a new position is reached in which the two products of current and flux density are again equal and balance is restored. The coil which now receives less current is compensated for its loss by greater flux density, and the coil which receives more current moves to a point of less flux density (see Fig. 14-18).

Each ratio of currents is associated with a definite ratio of flux densities in the two air gaps. Since these flux densities are permanently located along the air gaps, each current ratio turns the moving-coil system to a definite position which is different from those produced by other current ratios. A pointer attached to the system indicates the ratio of the two currents.

In reality a ratiometer does not look as simple as our description implies, but no matter how complicated its appearance, the principle of its operation is as we have stated it. A schematic circuit illustrating how a resistance bulb is connected in a ratiometer circuit is shown in Fig. 14-19. Note that the voltage furnished by a battery is divided between the circuits of the two coils by the fixed

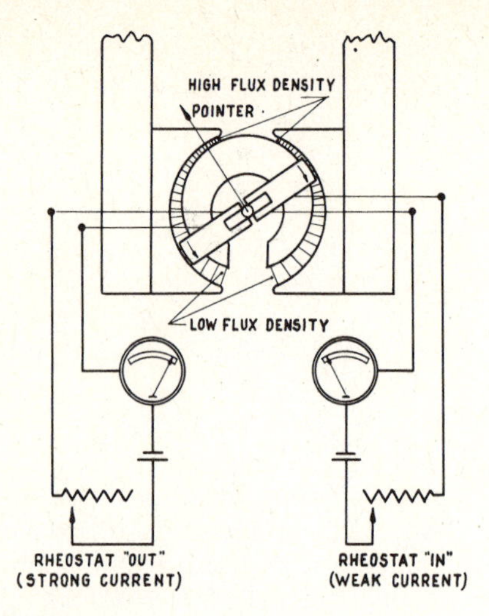

Fig. 14-18 Operation of a ratiometer circuit.

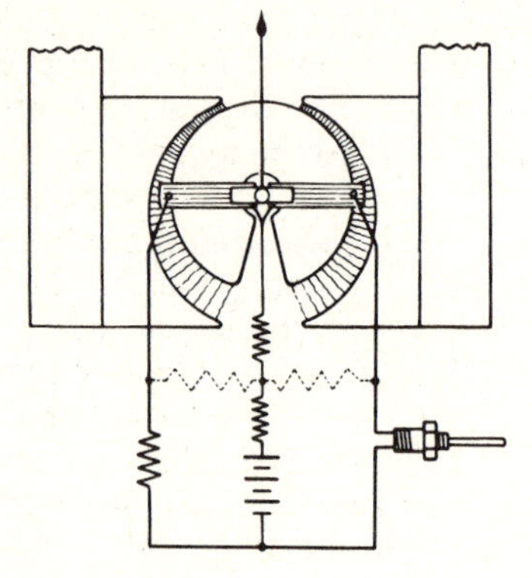

Fig. 14-19 Resistance bulb connected in a ratiometer circuit.

resistor in one side and the resistance bulb in the other. The series and shunt resistances shown are for the purpose of compensation and adjustment. It is obvious from the circuit that the current through the two sides of the circuit will be equal only when the resistance of the temperature bulb is equal to the resistance of the fixed resistor. At this point the moving coils assume positions in fields of equal flux density, as shown. Any change in the resistance of the resistance bulb will cause the ratio of the currents to change and the coils to shift position.

Ratiometer thermometers may be used for a variety of temperature indications, among which the most common are those of carburetor air, inlet air in a jet engine, cabin air, free air, and engine oil. A typical ratiometer indicator is shown in Fig. 14-20.

Fig. 14-20 Ratiometer temperature indicator.

SYNCHRO SYSTEMS

A synchro system is designed to measure an angular deflection at one point and reproduce this same deflection at a remote point. Synchro systems have been designed to employ both alternating current and direct current for power, but the present trend is to employ 400-Hz alternating current. These systems are used as position indicators, for remote indicating systems in radar systems, for autopilots, and for a wide variety of other remote-control and indicating systems used in aircraft, missiles, and spacecraft. Synchros have been designed and built under a variety of names, the most common being Selsyn (a General Electric trade name) and Autosyn (a Bendix trade name).

THE DC SELSYN

One of the early synchro systems developed by the General Electric Company was the dc Selsyn, often used as an indicator on aircraft with dc power systems to show the position of wing flaps and landing gear. The dc Selsyn instrument system consists of an indicator and a transmitter operating in synchronism; hence, the name *Selsyn* (self-synchronous).

The Selsyn indicator may display one or more indications, depending upon the needs of the airplane instrument system and the wishes of its designer. In some cases the indicator registers separate positions for the wing flaps, the nose gear, the left gear, and the right gear. This type of indicator, of course, requires four separate transmitters, one located in the vicinity of each part of the airplane from which a position indication is desired.

A schematic diagram of a single dc Selsyn system is shown in Fig. 14-21. The transmitter is merely a winding

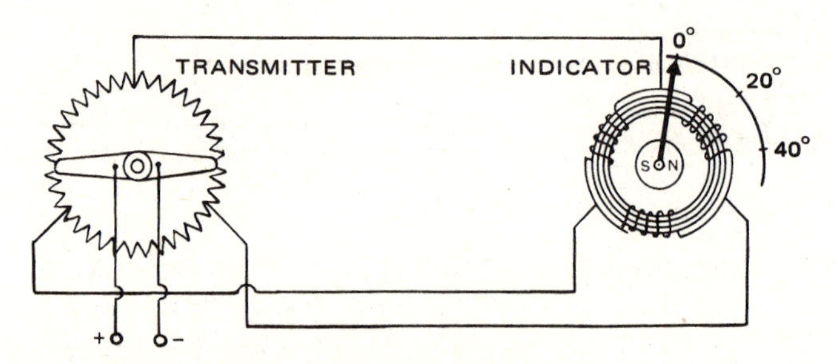

Fig. 14-21 Schematic diagram of a dc Selsyn circuit.

of fine resistance wire on a circular form with connections located at three equally spaced points around the winding. Dc power is fed to the ring winding at points 180° apart by means of wiper arms (see diagram). This arrangement is actually a special type of potentiometer, and when the wiper arms are rotated, the voltages appearing at the three connections will change with respect to one another. As shown in the diagram, the three connections to the transmitter are connected to three similar connections at the indicator. The indicator element consists of a laminated ring of ferromagnetic material on which three windings are equally spaced in a delta connection. When this unit is connected to the transmitter by means of three conductors as shown in the diagram, rotation of the wiper arms in the transmitter will vary the currents in the coils of the indicator in such a manner that the magnetic field of these coils will rotate also. The rotating element of the indicator is a permanent-magnet armature mounted on bearings so that it is free to turn with the rotation of the field. Thus, the indicating needle attached to the rotor shaft will follow the movement of the wiper arms at the transmitter. When the transmitter is linked to the flap-actuating mechanism, it will produce a signal which causes the indicator to show the position of the flaps. In like manner, it can be used to show the position of the landing gear or any other unit which moves through a range of various positions.

AUTOSYN INSTRUMENTS

The word *Autosyn* is the trade name applied by the Eclipse Pioneer Division of the Bendix Corporation to a system of self-synchronizing remote indicating instruments operated on alternating current. As with the Selsyn instruments, the Autosyn system activates indicators in the cockpit without using excessively long mechanical linkages or tubing. The indication is picked up by the transmitter near the engine, or at some other remote point, and is sent by electrical means to the indicator in the cockpit. This system has great value when used in airliners and other large aircraft.

An Autosyn synchro has the appearance of a small synchronous motor. In the Autosyn system, one synchro is employed as a transmitter and another as an indicator.

A schematic diagram of an Autosyn system is shown in Fig. 14-22. The system is basically an adaptation of the

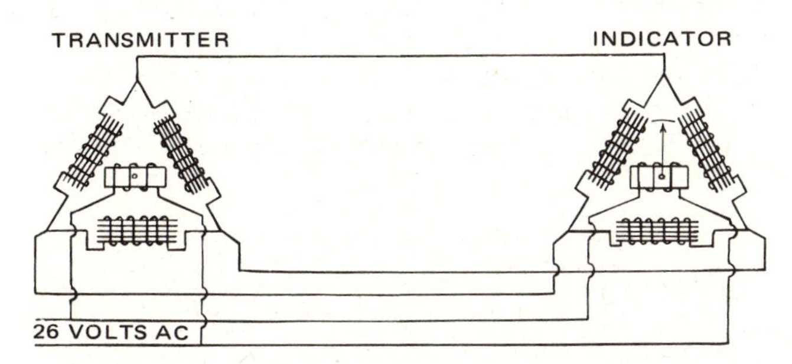

Fig. 14-22 Schematic diagram of an Autosyn synchro system.

self-synchronous motor principle, whereby two widely separated motors operate in exact synchronism; that is, the rotor of one motor spins at the same speed as the rotor of the other. When this principle is applied to the Autosyn system, however, the rotors neither spin nor produce power. Instead, the rotors of the two connected Autosyn units come into coincidence when they are energized by an alternating electric current, and thereafter the rotor of the first Autosyn moves only the distance necessary to match any movement of the rotor of a second Autosyn, no matter how slight that movement may be.

It must be understood that the transmitter and indicator of Autosyn units are essentially alike, both in electrical characteristics and in construction. Each has a rotor and a stator. When ac power is applied and a rotor is energized, the transformer action between the rotor and stator causes three distinct voltages to be induced in the rotor relative to the stator. For each tiny change in the position of the rotor, a new and completely different combination of three voltages is induced.

When two Autosyns are connected as shown in the diagram and the rotors of both units occupy exactly the same positions relative to their respective stators, both sets of induced voltages are equal and opposite. For this reason, no current flows in the interconnected leads, with the result that both rotors remain stationary. On the other hand, when the two rotors do not coincide in position, the combination of voltages of one stator is not like that of the other, and rotation takes place, continuing until the rotors are in identical positions. The induced voltages are then equal and opposite, and so there is no current flow in any of the three conductors; hence, the rotors will be in stationary and identical positions.

The Autosyn system may be used for a wide variety of indications on an airplane. Among these are manifold pressure, oil pressure, rpm (tachometer), remote compass indication, percent of power, and fuel pressure. Figure 14-23 shows exploded and cutaway views of the Autosyn transmitter for indicating oil pressure. The rear portion (not shown) is a standard bourdontube pressure-gage movement. Instead of the actuating center staff's being connected to an indicating needle, it is used to turn the rotor of an Autosyn unit. The position of the rotor is then duplicated in the indicating unit, in which the rotor is directly coupled to an indicating needle.

The power supply for an Autosyn system is usually 26 V alternating current at 400 Hz and is supplied to the rotors of both Autosyn units, as shown in Fig. 14-22.

Two functions may be shown on the same dial by means of a tandem indicator which has two synchro units, one behind the other. The shaft of the rear unit extends through a hollow shaft of the front unit, making it possible to obtain indications of any two similar functions at the same time on the same dial. When this plan is employed, a separate transmitter must be connected to the indicator for each indication to be shown. The purpose of dual instruments is to save instrument-panel space and weight and also to reduce the number of instruments which must be observed by the flight crew.

DIRECTION INDICATORS

Aircraft instrument systems utilize a number of different types of direction-indicating instruments, some of which are associated with electronic navigation systems. Among these are the magnetic compass, the magnetic direction indicator, the automatic direction finder (ADF), the radio magnetic compass (RMI), the horizontal situation indicator (HSI), and the flight director indicator (FDI).

The magnetic compass and the magnetic direction indicator are not strictly electric instruments, although they are controlled by magnetism. With other direction indicators, a remote device senses direction from the earth's magnetic field and transmits the indication to an instrument in the cockpit.

THE MAGNETIC COMPASS

Magnetic compasses are generally used to indicate the heading of the vehicle in which they are installed and to provide a means for determining position by cross bearings. In an airplane, the principal use of the magnetic compass is to indicate the direction in which the airplane is heading when flying a straight and level course, but it does not indicate correctly when turns are made at velocities which require any appreciable banking of the air-

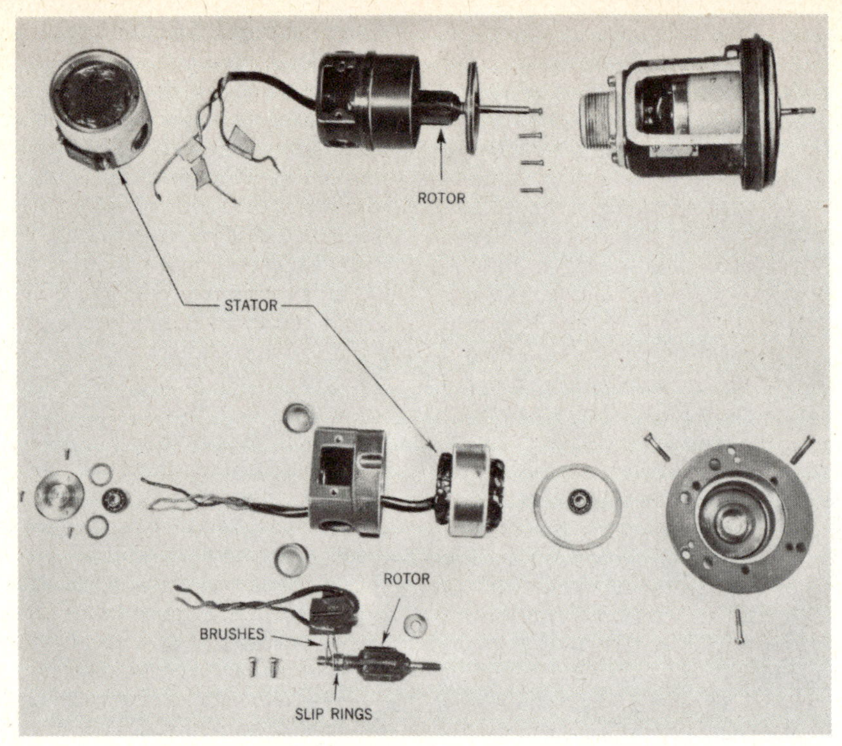

Fig. 14-23 Exploded views of Autosyn synchro units.

plane. Errors of this nature increase with an increase in the banking angle until the compass becomes worthless when the angle of bank is about 20°. These statements apply to all compasses which obtain their directive effort from the earth's magnetic field and their stabilization from the force of gravity.

The magnetic compass indicates direction with respect to the north magnetic pole; hence, to find true north a correction must be made. A compass is essentially a magnetized-needle arrangement mounted in a card element and suspended on a pivot in a metal bowl filled with a liquid. This construction permits the card to align itself with the magnetic lines of force surrounding the earth. The indications of the card are visible through a glass window in the side of the bowl, and the card is graduated in degrees of a circle. The cardinal headings, that is, north, east, south, and west, are shown by means of large letters on the card. North is indicated as 0 or 360°, east as 90°, south as 180°, and west as 270°. The fixed *lubber's line* is a vertical white line visible on the front of the instrument window, as shown in Fig. 14-24.

The interior of a typical compass is shown in Fig. 14-25. An expansion chamber is built into the compass to provide for the expansion and contraction of the liquid which result from changes of altitude and temperature. The liquid (*D* in Fig. 14-25) *damps* the oscillations of the card, thus preventing the card from turning violently and vibrating. An individual lighting system is usually provided in the compass for the illumination of the card during darkness.

In the illustration of Fig. 14-25, various parts of the compass assembly are indicated by letters. The magnets that rotate the card (*C*) are indicated by *A*. An air dome (*B*) keeps the card level. The jeweled bearing and pivot are shown at *E*. A diaphragm (*G*) separates the expansion chamber (*F*) from the air chamber. Compensating magnets for error correction are in the case at *H*. Illumination is provided by the lamp at *J*.

The earth acts as a great magnet with its magnetic north pole located near the north geographic pole and its mag-

Fig. 14-24 A magnetic compass.

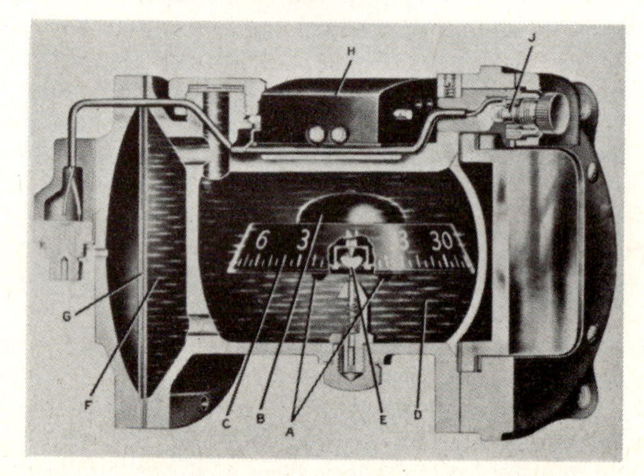

Fig. 14-25 Interior of a typical magnetic compass.

netic south pole near the geographic south pole. At the present time, the magnetic north pole is at latitude 70°N and longitude 96°W. The magnetic south pole is at latitude 73°S and longitude 156°E. These magnetic poles are not actual points on the earth, but centers of magnetic fields which slowly and constantly shift their positions.

It must be pointed out that the magnetic poles mentioned in the foregoing paragraph are not north and south poles in the conventional sense applied to permanent magnets or electromagnets. The north pole near the north geographic pole of the earth has the same polarity as the south pole of a common magnet. This was explained in a previous section, but is repeated here to avoid confusion. The north pole of a conventional compass needle is a north-seeking pole and is attracted to the magnetic pole near the north geographic pole of the earth. It follows, therefore, that the magnetic pole near the north pole of the earth must be a south-seeking or, simply, a south pole. For this reason, if a common bar magnet is suspended so that it is free to turn in the earth's magnetic field, the N, or north, end will point toward the magnetic pole near the geographic north pole.

The geographic north pole is called the *true north* and is so noted on all maps and charts used for ocean and air navigation. It is established by the axis of the earth's rotation. The magnetic north pole is about 800 mi [1287 km] from true north at the present time. All north-seeking needles, unless disturbed by local influences, point toward the magnetic north pole.

As previously explained, if a magnetized bar of steel (bar magnet) is so suspended that it can turn freely in any direction about its center of gravity and is not disturbed by surrounding magnetic influences, it will assume a position with one end pointing toward the north magnetic pole and the other pointing toward the south magnetic pole. The ends of the magnet are therefore known as the north-seeking, or N, end, and the south-seeking, or S, end, respectively. The magnetic force acting on the N end is equal and opposite to the magnetic force acting on the S end; hence, the magnetic effect on the N end is the only one considered in discussing a magnetic compass. Since the position assumed by a freely suspended bar magnet shows the direction of the magnetic force, that is, toward the magnetic north, the magnetic compass may be regarded as a direction-indicating instrument.

There are four principal causes of compass inaccuracies. These are (1) incorrect installation, (2) vibration, (3) magnetic disturbances, and (4) northerly turning error. **Deviation** and **variation** must be principally considered when a compass is used in an airplane or other vehicle.

Deviation is the angular difference, measured at the airplane, between the compass north and the magnetic north, assuming that the compass north is the direction shown by the compass needle or card. While an airplane is being built, the vibration and shock imparted to the various parts while they are being machined, forged, or fitted cause a certain amount of permanent magnetism to be induced in the airplane body by the earth's magnetic field. When the airplane is accepted and flown, this permanent magnetism acquired in various parts during construction is varied by the vibrations and shock received from landing, taking off, engine rotation, and other forces, with the result that the local magnetism of the airplane causes the compass to deviate from the magnetic north. There are also electric currents flowing in the airplane's electrical system and radio equipment which cause still more deviation. Furthermore, there are large steel masses in the engine and various parts of the airframes, for example, the landing gear, which cause additional disturbances. There are two general types of compensating devices for correcting the usual aircraft magnetic-compass deviation—the loose-magnet type and the screw type.

In the loose-magnet type of compensation, small cylindrical compensating magnets are placed in holes in a small drawer provided in the top of the compass. One series of holes extends in the airplane's fore-and-aft direction, and another series extends in a direction perpendicular to the longitudinal axis of the airplane. Compensating magnets of sufficient magnetic intensity are placed in either direction, or in both directions, to counteract any pull which the magnetized parts of the airplane may exert upon the compass.

The screw-type compensator has two screw-adjustable sets of magnets mounted on small rotatable pivots. The two adjusting screws are made accessible by removing a panel in front, and near the top, of the compass case. One screw is marked N-S for north-south and the other E-W for east-west. Error can be avoided by turning one or both screws with a nonmagnetic screwdriver.

After the compass has been compensated to eliminate the deviation error as much as possible, any remaining errors are noted on a correction card mounted near the compass in full view of the pilot or copilot, who takes the corrections so listed into consideration while flying his course.

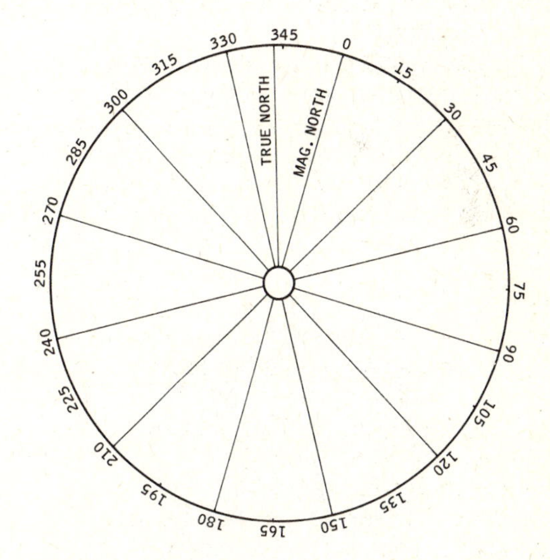

Fig. 14-26 A compass rose.

The process employed for determining deviation and making corrections in a compass is called *swinging the compass.* To swing a compass, it is necessary to use a compass *rose,* which is a circle marked on pavement with diameter lines marked for magnetic directions every 30° (see Fig. 14-26). The airplane is placed in a horizontal position (level flight) on the pavement and aligned with each direction line on the compass rose. This may be done by using plumb bobs suspended from the centerline of the airplane at the nose and near the tail. The engine of the airplane must be running, and all normal electric equipment must be operating when the compass readings are taken. A reading is taken for each 15° of the complete circle, and a deviation card is made showing the compass reading for each direction.

Variation is the angular difference, measured at the airplane, between magnetic north and true north. The amount of variation error affecting a compass may range from 0 to about 40°, either east or west, depending upon the geographical position of the airplane. Near the polar regions the variation can be as much as 180°, and the compass will probably fluctuate violently. The variation of a compass depends not only upon the geographical position but also upon the constant, progressive annual change which may not amount to much in any one year but changes appreciably over a period of many years.

All navigation charts show the amount of variation for different geographical positions on the earth by lines connecting points of equal magnetic variation, called **isogonic lines.** The line connecting all points of zero variation, found where the position of the compass is in line with the true and magnetic poles, is called the **agonic line.** Air navigation charts also show the amount of annual change of variation and the date on which the chart was prepared, so that anyone reading it can make allowances for changes since the chart was printed.

A magnetic compass should be installed where it may be read easily and where it is as far as possible from all sources of magnetic disturbance. Dc wires in the vicinity should be twisted so that they will not set up a magnetic field that will affect the compass. The lubber's line and the card pivot should be vertical and aligned with the fore-and-aft axis of the aircraft so the indication under the lubber line will show the magnetic heading of the aircraft.

VERTICAL-DIAL MAGNETIC DIRECTION INDICATOR

The vertical-dial direction indicator is actually a direct-reading magnetic compass; instead of the direction's being read from the swinging compass card, however, the reading is taken from a vertical dial as shown in Fig 14-27.

Fig. 14-27 Dial of a magnetic direction indicator.

The reference index on the vertical dial can be set at any desired heading by turning the knob at the bottom of the dial, and it is necessary only to match the indicator needle with the reference pointer to hold a desired course. The design of the dial provides easy reading of direction and also quick and positive indication of deviation from a selected heading. The compass liquid in the instrument is contained in a separate chamber, and the dial, instead of floating in the liquid as in other types of magnetic compasses, is completely dry. This makes the indicator easier to read and eliminates fluid leakage around the dial.

An important feature of the vertical-dial direction indicator is its stability in comparison to the conventional compass. Both the period and the swing of this indicator are less than half that of the ordinary magnetic compass.

The construction of the vertical-dial direction indicator is shown in Fig. 14-28. The float assembly (20), which contains the directive magnet (27), is located in a fluid-filled bowl. This assembly rides on a steel pivot (23), which in turn rests on a stud containing a cup jewel (24). The jewel stud rides in a guide and is supported on a spring (26) which absorbs external vibration. A change in the position of the instrument in relation to the directional magnet transmits this motion to the indicator needle (8) through the follower magnets (18), the bevel gears (16), and the horizontal bevel-gear shaft (11) onto which the indicator needle is pressed. The follower magnets and bevel gears are all contained in the housing (17). Both bevel-gear shafts are carefully balanced and ride on steel pivots in jeweled bearings.

A compensation system [polycompensator (13)] is mounted directly over the bevel-gear housing. The compensation-magnet gear assemblies are turned through gear trains to two slotted adjustment stem pinions (2 and 3). The reference lines (7) are set through a gear train by means of the adjustment knob (1).

As temperature varies, expansion of the compass liquid surrounding the float assembly takes place through a small hole in the baffle plate, and the level of the liquid in the expansion chamber above the baffle plate will rise and fall with changes in temperature. The normal level of the liquid is about even with the hole, which is used to fill the chamber.

The indicator is equipped with rim lighting which utilizes a special diffusing lens at the lamp socket to distribute the light evenly over the entire dial. Three-volt lamps are used and are accessible from the front to facilitate replacement.

REMOTE INDICATING COMPASS SYSTEMS

Because of the various errors peculiar to cockpit-mounted magnetic compasses, a number of remote indicating systems have been developed. By means of the gyrostabilization of sensing elements, the placing of sensing elements at a point free from local magnetic disturbances, and other refinements, most of the errors common to the magnetic compass have been eliminated or greatly reduced.

MAGNESYN COMPASS

The magnesyn compass may be described as a magnetic compass that is remotely located and whose indication is transmitted to the cockpit by means of a synchro system. The magnetic compass is located near the tip of the wing so it will be essentially free of electrical influences within the airplane. The transmitting synchro system operates as described previously in this section. The compass unit in the wing includes a transmitting synchro, and the indicator in the cockpit includes the receiving synchro.

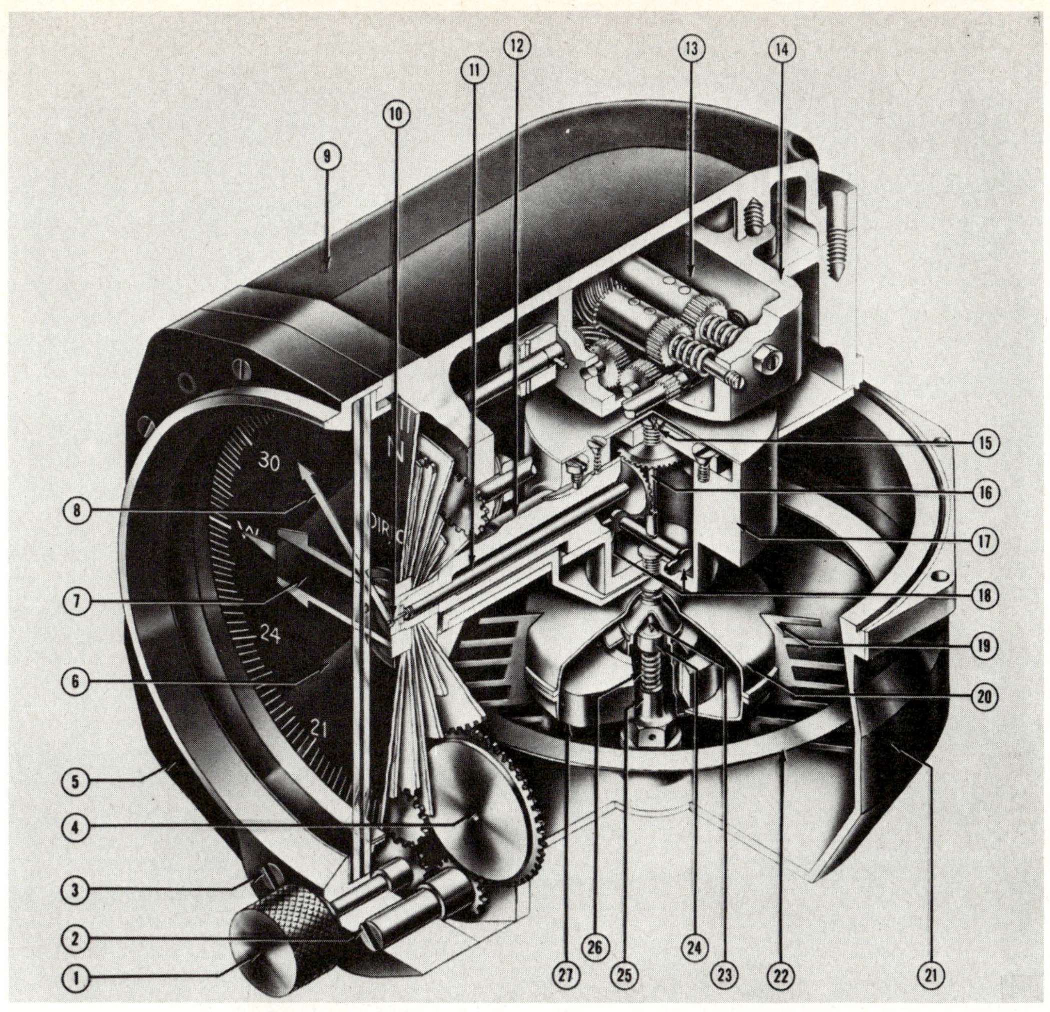

Fig. 14-28 Construction of a vertical-dial, magnetic direction indicator.

FLUX-GATE COMPASS SYSTEM

The Gyro Flux Gate compass system is one type of remote-indicating earth-inductor compass system. Its advantage is that it is comparatively free of the disadvantages of the standard magnetic compass. It consists of a flux gate (flux valve) transmitter, master direction indicator, amplifier, junction box, caging switch, and one or more compass repeaters. See Fig. 14-29.

The **flux gate** is a special three-section transformer which develops a signal whose characteristics are determined by the position of the unit with respect to the earth's magnetic field. The flux gate element consists of three highly permeable cores arranged in the form of an equilateral triangle with a primary and secondary winding on each core. The primary winding of the flux gate is energized by a single-phase 487.5-Hz 1.5-V power supply from an oscillator. This current saturates the three sections of the core twice each cycle. During the time that the core is not saturated, the earth's magnetic flux can enter the core and affect the induction of a voltage in the secondary. The effect of the primary excitation is to *gate* the earth's magnetic flux into and out of the core. This cycle occurs twice during each cycle of the excitation current; hence, the voltage induced in the secondary windings of the flux gate has twice the frequency of the primary current, because the core is saturated twice each cycle and has no excitation flux twice each cycle. The resulting voltage in the secondary windings has a frequency of 975 Hz.

The 975-Hz secondary signal is developed in the flux gate as illustrated in Fig. 14-30. At point 1 in the time cycle, the excitation voltage is zero, the earth's magnetic flux in the core of the flux gate is maximum, and the induced signal in the secondary is zero. There is no induced secondary signal because there is no flux change in the core at this instant. Between points 1 and 2, excitation voltage in the primary increases from zero to maximum. This saturates the core and causes the earth flux in the core to change from maximum to minimum, thus inducing a voltage in the secondary which rises to a maximum (*A*) in one direction and decreases to zero when there is no further change at point 2.

From points 2 to 3 the earth flux increases to maximum; hence, the secondary voltage at *B* is in a direction opposite that developed at *A*. Thus we see that there are two cycles of induced voltage in the secondary for each one cycle of primary excitation voltage.

The ratios of the output voltages in the three secondary coils are determined by the position of the flux gate element with respect to the earth's magnetic field, with only one possible position for any combination of voltages.

In the **transmitter**, shown in Fig. 14-31, the flux gate is held in a horizontal position by means of a gyro. This is

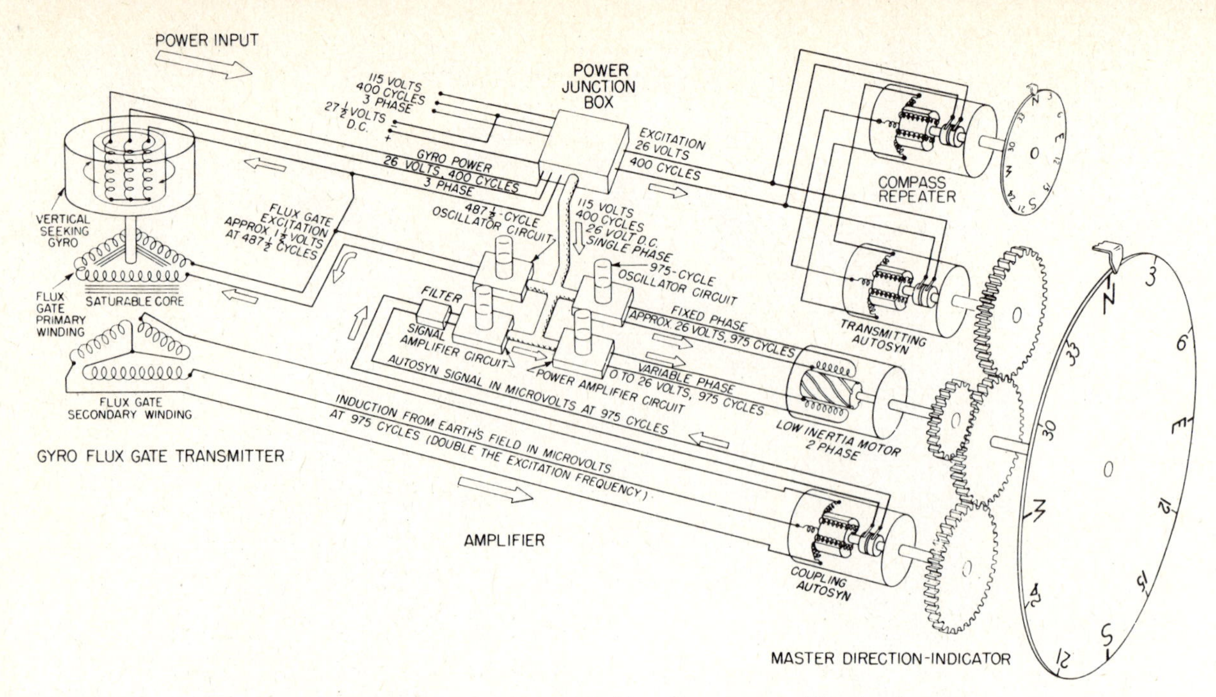

Fig. 14-29 Diagram of the Gyro Flux Gate system. (Bendix Corp.)

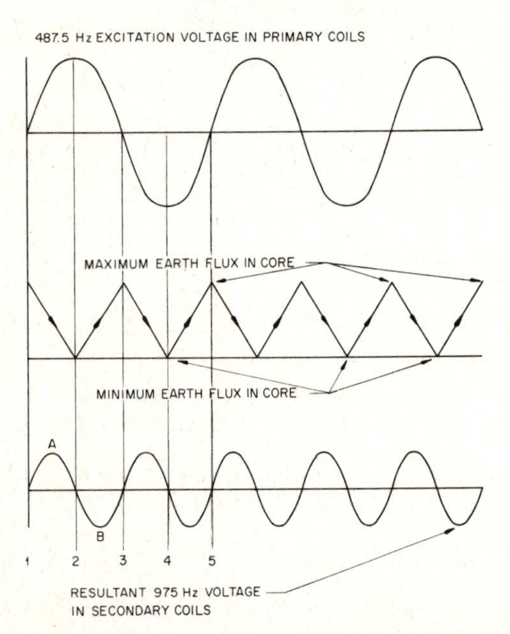

Fig. 14-30 Development of the signal in a flux gate. (Bendix Corp.)

Fig. 14-31 Flux gate transmitter. (Bendix Corp.)

necessary to provide a uniform signal from the flux gate element. If the gyro should be *tumbled,* that is, moved out of the horizontal position, it can be erected by means of the **caging switch.**

The 975-Hz output of the secondary windings of the flux gate is connected by three leads to the stator of a **coupling Autosyn** in the **master direction indicator.** The currents in these leads set up a magnetic field in the stator of the coupling Autosyn similar in direction and strength to that appearing in the flux gate element. This field shifts its position whenever any change occurs in the stator voltages as a result of a change in the heading of the aircraft.

The rotor of the coupling Autosyn normally lies in a null (neutral) position; in this position no signal is induced in it by the stator field. When the stator field shifts is a result of a change in aircraft heading, the rotor will no longer be in the null position, and a signal will be induced in the rotor winding. This signal is fed to the amplifier and then to the variable phase of a low-inertia two-phase motor in the master direction indicator. This causes a torque to be produced, and the motor turns in a direction determined by the phase of the current in the variable-phase winding. The motor shaft turns the indicator dial to provide a visual indication of the magnetic heading of the aircraft, and at the same time, it drives the rotor of the coupling Autosyn to a new null position. This reduces the rotor signal to zero, and the low-inertia motor stops turning until another change in the heading of the aircraft causes a new signal to be induced in the rotor of the coupling Autosyn.

The **master direction indicator,** shown in Fig. 14-32, is the indicating unit for the Gyro Flux Gate system. It contains the coupling Autosyn, the low-inertia two-phase motor, a transmitting Autosyn, and a course Autosyn synchro to supply a signal for the operation of the automatic pilot.

The **compass repeater** (Fig. 14-33) contains a single Autosyn which receives its signal from the transmitting

Fig. 14-32 Master direction indicator. (Bendix Corp.)

Fig. 14-33 Compass repeater. (Bendix Corp.)

Autosyn in the master direction indicator. The dial of the repeater shows the four cardinal headings as well as intermediate headings marked every 5° and numbered every 30°.

When it is desired to provide direction information at more than one location, the compass repeater is used. The repeater is a single Autosyn receiver having an indicating dial attached to the rotor shaft. As stated above, the repeater receives signals from the Autosyn transmitter in the master direction indicator and responds by reproducing the heading information on its dial.

The **amplifier,** shown in Fig. 14-34, serves to increase the power of the magnetic direction signal from the coupling Autosyn in the master direction indicator. It also furnishes a 487.5-Hz power supply for the excitation of

Fig. 14-34 Flux gate amplifier. (Bendix Corp.)

the primary windings of the flux gate transmitter and a 975-Hz power supply for the excitation of the fixed phase of the master-direction-indicator motor.

The caging switch shown in Fig. 14-35 is used to control the cage-uncage cycle of the gyro in the flux gate transmitter. The gyro is said to be *caged* when it is held in a fixed position by mechanical means; it is *uncaged* when it is allowed complete freedom of motion.

Fig. 14-35 Caging switch. (Bendix Corp.)

The need for caging a gyro may be easily understood. The gyro in the flux gate transmitter must spin on an axis perpendicular to the earth's surface to keep the flux gate in the correct position with respect to the earth's flux. Since the gyro is free to move away from the vertical axis with respect to the airplane in which it is installed, there are many times when it is not in the correct position for operation. At these times it is necessary to erect the gyro by means of the caging switch.

To erect the gyro, the MOMENTARY-CONTACT-SWITCH button on the switch box is depressed and released. This starts the cage-uncage cycle, which then continues automatically until the cycle is complete. An indicator light on the switch box comes on when the gyro is caged.

ELECTRONIC DIRECTION INDICATORS

In addition to the magnetic direction indicators described in the foregoing section, various types of direction indicators are associated with electronic systems. Among these are the radio magnetic indicator (RMI), the automatic direction finder (ADF), and the horizontal situation indicator (HSI). These and other direction indicators are discussed in the chapter on communication and navigation systems.

FUEL-QUANTITY INDICATORS

Fuel-quantity indicators for many general aircraft and all large, commercial aircraft are either electrically or electronically operated. The electrically operated indicating systems are usually of the variable-resistance type, and the electronically operated systems are of the capacitor type.

Electric fuel-quantity indicators utilize a variable resistance in the tank unit or sensor. The resistance is in the form of a rheostat or potentiometer, depending upon the method of indication. The fuel-quantity signal is provided as a float arm in the tank changes the resistance of the sensor as fuel level changes. A schematic diagram of a fuel-quantity-indicating system with a rheostat-type sensor is shown in Fig. 14-36. The variable resistance of the tank unit is connected in a bridge circuit with reference resistors in the indicator case. Refer to the section describing Wheatstone bridges in this text.

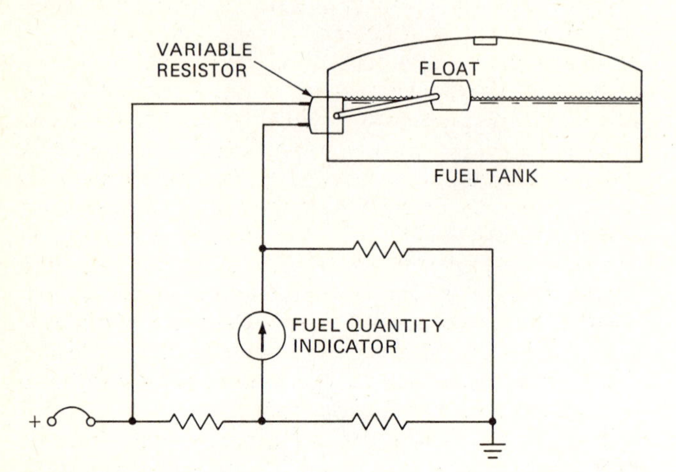

Fig. 14-36 Float-type fuel level indicator with variable resistor.

Electronic or capacitor-type fuel indicators utilize a variable capacitor as the sensor unit in the fuel tank. The capacitance of the sensor in the tank is changed as fuel rises and falls between the two electrodes (plates) in the tank unit. The change of capacitance is due to the difference between the dielectric strength of the fuel compared with that of air. Since fuel has a dielectric strength more than twice that of air, the capacitance of the sensor increases in accordance with the amount of fuel in the tank. The change in capacitance of the sensor unit is utilized in a bridge circuit to provide an amplified signal that rotates the pointer or actuates the digital display in the indicator. A schematic diagram to indicate the operation of a capacitance-type fuel-quantity-indicating system is shown in Fig. 14-37.

In an actual system in an aircraft, the fuel-quantity-indicating instrument may contain the amplifier or signal conditioner, thus eliminating the necessity for a separate unit installation. The indicating unit for fuel quantity may vary considerably from aircraft to aircraft. Formerly the majority of indicators utilized a needle-and-scale type indicator; however, with the availability of solid-state devices, many indicators are now of the digital type. The indication may be in pounds of fuel or in gallons or in both. Capacitance-type systems measure fuel quantity by weight (mass) rather than volume, because the dielectric strength of the fuel changes in accordance with density. Compensators are employed with the probes to assure an accurate indication. A fuel-quantity sensing probe (tank unit) is shown in Fig. 14-38.

Capacitance-type fuel-quantity-indicating systems require alternating current for operation. In large aircraft,

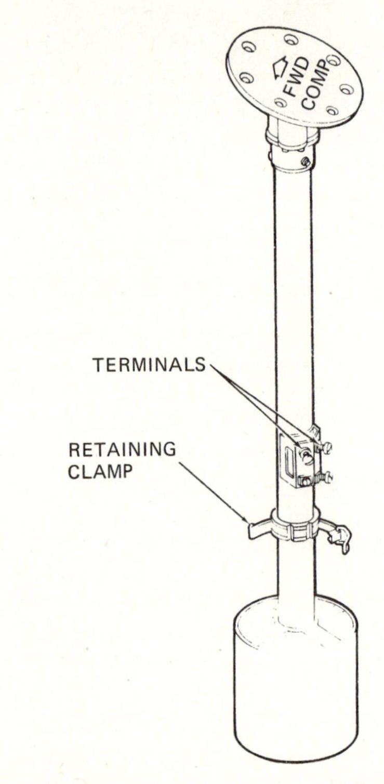

Fig. 14-38 Fuel-quantity sensing probe. (McDonnell Douglas)

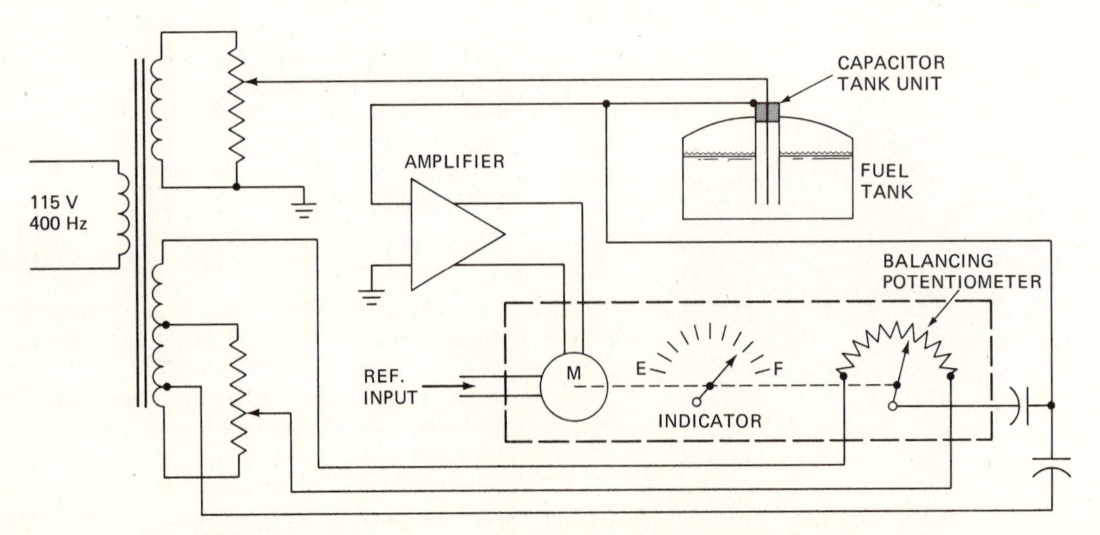

Fig. 14-37 Schematic diagram of capacitance-type fuel-quantity-indicating system.

this alternating current is taken directly from the 115-V 400-Hz power supply. In some systems, a back-up battery is provided in case of failure of the ac system. In this case, the direct current of the battery is fed through a solid-state inverter to supply 115-V 400-Hz current for the fuel-quantity-indicating system. In small aircraft with dc power systems, an inverter is required to furnish the necessary ac power.

FUEL-FLOW-INDICATING SYSTEMS

FLOW SENSORS

Flow sensors or transducers are devices that are installed in fluid lines to sense the volume of flow and produce a signal to an indicator that will provide an analog or digital readout for the use of the pilot or flight engineer in an airplane. A number of different devices can be made to serve this purpose. Among these are a restrictor that creates a difference of pressure proportional to fluid flow, a vane in a chamber having a variable cross section, and turbines. Fuel-flow systems utilizing vane-type and turbine-type sensors are described in the associated text, *Aircraft Maintenance and Repair,* Fourth Edition, in the chapter on fuel systems.

COMPUTERIZED FUEL SYSTEM

A system that provides fuel-flow and fuel-quantity indications is called *a computerized fuel system* (CFS). The indicator for the system provides fuel flow in gallons per hour or pounds per hour, gallons remaining, pounds remaining, time remaining for flight at the current power setting, and gallons used from initial start-up.

The transducer (sensor) for the CFS contains a neutrally buoyant rotor that spins at a rate proportional to the rate of fuel flow. The rotor is mounted in jewel bearings and incorporates notches that interrupt an infrared light beam from a light-emitting diode (LED). The light beam is aimed at a phototransistor, and the interruptions in the beam produce a series of pulses corresponding to the rate of fuel flow. The transducer is designed so the failure of the rotor cannot interrupt fuel flow. The transducer is mounted in a main fuel line and measures all fuel that flows through the line. A typical installation of a transducer is shown in the drawing of Fig. 14–39. In this case the transducer is installed in the main fuel line entering the fuel distribution manifold (fuel splitter) mounted on top of the engine. The transducer is mounted on a bracket with an Adel clamp.

The panel-mounted instrument contains computers that are designed to precisely count the number of pulses from the fuel-flow transducer and convert the count to gallons. A crystal-controlled clock provides the time reference needed to compute the rate of fuel flow and timer readout functions. The computer routinely calculates all the other displayed functions. The front panels of two types of fuel-indicating instruments are shown in Fig. 14-40. The top photograph is a standard round indicator, and the bottom photograph is the flatpack indicator. These instruments receive signals from two transducers as required for a twin-engine aircraft.

The basic program for the CFS is permanently "burned into" the computer chip. Variable data and intermediate computation values are stored by the computer in a separate memory bank. When the system is first installed in an aircraft, the aircraft's total usable fuel quantity is programmed into the separate memory circuit in accordance with installation instructions. This preprogramming is maintained in the computer memory bank through the use of three small dry batteries mounted in the rear of the instrument case. A warning indicator is provided to show when the battery power is getting low. The batteries will normally last up to a year and a half; however, it is recommended that they be changed during each annual inspection.

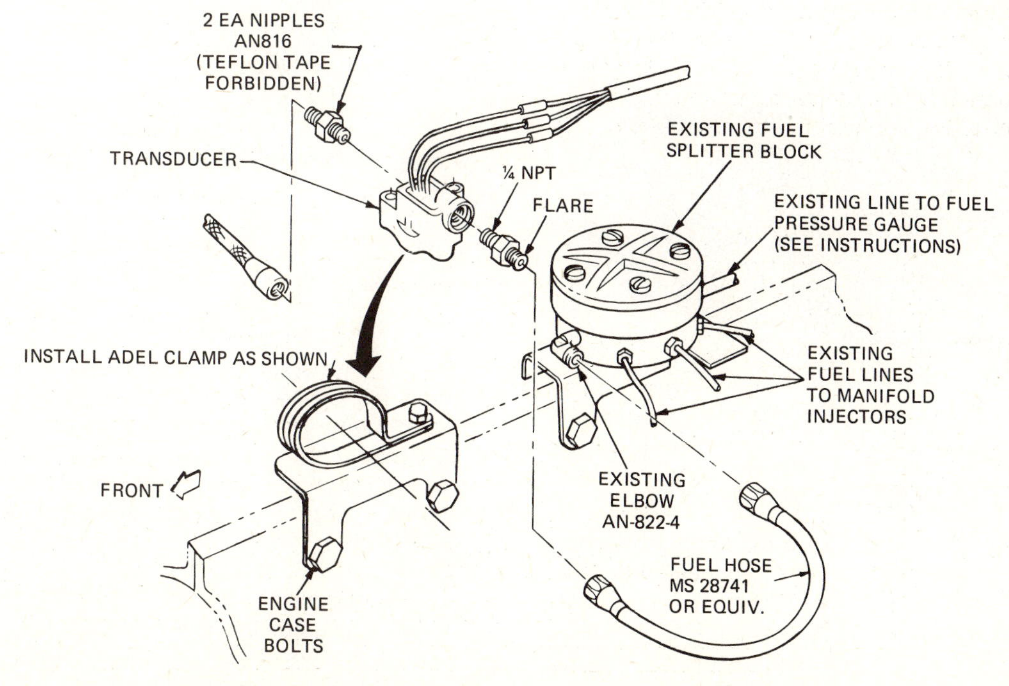

Fig. 14-39 Mounting of transducer for a computerized fuel system. (Symbolic Displays, Inc.)

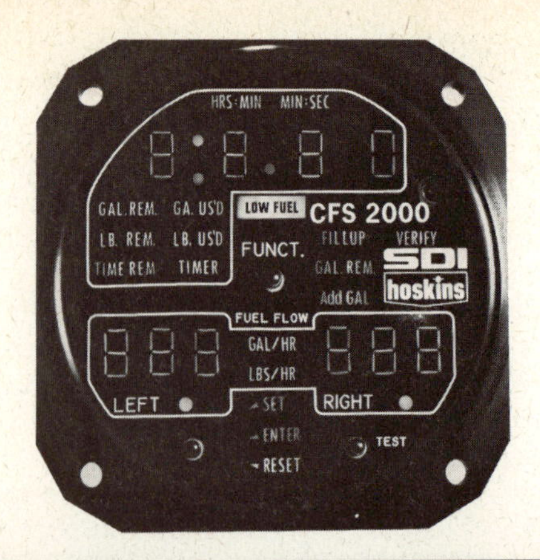

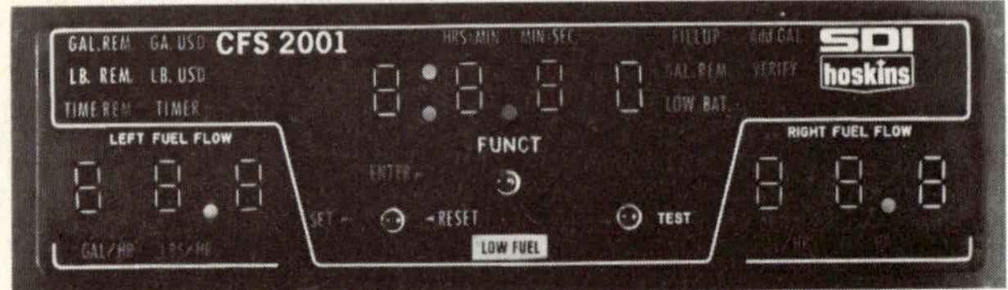

Fig. 14-40 Indicator panels for a computerized fuel system. (Symbolic Displays, Inc.)

At the beginning of a flight, the pilot must assure that the CFS is programmed with the correct information regarding quantity of fuel on board. If the tanks are filled to the normal full level, the "gallons-remaining" function should show the total usable fuel for the aircraft. During the flight, the computer will precisely count down from the initial fuel quantity. Instructions for programming the computer are provided by the manufacturer.

A **totalizer** is available for use with the CFS. This instrument reads the pulses from the transducers and continuously totals the fuel consumed. If not reset, the totalizer will add the fuel consumed for each flight to that previously totalized. When the unit is reset, it will start recording fuel consumption from the time of reset. The totalizer unit is shown in Fig. 14-41.

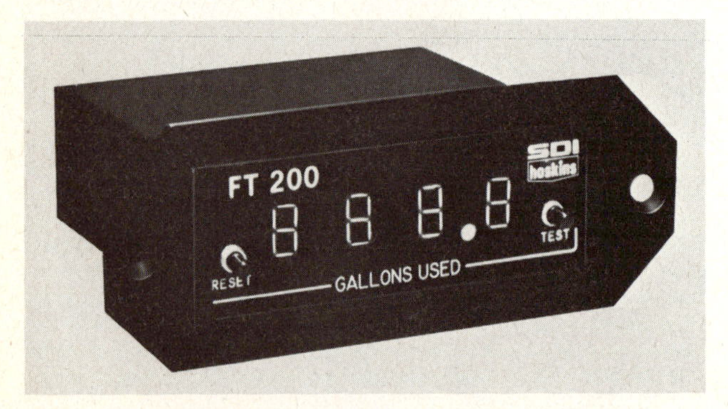

Fig. 14-41 Fuel totalizer indicator. (Symbolic Displays, Inc.)

VERTICAL-SCALE INSTRUMENTS

As aircraft have grown in size and complexity, the need for flight and engine instruments has also increased. This has resulted in an almost overwhelming array of instruments, particularly in large, transport-type aircraft. In addition to flying the airplane, the pilot must rapidly and accurately scan many instruments to assure that all conditions are correct for the mode in which the airplane is being flown. To assist the pilot and copilot of large aircraft, studies have been made on human factors in the reading of instruments. These studies have shown that a pilot can usually read a vertical-scale instrument more quickly and accurately than a round-dial instrument, especially when familiar with the use of vertical-scale instruments.

Vertical-scale instruments have been designed for both flight information and engine information, and some are installed in large aircraft such as the wide-bodied airliners. These instruments not only simplify the pilot's job, but also reduce the space required on the instrument panel. Figure 14-42 illustrates the advantage of the vertical-scale instrument as compared with the round-dial instrument for displaying the same information.

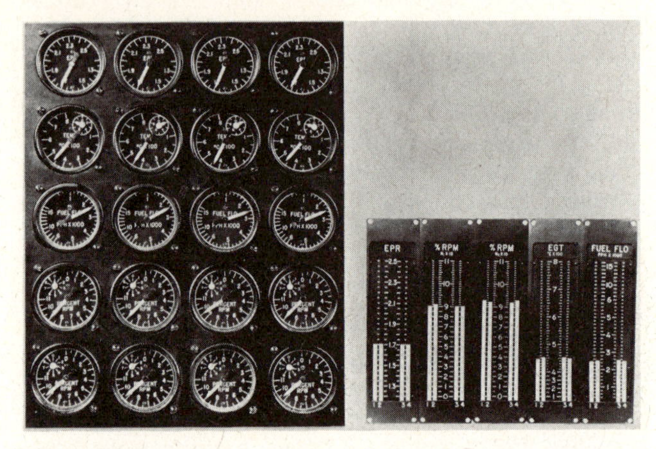

Fig. 14-42 Comparison of round-dial and vertical-scale instruments. (Bendix Corp.)

The advantage of the vertical-scale instrument in cockpit design is clearly shown in the photographs of Fig. 14-43. The instrument panels for two different four-engine jet aircraft are shown, one with vertical-scale instruments and the other with conventional, round-dial instruments. The same information is provided by each of the sets of instruments. The group of engine instruments is in the center of each of the instrument panels, and the advantage of the vertical-scale instrument with respect to space requirements is quite apparent. Furthermore, it can be seen that the pilots can quickly interpret a concentration of engine parameters (measurements) with a single, horizontal sweep of the eyes.

ENGINE INSTRUMENT GROUP

A modular grouping of primary engine instruments is shown in Fig. 14-44. From left to right, these instruments measure **engine-pressure ratio (EPR), percent of maximum rpm for the No. 1 compressor spool, exhaust-gas temperature (EGT), percent of maximum rpm for the No. 2 compressor spool, and fuel flow in pounds per hour (PPH).** This particular grouping of instruments is manufactured by the Bendix Corporation, Navigation and Control Division, for use in the Boeing 747 airplane.

From the illustration in Fig. 14-44, it can be seen that the instrument reading is presented by means of vertically moving column-like tapes which are read against numer-

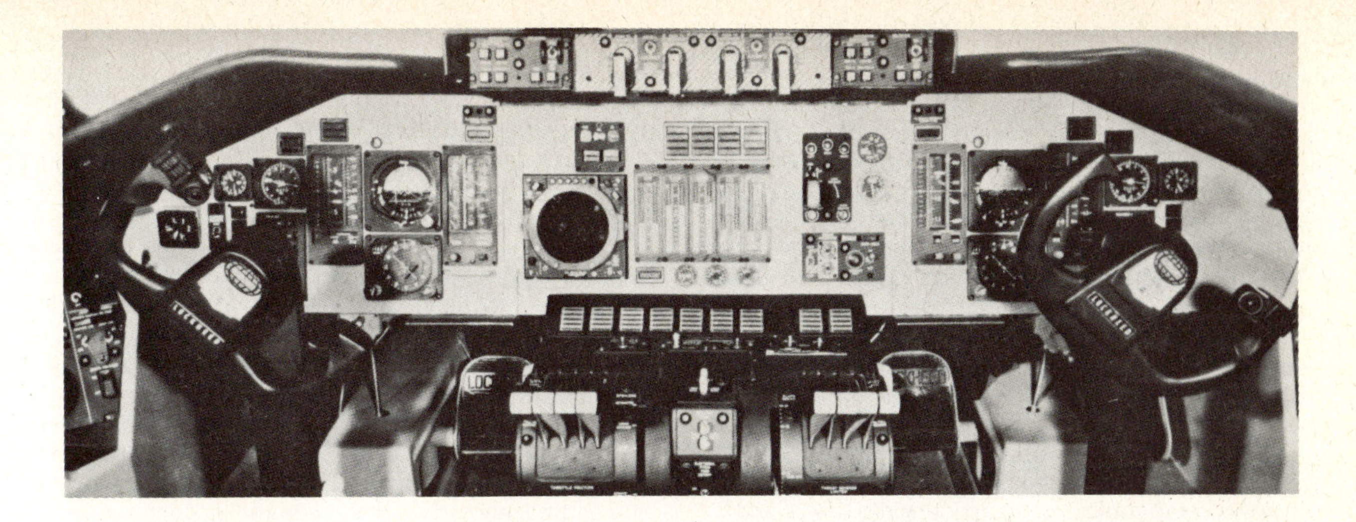

(*a*)

(*b*)

Fig. 14-43 Comparison of instrument panels when using vertical-scale and round-dial instruments. (Bendix Corp.)

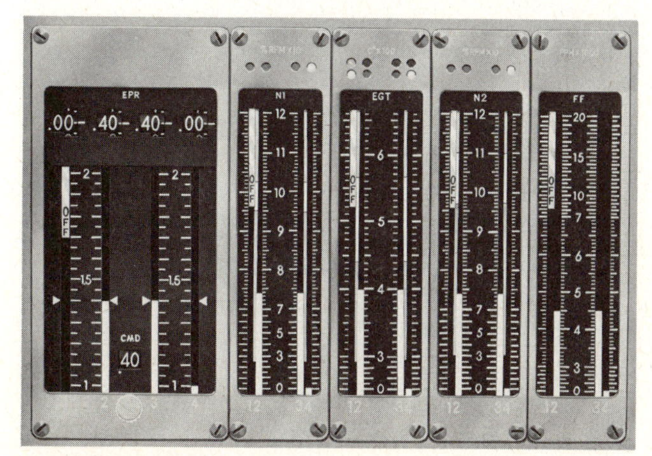

Fig. 14-44 Vertical-scale instruments for primary engine parameters. (Bendix Corp.)

als on fixed scales. Expanded graduations are provided for critical areas. The tapes are driven by roll-up mechanisms and are sprocketless to eliminate backlash and reduce wear. The tapes and scales are mounted at the same level to eliminate parallax errors and are illuminated from the back. The method of illumination provides balanced lighting and minimizes stray light. Provision is made for switching to backup lighting in the event of primary-lighting failure.

Each indicating tape has its own warning to show when it is inoperative. This is accomplished by means of a spring which drives the tape to a position where it displays the word *OFF* against a red and white diagonally striped background. An *OFF* warning will occur when there is a primary-power failure, loss of excitation to the transmitting servos, loss of gain in servo amplifiers, or open circuits in the synchro-control transformer rotors.

In the illustration of Fig. 14-44, the first instrument gives the **EPR.** This measurement is the ratio of the **turbine-discharge pressure** to the **engine-inlet air pressure.** In the illustration, the instrument is showing a ratio of 1.40 : 1 for engines 2 and 3. The **value counter** at the top of the instrument gives a digital reading equal to the indication given by the tapes. The tape of engine No. 1 instrument shows that this instrument is inoperative.

The **command bar** is used to place reference markings on the scales to give the pilot a quick indication of the value desired. This value is also shown on the **command counter** as a number. The **command knob** is used to set the desired value on the scales and on the counter. When the knob is pushed in, the command bar is set the same for all engines simultaneously, and the command counter shows the selected value. When the knob is pulled out, it will position the command bars for engines 2 and 3 only. The values for engines 1 and 4 will remain as previously set and will be in agreement with the command-counter reading.

The second and fourth instruments in the group shown in Fig. 14-44 are rpm indicators for the N_1 and N_2 turbines of the engine. This arrangement is necessary because the engines for which the instruments are designed have two main rotating assemblies. The low-speed turbine N_1 drives the forward section of the compressor through a shaft inside the hollow shaft of the high-speed turbine. The high-speed turbine N_2 drives the high-speed high-pressure section of the compressor. The instruments for turbine rpm are necessary to show the pilot whether the engines are operating at correct speed. The figure 10 on the scale indicates 100 percent of rated rpm, which also indicates 100 percent of rated power. Engine rpm limits are indicated by means of green, yellow, and red limit markings arranged vertically in the center of the instrument.

The EGT instrument is essential to the safe operation of a gas-turbine engine. Excessive temperatures may weaken the turbine blades to the extent that they will fracture; hence, the pilot must know when safe temperature is exceeded. If this occurs, fuel flow to the engine must be reduced immediately. If engine failure is indicated, the engine must be shut down. The EGT instrument includes warning lights which flash when the temperature becomes excessive. The lower light flashes when the temperature reaches the amber portion of the limit markings, and the upper light flashes when the temperatures rises into the red portion of the limit mark. This light then beams steadily until it is turned off.

The **fuel-flow indicator** provides a measurement of fuel flow to each engine in pounds per hour. The reading on the scale is multiplied by 1000 to give the correct reading. The rate of fuel flow to an engine is a good indicator of the satisfactory operation of the engine.

The electronic circuitry necessary for the operation of a typical vertical-scale rpm indicator is shown in the block diagram of Fig. 14-45. It will be noted that the sensing device is a tachometer generator which produces an alternating current with a frequency which is proportional to engine rpm. The ac from the tachometer generator is changed to a direct current with a voltage proportional to the ac frequency. This direct current is compared with a dc reference voltage, and the resulting signal is amplified and used to operate the servo motor which drives the indicating tape in the instrument. A **follow-up pot** (short terminology commonly used for potentiometers) produces a balanced signal which causes the servo motor to stop when the correct value is indicated on the instrument.

The electronic circuit modules and the drive modules for an EPR instrument are shown in Fig. 14-46. Note that this instrument provides EPR information from four engines.

SECONDARY PARAMETER INSTRUMENTS

A grouping of instruments for measuring secondary values associated with the engine and aircraft operation is shown in Fig. 14-47. The instrument group shown is designed for a twin-engine aircraft and operates in a manner similar to the primary group. The unit is 2.6 in [6.6 cm] high, 5.8 in [14.7 cm] wide, and 3.6 in [9.1 cm] deep. The advantage of this type of indicator is easily understood when one considers the space required for round-dial instruments necessary to provide the same information.

ENCODING ALTIMETER

Aircraft operating in controlled airspace and at certain altitudes must be equipped with a system that will trans-

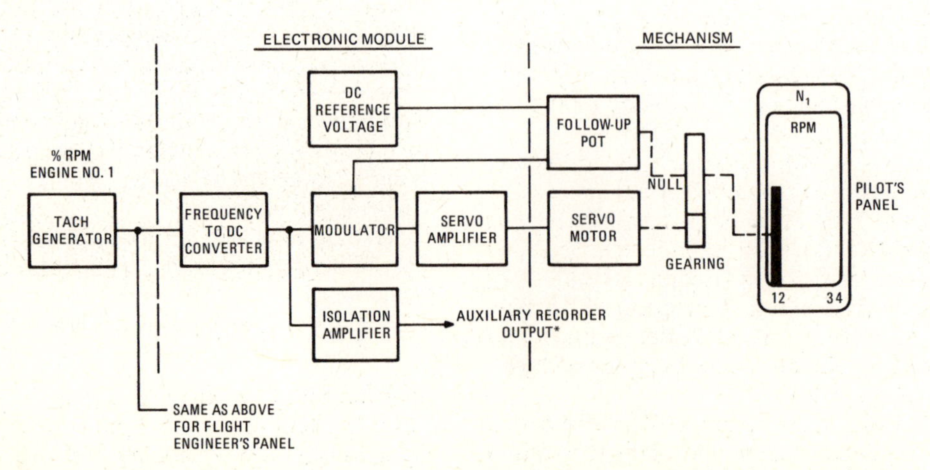

Fig. 14-45 Block diagram of electronic and electrical units in a vertical-scale rpm indicator.

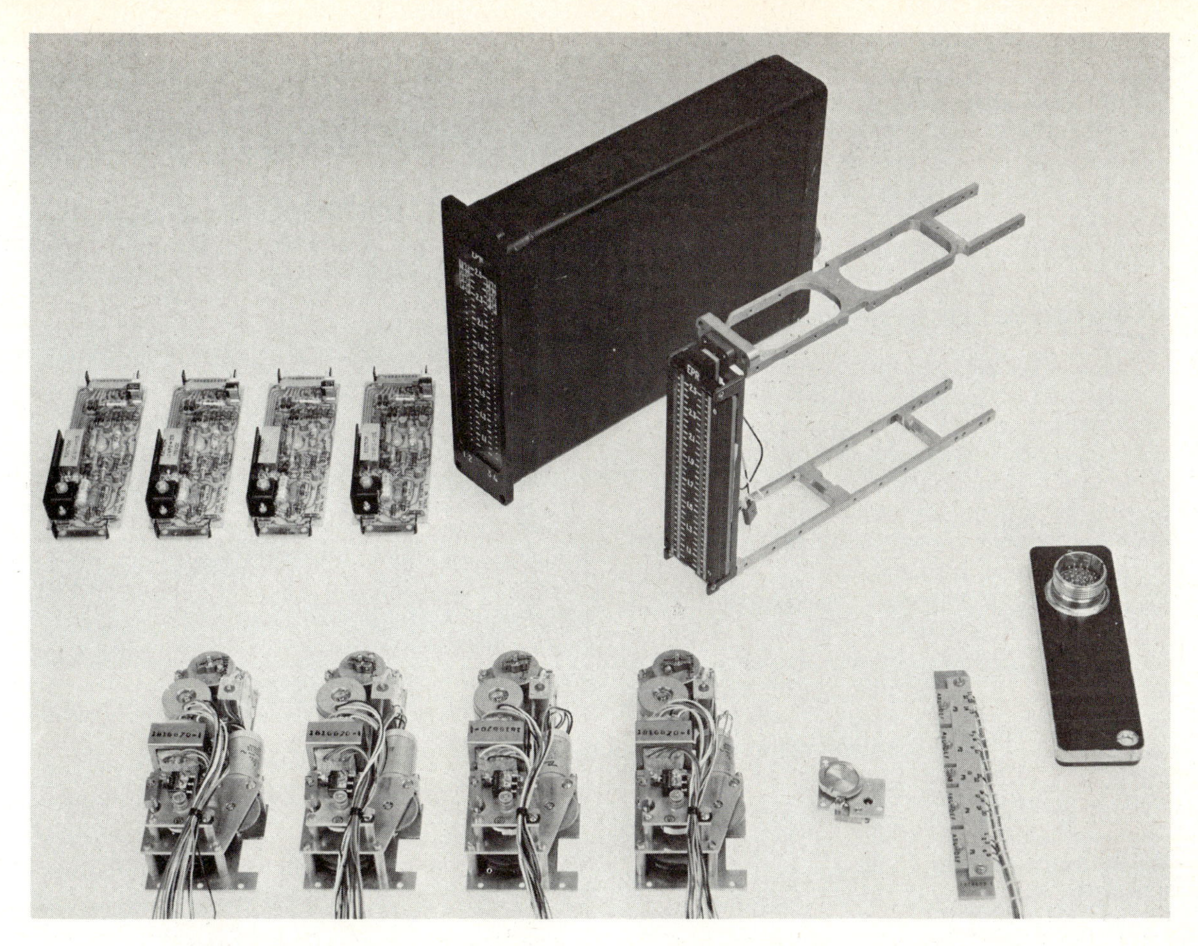

Fig. 14-46 Components of a vertical-scale instrument designed to serve four engines. (Bendix Corp.)

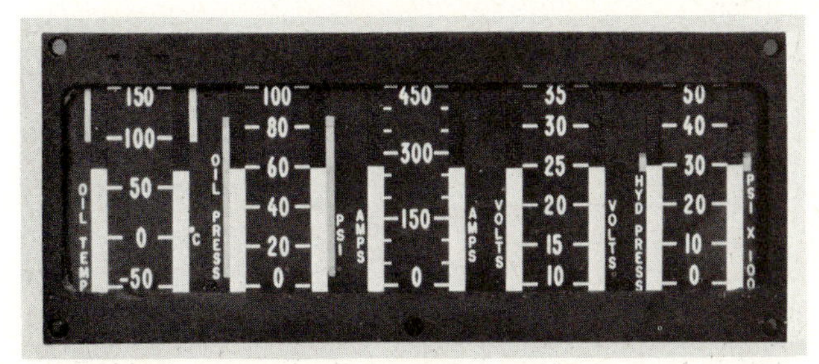

Fig. 14-47 Group of vertical-scale instruments for secondary engine parameters. (Bendix Corp.)

mit coded altitude information to the Air Traffic Control (ATC) center in the area. Such a system includes a standard **ATC transponder** electrically connected to the encoder of an encoding altimeter. ATC transponders are discussed in Chapter 17 of this text.

The encoding altimeter, also referred to as an altimeter/encoder, produces a binary code output that is sent by the transponder to the ATC radar. A decoder converts the signals to digital altitude information that is displayed on the radar screen to indicate the location and altitude of the aircraft involved. This information makes it possible for the air traffic controller to maintain both vertical and horizontal separation of aircraft. One type of encoding altimeter is shown in Fig. 14-48.

The encoding altimeter shown in Fig. 14-48 consists of a standard, panel-mounted, barometric altimeter with a built-in optical encoder that is driven by the altimeter

Fig. 14-48 An encoding altimeter. (Aero Mechanism, Inc.)

mechanism. The encoder utilizes light-emitting diodes (LED) as light sources and photodarlington transistors as light sensors. An encoding disk with nine concentric tracks of clear and opaque sections is positioned between the LEDs and the sensors. As the disk is rotated by the altimeter mechanism, the light passing through the clear sections of the disk tracks changes as altitude changes. For every 100 ft [30.5 m] of altitude a new combination of light patterns reaches the sensors and produces the coded altitude information. The signals from the light sensors are processed by solid-state logic circuitry before being sent to the ATC transponder.

RADIO ALTIMETER SYSTEM

A radio altimeter system, also termed a radar altimeter system, is designed to provide an indication of absolute altitude above the terrain. With this type of equipment in an aircraft, the pilot can tell exactly how high the aircraft is with respect to the ground rather than above sea level. This capability is of particular value when flying over strange territory or when barometric pressure may have changed during the course of a flight.

A radio altimeter system operates on the principle of radar as explained in Chapter 16 of this text. Pulses of energy are transmitted downward from the aircraft, and these pulse are reflected back to the aircraft where they are picked up by the receiver antenna. The time required for the pulses to reach the ground and return to the aircraft is a measure of the height above ground. A radar pulse travels at about 186 000 mps [300 000 000 m/s]. This is approximately 984 ft/Microsecond (μs) [300 m/μs]. Therefore, if it requires a time of 2μs for a pulse to travel from the airplane to the ground and return, the distance from the ground to the aircraft must be 984 ft.

A typical indicator for a radio altimeter system is shown in Fig. 14-49. Indicators for radio altimeter systems are manufactured in a variety of configurations. In many cases, the radio altimeter indicator is incorporated in the attitude director indicator (ADI) or similar instrument. In other cases, the indicator may be a rectangular instrument that provides an accurate indication of height above the terrain.

Fig. 14-49 A radio altimeter indicator. (Sperry Flight Systems, Avionics Division)

The transmitter of the radio altimeter system described here consists of a triode-cavity oscillator grid modulated by a solid-state modulator. The transmitted signal is detected at the cavity and applied to the range tracker as a pulse to start the timing sequence. The RF return signal from the

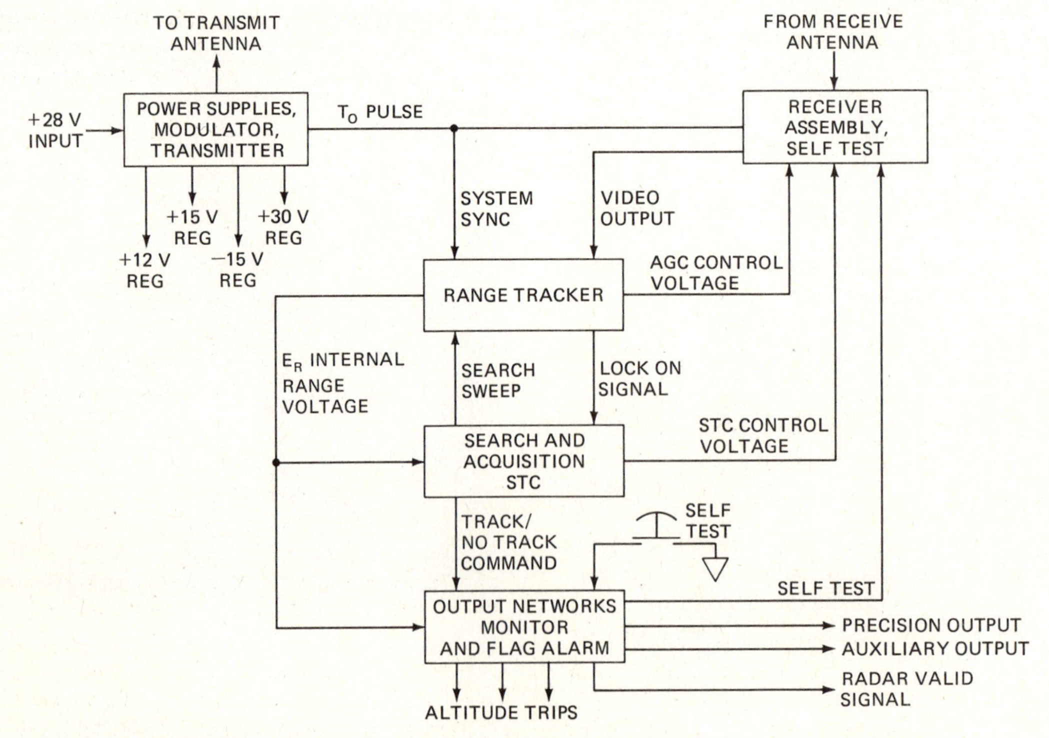

Fig. 14-50 Block diagram of a radio altimeter system. (Sperry Flight Systems, Avionics Division)

receiver antenna is input to the receiver system and heterodyned to produce an IF signal. The IF signal is amplified and detected to produce a video pulse. The video pulse is amplified and output to the range tracker. (See Fig. 14-50.)

Within the range tracker, the detected video pulse is compared to the system sync pulse, and the time difference between these pulses is determined. This time difference is converted to a dc analog voltage proportional to the shortest range to the ground. The internal range voltage is output to the search and acquisition circuit and output networks.

The search and acquisition circuits enable the radio altimeter system to initially acquire the ground return signal or reacquire the signal if tracking is interrupted. The search mode is initiated upon loss of the "lock-on" signal. A search-sweep signal is fed back to the range tracker which causes the system to search for the earliest return signal. This process continues until the track is reestablished and the lock-on signal is restored. A track/no track command is output to the output networks to cause the indicator needle to disappear from view during the search phase. A sensitivity time control (STC) voltage is fed back to the receiver to gain-program the receiver. Gain-programming prevents acquisition of the direct antenna leakage during the interval when the transmitter is on.

The internal range voltage is processed within the output networks to produce dc outputs for driving the indicator needle and other aircraft systems requiring radio altitude information. Three or four altitude trip outputs which supply a ground at or below the preset altitude are provided. The radar valid signal controls the indicator to provide warning of system malfunction.

A radio altimeter is essential for the approach and land functions of an automatic flight control system. The system provides annunciation of decision height (DH) and also signals the autothrottle system so engine power will be maintained at the correct value during landing.

The components of a radio altimeter system is shown in Fig. 14-51. The antennas are mounted on the bottom of the aircraft, the receiver/transmitter is located inside the fuselage as near to the antenna location as practical, and the indicator is mounted on the instrument panel in the cockpit or flight compartment.

The installation of the antennas is particularly important. The optimum installation is a situation where the two antennas are mounted in an area on the bottom of the aircraft which is entirely free of extraneous protrusions; the antennas are 36 in [91.44 cm] apart; they have a zero **squint angle;** and the plane of the antennas is parallel to the ground for all conditions of operation. The squint angle is the angle between the vertical axes of the antennas, and is a measure of their ability to illuminate the same area of the ground.

The area between the two antennas must be free of protrusions such as other antennas or drainage tubes. The area outside the antenna area should be as free of protrusions as possible. The antennas should be mounted so that no object is visible to either antenna within a 45° conic area below the aircraft.

The antennas must be mounted on a conductive surface for proper operation. Between the conductive surface and the antenna, a conductive gasket is placed. This assures that the base of the antenna is electrically bonded to the aircraft. The radomes of the antennas must not be painted.

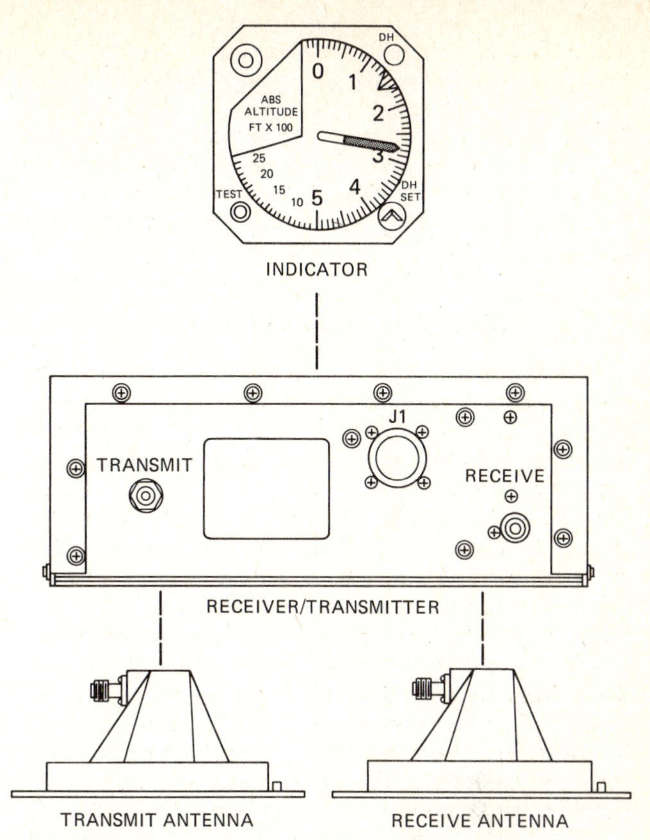

Fig. 14-51 Components of a radio altimeter system. (Sperry Flight Systems, Avionics Division)

The antennas must be mounted with H-plane coupling in order to maintain adequate isolation from one another. This means that the antenna connectors must be perpendicular to a line connecting the centers of the antennas, as illustrated in Fig. 14-52.

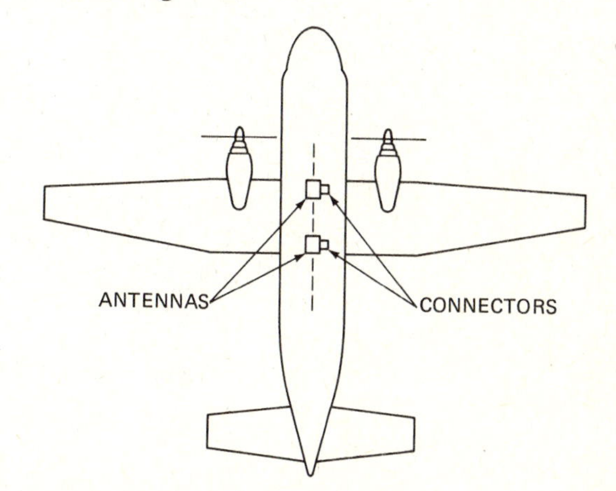

Fig. 14-52 Arrangement of antennas for a radio altimeter system. (Sperry Flight Systems, Avionics Div.)

Since there are a number of different configurations for radio altimeter systems, the technician installing such a system should follow closely the installation instructions provided by the manufacturer.

REVIEW QUESTIONS

1. List some of the parameters that are indicated by electric instruments in aircraft.
2. Why is it often necessary to utilize electric instruments rather than direct-driven instruments in an aircraft?
3. Why is a *tachometer* important in the operation of an aircraft engine?
4. Explain why an electric tachometer is more suitable than a mechanically driven tachometer in large aircraft.
5. Explain the operation of an ac tachometer system.
6. Describe the operation of a *thermocouple* temperature-indicating system.
7. Explain the important factors in the installation of a thermocouple system.
8. Explain the operation of a *Wheatstone bridge* temperature-indicating system.
9. Why is *ratiometer* system more satisfactory than a Wheatstone bridge system?
10. Describe the operation of a ratiometer temperature indicator.
11. What is a *synchro* system?
12. Explain the operation of a dc *Selsyn* system.
13. For what indications is a Selsyn system useful?
14. Describe an *Autosyn* system.
15. Give some uses for the Autosyn system.
16. Draw a schematic diagram for an Autosyn system.
17. What electric power is used for an Autosyn system?
18. Name several types of direction indicators.
19. Describe a *magnetic compass.*
20. What is the purpose of the compass fluid?
21. Name four causes for compass inaccuracies.
22. Explain *deviation* and *variation.*
23. What is meant by *swinging* a compass and how is it accomplished?
24. How is a magnetic compass compensated for error?
25. Describe the construction of a vertical-dial magnetic direction indicator.
26. Give three advantages of the vertical-dial magnetic direction indicator.
27. What is a *magnesyn compass* system?
28. Describe a *flux-gate* compass system.
29. Explain the operation of a flux-gate system.
30. Describe the *master direction indicator* in the flux-gate system.
31. What is the function of the *coupling Autosyn?*
32. What is meant by *caging* the gyro in a flux-gate system?
33. Name some electronic direction indicators.
34. Describe the operation of a float-type, electric fuel-quantity indicator.
35. Explain the operation of a capacitance-type fuel-quantity indicator.
36. Why does the capacitance of the fuel-level sensor change as fuel level changes?
37. What is required when a capacitance-type fuel-quantity-indicating system is installed in an aircraft that has a dc power system?
38. Describe three types of *fuel-flow sensors.*
39. Describe the fuel-flow sensor (transducer) for a *computerized fuel system.*
40. What information does the computerized fuel system provide for the pilot?
41. What information must be programmed into the CFS computer before the system can provide accurate information about the fuel quantity and flow?
42. What is the purpose of a *fuel totalizer?*
43. Give two advantages of *vertical-scale* instruments.
44. By what means is a measurement indicated on a vertical-scale instrument?
45. What components are required in a vertical-scale instrument system?
46. What is the function of an *encoding altimeter?*
47. Why is an encoding altimeter required in certain aircraft?
48. Describe the principal of a *radio altimeter?*
49. Explain installation requirements for the antennas of a radio altimeter.

CHAPTER 15

Electronic Control Systems

In this day of electronic "brains" or computers, automation, automatic flight, guided missiles, and spacecraft, electronic control has become an increasingly important factor in industrial, aircraft, and space technology. It is apparent that a detailed or complete coverage of electronic control systems cannot be given in one chapter of any text; however, it is possible to present certain principles and elements of control systems which will enable the student of electricity or electronics to obtain a basic understanding of the methods by which electronic control is accomplished.

It is the purpose of this chapter to explain some of the simpler types of control circuits and systems and point out how the more complex types are developed. It will be found that well-known laws of electricity operate in all systems, but because of the multitude of different combinations possible, it will be seen that there is no limit to the number of applications which may be produced.

BASIC ELECTRICAL CONTROL

The simplest type of electric control circuit is well known to almost everyone. Such a circuit is shown in Fig. 15-1, which is a schematic diagram of an electric lighting circuit. The control of this circuit is the manually operated switch by which a person may turn the light on or off. Such a circuit may be employed to operate an electric motor, an electric valve, or various other types of equipment. The circle marked *AC* represents a source of ac power.

Fig. 15-1 Control circuit for an electric light.

Another, slightly more complex, control circuit is one in which a relay is used for the remote control of a power circuit as shown in Fig. 15-2. In this case, a very small current is used to control a large current. The current supplied to the relay may be a fraction of 1 A, but the current through the power section of the relay and the load may be more than 1000 A. Such circuits as this are used in aircraft and other applications in which it is desired to keep heavy power cables as short as possible and still control the power from a distance.

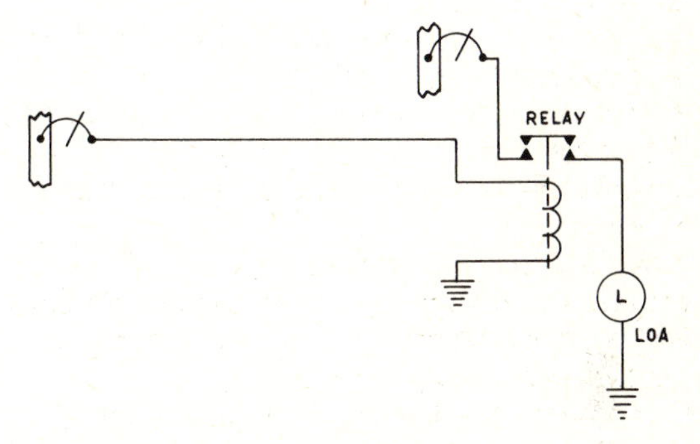

Fig. 15-2 Relay-operated control circuit.

An automatic control system has certain basic elements. The first of these is the **controlled item** or condition; control can only be applied to something that is controllable. Next, there must be a **corrector,** that is, a unit or system which applies the correction. A third item is the **sensor,** the device which determines that a correction is needed. The sensor can also be called an error detector, because it senses an error in the required condition. Finally, the system requires a device or devices working together to control the correction. This device or combination of devices may be called the **control equipment.**

Figure 15-3 is a simple schematic diagram of a thermostatic heat-control mechanism used to control temperature in a particular area. This system encompasses all the elements required for an automatic control system. The condition to be controlled is the temperature, the sensor is

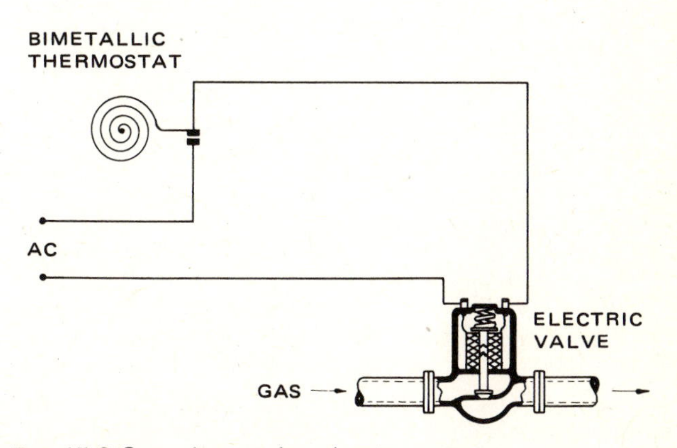

Fig. 15-3 Control circuit for a heating system.

the thermostatic switch, the control equipment is the electric valve and the gas heater to which it delivers gas, and the corrector is the system which produces and delivers heat to the area to be heat-controlled. When the temperature falls below the level for which the thermostat is set, the bimetallic element contracts and closes the switch contact, thus sending electric power to the valve. The valve opens, causing gas to be sent to the heater, with the result that the area under control receives a heat correction. When the temperature has reached the preset maximum, the thermostatic switch (sensor) opens the valve circuit and causes gas to be cut off from the heater.

BRIDGE CONTROL

A circuit commonly utilized in electronic control systems makes use of the bridge principle to apply correction to a moving unit. This principle is illustrated in the schematic diagram in Fig. 15-4, which is a circuit used for the operation of a flight control surface in an automatic-pilot system. In this circuit, the sensor is the gyro unit. A free vertical gyro installed in an airplane will maintain its vertical position with respect to the surface of the earth even though the airplane may pitch or roll many degrees from the horizontal position. This feature of a gyro makes it possible to attach the wiper (contact arm) of a potentiometer to the shaft of a lateral gimbal ring so that the pitching of the airplane will cause the wiper to move along the potentiometer resistance. Thus the gyro will be able to sense an error in the airplane's attitude and produce a correction signal.

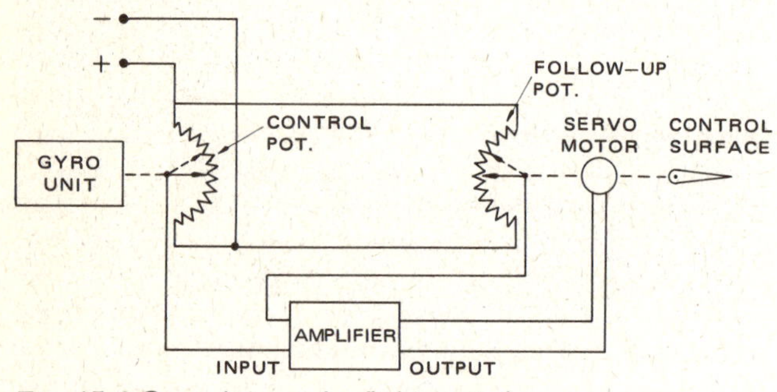

Fig. 15-4 Control circuit for flight control.

It will be noted that in the circuit in Fig. 15-4 two pots are connected in parallel to form a bridge circuit. One of these is the control pot, which is driven by the gyro unit, and the other is the follow-up or feedback pot, which is driven by the control-surface actuating mechanism. When a direct current is connected across both pots, as shown in the diagram, there will be no difference of electric potential between the wipers for as long as they are in the same relative position. Now, let us suppose that the airplane is caused to nose down by an air current. The gyro will hold its position with respect to the earth's surface, and so there will be a movement of the control-pot wiper along the resistance. Assume that the wiper moves to the position shown by the dotted arrow. This movement will produce a positive signal at the wiper with respect to the follow-up wiper, and the resulting voltage difference will be sensed by the amplifier. Within the amplifier is a discriminator circuit which develops one signal or another, depending upon the polarity of the input signal. In this case, the discriminator will produce a signal which energizes the control relay in the direction to produce an *up* correction at the control surface. The servo motor is designed so that it can move the control surface either up or down, depending on the nature of the energizing signal from the amplifier.

The follow-up circuit is an essential component of the bridge control system. If there were no follow-up, the control surface would move to the up position and would remain there until the airplane had passed through the horizontal position and moved to a nose-up position. Thus there would be a continual oscillation of the airplane up and down as the signal changed from up to down, and back to up. With follow-up, the follow-up wiper begins to move in a neutralizing direction as soon as the control surface begins to move. Thus, when the follow-up wiper reaches the same relative position as the control wiper, the control-surface movement stops. As the airplane attitude is corrected, the control wiper moves back toward neutral, producing a reverse signal which causes the follow-up wiper to move back to neutral also. In actual practice, there is a dead spot at the neutral position on each pot which prevents the continual *hunting* which would otherwise take place.

Further consideration of the circuit in Fig. 15-4 will reveal still another problem which would occur if there were no follow-up in the system. It is obvious that the servo motor will continue to drive the control surface in the up direction for as long as an up signal is coming from the amplifier, and that an up signal will come from the amplifier for as long as the airplane has a nose-down attitude. Hence, the control surface would still be moving toward the up position even when the airplane had corrected almost to the horizontal position; the surface would be in a maximum up position when the airplane had actually reached the horizontal position. These conditions, of course, could lead to very violent pitching of the airplane.

Follow-up or feedback is an essential part of all automatic control systems for airplanes and missiles, because the control equipment must "know" how much correction has been accomplished and how much the correcting activity must be reduced as the error diminishes.

PHOTOCONTROL SYSTEMS

In the study of control circuits, it is useful to consider the operation of devices that respond to the intensity and color of light to produce control signals. Some of these devices are the **phototransistor, phototube,** and **solar cell** and are also referred to as **optoelectronic** devices. The phototransistor and the phototube may be considered **photoconductive** units and the solar cell is a **photovoltaic** unit. Photoconductive units will conduct current when exposed to light, and photovoltaic units will convert light to electric energy.

The phototransistor acts as a light-operated switch. When it is exposed to a predetermined intensity of light, it becomes conductive and allows current to flow in the circuit in which it is connected. This current may be used to operate a low-power circuit, or it may be used to trigger a power transistor, a silicon-controlled rectifier (SCR), or a solid-state relay that will handle high-power circuits. A phototransistor and its symbol are shown in Fig. 15-5.

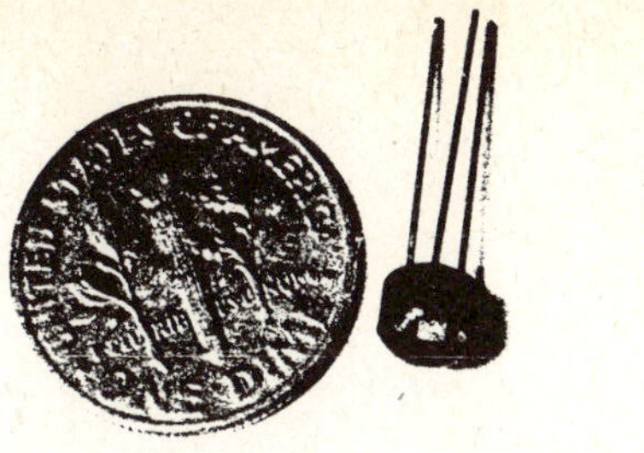

Fig. 15-5 Phototransistor and symbol.

A phototube was one of the first light-sensitive devices used extensively in applications where it was desired to use light to produce control signals. A drawing of such a tube is shown in Fig. 15-6. This unit may also be called a photoemissive cell inasmuch as light falling on the cathode causes the emission of electrons which provide a current flow through the tube. Phototubes were once used

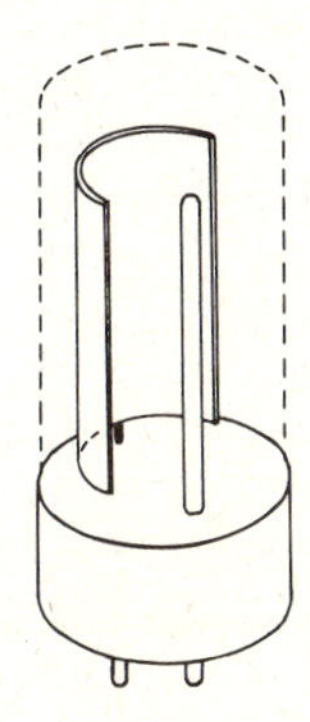

Fig. 15-6 Drawing of a phototube.

extensively for the operation of electric-eye warning systems, smoke detectors, counting systems, sorting machines, automatic switching circuits for street lights, sound reproduction from motion-picture films, and various other devices. Many of these systems are still in operation; however, phototransistors, photovoltaic cells, and other solid-state devices have largely replaced phototubes for systems responsive to light.

A photovoltaic cell or solar cell can be used to produce electric energy for a variety of purposes. If a large number of such cells are connected together in a solar panel, the power generated is limited only by the number of cells employed.

A silicon solar cell consists of a wafer of silicon that has been appropriately doped to make it a semiconductor and upon which a thin layer of boron has been diffused. The wafer is reinforced with metal and is provided with electric contacts to permit joining it with other cells to form modules of desired sizes. Modules are joined to form panels, and panels are joined to form arrays.

The construction and operation of a solar cell are illustrated in Fig. 15-7*a*, and solar cells joined to form a 16.01-V module are shown in Fig. 15-7*b*. This module will produce approximately 2.05 A or 32.8 W peak power, that is, when exposed to maximum sunlight.

Photovoltaic cells of various types have been used for many years to operate light meters for photographic light measuring and a variety of other devices. In one type of application, they are used to operate small motors which

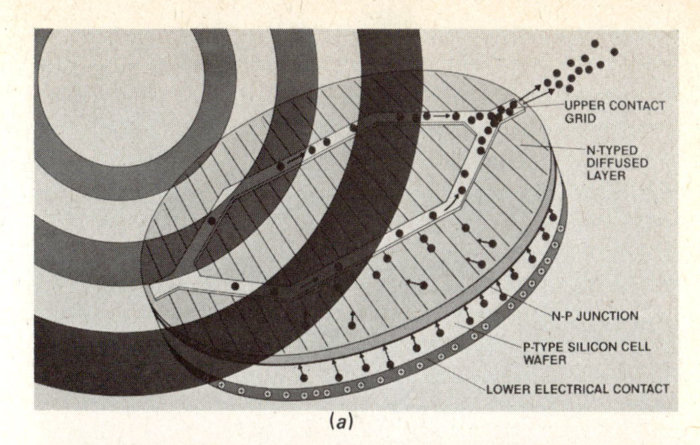

(*a*)

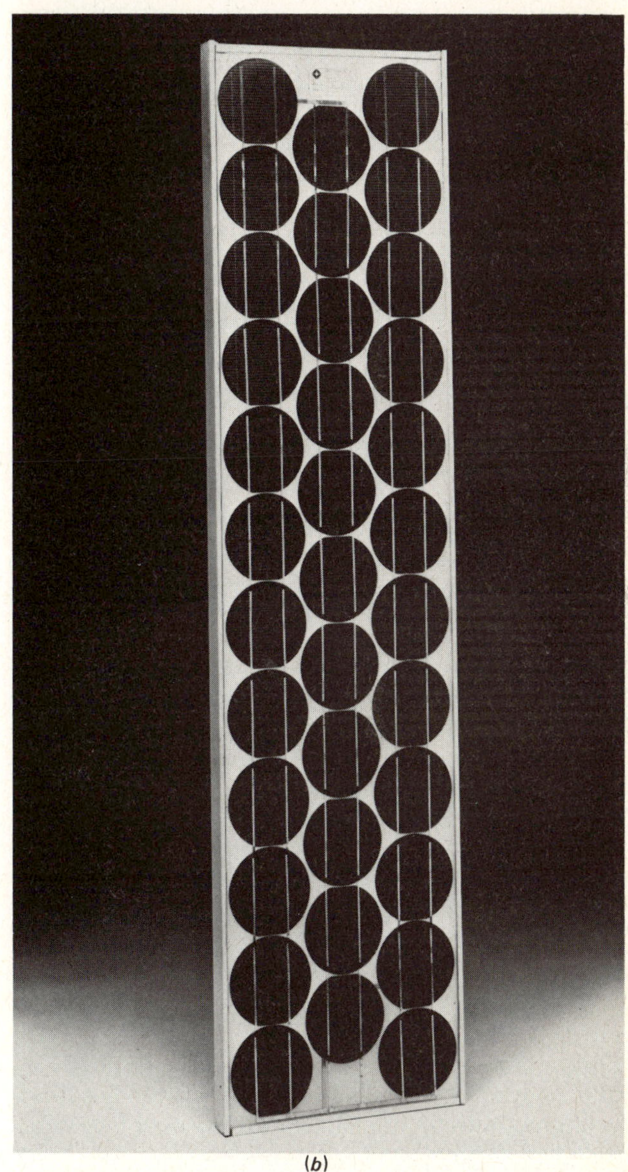

(*b*)

Fig. 15-7 Solar cells and panels. (Arco Solar, Inc.)

rotate the light irises for cameras to give automatic light adjustment as shown in Fig. 15-8. Light passing through the baffle strikes the photocell and generates a current which flows through the thermistor-resistor combination to the meter coil. The thermistor changes in resistance as temperature changes, thereby providing temperature

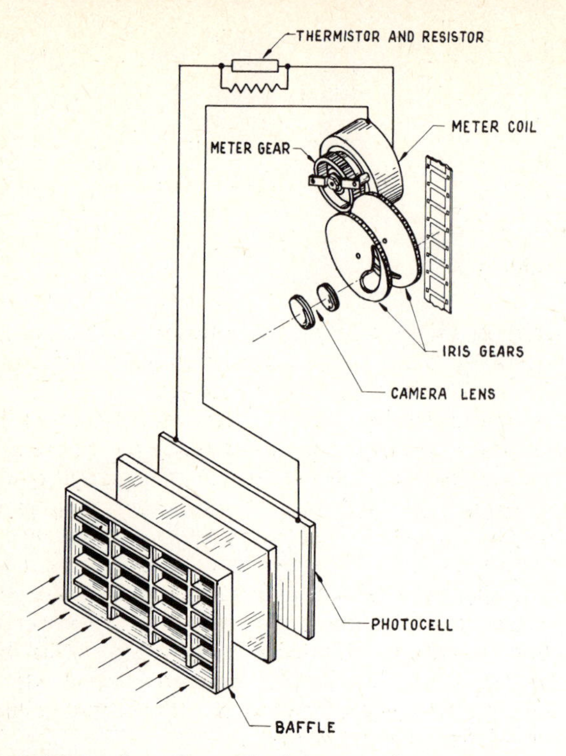

Fig. 15-8 A solar cell used for light control in a camera.

correction. The meter gear drives the iris gears which vary the incoming light by means of teardrop-shaped slits. As the light striking the photocell increases, current to the meter coil increases and causes the iris gears to move in a direction which will reduce the amount of light striking the film.

Both selenium and silicon cells are used in spacecraft and satellites to provide solar power to operate radio transmitters. Use of this source of power makes it possible for satellies to continue transmitting for many years.

For the person who likes the experiment with novel ideas, a solar cell or a phototransistor provides a wealth of possibilities. Any source of light, such as a flashlight, automobile headlights, a match, or any other lighting device which produces an appreciable quantity of light, may be used to control an electric circuit.

THE MULTIVIBRATOR

A multivibrator is one of the electronic circuits commonly found in electronic control systems to develop nonlinear signals of precise frequencies for the controlling of more complex systems. The multivibrator may be termed a **nonlinear oscillator** because it produces a signal which is oscillatory and also nonlinear. A nonlinear signal is one which is not sinusoidal; that is, its waveshape is not in the form of a sine wave.

A multivibrator circuit is shown in Fig. 15-9. Assume that both transistors are conducting when the circuit is started. One of the transistors will be carrying slightly more current than the other because it is not possible to create circuit elements that are perfectly balanced. It is assumed that resistor R_1 is carrying a slightly greater current than R_3 and that point X will be slightly more negative than point Y. The base of transistor Q_2 will then become biased to cutoff and Q_2 will stop conducting. When this occurs, transistor Q_1 will become biased to cutoff and Q_2

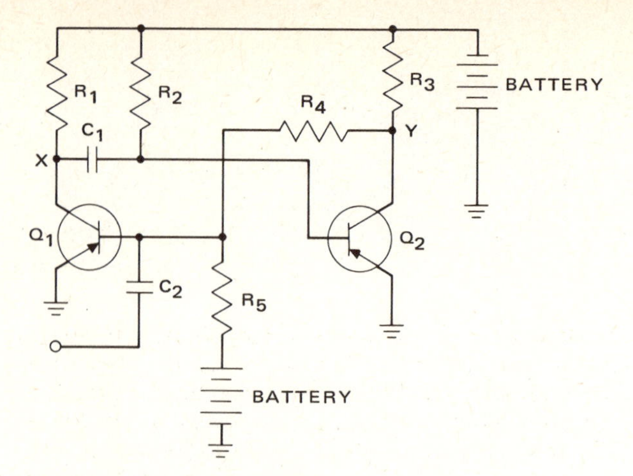

Fig. 15-9 A miltivibrator circuit.

will start conducting. Thus, the transistors will alternate in conducting and the output signal will be turned on and off at a rate depending upon the values of the capacitors and resistors involved.

SWITCHING CIRCUITS

One useful application of a multivibrator circuit is in the operation of an **electronic switch,** or **switching circuit.** One type of switching circuit is used to switch operation from one electronic circuit to another at a very high rate or to switch a circuit on and off rapidly. The rapid switching from one circuit to another makes it possible to use the switching circuit in conjuction with an oscilloscope to observe two signals at once on the oscilloscope screen. When the two signals are applied to the oscilloscope in very rapid succession, the image appearing on the screen will show both signals. In this way the two signals can be compared to determine whether they have the correct form.

The switching circuit in this application includes two transistors that are alternately made conductive and nonconductive by the signals from the multivibrator. The transistors conduct the two signals that are to be analyzed by the oscilloscope and deliver the signals alternately to the oscilloscope input.

Numerous types of switching circuits have been developed for a wide variety of purposes. The use of transistors, silicon-controlled rectifiers (SCR), and other solid-state devices makes it possible to design circuitry to accomplish any electrical or electronic function. Solid-state switching devices are used in place of electromechanical relays to control motor circuits and other circuits carrying high power. The advantage is that there are no moving parts to wear out, no contact points to arc and burn, and no emission of electromagnetic waves that would cause radio interference.

COUNTING CIRCUITS

It appears quite logical that an electronic computer will require a method for counting or storing up a quantity of information proportional to the duration of a signal, the number of pulses applied, or the frequency of the signal applied. Since computers are used to solve problems that require the sensing of possibly millions of impulses per second, it is necessary to incorporate into them counting

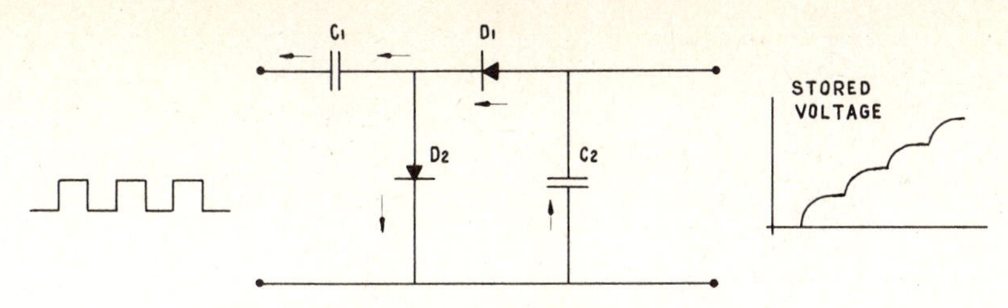

Fig. 15-10 A step-by-step counting circuit.

circuits suitable for summing up the number of impulses received. The two principal types of counting circuits are the step-by-step circuit and the frequency circuit.

A **step-by-step** counting circuit is shown in Fig. 15-10. When a square-wave signal is applied with positive pulse to C_1, current will flow through diode D_1, and capacitor C_2 will start and continue to charge for as long as the signal continues. Although this period will be only a fraction of a second, capacitor C_2 will hold the charge received. At the next pulse C_2 will increase its voltage and continue to do so for as long as the signal is applied. C_2 thus accumulates a charge proportional to the number of pulses received. During the time between waves or pulses, capacitor C_1 discharges through D_2 and prepares to receive the next pulse. In this way C_2 can continue to store the charges in response to the incoming pulses.

A diagram for a **frequency** counting circuit is shown in Fig. 15-11. When a positive pulse is applied to C_1, a current will flow for a short time in the direction 1-2-3-4.

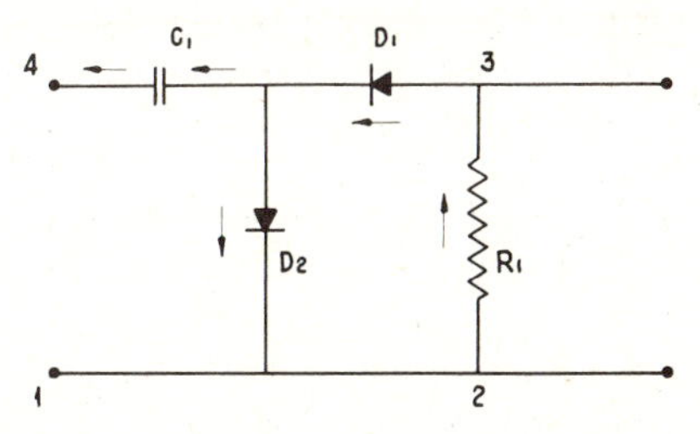

Fig. 15-11 A frequency-type counting circuit.

During this time current cannot flow through diode D_2 because of the negative charge on the line between 1 and 2. When the positive pulse ends, C_1 will discharge through D_2 and leave C_1 ready to receive another charge. The result is that current will flow through R_1 from point 2 to point 3 in pulses, with the average voltage depending upon the frequency of the pulses. It is, of course, essential that the pulses be of equal amplitude and duration if the voltage across R_1 is to give an accurate indication of the frequency.

ELECTRONIC DIFFERENTIATION AND INTEGRATION

Differentiation and integration are mathematical processes essential to the operation of electronic computers and other types of electonic circuits. Differentiation may be defined as the process of finding the differential, which is an infinitesimal difference between two consecutive values of a variable quantity. For example, if it is desired to find the differential at a given point on a sine curve, differentiation is employed. The result will show the rate of change taking place in the value at the particular time selected. In a sine curve, the rate of change is zero at the most positive value and at the most negative value and maximum when the sine curve is at zero.

An electronic circuit which performs the function of differentiation is shown in Fig. 15-12. Since the differential, also called the derivative, is the rate of change of the input voltage, the input must be constantly changing, or nothing will be obtained at the output. In this circuit the capacitor must have a very low capacitance so that it will become charged as rapidly as voltage is applied to it and will cause current to flow in the circuit only when there is a change in voltage. The result of this action will be that the voltage across R will be proportional to the rate of change of voltage at the input; hence, the output voltage will be a differential of the input.

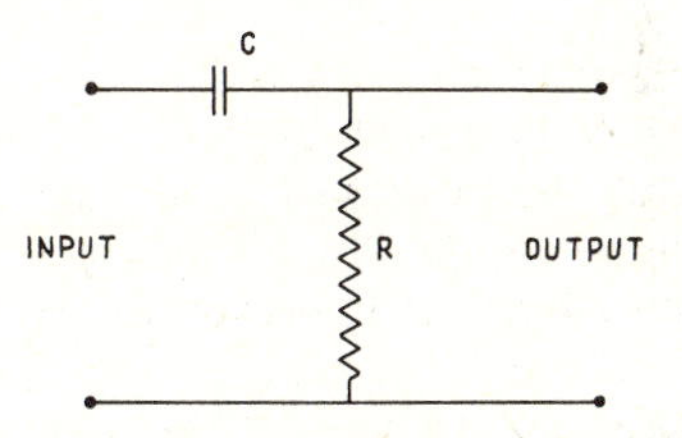

Fig. 15-12 A differentiating circuit.

Integration may be considered the reverse of differentiation. It is the process of adding many small quantities to obtain a total. An integrating circuit in electronics develops a voltage proportional to the average of the instantaneous voltages applied to the output. Such a circuit is shown in Fig. 15-13. In this circuit both the capacitance and the resistance are large and thus produce a long time constant. One important use of integrating circuits is in the analog computer, which utilizes them to solve very complex equations.

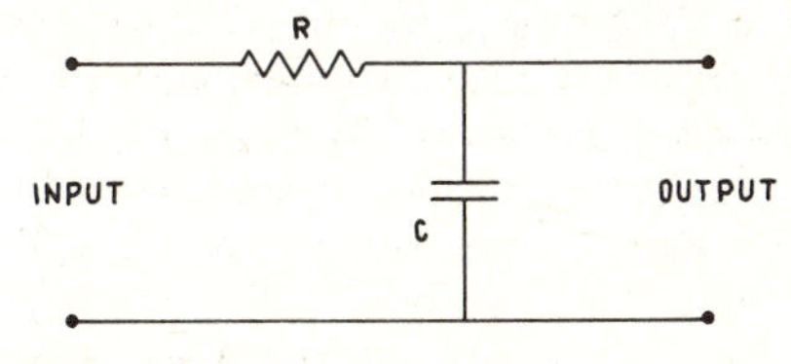

Fig. 15-13 An integrating circuit.

PULSE CIRCUITS

A wide variety of electronic circuits are used for the production of the electric pulses which are necessary in the operation of computers, radar, navigation, television, and various other electronic systems. One of the simplest methods for producing an electric pulse is to apply a square wave to a differentiating circuit. The diagrams in

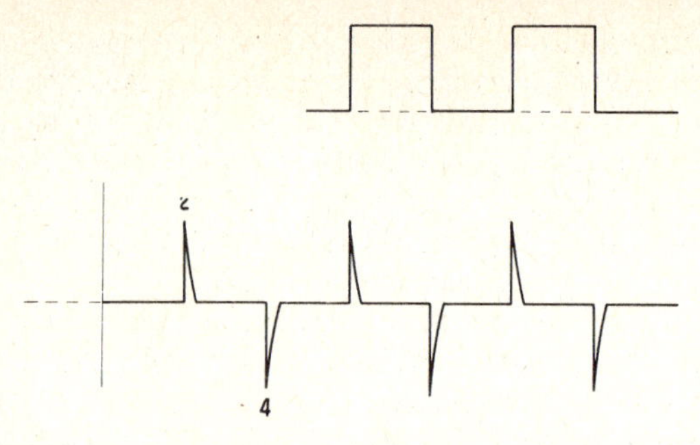

Fig. 15-14 Waveshapes in a pulse-forming circuit.

Fig. 15-14 show the input and output waveshapes. When the voltage rises to an instantaneous value, as at point 1 at the beginning of the square wave, the current flow to the capacitor will rise to a maximum and then immediately return to zero as the capacitor becomes charged. In this circuit the capacitor is of very low capacitance and so charges almost instantaneously. The sudden rise and fall of current to the capacitor will cause a sudden rise and fall of voltage across the resistor, point 2 on the curve (refer to Fig. 15-12 for the circuit used in this case). The rise and fall of voltage is the electric pulse. During the interval from point 1 to point 3 on the square wave, there is no change in voltage; hence, there can be no current flow, and the voltage across the resistor is zero. When the square-wave voltage returns to zero, the capacitor discharges and causes a voltage pulse in the opposite direction; this is indicated at point 4.

As mentioned previously, there are many methods by which pulses and other waveshapes are produced. The foregoing method is given to provide an understanding on the general theory of pulse formation.

In practice, some of the circuits described in this chapter are designed into solid-state devices, such as integrated circuits (ICs). As explained previously, such circuits are constructed in tiny chips of silicon from 0.05 to 0.24 in [1.25 to 6 mm] square, the size depending upon the number of elements included for the desired circuitry. Integrated-circuit chips may contain thousands of transistors, diodes, resistors, and capacitors.

LOGIC CIRCUITS

Logic circuits, used extensively in solid-state electronics, are simply control circuits designed to provide particular effects. The individual logic circuit is no more complex than any other electric or electronic circuit; however, when many such circuits are used together in a computer system, an almost infinite variety of results can be attained and it appears that the computer can actually think.

In aircraft control systems, logic circuits are used to sense certain conditions or events and to produce a desired response to them. These circuits may be very simple or quite complex, depending o the nature of the application.

A simple type of logic circuit is illustrated in Fig. 15-15. If power is available at buses *A* and *B* and switch *S* is closed (conducting), *R* will conduct, and a signal will be produced at lamp *L*. The switch *S* may be operated by the landing gear of an airplane in the UP position so that a red light will be on if the landing gear is not lowered before landing. To make the circuit more practical, two switches may be connected in series as shown in Fig. 15-16. One of the switches may be operated by the landing gear, and the other may be closed when the throttle is retarded as for landing. Now the circuit may be classed as an AND circuit because switches S_1 *and* S_2 must be closed and conducting before a signal will appear at *L*.

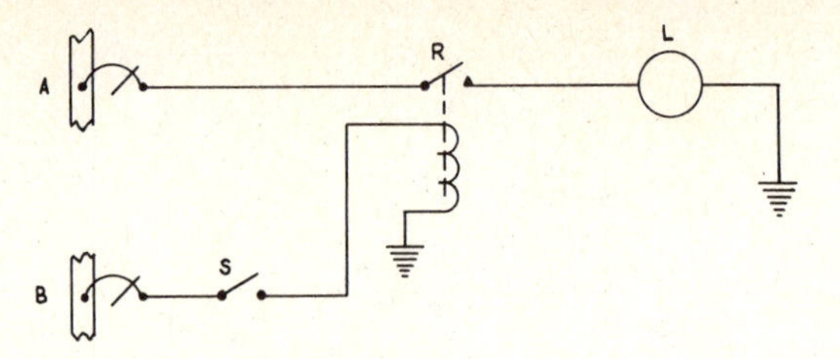

Fig. 15-15 A simple logic circuit.

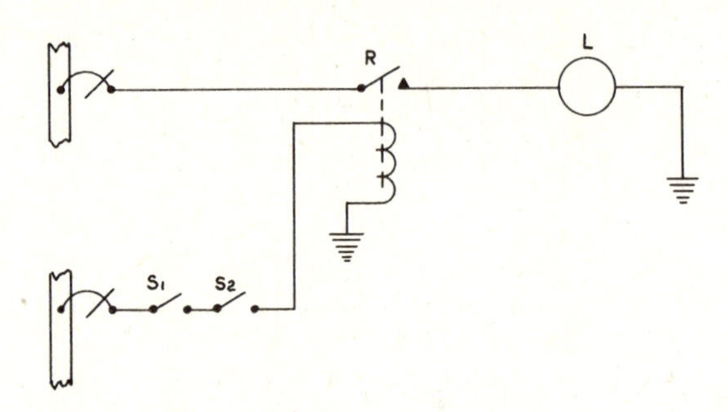

Fig. 15-16 AND-type logic circuit.

LOGIC GATES

In a logic system, control is attained through the use of electric circuits called **gates.** In the example of Fig. 15-16, the two switches constitute an AND gate because both switches must be closed to produce current in the circuit. The symbol for an AND gate and the associated **truth table** are shown in Fig. 15-17. The truth table makes it possible to see quickly what inputs are necessary to cause a gate to produce a desired result. A logic gate will have either a 0 output (nonconducting) or a 1 output (conducting). This makes it possible to utilize the **binary** digital system to indicate quantities by means of the outputs from a number of gates.

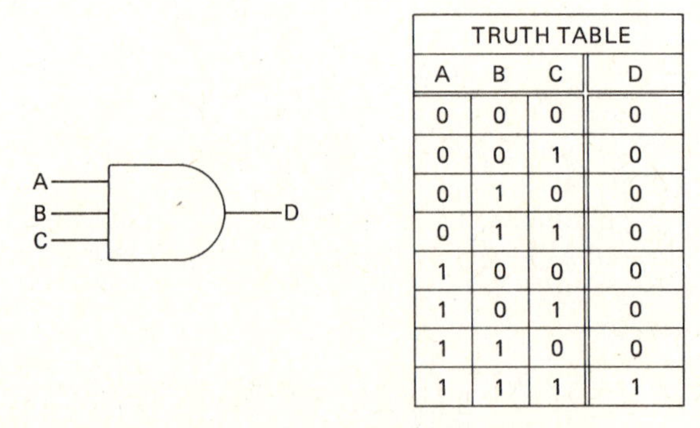

TRUTH TABLE

A	B	C	D
0	0	0	0
0	0	1	0
0	1	0	0
0	1	1	0
1	0	0	0
1	0	1	0
1	1	0	0
1	1	1	1

Fig. 15-17 Solar cell and module. (Arco Solar, Inc.)

The drawing of Fig. 15-18 illustrates one type of NOT gate. When voltage is applied at *A*, the transistor will con-

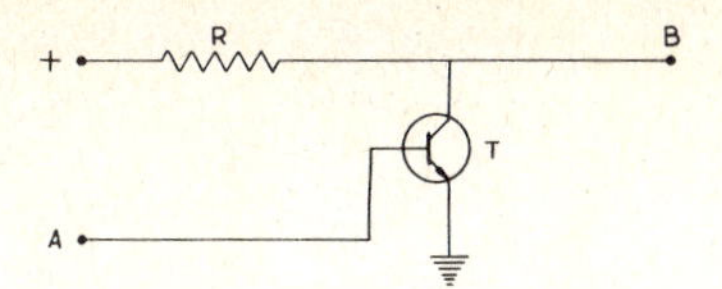

Fig. 15-18 A NOT-type logic circuit.

duct and this will leave the point *B* at approximately ground potential. Therefore, there will *not* be a voltage at *B*. If there is no voltage applied at *A*, the transistor will not conduct and voltage will appear at *B*. This circuit is called a NOT circuit or an **inverter.** It is an inverter because it makes a O output from a 1 input and a 1 output from a O input; that is, it inverts the input signal. The symbol for an inverter is shown in Fig. 15-19.

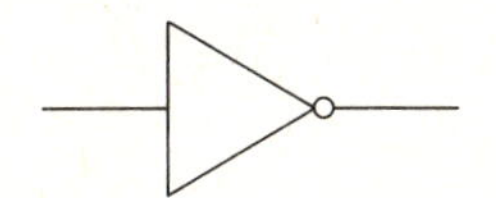

Fig. 15-19 An inverter or NOT gate symbol.

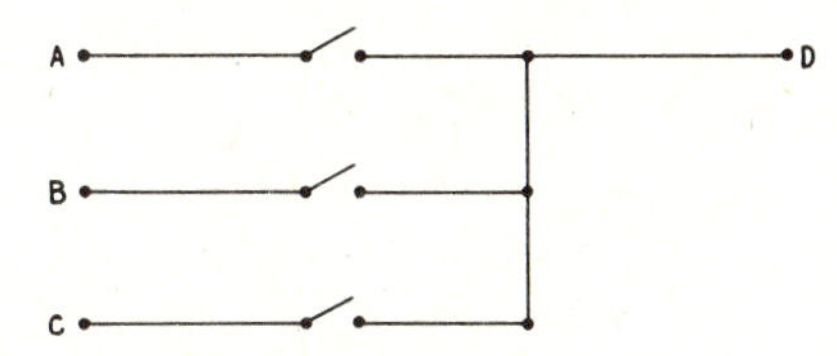

Fig. 15-20 An OR-type logic circuit.

The circuit shown in Fig. 15-20 is one type of OR gate. It is easily seen that if a voltage is applied to *A*, to *B*, *or* to *C*, a voltage will appear at *D*. Furthermore, if any one or all the switches are closed, a voltage will appear at *D*. The symbol and truth table for an OR gate are shown in Fig. 15-21.

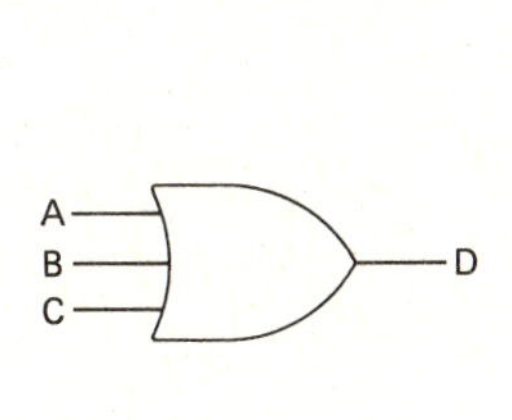

TRUTH TABLE

A	B	C	D
0	0	0	0
0	0	1	1
0	1	0	1
0	1	1	1
1	0	0	1
1	0	1	1
1	1	0	1
1	1	1	1

Fig. 15-21 Symbol and truth table for an OR gate.

Additional gates that will be found in logic circuits are *not and* (NAND) and *not or* (NOR) gates. These are made by inserting an inverter in the output circuit. The output of a NAND gate is the opposite of that of an AND gate for the same inputs, and the output of a NOR gate is the opposite of that of an OR gate with the same inputs. Symbols for these gates are given in the Appendix of this text.

BINARY DIGITAL SYSTEM

Since operating units in a digital electronic system have only two conditions or states, that is, conducting or nonconducting (on or off), it is necessary to employ a digital system that can utilize the on and off signals to produce letters, numbers, and symbols for communication. As mentioned previously, the *on* condition is given a 1 value and the *off* condition is given a 0 value. Through the binary system, these values can be used to indicate any required number, depending upon their timing or position in the output sequence of a particular unit.

When binary digits (bits) are written or printed, they are arranged in a row with the desired number of positions, and the number of 1 bits and their positions will determine the value of the actual number. The values are set with the lowest at the right. The value doubles for each digit position from right to left as follows:

12 11 10 9 8 7 6 5 4 3 2 1
2048 1024 512 256 128 64 32 16 8 4 2 1

15 14 13
16384 8192 4096

If all the values of the bits in all columns are 1s, the total figure indicated will be 32767. We can see, then, that with 15 columns of bits, we can indicate any quantity up to 32767. Suppose it is desired to indicate the quantity 9386 as a binary number. The 14-bit **word** would be arranged as follows:

1 0 0 1 0 0 1 0 1 0 1 0 1 0
8192 0 0 1024 0 0 128 0 32 0 8 0 2 0

When all the indicated values are added, the total of 9386 is obtained. The following will help the student to see how the numbers from 1 to 10 would be indicated:

1 0 0 0 0 1
2 0 0 0 1 0
3 0 0 0 1 1
4 0 0 1 0 0
5 0 0 1 0 1
6 0 0 1 1 0
7 0 0 1 1 1
8 0 1 0 0 0
9 0 1 0 0 1
10 0 1 0 1 0

Computers utilize binary digits for indicating words and symbols as well as numbers. This is accomplished by coding letters and symbols numerically. When the information is printed out or otherwise displayed, the coded numbers are revealed as letters or symbols.

REVIEW QUESTIONS

1. Give an example of a simple electric control system.
2. What basic elements are required for an automatic control system?
3. What is the function of a *sensor?*
4. Draw a bridge circuit and show how it may be used in a control system.
5. Why is a follow-up arrangement required with some control systems?
6. What is the difference between *photoconductive* and *photovoltaic* optoelectronic devices?

7. What is the difference between a *phototransistor* and other types of transistors?
8. Describe a *solar cell.*
9. Show how light may be used to operate a control system.
10. What is the purpose of a multivibrator circuit?
11. How is a multivibrator used in a switching circuit?
12. Discuss the use of solid-state devices in switching circuits.
13. Describe two types of *counting circuits.*
14. For what purposes are counting circuits employed?
15. Explain the values of *differentiating* and *integrating* circuits.
16. Explain the operation of a simple *pulse-forming* circuit.
17. Describe a simple *logic* circuit.
18. Describe a *gate* in a logic circuit.
19. What is the purpose of a *truth table?*
20. Compare an AND gate with an OR gate.
21. What is an *inverter* gate?
22. How does a NAND gate compare with an AND gate?
23. Describe the *binary* digital system.
24. How would you indicate the quantity 4536 in binary digits?
25. Why is the binary system important in solid-state digital circuitry?

CHAPTER 16

Weather Warning Systems

Since weather conditions along a flight route are of prime importance in determining whether a flight can be continued safely, means are needed to enable the pilot of an aircraft to "look" ahead of the airplane for many miles to see if dangerous weather exists. One of the principal means for accomplishing this purpose is **radar.** Another system developed to determine weather conditions ahead of an airplane is called a Stormscope. Stormscope is the trademark of a system that detects the electrical activity caused by storm conditions and displays the information on a cathode-ray tube screen.

It is the purpose of this chapter to examine both radar systems and the Stormscope and to describe some of the installations.

RADAR

The word *radar* is derived from the expression **radio detection and ranging.** This equipment was developed to a high level of performance by Great Britain and the United States during World War II for the detection of enemy aircraft and surface vessels. By the end of the war it was being used for many other purposes, including navigation, blind landing of aircraft, and bombing through overcast. Today, among numerous other functions, radar is used for weather mapping, terrain mapping, and air traffic control.

Radar systems have been developed for all types of aircraft from single-engine general aircraft to the largest airliners. Early radar systems were heavy and cumbersome and included many separate units. The extensive use of solid-state and microelectronics has brought about the development of small, compact systems weighing not more than 15 lb [6.8 kg]. The majority of radar systems operate on an echo principle in which high-energy radio waves in pulse form are directed in a beam toward a reflecting target. The beam of pulses is actually something like a stream of bullets from a machine gun, with a relatively long space between each pulse of energy. When the pulse of energy strikes the target, which may be a mountain, rain clouds, or an airplane, a portion of the pulse is reflected back to the receiving section of the radar (see Fig. 16-1).

In Fig. 16-1 at *A* the pulse has just been emitted from the airplane radar antenna. At this point a *pip* appears on the radar scope (cathode-ray tube). At *C* the pulse is striking a rain cloud. A portion of the pulse is reflected by the cloud and returns toward the airplane, as shown at *D*. When the reflected pulse reaches the airplane at *E*, a second, smaller pip appears on the radar screen. The dis-

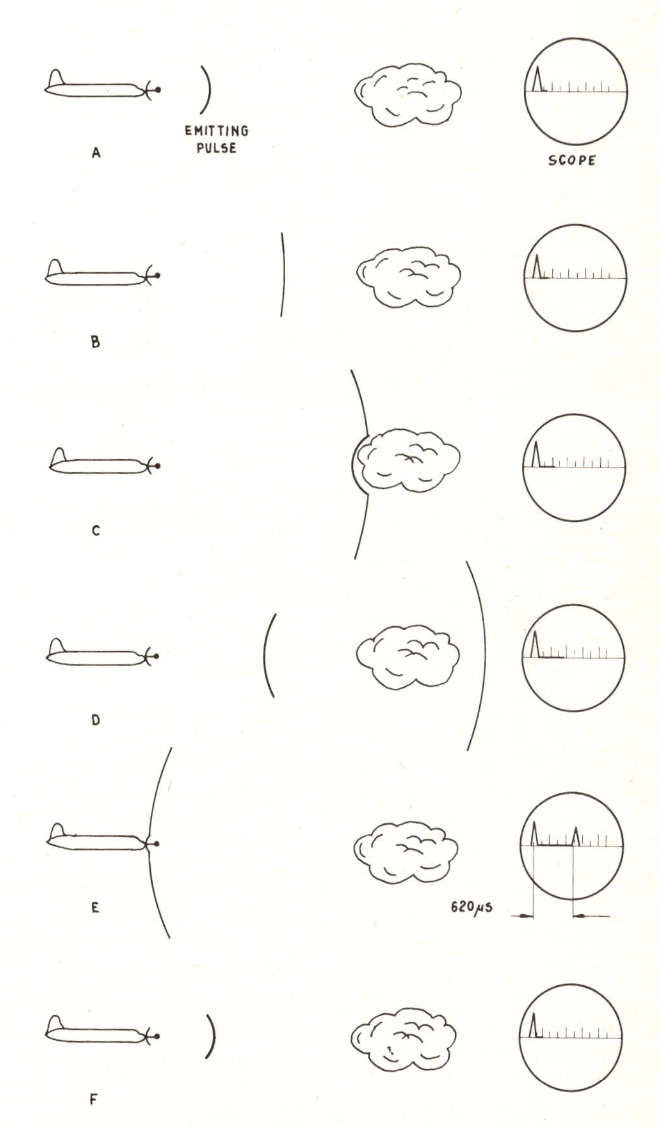

Fig. 16-1 Radar pulse transmission and reflection.

tance between the two pips indicates the distance from the airplane to the cloud. In Fig. 16-1 the time between the two pips is shown as 620 microseconds [μs], which represents a distance of approximately 50 mi [80 km]. At *F* another pulse is emitted from the radar antenna.

NATURE OF RADAR SIGNALS

A typical radar signal may consist of a carrier wave of 8000 MHz broken into pulses with a duration of 1 μs and spaced at intervals of $\frac{1}{400}$ s or 2500 μs. With the values stated, each pulse would have an actual length of about 300 m [483 km], and the distance between the pulses would be about 750 000 m or 466 mi [750 km]. This would be a ratio of roughly 2500 : 1 for the time of **no signal** to the time of **signal.**

It must be pointed out that the ratio of the length of a pulse to the time of no signal varies considerably with the frequency, which ranges from 150 to 30 000 MHz. The various bands are listed in Table 16-1.

TABLE 16-1

Wavelength	Identification letter	Frequency, MHz
2 m	P	150
30 cm	L	1 000
10 cm	S	3 000
5.6 cm	C	5 400
3 cm	X	10 000
1 cm	K	30 000

The length of the pulses of a radar signal may vary from 0.25 to 50 μs, depending on the requirements of the system. The pulse repetition frequency (PRF) also varies according to the distance over which the signals must travel. For very long distances, the pulse rate must be slow enough so that the return signal will be received before another pulse is transmitted. If this were not accomplished, it would be difficult to tell whether the pulse shown on the viewing screen (CRT or scope) was the one transmitted or the one received.

The use of a pulse system in radar makes it possible to transmit very powerful pulses. In effect, all the power is concentrated in the very short pulses. If the average power output if a transmitter is 10 W, the pulse power may be as high as 25 000 W.

In early types of radar systems, the display on the CRT was a horizontal scale and was called an A scan. The distance from the transmitted pulse to the received pulse indicates the distance of the target from the transmitter. With this type of scan, the direction of the target could not be determined except by noting the direction in which the antenna was pointed. To enable the radar to provide direction information, the P scan was developed. The P scan is described later. The A scan produces a horizontal line on the face of the CRT as the beam of electrons moves across the fluorescent surface. The horizontal line is used to establish the distance of the target from the radar transmitter and is indexed with range lines. For example, if the length of the line is to represent 20 nautical miles (nmi), the time in which the line is traced must be equal to the time that it takes a pulse to travel 20 nmi [37 km] and return. Since a radar signal travels at the rate of approximately 1 nmi [1.85 km] in 6.19 μs, it would take 12.38 μs for the radar signal to travel out 1 nmi and back 1 nmi. Therefore, if the horizontal line is to serve for a distance of 20 nmi, the time of scan must be 247.6 μs.

Figure 16-2 is a diagram of an A scope showing a scale on the face of the scope to provide a range reading. When the pulse is transmitted, the pip (*a*) is produced on the time base line. The reflected pulse, which has returned to the radar receiver, makes the pip at (*b*). If the full length of the line represents 20 nmi, we know that the target represented by (*b*) is approximately 10 nmi [18.5 km] from the radar set. Radar equipment is normally provided with adjustments so that the time base can be changed to represent different ranges. This is accomplished by changing the horizontal sweep circuit to increase or decrease the time required for the electron beam to sweep across the face of the scope. The sweep system is provided with a circuit which blanks out the electron beam in the CRT while it is shifting from the right-hand side of the scope back to the left-hand side, where it will start another visible sweep.

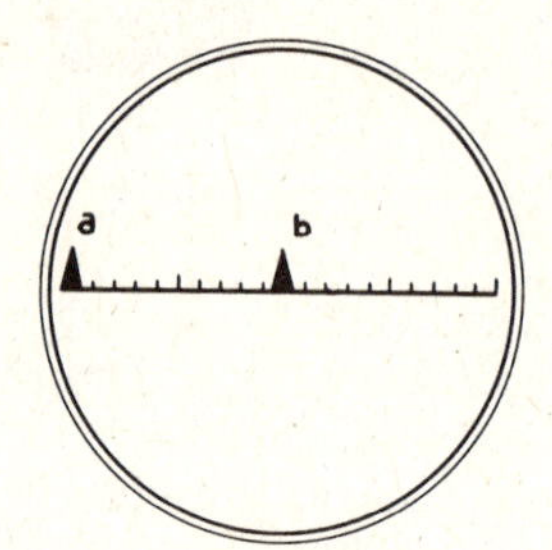

Fig. 16-2 Type-A scan.

The P scan is illustrated in Fig. 16-3. This type of radar scope may be called the **plan position indicator** (PPI), since it indicates both the distance and direction (azimuth) of the target. On the face of the PPI scope, the time trace starts at the same time that a pulse is transmitted from the radar antenna, and the reflected pulses cause bright spots along the trace line. The trace line is adjusted so that its intensity is very light or almost invisible except at the point that a target signal is received. The pulses are generated at such a frequency that the trace lines scan the entire face of the scope as the antenna makes a complete revolution; hence, as reflected signals appear on the screen, a picture appears in a shape similar to that of the object which reflects the signal. The fluorescent coating inside the face of the CRT is of a type which retains a fluorescent glow for several seconds after being activated by the electron beam. Thus the picture remains on the

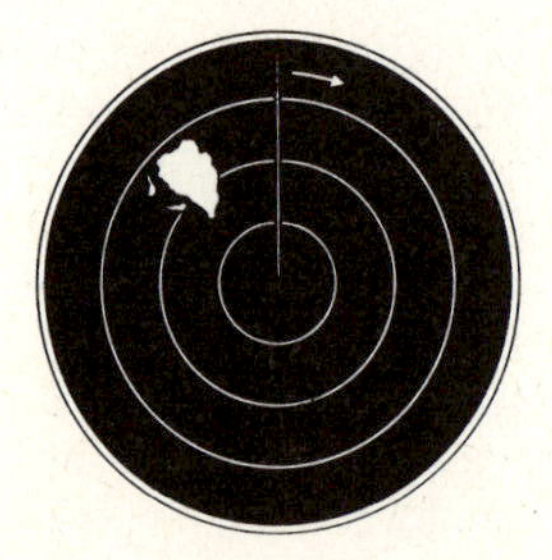

Fig. 16-3 Type-P scan.

screen and is reactivated each time that the time trace makes a complete circle.

OPERATION OF THE PPI

The operation of a PPI scope is illustrated in Fig. 16-4. The antenna rotates at 1 revolution per second (rps), more or less, as it searches a 360° area, and the time trace on the scope rotates at precisely the same rate. The time trace is rotated by means of a synchronized drive motor which rotates the deflecting coils of the CRT in synchronization with the antenna rotation.

Normally, when the PPI radar is installed in an airplane, the time trace will be vertical from the center of the scope to the top edge when the antenna is pointing directly ahead of the airplane. When the antenna is turned to the right, the time trace will be to the right (*see* Z in Fig. 16-4*c*).

Let us assume that the PPI scope in Fig. 16-4 is installed in an airplane and that the antenna is pointing directly forward. A pulse is transmitted, and a bright spot appears at *x*. If the length of the trace from the center of the scope to the edge represents a distance of 30 mi [48.3 km], the spot appearing at *x* indicates that the target (reflecting object) directly ahead of the airplane is at a distance of 20 mi [32.2 km]. Each pulse and each trace will continue to show the object for as long as the transmitted pulses strike it. This will add to the picture, providing an indication of the size of the target as well as its direction.

Figure 16-5 shows a map at (*a*) corresponding to the PPI indication at (*b*). Note that the airplane and the ship appear on the screen as white spots, the water as a black area, because it does not reflect the radar pulses, and the land as a mottled white area.

PRINCIPAL UNITS OF RADAR SYSTEMS

Figure 16-6 is a simplified block diagram of an aircraft radar system. This system consists of seven principal units, each serving a particular purpose in the system. The **synchronizer** provides the timing for the radar signal and synchronizes the transmitter, receiver, and indicator so that all operate together with correct timing. The timing and synchronizing are accomplished by trigger pulses generated in the synchronizer. These pulses originate in a multivibrator or similar pulse-generating circuit. In a system such as an airborne weather radar, the 400 Hz ac power supply is also sometimes used to synchronize the system and provide the 400 Hz PRF.

The **modulator** stores energy and supplies high-voltage pulses which are released by the trigger pulse from the synchronizer. During the interval between pulses, a network consisting of inductors and capacitors is charged to a high level. When the trigger pulse releases this energy, a high-voltage pulse, rectangular in shape, is delivered to the **transmitter.**

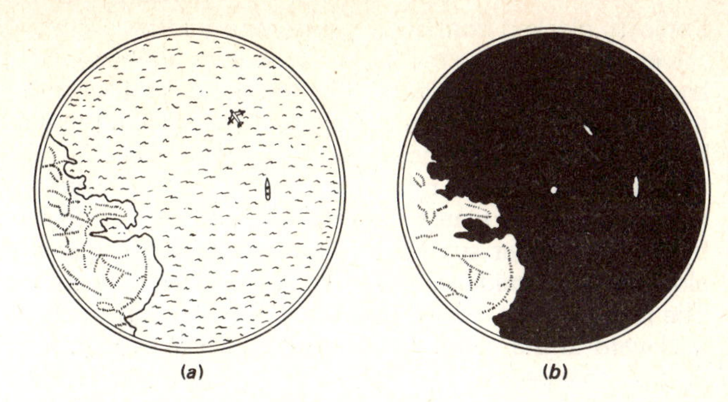

Fig. 16-5 Comparison of the actual area in view of a radar with the ppi indication.

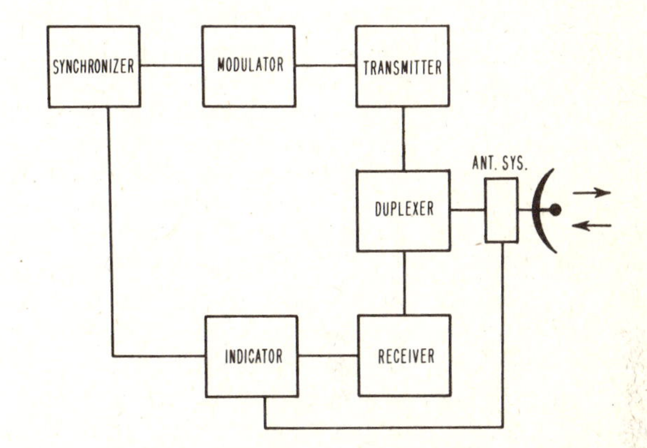

Fig. 16-6 Block diagram of a radar system.

The principal element of a radar transmitter is a **magnetron** tube. This tube receives the high-energy pulse from the modulator and converts it to an extremely high-frequency pulse which is sent on to the antenna system to be radiated into space. The magnetron makes use of **resonant cavities** to generate the correct frequency for the transmitted pulse. A resonant cavity may be compared to an empty shotgun shell: when a blast of air is blown across its open end, the shell will emit a whistle with a certain fixed frequency. In like manner, when a pulse of high-voltage electric energy is delivered to the magnetron cavities, a fixed-frequency radio wave is generated (5400 MHz in the case of a typical weather radar system).

The UHF pulse from the transmitter is carried by means of a waveguide section to the **duplexer** and from the duplexer through a waveguide to the antenna. The duplexer is an electronic switching device which alternately connects the transmitter and receiver to the antenna. When the pulse is emitted from the transmitter, it is elec-

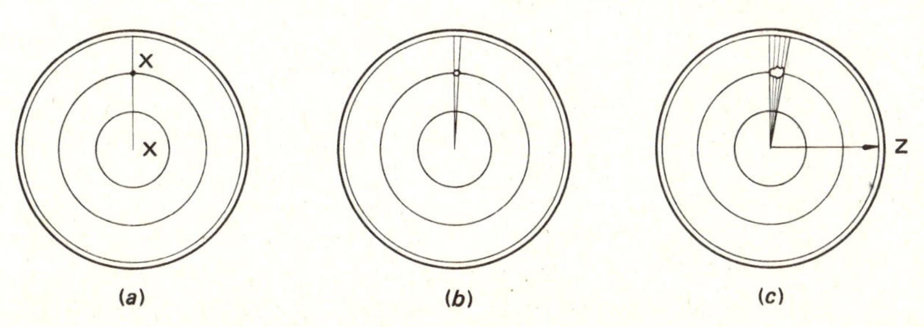

Fig. 16-4 Operation of the PPI.

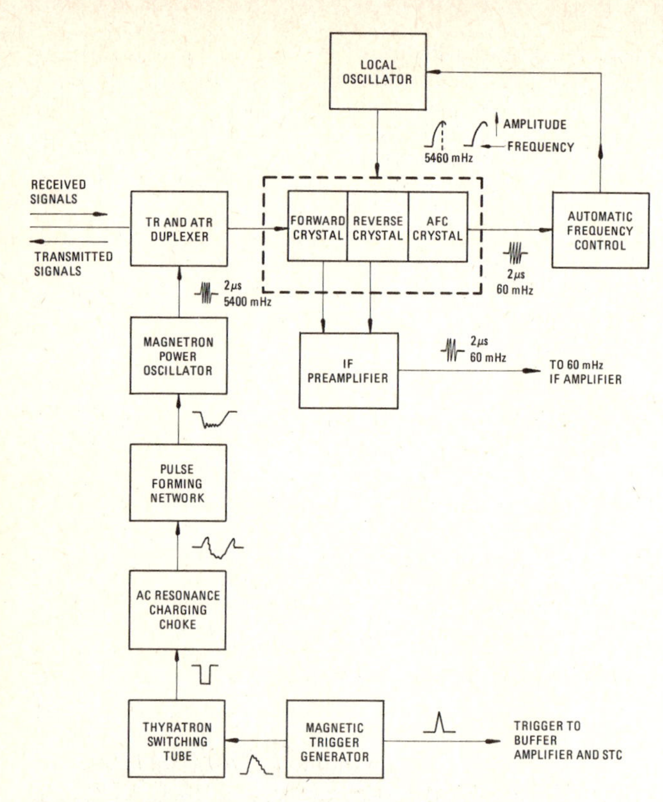

Fig. 16-11 Block diagram of a transmitter-receiver system.

rate of approximately 400 Hz. The PRF may vary somewhat from 400 Hz inasmuch as it is controlled by the line frequency of the 400-Hz power source. The trigger pulses are used in the transmitter to fire a **thyratron** switching tube which, in turn, releases the energy stored in the pulseforming network. This pulse, having a voltage of about 3200, is fed to the pulse transformer, which has a ratio of 1:5. The output of the transformer is, therefore, about 16 000 V, and this output is delivered to the cathode of the magnetron as one negative pulse. The peak power of this pulse is about 210 kW, which provides an RF output from the magnetron of over 75 peak kW at a 2-μs pulse width.

The magnetron tube is a package type with magnets permanently attached. The high-voltage pulse is fed to both sides of the filament at once, thus keeping the potential across the filament neutral so that no damage will be caused. The cathode is connected to the filament to receive the charge of the negative pulse, and filament voltage is fed from the magnetron filament-voltage transformer. When the negative pulse is applied, the flow of electrons creates a 5400-MHz oscillation in the cavity section of the magnetron, and this RF energy is coupled directly from the tube to the waveguide section.

The transmitter-receiver shown in Fig. 16-9 incorporates a **running-time meter** and a test meter with a switch to test eight metering positions, all on the front panel. Cooling is provided by means of blowers which circulate air through vents and louvers.

THE ACCESSORY UNIT

A complete accessory-unit assembly is shown in Fig. 16-12. This assembly includes power-supply, synchronizing, video, isocontour, intermediate-frequency, and stabilizing circuitry for the system. The unit also includes a test meter

Fig. 16-12 Radar accessory unit. (RCA)

and selector switch on the front panel, as shown in the illustration. Cooling is provided in the same manner as the cooling for the transmitter-receiver unit.

A block diagram of the accessory unit is shown in Fig. 16-13. When this diagram is compared with the diagram in Fig. 16-8, the corresponding sections can be identified easily. In Fig. 16-8 the **intermediate frequency** and **gain** are shown. In Fig. 16-13 the intermediate frequency is labeled **60-Hz IF amplifier,** and the gain is shown as being effective between the **sensitivity-time control** and the **IF amplifier.** The foregoing units are named according to their respective functions, that is, **gain** and **amplification.**

The **iso** and **iso switch** shown in Fig. 16-8 correspond to the following units in Fig. 16-12: **isocontour switch relay, isocontour clamp,** and two **isocontour ampli-**

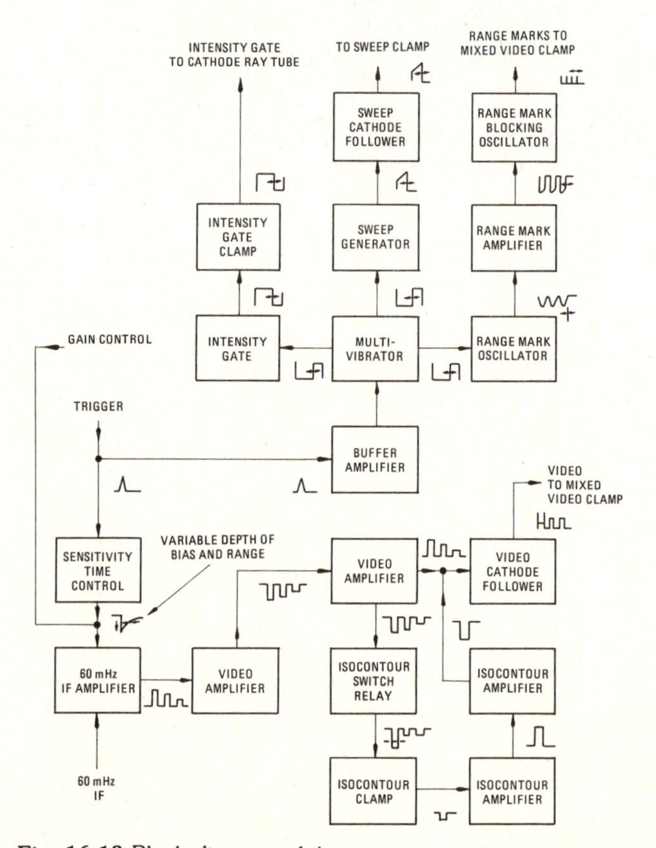

Fig. 16-13 Block diagram of the accessory unit.

fiers. The function of the isocontour system is to permit the operator to detect the hard-core or turbulent areas of the storm clouds. When the isocontour is turned on, the most intense portion of the picture on the indicator screen is inverted and, instead of showing as a bright area, shows black. Thus the hard core of a storm cloud shows as a black spot surrounded by a white area, which is the "soft" part of the cloud. The effect of picture inversion on a television screen may be seen on some sets by turning the **brightness** control to the extreme BRIGHT position. The normal dark portion of the picture will become light, and the light portions will become dark. In effect, a negative picture is produced on the screen.

In Fig. 16-8 the **video** section is shown as one block and in Fig. 16-12 as three blocks—two video-amplifier sections and a video-cathode follower. The function of the video circuits is to amplify further the detected signals from the IF amplifier section and also to permit the injection of the isocontour function.

The three circuits for the **intensity gate, sweep,** and **range** are all clearly indicated in both Figs. 16-8 and 16-12. The intensity gate turns on the electron beam in the CRT when the trigger signal starts the transmission of a pulse. The beam is on until it has completed the sweep from the center of the screen to the edge. At this time the intensity gate cuts the beam off until the next trigger signal starts the action again.

The sweep circuit provides a linear sawtooth signal which causes the electron beam to move steadily from the center of the tube screen to the rim. The sweep signal is started from the multivibrator along with all the signals sent to the other circuits which require timing with the transmitted pulse signal. The multivibrator is shown in Fig. 16-13 and is included in the **timer** shown in Fig. 16-8.

The **range-marks** system adds pulses to the electron beam at accurately timed intervals so that the beam will draw concentric circles on the full-circle screen. These circles represent an exact distance from the radar transmitter. For example, there are three circles evenly spaced from the center of the screen in Fig. 16-14. If these circles represent 10 nmi [18.5 km] each from the center of the screen, the target indication shown will be about 25 nmi [46.3 km] from the radar transmitter. Since the time for a pulse to travel 1 nmi [1.85 km] and return is approximately 12.38 μs, the distance between the range-mark rings will represent a sweep time of 123.8 μs, and the total sweep time from the center of the screen to the rim will be 371.4 μs. At 400 Hz the sweep signal interval is 2500 μs; hence, the shutoff time on the beam is 2500 —371.4, or 2128.6 μs, when the range of the indicator is set for 30 nmi [55.6 km]. This means that the electron beam is turned on for 371.4 μs while it sweeps from the center of the screen to the rim and is then turned off for 2128.6 μs until the next trigger appears.

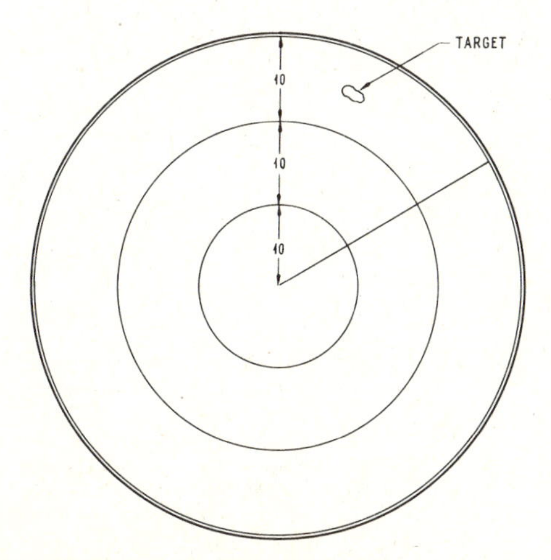

Fig. **16-14** Range markings on a ppi screen.

THE INDICATOR

The indicator of the radar system is shown in Fig. 16-15. The indicating tube is a PPI CRT with electrostatic focusing and magnetic deflection. The PPI indication is accomplished by intensity-modulating the radial-sweep beam while simultaneously rotating the beam in synchronism with the rotation of the antenna. The electron beam is modulated by means of pulse signals applied to the cathode. This modulation provides both range marks and video indication. Remember that the video signals or pulses "paint" the picture on the screen.

Fig. 16-15 A ppi indicator unit. (RCA)

On the front of the indicator case are four control knobs. At the upper left corner is the **range-marks** control, which is used to adjust the brillance of the range marks on the screen. The **lights** control at the upper right is used to adjust the brightness of the screen illumination and cursor lines. The **intensity** knob at the lower left adjusts the intensity, or brightness, of the reflected signals, thus permitting the operator to increase the intensity of a weak indication for better visibility. The **cursor** control at the lower right is used to rotate the cursor lines on the face of the scope to a desired azimuth position, making it possible to determine more accurately the direction of a target. The cursor lines may also be used to determine whether another airplane in the area is on a course which may result in a collison.

The appearance of a PPI scope for typical weather presentations is shown in Fig. 16-16. The bright areas indicate rain clouds and stormy areas. Any target lying in the area between the center of the scope and the top center is in the path of the airplane. When the antenna is at zero tilt, there are no indications in the bottom quadrant of the

Fig. 16-16 A weather indication on a full-circle ppi scope.

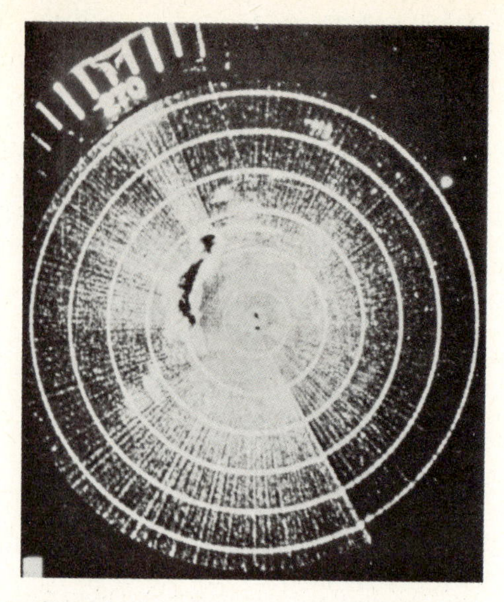

Fig. 16-17 Weather indication with isocontour on.

screen, because the airplane blocks out this section as the antenna turns back and sweeps across the fuselage area.

When it is desired to determine what areas in a storm system are producing heavy precipitation, the **contour** switch may be placed in the CONTOUR position. The most intense area of the signal will then be inverted, and the hard core of the storm will be shown as a black area on the screen as shown in Fig. 16-17.

As previously mentioned, the rotation of the radial-sweep beam is accomplished by rotating the deflection coils around the neck of the CRT (see Fig. 16-18). When the radar system is in operation, the **azimuth drive motor** rotates the antenna and through a gear train also rotates the **torque synchro generator.** When the **synchro relay** contacts are in the normally closed position, the **torque synchro motor** will also rotate and turn the deflection-coil yoke.

The **azimuth servo system** is self-synchronizing. On the antenna synchro slip ring there is an insulated 18° segment. The synchro ring at the CRT conducts on only a 6° segment. If the antenna and deflection coils are synchronized, the synchro relay will be deenergized, and the torque generator will produce a three-phase current which causes the torque synchro motor to rotate in synchronization. If the antenna and deflection coils are not in synchronization, the deflection coil will rotate with the antenna only until the 6° conducting segment comes in contact with the brushes. Since the antenna synchro ring will be conducting, the circuit will be closed through the synchro relay. This will disconnect and short-circuit two leads from the torque synchro motor, thus causing the synchro motor to stop with the deflection coils in the dead-ahead position. The antenna will continue to rotate, and the nonconducting segment of the antenna synchro ring will come into contact with the brushes, thus breaking the circuit to the synchro relay. The relay will then release the contact points and reestablish the circuit between the synchro generator and torque synchro

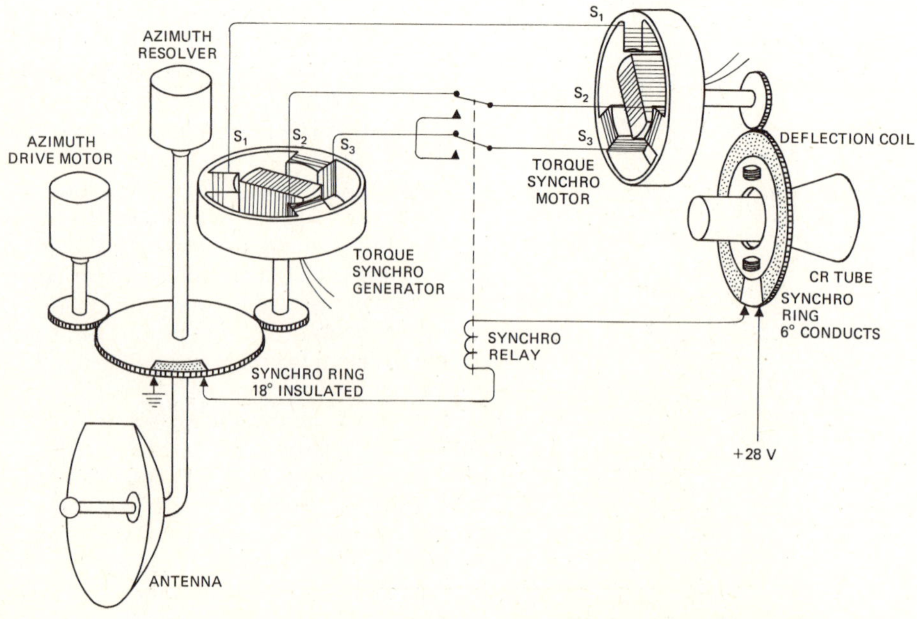

Fig. 16-18 Operation of antenna and indicator synchro system.

motor. At this time both the antenna and the deflection coil will continue to rotate together in synchronization.

ANTENNA ASSEMBLY

The antenna assembly of the radar system, illustrated in Fig. 16-19, is a rather complex device, largely because of its rotating and tilting mechanisms. The microwave RF signal is transmitted through a variety of waveguide sections and joints from the duplexer to the antenna reflector, and the reflected signal is returned through the same system. In each case the energy must be carried with a minimum of loss.

The antenna is mounted in the nose of the airplane with a Fiberglas or plastic nose fairing (radome) to cover it. This is necessary to protect the antenna and still allow the transmitted signals to be radiated into space. A metal nose would reflect the waves back; hence, there would be no radiation of energy.

In Fig. 16-19 the upper part, shown in the illustration, which is attached to the airplane structure, is called the **antenna base assembly.** All the power for rotation and tilt is brought into the base assembly by means of a plug connector. RF energy enters through the waveguide at the rear and passes down through the antenna frame assembly, which holds the reflector. In the base assembly are the azimuth drive motor and gear train, the synchro generator, and associated circuitry. Electric power is conducted from the stationary base to the rotating antenna by means of slip rings and brushes.

A simplified diagram of the **stabilization** system for the antenna is shown in Fig. 16-20. The purpose of the stabilization system is to enable the radar equipment to scan continuously in the same horizontal plane regardless of the pitch or roll of the airplane. From the diagram it can be seen that the pitch-and-roll signals originate with the flight gyro, which is a part of the automatic-pilot system. These signals are amplified and sent to the azimuth resolver. This resolver is connected through a resistor network to the elevation resolver and the tilt resolver. These resolvers furnish information to a computer system which

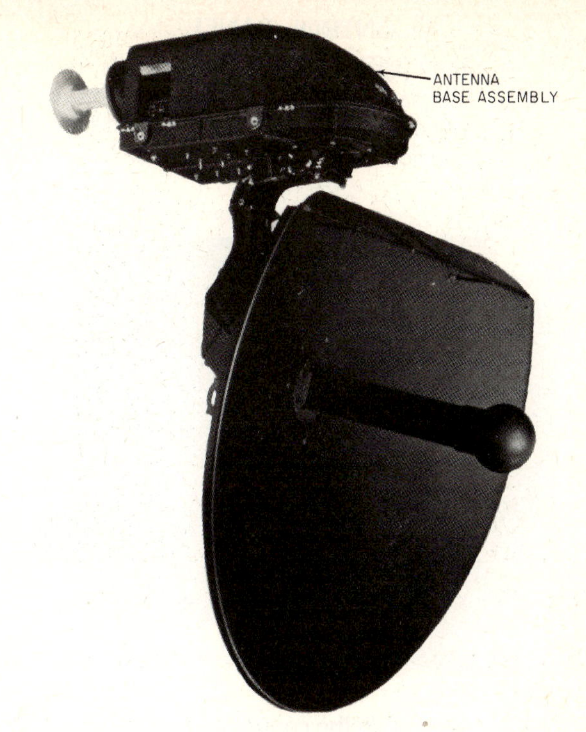

Fig. 16-19 Rotating antenna assembly. (RCA)

continuously solves the equation necessary to provide instructions for the elevation drive motor which performs the work of stabilization. The elevation drive motor actually positions the antenna reflector and feed to provide a continuous horizontal sweep, or rotation, of the antenna beam. The purpose of the tilt resolver is to inject a signal calling for a downward angle of sweep as desired by the operator. The tilt resolver is controlled by the TILT knob on the radar control panel.

The radar system can be operated either with or without antenna stabilization by means of the STABILIZER switch on the control panel. When the switch is placed in the stabilization position, the antenna becomes stabilized.

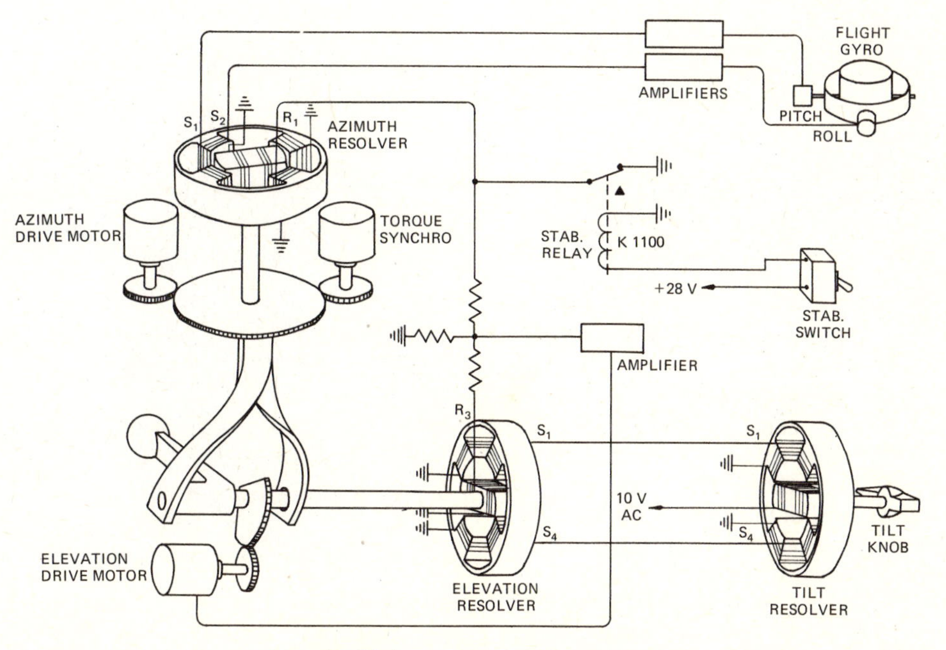

Fig. 16-20 Antenna stabilization system.

CHAPTER 17

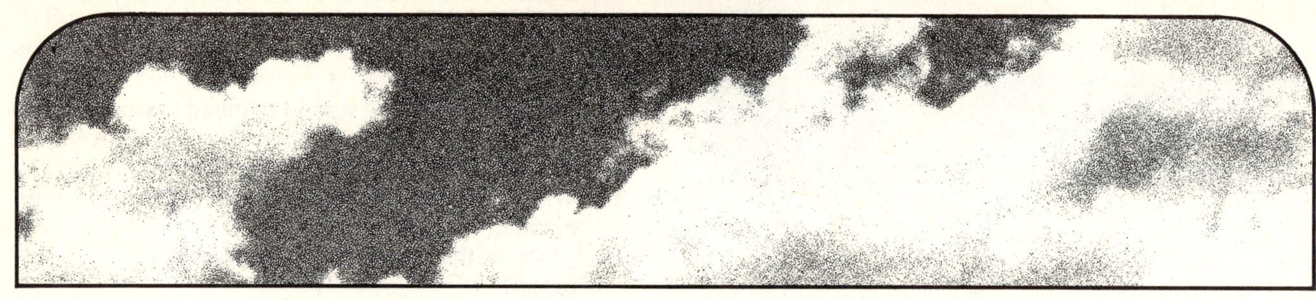

Communication and Navigation Systems

The use of radio equipment and avionics in general has increased markedly for all types of aircraft during the past thirty years. One of the reasons for this increase is the Federal Aviation Administration's requirement that all aircraft operating in high-traffic areas be equipped with two-way radio for communication with air traffic controllers and tower operators. The development of solid-state electronic technology has made it possible to install highly complex and sophisticated systems for communication, navigation, and automatic flight control in all types of aircraft. Previously, such systems could be utilized only in large aircraft because of the size and weight of the system components.

Avionic systems installed in aircraft can include communication radio (COM), navigation systems (NAV and RNAV), and automatic flight-control systems (AFCS). The navigation systems may include VHF omnirange receivers (VOR), distance-measuring equipment (DME), automatic direction finder (ADF), localizer receiver (LOC), glideslope receiver (GS), marker beacon receivers, identification transponder, radio altimeter, encoding altimeter, and numerous indicators. In some cases, particularly for small aircraft, many of these functions are built into one electronic package mounted in the instrument panel.

COMMUNICATIONS

Radio communications for aircraft are primarily for the purpose of air traffic control; however, commercial aircraft also utilize a range of high frequencies for communicating with ground stations and other aircraft for business and operational purposes. Communications for air traffic control are in the VHF band in the range between 118 and 135.975 MHz.

Communication systems for airliners involve interphone systems in the aircraft, passenger address systems, aircraft-to-aircraft communications, ground station communications, and air traffic control communications.

INTERPHONE

The interphone communications for one type of airliner is arranged in two separate systems which may operate together or independently. One system is used primarily for flight interphone operation, and the other, the service interphone system, is used for two-way communication between stations while the airplane is on the ground.

The flight interphone system allows crew members in the airplane during flight operation to communicate with each other by any one of the communication radio systems and also to monitor some of the navigation systems which produce audible signals. Both the pilot and copilot are provided with loudspeakers at their stations so that they may monitor incoming signals without using headsets. The flight interphone system is independent of the service interphone system and is operated by the use of control panels at the pilot's, copilot's, navigator's, engineer's, and radio-rack stations. A handset jack in the external power receptacle box is provided so that groundcrew members may communicate with anyone inside the airplane; the jack is connected to the pilot's interphone control panel. Such a control panel is illustrated in Fig. 17-1.

The service interphone system provides two-way communication between many external locations about the airplane. By using this system, maintenance personnel may communicate with each other at widely separated points on the airplane and also communicate with flight-crew members during preparations for flight. The attendant's interphone stations and the control-stand interphone station are equipped with telephone-type handsets. All other service interphone stations are equipped with a handset jack so that a handset or a combined handset and headset may be plugged in and operated at any time.

The general arrangement of the interphone system may be understood by examining the diagram in Fig. 17-2. This illustrates how the various sections of the system are interconnected and also indicates the locations of the different jacks into which headsets or headset combinations may be plugged. A switch on the engineer's panel is used to disconnect all service interphone stations during flight except those at the cabin attendant's panels, control stand, radio rack, and external power receptacle box.

Audio selector, or control, panels for the interphone system are located at the pilot's, copilot's, flight engineer's,

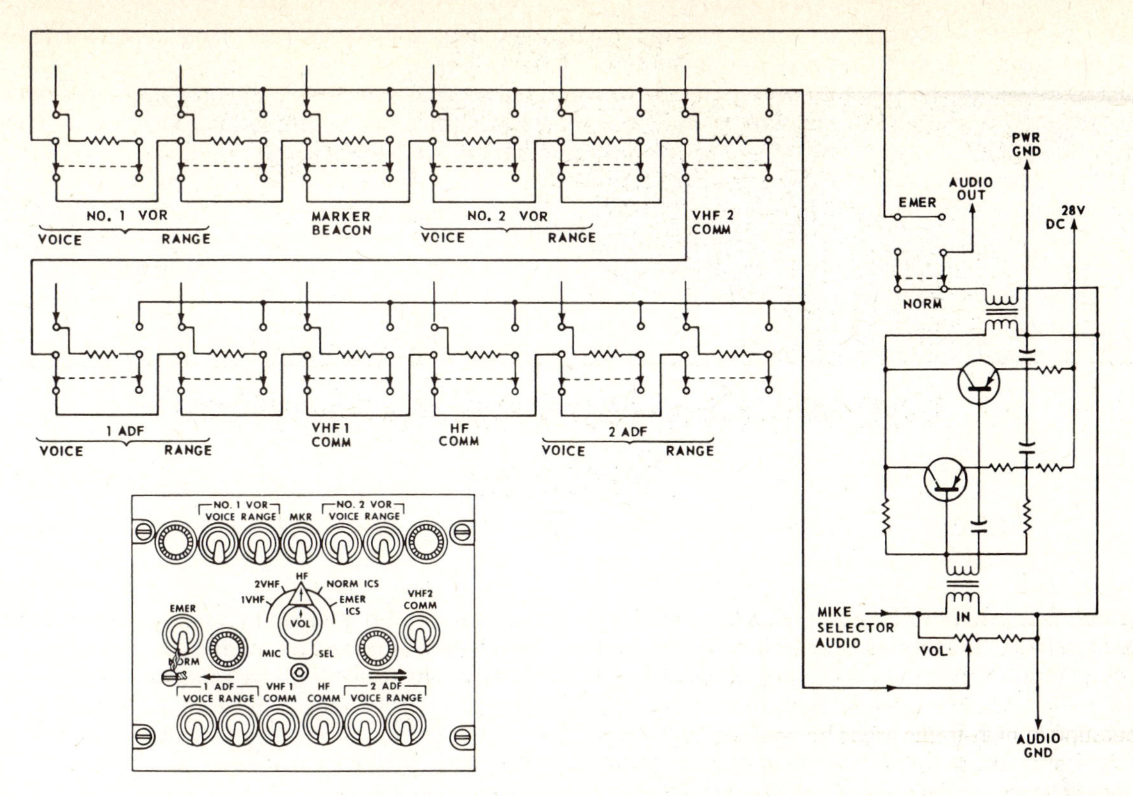

Fig. 17-1 Interphone control panel. (Boeing Co.)

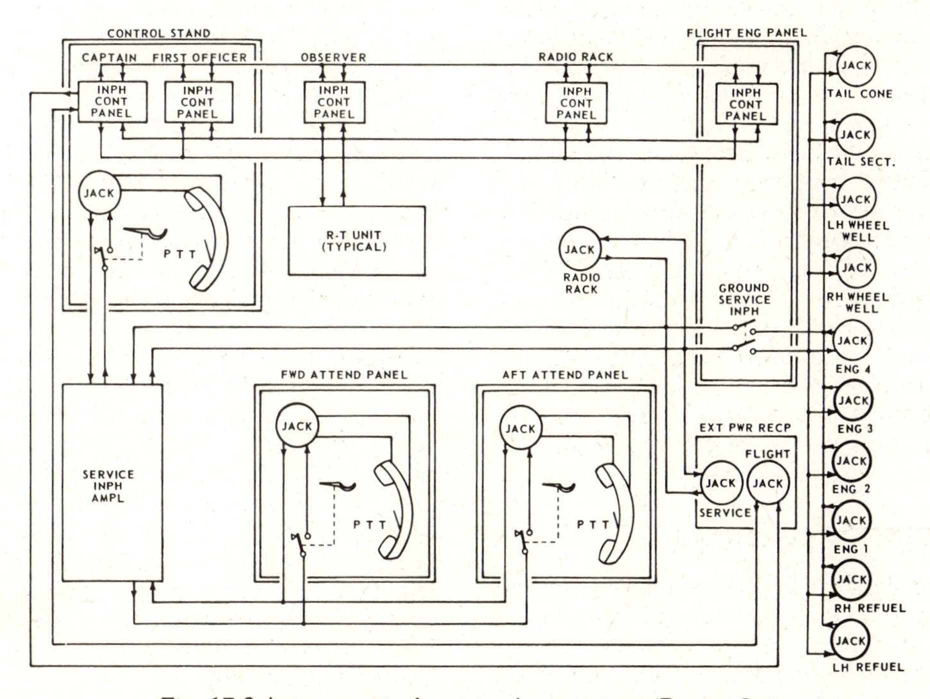

Fig. 17-2 Arrangement of an interphone system. (Boeing Co.)

navigator's, and radio-rack stations. Each panel includes signal-selection toggle switches, a microphone selector switch, and a transistorized audio amplifier with associated control and circuitry. The toggle switches are used to connect audio signals from the various radio-equipment units to the control-panel amplifier. Any one or any combination of audio facilities can be monitored.

One of the toggle switches on the audio selector panel permits emergency monitoring of signals in the event that the interphone control-panel audio amplifier is inoperative. Signals may be heard from any one facility when the EMER-NORM toggle switch is in the EMER position and the toggle switch for the desired facility is also on. Signal strength in the emergency position will be at the level produced by the particular receiver to which the interphone system is connected. The emergency monitoring circuit is arranged on a priority basis, and only one signal source may be monitored at a time.

A rotary-type microphone selector switch on the panel makes it convenient to transmit on any one communication facility. By rotating the selector switch to the desired position, microphone output may be directed to 2-VHF, 1-VHF, HF, NORM ICS, or EMER ICS. Selection of any facility directs microphone control and microphone audio output to the facility and also connects the selected facility's sidetone output to the audio amplifier in the control panel as shown in Fig. 17-3. In addition to the controls on the interphone audio selector panels, push-to-talk switches are provided at the flight interphone stations for use with oxygen-mask microphones. Selector switches on the pilot's and copilot's side panel are provided so that either a smoke-mask or an oxygen-mask microphone may be used when conditions so require, by actuating the switches to the desired position.

The forward and aft attendant's panels are equipped with telephone-type handsets connected to the service interphone system. The handsets are suspended on hook switches when not in use and must be removed from the hook to complete the receiver circuit. A separate push-to-talk switch on the handset must be actuated before the handset microphone is connected to the interphone system.

The **microphones** used with the interphone may be either carbon microphones or transistorized dynamic microphones. The interphone amplifier microphone input circuit is designed to provide polarizing voltage for carbon microphones. It should be remembered that a carbon microphone is not self-energizing and that it therefore requires a source of voltage before it can become effective. Hand-held microphones which incorporate microphone switches may be used at each of the interphone stations during normal flight operations. If the pilot, copilot, navigator, or flight engineer is wearing an oxygen mask, the microphone in the oxygen mask may be used when a separate push-to-talk switch at the crew-member's station is operated.

The pilot and copilot are provided with **loudspeakers** which make it possible for them to monitor communication equipment without wearing headsets. The loudspeakers operate on 28-V dc power which is obtained from the same source as the associated interphone and control panel. The loudspeakers contain transistorized amplifiers which provide a maximum power output of 3 W to the speaker voice coil. An ON-OFF-VOLUME knob on the front of the speaker assembly makes it possible to adjust loudspeaker output to the desired level. A muting relay in the loudspeaker assembly is energized whenever an associated microphone and control switch are closed. When the relay is energized, an additional resistance is placed in series with the volume control to reduce the output level of the loudspeaker. This arrangement is provided so the loudspeaker will not interfere with the transmission of messages while the operator is speaking on a microphone. Figure 17-4 illustrates the amplifier and circuitry for a loudspeaker system.

The extent of the dual interphone system on a large airliner is illustrated in the drawing of Fig. 17-5*a* and *b*. In

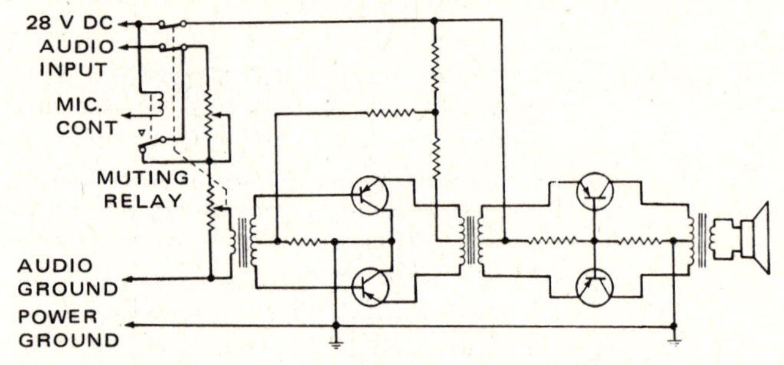

Fig. 17-4 Amplifier and circuitry for a loudspeaker system.

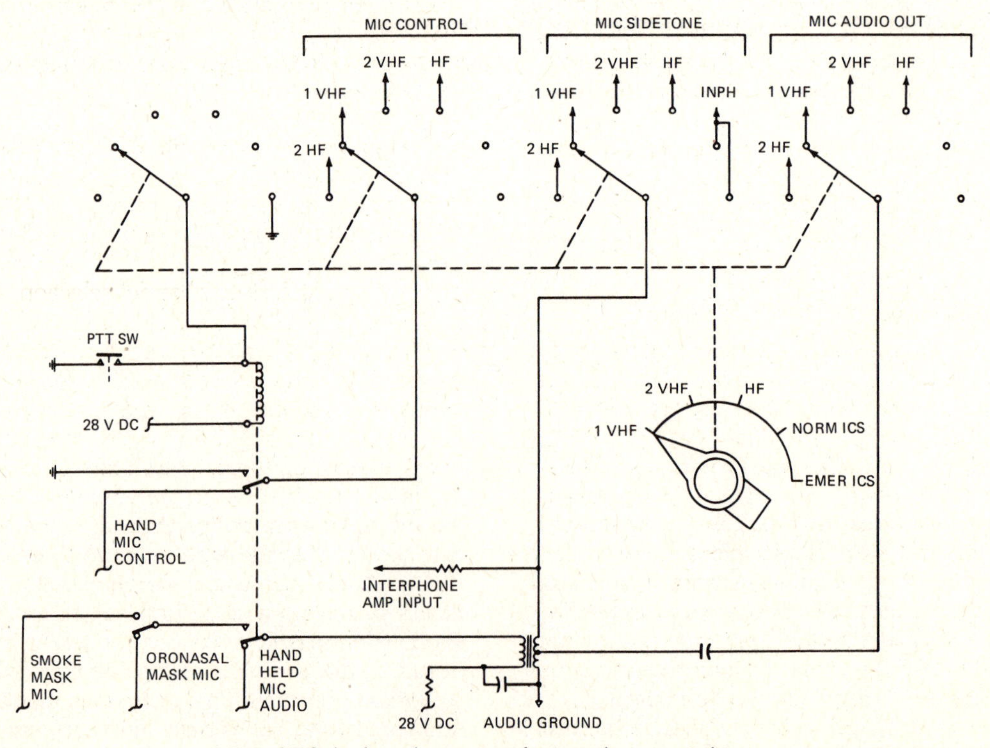

Fig. 17-3 Audio selector panel microphone switching.

this arrangement, the **flight interphone system** has two audio lines. One audio line, called *the cabin interphone*, provides a phone link between the flight station and ten attendant stations. These attendant stations are located close to each of the eight cabin doors and at the upper and lower levels of the galley lift. The second audio line interconnects four of the attendant stations to provide a separate line for communication between the galley and the principal service areas in the cabin. This is designated *the galley interphone system*, and it allows for peak-period galley operation, leaving the cabin line open for other use.

Headsets and hand-held microphones are located at each of the eleven interphone stations. The galley area has a unit with a built-in microphone and speaker, and it has an additional auxiliary microphone and speaker connected in parallel. There are also two speakers with integral amplifiers and volume controls installed in the flight station.

The **service interphone system** has twenty phone jacks located at major servicing areas throughout the aircraft to provide a communication link during ground service functions. This system also interconnects with the flight-crew audio selector panel positions and cabin-attendant handsets for increased service communication coverage.

The twenty jacks used solely for ground servicing and maintenance may be disconnected from the system by a switch in the flight station.

The pilot/mechanic call system provides an audible call facility in the flight station and nose wheel well. This is to alert the addressee that voice communication is required on the service telephone. The major components of the audible call facility are a horn and control switch located in the wheel well, a control switch on the flight station overhead panel, and an electronic chime unit above the flight station entry door.

A **cockpit voice recorder,** although not a component part of the interphone system, is a requirement for air transport aircraft and utilizes the principles of the intercom system. The voice recorder is installed in the rear part of the fuselage, where it is less likely to be damaged in case of a crash. The associated control unit, containing the area microphone, preamplifier, and the erase/test switch, is installed in the flight station to record cockpit conversations.

PASSENGER ADDRESS AND ENTERTAINMENT SYSTEM

The **passenger address and entertainment system** is provided so that flight-crew announcements and recorded music may be directed to the passengers while the airplane is in flight. The system consists of carefully placed loudspeakers which are driven from an amplifier in the radio rack. The passenger address system may be operated from a flight-deck microphone, from the forward or aft cabin-attendant's microphones, or from a magnetic-tape reproducer. Loudspeaker receptacles are located along each side of the passenger compartment and are all interconnected so that speakers may be connected at the most convenient locations to suit various seating arrangements. The loudspeakers, impedance-matching transformers, and speaker plugs are in the passenger service units.

Included in the passenger address and entertainment system for large, long-range aircraft are the passenger address system, passenger audio entertainment, passenger visual entertainment, and prerecorded announcement and boarding-music systems. These systems provide a continuous capability for passenger communication.

For one type of large, wide-bodied aircraft, passenger address speakers are installed in the flight station, galleys, cabin, and lavatories. Amplifiers are located in the forward electronics service center. These process inputs from microphones at the cabin hostess stations and flight station. Additional speakers are installed to sound chime announcements that direct attention to NO SMOKING and FASTEN SEAT BELT signs and passenger call signals.

The passenger address speakers are also used for boarding music and prerecorded announcements. Tape cartridges are played by a tape reproducer which is a part of the passenger audio entertainment system. Controls are placed at a cabin attendant's station.

Override switching for the passenger address system is set up with an established priority. Of highest priority are the flight station inputs, followed by the main-cabin attendant station inputs and, finally, by the boarding music or passenger entertainment. Automatic switching increases amplifier gain on the ground to overcome the increased noise level.

On some large aircraft, the passenger entertainment system is **multiplexed** with both audio functions and service functions interfaced. *Multiplexing* means that a number of different signals are conducted on the same wiring by means of different frequencies that are decoded at the receiving point. This practice greatly reduces the amount of wiring required for the various systems. Among the services that may be on a multiplexed system are stereophonic movie sound tracks, stereophonic or monaural music, passenger-to-attendant calls, remote-controlled reading lights, and remote-controlled individual air outlets.

A typical multiplex system is made up of the following components:

1. Four compartment controllers that provide the basic timing and synchronization for passenger service functions, as well as self-test capability.
2. One hundred and thirty-nine seat electric units that provide the encoding for passenger service functions and audio demultiplexing channel-selection logic for the audio entertainment.
3. One hundred and thirty-nine decoders that provide the decoding processing to operate the individual service functions, as well as the passenger oxygen test.
4. One main multiplexer to convert tape reproducer audio signals to a digital form suitable for multiplexing.
5. Three submultiplexers to distribute the multiplexed signals from the main multiplexer to the seat electronics units and convert the movie audio into a multiplexed type of signal. The passenger announcement override function is also channeled through the submultiplexers.
6. Power supply that provides the required dc voltages to all components of the system except the submultiplexers.
7. One self-test panel that provides a means for testing and monitoring the system for rapid fault isolation and service.

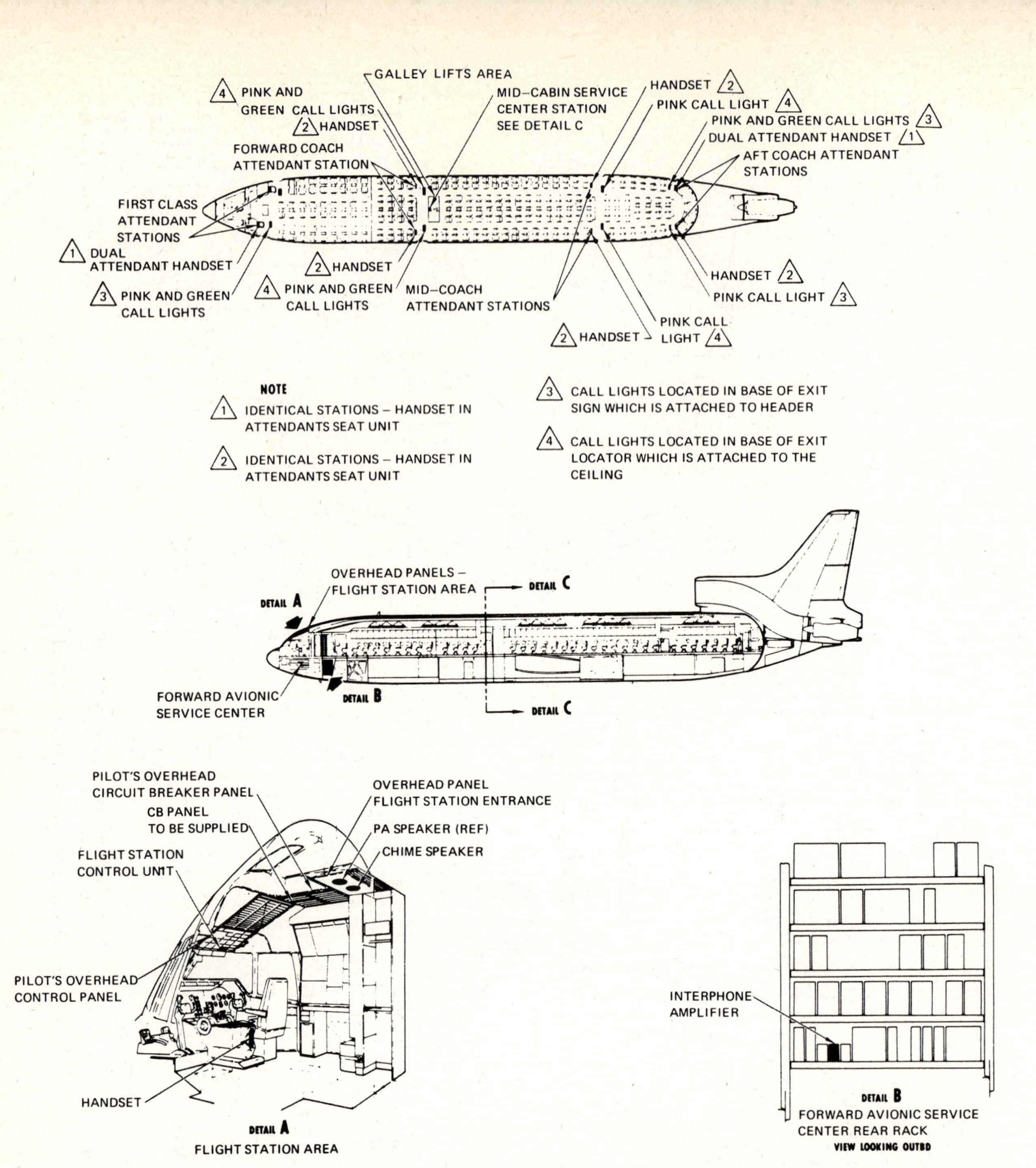

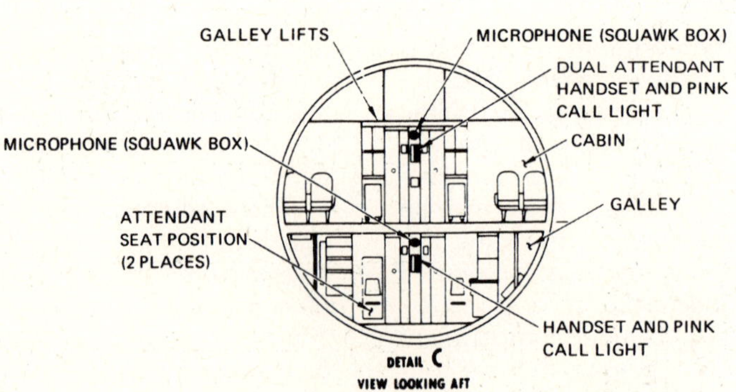

Fig. 17-5(a) Dual interphone system. (Lockheed California Co.)

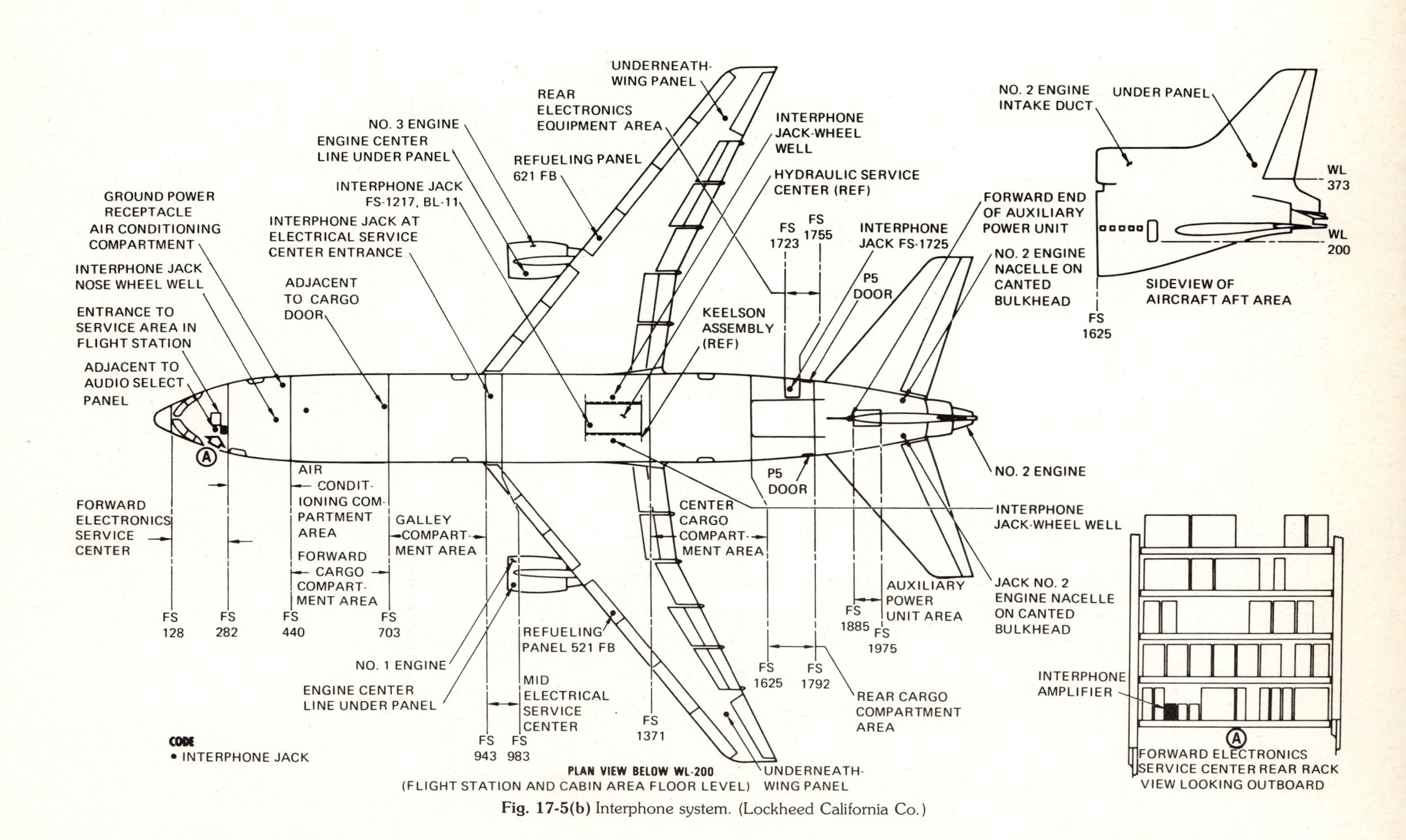

Fig. 17-5(b) Interphone system. (Lockheed California Co.)

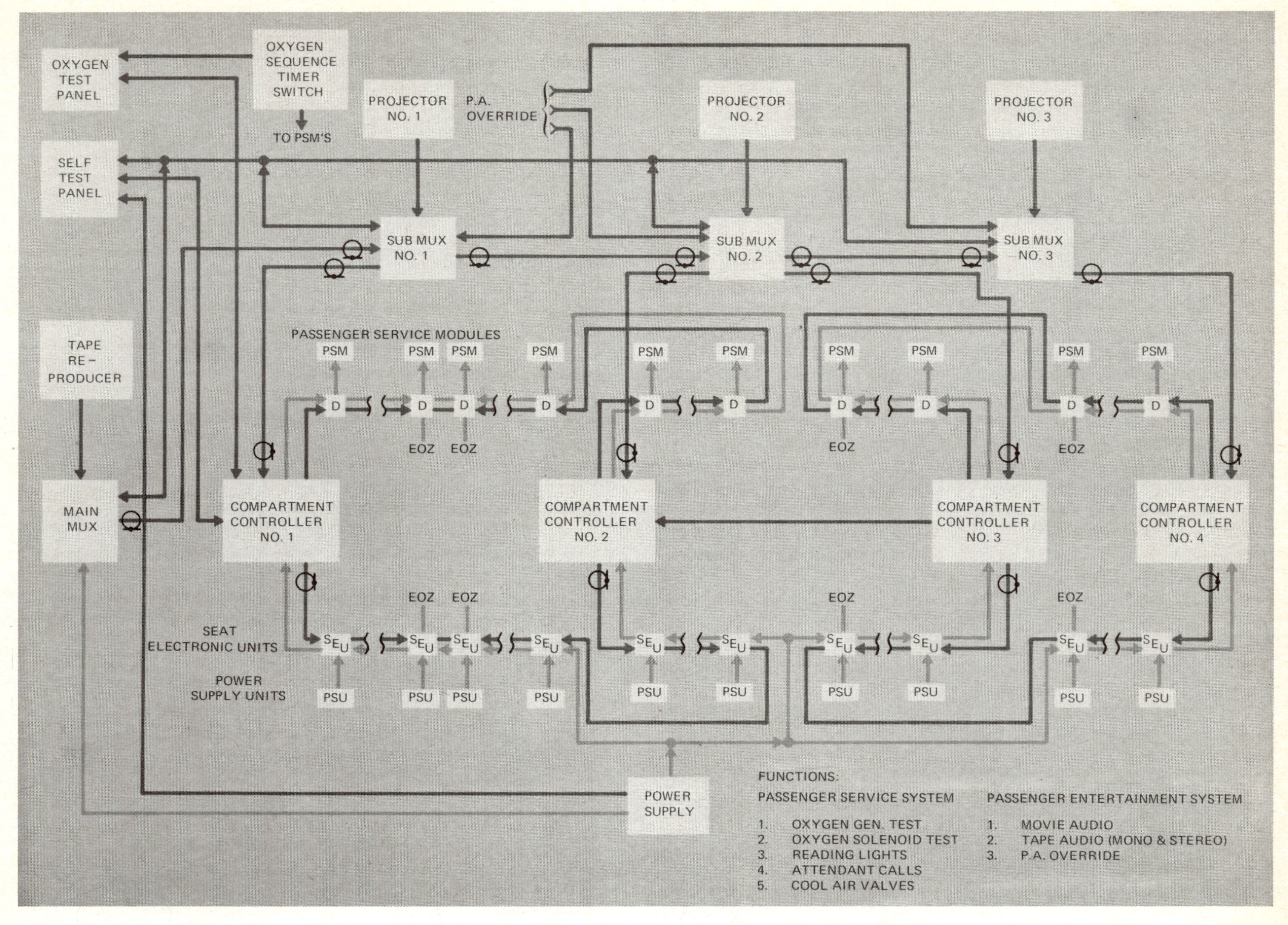

Fig. 17-6 Block diagram of a multiplex system. (Lockheed California Co.)

8. Two hundred and fifty-six passenger seat-control units that provide for each passenger the controls for selection of entertainment or service facilities.
9. One tape reproducer to provide audio signals for the system.

A block diagram of a multiplex system is shown in Fig. 17-6. It will be noted in the diagram that the system is regulated by four compartment controllers, each of these being associated with two of the eight service zones. The system is so designed that each pair of compartment controllers acts as a set. In the event that a controller should fail in one zone, the other would still provide full passenger service operation in that part of the cabin.

HIGH-FREQUENCY COMMUNICATION SYSTEM

The high-frequency (HF) communication system on a large airplane operates in the frequency range of 2.0 to 25 MHz. The HF range is actually a middle-frequency range inasmuch as it starts just above the standard broadcast band, which ends at approximately 1700 kHz. The HF system on an airplane is used to provide two-way voice communication with ground stations or with other aircraft. The output is also directed to Selcal decoders for air traffic control signaling.

The HF radio control panel is located where it is easily accessible to the pilot or copilot; it includes a frequency selector switch, a volume control, and a press-to-talk button for the Secal system. The volume control includes a switch for turning the equipment on and off.

The frequency selector switch is tuned as desired to select a particular communication channel. When the channel is selected, the number of the channel appears in a window directly above the switch knob. The volume control knob is located in the center of the frequency-selection dial. When the knob is turned to its maximum counterclockwise position, it turns the HF communications equipment off. A HF control panel is shown in Fig. 17-7.

Fig. 17-7 HF equipment control panel.

On the Boeing 707 airplane, the principal HF antenna consists of a probe extending forward from the upper tip of the vertical stabilizer. On other aircraft, a number of different designs are installed, including antennas within various components of the aircraft that are covered by plastic-type shields. The cover may be of fiberglas or a similar material that will allow electromagnetic waves to reach the antenna. The probe antenna is used for both receiving and transmitting and is matched to the transmission line at any frequency by means of a tuning unit. Figure 17-8 illustrates the method by which the probe antenna is attached to the fin cap.

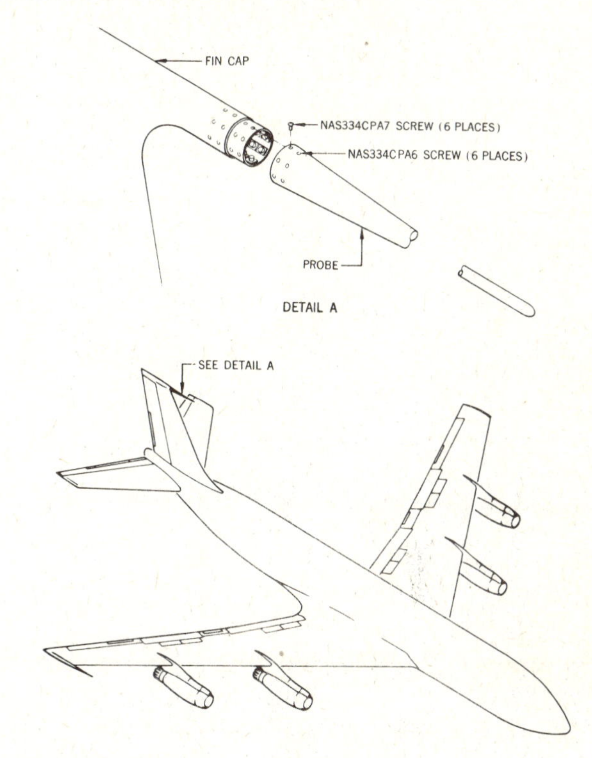

Fig. 17-8 Probe antenna installation on fin cap. (Boeing Co.)

The probe antenna has several advantages over long-wire antennas. It has essentially the same efficiency as a long-wire antenna, but is superior by reason of its light weight, ruggedness, and freedom from vibration.

Antenna tuning for all transmission frequencies is accomplished automatically by means of an **antenna-coupler system.** The complete antenna coupler system consists of a lightning-arrester assembly, a remote coupler unit, and coupler-unit control as shown in Fig. 17-9.

Initial adjustment of the antenna-coupler unit begins whenever a new frequency is selected for the transceiver, the combination transmitter and receiver used in the HF system. During the transceiver tuning cycle, a signal is sent to the antenna-coupler control which causes tuning circuits in the fin-tip antenna coupler to drive to a home position. When the transceiver tuning cycle is completed, a signal is sent to the coupler, which then tunes to provide a minimum standing wave on the antenna line. When the transmitter is turned on for transmitting, sensing circuits in the antenna coupler and control continuously measure the amount of reflected power and transmitted power. The sensing circuits cause antenna-tuning control circuits in the coupler to operate and adjust the impedance match between the antenna and the transmission line so that the amount of reflected power is at a minimum. The sensing and tuning circuits operate at all times that the

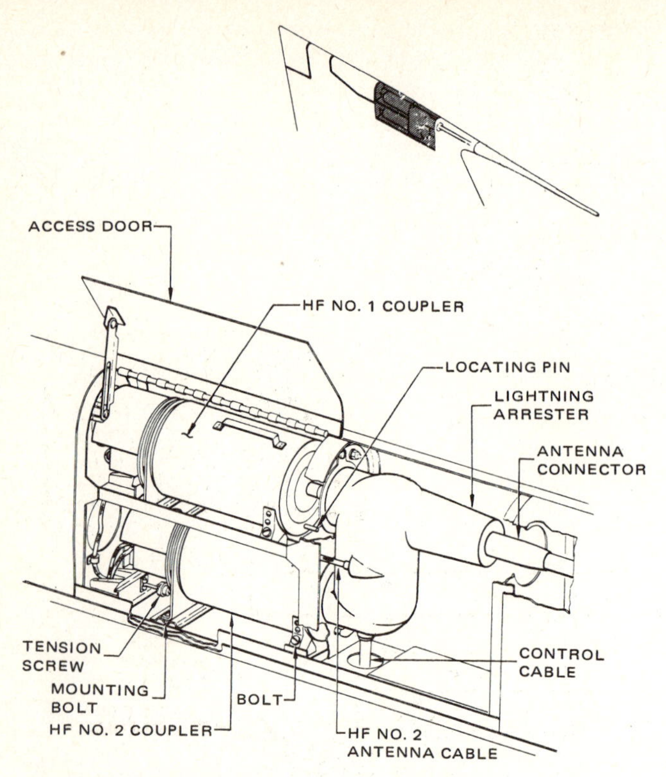

Fig. 17-9 Antenna coupler system. (Boeing Co.)

transmitter is operating and therefore continuously provides optimum match between antenna and line. The tuning unit will match any combination of antenna resistance and reactance to the characteristic impedance of the transmission line at any carrier frequency from 2.0 to 25 MHz. Completion of the tuning cycle occurs within 20 s after starting and requires an average time of 10 s. After final adjustment, the standing wave measured at the RF connector has a ratio of less than 2:1.

The HF transceiver is installed in an electronic equipment rack and is remotely controlled from the control unit on the flight deck. The unit consists of a transmitter, receiver, power supply, and transmitter modulator, all mounted in a single case.

The transmitter section operates in a frequency range of from 2.0 to 18.5 MHz. By means of the frequency selector switch on the control panel, any one of 10 channels, each having two available frequencies, may be selected. Each channel is crystal controlled, and the two available frequencies of each channel are within 1 percent of each other. The transmitter is capable of transmitting at 100-W output with frequencies up to 10.0 MHz. From 16.0 to 18.5 MHz, the transmitter is capable of 80-W output.

The receiver section operates at the same frequencies as the transmitter section and on the same channels. The receiver is a crystal-controlled superheterodyne and provides a minimum output of 50 milliwatts (mW).

The transceiver contains a maximum of 20 crystals, 10 for the transmitter and 10 for the receiver. All the power required for the transceiver unit is at 28 V direct current. The antenna-tuning unit utilizes 150-V 400-Hz power and 28-V dc power.

HF communications systems are not employed on all aircraft. Airlines may or may not utilize these systems, depending upon their particular requirements. HF systems are not usually found in light aircraft communication systems. Many airlines that employ HF communication systems do so because these systems provide for an extended range of communications between aircraft and from aircraft to ground stations. The arrangement of the HF communication system for one model of the Lockheed L-1011 airliner is shown in Fig. 17-10.

VHF COMMUNICATION SYSTEMS

As explained previously, VHF communication systems are employed largely for air traffic control. These systems are installed in all types of aircraft so the pilot may be given information and directions and may request information from control towers and flight service stations. On the approach to any airport with two-way radio facilities, the pilot of an aircraft calls the tower and requests information and landing instructions. In airline operations, the flight of an aircraft is continuously monitored by flight controllers and the crew is given instructions as necessary to maintain conditions of safe flight. The VHF communication system operates in the frequency range of 118 to 135.975 MHz. The nature of radio-wave propagation at these frequencies is such that communication is limited to line-of-sight distances. The advantage of VHF communication, however, is that these signals are not often distorted or rendered unintelligible by static and other types of interference.

VHF communication equipment for light aircraft may be separate units or they may be combined with VHF navigation (NAV) radio in the same case. A communication (COMM) radio unit is shown in Fig. 17-11. The photograph of Fig. 17-12 shows the interior arrangement of the system.

The transceiver shown in Figs. 17-11 and 17-12 is a solid-state, digital system that can receive or transmit on any one of the 720 channels in the COMM range of frequencies. The frequencies are spaced at 25 kHz intervals throughout the range. Frequencies are selected simultaneously for both the receiver and the transmitter by rotating the frequency selector knobs. The large, outer knob is used to change the megahertz portion of the frequency display, and the smaller, concentric knob changes the kilohertz portion. The small knob will change the frequency in 50-kHz increments when it is pushed in and in 25-kHz increments when it is pulled out.

To tune the transceiver to the desired operating frequency, the selected frequency must first be entered into the STANDBY display. It is then activated by pushing the transfer button, and the word USE will be displayed. Another frequency may then be entered in the STANDBY mode. To turn the transceiver on, it is necessary merely to rotate the ON/OFF/VOLUME knob out of the OFF position. It must be pointed out that the transceiver should not be turned on until after the engine is started. This is to prevent high transient voltages from the starting system from damaging the radio equipment. This practice should be followed with all avionics equipment.

Control panels for VHF communication systems vary in design, depending upon the manufacturer of the equipment and the requirements of the aircraft manufacturer. One type of control panel is shown in Fig. 17-13. This control panel contains frequency selector switches, vol-

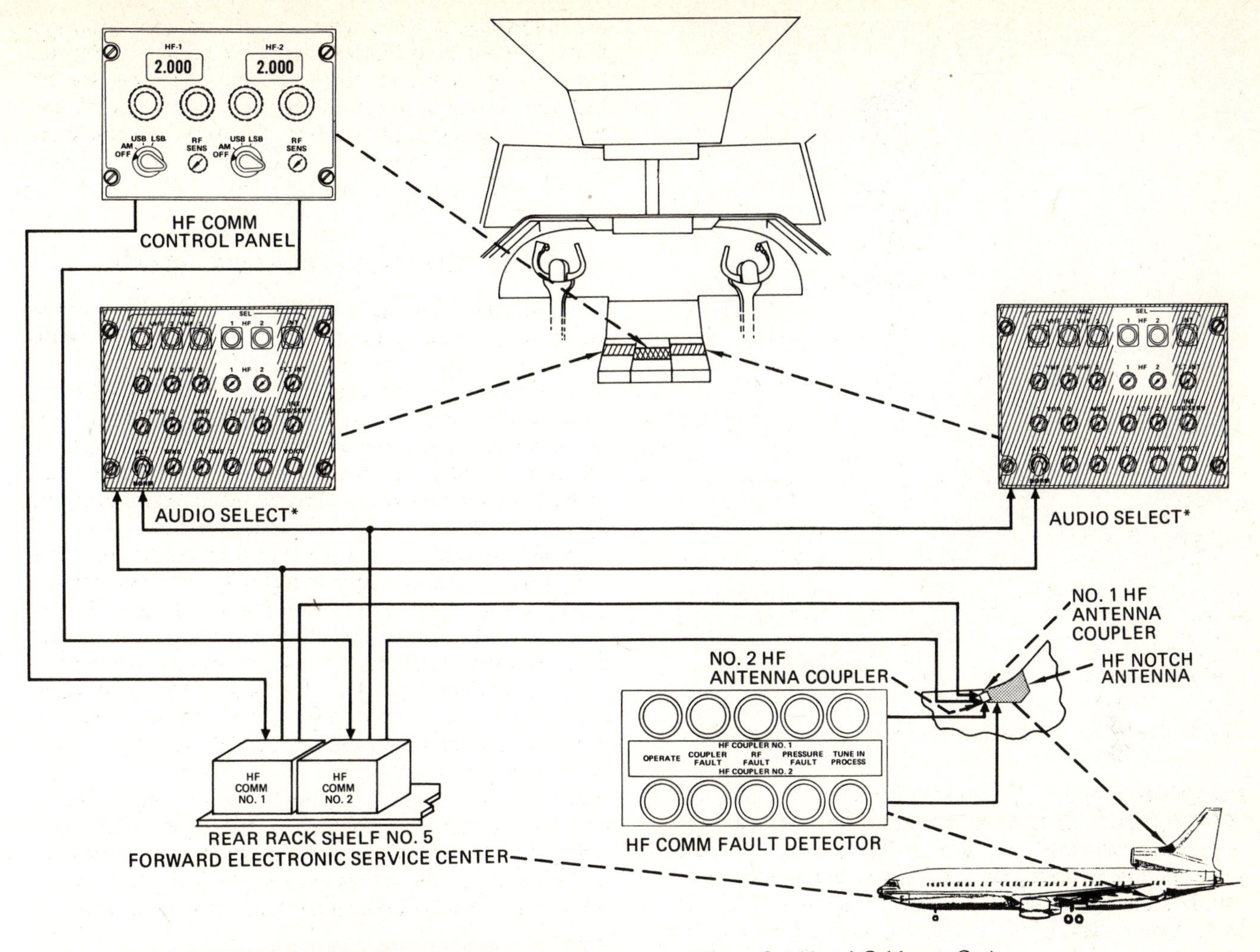

Fig. 17-10 An HF communications system for a large airliner. (Lockheed California Co.)

Fig. 17-11 A VHF communications radio. (King Radio Corp.)

Fig. 17-12 Interior arrangement of VHF COMM receiver. (King Radio Corp.)

Fig. 17-13 Control panel for a VHF system on an airliner.

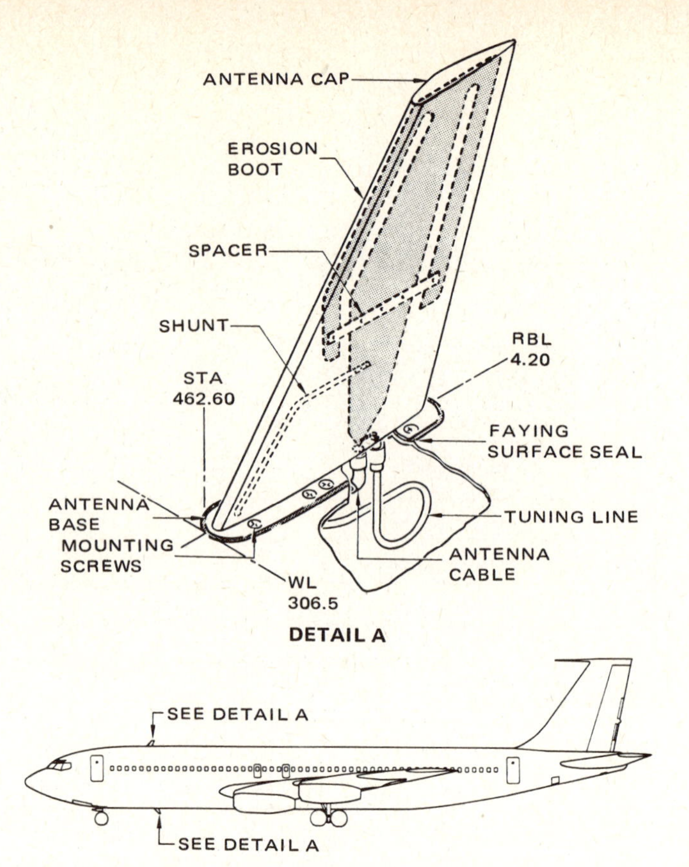

Fig. 17-14 VHF antenna configuration. (Boeing Co.)

ume control, squelch switch, and a selector switch to permit choice of double-channel or single-channel simplex operation. The volume control knob includes an ON-OFF switch which controls the operation of the transmitter and receiver through a 28-V relay. The double-channel—single-channel (DC-SC) switch is provided so that either double-channel simplex or single-channel simplex modes may be used. When the switch is placed in the DC position, transmissions are made at a frequency 6 MHz above the receiving frequency. When the switch is placed in the SC position, transmissions are made at the same frequency as the receiving frequency. In each case, the receiver is disconnected from the antenna.

Antennas for VHF systems are low-drag stub units extending from the top and bottom centerlines of the airplane. These antennas are matched to their respective transmission lines by means of carefully measured lengths of tuning line. The antennas are used for both transmitting and receiving. Typical VHF antenna configurations are shown in Fig. 17-14. A typical VHF system employs a superheterodyne receiver to receive signals with the frequency range of 118.0 to 135.95 MHz. A maximum of 360 channels, spaced 50 kHz apart, are available. The internal power supply requires 115-V alternating current, and the control circuits use 28-V dc power.

All frequency selection in the VHF receiver is remotely controlled and is accomplished by the selection of various crystals in crystal banks for the receiver oscillator and by mechanical adjustment of the tuned circuits. Under normal operating conditions frequency-selection cycling time does not exceed 5 s. The audio output of the VHF receiver is fed through the remote volume control and a muting relay in the transmitter to the interphone system.

VHF communication transmitters provide AM voice-communication transmission between aircraft and ground stations or between aircraft. Transmission is on the same number of channels and frequencies as provided in the receiver. Because of the nature of VHF radio signals, the average communicating distance from aircraft to ground is approximately 30 mi [48 km] when the airplane is flying at 1000 ft [305 m] and approximately 135 mi [217 km] when the airplane is at 10 000 ft [3048 m]. Transmitting frequency is determined by the position of the selector switches on the VHF control panel. The transmitter is tuned at the same time and to the same frequency as the receiver.

On many aircraft, one control panel contains the controls and switches for a number of different radio systems. The drawing on Fig. 17-15 illustrates the system used on one model of the Lockheed L-1011 aircraft. This system includes multipurpose control panels called audio selectors. These panels contain controls for VHF, HF, VOR, ADF, DME, marker beacon, interphone, and others.

SELCAL DECODER

The word *Selcal* is derived from the term *selective calling,* and the **Selcal decoder** is an instrument designed to relieve the pilot and copilot from continuously monitoring the aircraft radio receivers. The Selcal decoder is in effect an automatic monitor which listens for a particular combination of tones which are assigned to the individual aircraft. Whenever a properly coded transmission is received from a ground station, the signal is decoded by the Selcal unit, which then gives a signal to the pilot indicating that a radio transmission is being directed to the aircraft. The pilot can then listen to the receiver and hear the message.

Ground stations equipped with tone-transmitting equipment call individual aircraft by transmitting two pairs of tones which will key only an airborne decoder set to respond to the particular combination of tones. When the

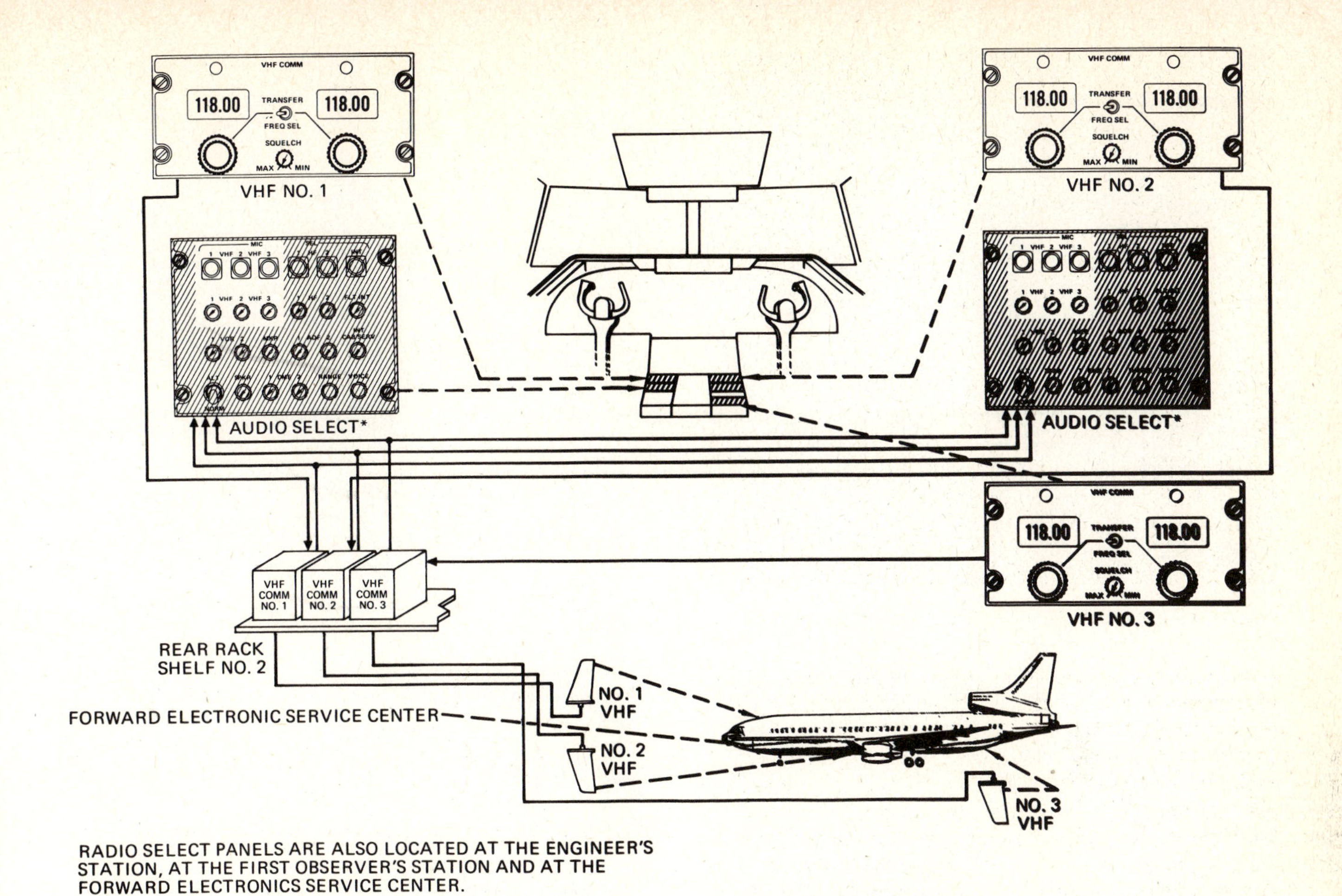

Fig. 17-15 VHF communications system. (Lockheed California Co.)

proper tones are received, the decoder operates an external alarm circuit to produce a chime, light, buzz, or a combination of such signals.

A ground operator who wishes to contact a particular aircraft by means of the Selcal unit selects the four-tone code which has been assigned to the aircraft. The tone code is transmitted by a radio-frequency wave, and the signal can be picked up by all receivers tuned to the frequency used by the transmitter. The only receiver which can respond to the signal and produce the alert signal for the pilot is the receiver and decoder system which has been set for the particular combination of tone frequencies.

A typical Selcal decoder is shown in Fig. 17-16. This unit consists of a single one-half Air Transport Rack (ATR) standard-size frame and housing assembly which contains two identical decoder chassis. The electronic circuits for the unit are completely transistorized, and transistor switches are used in place of relays to open and close circuits.

Each decoder chassis requires four **Vibrasponder** reeds to establish the code for the unit. The reeds respond to audio frequencies and cause the generation of a voltage which is processed by the unit. The resulting voltage is used to operate the alarm circuits. Vibrasponder reeds are manufactured with tone frequencies from 312.6 to 977.2 Hz. Twelve separate tones are available.

The arrangement of a Selcal decoder system in one type of large airliner is shown in Fig. 17-17. The decoder is a

Fig. 17-16 A Selcal decoder unit. (Motorola, Inc.)

solid-state unit that can be used to monitor any two of five receivers. Receivers to be monitored are selected by means of the rotary selector switches on the Selcal control panel.

SATCOM COMMUNICATIONS

Since communications satellites have been placed in orbit around the earth, some large aircraft have been

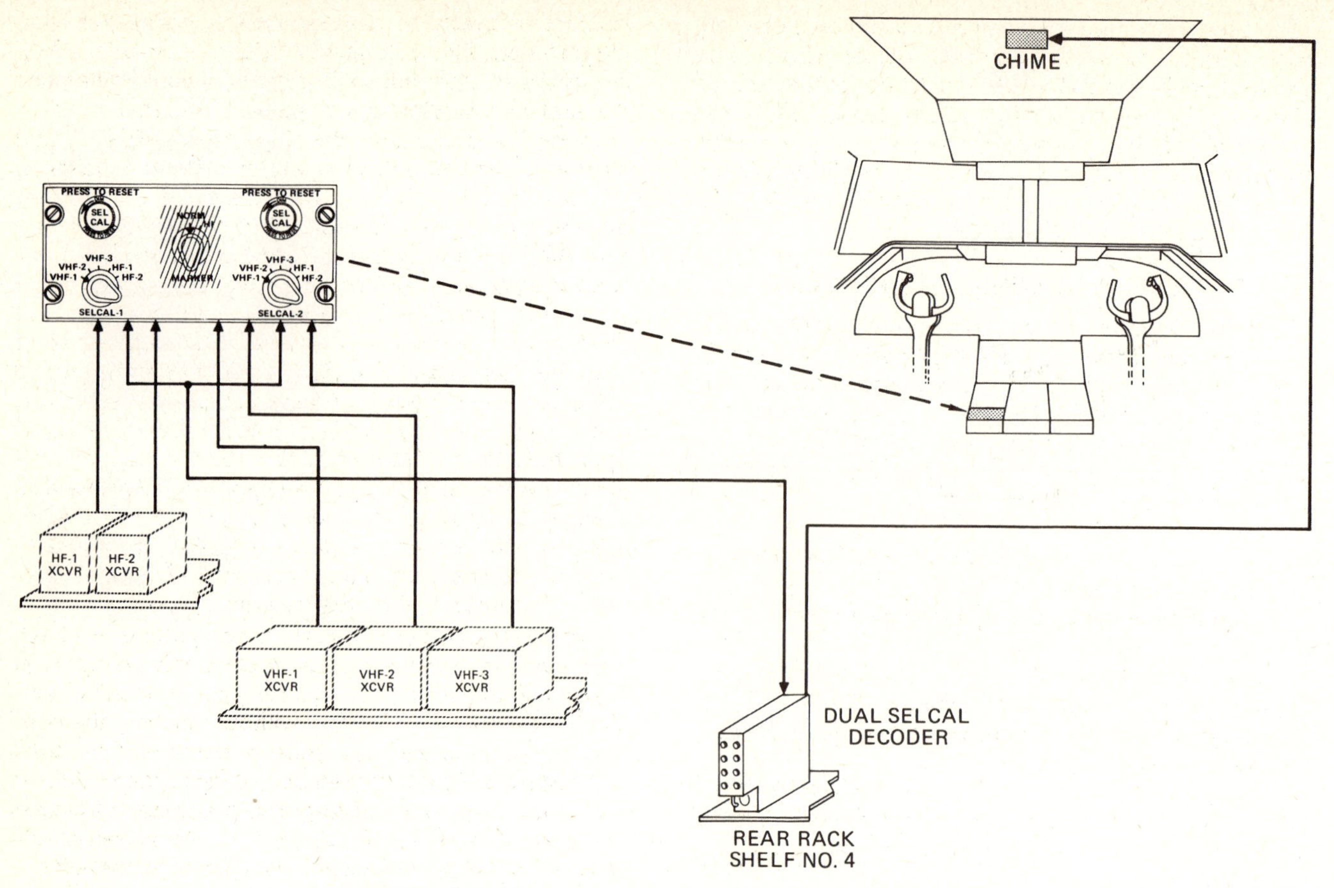

Fig. 17-17 Selcal system in an airliner. (Lockheed California Co.)

equipped with receivers and transmitters that are capable of utilizing the benefits of the satellites. SATCOM equipment extends the line-of-sight range of VHF to an over-the-horizon range.

FEDERAL COMMUNICATIONS COMMISSION REGULATIONS

Because of the very nature of radio waves and their effect upon many activities of modern life, all electromagnetic emissions are controlled by a single government agency. In the United States this agency is the Federal Communications Commission. It is the responsibility of this body to supervise all radio transmission in the United States, its territories, and possessions. The FCC licenses radio operators, technicians, amateur stations and operators, commercial radio stations, marine radio stations, television stations, and various special radio or television operations. Furthermore, the FCC assigns frequency ranges for different types of operations and assigns specific frequencies to individual stations. The agency also cooperates with international agencies to work out agreements to prevent, as much as possible, interference between stations of different nationalities.

Some typical FCC regulations are listed below, but not necessarily in the actual language of the law:

1. All radio transmitters installed in operating aircraft must be licensed.
2. Distress calls or messages have priority over all others.
3. The distress call for radiotelephone is *Mayday*. The distress call for radiotelegraph is --- ——— ---, which may be interpreted S O S.
4. The penalty for willfully violating the Communications Act is a $10 000 fine or imprisonment for a term of not more than two years or both.
5. No obscene, indecent, or profane language shall be transmitted by radio.
6. No fraudulent signals shall be transmitted.
7. Information received by radio and not intended for the person receiving such information shall not be divulged to any person other than the one for whom it is intended; neither shall the existence of the information be divulged.
8. No unnecessary communications shall be transmitted.
9. Noncitizens may not be issued radio-operator's licenses or permits.
10. No operator of a radio station shall violate the provisions of any treaty to which the United States is one of the parties.
11. A person holding a third-class or restricted license shall not adjust the frequency of a radio transmitter.
12. The operating power of a radio station may be permitted to vary from 5 percent above the assigned power to 10 percent below the assigned power.
13. A person's radio license may be suspended or revoked for violation of FCC regulations.
14. The radiotelephony urgency signal is *pan*.
15. The radiotelephony safety signal is *security*.

The above laws are only a few of the most important relating to the operation of radio. A person who is to be involved in the operation of a radio transmitter should obtain all the necessary information from the FCC and then apply for the license appropriate to the operation concerned. The operation of a radio transmitter in an airplane usually does not require more than a **restricted** or **third-class** radiotelephony license.

TESTING COMMUNICATIONS RADIO

Testing of communications radio in small aircraft used in general aviation may be accomplished in accordance with procedures appropriate for the airport and area in which the test is made. The testing operator must have the appropriate FCC license. Testing on the ground with the engine or engines not operating can be accomplished as follows:

1. Turn on the aircraft power with the master switch.
2. Turn on the transceiver.
3. Select the frequency of the station with which the test is to be conducted.
4. Listen to the receiver to be sure there is no radio traffic in progress at the selected frequency.
5. If no radio traffic is heard, press the transmit switch and call the selected station.
6. Upon receipt of a reply, request a radio check.
7. If reception and transmission are satisfactory, turn the transceiver off. Be sure to do this before turning off the master switch.

NAVIGATION SYSTEMS

The development of navigation instruments and systems for aircraft is almost unbelievable when one considers what is possible today as compared with the early days of aviation. For example, it is now possible for a properly equipped aircraft to fly anywhere over the surface of the earth, and the crew will know its precise position at all times. The fact that we are now able to place a space vehicle upon the surface of the moon at a predetermined point and time is evidence that the art and science of aerospace navigation has advanced to an extremely high level of precision and capability.

For a number of years, the size, weight, and cost of electronic navigation systems made it impractical to use complex systems for small aircraft navigation. However, owing to the development of solid-state technology and microelectronics, the most sophisticated systems can now be used in small aircraft.

In the early days of airplane operation, navigational instruments either did not exist or, at most, consisted of a magnetic compass and an airspeed indicator. When flying by visual reference, the early pilot would usually navigate from one landmark to another, following roads and railroads or rivers and valleys. Flights were made at comparatively low altitudes providing a view of the ground that was usually good enough for the pilot to clearly identify objects there. Under the flying conditions that existed when the airplane was considered a novelty, complex navigation instruments and systems were not in great demand. As the use of airplanes increased and flights were made at higher altitudes, above the clouds and at night, it became necessary to develop reliable navigation techniques along with instruments indicating attitude, heading, airspeed, and drift so that the pilot could determine the airplane's position by computation and map plotting.

Among the instruments developed for navigation and flight were the magnetic compass, the airspeed indicator, the rate-of-climb indicator, the bank-and-turn indicator, the directional gyro (gyrocompass), and the artificial horizon. All these are mechanical instruments, with the exception of the magnetic compass, which responds to the earth's magnetic field. The airspeed and rate-of-climb indicators are operated by means of pitot static pressure from the pitot system. The gyro instruments are energized by means of venturi suction or a vacuum-pump system, which causes air jets inside the instrument to impinge upon buckets on the rims of the gyros. This spins the gyros at a high speed, thus enabling them to respond to changes of airplane attitude and, through mechanical linkages, produce indications for the pilot's information. Many gyro instruments today are energized by small electric motors instead of by the vacuum system.

From the 1930s to the present time, great strides have been made in the development of electronic navigational systems. Today a pilot can fly an airplane across the continent from takeoff to landing without touching the controls, all the navigation and pilotage being accomplished electronically. Completely automatic flying may be done commercially, but the application of such systems is most useful in the military services for flying drone target aircraft and missiles and for recovering such devices after a mission is completed. By this means, certain types of missiles are flown many times for testing and target work and then are brought back and landed safely at the home base.

It is the purpose of this section to describe and explain some of the electronic navigation systems and equipment on modern aircraft. It is beyond the scope of the text to describe the details of circuitry and all the electronic principles employed, because to do so would require far more space than is available; however, the general principles of operation and the individual components will be explained.

AUTOMATIC DIRECTION-FINDER SYSTEMS

Since before World War II ADF equipment has been used on aircraft as an aid to navigation. The function of an ADF system is to enable the pilot to determine the headings, or direction, of the radio stations being received. The ADF system operates on a frequency range of 90 to 1800 kHz, a range which makes it possible for the system to receive radio-range stations in the LF band and standard broadcast stations. By use of the ADF system, a pilot can determine the aircraft's position, or the pilot can "home" on a radio broadcast station or a radio beacon station by flying directly toward that station using the indication of the radio compass or radio magnetic indicator. To find the aircraft's position, the pilot or the navigator determines the headings of two different radio stations and then plots the headings on a navigation chart. The point at which the heading lines cross will be the location of the aircraft.

ADF systems utilize the directional characteristics of a loop antenna to determine the direction of a radio station. A simple direction finder which is not automatic may be made by using a loop antenna with an ordinary radio receiver. By rotating the antenna, the strongest reception can be determined and also the point at which the signal fades out. This point is called the **null position,** and from it a fairly accurate indication of the station direction can be determined.

On modern airliners it is common practice to utilize two sets of ADF equipment. The two units may be tuned in to two different radio stations and an immediate fix determined by plotting the lines of position for the two radio stations on the navigation chart. A further value of having an airplane equipped with two ADF systems is that if one system fails, the other is still available for direction finding. ADF equipment is especially valuable in areas of the world where special navigational aids are not available but where the pilot may tune in on a standard broadcast station.

THEORY OF OPERATION

As previously explained in this text, radio waves are propagated in the form of electromagnetic and electrostatic lines of force which travel at a speed of approximately 186 000 mph [299 300 km/h] from the radio transmitter. When these lines of force cut across the radio antenna, a voltage is induced in the antenna. This voltage is amplified and demodulated so that the intelligence contained on the radio wave may be determined. If a loop antenna is placed in such a position that it is at 90° to the direction of wave travel, equal and opposite voltages will be induced in the sides of the antenna as shown in Fig. 17-18. The voltages thus induced in the loop will cancel each other, with the result that the loop will have no output. If the loop is connected to a radio receiver, the signal will disappear at this point. If the loop is turned either one way or the other, a voltage will be induced in one side slightly before it is induced in the other, with the result that there will be a difference between the two which will provide a signal that may be fed to the receiver. When the plane of the loop is parallel to the direction of wave propagation, the strongest signal will be developed.

In determining the direction of a radio station with the loop antenna, it is far more accurate to use the null position of the antenna than it is to attempt to use the position at which the strongest signal is received. This is because the null point is very narrow, whereas the point of strongest signal is several degrees wide.

ADF COMPONENTS

The principal units of an ADF system are **a radio receiver,** which includes the amplifiers and various other electronic components, **a loop antenna, a sense antenna, a radio magnetic indicator (RMI),** and a remote-control unit, or **control panel.** The loop antenna in older systems is rotated electrically and is connected by a synchro system to the RMI. Hence, as the loop is rotated, the pointer of the RMI also rotates. When the system is operating in a fully automatic mode and a station is tuned in, the loop antenna will automatically turn to a direction which provides the correct heading of the station on the RMI.

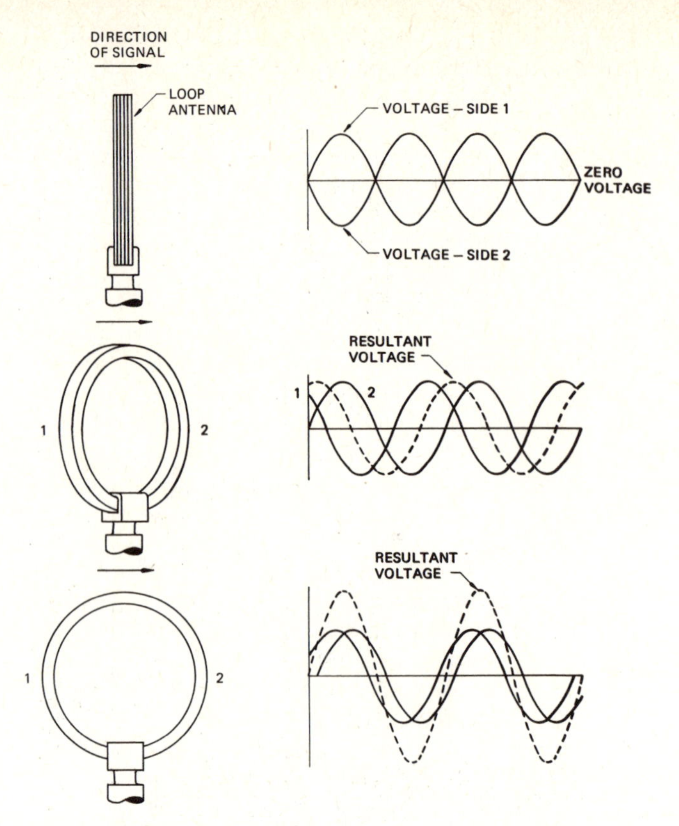

Fig. 17-18 Operation of ADF loop antenna.

Because of inherent problems with any moving device, electronic engineers designed a loop antenna that is "rotated" electronically. Actually, the antenna does not move; however, its output is equivalent to a rotating loop.

The function of the sense antenna is to provide an input signal which is out of phase with the signal received with the loop antenna. This is necessary to provide a correct heading indication for a radio station. If a sense antenna were not used, the indicator might show the heading of the station or it might point to the reciprocal of the heading, that is, to a direction 180° away from the heading of the radio station being received.

Figure 17-19 is a block diagram showing the general arrangement of one of the ADF systems installed in one model of airliner. In this diagram the coordination between the loop antenna, the synchro transmitters, the loop drive motor, and the indicator is clearly seen.

ADF CONTROLS

The control panel for an ADF system is shown in Fig. 17-20. The control panel contains a **gain control, a beat-frequency-oscillator (BFO) switch, a selectivity switch, a tuning meter, a frequency indicator, frequency selector knobs, a loop-control switch, and a function-selector switch.** The gain control is used to adjust the level of the signal being received, and the BFO switch is used to control the beat-frequency oscillator which produces a beat frequency to be mixed with an incoming continuous-wave (CW) signal. By using the beat frequency, a 1020-Hz tone is produced to make the CW signal audible.

The function selector switch is a four-position rotary switch having OFF, ADF, ANT (antenna), and LOOP posi-

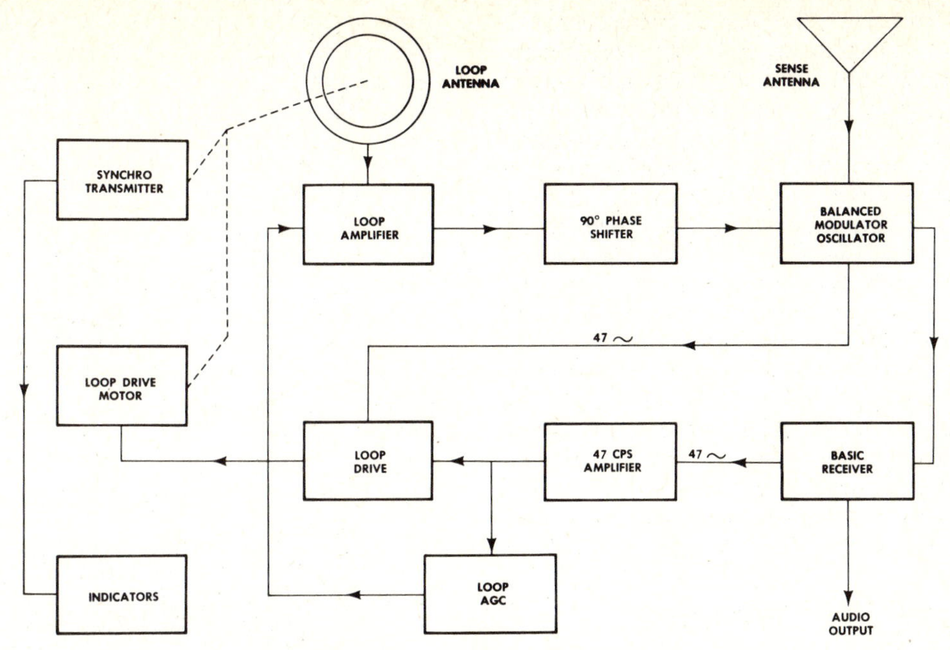

Fig. 17-19 Block diagram of ADF system.

Fig. 17-20 One type of control panel for an ADF system.

tion. When the switch is placed in the ADF position, the receiver is energized, and circuits are selected to determine automatically the bearing to the station being received. During ADF operation, both the sense- and loop-antenna circuits are in operation, and bearing information is shown on the RMIs. This means that the pilot or navigator can read the indicators to determine the bearing of the station being received. In some systems, an instrument called a **radio compass** is used in place of the RMI. The radio compass has a dial indexed for a complete circle of 360°; the top of the dial represents the position of the front of the airplane. If the pointer is pointing directly to the top of the dial, which is the zero position, the airplane is heading directly toward the station being received.

When the ANT function is selected, the sense-antenna circuits only are utilized, and the receiver is used for reception of audio signals. The ANT function is used for reception of weather broadcasts and radio-range signals without any attempt to locate the bearing of the received station.

The LOOP function of the selector switch is used when it is desired to determine by manual operation of the loop the bearing to the station being received. In this function, the loop antenna is manually controlled through the **loop-control** switch, and a null in reception is determined by monitoring the tuning meter. When the null position is located, the tuning-meter reading will drop to zero. The LOOP function may also be used for audio reception under conditions of severe precipitation static, since the loop is shielded electrostatically and may reduce interference. The LOOP function should not be used on radio ranges, since it may give confusing and unreliable reception of range signals.

The loop-control switch on the control panel is used only when the LOOP function is selected by the **function selector switch.** Rotating the loop-control switch to the L or R position will cause the loop to rotate counterclockwise or clockwise, respectively. A mark at the midpoint to either extreme switch position indicates the correct position for slow rotation of the loop. If the loop switch is rotated to the extreme position, the loop antenna will rotate at high speed in the selected direction. Manual control of the loop antenna is used to determine **aural nulls.** Determination of the null point is achieved by watching the tuning meter during the rotation of the loop. When the meter deflection is at a minimum, the loop is tuned to the null point, and the indicator will be pointing to the bearing of the station being received.

To select a desired frequency range for the ADF receiver, the frequency selector knobs are used. Three of these selector knobs are located on the front of the control panel. The left knob selects hundreds (100 to 999) of kilohertz, the center knob selects tens (10 to 99) of kilohertz, and the right knob selects units (1 to 9) of kilohertz. As the knobs are rotated, the selected frequency appears on the indicator dials above the knobs. To tune between 90 and 99 kHz, the hundreds knob is turned one position left of the 100 kHz position. At this time a mask will cover the hundreds and tens indicator dials, and the number 9 will appear over the tens indicator dial. The units selector knob may then be rotated to select the desired frequency. While the receiver is being tuned, it is best to have the function selector switch in the ANT position, so that the tuning meter may be observed to determine the point of maximum signal strength.

The selectivity of the receiver, that is, the ability of the receiver to separate the desired signals from adjacent signal sources, is determined by a mechanical magnetostrictive filter. A bandpass filter produces a selectivity curve having a pass band 3.1 kHz in width with the center of the band at 455 kHz. A narrow, or *sharp,* bandpass filter produces a selectivity curve only 500 Hz wide with the center at 455.7 kHz. Selection of either BROAD bandpass or SHARP bandpass is made by a selector switch on the control panel. Under normal conditions, the BROAD position is used, but if interference from adjacent stations cannot be satisfactorily tuned out, the SHARP position is used. The extremely narrow bandwidth characteristic of the sharp bandpass filter enables sharp separation and rejection of undesired signals.

RADIO MAGNETIC INDICATOR

The RMI shown in Fig. 17-21 is the instrument designed to provide visual information to the pilot and copilot concerning the data received by the ADF equipment. This instrument makes it possible for the pilot to navigate the aircraft without the necessity of numerical or graphical calculations. The instrument displays the magnetic heading of the aircraft and the magnetic bearing of two radio stations. Magnetic-heading information for the instrument

Fig. 17-21 A radio magnetic indicator.

is provided by a flux gate compass or other system. The bearings of the two radio stations are provided by the two separate ADF receivers operating with the loop antenna. The magnetic heading of the aircraft is indicated on a rotating disk-type dial, and the magnetic bearings of the two radio stations are shown under the two pointers.

The face of the RMI consists of a fixed outer dial with 45° markings through 360°, an inner rotating compass dial graduated from 0 to 360° clockwise in 2° increments, a wide pointer with parallel grids at the outer edge, and a narrow pointer mounted concentrically with the wide pointer and compass dial. The two pointers provide radio-bearing indications, that is, the bearings of two radio stations being received. The indications of the pointers are read in reference to the compass dial, or card, and thus provide the magnetic bearings of the radio stations on the navigation chart and immediately locate the airplane at the position where the two lines cross.

ADF WITH DIGITAL TUNING

An ADF system with digital tuning provides a more convenient and more rapid tuning than that possible with the ADF system previously described. In order to tune to a station in the 190- to 1750-kHz range, the pilot selects the frequency desired on the digital control unit. The dual configuration of the control unit is shown in Fig. 17-22. The system automatically switches to the selected band and tunes the receiver to the selected frequency within 0.5 kHz.

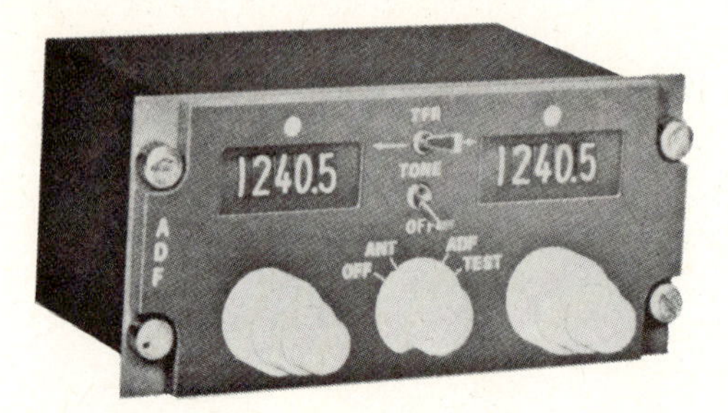

Fig. 17-22 Control panel for dual ADF. (Collins Air Transport Division, Rockwell International.)

The ADF receiver contains a phase-lock fine-tuning system which automatically locks onto the strongest portion of the received signal. It eliminates the need for manual fine-tuning controls and tuning meters, and it obtains high signal selectivity for reliable ADF indications under adverse reception conditions.

The system is composed of solid-state circuitry and makes extensive use of integrated circuits. The only moving parts in the entire system are in the RMI. The absence of moving parts makes the system almost completely troublefree and maintenance free.

Courses to radio stations are displayed as magnetic bearings on the RMI. Instrument-mounted selector knobs provide the pilot with flexible combinations of dual ADF/VOR displays.

Magnetic headings displayed by the RMI must be supplied from a remote, gyrostabilized compass or other system.

The antennas for the system are of the series, fixed-loop type. This means that there are no moving parts to wear or come out of adjustment; hence, there is little or no need for service or maintenance.

Currently manufactured ADF systems are all of the solid-state type with digital readouts in addition to the RMI. Many are designed to interface with other components in an integrated flight system. These systems are described later in this section. Figure 17-23 shows how ADF equipment is arranged in a large airliner.

INSTRUMENT-LANDING SYSTEM

The instrument-landing system (ILS) was developed as a result of the need for a method whereby a pilot flying an airplane could locate and fly to an airport even when visi-

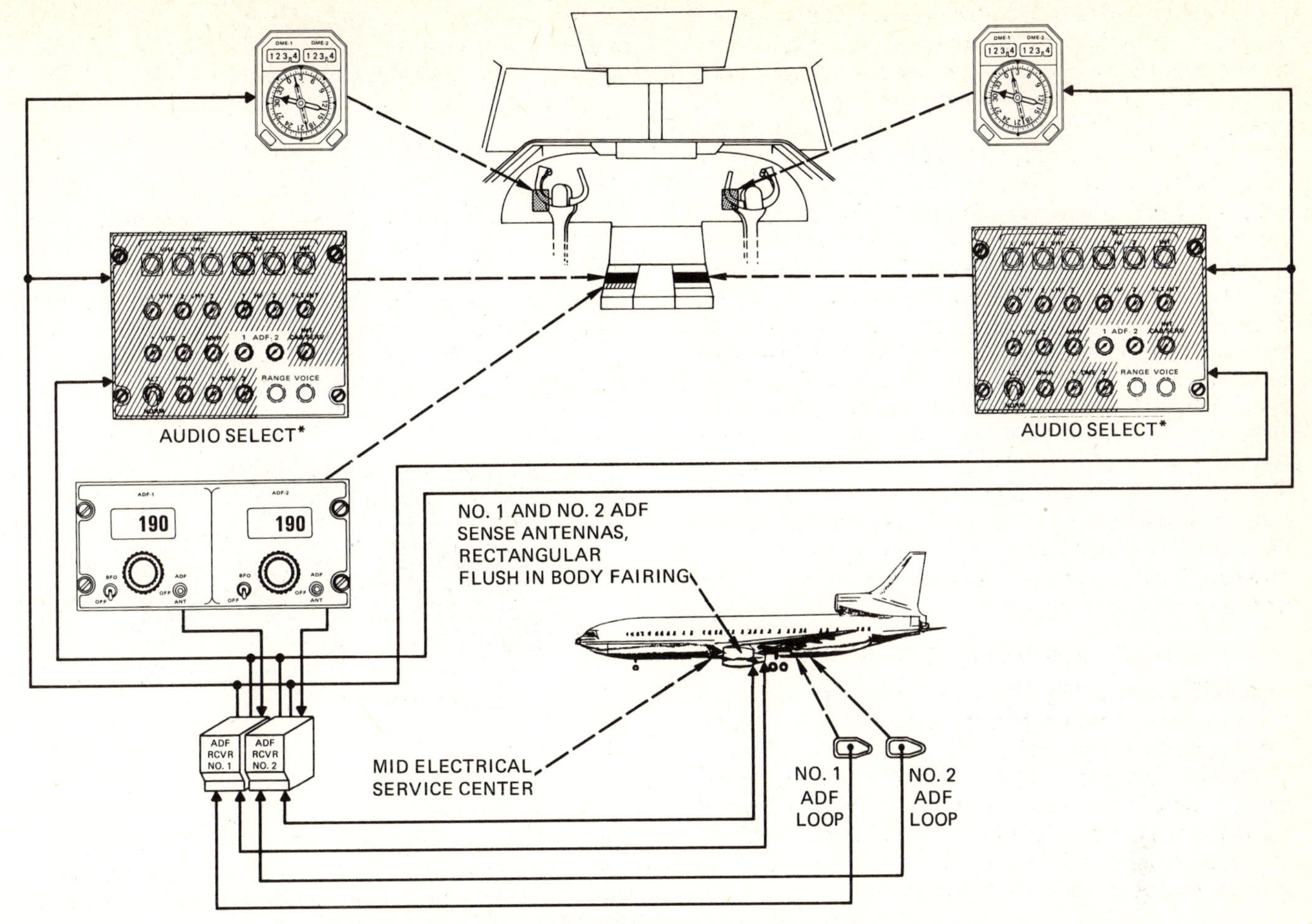

Fig. 17-23 ADF system for an airliner. (Lockheed California Co.)

bility was very poor. The systems now being operated by the FAA are the result of development work which has been conducted since 1928. The present system, except for the glideslope, was demonstrated in commercial form in Indianapolis in 1939 and 1940. Commercial use of the ILS was greatly delayed by the war, however, because of manpower shortages and because it was not possible to obtain the necessary equipment.

The ILS provides a horizontal directional reference and a vertical reference called the **glideslope.** The directional reference signal is produced by the runway **localizer** transmitter installed approximately 1000 ft [305 m] from the end of the runway and operating at frequencies of 108 to 112 MHz. The glideslope signal is produced by the glideslope transmitter, which is located near the side of the runway on a line perpendicular to the runway centerline at the point where airplane touchdown occurs. This point is generally about 15 percent of the runway length from the approach end of the runway. The glideslope transmitter operates on a frequency of 328.6 to 335.4 MHz.

THE LOCALIZER

The localizer consists, essentially, of two RF transmitters and an eight-loop antenna array. The transmitters broadcast a complete system of radiation patterns which produce a null signal along the center of the runway. The radiation pattern is such that when an airplane is approaching the runway for a landing, the signal to the right of the localizer path will be modulated with 150 Hz and the signal to the left of the localizer path will be modulated with 90 Hz. The localizer receiver on board an airplane is able to discriminate between the 90- and 150-Hz signals. The output of the receiver is fed to the vertical needle of a **course deviation indicator** (CDI) (shown in Fig. 17-24) or to another type of instrument such as a **flight-director indicator.** If the airplane is to the right of the localizer centerline, the 150-Hz modulation signal will predominate, and the vertical needle of the indicator will point to the left of the centerline, indicating that the pilot should fly left in order to return to the centerline of the localizer beam.

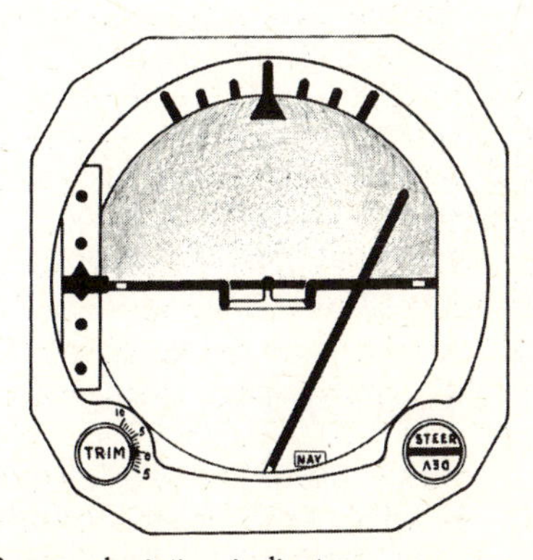

Fig. 17-24 Course deviation indicator.

Fig. 17-25 Radiation pattern from the glideslope transmitter.

THE GLIDESLOPE

The glideslope transmitter operates on a principle similar to that of the localizer. As previously mentioned, the glideslope transmitter is located at a distance from the approach end of the runway approximately 15 percent of the length of the runway. A schematic diagram illustrating the radiation pattern from the glideslope transmitter is shown in Fig. 17-25. In an airplane is approaching the runway and is above the glide path, the 90-Hz signal will predominate; and if the airplane is below the glide path, the 150-Hz signal will predominate. The glideslope receiver will provide an output to the crosspointer indicator in such a way that the pilot will have a visual indication of the airplane position with respect to the glide path. If the horizontal pointer is above the center of the indicator, the airplane is below the glide path.

A diagram of the beam provided by the combination of localizer and glideslope transmitter is shown in Fig. 17-26. The beam is electronically exact and provides a precise path by which an airplane may approach a runway and reach the point of touchdown. This provides a most valuable aid for conditions of poor visibility in the vicinity of an airport.

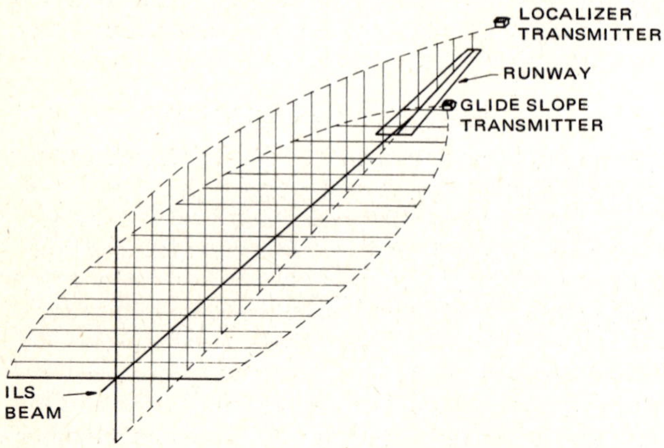

Fig. 17-26 ILS beam formed by localizer and glideslope transmitters.

Receivers for the ILS are usually combined with the receiver for VHF omnirange (VOR) and are often designated as VOR/LOC receivers. The indicators for the system are often combined with other indicators that provide a number of navigation indications. Among these are the **attitude director indicator (ADI)** shown in Fig. 17-27 and the **horizontal situation indicator (HSI)** shown in Fig. 17-28.

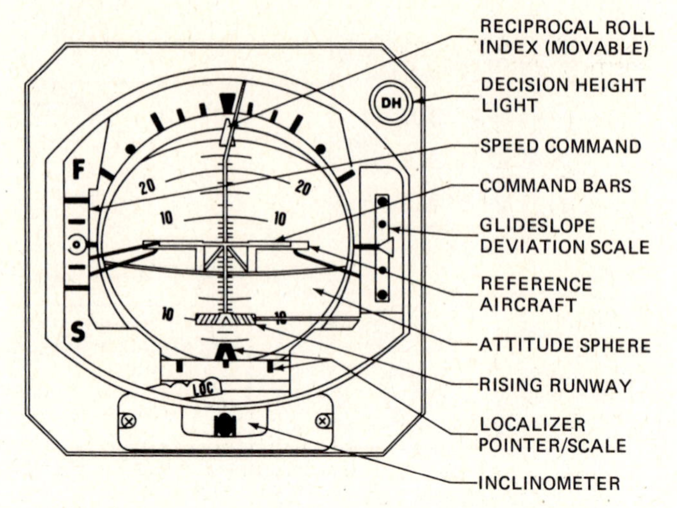

Fig. 17-27 Attitude director indicator. (Lockheed Calif. Co.)

MARKER BEACONS

In order to provide pilots with an indication of their distance from the runway, marker-beacon transmitters are installed with the outer-marker transmitter at approximately 5 mi [8 km] from the runway and the midmarker approximately $\frac{2}{3}$ mi [1 km] from the end of the runway. The marker-beacon transmitter operates at a frequency of 75 MHz and produces both aural and visual signals. The outer-marker transmitter produces a 400-Hz intermittent signal which causes a blue indicator light on the instrument panel to glow intermittently. The midmarker transmitter produces a signal modulated at 1300 Hz which causes the amber marker-beacon light on the instrument panel to glow. Thus, when an airplane is approaching the runway and is approximately 5 mi from its end, the blue light will flash. A short time later, when the airplane is within $\frac{2}{3}$ mi of the runway, the amber light will flash. This system provides an excellent indication to the pilot of the plane's distance from the runway.

THE MARKER-BEACON RECEIVER

The marker-beacon receiver for a typical large-aircraft navigation system is a fixed-frequency superheterodyne designed to operate only on a frequency of 75 MHz. The receiver is equipped with output circuits which enable it to deliver both aural and visual signals to the flight crew. A portion of the output signal is fed through a transformer to audio filters tuned to 400, 1300, and 3000 Hz. The 75 MHz-signal from the marker-beacon transmitters is modulated with the three different audio tones, depending upon whether the transmitter is a midmarker, an outer marker, or an airways Z, or fan, marker. Each of the audio filters is designed to select one of the frequencies and with this signal activate a switching circuit which causes the appropriate signal light on the instrument panel to flash. The indicator lights are white, amber, and blue, thus making it possible for the pilot to know what type of marker the airplane is passing over. For example, a flashing blue light indicates that the pilot is passing over the outer marker.

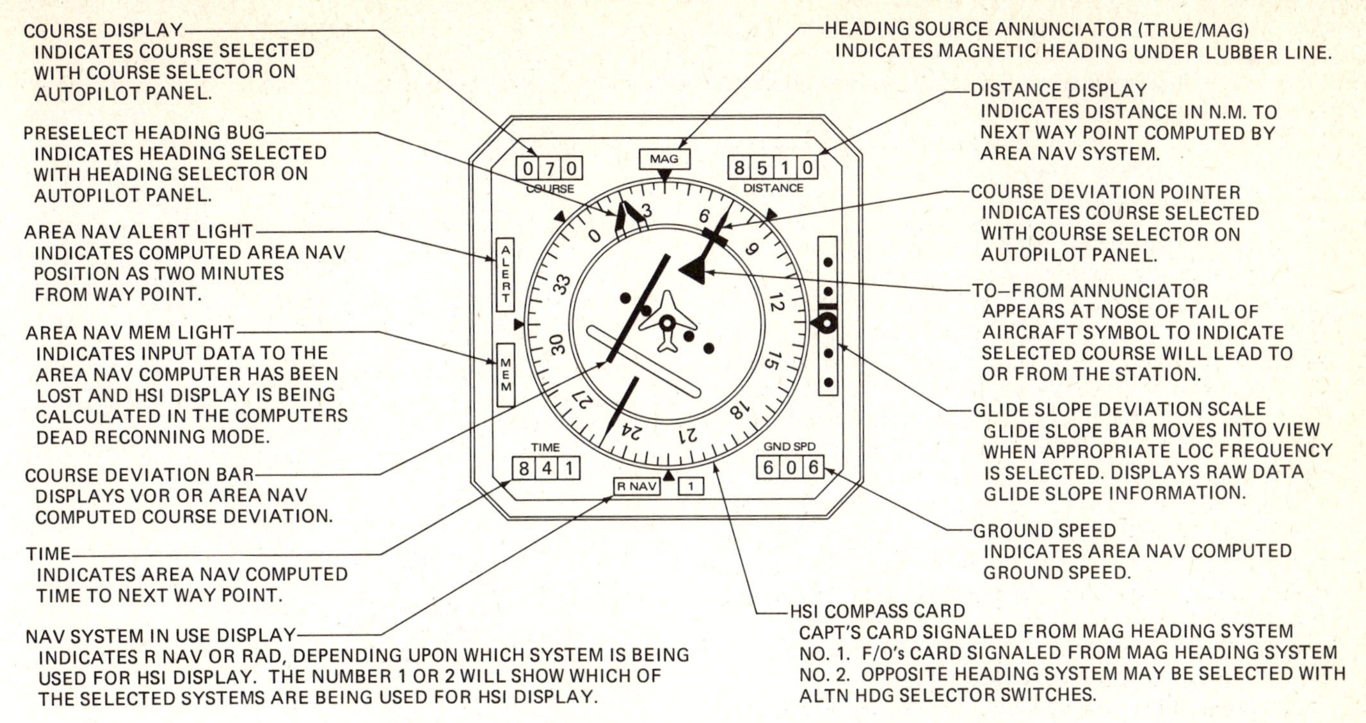

Fig. 17-28 Horizontal situation indicator. (Lockheed California Co.)

MICROWAVE LANDING SYSTEM

The microwave landing system (MLS) has been developed to overcome some of the problems and limitations associated with ILS system. The MLS system currently being deployed is called a **time-reference scanning-beam microwave landing system,** or TRSB. The ILS provides one narrow flight path and operates at VHF/UHF frequencies, whereas the MLS provides a wide range of flexible flight paths on an approach to an airport. In addition, the MLS has the advantages inherent with operating at microwave frequencies (3 to 30 GHz). Among the benefits of the microwave frequencies are a much larger number of frequency channels, fewer problems with finding suitable sites for ground components, and elimination of severe multipath interference caused by signal reflections from buildings, hills, and other objects. With MLS, aircraft can approach a runway from a wide variety of angles rather than being required to be aligned with the runway for many miles on the approach.

As explained previously, the ILS transmits signals that, when combined, provide a narrow beam rising from a point on the runway and extending for an indefinite distance along the approach to a runway. This situation requires that all aircraft approaching a particular runway be "funneled" into the one approach path. With MLS, aircraft can approach the runway from many different angles, thus making it possible to accommodate more flights and shorten flight paths.

PRINCIPLE OF OPERATION

The principle of operation for the TRSB microwave landing system may be illustrated as shown in Fig. 17-29. Two transmitters, one for azimuth and one for elevation, transmit fan scanning beams toward approaching aircraft. The precise timing of the scanning beams provides exact information for the pilot regarding the position of the aircraft. Beams are scanned rapidly "to" and "fro" throughout the area shown in the drawing. In each complete scan cycle, two pulses are received by the aircraft. One pulse is received during the "to" scan and the other during the "fro" scan. The aircraft receiver derives its position angle directly from the measurement of the time difference between the two pulses. The receiver proces-

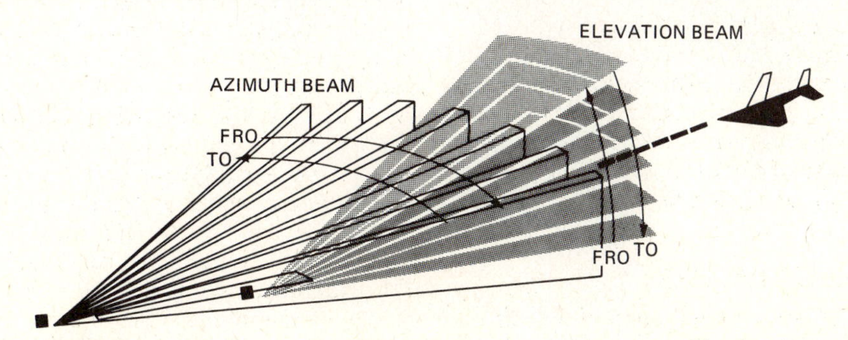

Fig. 17-29 Scanning beams of the TRSB microwave landing system. (Bendix, Aerospace-Electronics Group, Communications Div.)

sor computes the information and prepares it for display on a conventional course deviation indicator (CDI). In addition, a digital display of the information is presented on the control panel.

Distance information for the system is derived from conventional distance-measuring equipment (DME). DME is explained later in this chapter.

AIRBORNE MLS EQUIPMENT

The avionics equipment needed for MLS on an airliner is shown in Fig. 17-30. The units shown are (1) a precision DME unit, (2) the angle guidance receiver/processor, (3) the DME display, (4) the control panel, and (5) an auxiliary data display. MLS equipment for light aircraft is similar but smaller and lighter.

With the equipment shown and associated ground equipment, a pilot approaching the runway of an airport can always tell exactly the aircraft location. Bearings from the airport are provided by the azimuth beam, elevation is provided by the elevation beam, and distance is provided by the DME.

VHF OMNIRANGE

VOR is an electronic navigation system which enables the pilot to determine the bearings of the VOR transmitter from any position in its service area. This is possible because the VOR ground station, or transmitter, continually broadcasts an infinite number of directional radio beams or radials. The VOR signal received in an airplane is used to operate a visual indicator from which the pilot determines the bearings of the VOR station with respect to the airplane.

Figure 17-31 shows diagrams illustrating both the VOR transmitter and the VOR receiver systems; the receiver system is of the type used on one model of airliner. The diagram for the VOR ground station, or transmitter, shows a five-unit antenna array. The center loop of the antenna array continuously broadcasts the reference phase signal, which is modulated at 30 Hz. Two outputs are radiated by the diagonal pairs of corner antennas, and the signals radiated from these pairs are modulated by 30 Hz and differ in phase by 90°. Each pair of antennas radiates a figure-eight pattern, each pattern being displaced from the other by 90° both in space and in time phase. The resulting pattern is the sum of the two figure-eight patterns and consists of a rotating field turning at 1800 rpm, or 30 Hz.

The total effect of the radiation from the VOR transmitter is to produce two signals whose phase characteristics vary in accordance with the direction (bearing) of the transmitter from the receiver. The two signals radiated due south (magnetic) of the transmitter are exactly in phase; hence, an airplane flying magnetic north directly toward the VOR transmitter will show an indicated bearing of 0° to the VOR station. The TO-FROM indicator will show that the airplane is flying to the station.

In a clockwise direction around the VOR station, the radiated signals become increasingly out of phase. At 90° clockwise from the due south direction, the signals are 90° out of phase, at 180° they are 180° out of phase, at 270° they are 270° out of phase, and at 360° (0°) they are back in phase. The phase difference of the two signals makes it possible for the receiver to establish the bearings of the ground station. The directional bearings of VOR stations are set up in accordance with the earth's magnetic field so that they may be compared directly with magnetic-compass indications on the airplane.

During the operation of VOR equipment on a particular heading, an airplane flying toward the VOR station will show a TO indication on the omniindicator. After the airplane passes the station, the indicator will show FROM,

Fig. 17-30 Airborne avionics equipment for MLS. (Bendix, Aerospace Electronics Group, Communications Div.)

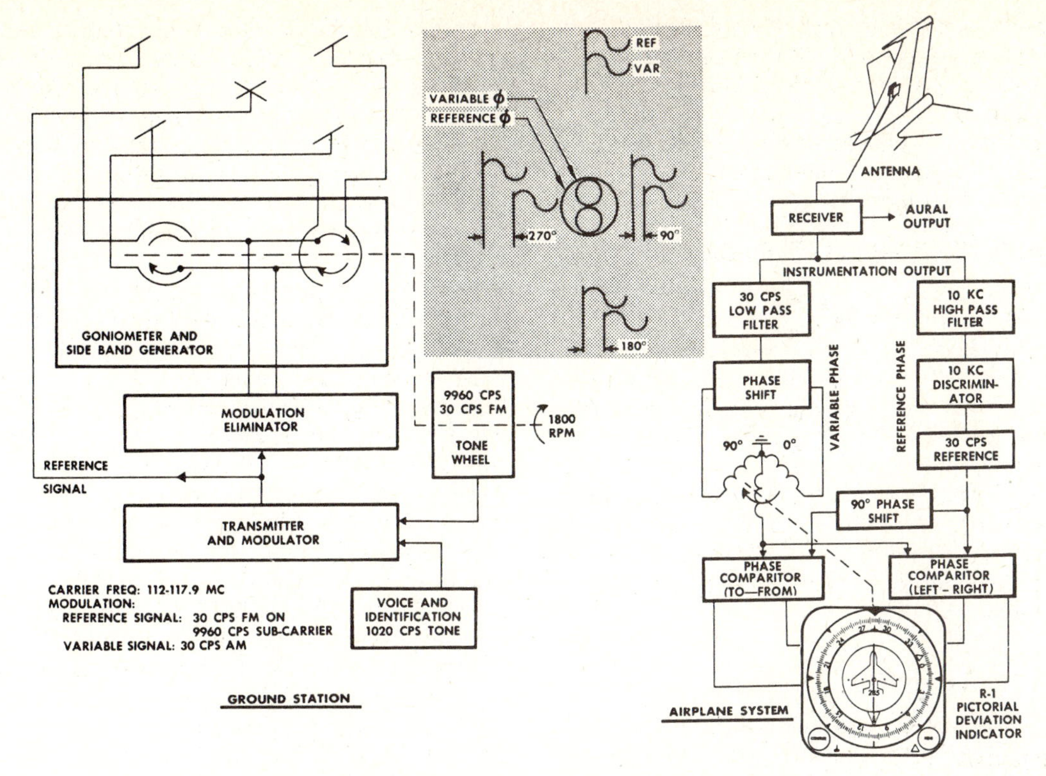

Fig. 17-31 Block diagrams of VOR and receiver systems.

and the heading information will remain the same as it was. For example, if an airplane is flying toward a VOR station having a bearing of 200°, the omnibearing indicator will show 200° TO. After the airplane passes over the VOR station, the indicator will show 200° FROM.

The carrier frequency of the VOR station is in the VHF range between 112 and 118 MHz. A modulation of 9960 Hz is placed on the carrier of the reference signal to provide a subcarrier which is modulated by a 30-Hz signal. The 9960-Hz modulation on the original carrier wave is AM, and the 30-Hz signal on the subcarrier is FM. The carrier wave for the variable-phase signal is amplitude-modulated by a 30-Hz signal. The VOR receiver mounted in an airplane may be either an independent VOR receiver or a special VOR instrumentation unit operated in connection with a standard VHF navigation receiver. The VOR receiver receives both components of the VOR signal transmitted from the ground station and from these signals produces two 30-Hz signals, one being the reference phase and the other being the variable phase. The angular distance between the two phases is applied to the omnirange indicating unit, by which it is translated into usable heading information.

The omnirange indicator includes an azimuth dial, a LEFT-RIGHT deviation needle, and a TO-FROM indicator. When the VOR receiver on an airplane is tuned to a VOR ground station, the LEFT-RIGHT indicating needle will be deflected either to the right or to the left unless the selected course on the omnirange indicator is in agreement with the bearing of the VOR ground station. If the LEFT-RIGHT needle is deflected, the pilot may rotate the course selector until the needle is centered. At this point, the heading shown on the indicator will be the actual magnetic heading of the VOR station with respect to the airplane. On some omnirange indicators, the bearing information is shown as a digital display and also on the azimuth dial. In addition to the indications already mentioned, the omnirange indicator shows whether the airplane is flying to or from an omnirange station.

Figure 17-32 is one type of **course-deviation indicator** used in connection with the VOR and other navigation equipment. (See also Fig. 17-28.) The face of this instrument consists of fixed lubber-line markings and 45 and 90° reference markings. These markings are shown as small white triangles at the outer perimeter of the azimuth dial. The azimuth dial is a rotating servoed-compass repeater dial graduated from 0 to 360° clockwise in 2° increments. The face of the dial also includes an open triangular cursor which travels around the compass dial and is positioned by the heading (HDG) input knob, an inner rotating carriage which carries the course-deviation bar, a course-deviation warning flag, solid triangular flags to show if the selected course is to or from the VOR sta-

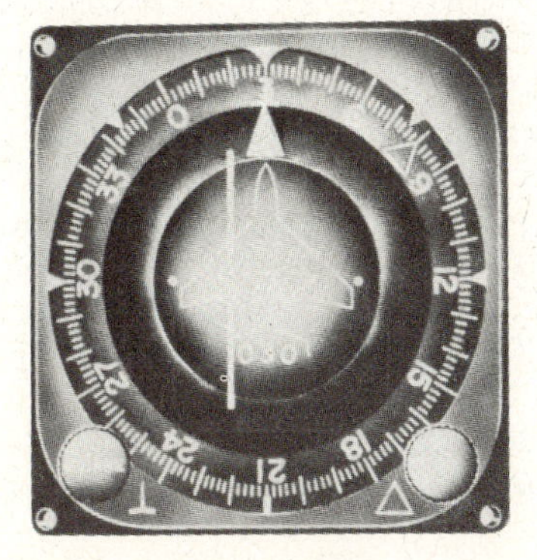

Fig. 17-32 One type of course deviation indicator.

tion, and a cursor shaped like an inverted T to show the selected course. On some instruments an omnirange bearing selector (OBS) knob is used instead of the HDG knob. A central pedestal is marked with the figure of an airplane aligned with the lubber line and contains a digital presentation of the selected course. The course input knob shown at the lower left-hand corner of the instrument face has two positions. When it is in the NORMAL position, the pilot is able to select any desired course line, which is then displayed in digital form on the central pedestal and is also indicated by the position of the inverted-T cursor against the compass dial. Internal electric components positioned in accordance with a selected course transmit course-line information to the VOR navigation receiver and to the autopilot, computer, or coupler. The EMERGENCY position of the course knob is provided to enable the course-deviation components to remain usable in the event of compass-system malfunction. To place the knob in the EMERGENCY position, it is pulled out from the case front. In this position, course-selection features are disengaged and the compass-dial servo is deenergized. This makes it possible for the pilot to orient the course-deviation bar parallel to the instrument's vertical axis without disturbing the selected course information.

The heading input knob positions an internal preset heading-control transformer in accordance with the selected position of the triangular cursor. The output of this transformer is an electric signal available for autopilot, coupler, or computer use. It must be remembered that all these circuits and components associated with the HSI indicator shown in Fig. 17-28 are not necessarily a part of the VOR system.

DISTANCE-MEASURING EQUIPMENT

To make it possible for a pilot to determine the distance of the aircraft from a particular VOR/DME or VORTAC station, distance-measuring equipment (DME) was developed. VORTAC indicates a station that includes both VOR and TACAN equipment. TACAN is described later in this section.

Through the use of the combined facilities of DME and VOR, a pilot is given a constant visual indication of the slant distance from a particular VOR station and also the bearing of the station. It is therefore possible at all times to establish the exact location of the airplane.

The operation of DME units is similar to that of radar beacons. That is, the communication between the airborne unit, called the **interrogator,** and the ground station is by means of pulses similar to those utilized in radar. The ground DME unit is called a **transponder.** During operation of VOR, when the pilot selects a particular ground-station frequency by means of the control, the coded pulse is automatically selected in the DME interrogator associated with the VOR. The interrogator transmits pairs of coded pulses to the transponder, and if the code is correct, the transponder will send a reply back to the interrogator. The time interval between the signal transmitted by the interrogator and the signal received from the transponder determines the distance of the airplane from the ground station. Remember that it requires approximately 6.19 μs for a radio wave to travel 1 nmi [1.8 km].

The DME challenge sent by an aircraft, when a particular VOR station is selected, consists of spaced pulses in the frequency range of approximately 987 to 1213 MHz. The ground-station transponder will accept only signals which are spaced correctly and have the correct frequency.

The DME equipment mounted in an airplane consists of **timing** circuits, **search** and **tracking** circuits, and the **indicator.** The timing circuits measure the time intervals between the interrogation and the reply, thus establishing the distance of the ground station from the airplane. The search circuits cause the airborne equipment to seek a reply after each challenge, a function accomplished by triggering the receiver into operation after each interrogation. When the receiver picks up a reply, the tracking circuits operate and enable the receiver to hold the signal which it has found. The time interval is measured and converted into a distance reading, which is then shown on the indicator. DME distance indications are displayed digitally on one or more panels or instruments. Figure 17-33 shows how an RMI has been combined with DME indicators in an instrument called a radio digital distance magnetic indicator. DME indications are also displayed on other instruments and panels. (See Fig. 17-28.)

As an airplane equipped with DME is approaching a DME station and is receiving DME information, the distance readout will continue to change as the distance from the station changes. The rate of change is fed to a computer that produces a ground-speed indication. In many of the advanced navigation systems, the time required to reach a given station or waypoint is also displayed. This is shown in the photograph of a DME unit and indicators in Fig. 17-34.

Airborne receivers for DME are provided with an audio system that receives identification codes from DME stations. This makes it possible for the pilot to identify positively the station that the DME has locked onto. In the majority of VOR/DME receivers, when a particular VOR frequency is selected, the associated DME frequency is automatically selected for that station.

In the installation of DME equipment in an aircraft, the location of the antenna is critical. The antenna is a short stub approximately 2.5 in [6.35 cm] in length, usually mounted on the bottom of the fuselage. Care must be taken in locating the antenna, because it can be blanked out easily by obstructions such as landing gear or other antennas nearby. It is recommended that manufacturer's instructions for installations in similar aircraft be observed when making a new installation.

TACAN

A distance-measuring and bearing-indicating system similar to the VOR/DME system described earlier in this chapter is called **TACAN.** This system was developed by the Navy for use on aircraft carriers and other Navy air installations. The word *TACAN* is a short version of the descriptive term **tactical air navigation.**

The TACAN distance-measuring facility is now ultilized for civilian air navigation, as well as for the military. The combination of VOR and TACAN to give both bearing and distance information is called **VORTAC.** To utilize VORTAC, an aircraft must be equipped with UHF radio

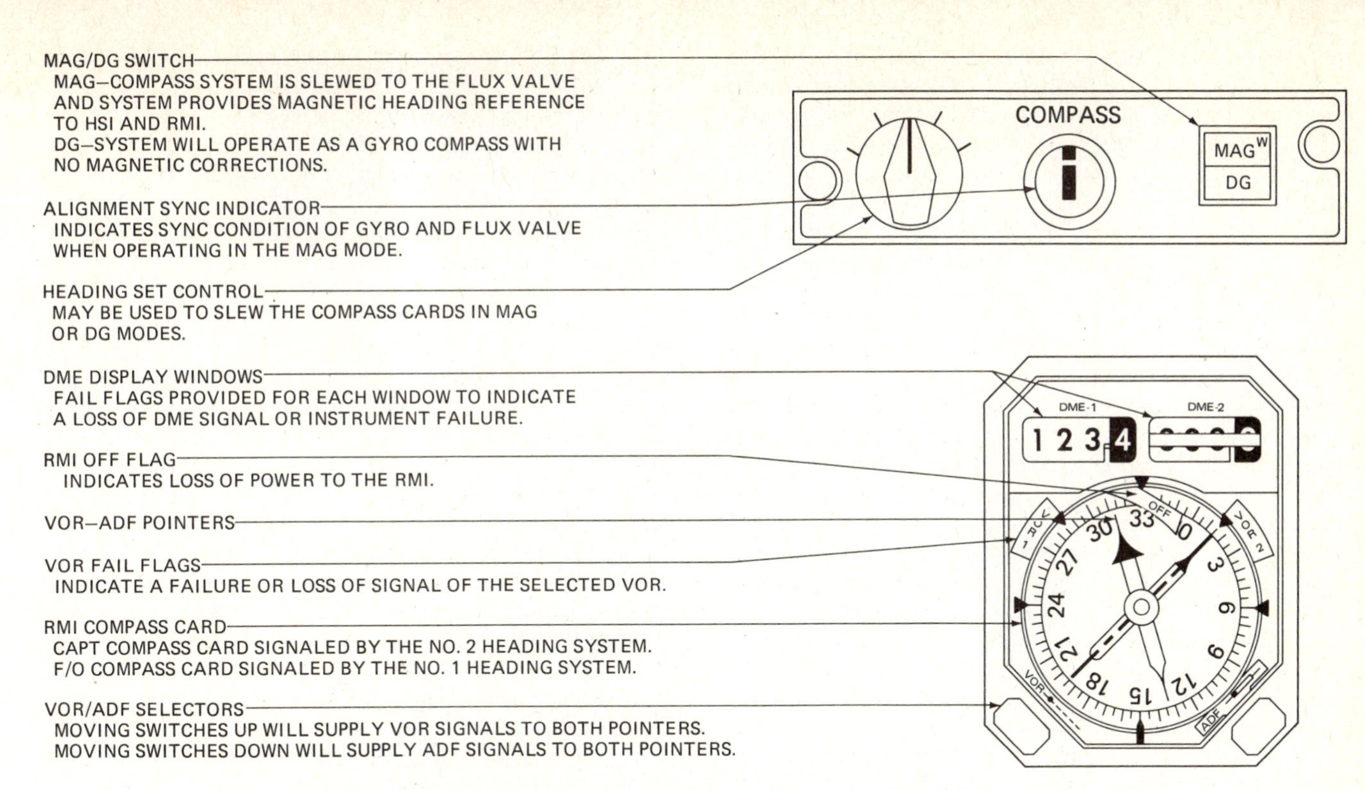

Fig. 17-33 A radio digital distance magnetic indicator. (Lockheed California Co.)

Fig. 17-34 DME remote unit and indicators. (King Radio Corp.)

units which can operate on the TACAN frequencies. The low TACAN band has receiving frequencies from 1025 to 1087 MHz and transmitting frequencies from 962 to 1024 MHz. The high TACAN band has receiving frequencies from 1088 to 1140 MHz and transmitting frequencies from 1115 to 1213 MHz.

NAVIGATION EQUIPMENT

RECEIVERS

A basic navigation (NAV) receiver is designed to receive VOR signals and display course, bearing and heading in-

NAV RECEIVER WITH OPTIONAL GS RECEIVER

(DISPLAYS BOTH ACTIVE AND STANDBY FREQUENCIES)

Fig. 17-35 A navigation receiver. (King Radio Corp.)

formation on an RMI, HSI, or other instrument. If the receiver is equipped to receive LOC and glideslope (GS) signals, the system will include a CDI or similar instrument to show the pilot whether the aircraft is tracking the ILS beam as it approaches a runway.

A typical NAV receiver is shown in Fig. 17-35. This receiver is equipped with a standby frequency provision so a frequency may be preselected and held in readiness for use when needed. The frequencies are selected by means of the frequency selector knobs at the right of the panel. The outer knob selects the megahertz portion of the frequency, and the inner knob selects the kilohertz portion. Some NAV receivers use pushbutton panels for selecting frequencies.

As mentioned previously, NAV receivers and COM transceivers are often combined into one unit of equipment. This saves weight and space and simplifies the installation. A unit of this type is called a NAV/COM unit and may include ILS receivers for LOC and GS.

INTEGRATED NAVIGATION SYSTEM

Area navigation (RNAV) is a system that makes it possible to use the information from VOR/DME or VORTAC stations and fly a direct route from a point of departure to a destination without following a dogleg course, which would result if NAV were used only with VOR/DME information. This is accomplished by using a microprocessor to continuously receive and process data originally entered plus new data that is supplied by the DME during flight.

The use of RNAV with other electronic navigation components makes possible a completely **integrated navigation system.** Such a system includes VOR, LOC, GS, DME, NAV, RNAV, and associated circuitry. The principal elements of an integrated navigation system are shown in Fig. 17-36. In addition to the equipment shown, a CDI, HSI, or similar instrument is required to provide visual course information to the pilot. Controls for programming data into the system are on the front of the unit. The unit shown in Fig. 17-36 utilizes rotary knobs to enter frequencies, radials or bearings, and distances into the system. Other types of equipment may employ pushbutton controls as shown in Fig. 17-37.

PROGRAMMING A FLIGHT

The RNAV system, in effect, makes it possible to "move" a VOR/DME station electronically from its actual location to a location on the proposed flight route. The mathematics of this operation is handled by the large-scale integrated (LSI) circuitry of the microprocessor. Figure 17-38 is a drawing of a proposed flight route showing how three VOR/DME stations are employed to produce four waypoints along the route. To set up these waypoints the pilot or other operator uses the control

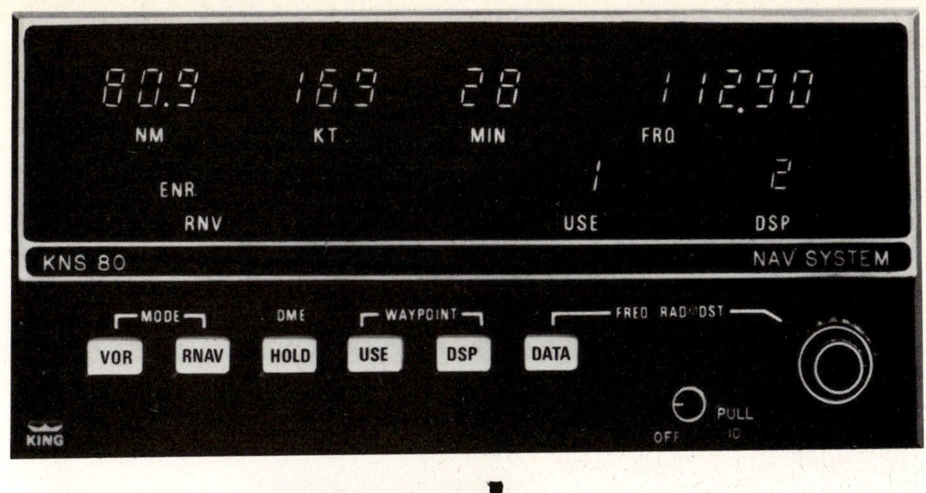

Fig. 17-36 An integrated navigation system showing internal electronic circuitry. (King Radio Corp.)

Fig. 17-37 Pushbutton control panel for a navigation system. (King Radio Corp.)

panel such as that shown in Fig. 17-36. Waypoints are loaded and stored in the system in the following steps:

1. After the master switch is turned on and the engine is started, turn on the equipment by rotating the ON/OFF knob. The display will immediately show the last information being displayed before the previous turn-off.

2. Put waypoint 1 in the DSP window by depressing the DSP button one, two, or three times until the figure 1 is displayed over DSP.

3. If the letters FRQ are not displayed, press the DATA button until the data is cycled through RAD and DST to display FRQ.

4. Rotate the outer data input control knob until the megahertz portion of the frequency for waypoint 1 (116) appears in the display. Then rotate the inner control knob until the kilohertz portion of the frequency (.90) appears in the display. The frequency shown should then be 116.90 MHz.

5. Select the waypoint 1 radial (bearing 275°) by depressing the DATA button. This will cause the radial for the previous waypoint 1 to appear in the data display over the annunciation RAD. Select the radial 275° by rotating the data input controls. The outer knob controls the 10° and 100° digits and the center knob controls the 1° digits when in the "in" position. If a 0.1 digit is required, this can be obtained by pulling the center knob out and rotating it.

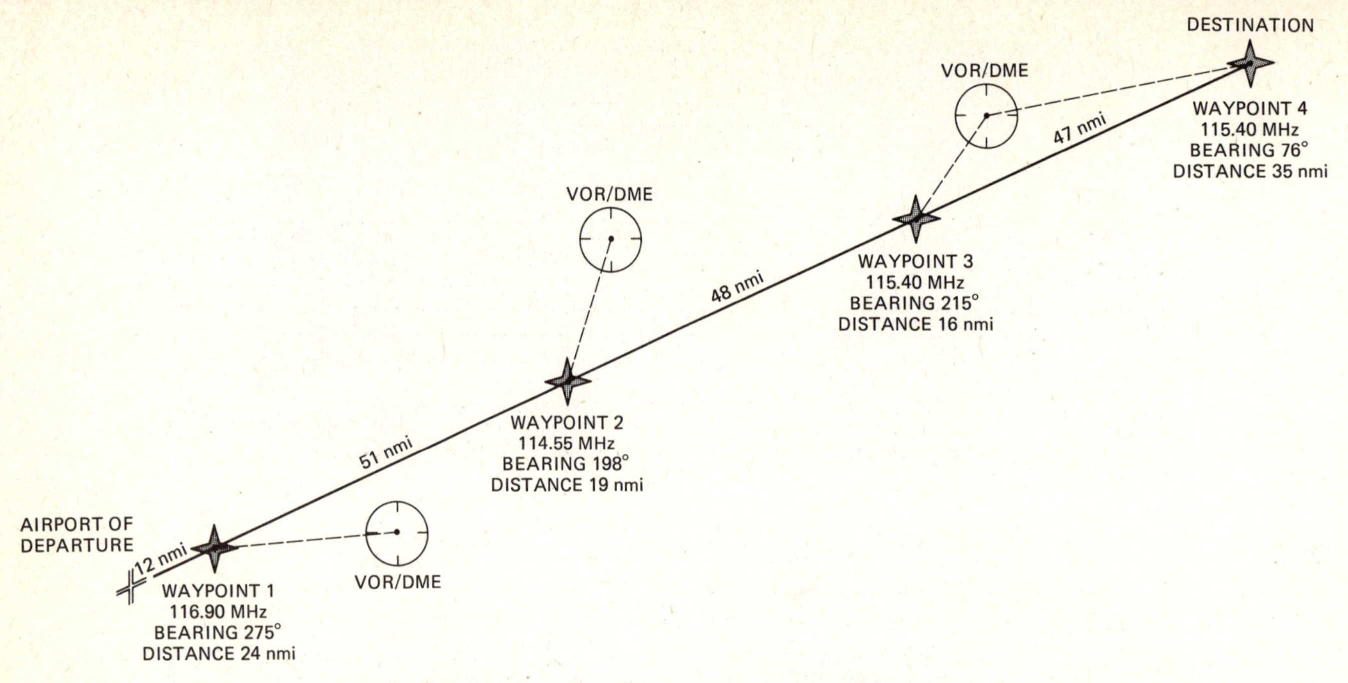

Fig. 17-38 Direct flight by RNAV utilizing VOR/DME stations.

6. Select the waypoint 1 distance by depressing the DATA button. DST will appear in the display, and the previous waypoint 1 distance will be displayed. Select the correct waypoint 1 distance by rotating the data input knobs. The outer knob controls the 10 nautical-mile (nmi) digits, and the inner knob controls the 1-nmi digits. In the "out" position, the inner knob provides 0.1 digits. Throughout the loading sequence, the figure 1 over the DSP annunciation will blink. It will stop blinking and remain steady only when the waypoint number in DSP is the same as the waypoint number in USE. This is to alert the pilot that the equipment is utilizing data other than that displayed.

7. Load waypoint 2 data by pressing the DSP button to place the figure 2 in the DSP position. Then proceed as with waypoint 1, loading frequency, radial, and distance in that order. Load data for waypoints 3 and 4 in the same manner.

8. After the four waypoints are loaded, depress the RNAV button until ENR/RNV is annunciated. Then depress the USE and DSP buttons until the figure 1 is displayed at both points. At this time, the equipment is set to operate on the data displayed for waypoint 1.

After takeoff and climb to a line-of-sight altitude to the DME, the DME will lock on and display distance to waypoint 1 over the annunciation NM. After a short time of flying the correct course, the ground speed and time to waypoint 1 will be accurately displayed.

In Fig. 17-38, the bearing of the destination from waypoint 1 is 65°. Therefore, the omni bearing selector knob (OBS) on the CDI should be rotated to 65°. The airplane is then flown with the needle centered to follow the correct course.

When the airplane passes over waypoint 1, the omni indicator on the CDI will shift from TO to FROM. The flight may be continued on the basis of waypoint 1 data until it is desired to shift to waypoint 2. This is accomplished by pressing the DSP and USE buttons to display the figure 2. Waypoint 2 data will be displayed in accordance with the data selected. The DME will shift to the waypoint 2 data and will lock onto the new station. Distance, ground speed, and time to waypoint 2 will be displayed.

The integrated navigation system for a large airliner is illustrated in the block diagram of Fig. 17-39. This system includes ADF, radio altimeters, and gyros to provide complete data for the captain and first officer. Dual systems provide safety in case of failure of any subsystem. By coupling this system with the automatic pilot, fully automatic flight is possible from takeoff to landing and rollout.

LONG-RANGE NAVIGATION SYSTEMS

Because of the need for accurate and effective navigation on overseas routes, polar routes, and other areas where conventional navigational aids are not available, a variety of long-range systems have been developed. Among these are LORAN, decca, and VLF/OMEGA. These are effective systems and have been in use in various forms for many years.

LORAN (LOng-RAnge Navigation) is a system operating in the LF range (30 to 300 kHz) and utilizing pulse-transmitting stations to provide the signals necessary for navigational computation. A master station and one or more slave stations send out synchronized pulses that are received by equipment in the airplane. If the airplane receives pulses from the master station and slave station at the same time, the airplane crew knows that the airplane is equidistant from the two stations and it is located somewhere on the perpendicular bisector of a line between the stations. If the pulses are not received at the same time, the airplane will be located on a hyperbolic line representing the difference in distances from the airplane to the two stations.

The drawing of Fig. 17-40 represents a master station, M, and two slave stations, S_1 and S_2. Any point on line AB is equidistant from the master station and S_1, and any

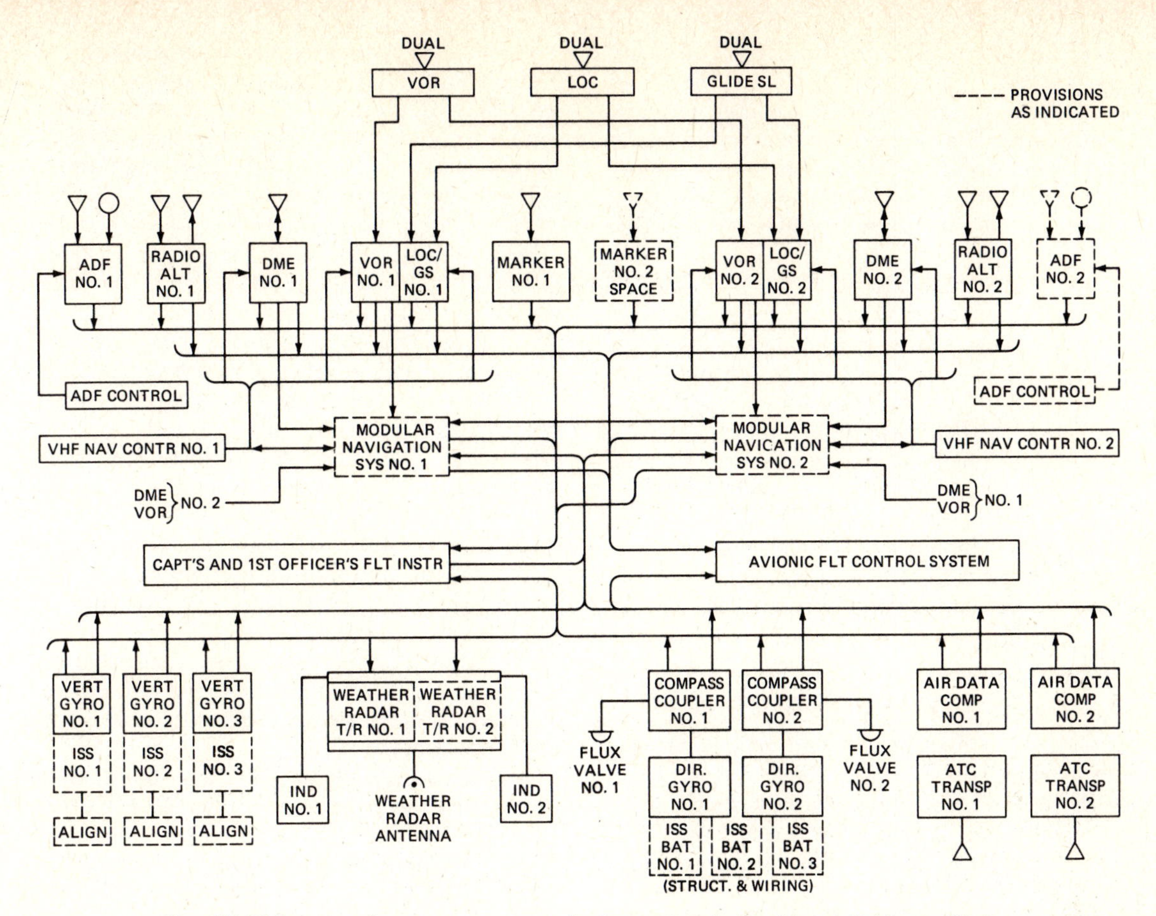

Fig. 17-39 Integrated navigation system for large aircraft. (Lockheed California Co.)

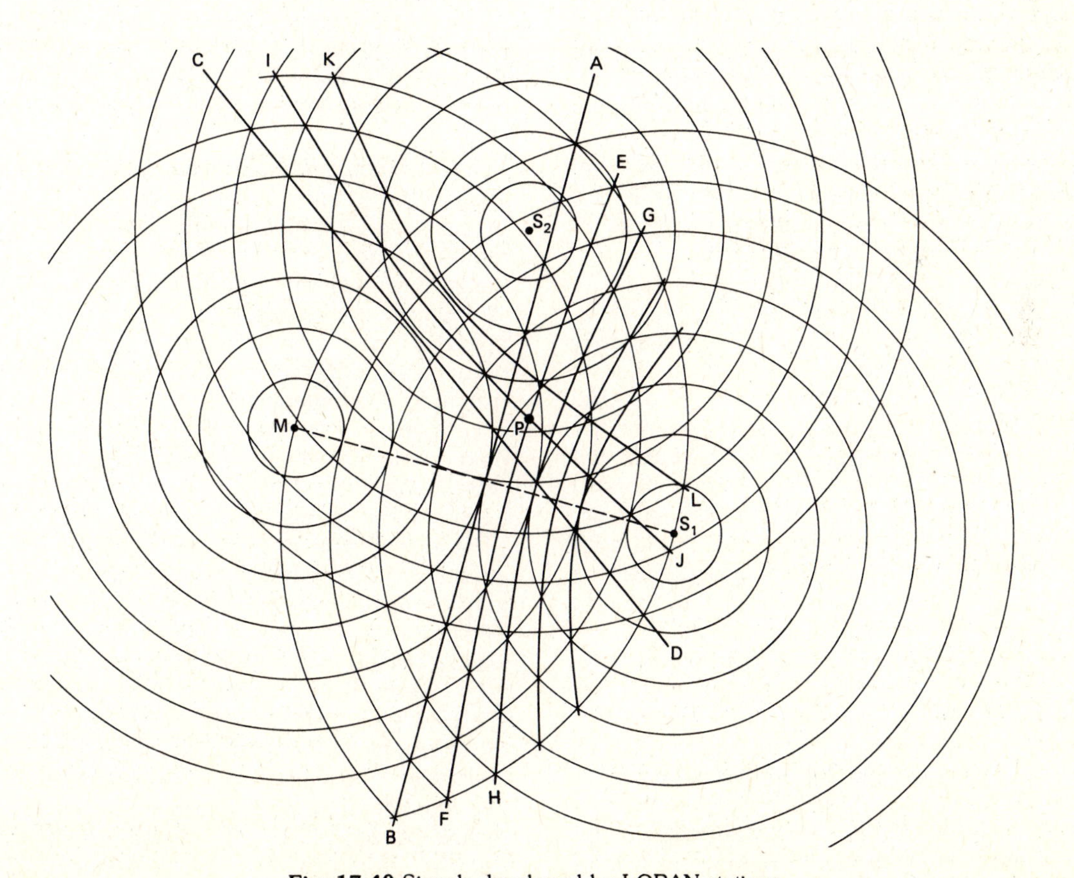

Fig. 17-40 Signals developed by LORAN stations.

point on line CD is equidistant from M and S_2. The space between each circle, in this case, is given a value of 150 μs or a distance of approximately 24 nmi. The hyperbolic line EF is the line where any point on the line is 150 μs closer to S_1 than to M. Any point on the hyperbolic line GH is 300 μs closer to S_1 than to M. The other hyperbolic lines represent increased differences in distance.

In order to establish an exact location, it is necessary to utilize another slave station or another combination of a master station and slave station to provide a cross reference. In the drawing of Fig. 17-40, this is accomplished with slave station S_2. The hyperbolic line IJ is 150 μs closer to S_2 than to M. Therefore, if an airplane receives signals that indicate that it is 150 μs closer to S_1 and 150 μs closer to S_2 than to M, then it can compute that it is at position P on the chart. In current LORAN systems, this computation is accomplished by electronic computers, and the signals received are translated to coordinates of longitude and latitude and displayed digitally on an indicator.

VLF/OMEGA NAVIGATION SYSTEM

The VLF/OMEGA navigation system utilizes worldwide stations established by the U.S. Navy and the U.S. Coast Guard. The VLF communications network is operated and maintained primarily by the U.S. Navy, and the OMEGA navigation network is operated and maintained by the U.S. Coast Guard. VLF communications stations are located at Annapolis, Maryland; Anthorne, England; Northwest Cape, Australia; Rugby, England; Lualualei, Hawaii; Yosami, Japan; Cutler, Maine; and Jim Creek, Washington. The Omega network stations are located at Gulfo Nueva, Argentine; Haiku, Hawaii; Tsushima, Japan; Reunion Island, Indian Ocean; Liberia, Africa; La Moure, North Dakota; Aldra, Norway; and Australia. The locations of the stations and the fact that they utilize VLF (10 to 30 kHz) frequencies result in a global coverage such that the signals can be employed for navigation anywhere on the face of the earth.

Each communications (COMM) station in the network transmits on an assigned frequency between 14 and 24 kHz. These stations transmit continuously on individually assigned frequencies. The OMEGA navigation stations utilize four frequencies: 10.2 kHz, 11.05 kHz, 11.33 kHz, and 13.6 kHz, but no two stations are transmitting on the same frequency at the same time. The entire segmented format is repeated every 10 s. The station in Norway transmits at 10.2 kHz for 0.9 s in the first segment, 13.6 kHz for 1.0 s in the second segment, 11.33 kHz for 1.1 s in the third segment, and 11.05 kHz for 1.0 s in the sixth segment. Then it does not transmit in the fourth, fifth, seventh, or eighth segments. The station in Liberia transmits with the same order of frequencies in the second, third, fourth, and seventh segments. The other stations are assigned segments in a similar manner such that a computer in a receiving system can sort out the stations it is receiving by a process called **commutation.** After the computer has determined what stations it is receiving, it performs the mathematics necessary to determine its location with respect to the position information it contained at the time it started receiving and computing position. The computer must receive signals from at least two stations in order to determine its change in position as a flight progresses. If the computer is given the correct position coordinates at its starting point, it will provide accurate position data digitally as latitude and longitude.

The long-range capabilities of the VLF/OMEGA system is made possible by the use of the VLF frequencies. These waves can be effective for distances up to 10 000 mi [16 098 km] because they are reflected back to the earth by the ionosphere. Thus each station can blanket half the earth's surface, and there is no location where a suitable number of stations cannot be received.

A number of manufacturers have produced equipment for VLF/OMEGA navigation. For the purposes of this discussion, the GNS-500A global navigation system manufactured by Global Navigation, Inc. is selected. The principal components of this system are the **control display unit (CDU),** the **receiver computer unit (RCU), antenna,** the **optional equipment unit (OEU),** an optional **CDI,** and an optional **pilot numerical display (PND).**

The control display unit is located in the cockpit and serves as the link between the pilot and the computer. It contains all necessary controls and displays for operation of the system.

The receiver computer unit is the "brain" of the system. It contains the power supply, the VLF/OMEGA receivers, the computer, and the required interfacing circuits. This unit is remotely mounted and contains no operational control.

Any one of three different types of antennas may be used with the VLF/OMEGA equipment. These are an E-field blade-type antenna with a preamplifier built into the base; an H-field double-loop antenna designed to reduce 400-Hz noise; or an ADF loop and sense antenna used in conjunction with an optional ADF/VLF antenna coupler.

The OEU contains the rubidium frequency standard that enables the system to function satisfactorily when receiving only two VLF stations rather than three as is otherwise required. The unit is comprised of three major components: the ribidium frequency standard, a standby battery, and battery-charging circuitry. The standby battery operates the system through power interruptions that occur when engines are started or when power buses are switched. The battery is a 26-V nickel-cadmium unit that is kept charged by the aircraft electric system. Optional plug-in interface boards are available with the OEU to provide displays of desired course, distance-to-go, and bearing to waypoint on the aircraft flight instruments.

The system provides standard outputs for the aircraft autopilot and HSI; however, a small panel-mounted CDI may be installed where a separate indicator is desired. The CDI provides an analog indication of the selected leg with respect to the position of the aircraft. The CDI sensitivity when in the enroute (EN) position is 7.5 mi [12.1 km] full-scale deflection from center, and when in the approach (AP) position it is 1.5 mi [2.4 km] full-scale deflection from center. A TO/FROM/OFF flag is located near the center of the instrument to show the aircraft direction of movement with respect to a particular waypoint. An SX annunciator light is located on the right side of the instrument to indicate that the course guidance displayed is with respect to an offset parallel leg selected by the pilot.

The pilot numerical display (PND) is mounted on the aircraft instrument panel and provides a remote readout of groundspeed and distance-to-go. It contains a three-digit, numerical readout with a dim knob for illumination control and a switch to select either ground speed or distance for display. This unit is not needed for operation of the system because the information may be obtained from the CDU; however, it is useful in that it can display speed and distance data when the CDU is displaying other desired data. Distance is displayed in nautical miles when the instrument is in the EN position and in nautical miles and tenths of nautical miles when in the AP position.

A drawing of the CDU for the system is shown in Fig. 17-41. A description of the components of this unit will clarify and explain many of the functions of the system.

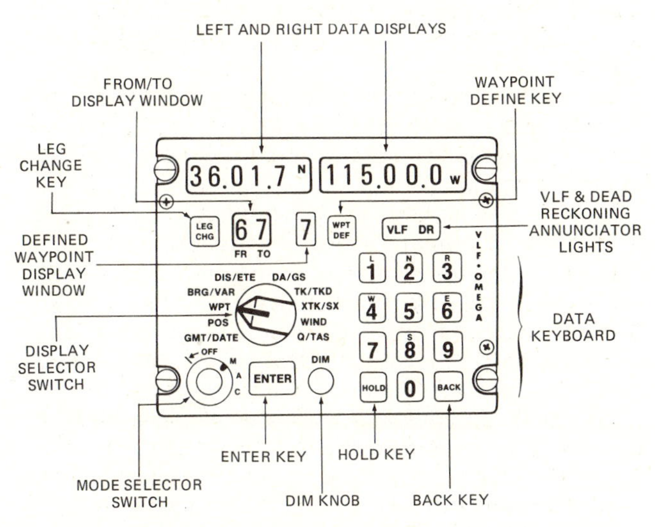

Fig. 17-41 Control display unit for a VLF/OMEGA navigation system. (Global Navigation, Inc.)

The **left and right data displays** at the top of the panel provide for a visual readout of selected navigational data from the computer and allow verification of data that is to be entered into the computer memory. The characters are comprised of seven-segment neon lights (gas-discharge tubes). Characters displayed include numerals, decimal points, dashes, and the six letters *L*, *R*, *N*, *S*, *E*, and *W*.

The **leg change (LEG CHG) key** and **TO/FROM display window** are used for manual leg change selection. Pilots use these to program the system for the Great Circle route between the two waypoints of their choice. In the drawing of Fig. 17-41, the display shows that the leg selected is waypoint 6 to waypoint 7. If either of the waypoints selected have not been defined, the letter *E* will appear in its place indicating that the computer is empty of that information.

The **waypoint define (WPT DEF) key** and the **defined waypoint display window** are used for pairing each waypoint's geographical coordinates (latitude and longitude) with a convenient reference number. They are used only in conjunction with the WPT position of the display selector switch.

The **VLF and DR annunciator lights** tell the pilot whether the system is operating and in what navigational mode it is functioning. If the system is not receiving signals of adequate quality and sufficient number to operate in the VLF mode, it will switch to dead reckoning and DR will be annunciated. The system is operating if either of the two lights is illuminated.

The **data keyboard** is used by the pilot or other members of the flight crew to key the various characters into the left and right data displays, the TO/FROM display window, and the defined waypoint display window. It is always used with either the display selector switch or the leg change key for proper character placement and distinction between its numbers and letters. The **HOLD key** is used for **present position display** updates and to place the system into the VLF mode. The **BACK key** has three functions. It may be used to erase individual incorrect digits during programming, test all display lights on the face of the CDU, and erase previously stored waypoint coordinates.

The **DIM knob** is used to regulate the illumination of the characters on the left and right data displays, the FROM/TO display window, the VLF and DR annunciator lights, the ENTER light, and waypoint window. The backlighting of the front panel and the data keyboard of the CDU is normally controlled with the same rheostat that controls the aircraft panel lighting.

When illuminated, the **ENTER key** is used to insert into the computer memory the information displayed at that time in the left and/or right data displays. The word ENTER will be steadily illuminated whenever the computer is about to receive some new information. The light will go out after the computer has accepted the information. The light will blink if the computer needs to have the information just entered verified by the pilot. If an error is noticed after information has been entered, the pilot or other operator should, in most cases, reenter the correct data. Totally erroneous information such as a latitude over 90° or a GMT (Greenwich mean time) greater than 2400 will not be accepted by the computer. The blinking ENTER light will not extinguish when information of this type is input.

The **mode selector switch** is used to control the operating mode of the system. It is a four-position switch with functions of OFF, **manual (M)**, **automatic (A)**, and **computer access (C)**. In the manual position the pilot or other crew member must make the selections of legs, whereas in the automatic position of the mode switch, the computer automatically makes sequential leg updates as it nears each waypoint. For example, just before reaching waypoint 4, the computer will reprogram itself for the leg from waypoint 4 to waypoint 5. The WAYPOINT ALERT annunciator is illuminated for 2 min before reaching the waypoint in the A position only.

The **computer access** mode provides for access to the computer for secondary functions such as preflight planning. Use of this mode enroute does not interfere with the computer's primary task of navigation. Other functions of this mode are its use in facilitating system troubleshooting and display of data from other systems interfacing with this system.

The **display selector switch** is used to call from the computer the information required or to inform the computer of the type of information to be entered on the data keyboard. The data selector switch has ten operating positions. Greenwich mean time and the data are displayed in the left and right data display when the display selector switch is placed in the GMT/DATE position. GMT is shown in hours, minutes, and tenths of minutes. The day, month, and year are displayed in the right data display. This information is initially provided by the continuously running clock and needs only to be checked by the operator for accuracy and verified to the computer. It is used by the computer for automatic **diurnal shift compensation.** Diurnal shift is the change in altitude of the ionosphere that takes place from night to day and during different seasons of the year.

The **present position** (POS) of the display selector switch displays the present position of the aircraft in terms of its latitude and longitude. Latitude is displayed in the left display, and longitude is displayed in the right display. The coordinates are presented in degrees, minutes, and tenths of minutes. The CDU shown in Fig. 17-41 displays 36° 1.7′ north (N) latitude and 115° west (W) longitude.

The **waypoint position (WPT)** of the selector switch is used to display the geographical coordinates of the waypoint identified in the defined waypoint display window. These coordinates will be displayed in the left and right data displays. Any waypoint reference number may be entered into the defined waypoint display by depressing that number on the data keyboard. If coordinates have not yet been entered for the waypoint selected, a series of dashes will appear in the left and right data displays. When the system is turned off, all waypoint coordinates defined in the waypoint storage locations 1 through 9 will remain stored in memory.

The position of the display selector switch marked BRG/VAR (bearing/variation) is used to display the bearing from the present location of the aircraft to the TO waypoint selected in the FROM/TO display window. The bearing is displayed to the nearest degree in the left data display. At the same time, the local magnetic variation will be computed and displayed in the right data display if the aircraft position is between N70° and S60°. The bearing is magnetic; however, true north can be called up when the operator manually cancels the automatic variation by entering E (east) or W (west) and 0 into the system. This will be indicated in the left data display by the letter *N* following the bearing display. Outside latitudes between N70° and S60°, bearing and variation will not be displayed until correct local magnetic variation is manually input to the system.

Distance and estimated enroute are displayed when the data selector switch is placed in the DIS/ETE position. Distance-to-go is in the left data display, and estimated time enroute is in the right data display. The distance is measured from the aircraft's present position to the TO waypoint shown in the TO/FROM display window and is given to the nearest tenth of a nautical mile. Estimated time enroute for this same segment is given in hours, minutes, and tenths of minutes. Dashes will appear in the displays if TO/FROM waypoint data have not been entered. Depressing the 0 key in this position will cause the right data display to show the **estimated time of arrival (ETA)** for this segment in hours, minutes, and tenths of minutes GMT. This information will be correct only if the correct GMT was entered in the system at the start of the flight. Releasing the 0 key will cause the display to revert to ETE indication.

Drift angle and ground speed are displayed by means of the DA/GS position of the display selector switch. Drift angle is shown in the left data display, and ground speed is shown in the right data display. The *R* and *L* preceding the drift angle display indicates whether the track of the aircraft is to the right or left of the heading. Aircraft heading and magnetic variation are required to compute the drift angle. Dashes will appear in both the left and right data displays when the actual ground speed is below 50 knots (kn) [92.7 km/h].

With the TK/TKD position selected, the ground **track angle** being made good is shown to the nearest degree on the left data display with respect to either true or magnetic north. An *N* will appear in the left data display if the track is relative to true north. Track angle information will not be provided if the ground speed is less than 50 kn or if magnetic variation is not defined. **Track angle deviation** to the nearest degree will be displayed on the right data display. The *R* or *L* preceding this readout indicates that the present track is to the right or left of the bearing to the TO waypoint. TKD is available only after a leg has been selected in the TO/FROM display window.

Crosstrack is the distance of an aircraft from an originally selected course. **Selected crosstrack** is a track parallel to a specified track and a selected distance from it. To display **cross-track distance** and **selected cross-track distance,** the display selector switch is placed in the XTK/SK position. Cross-track distance will be displayed to the nearest tenth of a nautical mile in the left data display. The *R* or *L* preceding the readout shows whether the aircraft is to the right or left of the selected Great Circle route. XTK is only available after a leg has been selected in the FROM/TO display window. A selected cross-track distance, SX, may be entered by the operator in the right data display to produce parallel course information. After SX has been entered, the CDI will be centered when the aircraft is positioned on the selected parallel leg. The SX can be canceled manually by entering a zero SX and will be canceled automatically upon making any leg change. The SX annunciator on the CDI illuminates whenever a parallel course is selected.

When the display selector switch is placed in the WIND position, the **wind direction and speed** will appear in the left and right data displays, respectively, provided that the proper data have been input to the computer. Wind direction is given to the nearest degree referenced to true north, and speed is given to the nearest knot.

Wind data will be displayed only if the following conditions are met:

1. The aircraft heading is automatically input to the computer.
2. The true airspeed is input to the computer, either automatically from an air data computer or similar source, or manually by the operator.
3. The ground speed is in excess of 50 kn.
4. The wind speed is greater than 4 kn [7.4 km/h].
5. Magnetic variation is valid and entered either automatically or manually.

When the display selector switch is placed in the Q/TAS position, the system will display two numbers in the left data display relating to the general quality of navigation to be expected. The state of the ribidium frequency standard in the OEU is shown on the far left of the left data display as either a blank, when the frequency standard is unstable, or a 1 when the frequency standard is stable. The right side of the left data display shows the navigational **quality factor.** This is determined by the number of VLF/OMEGA stations being received, the signal strength and stability, the reception angularity, and the status of the frequency standard. A low number such as between 2 and 4 is optimum, while 8 or above indicates that the system is in dead reckoning.

The right data display presents the true airspeed in knots. Those systems that receive a TAS input from an air data computer will present TAS for display. TAS can be input manually on the data keyboard when no automatic input exists. This should be updated whenever any significant TAS changes take place.

OPERATION OF THE SYSTEM

The operation of a VLF/OMEGA equipment on an airplane should follow the instructions provided in the operator's manual supplied by the manufacturer. The procedures involve preflight planning which includes selection of the route, establishment of waypoints, and entering correct data in the computer. Waypoint 1 should be the point of departure, and the geographical coordinates for the end of the runway should be entered. The computer receives position information in degrees, minutes, and tenths of minutes. If the information from the navigation chart is in degrees, minutes, and seconds, the seconds may be converted to tenths of minutes by dividing by six. For example, if the coordinates to be entered are given as N33°24′31″ and W117°40′43″, these data must be entered as N33.24.5 and W117.40.7.

Unless otherwise advised by the manufacturer, the system should be turned on only after the engine or engines have been started to avoid voltage transients that could damage the equipment. To turn the equipment on, the mode selector switch is rotated out of the OFF position to M or A as desired. The system will self-test, and all displays can be tested by use of the BACK key on the data keyboard. Data can now be entered for all waypoints as selected. Entry of data should be in accordance with manufacturer's instructions. In addition to waypoint data, the GMT and DATE and present position should be checked for accuracy. If not correct, the proper data should be entered.

After all the required data is stored in the computer and the aircraft is in flight, the computer will continuously compute the position of the aircraft, based on the position information it was given at the start of the flight. This is accomplished by comparing the VLF signals received with a reference signal produced in the system. Rotation of the display selector switch will cause the display of any available information desired by the pilot and crew.

INERTIAL NAVIGATION SYSTEM

An inertial navigation system (INS) provides information similar to the VLF/OMEGA system and doppler system. Information as desired is displayed in the flight compartment on a digital display panel in response to selection by the pilot or a crew member.

An inertial system utilizes extremely sensitive gyros and accelerometers to develop signals from which navigation information is computed. Gyros sense pitch, roll, and yaw, while accelerometers sense accelerations and decelerations.

The heart of a typical inertial system is the **inertial sensor unit (ISU).** The ISU contains the **inertial measurement unit (IMU)** and may include the system power supply. It contains a stabilized platform on which are mounted gyros and accelerometers. The stable platform is isolated from angular motions in the aircraft by a shock-mounted gimbal system providing isolation from aircraft vibration. The IMU is contained within a fixed outer frame that serves to mount the platform, the IMU connectors, and the electronics for instrument compensation and electrical adjustments. Aircraft pitch, roll, and heading data is provided by means of synchros mounted on the platform-gimbal axis.

The gyros in the IMU sense pitch, roll, and yaw and produce signals which are converted to usable information by the computer. Acceleration in any direction is sensed by the accelerometers, and this information is also employed by the computer to provide navigation information. The information developed by the computer is displayed on the control/display panel. The following information is available: track angle and ground speed, true heading and drift angle, cross-track distance and track-angle error, present position in latitude and longitude, latitude and longitude of any of the nine waypoints stored in the system, wind speed and direction, distance and time to the next waypoint, desired track **from** or **to** a waypoint, waypoint alert, system-failure warning, battery mode of operation annunciation, and attitude reference-mode annunciation. Desired information is displayed when the selector switch on the control/display panel is rotated to the appropriate position.

It must be noted that the INS discussed here is only one of a number of systems in use by long-range aircraft. The technician will encounter different types of systems; however, the basic principles of operation are the same.

DOPPLER NAVIGATION SYSTEM

The doppler navigation system is so named because it utilizes the **doppler shift** principle. The doppler shift is the difference in frequency which occurs between a radar signal emitted from an aircraft radar antenna and the signal returned to the aircraft. If the signal is sent forward from an aircraft in flight, the returning signal will be at a higher frequency than the signal emitted. The difference in the frequencies makes it possible to measure speed and direction of movement of the aircraft, thus providing information which can be computed to give the exact position of the aircraft at all times with respect to a particular reference point and the selected course.

In the doppler navigation system, flight information is obtained by sending four radar beams of continuous-wave 8800-MHz energy from the aircraft to the ground and measuring the changes in frequencies of the energy returned to the aircraft. The change in frequency for any one beam signal is proportional to the speed of the aircraft in the direction of the beam. The radar beams are pointed forward and down at an angle of approximately 45° to the right and left of the center of the aircraft and rearward and down at a similar angle. When the airplane is flying with no drift, the forward signals will be equal. The rearward signals will be equal to the forward signals, but opposite in value. The difference between the frequencies of the forward and rearward signals will be proportional to the ground speed; hence, this difference is used to compute the ground speed and display the value on the **doppler indicator.** If the airplane drifts, there will be differences in the frequencies between the right and left beam signals, and these differences are translated into drift angle and displayed on the doppler indicator. Figure 17-42 is a drawing showing how the radar beams are aimed with respect to the aircraft.

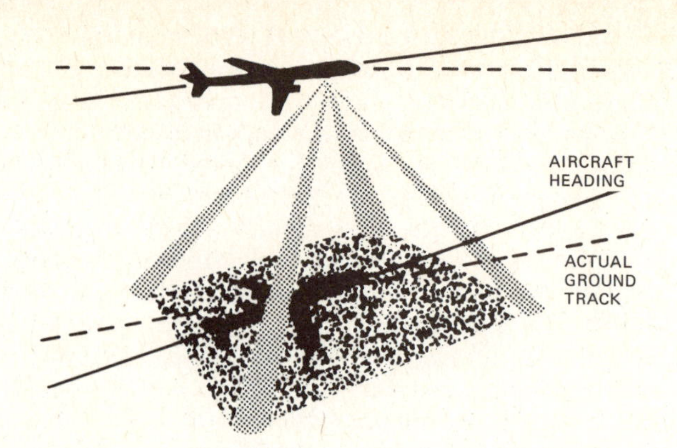

Fig. 17-42 Drawing to show radiation of doppler radar beams from an aircraft.

The advantage of a doppler system is that it is completely contained in the aircraft and requires no external signals. At the start of a flight, the course or courses to be flown are programmed into the system. Therefore, continuous information regarding the position of the aircraft will be displayed on the doppler indicator and the computer controller.

INTEGRATED NAVIGATION AND FLIGHT SYSTEMS

When all or most of the conditions affecting the flight of an airplane are brought together and sensed by a system that is able to present information regarding the conditions to the pilot, the total system may be termed an integrated navigation and flight system or simply an **integrated flight system.**

A completely integrated navigation and flight system includes flight instrumentation, navigation systems, communications systems, and the automatic flight system. The instrumentation associated with such an integrated sytem is shown in Fig. 17-43. This is an arrangement for one particular airline on a large passenger aircraft. At the top of the panel are warning annunciators to indicate that a segment of the system is inoperative. The AFCS MODES

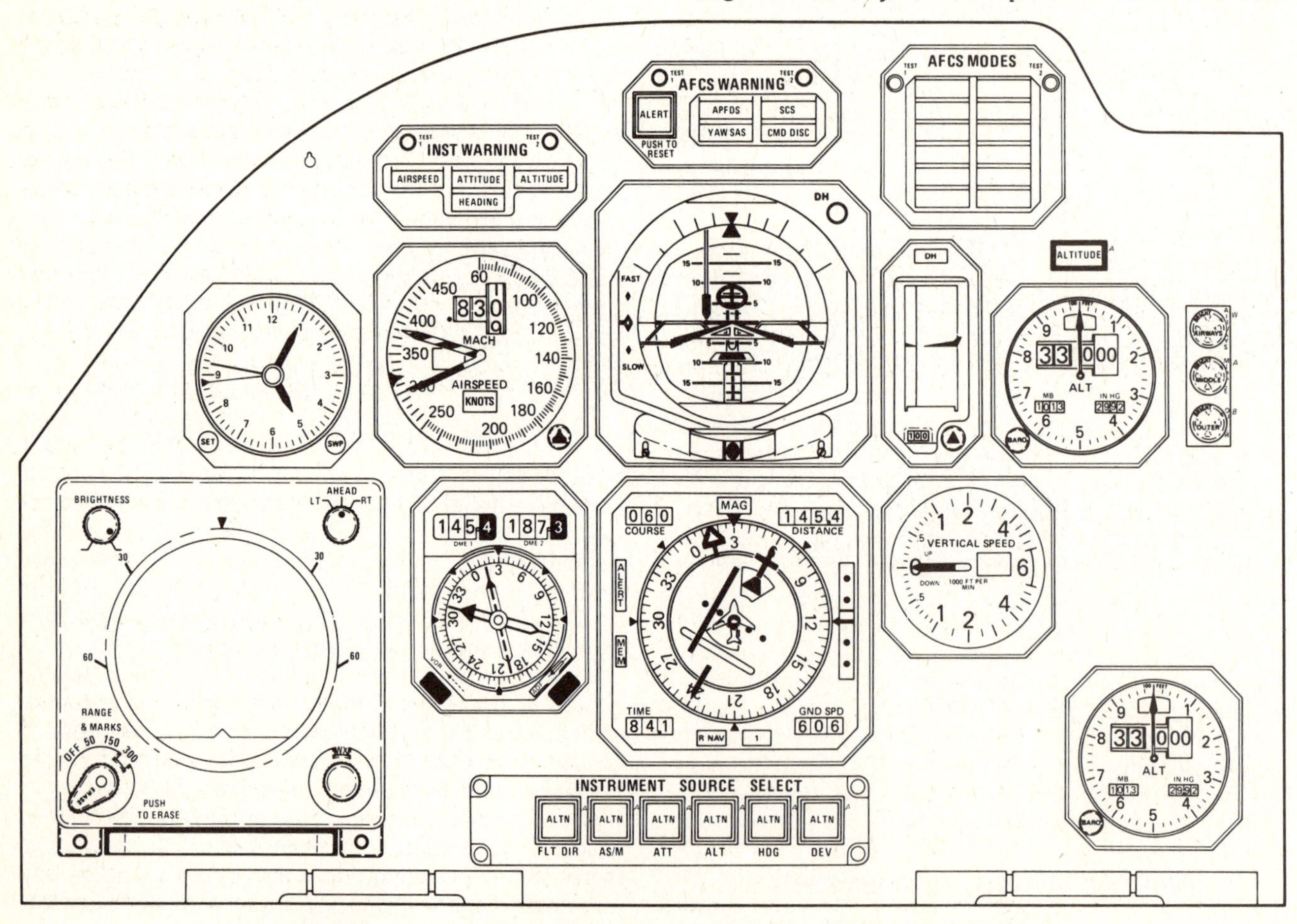

Fig. 17-43 Instrumentation associated with an integrated flight system. (Lockheed California Co.)

annunciator shows in what mode the avionics flight control system is operating, that is, what signals and control elements are in effect at a given time. The upper row of instruments includes the clock, Mach and airspeed indicator, ADI, decision height annunciator (DH) (which is a radio altimeter), and an altimeter. In the bottom row of instruments are the radar scope, a radio digital distance magnetic indicator, an HSI, a vertical speed indicator, and, in the lower left corner, a backup altimeter. Adjacent to these instruments and located in the center panel are a total air temperature indicator, an altitude indicator, and a true airspeed indicator. These instruments receive data from the central air data computer.

The instruments shown in Fig. 17-43 are at the captain's station; however, a duplicate set of the instruments are at the first officer's station. Thus, both pilots are continuously informed regarding all flight conditions and situations regardless of whether the aircraft is being flown manually or in an automatic mode.

ATC TRANSPONDER

Because of the difficulty that flight controllers had in identifying aircraft on radar scopes in tower stations and control centers, radar devices called ATC (Air Traffic Control) **transponders** were developed. In general, a transponder is an automatic receiver and transmitter which can receive a signal (be interrogated) by a ground station and then send a reply back to the station. For the purpose of this discussion, we shall consider the type of transponder which receives an interrogation from a ground radar station and sends a reply signal for identification. The reply shows on the radar scope as a double slash. A controller who wishes to obtain positive identification of an aircraft will request that an "IDENT" signal be returned from the transponder. The pilot of the aircraft will then press the IDENT button on the transponder control panel to send a special-image signal which the controller will recognize for identification.

One type of transponder for light aircraft is shown in Fig. 17-44. At the left of the transponder panel is the **function selector switch.** This is a five-position rotary selector which determines the operating mode of the unit.

Fig. 17-44 Operator's panel for a transponder. (NARCO Avionics)

In the OFF position of the function selector switch, all power is off and the transponder is inoperative. In the ON position, the switch places the unit in the operating mode for normal operation. This is called mode *A*. The transponder is ready to reply to interrogations from a ground station after a 1-min warm-up period.

The ST BY position of the selector turns the transponder power on and applies power to the transmitting system. ST BY is used at the request of the ground controller to selectively clear the radar scope of traffic. Turning to ST BY will keep the transponder from replying to interrogations, but will allow instant return to the operating mode when switched to ON.

The ALT position of the selector switch activates mode *C*, the altitude-reporting capability of the transponder. When used with an **altitude digitizer** or an **encoding altimeter,** the unit will automatically transmit altitude information. The altitude is given as standard-pressure altitude which is converted to real altitude by ground computers.

The TEST position of the selector switch is used to self-test the operation of the unit. It may be used at any time, as it does not interfere with the normal operation. By turning the selector switch to TEST and holding it there, a test signal is developed to interrogate all internal circuitry of the transponder except the receiver. If the transponder is working in a normal manner, the REPLY lamp will remain on as long as the switch is in the TEST position.

The REPLY lamp and PUSH IDENT button are contained within a single assembly. The REPLY lamp automatically goes on when the transponder is replying to ground interrogation or when the function selector switch is placed in the TEST position. The PUSH IDENT button is used to send the special position identification pulse (spip). When the pilot is asked by the ground controller for an "IDENT," the pilot presses the button and activates a special signal which "paints" an instantly identifiable and unmistakable image on the controller's radar scope. This signal must be used only when requested by the controller, because use at any other time could interfere with another aircraft sending a spip. It is not necessary to hold the IDENT button down, because the reply will last about 20 s after release. This is sufficient time for the ground controller to make a good identification and determine position.

The **code selector** comprises four 8-position rotary switches providing a total of 4096 active settings available for selection of the identification code. The code selector sets up the number of spacing of the pulses that are transmitted at the transponder frequency of 1090 MHz.

During operation, the pilot must set in the transponder code requested by the ground controller. When the unit responds to an interrogation from the ground station, the REPLY lamp will light, thus telling the pilot that the code is correct.

Power for the transponder and for other radio equipment is usually supplied by the aircraft generator or alternator. When additional equipment is installed, the person responsible must make sure that the electric system of the aircraft has sufficient capacity to supply all the aircraft requirements.

INSTALLATION AND MOUNTING OF AVIONICS EQUIPMENT

Plans for mounting avionics equipment in aircraft should include careful consideration of location, strength of mounting structures, reduction of vibration and shock, bonding and shielding, and serviceability. Hazards to personnel and to the aircraft must be avoided, since high voltages are developed in some types of equipment and since some units may develop sufficient heat to ignite any particularly flammable material in the immediate vicinity. The manufacturer provides complete information for the installation of avionics equipment.

Avionics equipment, controls, and indicators should be located in the positions most convenient to those who must operate them; in light aircraft, the controls and indicators should be easily accessible to the pilot. Sufficient ventilation should be provided for equipment subject to heating so that it will not exceed its normal operating temperature. To avoid the danger of fire, equipment which naturally operates at high temperatures must be sufficiently removed from flammable materials.

The actual attachment of avionics-equipment units to the aircraft must be such that there is no danger of a unit's becoming loose because of vibration. Fastening devices include standard bolts, nuts, and screws with effective locking devices such as self-locking nuts, lock washers, safety wire, and cotter pins. Self-locking hold-down clamps and snap slides are hold-down devices specially designed for radio equipment.

Radio units in light aircraft may be mounted on brackets attached to the rear of shock-mounted instrument panels, or they may be secured on shock-mounted brackets or racks attached to a solid structure of the airplane. In any event, shock mountings must be placed between the actual radio equipment and the basic aircraft structure. In some cases, shock-mounting bases designed especially to fit particular units are attached directly to the airplane. One type of shock mounting is shown in Fig. 17-45.

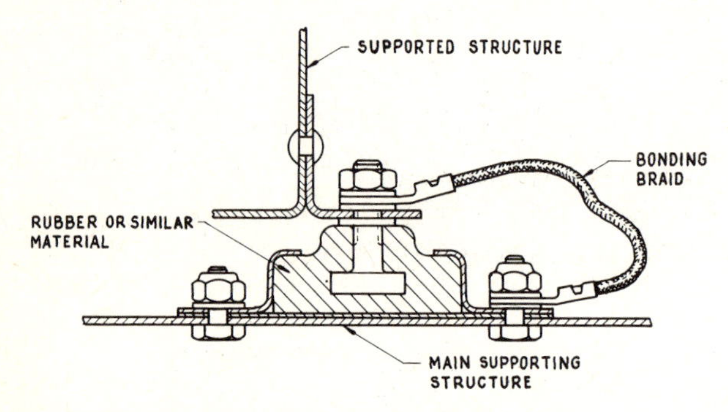

Fig. 17-45 One type of shock mounting.

Because shock mountings utilize rubber, synthetic rubber, plastic, or some other insulating material as the shock-absorbing agent, it is essential that grounding or bonding jumpers be connected from the aircraft structure to the avionics-unit case. These serve as a part of the ground circuit for the equipment and also help to reduce noise from static and other types of interference. Bonding and shielding information is given in Chapter 11 of this text.

In the installation of shock-mounted units, adequate clearance must be provided to prevent any contact between the mounted unit and the adjacent structure under conditions of violent shock or vibration. Electric cables and control cables connected or attached to radio equipment must have sufficient play and be mounted in such a manner that the vibration and sway of the radio equipment will have no adverse effect on them or cause undue wear. The strength of the mounting for radio equipment should be at least such that the equipment can withstand the ultimate accelerations for which the airplane structure is designed.

Avionics mounting racks are usually designed to ARINC (Aircraft Radio, Inc.) standards, and equipment cases are designed to fit such racks. This is particularly true of large, commercial aircraft; however, mounting racks and avionics equipment for smaller aircraft are also being designed according to ARINC standards. Technicians installing avionics equipment should ensure that equipment and racks are compatible.

ANTENNAS

The performance of radio systems on aircraft is profoundly affected by the design and placement of antennas. This is particularly true of antennas for transmitters, since the antenna system is a tuned circuit, and its ability to radiate energy into space is determined by its length in relation to the frequency to be transmitted. In general, the higher the frequency, the shorter the antenna. In practice it is possible to adjust the length of an antenna electronically by means of an inductance coil in series with the antenna. The inductance coil is provided with taps or some other means of adjustment to vary the impedance of the antenna.

It is common practice to use one antenna for both transmitting and receiving if the radio equipment is to be used only for communications, provided that the length of the antenna is such that it will accommodate the frequencies to be transmitted and received. When a single antenna is used, it is normally connected to the receiver and switched to the transmitter for broadcasting by means of a relay and a push-to-talk switch on the microphone.

Navigation and communications antennas are manufactured in many sizes and shapes, depending upon their particular functions. As explained earlier, antenna length or size is determined by the frequency range in which it is intended to operate. Special designs such as loops and dipoles are used for certain types of signals and provide directional references. A few typical antennas are shown in Fig. 17-46. The antennas numbered 1 and 2 are designed to receive VOR navigation signals to provide bearing information. An antenna coupler is shown between the two sections of number 2. The antennas identified by the numbers 3, 4, and 5 are VHF communications antennas. Number 6 is a DME and transponder antenna, and number 7 is a marker beacon antenna.

ANTENNAS FOR LOW AND MEDIUM FREQUENCIES

The antennas for low and medium frequencies are greater in length than those for the higher frequencies. Nondirectional beacons (NDB) operate in these ranges. Antennas for these frequencies may be T, L, or V types mounted on the top or the bottom of the fuselage (see Fig. 17-47). Clearance between the antenna and the fuselage or other structures should not be less than 1 ft [30 cm] in any case, and the main leg of the antenna should be not less than 6 ft [183 cm] in length. A whip antenna may be used provided that tests show satisfactory performance with such an installation.

MAST AND WHIP ANTENNAS

When a mast or a whip antenna is installed on the fuselage or any other part of an aircraft structure, it is necessary to make sure that the structure of the airplane is sufficiently strong to support the unit under all conditions

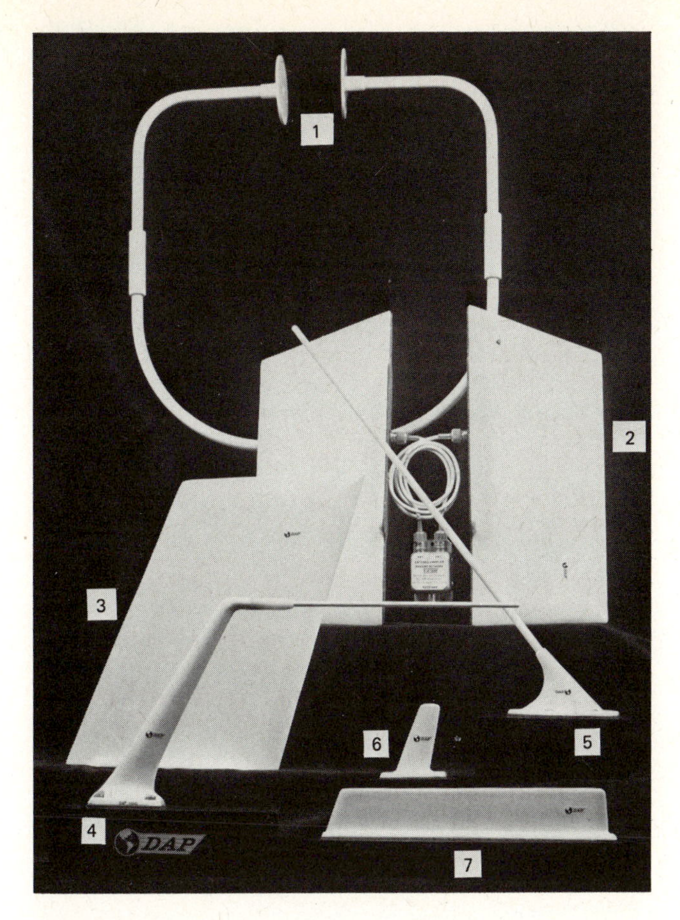

Fig. 17-46 Typical avionics antennas. (Dayton Aircraft Products)

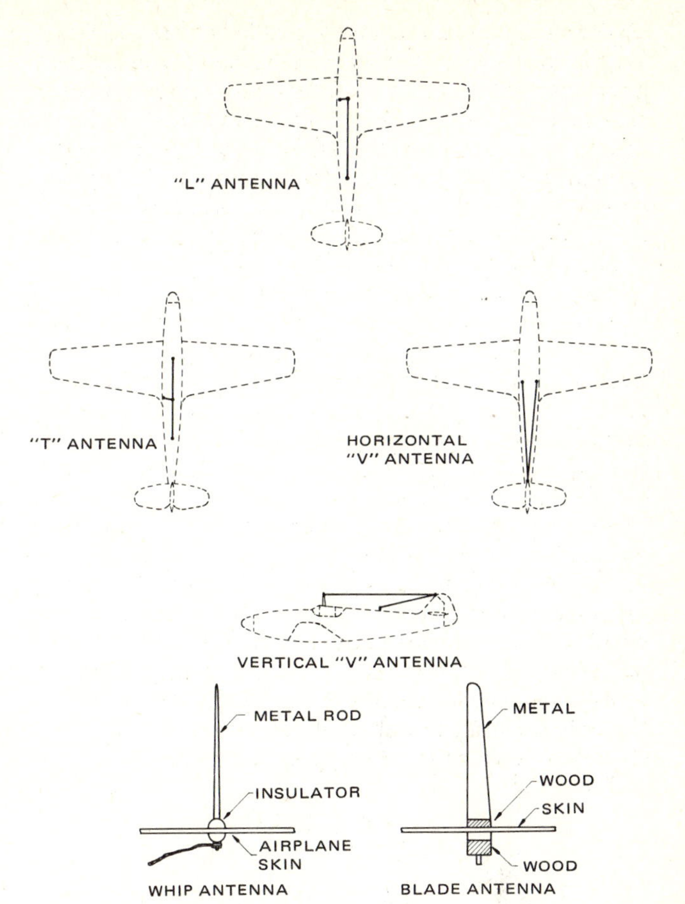

Fig. 17-47 Antenna installations.

of shock, vibration, or continued movement. This often requires that a doubler plate be installed around the point at which the antenna is attached as shown in Fig. 17-48.

LOOP ANTENNA

Loop antennas are used for both communications and navigation and were at one time rotated manually or automatically by means of a servo motor. Currently manufactured loop antennas are of the fixed type. The effect of rotation is produced electronically. Usually loop antennas are designed for operation with specific types of ADFs. The antenna employed with a particular make and model of equipment should be the antenna designed for that equipment.

Loops are usually installed on the bottom of the fuselage and are enclosed in nonmetallic, streamlined housings. Particular attention must be given to locating the antenna in an area which will not be affected by nearby metallic structures. Improper location may cause certain signals to be blanked out or may produce a distortion of the signal so that a true directional reading cannot be obtained. Equipment manufacturers often supply information regarding loop location.

The outstanding characteristic of loop antenna is its directional sensitivity, which makes it useful as a navigational device. At one point in the rotation of the loop, there will be a complete fade-out of the signal being received. This point is called the *null* and is used to pinpoint the direction of the station from which the signal is being received. The radio magnetic indicator or ADF display

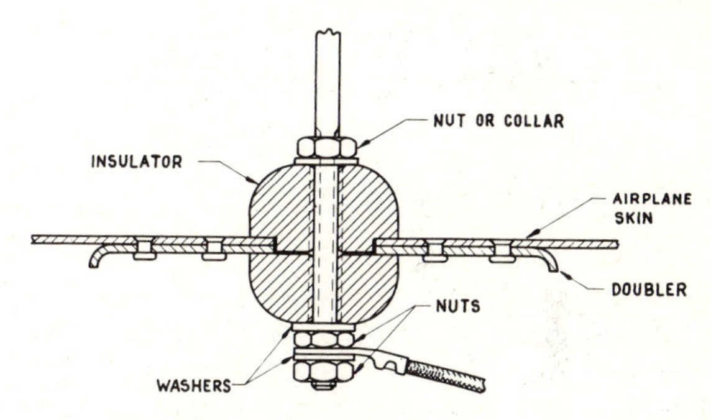

Fig. 17-48 One type of mounting for a whip antenna.

then continuously indicates the bearing of the station with respect to the airplane.

In determining the best location for a loop antenna, consideration should be given the space necessary both inside and outside the fuselage, structural requirements, length of cables, location of the receiving equipment, and the effect on operation and maintenance of the aircraft. Furthermore, the location should be selected to provide a balanced **quadrantal error.**

Quadrantal error is the installation error caused by metal fuselage, wings, and other parts of the aircraft. This metal distorts the electromagnetic field of a received signal and produces azimuth-reading inaccuracies, which are greatest between the four cardinal points with respect to the centerline of the aircraft. When a loop antenna has been installed, it is necessary to check the direction of the radio bearings every 45° from the longitudinal axis of the aircraft in order to determine the deviations caused by distortion of the radio field pattern because of the effects of the metal structures of the aircraft. If the loop is of the type which includes compensating adjustments, it is important that no compensation be present in the loop at the time that the calibration is made. If the loop has no provision for the adjustment of quadrantal error, the calibration data should be used for the preparation of a correction card to be mounted in the cockpit near the indicator. This card will provide the pilot with corrected bearing information.

The installation and calibration of antennas in particular types of equipment should be made according to the manufacturer's instructions. All manufacturers of radio equipment furnish manuals describing the installation and operation of their equipment. The instructions contained in these manuals should be followed carefully if satisfactory service is to be expected.

HIGH-FREQUENCY ANTENNAS

As explained previously, the wavelength of any radio wave may be found by using the proper formula. For example, if it is desired to find the wavelength of a 100-MHz wave, we find

$$\lambda = \frac{300\ 000\ 000}{100\ 000\ 000} = 3 \text{ m} \qquad \text{or } 9.84 \text{ ft}$$

The length of an antenna for HF transmission must be determined as an exact fraction of the wavelength of the signal to be transmitted. For example, when a Hertz antenna such as that shown in Fig. 17-49 is to be used, the

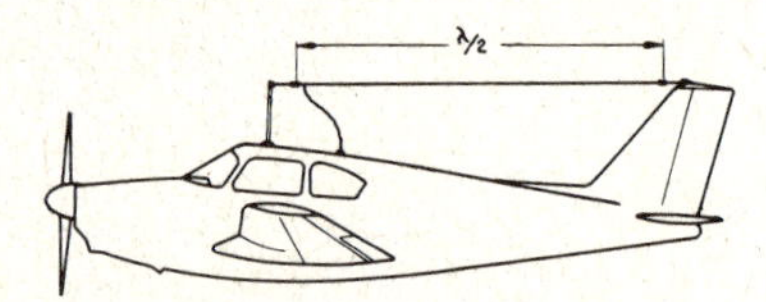

Fig. 17-49 Installation of a Hertz antenna.

total length of the antenna should be one-half the wavelength. Hence, if we wish to find the proper length of the Hertz antenna in feet, we use the following formula:

$$\text{Antenna length (in feet)} = \frac{300\ 000\ 000 \times 3.28}{2 \times f}$$

where 3.28 = conversion factor for changing meters to feet
f = frequency of the transmitter, Hz

For a Marconi antenna the extended section should be one-quarter the wavelength. The correct antenna length may be determined by using the foregoing formula and dividing the result by 2 or by making the denominator of the fraction $4 \times f$ instead of $2 \times f$. It has been found in actual practice that the current in an antenna travels at about 5 percent less velocity than it does in free air; hence, the actual length of an antenna should be about 5 percent less than that computed with the foregoing formulas. In view of this fact, we may use a simplified formula as follows to determine the actual length of a half-wave antenna:

$$\text{Length (in feet)} = \frac{468}{f}$$

where f = frequency, MHZ
468 = 95 percent of 492, which is half the number of millions of feet in 300 000 000 m

On fabric-covered or wood aircraft, it is necessary to provide a ground plane counterpoise for a vertical Marconi antenna. This is accomplished by placing a number of metal-foil strips in a position radial from the antenna base and securing them under the fabric or wood skin (see Fig. 17-50). The length of the strips should be no more than the length of the antenna.

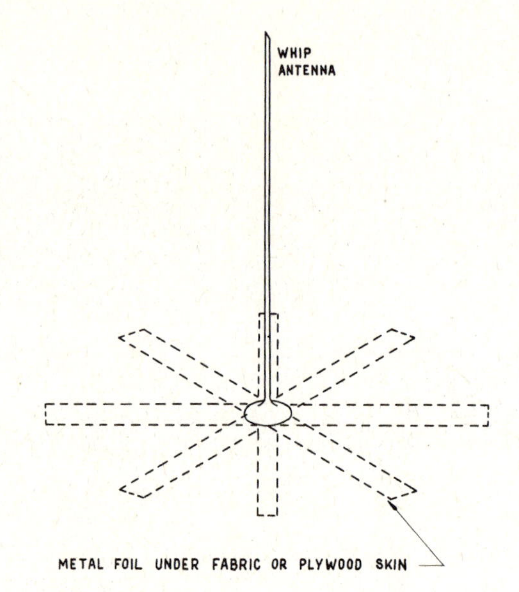

Fig. 17-50 Ground plane for a whip antenna.

ANTENNAS FOR VHF NAVIGATION EQUIPMENT

Particular attention must be given to the proper location of antennas which operate at very high frequencies for navigation equipment. For example, the horizontal V antenna for VOR and localizer is most effective on small airplanes when mounted over the forward part of the cabin. The apex of the V should be pointing forward, and the plane of the V should be horizontal when the airplane is in level flight. Figure 17-51 shows suitable locations for a VOR-localizer antenna.

The glideslope antenna is usually a small dipole mounted on the forward part of the airplane, but it is sometimes mounted on the same mast as the VOR-localizer V antenna.

The installation of any antenna for any purpose should follow closely the instructions of either the radio manufacturer or the airplane manufacturer. All approved installations have been carefully engineered for best performance, and any deviation from the approved installation is likely to give inferior performance.

TYPICAL ANTENNA INSTALLATIONS

Observation of aircraft will reveal many different types and shapes of antennas and installations. Figure 17-52

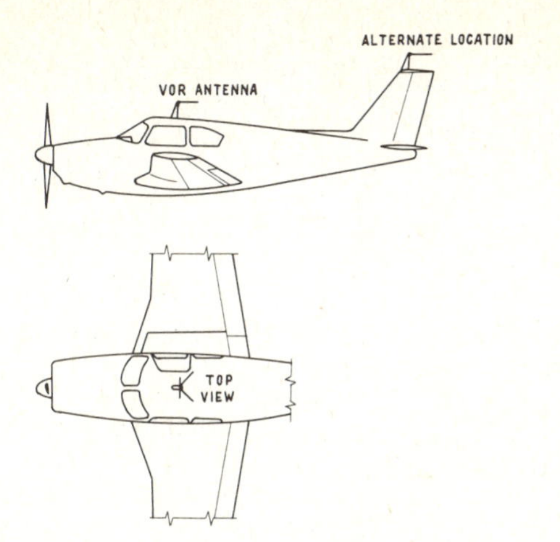

Fig. 17-51 Installation of a VOR antenna.

shows six aircraft to illustrate various antennas. Number 1 shows a Hertz antenna for low- and medium-frequency wave reception and two VHF antennas. Number 2 is a VOR antenna streamlined with a plastic cover. Number 3 shown a fixed-loop ADF antenna with a sense antenna and a DME/transponder antenna in the background. Number 4 is an open antenna for a marker beacon receiver, and number 5 shows an enclosed marker beacon receiver antenna. Number 6 is an open DME/transponder antenna which has been mounted on the bottom of the airplane.

ELIMINATION OF NOISE

Noise in avionic systems includes not only the audible sounds heard in receiver earphones or loudspeakers but

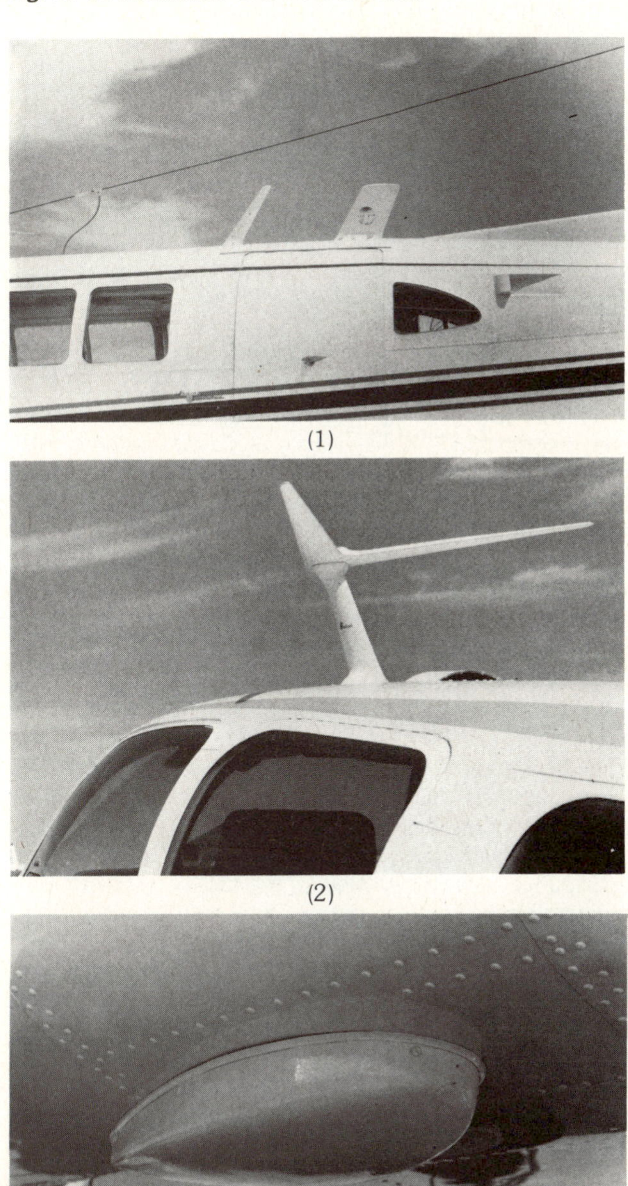

(1)

(2)

(3)

(4)

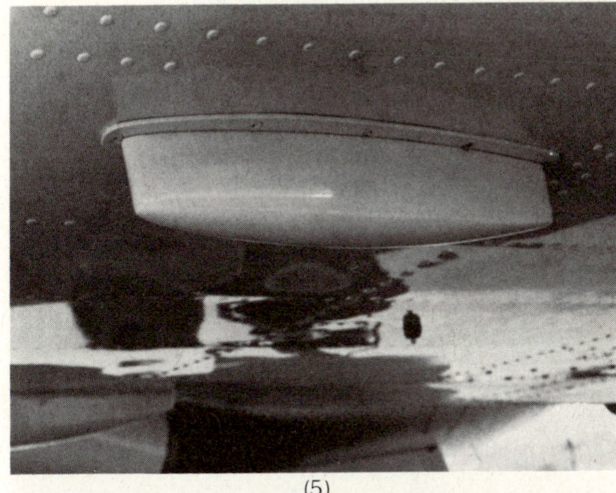

(5)

(6)

Fig. 17-52 Typical antenna installations for avionic equipment.

also the effects that show up in the outputs of navigation receivers, radar, and other avionic equipment. Some noise is generated within the equipment itself, but the most troublesome noise is that which is caused by various stray electromagnetic waves that emanate from electric equipment, such as ignition, generators, and relays, and by precipitation (p) static.

The effects of noise in audio receivers can be reduced to some extent by the use of **squelch** circuits in the radio. If the noise is too intense, however, a squelch will not be effective.

The most effective method of eliminating engine noise is to shield the ignition system. This means that the entire system, including the magnetos, low-tension wires, high-tension wires, spark plugs, and ignition switch, are completely enclosed in metal covers or metal braid. When a system of this type is employed, all connections must be clean and tight. If any of the coupling nuts between the wires and the magneto or spark plugs come loose, there will be an immediate increase of noise in the radio.

If it is found that the shielded ignition system does not adequately reduce the ignition noise, it may be necessary to install a filter between the magneto and magneto switch. This filter may consists of a single bypass capacitor or a combination of capacitors and choke coils.

Other sources of noise may be the generator, voltage regulator, or various actuator motors. These are usually quieted by connecting filter capacitors in parallel with the contact points or brushes causing the disturbance. It is important to keep the brushes and commutators of generators and motors clean and smooth to prevent arcing, since arcing is the primary cause of interference from these units.

The most effective means for eliminating the effects of p static are to ensure that all parts of the aircraft are electrically bonded as explained previously and to install static-discharge devices at effective points on the aircraft. The use of static dischargers was discussed in Chapter 1 of this text.

Extensive research has been performed on the most effective use of static dischargers for all types of aircraft. A comparatively slow—100 to 150 kn [185 to 278 km/h]—aircraft that does not fly in bad weather conditions may have little need for static discharges because the generation of p static in such conditions is relatively small. However, high-speed all-weather aircraft must be equipped with such dischargers if satisfactory avionic operation is to be expected.

The location and number of dischargers are particularly important. Manufacturers of the units can recommend the type of installation needed for an aircraft, based on the many tests that have been conducted. The dischargers are attached by riveting or bonding at selected locations near the tips of wings, on outboard ailerons, on rudders, and on elevators. The installations should be made in accordance with the manufacturer's installation instructions.

A null-field static discharger is shown in Fig. 17-53. This type of discharger is particularly effective in preventing electromagnetic coupling of the static discharge impulses with the antennas on the aircraft. The unit shown is designed to be attached by means of a conductive, epoxy cement.

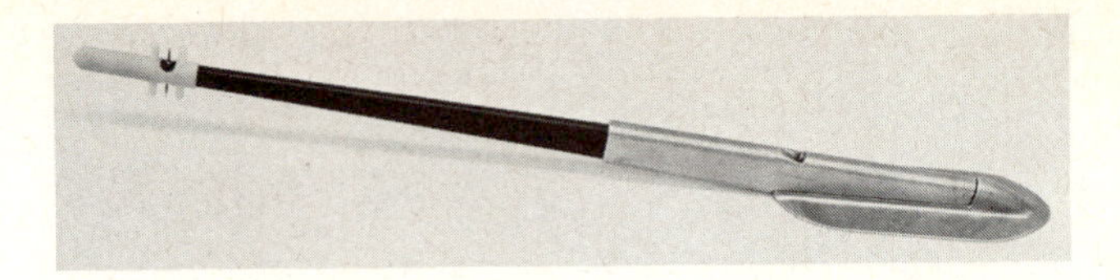

Fig. 17-53 A null-field static discharger. (Dayton Aircraft Products)

EMERGENCY LOCATOR TRANSMITTER

An **emergency locator transmitter (ELT),** also referred to as a **locator beacon,** is required on aircraft to provide a signal or signals that will enable search aircraft or ground stations to find aircraft that have made crash landings in remote or mountainous areas. Even though the ELT is not strictly a communications or navigation device, it has elements of both and so it is described in this section.

A typical ELT consists of a self-contained dual-frequency radio transmitter and battery power supply with a suitable whip antenna. When armed, it will be activated by an impact force of 5 *g* or more, as may be experienced in a crash landing. The ELT emits an omnidirectional signal on the international distress frequencies of 121.5 and 243.0 MHz. General-aviation aircraft, commercial aircraft, the FAA and the CAP (Civil Air Patrol) monitor 121.5 MHz, and 243.0 MHz is monitored by the military services.

After a crash landing, the ELT will provide line-of-sight transmission up to 100 mi [161 km] at a receiver altitude of 10 000 ft [3050 m]. The ELT transmits on both distress frequencies simultaneously at 75-mW rated power output for 50 continuous hours in the temperature range of −4°F to 131°F [−20°C to +55°C].

The fixed ELT must be installed securely in the aircraft at a location where crash damage will be minimal. The location selected is usually in the area of the tail cone; however, in some cabin-type aircraft, the unit is installed in the aft, top part of the cabin. Access is provided in either case so the unit can be controlled manually.

TESTING THE ELT

The ELT control panel on the unit contains a switch with three positions: AUTO, OFF, and ON. The unit may be tested by tuning the VHF COMM receiver to 121.5 MHz and then placing the VLF switch in the ON position. The emergency tone will be heard if the ELT is operating. Immediately after the test, the switch should be returned to the AUTO position.

If the ELT is inadvertently turned on, owing to a lightning strike or an exceptionally hard landing, the control switch should be placed in the OFF position to stop the transmission, and then the switch should be returned to the AUTO position to arm the unit.

SERVICE FOR THE ELT

The ELT requires a minimum of service; however, certain procedures are necessary to assure operation. The battery pack must be changed in accordance with the date stamped on the unit. The new battery pack installed must be of the type specified by the manufacturer. If the unit contains a rechargeable battery pack, charging must be accomplished in accordance with the established schedule. The ELT should be tested regularly to assure satisfactory operation. An inspection of the ELT mounting and antenna should be made periodically to ensure firm attachment to the aircraft.

Regulations regarding the use of operation of ELT equipment is set forth in FAR Part 91.52. Technicians involved with the installation and service of ELTs should be familiar with these regulations and manufacturer's data.

REVIEW QUESTIONS

1. List subsystems that may be included in a complete navigation system for an aircraft.
2. What is the frequency range employed for air traffic control communications?
3. Describe an *interphone system* for a large passenger airplane.
4. What functions do *flight interphone systems* and *service interphone systems* have?
5. What is the purpose of a cockpit voice recorder?
6. Give a general description of a *passenger address and entertainment system.*
7. Describe the purpose of the *antenna coupler system* for a probe antenna.
8. For what purposes are VHF communications employed?
9. What frequency range is utilized for VHF communications systems?
10. Describe a typical VHF communications transceiver.
11. What is the difference between the USE and STANDBY display in a VHF COMM transceiver?
12. Describe antennas for VHF systems.
13. What is the function of a *Selcal decoder?*
14. What is the advantage of SATCOM communications?
15. Explain the importance of FCC regulations.
16. Describe the procedure for testing a COMM radio unit.
17. What technology has made possible the use of complex avionic navigation systems for small aircraft?
18. Describe the operation of an automatic direction finding (ADF) system.
19. What device enables an ADF receiver to determine direction?
20. What is the purpose of the BFO in an ADF system?
21. Describe a *radio magnetic indicator* (RHI).
22. What signals are generated by an instrument landing system (ILS)?
23. Briefly describe the operation of the *localizer.*
24. Compare the glideslope function of the ILS with that of the localizer.
25. What is the purpose of the *marker-beacon system?*
26. What frequency is utilized by the marker-beacon system?
27. What are the indicator lights associated with the marker beacon?
28. Compare the *microwave landing system* (MLS) with the ILS.
29. What are the advantages of the MLS?
30. Describe the principle of operation of the MLS.
31. List typical avionics items needed in the aircraft to operate with the MLS.
32. What type of signal is radiated by a *VHF omnirange* ground station?
33. In what ways may VOR information be displayed to the pilot of an aircraft?
34. Describe a *course deviation indicator* (CDI).
35. Explain the principle of *distance-measuring equipment* (DME).
36. What is an important consideration with respect to a DME antenna installation?
37. What is the meaning of the term *TACAN?*
38. What are the functions of a *basic navigation* (NAV) receiver?
39. Describe the controls on a typical NAV receiver.
40. What is the advantage of an *area navigation* (RNAV) system?
41. What subsystems are usually included in an integrated navigation system?
42. Describe the controls for an RNAV system.
43. How would a flight utilizing an RNAV system be programmed?
44. What is meant by a *waypoint?*
45. Explain how a LORAN system establishes location.
46. Explain the principle of the VLF/OMEGA navigation system.
47. What is the possible range of a VLF navigation station?
48. Give some advantages of the VLF/OMEGA navigation system.
49. Describe the *control display unit* (CDU) for the VLF/OMEGA system.
50. What is meant by *crosstrack?*
51. Why should avionics equipment be turned on only after the master switch is on and the engines have been started?
52. Explain the principle of an inertial navigation system (INS).
53. What is the principle employed in a *doppler* navigation system?
54. Describe an *integrated flight system.*
55. What is the function of an *ATC transponder?*
56. What is the value of an *encoding altimeter* when used in connection with an ATC transponder?
57. Describe key points in the installation and mounting of avionics equipment.
58. What is the primary governing factor in establishing the size or length of a radio antenna?
59. What considerations must be taken into account when installing a mast or whip antenna on an aircraft?
60. What is meant by *quadrantal error* in respect to an ADF loop antenna?
61. What are the causes of *noise* in avionic equipment?
62. How is the noise from the ignition system of a reciprocating engine reduced?
63. What methods are employed to reduce the noise caused by *p-static?*
64. What important factors must be considered in the installation of static dischargers?
65. What is the purpose of an *emergency locator transmitter* (ELT)?
66. What frequencies are used by an ELT?
67. How can an ELT be tested?
68. What services should be performed on the ELT?

CHAPTER 18

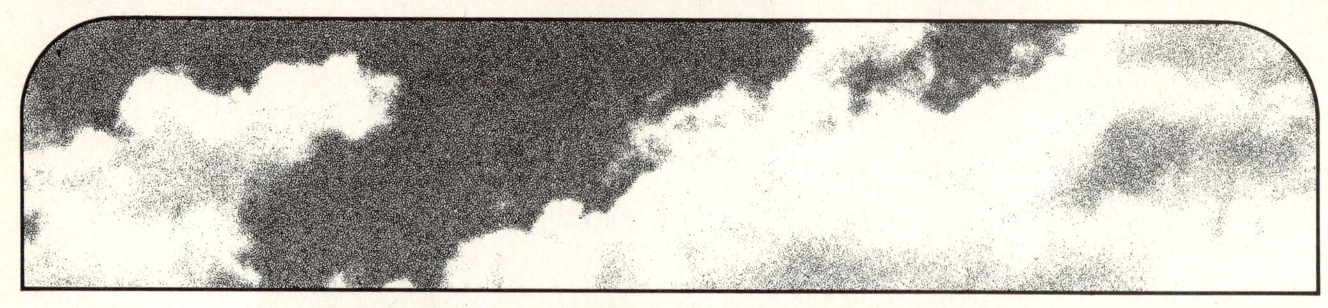

Automatic Pilots and Landing Systems

For many years large aircraft such as airliners, transport aircraft, and military aircraft have been equipped with automatic pilots (automatic flight-control systems) to relieve the human pilot and other members of the flight crew of the tedious duty of keeping the aircraft on course for periods of many hours. Today automatic pilots are installed on all large aircraft of both commercial and military types and on many light aircraft. The use of autopilots for light aircraft has greatly increased the convenience and utility of these vehicles and has made it considerably more desirable for owners and operators to use such aircraft much more extensively than was common in the past.

It is the purpose of this chapter to examine the principles of operation for automatic-pilot systems and to explore in some detail typical systems being commonly used. The basic principles are quite simple; however, many new and complex capabilities have been added to the basic systems.

The original automatic pilots for aircraft were largely mechanical in function. For example, one early system used extensively by both military and commercial aircraft utilized gyros to actuate hydraulic valves, which, in turn, directed fluid under pressure to hydraulic cylinders. These actuating cylinders moved the control surfaces of the aircraft through cables by which the cylinders and control surfaces were connected. The movement of the control surfaces made the required corrections in the flight attitude of the aircraft.

BASIC THEORY OF AUTOMATIC PILOTS

The use of gyros to develop electric signals for the operation of control surfaces is illustrated in Fig. 18-1. The fact that a gyro will hold a constant position in space makes it possible to develop a relative motion between the gyro and its supporting structure when the supporting structure changes position. When the gyro is supported with gimbal rings and is installed in an airplane as shown in Fig. 18-1, any pitch or roll of the airplane will cause a relative movement between the gyro and the airplane. This movement can be used to operate contact points, move the wiper of a potentiometer, or operate hydraulic valves, thus providing a means for operating aircraft-control surfaces in response to the movement. In Fig. 18-1, the two upper drawings show the gyro moving the wiper of a potentiometer to develop a proportional electric signal which may be amplified and used to apply a flight correction. In the two lower drawings of the illustration, the gyro is shown operating a set of double-throw contact points. In this case the closed contacts direct a signal to a relay, thus sending power to a servo motor which will move the control surfaces in the appropriate direction to level the aircraft.

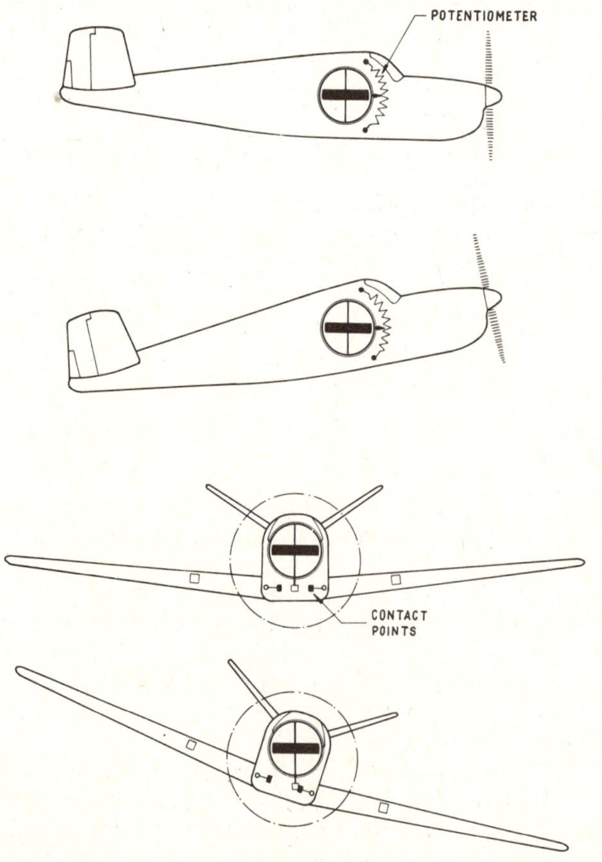

Fig. 18-1 Operation of gyro sensing units.

ELECTRICAL PRINCIPLES

A number of electromechanical and electronic autopilots for single-engine and twin-engine aircraft have been designed and manufactured, and all such systems utilize gyro outputs to provide correction signals for aircraft in pitch, roll, and yaw. The methods by which these signals are utilized vary according to the complexity of the systems.

ESSENTIAL FEATURES

In any autopilot system, it must be remembered that a reversal of the corrective signal must take place before the aircraft has returned to its corrected attitude. For example, if the rudder of an aircraft is moved to the right by the autopilot to correct the aircraft heading, it must be moved back to neutral by the time that the aircraft has reached its corrected position. Otherwise the system will overshoot and cause the aircraft to oscillate violently. To accomplish this purpose, **feedback** systems are employed.

In electronic autopilot systems, the gyros which sense flight deviations of the aircraft develop corrective signals by means of transformer action or a similar principle. In this way no friction is developed, and very little mechanical force is evident to restrict the movement of the gyro. One type of signal-generating device is called an EI pickoff and is diagrammed in Fig. 18-2. The pickoff consists of three coils mounted on an E-shape piece of laminated steel. The coils wound on the outer legs of the E are connected together in such a phase relationship that the voltages induced from the center coil cancel. A movable steel

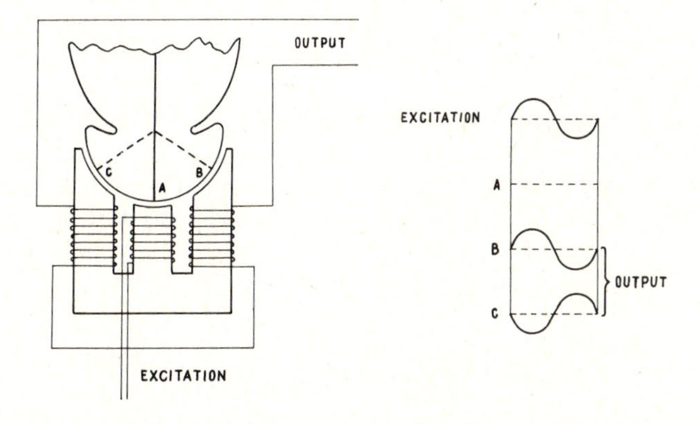

Fig. 18-2 Electromagnetic signal-generating device for a gyro sensing unit.

armature, called the I member, provides a low-reluctance path for the magnetic field. When this armature is moved relative to the E section, it changes the ratio of coupling between the secondary windings and the primary, and this changes the value of the induced voltage in the outside legs so that they are no longer equal. The result is a voltage that is either in phase or out of phase with the excitation voltage. Note that the excitation voltage is applied to the center leg of the E section.

The electric voltage output from the secondary coil of the EI pickoff assembly will be proportional to the displacement of the elements away from the null position. The null is the mechanical position which results in zero electric output, and is indicative of the fact that the aircraft

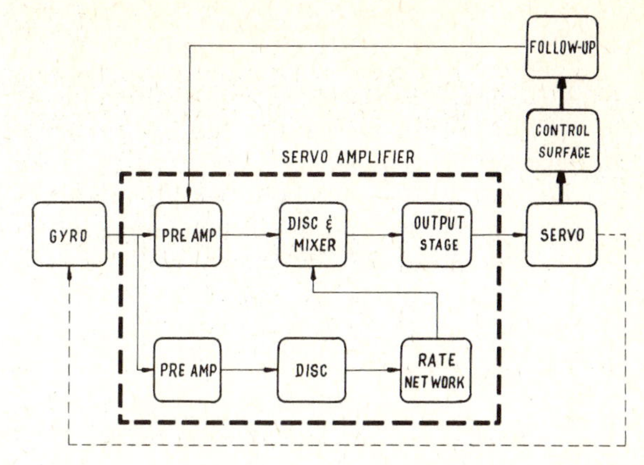

Fig. 18-3 Block diagram of an autopilot system.

is in its proper flight position in relation to the reference established by the gyro unit. Movement of the aircraft away from the established flight reference causes relative mechanical displacement of the pickoff elements away from the electrical null, thus producing an electric output voltage which is the signal to the system calling for a corrective action.

It should be noted by referring to Fig. 18-2 that displacement of the armature from the centered, or null, position results in an output signal that is proportional to the displacement and that the direction of displacement will determine the nature of the signal produced. In one direction of armature movement, the secondary voltage will be out of phase with the primary, and in the other direction, the secondary voltage will be in phase with the primary. Thus, there are several kinds of information which may be gained from an examination of the output voltage of the electrical pickoff system. Among them are the following:

1. If the voltage is at its null value, the aircraft is in the proper flight position.
2. If the magnitude of the signal voltage increases, it is an indication of an angular deviation from the established reference course.
3. The relative phasing of the signal voltage with respect to the primary excitation voltage is indicative of the direction of departure from the established reference course.
4. The rate of change of the magnitude of the signal voltage is indicative of the angular velocity with which the aircraft is changing its course.

The information gained from the signal system provides the means whereby the control surfaces of the aircraft are moved to cause corrective action and to return the aircraft to the proper flight position. The extremely small amount of power represented by the signal output cannot, in itself, operate the controls on the aircraft and, therefore, must be suitably amplified. This may be accomplished by means of a servo amplifier containing a discriminator system or by a computer amplifier.

On some currently manfactured gyros, the signal is produced optically by means of light-emitting diodes (LED) and photo transistors. Movement of the gyro varies the amount of light received by the photo transistor, thereby changing the output.

BASIC ELECTRONIC AUTOPILOT SYSTEM

A block diagram of a basic electronic autopilot system is shown in Fig. 18-3. It will be noted in the diagram that the system includes a follow-up feedback from the control surface to the preamplifier, a feedback signal from the servo motor to the sensing gyro, and a rate network which tells the system how rapidly a change is taking place.

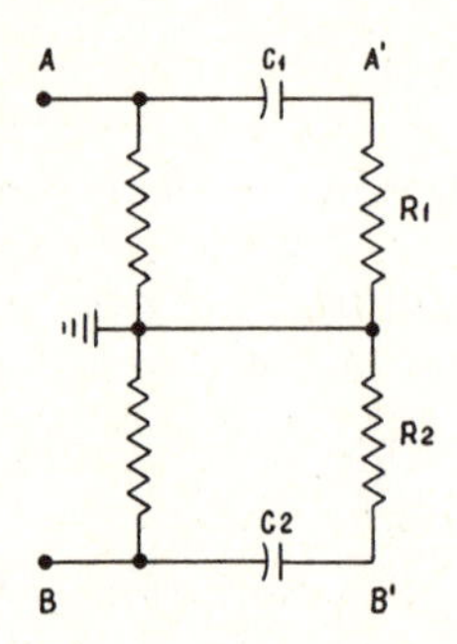

Fig. 18-4 A rate network for an autopilot.

The follow-up, or feedback, system consists of an electrical pickoff similar to those used with a gyro, but connected through linkage to the control surface of the aircraft. An electric signal, produced by the displacement of the control surface away from its neutral position, is fed back into the servo amplifier in such a manner that the phasing of the signal is opposite to that of the gyro signal which produces the control motion. This negative-feedback arrangement produces a closed-loop servo system which provides more positive positioning of the control surface than would be possible by depending solely upon aerodynamic resistance to neutralize the surface when control force is no longer needed.

To provide an extremely sensitive control so that deviations from course will be small and the aircraft will be returned without overshooting the course, a further addition is made to the servo control system in the form of a rate circuit which modifies the servo-amplifier output signal. The gyro signal is fed through a separate preamplifier and discriminator and then into a rate network. A simple schematic diagram of such a network is shown in Fig. 18-4.

Points A and B are the output of the discriminator circuit. Since the voltage existing at points A and B is direct current, it is obvious that a voltage will be developed across the resistors R_1 and R_2 only when a current flows to charge or discharge capacitors C_1 and C_2. Thus a voltage will be apparent at points A' and B' only when the voltage at A and B is changing. For any given value of C and R, the magnitude of the voltage appearing across points A' and B' will be proportional to the rate of change of the voltage across A and B or, in effect, a measure of the velocity of the aircraft toward or away from its correct flight attitude. This signal from the rate circuits is fed into the same discriminator that detects the displacement signal, and the combined output is applied to the output stage.

The total effect of these signals may best be understood if we analyze their effects while following the aircraft through one complete cycle of deviation from its course. For this discussion we shall assume that we are concerned with the yaw axis, or rudder control, and that the aircraft has turned (yawed) to the right from its prescribed course. This produces an electric signal from the directional gyro which causes the servo amplifier to direct the servo system to apply left rudder. The amount of rudder will be proportional to the angular deviation of the plane from its original heading.

The rate of change of the signal from the gyro causes the rate circuits to produce additional left-rudder control, which acts as a brake to reduce the velocity of the angular deviation. This means that a very strong application of left rudder is applied if the yaw movement was very rapid. The corrective rudder action causes the airplane to stop its movement to the right, and this arresting movement stops the change in the gyro signal. Since the gyro signal is no longer changing, the rate signal disappears, but the displacement signal continues to give corrective control for as long as the aircraft has not returned to its original heading. The aircraft then swings back to its original heading, and the rate signal is the reverse of that which occurred when the gyro was increasing the correction signal. Since the rate signal has reversed, it will call for reduction of left rudder; that is, it will tend to return the rudder to neutral. This reduction in corrective control is proportional to the speed of return of the aircraft to its course, and if the rate of return is high enough, it may even result in opposite control in order to prevent the aircraft from overshooting the correct heading.

With the proper values in the rate circuits, it is possible to adjust the response of the autopilot so that each correction is rapid and smooth.

In the correction cycle described, the follow-up (feedback) signal appears at all times when the rudder is displaced from the streamline position. The follow-up signal acts as "synthetic springs" which overcome static friction forces in order to position the control surface accurately.

The basic system of flight control outlined here is used on all three axes of control on the aircraft, although various features are added to each axis to solve the control problems peculiar to each.

TYPICAL AUTOMATIC PILOT AND FLIGHT-CONTROL SYSTEM

Automatic pilots are manufactured in many configurations by a number of companies. Some systems are comparatively simple, while others become complex, especially when integrated with navigation systems. As explained previously, all systems utilize gyros to sense aircraft attitude and provide signals for correction.

For the purpose of this section, the Bendix M-4D automatic pilot and flight-control system has been selected for description. This system not only includes the basic elements of autopilot operation but also includes components that give it all the capabilities of an automatic flight-control system. A number of systems by other manufacturers utilize similar principles and perform the same functions.

The system described here can be programmed to fly a predetermined course, either NAV or RNAV, maintain a selected altitude, capture a VOR radial or ILS beam from any angle, make back-course approaches, and perform other functions. The system also includes automatic pitch trim, pitch synchronization, pitch integration, and altitude control. The computer portion of the system can also be used to display computed command data on a director-horizon indicator (flight director) and give directional data on a horizontal-situation indicator. Thus the aircraft is provided with a fully integrated flight-control system.

SYSTEM COMPONENTS

The arrangement of the principal components of the M-4D automatic flight-control system in an aircraft is shown in Fig. 18-5. The basic autopilot is comprised of the controller, gyros, servos, and the section of the computer amplifier that accepts signals from the gyros and converts them to flight commands for the servos.

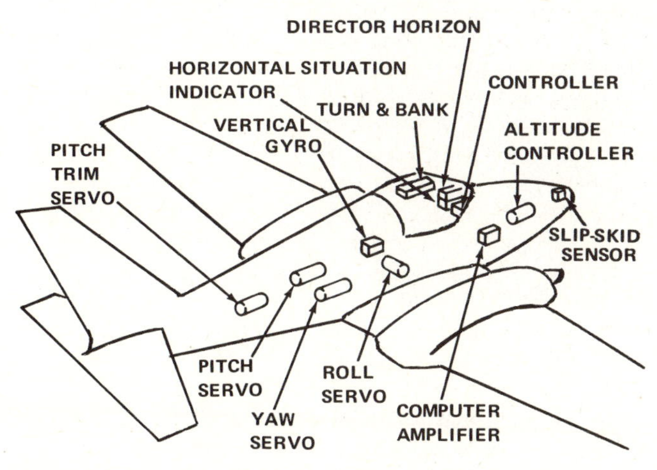

Fig. 18-5 Arrangement of the components of an autopilot system in an aircraft. (Bendix Avionics Division)

FLIGHT CONTROLLER

The flight controller, shown in Fig. 18-6, is the unit by which the human pilot controls the operation of the autopilot. The autopilot is engaged by pressing the AP button. An annunciator on the lower part of the AP button will illuminate when the autopilot is engaged. The YAW button provides for automatic control of rudder trim during changes in airspeed.

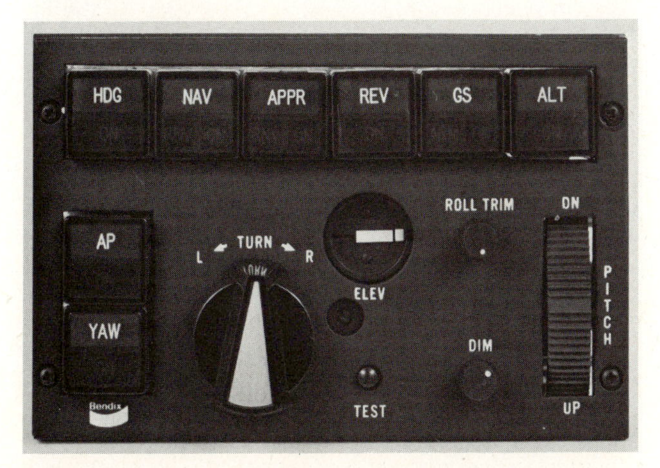

Fig. 18-6 Flight controller. (Bendix Avionics Division)

The HDG and NAV buttons command the autopilot to follow a preselected heading, couple with a VOR radial, or fly a preselected RNAV course. The APPR button commands the autopilot to capture a localizer beam on the approach to an airport or to approach an airport using a VOR radial or RNAV data. If the aircraft must approach from the opposite direction of the normal course signal, the REV button is pressed to provide the back course information. The GS button commands capture and tracking of the glideslope signal, and the ALT button commands the autopilot to maintain a certain selected barometric altitude. All selector buttons are provided with annunciators that illuminate when each mode is active. The NAV annunciator will show RNAV or ON, depending upon whether the system is operating on data from the RNAV computer or is tracking a VOR radial. The GS annunciator will show ARM in the APPR mode if the airplane is below the glideslope beam and the signal has been present for approximately 20 s. When the aircraft reaches the beam center, the annunciator will show ON. At this time the ALT and ARM annunciators will be off.

The TURN control is used to initiate either a 12° banked turn or a 24° banked turn to the right or left. Other lateral modes will be released during operations of the turn control. The TURN control must be in the center position when the autopilot is first engaged. The PITCH control is spring-loaded in the center position. When moved up or down, it commands pitch changes up to ±20°. During this operation, other longitudinal modes such as GS, ALT, or GA are disengaged. The GA button is on the pilot's control wheel and is used to command a go-around in case of a missed approach.

The ELEV indicator shows whether the elevator is in the neutral position. Before autopilot engagement, this meter shows the status of the autopilot signal relative to the pitch axis of the aircraft. It should always be near center because the system has automatic pitch synchronization. After engagement, the meter shows the force required by the primary servo to hold the aircraft in the desired pitch attitude.

The ROLL TRIM is used to trim the ailerons so the aircraft is in a level attitude. It should not be used with any of the lateral modes engaged or with the TURN control out of the center position.

TURN-AND-BANK INDICATOR

The **turn-and-bank indicator** is an electrically driven gyro unit which provides a visual indication of the rate of turn and at the same time provides an electric signal to the autopilot with the same information. This signal is used by the computer amplifier in the development of turn signals for the aircraft.

The ball inclinometer in the turn-and-bank indicator shows the pilot whether the aircraft is being flown with properly coordinated rudder and aileron control.

DIRECTOR-HORIZON INDICATOR

The **director-horizon indicator** is similar to instruments described as flight-director indicators or attitude-director indicators. This instrument, shown in Fig. 18-7, serves several functions. It is an attitude indicator and an attitude sensor. That is, it senses attitude and develops electric sig-

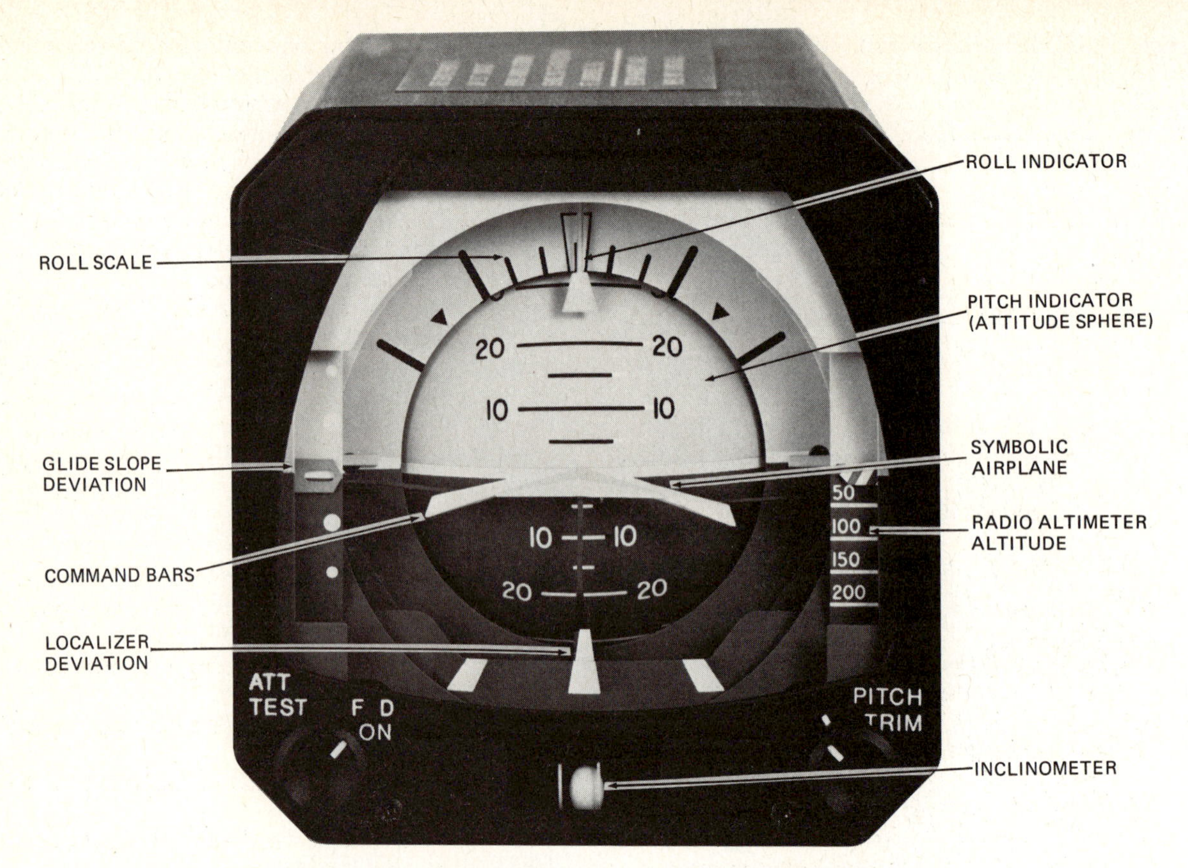

Fig. 18-7 Director-horizon indicator. (Bendix Avionics Div.)

nals that are sent to the computer amplifier. These signals are amplified and sent to the primary servos commanding control-surface movement for control of pitch and roll. At the same time, the instrument is indicating the degree of pitch and roll for the information of the pilot.

The vertical gyro in the instrument may be driven either by a vacuum system or electrically, depending upon the particular instrument selected for the system. Since the gyro remains vertical with respect to the surface of the earth, the pitch and roll movements of the airplane cause relative movements between the gyro and the instrument case, thus producing signals for the autopilot and indications on the instrument.

The **localizer deviation** is displayed by a meter-driven pointer read against a fixed horizontal scale. The meter is controlled by signals from the navigation receiver.

The V-type **command bars** show the pilot how the airplane should be flown to satisfy flight commands. If the symbolic airplane is nested in the V, the flight commands are being satisfied. If the command bars should tilt upward to the right, the airplane should be flown upward and to the right to meet the command bars. The command bars are light yellow in color, and the airplane symbol is orange to provide visible contrast.

Glideslope deviation is displayed at the left of the instrument by a meter-driven pointer read against a fixed vertical scale. The meter is controlled by signals from the glideslope receiver.

The **pitch trim knob** is used to adjust manually the attitude sphere to compensate for variations in pilot eye level and for changes in aircraft cruise attitude.

The **inclinometer** is used to indicate aircraft slip or skid. It is adjustable to compensate for installation tilt.

To warn the pilot if some elements of the instrument are not providing valid information, warning flags appear. For example, if the ATT flag appears, it means that there is a malfunction in the pitch or roll servos or in the vertical gyro.

SERVOS

Three primary servos control pitch, yaw, and roll by moving the elevators, rudder, and ailerons in response to commands from the autopilot. These servos are small electric motors that drive capstans through magnetic clutches. The capstan in each servo contains an adjustable slip clutch which is preset for each axis by the installer. These clutches make it possible for the pilot to override the automatic pilot if necessary.

The **trim servo** is used to activate an elevator trim tab control to relieve long-term aerodynamic loading and generally assist in smoother operation of the elevator surface without requiring large amounts of power from the primary servo. Its operation is either automatic, using the composite pitch error signal from the computer-amplifier, or both automatic and manual if an optional manual electric trim adapter is utilized.

The location and installation procedures for all the servos depend upon the type of aircraft in which they are being installed. Detailed instructions concerning cable tensions, clutch settings, and other pertinent data are included in the installation kit for the particular aircraft involved.

COMPUTER-AMPLIFIER

The photograph of Fig. 18-8 illustrates the computer-amplifier for the Bendix M-4D automatic flight-control system. The computer-amplifier is the heart and brain of the system. It consists of nine plug-in modules, one of which is peculiar to a particular aircraft type installation and another of which relates to a specific type of flight-director installation. The modules can be removed and replaced easily in case of malfunction.

The computer-amplifier combines inputs from the gyro sensors, control panel, heading and radio sources (NAV and RNAV) to compute and deliver appropriate electrical commands to the control surface servos and to the flight-director instrumentation.

Fig. 18-8 Computer-amplifier. (Bendix Avionics Division)

The computer-amplifier is mounted on a shock mounting especially designed to receive it. The mounting is shown in the drawing of Fig. 18-9. The location of the installation depends upon the type of aircraft in which the equipment is being installed. The installation kit for the particular aircraft gives detailed instructions.

HORIZONTAL SITUATION DISPLAY

The horizontal situation display (HSD) performs as a typical HSI but includes features and capabilities in addition. The HSD is not an essential component of the autopilot system; however, it does give the pilot useful visual information that aids in flying the aircraft more precisely.

The HSD is shown in Fig. 18-10. Its functions are described as follows:

Heading card: Graphically illustrates magnetic headings.

Lubber line: Indexes the actual magnetic heading from the heading card.

Heading bug: Indexes the heading to be maintained by the flight-control system

Heading select knob: Positions the heading bug with respect to the heading card.

Heading flag: Indicates a malfunction in the directional gyro.

Function switch: Selects the desired mode—ADF, standard HSI, or automatic VOR.

RNAV annunciator: Indicates that the HSD is operating in conjunction with an RNAV programmer.

Course pointer: Indicates the selected VOR or LOC course in the HSI mode, or bearing to the station in the VOR or ADF mode.

Course select knob: Positions the course pointer in the HSI mode only.

Course deviation bar: Indicates the relative position of the selected VOR radial or localizer beam from the aircraft.

Course deviation scale: Indicates the relative position 10° full scale for VOR and 2.5° at the third dot for LOC.

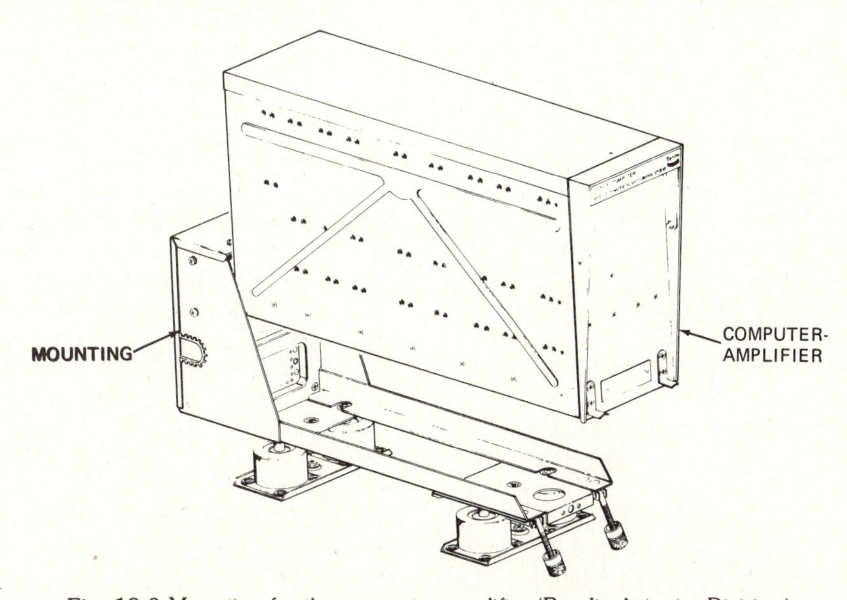

Fig. 18-9 Mounting for the computer-amplifier (Bendix Avionics Division)

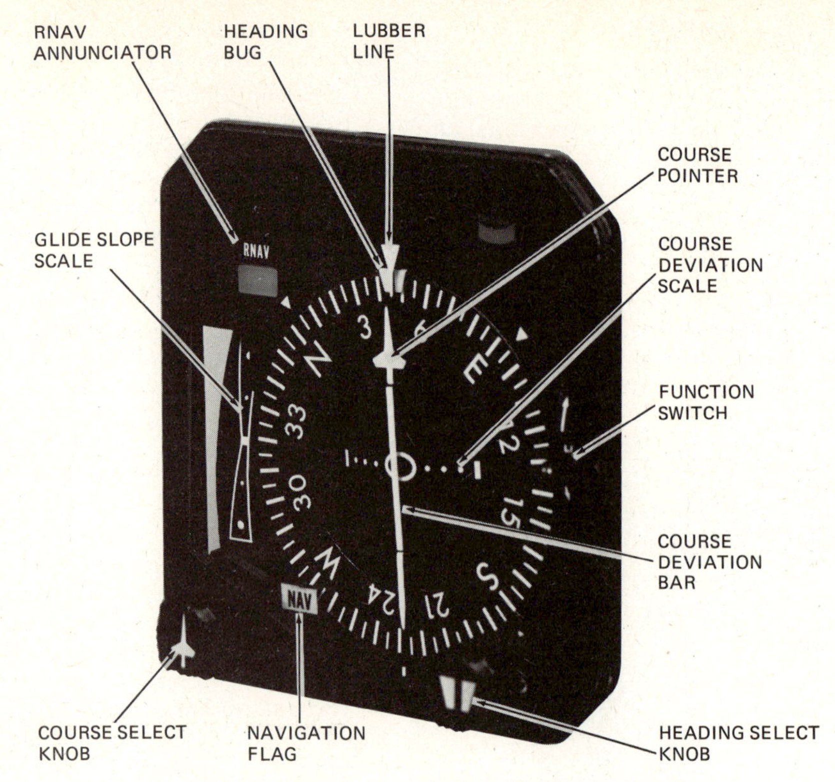

Fig. 18-10 Horizontal situation display. (Bendix Avionics Div.)

Navigation flag: Indicates that navigation signals being received are not usable.

The HSD heading system is operable in any of three modes: **slaved, magnetic,** or **free gyro,** as determined by the selector on the **slaving control.** The slaving control is shown in Fig. 18-11. The heading is indicated by the heading card as indexed by the lubber line. The HDG knob slews (turns) the heading bug to the desired reference on the heading card. The heading bug moves with the heading card as heading changes, and it provides an output to the autopilot proportional to the angular distance between the lubber line and the heading bug. When the heading bug is aligned with the lubber line, the aircraft is on the selected heading.

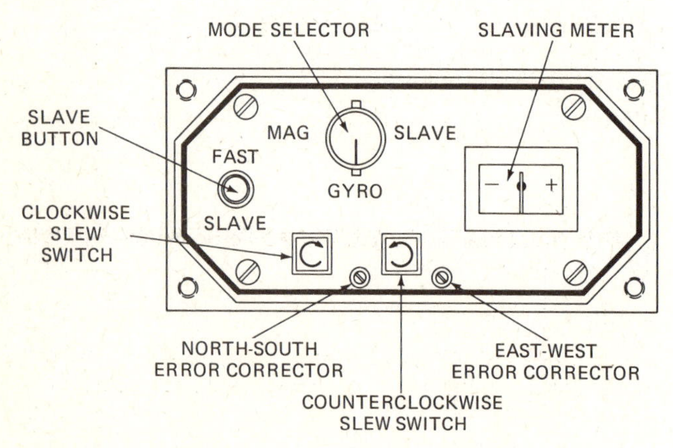

Fig. 18-11 Slaving control. (Bendix Avionics Division)

The **slaved mode** combines the accuracy and long-term stability of a flux sensor (flux valve or flux gate) with the short-term stability of a gyro. The gyro provides smooth, accurate heading information, while the microprocessor in the HSD implements a slow slew signal to correct for error signals. The microprocessor develops the slew signal by comparing the flux sensor signal to the gyro output. Any differences noted will be removed by a microprocessor slew signal.

The **magnetic mode** couples the flux sensor to the heading card in a fast slew configuration. The magnetic output, without the stabilizing effect of the gyro, allows the HSD to operate as a magnetic compass. This mode is normally used only if the gyro has failed.

In the **free gyro mode,** the directional gyro provides the only heading information in the HSD. The heading card must be periodically aligned with correct magnetic heading information from a separate source by depressing the manual slew buttons on the slaving control. This would be necessary only in the case that the flux sensor is inoperative.

The **slaving meter** on the slaving control is an indicator to show when the flux sensor is sending a correction signal to the directional gyro.

SLAVED DIRECTIONAL GYRO

The slaved directional gyro operates in connection with the flux sensor to provide accurate directional information for the HSD. As explained previously, the microprocessor in the HSD develops correctional signals that keep the directional gyro in agreement with the magnetic heading sensed by the flux sensor. A photograph of a slaved directional gyro is shown in Fig. 18-12.

Fig. 18-12 Slaved remote directional gyro. (Bendix Avionics Div.)

FLUX SENSOR

The flux sensor is a flux valve system that detects the magnetic lines of force in the earth's magnetic field and produces magnetic heading information for display on the HSD, HSI, or other indicator as required for navigation purposes. The system operates on a principle similar to that of the flux gate, described previously.

ALTITUDE CONTROLLER

The **altitude controller** is an aneroid-type instrument which constantly senses the pressure altitude of the aircraft. When the flight controller is in the ALT mode, the altitude controller sends signals to the computer amplifier to call for corrective action if the aircraft is not at the preselected altitude. The signals call for movement of the elevator to accomplish the required correction.

SLIP-SKID SENSOR

The **slip-skid sensor** provides signals to the yaw channel of the computer amplifier to provide the autopilot with information which enables it to perform coordinated turns and keep the airplane under stable control, even if one of the engines becomes inoperative.

AUTOMATIC FLIGHT AND LANDING SYSTEMS

It is the purpose of this section to give the student a general understanding of the concept of a completely integrated flight-control system that enables an aircraft to take off, fly a prescribed route, and land at a designated airport withoug the aid of a human pilot. The system described here is one configuration designed for the Lockhead L-1011 airplane. The complete system includes the elements described in the first part of this chapter plus other systems and subsystems that provide the capability of automatic takeoff and landing.

A number of automatic flight-control systems have been developed, and even though they may vary in some respects, they must have the ability to track ILS and VOR signals; maintain prescribed altitudes; adjust power for takeoff, climb, cruise, and landing; operate flaps, spoilers, and landing gear as required; and maintain proper flight attitudes. In each such system, the autopilot flies the airplane as it responds to commands from attitude sensors, navigation systems, and flight-control units. Power is controlled through the engine throttles moved by throttle servos responding to commands from throttle computers. Flaps, landing gear, and spoilers are operated by an electronic system that is programmed to provide proper flap extension for takeoff and landing, landing gear retraction and extension at the proper times, and spoiler deployment as necessary to reduce lift and act as an air brake.

SUBSYSTEMS

The complete system described here is designated as the **avionics flight-control system (AFCS)** and could also be called an automatic flight-control system. It provides manual or automatic modes of control throughout the entire flight envelope from takeoff to landing and rollout. This is achieved in an integrated system comprised of the **autopilot/flight director system (APFDS),** the **stability augmentation system (SAS),** the **speed-control system (SCS),** and the primary **flight-control electronic system (FCES).** Figure 18-13 is a block diagram showing how the components of these systems are interrelated. All the subsystems of the AFCS are fully integrated and have levels of redundancy to achieve a high level of reliability. Redundancy is accomplished by providing two or more systems of each type so a failure of one system will not affect the operation of the complete system.

AUTOPILOT/FLIGHT-DIRECTOR SYSTEM

The APFD, through the autopilot, provides automatic pitch and roll control to stabilize the aircraft and maintain selected altitude, attitude, and heading in flight. In this fully automatic mode, the flight director may be used in a monitoring capacity; that is, the human pilot may watch the flight director (ADI) to observe the operation of the autopilot. In other modes, the flight director may be used for flight guidance.

The APFDS operates in the modes listed below.

Altitude Select and Hold

In this mode, the autopilot maintains the altitude that has been programmed.

Vertical Speed Select and Hold

This mode causes the autopilot to control rates of climb and descent in accordance with the rates that have been programmed.

Heading Select and Hold

The autopilot controls the airplane to maintain selected headings when this mode is in operation.

Control Wheel Steering (CWS)

This mode allows the human pilot to change commands to the autopilot by moving the control wheel. After a change is made, the pilot releases the wheel and the autopilot will follow the new command.

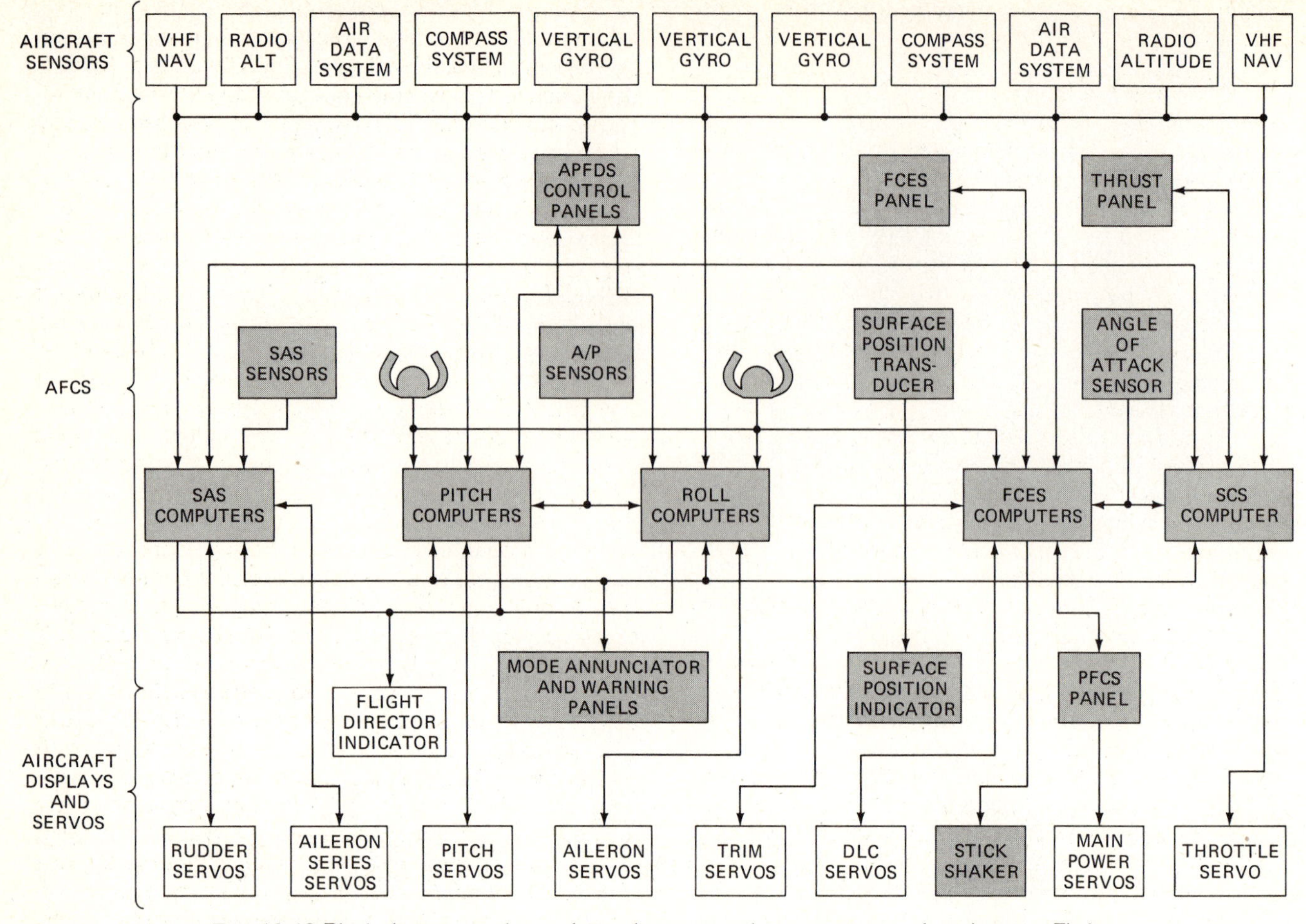

Fig. 18-13 Block diagram to show relationship among the components of an Avionics Flight Control System. (Lockheed California Co.)

IAS Select and Hold

In this mode, the human pilot selects the desired indicated airspeed, and the autopilot together with the speed-control system will maintain the selected airspeed.

Mach Hold

In this mode, the system uses Mach number rather than IAS as the speed reference.

Localizer

This is an approach mode wherein the aircraft tracks the localizer beam of the ILS system.

VOR and Area Navigation (RNAV)

When in this mode, the system navigates automatically, utilizing VOR data or commands from an RNAV computer.

Approach/Land

As the aircraft approaches its destination, the speed-control system reduces power as necessary for the correct rate of descent and the navigation system locks onto the ILS beam for guidance to the runway. The radio altimeter provides altitude signals for landing, flare, and touchdown. Upon touchdown, the power is reduced to idle and the aircraft tracks the localizer beam to roll out along the center of the runway.

CENTRAL AIR-DATA COMPUTERS

Two central air-data computers (CADCs) are employed to provide information for both manual and automatic flight. They receive pneumatic inputs from the pitot-static system and electric signals from the total air temperatures (TAT) probes to generate altitude, airspeed, and temperature information. This information is then used to provide outputs for the air data instruments and recorders as well as for the automatic flight-control systems, the stability augmentation system, the Mach trim/feel systems, and for various other purposes.

The incorporation of CADCs permits the use of electrically servoed instrumentation, including primary flight instruments, flight data recorders, and the optional true-airspeed (TAS) and static air temperature (SAT) indicators. This use of electrically driven instruments reduces the amount of pneumatic plumbing required behind the instrument panels to only those lines connected to the standby airspeed indicator and altimeter. The use of servoed instruments also makes it possible to incorporate Mach and overspeed warning switches into the electric flight instruments and permits the use of electric switching between the normal and alternate air data systems. This eliminates the need for troublesome pitot-static system selector switches.

STABILITY AUGMENTATION SYSTEM

Inflight stability and control of the aircraft are augmented by the stability augmentation system (SAS) which provides yaw damping. Yaw damping is the process of limiting the rate and degree of yaw to a safe and controllable level. Two computers are used for improved reliability, and limited averaging improves tracking of the servos.

With either of the two dual-channel computers engaged, the SAS provides dutch-roll damping and turn coordination during all phases of flight. Both computers are engaged for the runway alignment and roll-out functions. This provides the fail-operational requirement that is specified for autoland. *Fail-operational* means that the system will continue to function even though there are failures in some parts of the system. The system, therefore, meets the requirements for Category III fully automatic blind-landing capability.

During autoland, the yaw computers start monitoring the availability of runway alignment and roll-out and the results are displayed on each AFCS mode annunciator during the approach/land sequence if the system is functioning normally. Failures are annunciated on the AFCS warning indicators.

All the computer channels receive a yaw-rate signal generated from one of three yaw-rate gyros. In the basic yaw SAS mode, turn coordination is achieved by processing signals from four aileron position transducers. In LOC track or autoland modes the aileron position signals are switched out.

The runway alignment is generated as a function of altitude and alignment/roll-out logic. Damping is provided by mixing heading and heading-rate signals with yaw-rate signals. The alignment scheme is a limited forward slip maneuver in which up to 8° of initial crab angle is removed by lowering a wing and slipping the aircraft. This is to align the aircraft with the runway in case of a crosswind.

The roll-out mode is initiated at touchdown as a function of altitude. The roll-out guidance system utilizes the LOC beam to track the center of the runway until speed is reduced to taxiing speed.

SPEED-CONTROL SYSTEM

The speed-control system (SCS) provides an **airspeed autothrottle** mode and an **angle-of-attack autothrottle** mode. The airspeed mode is used for all flight conditions through initial approach, and the angle-of-attack mode is used in the final approach and landing. The SCS also provides the go-around command for both manual and automatic go-around, and the takeoff command for manual takeoff guidance.

The SCS consists of a single computer with two identical computation channels, a single, monitored autothrottle servo, and dual sensor inputs. The autothrottle function controls the engine throttles through the autothrottle servo to maintain a selected airspeed or a precomputed angle of attack which corresponds to a "stall margin" airspeed.

The normal mode of operation for the autothrottle with the airplane in the cruise configuration is **airspeed select.** In the approach and landing configuration, the autothrottle is operated in the **stall-margin** mode. At the flare initiation altitude, the SCS switches to the **flare** mode, which provides automatic closed-loop throttle retard before touchdown. At touchdown, the system switches to the **touchdown** mode and the throttles are driven to idle, whereupon the autothrottle system automatically disengages. The takeoff/go-around computation is sent to the APFDS for use in controlling the pitch axis during takeoff and go-around maneuvers. The thrust panel by which the aircraft speed is controlled is accessible to both pilots. It consists of an autothrottle engage switch, an airspeed select control, a selected airspeed digital readout, and a stall-margin mode annunciator. In addition, the thrust panel contains the necessary elements to generate the airspeed error signal used by the SCS computer when in the airspeed-select mode.

A single monitored servo drives the engine throttles at a rate proportional to the amplitude of the command signal. A dual-input bidirectional sprag clutch is continuously engaged except when a manual override force is applied to the throttle levers. With the autothrottle engaged or disengaged, there is no substantial difference in the force required to manually operate the throttle levers.

The autothrottle operating range is defined by limit switches located at the maximum thrust limit, the minimum thrust limit, and the idle position. Actuation of the maximum- or minimum-thrust-limit switches will open the control phase of the servo motor, thus stopping further throttle motion in that direction. With the airplane on the ground, as sensed by the main-landing-gear strut compression switches, actuation of the idle-disconnect switch will result in total disengagement of the autothrottle.

The autothrottle servo is engaged by moving the engage switch on the thrust panel to the ATS position. In this configuration, the servo will drive the engine throttle levers to satisfy the computed throttle command. Disengagement of the autothrottle servo is accomplished by moving the engage switch to OFF; actuation of either of the two throttle-lever-mounted disconnect switches; detection of a failure in either of the autothrottle computation channels or the servo loop; selection of the IAS HOLD, MACH HOLD, or TURBULENCE mode of the APFDS; or engagement of the takeoff/go-around mode. The go-around mode is selected by means of the go-around switch on either the captain's or first officer's control wheel. Alert lights will flash on the AFCS warning indicator when the autothrottle is disengaged.

PRIMARY FLIGHT-CONTROL ELECTRONIC SYSTEM

The primary flight-control electronic system (PFCES) consists of various automatic control, warning, and indicating subsystems which are principally concerned with manual aircraft control. These subsystems provide means for operating and monitoring the many control surfaces that are necessary for the safe and efficient flight of a large airplane. The systems operate for both automatic and manual flight.

The positions of flight-control surfaces must be known to the pilots, and these positions are displayed by means of the **surface-position indicator.** This assists the pilots to verify proper operation of the flight-control surfaces, primarily during ground operation. An autopilot mistrim display is also included. Control and trim positions are provided for spoilers, rudder, stabilizer trim, and aileron trim.

A **primary flight-control monitoring system (PFCS)** detects and displays to pilots the means for alleviation of opens and jams in the pitch-axis and jams in the roll-axis control systems. Two independent sensors and monitoring channels are used in each instance, and warnings can result from either or both channels.

The **rudder-control limiting system** automatically restricts the rudder authority and limits rudder hydraulic power capability during high-speed flight. The system mechanically limits rudder deflection as a function of airspeed when the flaps are retracted with less than 4° of flap deflection.

The **spoiler mode-control system** automatically changes the configuration of the roll and speedbrake inputs to the spoilers to optimize roll, direct lift control, and speedbrake control characteristics for low- and high-speed flight. Since the spoilers are used to control lift, speed, and roll, the selection of the spoilers to be employed in each mode is critical. The spoilers operate in conjunction with the **direct lift control (DLC)/automatic ground speedbrake (AGSB)** system to respond automatically during approach and landing through the use of the DLC and to deploy spoilers automatically for braking after landing and a rejected takeoff.

The **stall warning system** artificially vibrates the control columns to warn of an impending stall. Stall warning computations are based on angle-of-attack measurements as modified by position measurements of flaps and leading-edge slats to obtain an unmistakable warning at a minimum of 7 percent above stalling speed.

The **electric pitch trim system** permits electrical control of pitch-axis trim by the pilots through the use of thumbwheels on the control wheels. The system is operated automatically during automatic flight.

Mach trim, as required by FAR, is used to control the average gradient of the stable slope of the stick-force versus-speed curve to less than 1 lb [0.45 kg] for a change in speed of 6 kn [11 km/s]. Trim changes are scheduled from Mach data derived from the two CADCs. The system normally controls only one of two electric motors in the electrical trim/feel mechanism. The data is processed by the trim augmentation computer.

The **Mach feel system** automatically adjusts the pitch-axis feel-force gradient for all flight conditions. This is a two-channel active-standby system that operates together with stabilizer trim angle and Mach number to provide proper pitch-control-force characteristics as established by FAR requirements; that is, that the pitch control force at limit load be controlled within 50 to 100 lb [22.7 to 45.4 kg].

The **pitch-trim disconnect system** is a means of alleviating mechanical jams in the pitch-trim system. When disconnected, pitch-trim control can then be continued, either by use of the electrical thumbwheel controls or through the mechanical trim wheels. If only electrical trim is operative, there is a loss of series trim output. If only mechanical trim is operative, there is no effect on performance of the trim system.

The **altitude alert system** indicates by visual and aural warnings of approaches to and deviations from selected altitudes. It also provides for automatic capture of selected altitudes through the autopilot/flight director system. The system employs dual redundant barometric altimeters, computational channels, and annunciators. Each channel is completely independent, inclusive of electric power source, so that warnings will result from both channels.

The altitude alert system is integrated with the altitude-capture and altitude-hold functions of the APFDS and utilizes the same controls on the ALT SELECT panel as those used for the altitude-select function of the APFDS. However, the altitude-alert system operates independently of the autopilot.

The major elements of the PFCES, excluding interfacing components, are a flight-control-electronic-system (FCES) computer, a trim-augmentation computer, a trailing-edge flap-load-relieving-system computer, angle-of-attack sensors, stick shakers, control-surface-position transducers, and associated flight-station control panels and indicators. Functions of the FCES computer are AFSC monitoring, rudder limiting, direct-lift control, automatic ground spoiler control, stall warning, altitude alert, and fault-monitoring indication.

The trim-augmentation computer's functions are manual and automatic pitch trim, and Mach-trim and Mach-feel compensation. The fault-isolation monitoring indication employs a single computer channel and the direct-lift control has a fail-operational capability to meet autoland requirements. All the other functions have dual computer channels with a fail passive capability.

FLIGHT STATION EQUIPMENT

Equipment for control and operation of the autopilot/flight director system (APFDS) is located to be easily accessible to the pilots. The flight station equipment is shown in Fig. 18-14. As will be noted in the drawing, the controls permit selection of any mode of APFDS operation desired plus the selection of the parameters for the flight. When fully automatic flight is programmed, additional data must be entered through other controls.

The AFCS mode annunciation unit is shown in Fig. 18-15. In the drawing, all the modes are illuminated as when the test 1 button is pressed. During normal flight, the only modes illuminated are those in actual use.

REVIEW QUESTIONS

1. What device is used to produce aircraft-attitude signals for an autopilot?
2. Why is there a need for a *feedback* system for each correction command signal in an autopilot?
3. Describe a signal pickoff device that may be used with a gyro to produce correction signals.
4. What is meant by an *optical* signal device used with a gyro?
5. What are the principal components of a basic electronic autopilot?

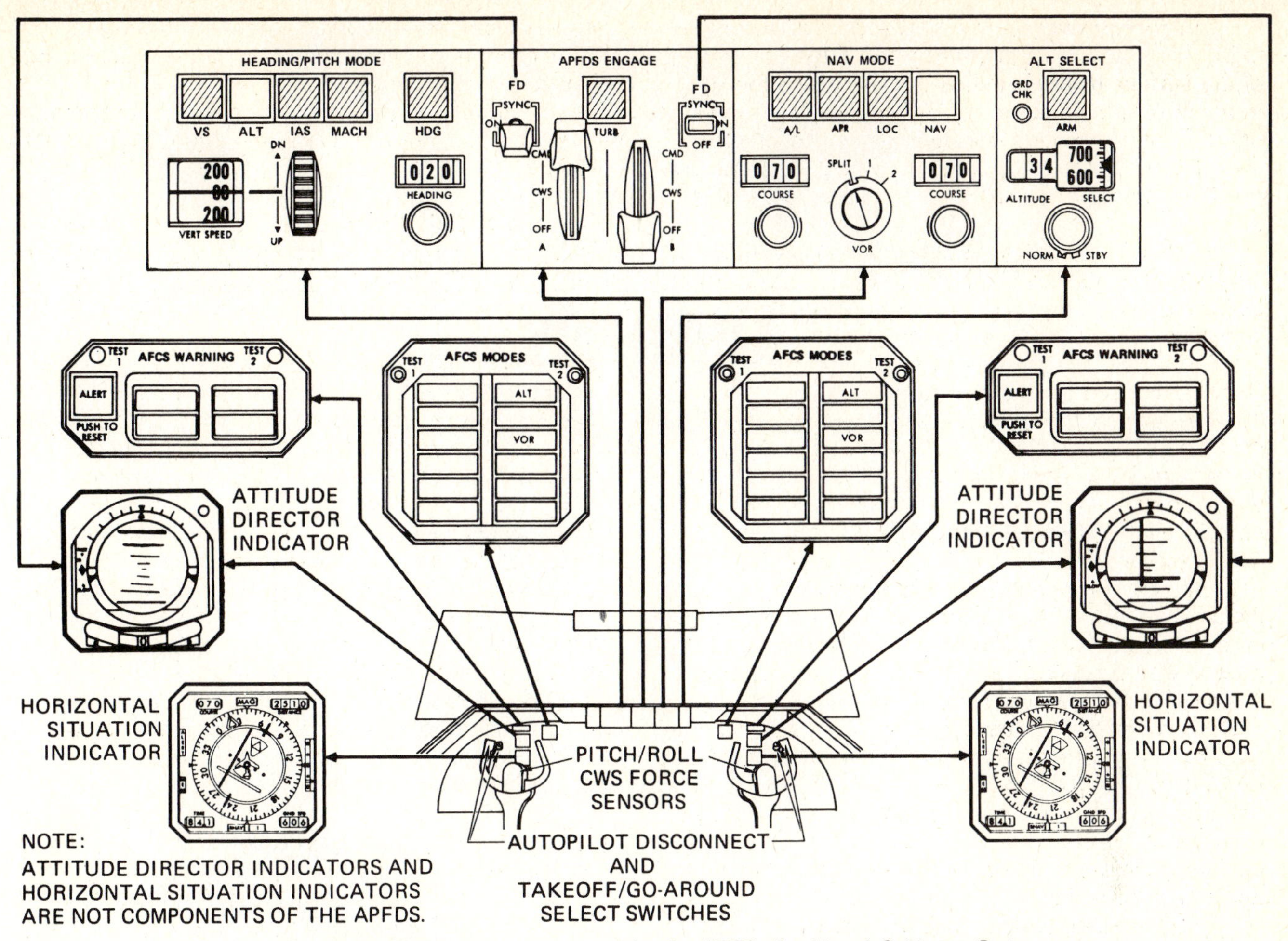

Fig. 18-14 Flight station equipment for the AFCS. (Lockheed California Co.)

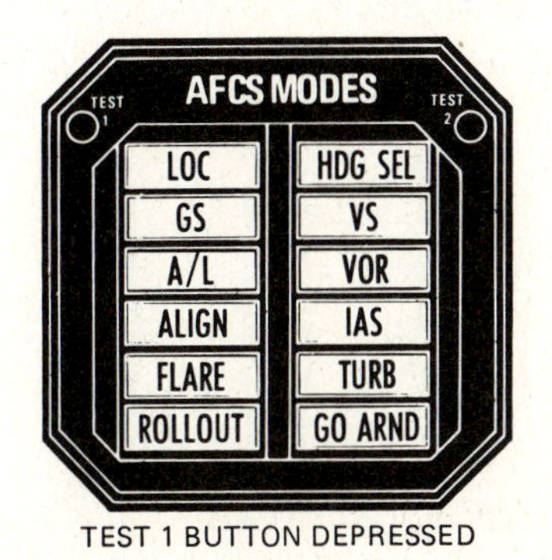

Fig. 18-15 AFCS mode-annunciation unit. (Lockheed California Co.)

6. What is the purpose of a *rate network* in an autopilot system?
7. Describe the *flight controller* for a Bendix M-4D automatic flight-control system.
8. What guidance is employed by the autopilot when the RNAV annunciator is illuminated?
9. What is the purpose of the GA button on the pilot's control wheel?
10. What electric signal is produced by the turn-and-bank indicator?
11. List the functions of the *director-horizon indicator.*
12. Describe a *primary servo unit.*
13. What is the purpose of a *slip clutch* on a servo?
14. Give the functions of the *computer-amplifier.*
15. Describe the *horizontal situation display* (HSD).
16. Describe the *slaving control* and explain its function.
17. What is the purpose of the *slaved directional gyro?*
18. What is a *flux sensor?*
19. In what modes does the *autopilot/flight-director system* (APFDS) operate?
20. What is the purpose of an air-data computer (CADC)?
21. Describe the function of the *yaw stability augmentation system* (SAS).
22. By what means does the *speed-control system* (SCS) control the speed of the aircraft?
23. In what modes does the SCS operate?
24. By what means is the autothrottle system prevented from exceeding limits of power?
25. What are the functions of the *primary flight-control electronic system?*
26. How does the pilot determine the position of flight-control surfaces?
27. How does the pilot know when the aircraft is approaching stalling speed?
28. Describe the control panel for an automatic flight-control and landing system.

APPENDIX

USEFUL FORMULAS

Ohm's law

$$I = \frac{E}{R} \qquad R = \frac{E}{I} \qquad E = IR$$

where I = current (intensity of current flow)
E = voltage (emf)
R = resistance

Resistances in series

$$R_t = R_1 + R_2 + R_3 \cdots$$

Resistances in parallel

$$R_t = \frac{1}{1/R_1 + 1/R_2 + 1/R_3 \cdots}$$

or

$$\frac{1}{R_t} = \frac{1}{R_1} + \frac{1}{R_2} + \frac{1}{R_3} \cdots$$

Two resistances in parallel

$$R_t = \frac{R_1 \times R_2}{R_1 + R_2}$$

Capacitances in series

$$C_t = \frac{1}{1/C_1 + 1/C_2 + 1/C_3 \cdots}$$

or

$$\frac{1}{C_t} = \frac{1}{C_1} + \frac{1}{C_2} + \frac{1}{C_3} \cdots$$

Capacitances in parallel

$$C_t = C_1 + C_2 + C_3 \cdots$$

Electric power

$$P = EI \qquad \text{or} \qquad P = I^2R \qquad \text{or} \qquad P = \frac{E^2}{R}$$

where P = power, W
1 hp = 550 ft·lb/s = 746 W
1 J = 1 W/S

Frequency and wavelength

$$f = \frac{300\,000\,000}{\lambda}$$

$$\lambda = \frac{300\,000\,000}{f}$$

where f = frequency, Hz
λ = wavelength, m

Capacitive reactance

$$X_C = \frac{1}{2\pi fC}$$

where X_C = capacitance reactance, Ω
f = frequency, Hz
C = capacitance, F

Inductive reactance

$$X_L = 2\pi fL$$

where X_L = inductive reactance, Ω
f = frequency, Hz
L = inductance, H

Resonant frequency

$$f = \frac{1}{2\pi\sqrt{LC}} \qquad \text{or} \qquad f = \frac{0.159\,155}{\sqrt{LC}}$$

Impedance: series circuit

$$Z = \sqrt{(X_L - X_C)^2 + R^2}$$

where Z = impedance, Ω
X_L = inductive reactance, Ω
X_C = capacitance reactance, Ω
R = resistance, Ω

Impedace: parallel (tank) circuit

$$Z_{\text{par}} = \frac{L}{RC}$$

Inductance of a coil

$$L = \frac{1.26N^2\mu A}{10^8 l}$$

where L = inductance, H
N = number of turns of wire on the coil
μ = permeability of core material
A = cross-sectional area of the coil
l = length of the coil, cm

Inductances in series, no magnetic coupling

$$L_t = L_1 + L_2 + L_3 \ldots$$

Inductances in parallel, no coupling

$$\frac{1}{L_t} = \frac{1}{L_1} + \frac{1}{L_2} + \frac{1}{L_3} \cdots$$

Figure of merit for a coil

$$Q = \frac{X_L}{R} \quad \text{or} \quad Q = \frac{2\pi f L}{R}$$

Amplification factor: vacuum tube

$$\mu = \frac{\Delta E_p}{\Delta E_g} \quad (I_p \text{ constant})$$

where ΔE_p = change in plate voltage, V
ΔE_g = change in control-grid voltage, V

Plate resistance

$$R_p = \frac{\Delta E_p}{\Delta I_p} \quad (E_g \text{ constant})$$

where R_p = plate resistance, Ω
ΔE_p = change in plate voltage, V
ΔI_p = change in plate current, A

Transconductance

$$G_m = \frac{\Delta I_p}{\Delta E_g} \quad (E_p \text{ constant})$$

where G_m = transconductance, mhos

ABBREVIATIONS

ampere	A
ampere-hour	Ah or A · h
billion electronvolts	GeV
circular mil	cmil
coulomb	C
decibel	dB
dyne	dyn
electronvolt	eV
erg	erg
farad	F
gauss	G
gilbert	Gb
henry	H
hertz	Hz
horsepower	hp
joule	J
kilowatthour	kWh or kW · h
mho (see siemens)	
oersted	Oe
ohm	Ω
siemens	S
var	var
volt	V
voltampere	VA or V · A
watt	W
watthour	Wh or W · h
weber	Wb

Greek Alphabet

Name	Capital	Lowercase	Use in Electronics
Alpha	Α	α	Angles, area, coefficients
Beta	Β	β	Angles, flux density, coefficients
Gamma	Γ	γ	Conductivity
Delta	Δ	δ	Variation, density
Epsilon	Ε	ϵ	
Zeta	Ζ	ζ	Impedance, coefficients, coordinates
Eta	Η	η	Hysteresis coefficient, efficiency
Theta	Θ	θ	Temperature, phase angle
Iota	Ι	ι	Current
Kappa	Κ	κ	Dielectric constant
Lambda	Λ	λ	Wavelength
Mu	Μ	μ	Micro, amplification factor, permeability
Nu	Ν	ν	Reluctivity
Xi	Ξ	ξ	
Omicron	Ο	ο	
Pi	Π	π	Ratio of circumference to diameter (3.1416)
Rho	Ρ	ρ	Resistivity, density
Sigma	Σ	σ	Sign of summation
Tau	Τ	τ	Time constant, time phase displacement
Upsilon	Υ	υ	
Phi	Φ	φ	Magnetic flux, angles
Chi	Χ	χ	
Psi	Ψ	ψ	Dielectric flux, phase difference
Omega	Ω	ω	Capital, ohms; lowercase angular velocity

Copper Wire
Single Strand, American Wire Gage

Gage	Diameter mils	Cross section, cir mils	Resistance, Ω/1000 ft (25°C)	Weight, lb/1000 ft
0000	460.0	211 600.0	0.0500	641.0
000	410.0	167 800.0	0.0630	508.0
00	365.0	133 100.0	0.0795	403.0
0	325.0	105 500.0	0.1000	319.0
1	289.3	83 690.0	0.126	253.0
2	258.0	66 370.0	0.159	201.0
3	229.0	52 640.0	0.201	159.0
4	204.0	41 740.0	0.253	126.0
5	182.0	33 100.0	0.319	100.0
6	162.0	26 250.0	0.403	79.5
7	144.3	20 820.0	0.508	63.0
8	128.5	16 510.0	0.641	50.0
9	114.4	13 090.0	0.808	39.6
10	102.0	10 380.0	1.02	31.4
11	91.0	8 234.0	1.28	24.9
12	81.0	6 530.0	1.62	19.8
13	72.0	5 178.0	2.04	15.7
14	64.1	4 107.0	2.58	12.4
15	57.1	3 257.0	3.25	9.9
16	50.8	2 583.0	4.09	7.8
17	45.3	2 048.0	5.16	6.2
18	40.3	1 624.0	6.51	4.9
19	35.9	1 288.0	8.21	3.9
20	32.0	1 022.0	10.4	3.09
21	28.5	810.0	13.1	2.45
22	25.3	642.4	16.5	1.95
23	22.6	509.0	20.8	1.54
24	20.1	404.0	26.2	1.22
25	17.9	320.0	26.2	0.97
26	15.9	254.0	41.6	0.769
27	14.2	202.0	52.5	0.610
28	12.6	160.0	66.2	0.484
29	11.3	127.0	83.4	0.384
30	10.0	100.5	105.2	0.304
31	8.93	79.70	132.7	0.241
32	7.95	63.21	167.3	0.191
33	7.08	50.13	211.0	0.152
34	6.31	39.75	266.0	0.120
35	5.62	31.52	335.5	0.095
36	5.00	25.00	423.0	0.0757
37	4.45	19.83	533.4	0.0600
38	3.96	15.72	672.6	0.0476
39	3.53	12.47	848.1	0.0377
40	3.14	9.98	070.0	0.0299

ELECTRICAL AND ELECTRONIC SYMBOLS

The symbols shown here are those that are likely to be encountered by the aviation maintenance technician. Only the primary symbols are provided in this section. For the additional symbols representing variations of the primary symbols, the technician should consult the document "Graphic Symbols for Electrical and Electronics Diagrams" published by the Institute of Electrical and Electronic Engineers (IEEE), IEEE Std 315-1975, or ANSI Y32.2-1975 furnished by the American National Standards Institute. Symbols shown usually comply with International Electronical Commission (IEC) Publication 117.

QUALIFYING SYMBOLS

Qualifying symbols are applied to standard symbols to provide an indication of the special characteristics of the symbols as they are employed in specific circuits.

ADJUSTABILITY OR VARIABILITY

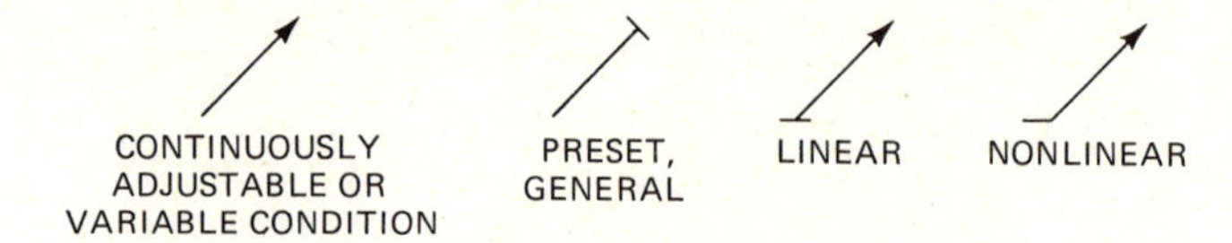

SPECIAL PROPERTY INDICATORS

t°
TEMPERATURE DEPENDENCE

x
MAGNETIC FIELD DEPENDENCE

τ
(GREEK LETTER TAU) STORAGE

SATURABLE PROPERTIES

RADIATION INDICATORS

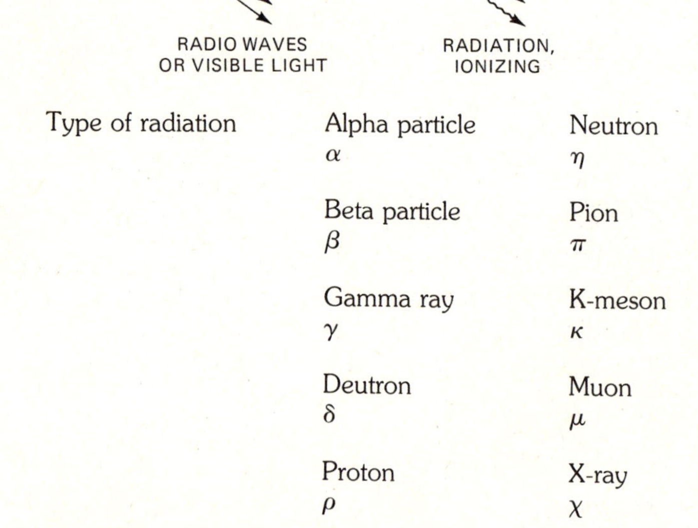

Type of radiation

Alpha particle α	Neutron η
Beta particle β	Pion π
Gamma ray γ	K-meson κ
Deutron δ	Muon μ
Proton ρ	X-ray χ

PHYSICAL-STATE RECOGNITION SYMBOLS

GAS, AIR, OR PNEUMATIC
LIQUID
SOLID
ELECTRET MATERIAL

TEST-POINT RECOGNITION SYMBOL

● OR

TEST POINT FOR CIRCUIT TERMINAL

DIRECTION OF FLOW OF POWER, SIGNAL, OR INFORMATION

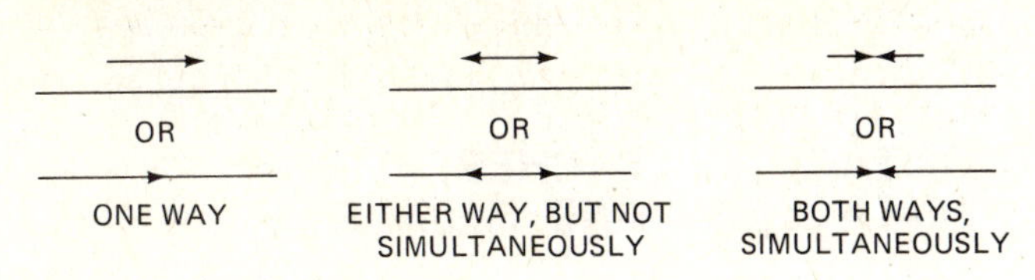

OR — ONE WAY

OR — EITHER WAY, BUT NOT SIMULTANEOUSLY

OR — BOTH WAYS, SIMULTANEOUSLY

KIND OF CURRENT (GENERAL)

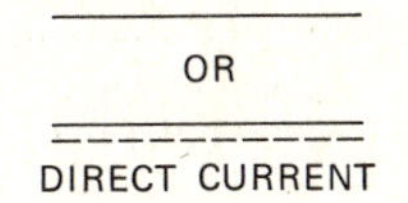

OR

DIRECT CURRENT

ALTERNATING CURRENT

CONNECTION SYMBOLS

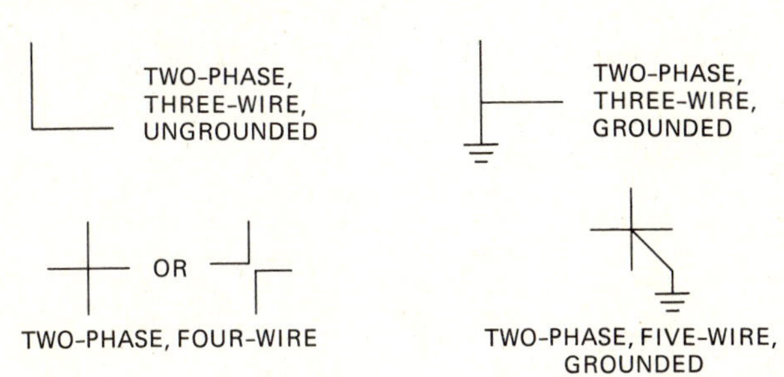

TWO-PHASE, THREE-WIRE, UNGROUNDED

TWO-PHASE, THREE-WIRE, GROUNDED

OR

TWO-PHASE, FOUR-WIRE

TWO-PHASE, FIVE-WIRE, GROUNDED

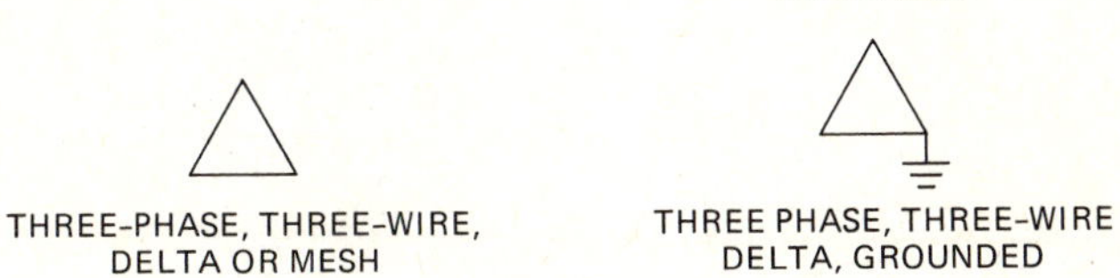

THREE-PHASE, THREE-WIRE, DELTA OR MESH

THREE PHASE, THREE-WIRE DELTA, GROUNDED

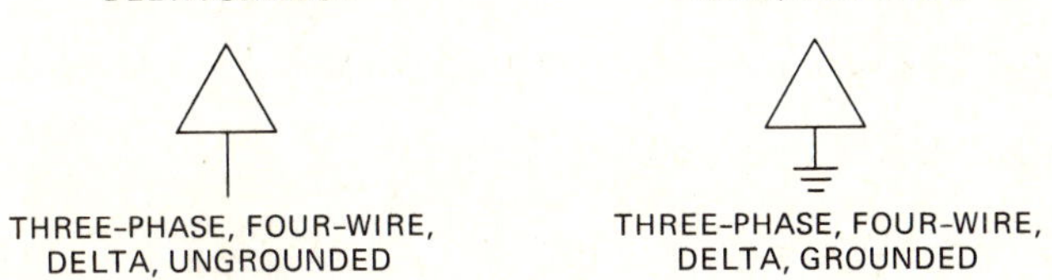

THREE-PHASE, FOUR-WIRE, DELTA, UNGROUNDED

THREE-PHASE, FOUR-WIRE, DELTA, GROUNDED

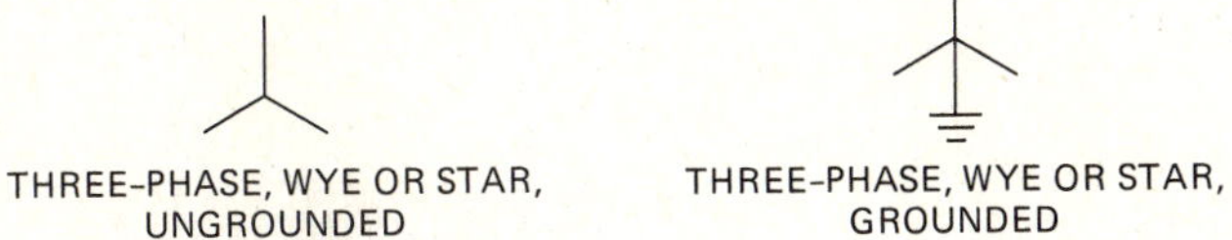

THREE-PHASE, WYE OR STAR, UNGROUNDED

THREE-PHASE, WYE OR STAR, GROUNDED

FUNDAMENTAL ITEMS

RESISTOR

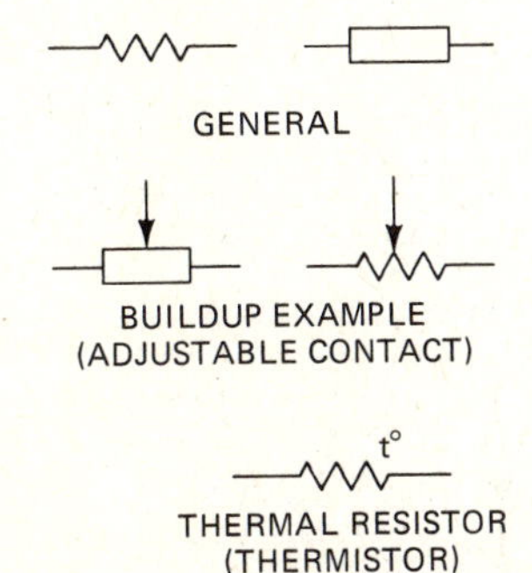

GENERAL

BUILDUP EXAMPLE (ADJUSTABLE CONTACT)

THERMAL RESISTOR (THERMISTOR)

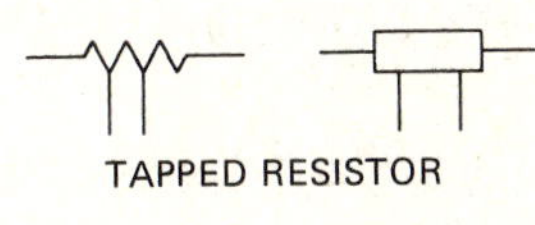

TAPPED RESISTOR

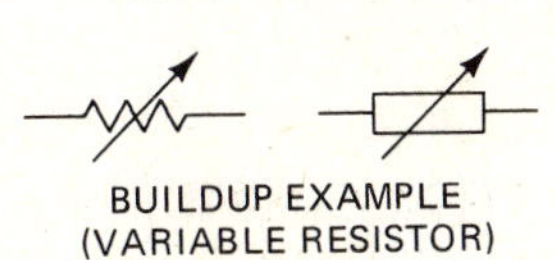

BUILDUP EXAMPLE (VARIABLE RESISTOR)

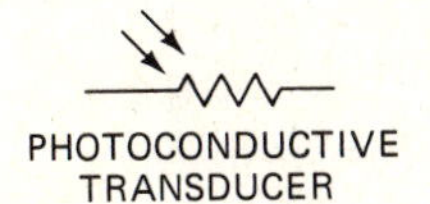

PHOTOCONDUCTIVE TRANSDUCER

CAPACITOR

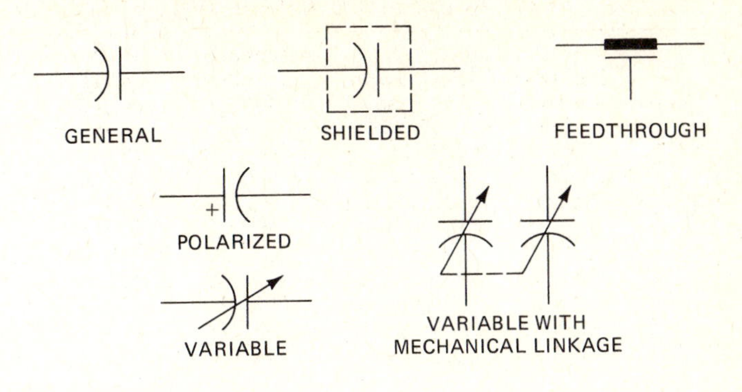

GENERAL

SHIELDED

FEEDTHROUGH

POLARIZED

VARIABLE

VARIABLE WITH MECHANICAL LINKAGE

ANTENNA

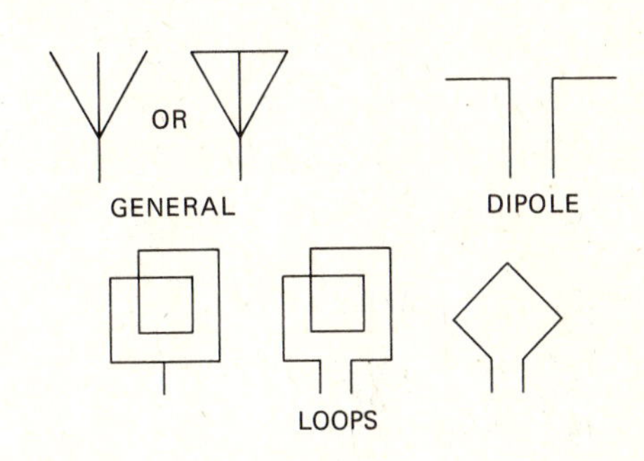

GENERAL

DIPOLE

LOOPS

BATTERY

ONE-CELL

MULTICELL

ALTERNATING-CURRENT SOURCE

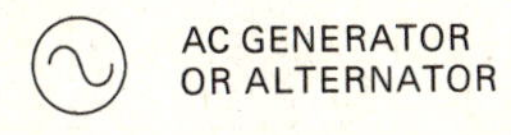

AC GENERATOR OR ALTERNATOR

PERMANENT MAGNET

PICKUP HEAD

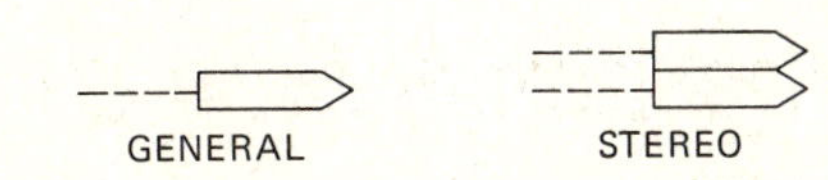

GENERAL

STEREO

PIEZOELECTRIC CRYSTAL UNIT

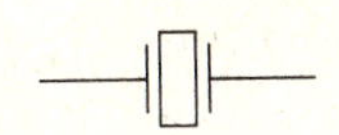

THERMOCOUPLES

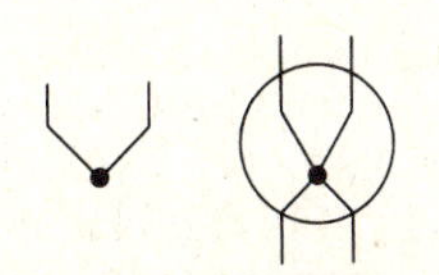

TEMPERATURE-MEASURING THERMOCOUPLE WITH INTEGRAL HEATER

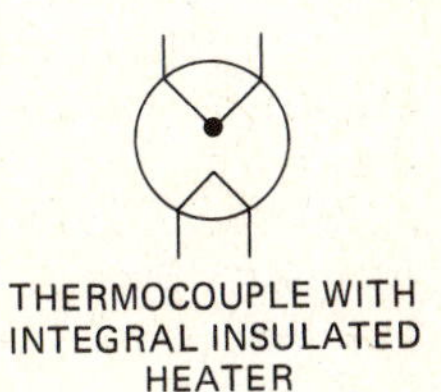

THERMOCOUPLE WITH INTEGRAL INSULATED HEATER

THERMOMECHANICAL TRANSDUCERS

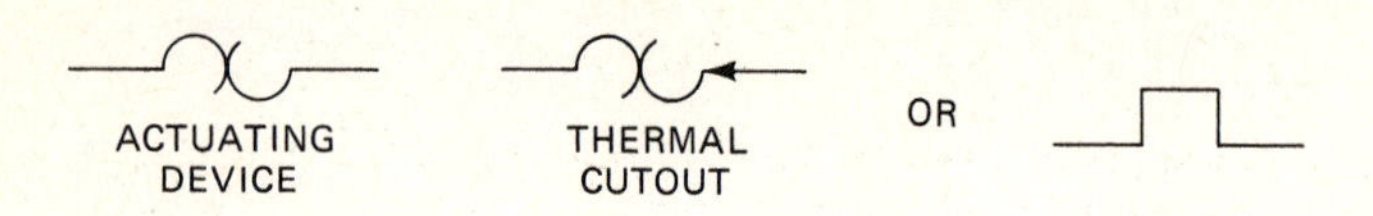

IGNITOR PLUG

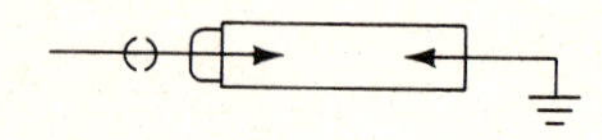

TRANSMISSION PATH

CONDUCTOR, CABLE, WIRING

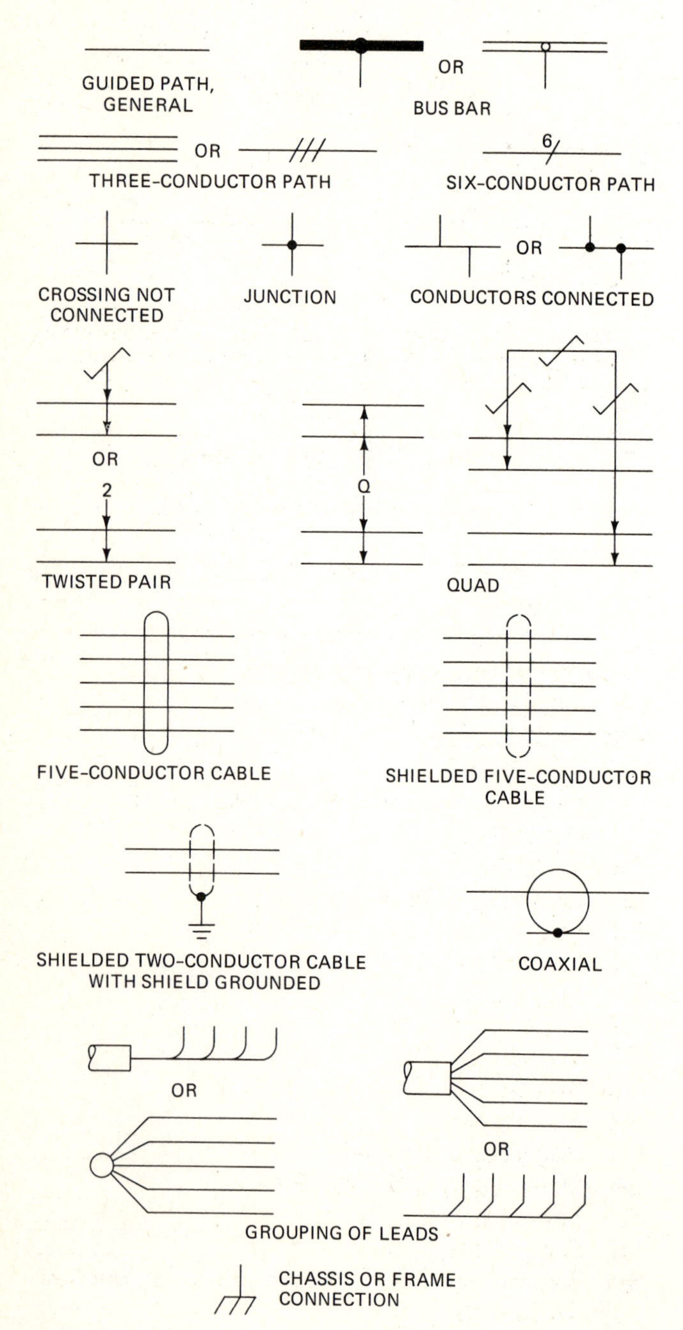

CHASSIS OR FRAME CONNECTION

WAVEGUIDES

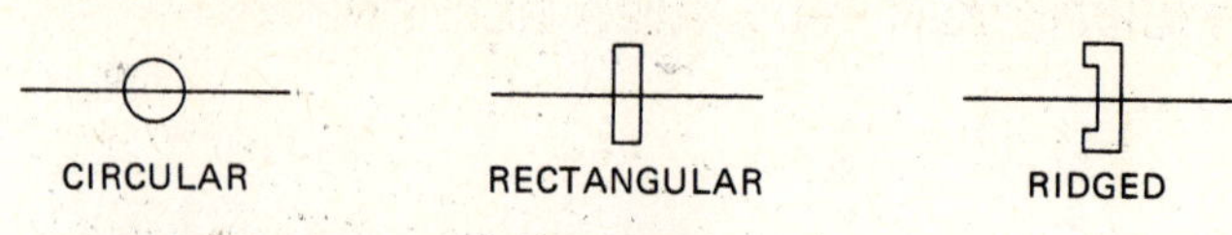

CONTACTS, SWITCHES, CONTACTORS AND RELAYS

SWITCHING FUNCTION

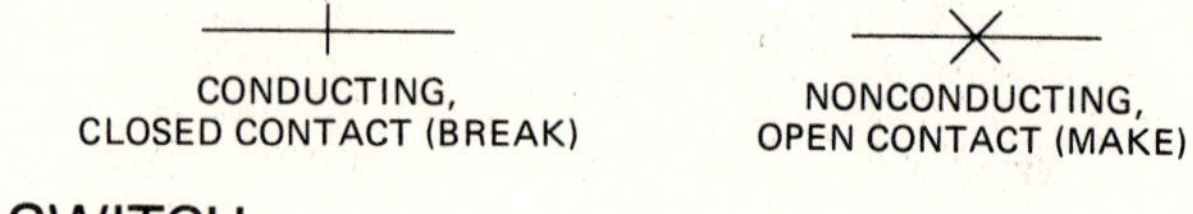

SWITCH

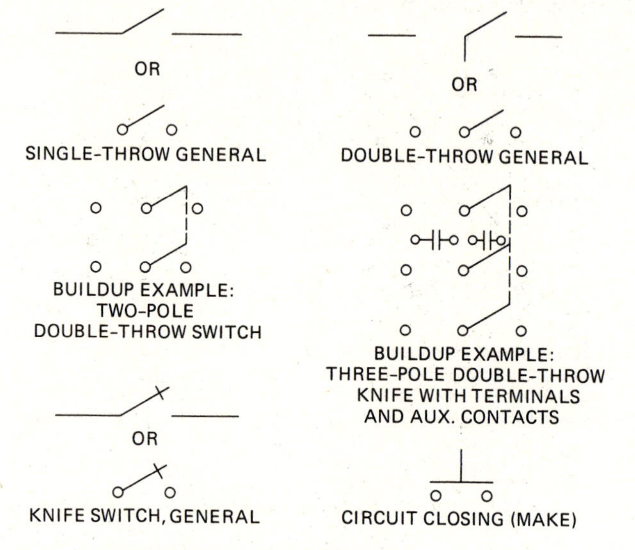

MULTIWAY TRANSFER SWITCH

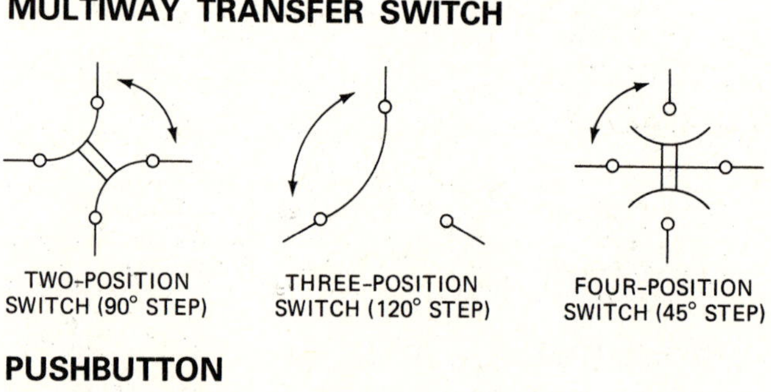

PUSHBUTTON

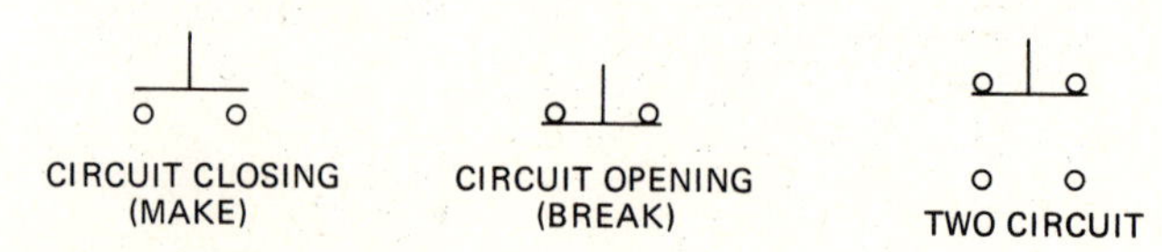

LOCKING SWITCH

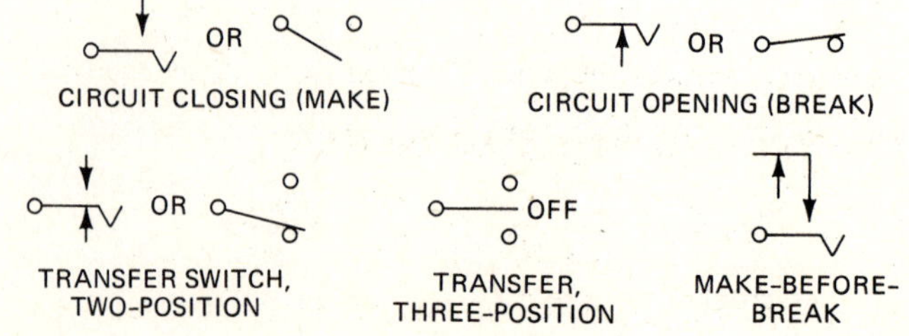

NONLOCKING SWITCH, MOMENTARY OR SPRING RETURN

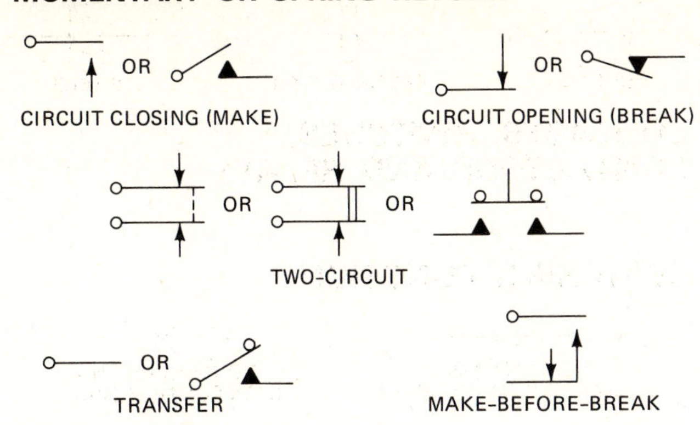

SELECTOR OR MULTIPOSITION SWITCH

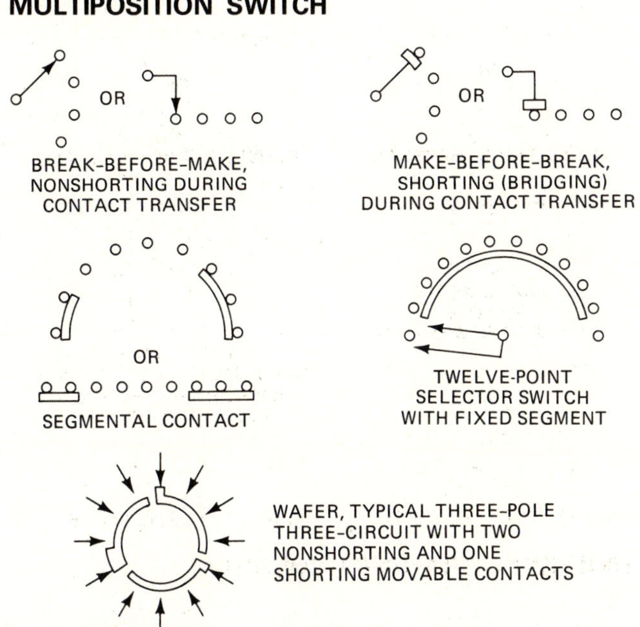

LIMIT SWITCH

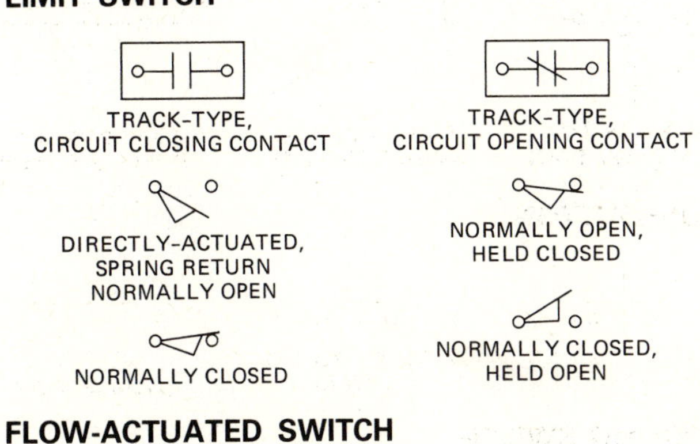

FLOW-ACTUATED SWITCH

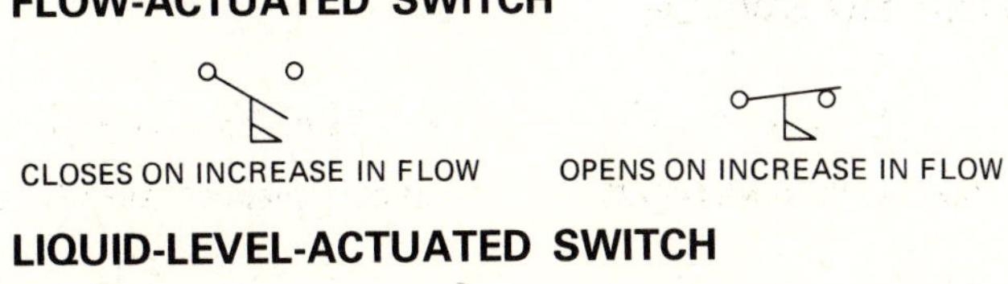

LIQUID-LEVEL-ACTUATED SWITCH

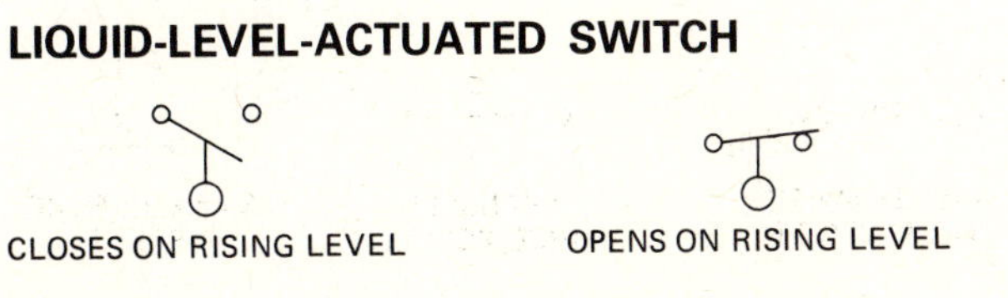

PRESSURE OR VACUUM-ACTUATED SWITCH

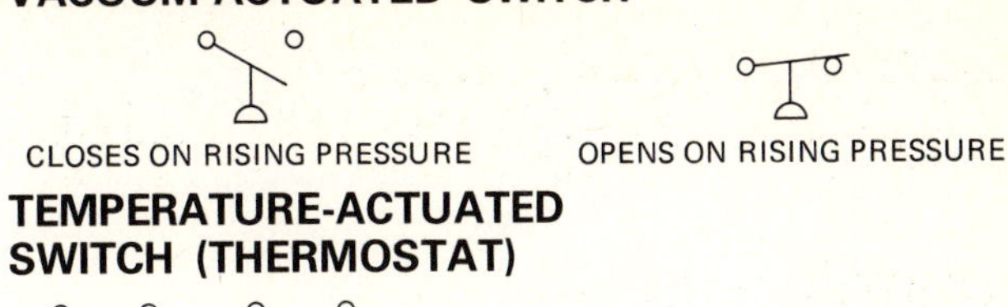

TEMPERATURE-ACTUATED SWITCH (THERMOSTAT)

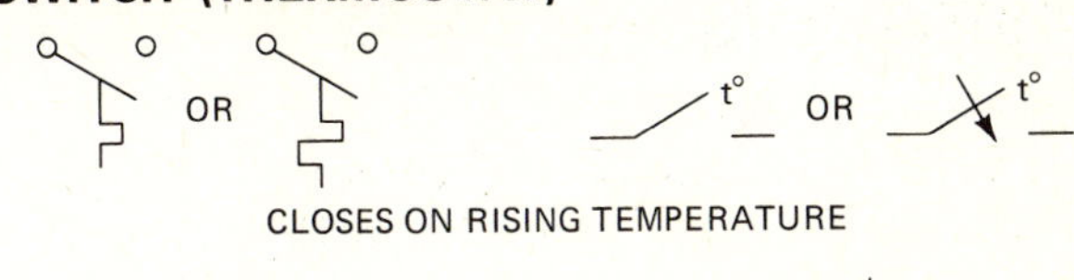

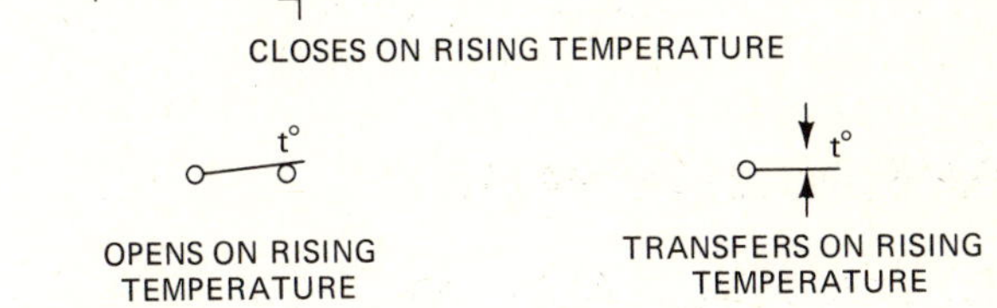

FLASHER

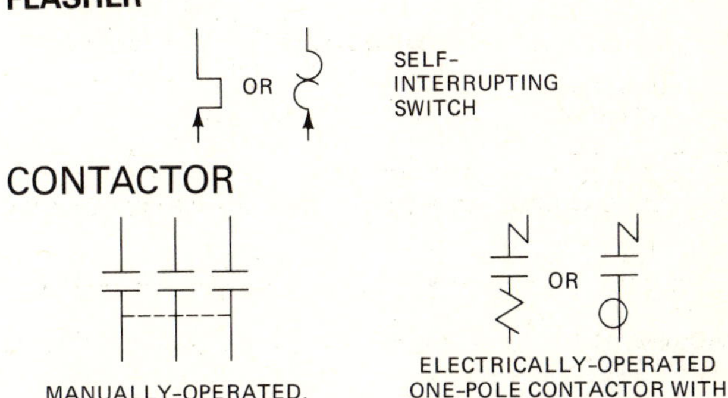

RELAYS

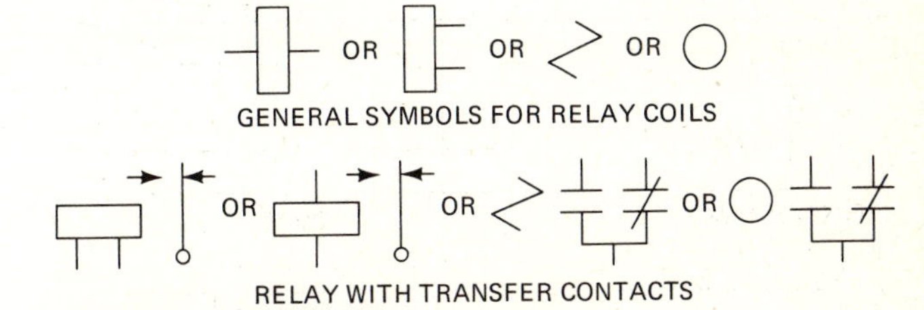

LETTER COMBINATIONS THAT MAY BE USED WITH RELAY SYMBOLS

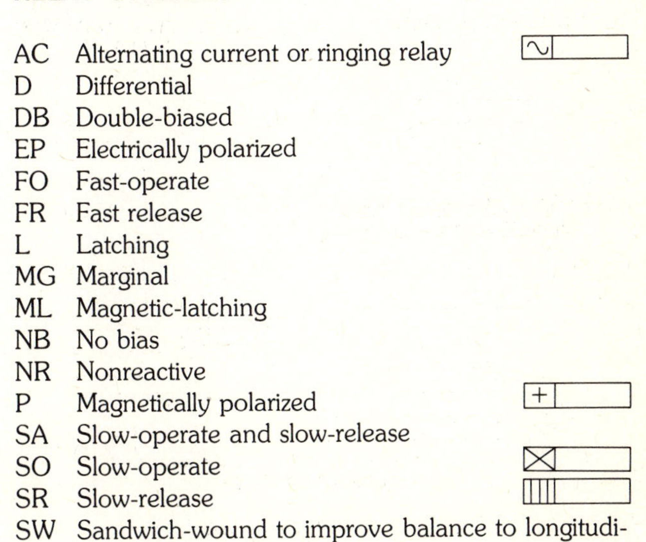

AC Alternating current or ringing relay
D Differential
DB Double-biased
EP Electrically polarized
FO Fast-operate
FR Fast release
L Latching
MG Marginal
ML Magnetic-latching
NB No bias
NR Nonreactive
P Magnetically polarized
SA Slow-operate and slow-release
SO Slow-operate
SR Slow-release
SW Sandwich-wound to improve balance to longitudinal currents

TERMINALS AND CONNECTORS

TERMINALS

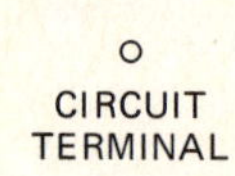

CIRCUIT TERMINAL

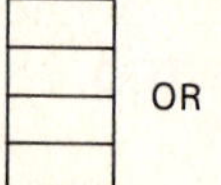

OR

TERMINAL STRIP WITH FOUR TERMINALS

TERMINALS FOR ELECTRON TUBES

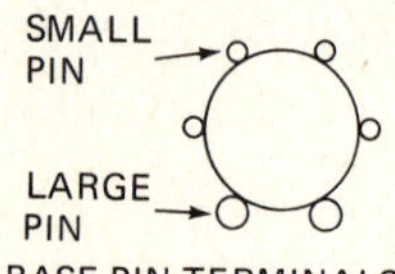

BASE PIN TERMINALS

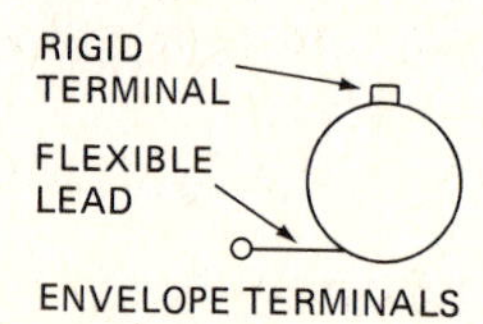

ENVELOPE TERMINALS

DEVICE WITH BASE-ORIENTATION KEY

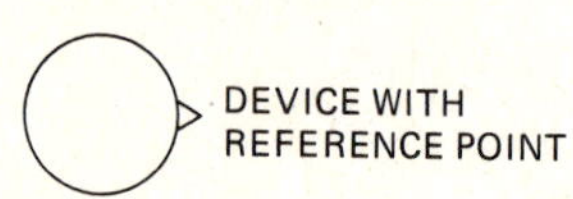

DEVICE WITH REFERENCE POINT

CABLE TERMINATION

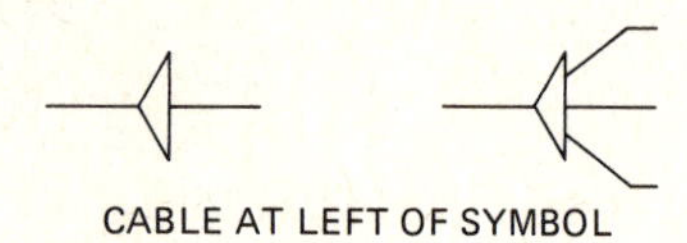

CABLE AT LEFT OF SYMBOL

CONNECTORS

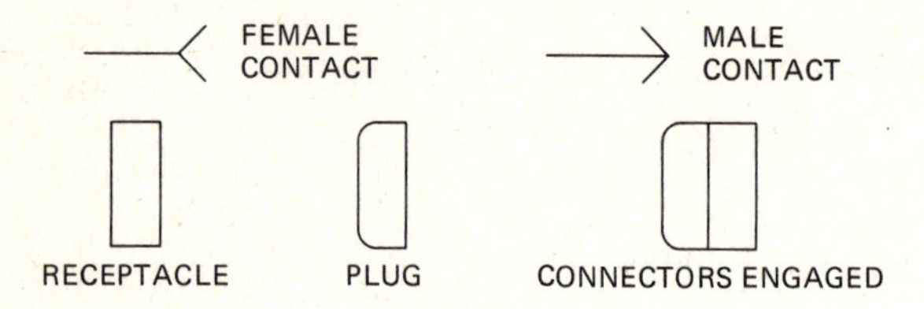

RECEPTACLE PLUG CONNECTORS ENGAGED

(Type of contacts in connectors are indicated as male or female.)

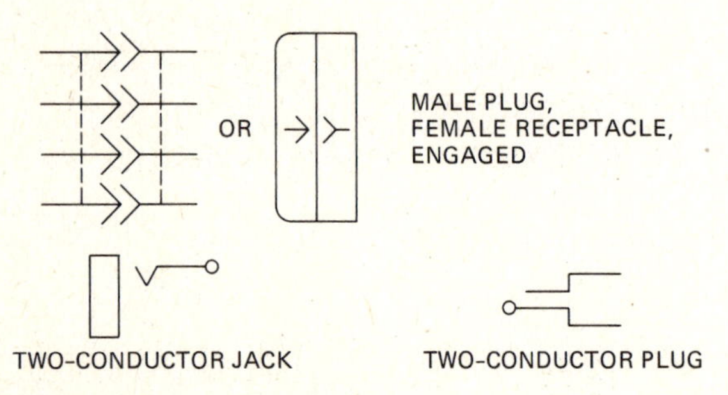

MALE PLUG, FEMALE RECEPTACLE, ENGAGED

TWO-CONDUCTOR JACK

TWO-CONDUCTOR PLUG

POWER SUPPLY CONNECTORS

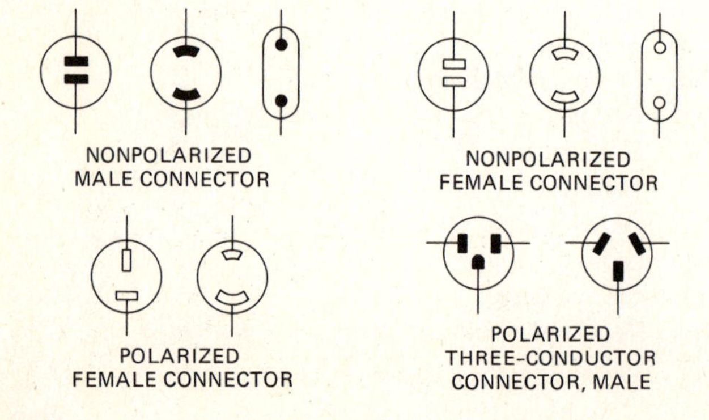

NONPOLARIZED MALE CONNECTOR

NONPOLARIZED FEMALE CONNECTOR

POLARIZED FEMALE CONNECTOR

POLARIZED THREE-CONDUCTOR CONNECTOR, MALE

COAXIAL CONNECTOR

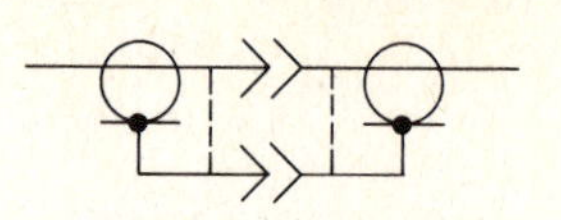

COAXIAL WITH OUTSIDE CONDUCTOR CARRIED THROUGH

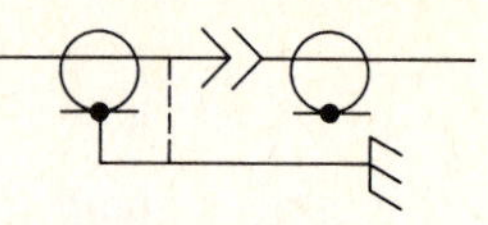

COAXIAL WITH OUTSIDE CONDUCTOR TERMINATED ON CHASSIS

TRANSFORMERS, INDUCTORS AND WINDINGS

CORE SYMBOLS

No symbol is used for an air core

MAGNETIC CORE OF INDUCTOR OR TRANSFORMER

CORE OF MAGNET

INDUCTOR

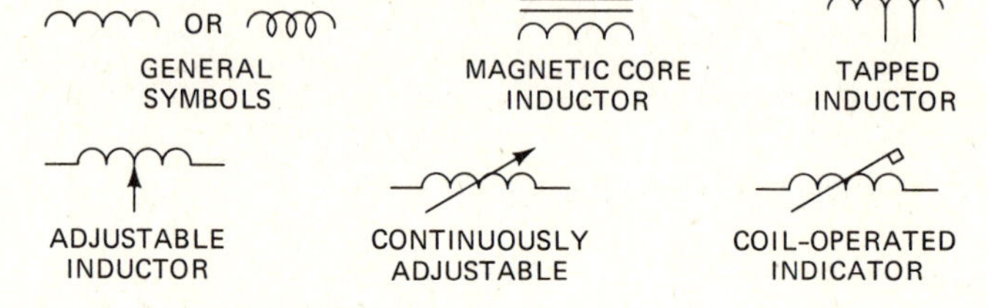

GENERAL SYMBOLS

MAGNETIC CORE INDUCTOR

TAPPED INDUCTOR

ADJUSTABLE INDUCTOR

CONTINUOUSLY ADJUSTABLE

COIL-OPERATED INDICATOR

TRANSDUCTOR

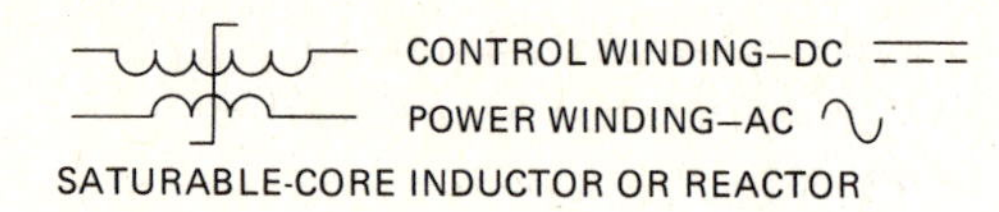

SATURABLE-CORE INDUCTOR OR REACTOR

TRANSFORMER

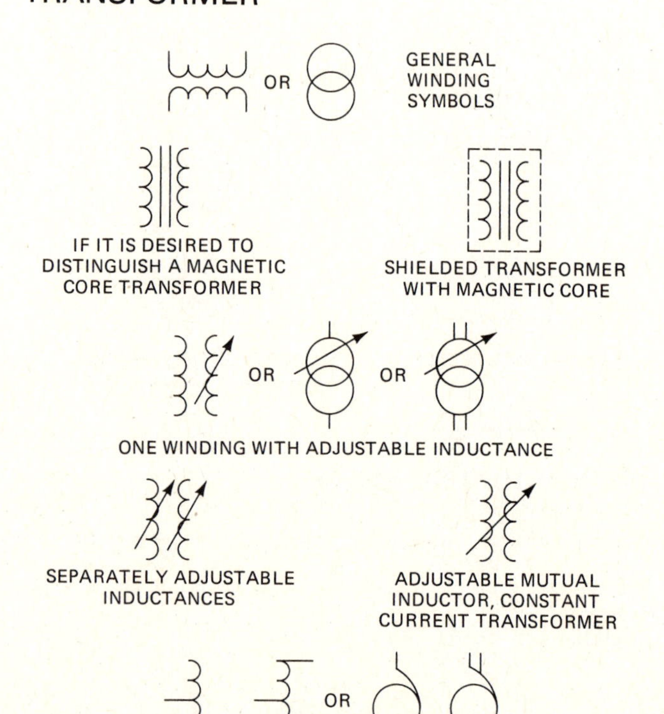

GENERAL WINDING SYMBOLS

IF IT IS DESIRED TO DISTINGUISH A MAGNETIC CORE TRANSFORMER

SHIELDED TRANSFORMER WITH MAGNETIC CORE

ONE WINDING WITH ADJUSTABLE INDUCTANCE

SEPARATELY ADJUSTABLE INDUCTANCES

ADJUSTABLE MUTUAL INDUCTOR, CONSTANT CURRENT TRANSFORMER

AUTOTRANSFORMER, ONE-PHASE

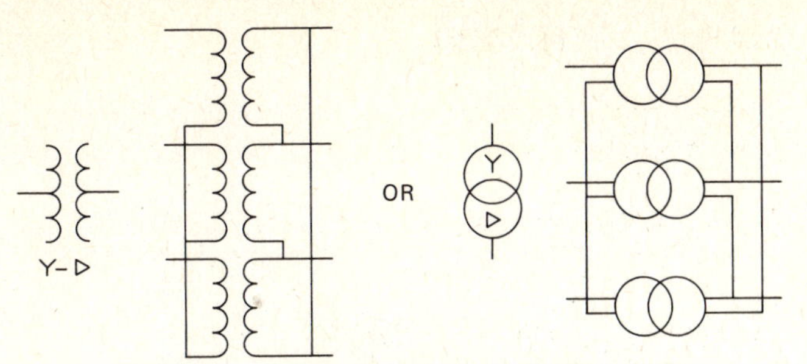

THREE-PHASE BANK OF ONE-PHASE, TWO-WINDING TRANSFORMERS WITH WYE-DELTA CONNECTIONS

ELECTRON TUBES AND RELATED DEVICES

ELECTRON TUBE

EMITTING ELECTRODE

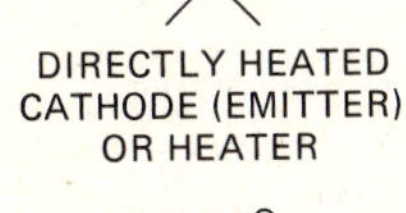

DIRECTLY HEATED CATHODE (EMITTER) OR HEATER

INDIRECTLY HEATED CATHODE

COLD CATHODE

PHOTOCATHODE

CONTROLLING ELECTRODE

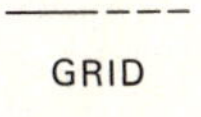

GRID

DEFLECTING ELECTRODES (USED IN PAIRS)

IGNITOR (FOR POOL TUBES)

EXCITOR (CONTACTOR TYPE)

COLLECTING ELECTRODE

ANODE OR PLATE

TARGET OR X-RAY ANODE

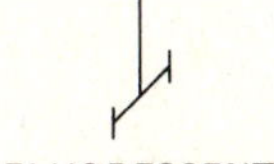

FLUORESCENT TARGET

COLLECTOR

TUBE SYMBOLS

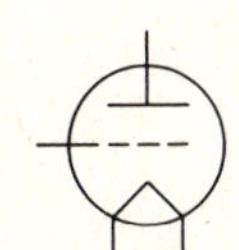

TRIODE WITH DIRECTLY HEATED CATHODE

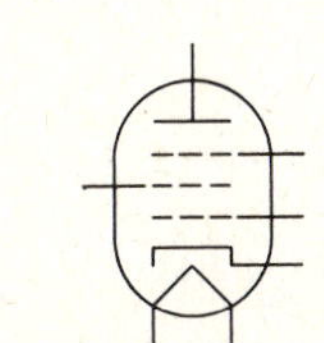

EQUIPOTENTIAL-PENTODE INDIRECTLY HEATED CATHODE

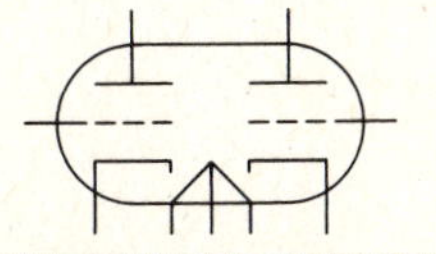

EQUIPOTENTIAL-CATHODE TWIN TRIODE

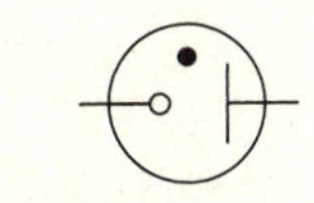

COLD-CATHODE, GAS-FILLED RECTIFIER

X-RAY TUBE

WITH FILAMENTARY CATHODE AND FOCUSING GRID (CUP)

CATHODE-RAY TUBE

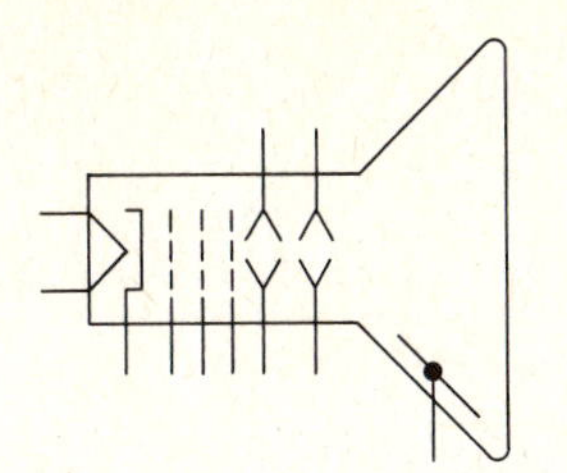

CRT WITH ELECTRIC FIELD DEFLECTION

SEMICONDUCTOR DEVICES

TRANSISTORS AND DIODES

ELEMENT SYMBOLS

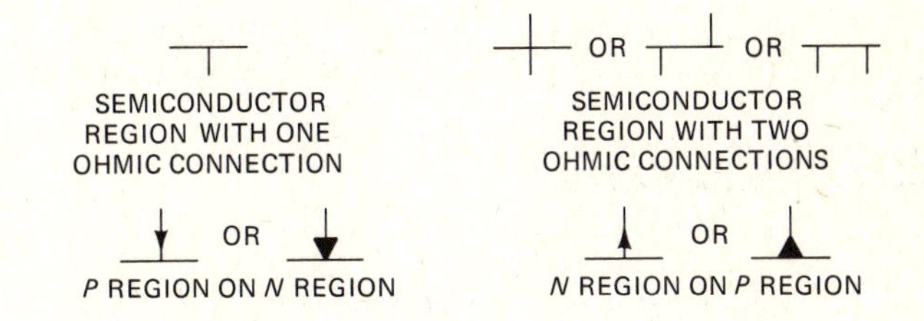

SEMICONDUCTOR REGION WITH ONE OHMIC CONNECTION

SEMICONDUCTOR REGION WITH TWO OHMIC CONNECTIONS

P REGION ON *N* REGION

N REGION ON *P* REGION

(Arrow points opposite electron flow)

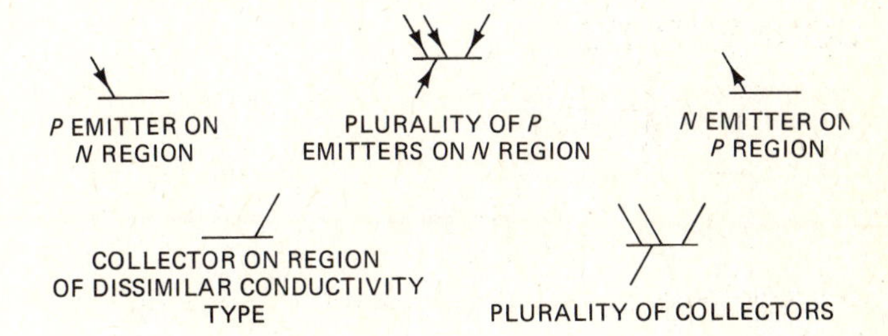

P EMITTER ON *N* REGION

PLURALITY OF *P* EMITTERS ON *N* REGION

N EMITTER ON *P* REGION

COLLECTOR ON REGION OF DISSIMILAR CONDUCTIVITY TYPE

PLURALITY OF COLLECTORS

TWO-TERMINAL DEVICES

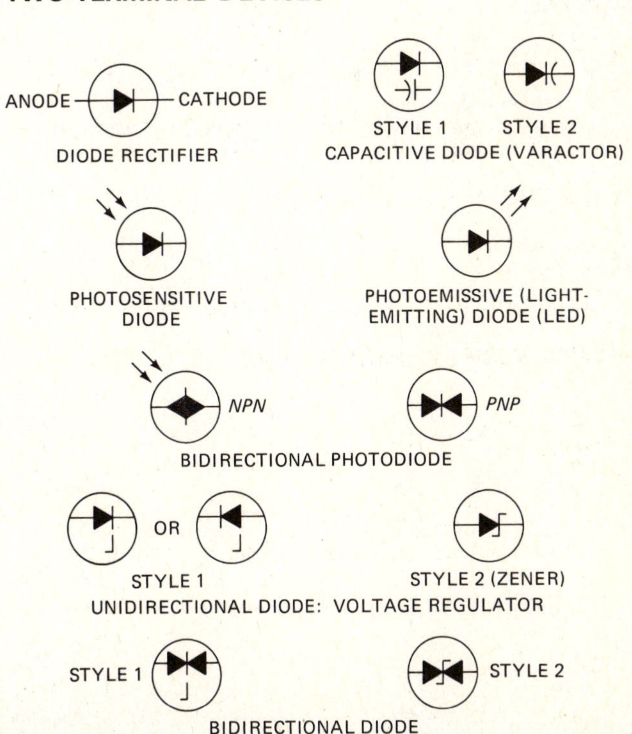

DIODE RECTIFIER

CAPACITIVE DIODE (VARACTOR)

PHOTOSENSITIVE DIODE

PHOTOEMISSIVE (LIGHT-EMITTING) DIODE (LED)

BIDIRECTIONAL PHOTODIODE

UNIDIRECTIONAL DIODE: VOLTAGE REGULATOR

BIDIRECTIONAL DIODE

TWO-TERMINAL DEVICES (*Cont.*):

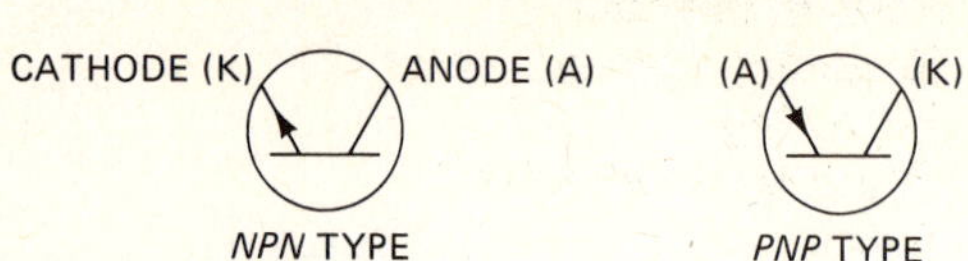

NPN TYPE *PNP* TYPE

UNIDIRECTIONAL NEGATIVE-RESISTANCE BREAKDOWN DIODE; TRIGGER DIAC

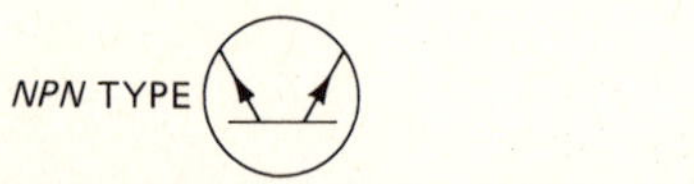
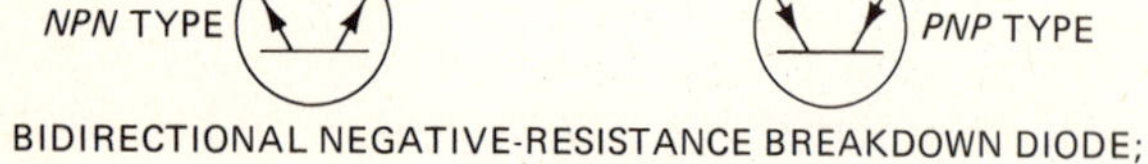

BIDIRECTIONAL NEGATIVE-RESISTANCE BREAKDOWN DIODE; TRIGGER DIAC

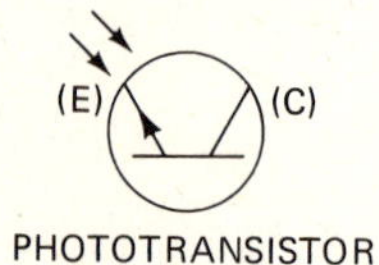

PHOTOTRANSISTOR

CURRENT REGULATOR

THREE-OR-MORE-TERMINAL DEVICES

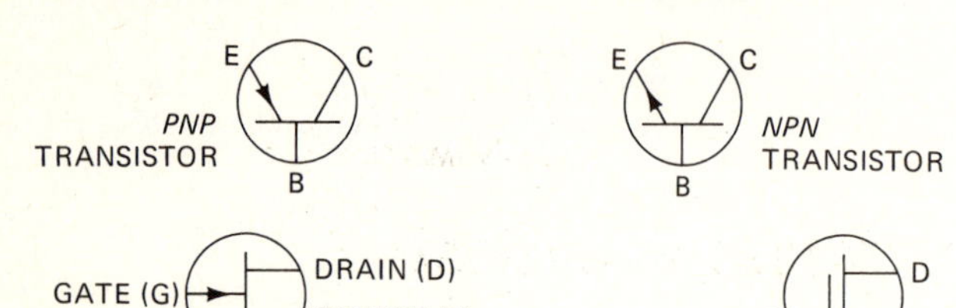

PNP TRANSISTOR *NPN* TRANSISTOR

FIELD EFFECT TRANSISTOR (FET) WITH *N*-CHANNEL JUNCTION GATE

FET WITH INSULATED GATE

A K G

THYRISTOR, STYLE 1

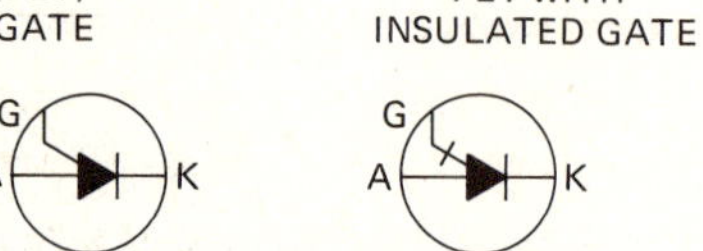

THYRISTOR, STYLE 2

GATE TURN-OFF TYPE

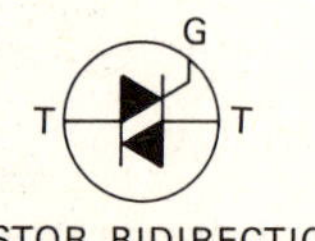

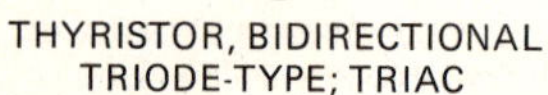

THYRISTOR, BIDIRECTIONAL TRIODE-TYPE; TRIAC

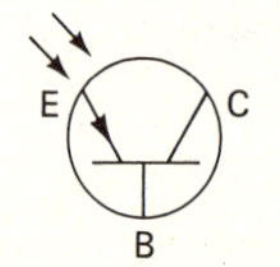

PHOTOTRANSISTOR

CIRCUIT PROTECTORS

FUSES

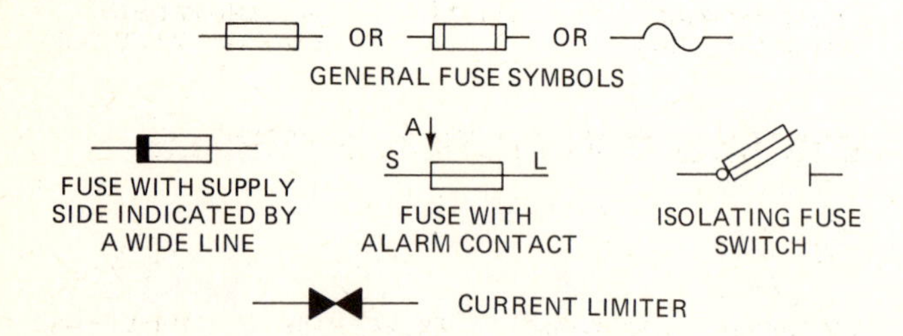

GENERAL FUSE SYMBOLS

FUSE WITH SUPPLY SIDE INDICATED BY A WIDE LINE

FUSE WITH ALARM CONTACT

ISOLATING FUSE SWITCH

CURRENT LIMITER

LIGHTNING ARRESTOR

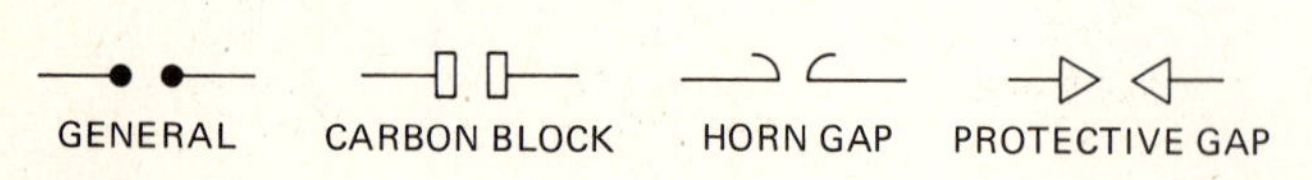

GENERAL CARBON BLOCK HORN GAP PROTECTIVE GAP

CIRCUIT BREAKER

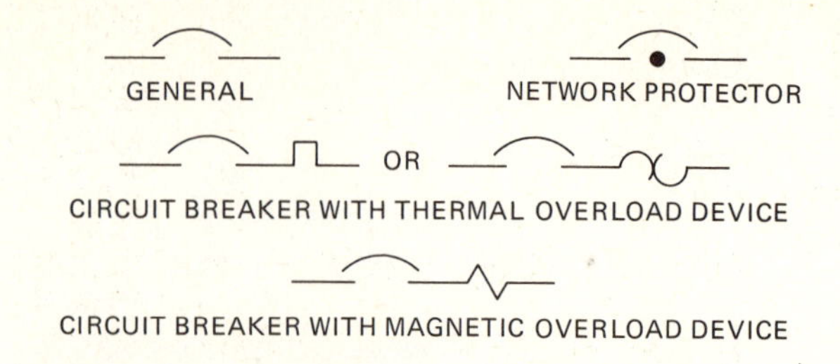

GENERAL NETWORK PROTECTOR

CIRCUIT BREAKER WITH THERMAL OVERLOAD DEVICE

CIRCUIT BREAKER WITH MAGNETIC OVERLOAD DEVICE

ACOUSTIC DEVICES

AUDIBLE SIGNALING DEVICE

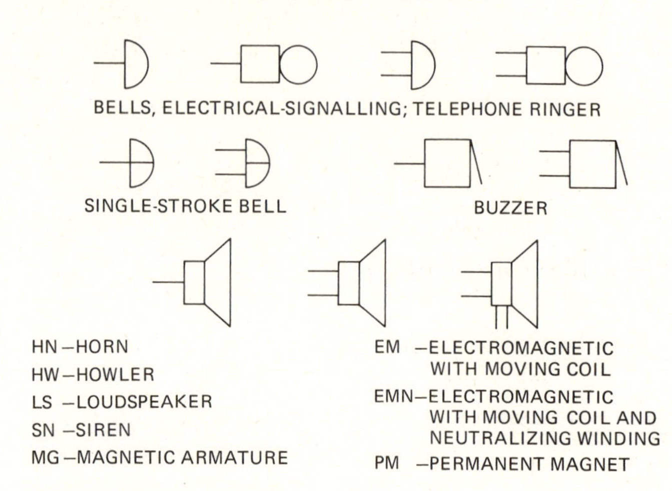

BELLS, ELECTRICAL-SIGNALLING; TELEPHONE RINGER

SINGLE-STROKE BELL BUZZER

HN—HORN
HW—HOWLER
LS —LOUDSPEAKER
SN —SIREN
MG—MAGNETIC ARMATURE

EM —ELECTROMAGNETIC WITH MOVING COIL
EMN—ELECTROMAGNETIC WITH MOVING COIL AND NEUTRALIZING WINDING
PM —PERMANENT MAGNET

MICROPHONE, TELEPHONE TRANSMITTER

MICROPHONE, GENERAL

HANDSET

GENERAL WITH PUSH-TO-TALK SWITCH

TELEPHONE RECEIVER

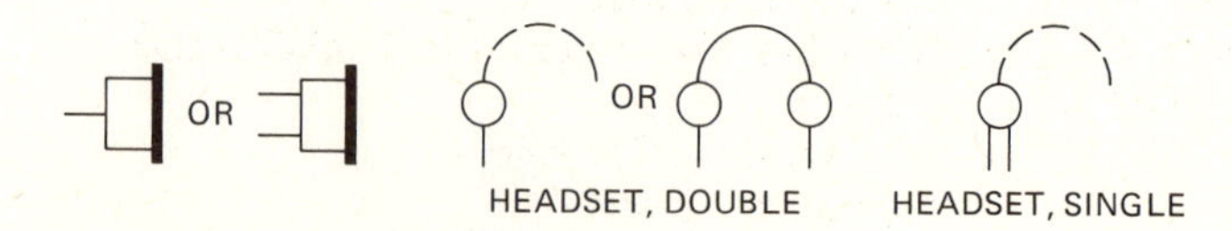

HEADSET, DOUBLE HEADSET, SINGLE

LAMPS AND VISUAL-SIGNALING DEVICES

LAMP

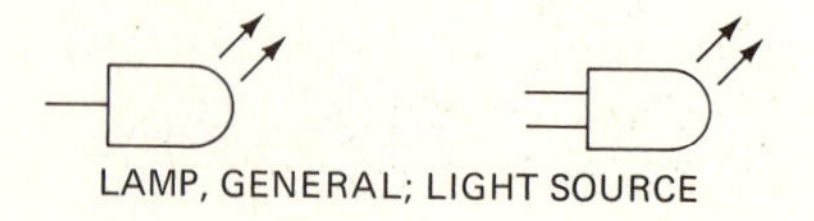

LAMP, GENERAL; LIGHT SOURCE

A	AMBER
B	BLUE
C	CLEAR
G	GREEN
O	ORANGE
ARC	ARC
EL	ELECTROLUMINESCENT
FL	FLUORESCENT
HG	MERCURY VAPOR
IN	INCANDESCENT
OP	OPALESCENT
P	PURPLE
R	RED
W	WHITE
Y	YELLOW
IR	INFRARED
NA	SODIUM VAPOR
NE	NEON
UV	ULTRAVIOLET
XE	XENON
LED	LIGHT-EMITTING DIODE

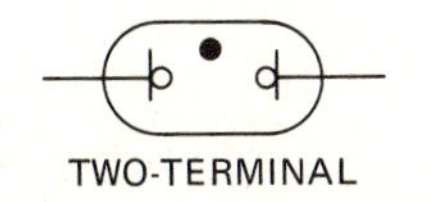
TWO-TERMINAL

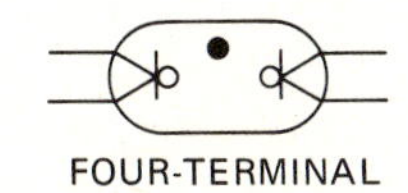
FOUR-TERMINAL

FLUORESCENT LAMPS

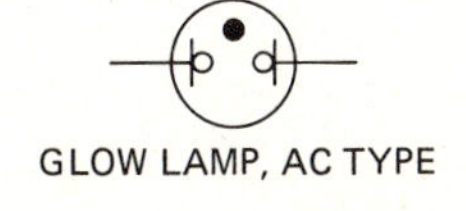
GLOW LAMP, AC TYPE

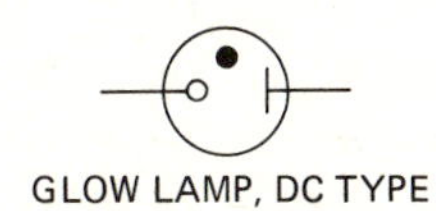
GLOW LAMP, DC TYPE

COMMUNICATION-SWITCHBOARD LAMP

READOUT DEVICES

METER INSTRUMENT

GENERAL

A	Ammeter
AH	Ampere-hour meter
C	Coulombmeter
CMA	Contact-making (or breaking) ammeter
CMC	Contact-making (or breaking) clock
CMV	Contact-making (or breaking) voltmeter
CRO	Cathode-ray oscilloscope
DB	DB (Decibel) meter
DBM	DBM (Decibels referred to 1 mW) meter
DM	Demand meter
DTR	Demand-totalizing relay
F	Frequency meter
GD	Ground detector
I	Indicating meter
INT	Integrating meter
uA or UA	Microammeter
MA	Milliammeter
NM	Noise meter
OHM	Ohmmeter
OP	Oil pressure
OSCG	Oscillograph, string
PF	Power-factor meter

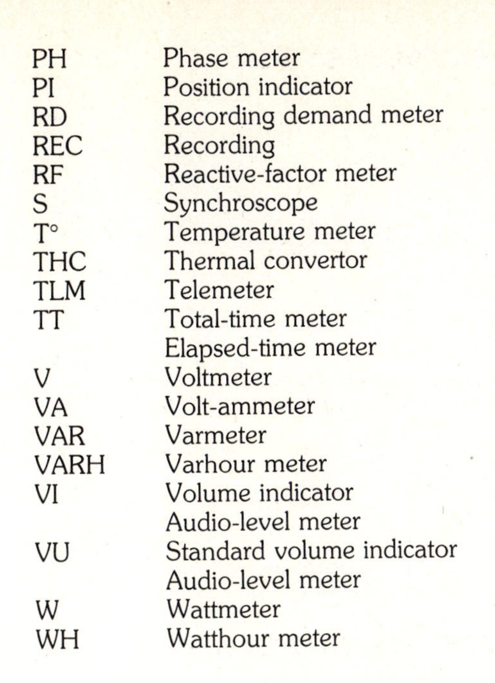

PH	Phase meter
PI	Position indicator
RD	Recording demand meter
REC	Recording
RF	Reactive-factor meter
S	Synchroscope
T°	Temperature meter
THC	Thermal convertor
TLM	Telemeter
TT	Total-time meter Elapsed-time meter
V	Voltmeter
VA	Volt-ammeter
VAR	Varmeter
VARH	Varhour meter
VI	Volume indicator Audio-level meter
VU	Standard volume indicator Audio-level meter
W	Wattmeter
WH	Watthour meter

OR G

GALVANOMETER

ROTATING MACHINERY

ROTATING MACHINE

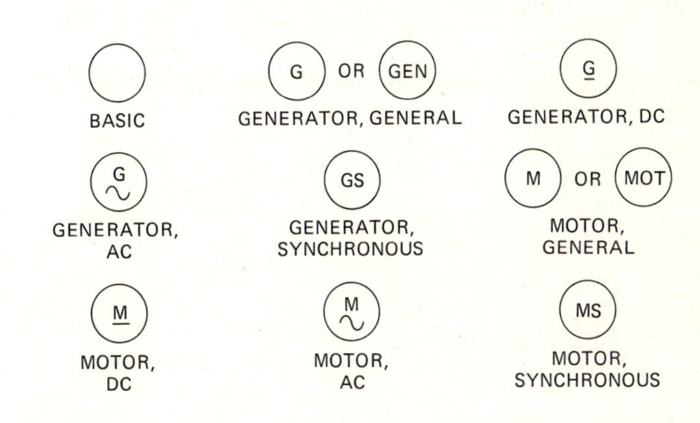

BASIC
GENERATOR, GENERAL
GENERATOR, DC
GENERATOR, AC
GENERATOR, SYNCHRONOUS
MOTOR, GENERAL
MOTOR, DC
MOTOR, AC
MOTOR, SYNCHRONOUS

FIELD, GENERATOR OR MOTOR

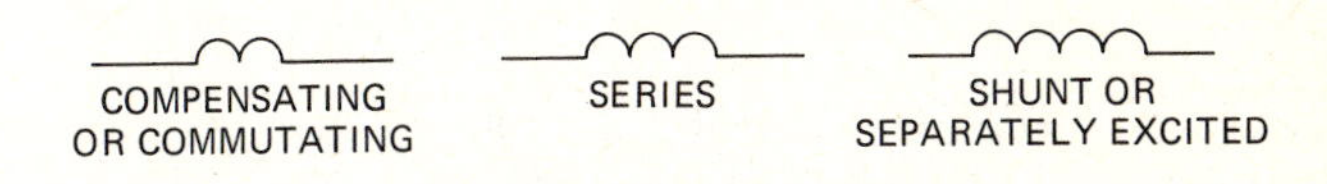
COMPENSATING OR COMMUTATING
SERIES
SHUNT OR SEPARATELY EXCITED

WINDING CONNECTION SYMBOLS

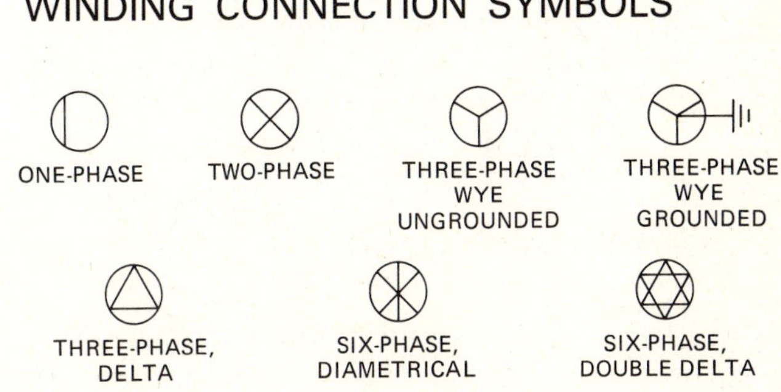
ONE-PHASE
TWO-PHASE
THREE-PHASE WYE UNGROUNDED
THREE-PHASE WYE GROUNDED
THREE-PHASE, DELTA
SIX-PHASE, DIAMETRICAL
SIX-PHASE, DOUBLE DELTA

DIRECT-CURRENT MACHINES

SEPARATELY EXCITED DC GENERATOR OR MOTOR WITH COMMUTATING FIELD WINDING

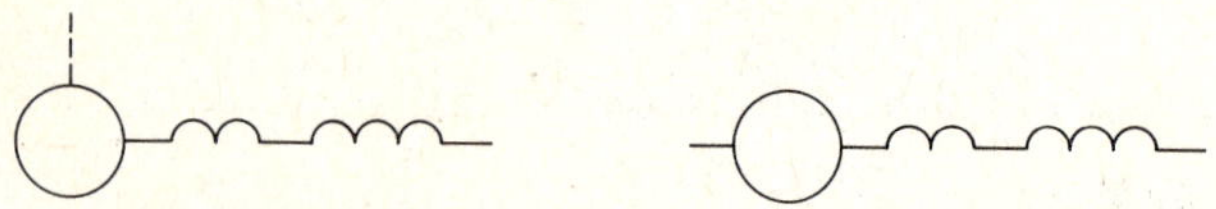

SERIES MOTOR OR TWO-WIRE DC GENERATOR WITH COMMUTATING FIELD WINDING OR BOTH

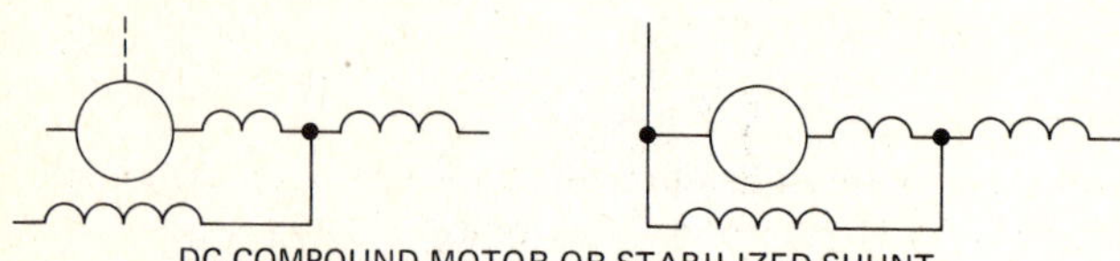

DC COMPOUND MOTOR OR STABILIZED SHUNT MOTOR WITH COMMUTATING FIELD WINDING

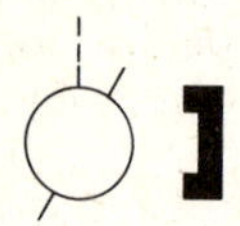
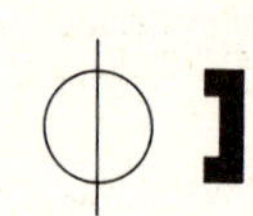

DC, PERMANENT-MAGNET FIELD GENERATOR OR MOTOR

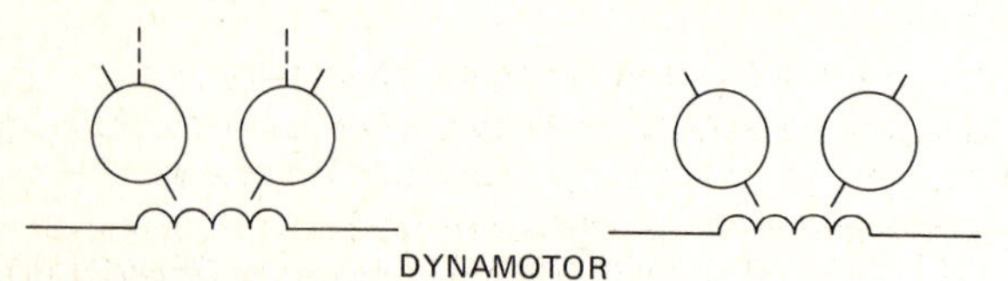

DYNAMOTOR

ALTERNATING-CURRENT MACHINES

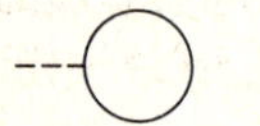
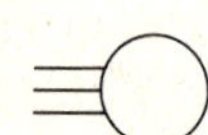

SQUIRREL-CAGE INDUCTION MOTOR OR GENERATOR, SYNCHRONOUS MOTOR OR GENERATOR, SPLIT-PHASE INDUCTION MOTOR OR GENERATOR, ROTARY-PHASE CONVERTOR, OR REPULSION MOTOR

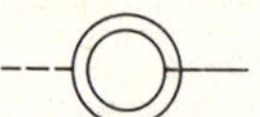
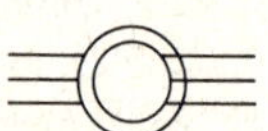

WOUND-ROTOR INDUCTION MOTOR, SYNCHRONOUS INDUCTION MOTOR, INDUCTION GENERATOR, OR INDUCTION FREQUENCY CONVERTOR

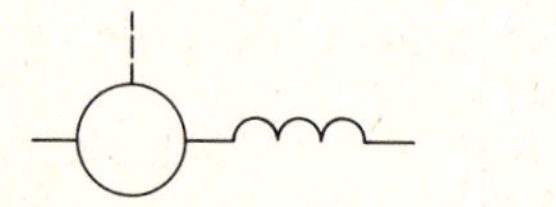
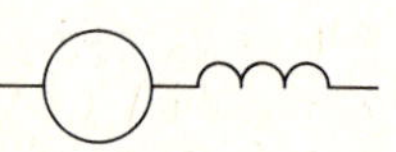

AC SERIES MOTOR

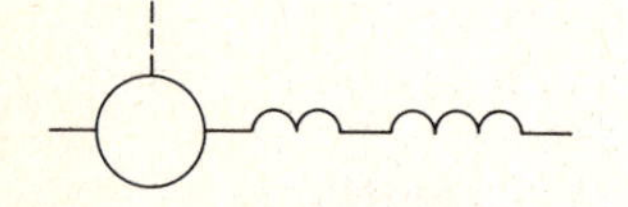
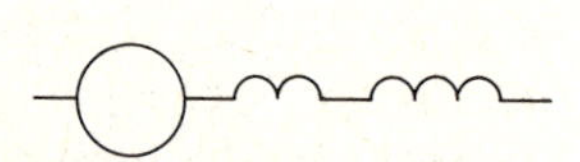

AC SERIES MOTOR WITH COMMUTATING OR COMPENSATING FIELD WINDING OR BOTH

ONE-PHASE, SHADED-POLE MOTOR

SYNCHRONOUS MOTOR OR GENERATOR WITH DC FIELD EXCITATION

SYMBOLS USED IN LOGIC CIRCUITS AND DIAGRAMS

(Not included in ANSI Y32.2)

AMPLIFIERS

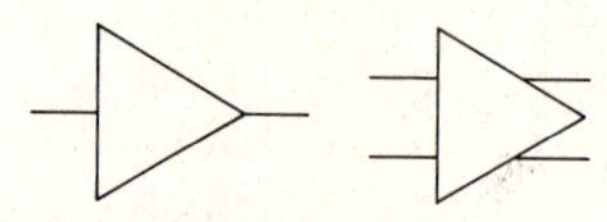

GENERAL SYMBOLS

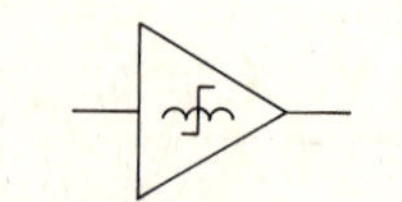

MAGNETIC AMPLIFIER

LOGIC GATES

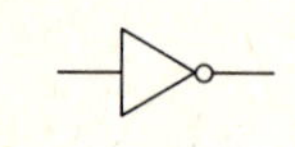

INVERTOR OR **NOT** GATE

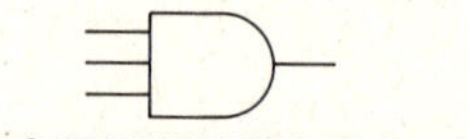

AND GATE WITH THREE INPUTS

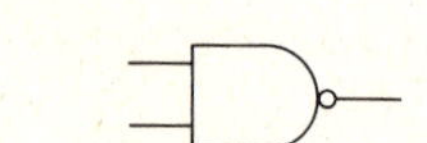

NAND (NOT AND) GATE

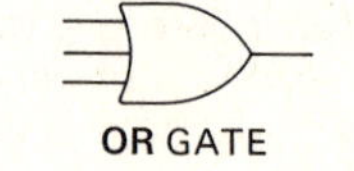

OR GATE

NOR (NOT OR) GATE

GLOSSARY

accelerate: To change velocity, that is, increase or decrease speed.

accelerometer: A device for sensing or measuring acceleration and converting it to an electric signal.

acceptor: An *impurity* atom in a semiconductor material which will receive or *accept* electrons. Germanium with an acceptor impurity is called *p*-type germanium because it has a positive nature.

actuator: A hydraulic, electric, or pneumatic device used to operate a mechanism by remote control on a plane or missile.

alignment, electric: The tuning of electronic components in a particular circuit so that all portions of the circuit will respond to the correct frequency.

alternating current (ac as adjective): An electric current which periodically changes in direction and constantly changes in magnitude.

alternation: The part of an ac cycle during which current is flowing in one direction; one-half a cycle.

alternator: An electric generator designed to produce alternating current.

ammeter: An instrument used to measure current flow.

ampere (A): The basic unit of current flow. One ampere is the amount of current which flows when an emf of 1 V is applied to a circuit with a resistance of 1 Ω. One coulomb per second.

ampere-hour (Ah): The quantity of electricity which has passed through a circuit when a current of 1 A has flowed for 1 h. Current (in amperes) × time (in hours) = ampere-hours. One Ah is equal to 3600 C.

ampere-turn: The magnetizing force produced by a current of 1 A flowing through one turn of a coil. Ampere-turns = amperes × number of turns of wire in the coil.

amplication: The increase of power, current, or voltage in an electronic circuit.

amplication factor: The ratio of a small change in plate voltage to a small change in grid voltage when the plate circuit is operating through a load. It is denoted by the Greek letter mu (μ). Amplification factor for a transistor is the ratio of change in emitter current to a change in base current.

amplifier: An electronic circuit designed to produce amplification.

amplitude modulation (AM): Modulation of a carrier wave in which the modulating signal changes the amplitude of the carrier in proportion to the strength of the modulating signal.

angular velocity: Time rate of change of an angle rotated around an axis in degrees per second or degrees per minute.

anode: Positive electrode of a battery; the electrode of an electron tube, diode, or electroplating cell to which a positive voltage is applied.

antenna: A device designed to radiate or intercept electromagnetic waves.

armature: In a dc generator or motor, the rotating member. In an ac generator the armature is stationary and is acted upon by the rotating field produced by the rotor. The moving element acted upon by the magnetic field in a relay is also called the armature.

armature reaction: The interaction of the armature field upon the main field of a generator or motor, resulting in distortion of the main field.

atom: The smallest possible particle of an element.

attenuation: A reduction in the strength of a signal, the flow of current, flux, or other energy in an electronic system.

audio frequency (AF as adjective): A frequency in the audible range, generally considered to be from about 35 to 20 000 Hz.

automatic direction finder (ADF): A radio receiver utilizing a directional loop antenna which enables the receiver to indicate the direction from which a radio signal is being received; also called radio compass.

automatic flight-control system (AFCS): A flight-control system incorporating an automatic pilot with additional systems such as a VOR coupler, an ILS approach coupler, and an internal navigation system which is fully automatic, so the aircraft can be flown in a completely automatic mode.

automatic frequency control (AFC): A circuit arrangement which maintains the frequency of the system within specified limits.

automatic pilot: A system installed in an airplane or missile which senses deviations in the flight path and moves the control surfaces to maintain the selected flight path.

automatic volume control (AVC): A circuit arrangement in which the dc component of the detector output in a radio receiver controls the bias of the rf tubes, thus regulating their output to maintain a reasonably constant volume.

Autosyn: A trade name of the Bendix Corporation used to designate certain types of synchro devices.

azimuth: Angular distance measured on a horizontal circle in a clockwise direction from either north or south.

ballast: A circuit element designed to stabilize current flow.

band: A range of frequencies.

bandpass filter: A filter circuit designed to pass frequencies within a specific band and attenuate frequencies outside the band.

bandwidth: The difference between the maximum and minimum frequencies in a band.

base: The terminal of a transistor to which the controlling current is applied.

battery: A group of voltaic cells connected together to produce a desired voltage and current capacity. Typical batteries utilize primary cells, secondary cells, and photovoltaic cells.

beam power tube: An electron tube which utilizes directed electron beams to add to its power-handling capability.

beat frequency oscillator (BFO): An oscillator designed to produce a signal frequency which is mixed with another frequency in order to develop an intermediate frequency or an audio frequency.

bel: A unit used to express the ratio of two values of power. The number of bels is the logarithm to the base 10 of the power ratio.

bias: A voltage applied to the control grid of an electron tube or the control element of a transistor to establish the correct operating point.

binary system: A numbering system using only two symbols 0 and 1 and having 2 as a base. In the decimal system, 10 symbols are used and the base is 10.

bit: One unit of a binary number.

bleeder resistance: A permanently connected resistor connected across the output of a power supply and designed to "bleed-off" a small portion of the current.

bonding: The connecting together of metal structures with electric conductors, thus establishing a uniform electric potential among all the parts bonded together.

breakdown voltage: Breadkdown voltage in a capacitor is that voltage at which the dielectric is ruptured, or the voltage level in a gas tube at which the gas becomes ionized and starts to conduct.

brush: A device designed to provide an electrical contact between a stationary conductor and a rotating element.

buffer amplifier: An amplifier in a transmitter circuit designed to isolate the oscillator section from the power section, thus preventing a frequency shift.

bus bar: A power distribution point to which a number of circuits may be connected. It oftens consists of a solid metal strip in which a number of terminals are installed.

cable: A group of insulated electric conductors, usually covered with rubber or plastic to form a flexible transmission line.

capacitance: The property enabling two adjacent conductors separated by an insulating medium to store an electric charge. The unit of capacitance is the *farad.*

capacitive reactance: The reactive effect of capacitance in an ac circuit. The formula is $X_c = \frac{1}{2}\pi f C$, where X_c is capacitive reactance in ohms, f is frequency in hertz, and C is the capacitance in farads.

capacitor: A device consisting of conducting plates separated by a dielectric and used to introduce capacitance into a circuit.

carrier wave: A radio-frequency electromagnetic wave used to convey intelligence impressed upon it by modulation.

cathode: (1) The negative electrode of a battery; (2) the element in an electron tube which emits electrons into the interelectrode space; (3) the negative terminal of a diode or electroplating cell.

cathode-ray tube (CRT): A special type of electron tube in which a stream of electrons from an *electron gun* impinges upon a fluorescent screen, thus producing a bright spot on the screen. The electron beam is deflected electrically or magnetically to produce patterns on the screen.

characteristic curve: A graph which shows the performance of an electron tube or a transistor under various operating conditions.

charge: A quantity of electricity. A charge is negative when it consists of a number of electrons greater than the number normally held by the charged material in a neutral condition. The charge is positive when there is a deficiency of electrons.

choke coil: An inductance coil designed to provide a high reactance to certain frequencies and generally used to block or reduce currents at these frequencies.

circuit: Conductors connected together to provide one or more complete electrical paths.

circuit breaker: A device which automatically opens a circuit if the current flow increases beyond an established limit.

circuit protection: The provision of devices in an electric circuit to prevent excessive current flow. These devices may be fuses, circuit breakers, current limiters, or sensing relays.

circular mil (cmil): The cross-sectional area of a circle having a diameter of 1 mil (0.001 in). The circular mil is used to indicate the size of electric wire.

clutch: A mechanical device used to connect or disconnect a motor or other driving unit from the driven device.

coaxial cables: A pair of concentric conductors. The inner conductor is supported by insulation which holds it in the center of the outer conductor. A coaxial cable is normally used to conduct HF currents.

coil: One or more turns of a conductor designed for use in a circuit to produce inductance or an electromagnetic field.

collector: The section of a transistor corresponding to the plate in an electron tube.

collector ring: A rotating electrical contact used with a brush to transfer electric current from a rotating unit to a stationary unit or vice versa.

color code: A system of colors used to indicate component values, or identify wires and terminals.

commutator: A rotating contact device in the armature of a dc generator or motor, which in effect changes the ac current flowing in the armature windings to a dc current in the external circuit.

compass: A device used to determine direction on the earth's surface. A magnetic compass utilizes the earth's magnetic field to establish direction.

compound winding: A combination of series and parallel or shunt windings to provide the magnetic field for a generator or motor.

conductance: The reciprocal of resistance.

conductor: A material through which an electric current can pass easily.

conduit: A metallic tubular sheath through which insulated conductors are run. The conduit provides mechanical protection and electric or magnetic shielding for the conductors.

continuity tester: A device designed to test the electrical continuity of a conductor or circuit. A battery and light, or other indicating unit, connected in series, or an ohm-meter may serve as continuity testers.

continuous wave (CW): An RF carrier wave whose successive oscillations are identical in magnitude and frequency.

control circuit: Any one of a variety of circuits designed to exercise control of an operating device, to perform counting, timing, switching, and other operations.

control grid: The grid nearest the cathode in an electron tube. The charge on the control grid governs the flow of electrons from cathode to plate.

corona loss: Power loss due to the ionization of gas adjacent to a high-potential conductor.

cosine: The ratio of the side adjacent to an acute angle of a right triangle to the hypotenuse.

coulomb (C): The international coulomb is a unit of electric charge consisting of approximately 6.28×10^{18} electrons. The absolute coulomb is slightly greater than the international coulomb; that is, 1 absolute coulomb = 1.000165 international coulomb.

counter electromotive force (cemf): A voltage developed in the armature of a motor which opposes the applied emf. The same principle applied to any inductance through which an alternating current is flowing.

counterpoise: One or more conductors used under certain types of antennas to take the place of the usual ground circuit.

coupling: Energy transfer between elements or circuits of an electronic system.

cross modulation: The modulation of a desired signal by an unwanted signal resulting in two signals in the output.

crystal: A solid body with symmetrically arranged plane surfaces. In electronic systems, crystals are used as rectifiers, semiconductors, transistors, and frequency controllers and to produce oscillatory voltages.

crystal diode: A diode constructed from a crystal semiconductor material such as silicon or germanium.

current: The movement of electricity through a conductor, i.e., the flow of electrons through a conductor.

current limiter: A device installed in a circuit to prevent current from increasing above a specified limit.

cutoff: The point at which an operation stops because a cutoff condition has been reached. In electron tubes, the point at which grid bias voltage stops the flow of electrons from cathode to plate.

cycle: A complete sequence of events in a recurrent series of similar periods.

damping: The decay in amplitude or strength of an oscillatory current when energy is not introduced to replace that lost through circuit resistance.

d'Arsonval meter movement: A meter movement consisting of a movable coil suspended on pivots between the poles of a permanent magnet.

decades: A series of quantities in multiples of 10, for example, 10, 100, 1000, 10 000.

decibel (db): One-tenth of a bel.

decoupling: The process of eliminating electrical or magnetic coupling between units in an electronic system.

deflection: The movement of an electron beam up and down or sideways in response to an electric or magnetic field in a cathode-ray tube.

degeneration: Feedback of a portion of the output of a circuit to the input in such a direction that it reduces the magnitude of the input; also called negative feedback. Degeneration reduces distortion, increases stability, and improves frequency response.

delta connection: A method of connecting three components to form a three-sided circuit, usually drawn as a triangle, hence the term *delta.* Delta (Δ) is the Greek letter corresponding to the English D.

demodulation: The recovery of the AF signal from an RF carrier wave. Also called *detection.*

dectector: That portion of an electronic circuit which demodulates or detects the signal.

deviation, compass: The error in a magnetic compass due to construction, installation, and nearby magnetic materials.

diac: A negative-resistance breakdown diode, constructed in both unidirectional and bidirectional forms.

dielectric: An insulating material used to separate the plates of a capacitor.

dielectric constant: A measure of the effectiveness of a dielectric for holding a charge in a capacitor. Air is given a dielectric constant of 1; mica has a dielectric constant of 5.5; hence, a capacitor having mica as a dielectric will have a capacitance 5.5 times as great as the same capacitor with a dielectric of air.

differentiating circuit: A circuit which produces an output voltage proportional to the rate of change of the input.

diode: A semiconductor device or an electron tube with only a cathode and anode, usually used as rectifiers and detectors.

dipole antenna: An antenna consisting of two equal lengths of wire or other conductor extending in opposite directions from the input point. Each section of the dipole is approximately one-quarter wavelength.

direct current (dc as adjective): An electric current which flows continuously in one direction.

directional gyro: A direction-indicating instrument which utilizes a gyroscope to hold the moving element in a fixed position relative to a directional reference.

discriminator: A circuit whose output polarity and magnitude are determined by the variations of the input phase or frequency.

distance-measuring equipment (DME): An electronic system used with radio navigation equipment to provide an indication of the distance to a specific point.

distortion: Undesirable change in the waveform of the output of a circuit compared with the input.

donor: An impurity used in a semiconductor to provide free electrons as current carriers. A semiconductor with a donor impurity is of the *n* type.

doppler effect: The effect noted as one moves toward or away from a source of a sound-wave or electromagnetic-wave propagation. Moving toward the source results in receiving a higher-frequency sound or signal than the source is emitting, and moving away from the source results in receiving a lower-frequency sound or signal.

duplexer: A circuit which makes it possible to use the same antenna for both transmitting and receiving without allowing excessive power to flow to the receiver.

dynamotor: An electric rotating machine with a double armature, usually designed to produce a high dc voltage for plate circuits in radio transmitters and receivers. One end of the armature serves a low-voltage dc motor, and the other end is wound for a high-voltage dc generator.

dynatron effect: The area of operation in a tetrode electron tube where plate current decreases as plate voltage increases. This effect is caused by secondary emission from the plate and by the attraction of the secondary electrons to the screen grid.

dynode: The elements in a multiplier tube which emit secondary electrons.

eddy currents: Currents induced in the cores of coils, transformers, and armatures by the changing magnetic fields associated with their operation. These currents cause great losses of energy. For this reason such cores are composed of insulated laminations which limit the currents paths.

Edison effect: The discovery of Thomas A. Edison in 1883 that a heated filament placed in an evacuated tube with another electrode will emit electrons.

effective value: A term used to indicate the actual working value of an alternating current based upon its heating effect. Also called the root-mean-square (rms) value and is equal to $1/\sqrt{2}$ times the maximum value in a sinusoidal current.

electret: A dielectric body in which a permanent state of electric polarization has been set up. Also, the material of which an electret is composed.

electricity: In general terms electricity may be said to consist of positive or negative charges at rest or in motion.

electrode: A terminal element in an electric device or circuit. Some typical electrodes include the plates in a storage battery, the elements in an electron tube, and the carbon rods in an arc light.

electrolysis: The process of decomposing a chemical compound by means of an electric current.

electrolyte: Any solution which conducts an electric current.

electromagnet: A magnet formed when an iron core is placed in a current-carrying coil.

electromagnetic induction: The transfer of electric energy from one conductor to another by means of a moving electromagnetic field. A voltage is produced in a conductor as the magnetic lines of force cut or link with the conductor. The value of the voltage produced by electromagnetic induction is proportional to the number of lines of force cut per second. When 100 000 000 lines of force are cut per second, an emf of 1 V will be induced.

electromagnetism: The magnetism produced by the flow of electric current.

electromotive force (emf): The force which causes current to move through a conductor. The unit of measurement for emf is the volt; hence, emf is often called *voltage*.

electron: A negatively charged nonnuclear particle which orbits around the nucleus of an atom. Generally speaking, the electron may be considered the carrier of electric current through a conductor. An electron at rest has a mass of 9.107×10^{-28} g and a charge of 1.6×10^{-19} C.

electron gun: The combination of an electron-emitting cathode together with accelerating anodes and beam-forming electrodes to produce the electron beam in a CRT.

electron tube: A device consisting of an evacuated or gasfilled envelope containing electrodes for the purpose of controlling electron flow. The electrodes are usually a cathode (electron emitter), a plate (anode), and one or more grids.

electrostatic field: The field of electric force existing in the area around and between any two oppositely charged bodies.

elements: Any substance which cannot be changed to another substance except by nuclear disintegration. There are more than 100 known elements.

emission, electronic: The freeing of electrons from the surface of a material, usually produced by heat.

emitter: The electrode in an electron tube which emits electrons, that is, the cathode; also the electrode in a transistor from which current carriers enter the interelectrode region.

equalizer circuit: A circuit in a multiple-generator voltage-regulator system which tends to equalize the current output of the generators by controlling the field currents of the several generators.

excitation: The application of electric current to the field windings of a generator to produce a magnetic field; also the input signal to an electron tube.

fading: A decrease in strength of a received radio signal.

farad: The unit of capacitance; the capacitance of a capacitor which will store 1 C of electricity when an emf of 1 V is applied.

feedback: A portion of the output signal of a circuit returned to the input. Positive feedback occurs when the feedback signal is in phase with the input signal. Negative feedback occurs when the feedback signal is 180° out of phase with the input signal.

ferromagnetic materials: Magnetic materials composed largely of iron.

fidelity: The degree of similarity between the input and output waveforms of an electronic circuit.

field: A space in which magnetic or electric lines of force exist.

field coil: A winding or coil used to produce a magnetic field.

field frame: The main structure of a generator or motor within which are mounted the field poles and windings.

filament: A resistive element in an electron tube which supplies the heat necessary for thermionic emission of electrons; also the heated element in an electric light bulb.

filter: A circuit arranged to pass certain frequencies while attenuating all others. A high-pass filter passes high frequencies and attenuates low frequencies; a low-pass filter passes low frequencies and attenuates high frequencies.

flux: Electrostatic or magnetic lines of force.

flux gate: An electromagnetic sensing device designed to determine the direction of the earth's magnetic field and thus produce magnetic-direction information for navigation systems.

flywheel effect: The characteristic of a parallel *LC* circuit which permits a continuing flow of current even though only small pulses of energy are applied to the circuit.

free electrons: Those electrons so loosely bound in the outer shells of some atoms that they are able to move from atom to atom when an emf is applied to the material.

frequency: The number of complete cycles of a periodic process per second. In electricity the unit of frequency is the hertz.

frequency modulation (FM): Modulation of a carrier by causing changes in carrier frequency proportional to the amplitude of the modulating signal.

frequency multiplier: A circuit designed to double, triple, or quadruple the frequency of a signal by harmonic conversion.

fuse: A metal link which melts when overheated by excess current, used to break an electric circuit whenever the load becomes excessive.

gain: The increase in signal power through a circuit.

galvanometer: A device for measuring electric currents. It usually consists of a current-carrying coil which produces a field to react with the field of a permanent magnet.

ganged tuning: A mechanical arrangement to permit the simultaneous tuning of two or more circuits.

gas-filled tube: An electron tube with gas introduced into the envelope to produce certain desired operating characteristics (see **thyraton tube**).

gate: An electronic switching circuit commonly employed in digital electronics to produce required outputs in response to particular inputs. The outputs are either "on" or "off" to produce the binary digits 1 or 0. Also, the control circuit built into various semiconductor devices.

gauss (G): The unit of magnetic flux density equal to 1 Mx (line of force) per square centimeter.

generator: A rotating machine designed to produce a certain type and quantity of voltage and current.

gilbert (Gb): The unit of magnetomotive force; it is equal to approximately 0.768 ampere-turns.

gimbal: A mechanism consisting of a pair of rings, one ring pivoted within the other and the outer ring supported on pivots 90 from the inner-ring pivots. A gyroscope pivoted in the inner ring at right angles to the inner-ring pivots will be free to precess in response to applied external forces.

glideslope: A directed radio beam emanating from a glideslope transmitter located near the runway of an instrumented airport to provide a reference for guiding an airplane vertically to the runway.

grid: An element in an electron tube used to regulate or control the flow of electrons from the cathode to the plate.

ground: (1) An electrical connection to the earth; (2) a common connecting device for the *zero-potential* side of the circuits in an elecrical or electronic system; (3) the accidental connection of a hot conductor to the ground (a hot conductor is one whose potential differs from ground potential).

ground wave: That portion of a radio wave which travels to the receiver along the surface of the earth.

growler: An electromagnetic device which develops a strong alternating field by which armatures may be tested.

guidance: The control of missiles or aircraft in flight.

gyroscope: A comparatively heavy wheel mounted on a spinning axis which is free to rotate about one or both of two axes perpendicular to each other and to the spinning axis. The gyroscope is used to sense directional changes and to develop signals for operating automatic pilots and inertial guidance systems.

harmonics: Multiples of a base frequency.

henry (H): The unit of inductance. It is the amount of inductance in a coil which will induce an emf of 1 V in the coil when the current flow is changing at the rate of 1 A/s.

hertz (Hz): The unit of frequency. One Hertz is equal to 1 c/s.

heterodyne: The process of mixing two frequencies to produce both sum and difference frequencies. The principle is used in superheterodyne receivers.

hexode: An electron tube having six active elements.

HIG: *Hermetically sealed integrating gyro.* A gyro mounted in a sealed case with a viscous damping medium. The output is therefore an indication of the total amount of angular displacement of the vehicle in which the gyro is installed, rather than the rate of angular displacement.

horizontal situation indicator (HSI): A flight instrument that provides the pilot with information regarding heading, course, glideslope deviation, course deviation, and other data regarding aircraft position.

horsepower (hp): A common unit of mechanical power. The time rate of work which will raise 550 lb through a vertical distance of 1 ft in 1 s; also 33 000 ft·lb/min. One horsepower is equal to 746 W of electric power.

hot-wire meter: An electric instrument for measuring alternating currents. A wire is heated by the current flow, and the expansion of the wire is used to provide movement for the indicating needle.

hydrometer: A calibrated float used to determine the specific gravity of a liquid.

hypotenuse: The side of a right triangle opposite the right angle.

hysteresis: The ability of a magnetic material to withstand changes in its magnetic state. When a magnetomotive force (mmf) is applied to such a material, the magnetization lags the mmf because of a resistance to change in orientation of the particles involved.

ignition: Pertaining to engines, the introduction of an electric spark into a combustion chamber to fire the fuel-air mixture.

image frequency: The heterodyne action of an oscillator in a superheterodyne receiver. An image frequency is produced when an unwanted signal is of such a frequency that when mixed with the oscillator frequency, it produces a differenece frequency equal to the intermediate frequency of the receiver.

impedance (Z): The combined effect of resistance, capacitive reactance, and inductive reactance in an ac circuit. Z is measured in ohms.

inductance: The ability of a coil or conductor to oppose a change in current flow (see **henry**).

inductance coil: A coil designed to introduce inductance into a circuit.

induction motor: An ac motor in which the rotating field produced by the stator induces currents and opposing fields in the rotor. The reaction of the fields creates the rotation force.

inductive reactance (X_L): The effect of inductance in an ac circuit. The formula for inductive reactance is $X_L = 2\pi fL$. X_L is measured in ohms.

inductor: An inductance coil.

inertia: The tendency of a mass to remain at rest or to continue in motion in the same direction.

inertial guidance: The guidance of a missile or airplane by means of a device which senses changes of direction or acceleration, and automatically corrects deviations in planned course.

instrument landing system (ILS): A radio guidance and communications system designed to guide aircraft through approaches, letdowns, and landings under instrument flying conditions.

insulator: A material which will not conduct current to an appreciable degree.

integrated circuit: A microminiature circuit incorporated on a very small chip of semiconductor material through solid-state technology. A number of circuit elements such as transistors, diodes, resistors, and capacitors are built into the semiconductor chip by means of photography, etching, and diffusion.

integrating circuit: A network circuit whose output is proportional to the sum of its instantaneous inputs.

interelectrode capacitance: The capacitance existing between the electrodes of an electron tube.

interpoles: Small magnetic poles inserted between the main field poles of a generator or motor in series with the load circuit to compensate for the effect of armature reaction.

inverter: A mechanical or electronic device which converts direct current to alternating current. Also, a binary digital circuit element or circuit with one input and one output. The output state is always the inverse (opposite) the input state.

ion: An atom or molecule which has lost one or more electrons (positive ion) or one which has one or more extra electrons (negative ion).

ionization: The process of creating ions by either chemical or electrical means.

iron-vane movement: An ac electric measuring instrument which depends upon a soft-iron vane or movable core operating with a coil to produce an indication of ac current flow.

joule (J): A unit of electric energy or work equivalent to the work done in maintaining a current of 1 A against a resistance of 1 Ω for 1 s; 1 J = 0.73732 ft. lb.

jumper: A short conductor usually used to make a temporary connection between two terminals.

junction box: An enclosure used to house and protect terminal strips and other circuit components.

junction transistor: A transistor consisting of a single crystal of *p*- or *n*-type germanium between two electrodes of the opposite type. The center layer is the base and forms junctions with the emitter and collector.

Kennelly-Heaviside layer: An ionized layer in the upper atmosphere which reflects radio waves to earth; also called E layer or ionosphere.

keying: The process of modulating a CW carrier wave with a key circuit to provide interruptions in the carrier in the form of dots and dashes for code transmission.

kilo: A prefix meaning 1000; e.g., kilocycle, kilovolt, kilowatt, etc.

kinetic energy: The energy which a body possesses as a result of its motion. It is equal to $\frac{1}{2}MV^2$ where M is mass, and V is velocity.

klystron tube: A special electron tube for UHF circuits in which modulation is accomplished by varying the velocity of the electrons flowing through the tube.

***LC* circuit:** A circuit network containing inductance and capacitance.

Lenz's law: A law stated by H. F. E. Lenz in 1833 to the effect that an induced current in a conductor is always in such a direction that its field opposes the change in the field causing the induced current.

light-emitting diode (LED): A semiconductor that utilizes a light-producing material such as gallium phosphide. The material produces light when an electric current is passed through it in a certain direction. LEDs are often used for digital displays.

limit switch: A switch designed to stop an actuator at the limit of its movement.

load factor: The ratio of average load to greatest load.

local oscillator: The internal-oscillator section of a superheterodyne circuit.

localizer: That section of an ILS which produces the directional reference beam.

logic circuit: A circuit designed to operate according to the fundamental laws of logic.

loop: A control circuit consisting of a sensor, a controller, an actuator, a controlled unit, and a follow-up or feedback to the sensor; also, any closed electronic circuit including a feedback signal which is compared with the reference signal to maintain a desired condition.

loop antenna: A bidirectional antenna consisting of one or more complete turns of wire in a coil.

loopstick: A loop antenna consisting of a large number of turns of wire wound on a powdered iron (ferrite) rod. Loopsticks are particularly useful in small portable radio receivers.

LORAN (LOng-RAnge Navigation): A radio navigation system utilizing master and slave stations transmitting timed pulses. The time difference in reception of pulses from several stations establishes a hyperbolic line of position which may be identified on a LORAN chart. By utilizing signals from two pairs of stations, a fix in position is obtained.

low-pass filter: A filter circuit designed to pass LF signals and attenuate HF signals.

Mach number: The ratio of actual speed to the speed of sound. An object moving at the speed of sound has a Mach number of 1.

Machmeter: An instrument for indicating the speed of a vehicle in terms of Mach number.

magamp: A contraction of *magnetic amplifier*. An amplifier system using saturable reactors to control an output to obtain amplification.

magnet: A solid material which has the property of attracting magnetic substances.

magnetic field: A space where magnetic lines of force exist.

magneto: A special type of electric generator having a permanent magnet or magnets to provide the field.

magnetomotive force (mmf): Magnetizing force, measured in gilberts or ampere-turns.

magnetron tube: A special electron tube for use in microwave systems. It uses strong magnetic and electric fields and tuned cavities to produce microwave amplification.

marker beacon: A radio navigation aid used in the approach zone of an instrumented airport. As the airplane crosses over the marker-beacon transmitter, the pilot receives an accurate indication of his distance from the runway through the medium of a flashing light and an aural signal.

master switch: A switch designed to control all electric power to all circuits in a system.

matter: That which has substance and occupies space; material.

maxwell (Mx): A unit of magnetic flux; one magnetic line of force.

mega: A prefix denoting one million, e.g., megahertz, megohm, etc.

mercury-vapor rectifier: A rectifier tube containing mercury which vaporizes during operation and increases the current-carrying capacity of the tube.

mho: A unit of conductance, the reciprocal of ohm.

microfarad (μF): One-millionth of a farad.

microphone: A device for converting sound waves to electric impulses.

miscrosecond (μs): One-millionth of a second.

microprocessor: An integrated circuit (*IC*) that can be programmed to perform a variety of desired functions. The circuit contains an arithmetic and logic unit, a controller, some registers, and possibly other elements.

microwave: An electromagnetic wave with a length of less than 10 m; i.e., it has a frequency of 30 MHz or more.

microwave landing system (MLS): A radio landing system for aircraft that utilizes microwave frequencies for the transmission of guidance and control signals.

mil: One-thousandth of an inch.

milli: A prefix meaning one-thousandth; e.g., milliammeter, milliampere, millihenry, etc.

mixer: A circuit in which two frequencies are combined to produce sum and difference frequencies (see **heterodyne** and **beat frequency oscillator**).

modulation: The impressing of an information signal on a carrier wave.

modulator: That portion of a transmitter circuit which modulates the carrier wave.

molecule: The smallest particle of a substance which can exist in a free state and maintain its chemical properties.

motor, electric: A rotating device for converting electric energy to mechanical energy.

multimeter: A combination instrument designed to measure a variety of electrical quantities.

multiplier tube: An electron tube designed to amplify or multiply very weak electron currents by means of secondary emission.

multivibrator: A special type of relaxation oscillator circuit designed to produce nonlinear signals such as square waves and sawtooth waves.

mutual inductance: The inductance of a voltage in one coil due to the field produced by an adjacent coil. Inductive coupling is accomplished through the mutual inductance of two adjacent coils.

neutralization: The use of circuits external to electron tubes to cancel the effects of interelectrode capacitance.

neutron: A neutral particle found in the nucleus of an atom.

north pole: The north-seeking pole of a magnet.

nucleus: The core or center particle of an atom.

null: An indicated low or zero point in a radio signal.

oersted: An mmf of 1 Gb/cm^2.

ohm (Ω): The unit of resistance which limits the current to 1 A when an emf of 1 V is applied.

ohmmeter: An electric measuring instrument designed to measure resistance in ohms.

Ohm's law: A law of current flow stated by George S. Ohm as follows: One volt of electrical pressure is required to force 1 A of current through 1 Ω of resistance; also, the current in a circuit is directly proportional to the voltage and inversely proportional to the resistance. The formula for Ohm's law may be expressed $I = E/R$, $R = E/I$, or $E = IR$.

optoelectronics: Electronic systems that utilize light-emitting and light-sensitive devices such as light-emitting diodes (LED) and phototransistors for control and operation.

oscillator: An electronic circuit which produces alternating currents with frequencies determined by the inductance and capacitance in the circuit.

oscillograph: A device for producing a graphical representation of an electric signal mechanically or photographically.

oscilloscope: An electronic device utilizing a CRT for observing electric signals.

parallel circuits: Two or more complete circuits connected to the same two power terminals.

peak inverse voltage (PIV): The maximum voltage which may be applied safely to an electron tube or semiconductor device in the direction inverse to normal current flow.

peak voltage: The maximum level of a variable voltage.

pentagrid converter: A five-grid electron tube which serves as a mixer, local oscillator, and first detector in a superheterodyne radio receiver.

pentode: An electron tube containing five electrodes—cathode, plate, control grid, suppressor grid, and screen grid.

permeability (μ): The property of a magnetic substance determining the flux density produced in the substance by a magnetic field of a given intensity. The formula is $\mu = B/H$, where B is flux density in gauss, and H is the field intensity in oersteds. The permeability of air is 1.

phase angle: The angular difference between two sinusoidal waveforms. When the voltage of an ac signal leads the current by 10°, there is a phase angle of 10° between the voltage and current.

phase inverter: An electronic circuit whose output is 180° out of phase with the input.

photo cell: An electronic device which becomes conductive or produces a voltage when struck by light.

phototransistor: A transistor in which light is used to control the collector-emitter current.

picofarad (pF): One-millionth of a microfarad.

piezoelectric effect: The property of certain crystals enabling them to generate an electrostatic voltage between opposite faces when subjected to mechanical pressure. Conversely, the crystal will expand or contract if subjected to a strong electrical potential.

pitch: The rotation of an airplane or missile about its lateral axis.

plan position indicator (PPI): A radar system component for presenting a maplike display of the search area on the screen of a CRT.

plate, electron tube: The anode of an electron tube.

plate resistance: In an electron tube, the ratio of a change in plate voltage to a change in plate current with grid voltage constant; expressed $R_p = \Delta E_p/\Delta I_p$.

plate saturation: The condition in an electron tube when the plate will no longer attract electrons as fast as they are emitted by the cathode.

polarity: (1) The nature of the electric charge on each of two terminals between which there is a potential difference; (2) the difference in the nature of the magnetic effect exhibited by the two poles of a magnet.

potential difference (PD): The voltage existing between two terminals or two points of differing potential.

potentiometer: A variable resistor often used as a voltage divider.

power: The rate of doing work (see **horsepower**).

power factor: In ac circuits, a multiplier equal to the cosine of the phase angle (θ) between the current and voltage. The power of an ac circuit in watts is equal to $EI \cos \theta$.

power supply: The part of an electronic circuit which supplies the filament and plate voltages for the operation of the circuit.

primary cell: A voltaic cell whose chemical action destroys some of the active elements in the cell, thus making it impossible or impractical to recharge.

primary winding: The input winding of a transformer.

proton: A positively charged particle found in the nucleus of an atom.

pulse generator: An electronic circuit designed to produce sharp pulses of voltage.

Q factor: The "figure of merit" or "quality" of an inductance coil. The formula for the Q of a coil is $Q = X_L/R = 2\pi fL/R$.

radar (*ra*dio *d*etecting *a*nd *r*anging): Radio equipment which utilizes reflected pulse signals to locate and determine the distance to any reflecting object within its range.

radar mile: The time required for a radar pulse to travel a distance of 1 nmi and return to the radar receiver; approximately 12.4 μs.

radio frequency (RF as adjective): All frequencies above the audible range, usually above 20 000 Hz.

rate gyro: A gyro unit whose output is proportional to the rate of changing direction.

rate signal: Any signal proportional to a rate of change.

ratiometer: A measuring instrument in which the movement of the indicator is proportional to the ratio of two currents.

***RC* circuit:** A circuit containing both resistance and capacitance.

RC time constant: The time required to charge a capacitor to 63.2 percent of its full-charge state through a given resistance.

rectification: The conversation of alternating current to direct current by means of a rectifier.

rectifier: A device which permits current to flow in one direction only.

regeneration: Positive feedback of an output signal to the input of an electronic-tube to increase the power of a signal.

relaxation oscillator: An oscillator circuit in which an *RC* circuit determines frequency of oscillation. The output is a sawtooth or rectangular wave.

relay: An electric switch operated by an electromagnet or solenoid.

reluctance: The property of a material which opposes the passage of magnetic flux lines through it.

resistor: A circuit element possessing a finite amount of resistance.

resonance: A condition in an *LC* circuit in which capacitive reactance and inductive reactance are equal.

reverse-current relay: A relay incorporated into a generator circuit to disconnect the generator from the battery when battery voltage is greater than generator voltage.

rheostat: A variable resistor.

ripple: A small periodic variation in the voltage level of a dc power supply.

roll: The rotation of an airplane or missile about its longitudinal axis.

rotor: A rotating part of an electric machine.

sawtooth wave: The output of a relaxation oscillator, rising slowly and then dropping sharply to zero to form waveshapes resembling sawteeth.

scope: A contraction of *oscilloscope.* Also used to designate the CRT used in radar.

screen grid: A grid constructed of fine wire mesh placed between the control grid and plate in an electron tube to reduce the effects of grid-plate capacitance.

secondary cell: An electrolytic voltaic cell capable of being repeatedly charged and discharged.

secondary coil: The output winding of a transformer.

secondary emission: The emission of electrons from a surface when struck by high-velocity electrons from the cathode.

Selcal: A contraction of *selective calling* referring to an automatic signaling system used in aircraft to notify the pilot that the aircraft is receiving a call.

selectivity: The ability of a radio receiver to tune in desired signals and tune out undesired signals.

selenium rectifier: A rectifier using a thin coating of selenium on an iron disk to develop a unidirectional current-carrying characteristic. Electrons flow easily from the iron to the selenium but encounter high resistance in the opposite direction. A metal alloy is used in order to form the electrical connection with the selenium.

self-inductance: The property of a single conductor or a coil causing it to induce a voltage in itself whenever there is a change of current flow.

Selsyn: A trade name of the General Electric Company applied to self-synchronizing units or synchros.

semiconductor controlled rectifier (SCR): A semiconductor rectifier that is controlled by means of a gate signal.

sensitivity: A measure of the ability of a radio receiver to receive very weak signals.

sensor: A sensing unit used to actuate signal-producing devices in response to changes in physical conditions.

series circuit: A circuit in which the current flows through all the circuit elements via a single path.

servo: An actuating device which feeds back an indication of its output or movement to the controlling unit, where it is compared with a reference at the input. Any difference between the input and output is used to produce the required control.

shielding: Metal covers placed around electric and electronic devices to prevent the intrusion of external electrostatic and electromagnetic fields.

shunt: A calibrated resistor connected across an electric device to bypass a portion of the current.

side bands: The bands of frequencies on each side of carrier frequency produced by modulation.

signal: The electric current, voltage, or waves constituting the inputs and outputs of electric or electronic circuits or devices. A signal may be the electric energy carrying information or may be the information itself.

signal generator: A test unit designed to produce reference electric signals which may be applied to electronic circuits for testing purposes.

sine curve or wave: A graphical representation of a wave proportional in magnitude to the sine of its angular displacement; hence, the sine wave is most useful in representing ac values.

skin effect: The tendency of HF alternating currents to flow in the outer portion of a conductor.

skip distance: The distance from a transmitter to the point where the reflected sky wave first reaches the earth.

sky wave: That portion of a radio wave which is reflected from the ionosphere.

slip rings: Conducting rings used with brushes to conduct electric current to or from a rotating unit.

solenoid: An electromagnetic device having a movable core.

space charge: The electric charge carried by the cloud of electrons in the space between electrodes of an electron tube.

split-phase motor: An ac motor which utilizes an inductor or capacitor to shift the phase of the current in one of two field windings. This causes the resultant field to have a rotational effect.

square mil (mil^2): An area equivalent to a square having sides 1 mil (0.001 in) in length.

square wave: An electric wave having a square shape.

standing waves: Stationary waves occurring on an antenna or transmission line as a result of two waves, identical in amplitude and frequency, traveling in opposite directions along the conductor.

stator: The stationary winding of a rotating ac machine.

substrate: The semiconductor material upon which diffused and epitaxially deposited regions are formed to construct diodes, transistors and similar devices.

superheterodyne: A radio receiver using the heterodyne principle to produce an intermediate frequency (IF).

suppressor grid: A grid placed between the screen grid and the plate in an electron tube to cause secondary electrons to return to the plate.

susceptance: The ratio of the effective current to the effective voltage in an ac circuit multiplied by the sine of the phase difference between the current and voltage.

sweep: The horizontal deflection of the electron beam in a CRT.

switch: A device for opening and closing electric circuits.

synchro: A device for transmitting indications of angular position from one point to another.

synchronous motor: An ac motor whose rotor is synchronized with the rotating field produced by the stator. The speed of rotation is always in time with the frequency of the applied alternating current.

synchroscope: An instrument designed to show whether two rotating elements are in synchronization.

tachometer: An instrument designed to indicate the rpm of a rotating device.

tank circuit: A parallel resonant circuit including an inductance and a capacitance.

telemetering: A system of sending measurements over great distances by radio.

terminal: A connecting fitting attached to the end of a circuit element.

terminal strip: An insulated strip with terminal posts to provide a convenient junction point for a group of separate circuits.

tetrode: An electron tube or semiconductor device containing four active electrodes.

thermionic: A term describing electron emission caused by heat.

thermocouple: A junction of two dissimilar metals which generates a small current when exposed to heat.

three-phase system: An ac electric system consisting of three conductors, each carrying a current 120° out of phase with each other. Three-phase systems are used extensively in modern electric and electronic actuating systems.

thyratron tube: A triode tube into which a gas has been introduced to change its operating characteristics. The control grid in a thyratron tube is used to start conduction when the correct potential difference exists between the cathode and plate. After the tube starts to conduct, the control grid is no longer effective, and the tube will continue to conduct even though the cathode-plate voltage drops very low.

thyristor: A four-layer (pnpn) semiconductor device with two, three, or four external terminals. Current flow through a thyristor may be controlled by one or more gates, by light, or by voltage applied between the two main terminals.

transconductance (G_M): The ratio of a small change in plate current to a small change in grid voltage. The formula is $G_M = I_p/E_g$ (plate voltage constant).

transformer: A device used to couple electric energy between circuits by means of mutual inductance.

transistor: A semiconductor device, usually made of a germanium or silicon crystal, used to rectify or amplify an electric signal.

transmission line: A conductor for radio waves, usually used to conduct RF energy from the output of a transmitter to the antenna.

transmitter: An electronic system designed to produce modulated RF carrier waves to be radiated by the antenna; also, electric devices used to collect quantitative information at one point and send it to a remote indicator electrically.

triac: A thyristor that provides bilateral operation. It is equivalent to two silicon-controlled rectifiers in inverse parallel connection. It is described as a bidirectional triode thyristor and is controlled by a gate circuit.

trigger pulse: An electric pulse applied to certain electronic circuit elements to start an operation.

trimmer capacitor: A low-capacity, adjustable capacitor connected in parallel with a large capacitor to provide fine tuning adjustments.

triode: An electron tube or semiconductor device with three active electrodes.

tuned radio frequency (TRF) receiver: A radio receiver in which tuning and amplification are accomplished in the RF section before the signal reaches the detector. After the detector one or more stages of AF amplification are employed to increase the output sufficiently to operate a loudspeaker.

tungar rectifier: A high-capacity diode rectifier tube having a heated cathode and a graphite plate in an envelope filled with argon gas.

tuning: The process of adjusting circuits to resonance at a particular frequency.

turn-and-bank indicator: A gyro-operated instrument designed to show the pilot of an airplane the rate of turn. It also has a curved tube containing a ball to show whether the airplane is correctly banked.

ultrahigh frequency (UHF): Radio frequencies between 300 and 3000 MHz.

vacuum tube: An electron tube with an evacuated envelope.

vacuum-tube voltmeter (VTVM): An electronic voltage-measuring instrument used for electronic circuit testing. Its very high input impedance prevents it from drawing appreciable power from the circuit being tested.

variable-mu (μ) tube: An electron tube having a control grid in which the grid wires are spaced less closely at the center than at the ends. This causes the amplification factor to change as grid bias is changed. Also called a remote-cutoff tube.

vector: A quantity having both magnitude and direction.

velocity: A measure of speed with direction.

very high frequency (VHF): The frequency range between 30 and 300 MHz.

very low frequency (VLF): The frequency range between 3 and 30 kHz.

VHF omnirange (VOR): An electronic air navigation system which provides accurate direction information in relation to a certain ground station.

video: A term describing electronic circuit components controlling or producing the visual signals displayed on a CRT.

volt: The unit of emf or voltage.

volt-amperes: The product of the voltage and current in a circuit.

voltage divider: A resistance arranged with connections (taps) to provide for the removal of voltages of any desired level. A potentiometer is often used as a variable voltage divider.

voltage regulator: A circuit which maintains a constant level voltage supply despite changes in input voltage or load.

voltmeter: A voltage-measuring instrument.

volume control: The circuit in a receiver or amplifier which varies loudness.

watt (W): The unit of electric power. In a dc circuit, power (in watts) = volts × amperes, or $P(W) = EI$.

watthour (Wh): The commercial unit of electric energy; watthours = watts × hours.

wattmeter: An instrument designed to measure electric power.

waveguide: A hollow metal tube designed to carry electromagnetic energy at extremely high frequencies.

wavelength (λ): The distance between points of identical phase in a radio wave. The formula for wavelength is λ (lambda) = $300\ 000\ 000/f$ where λ is wavelength in meters, and f is frequency in hertz.

Weston meter movement: A moving-coil instrument movement.

Wheatstone bridge: A bridge circuit consisting of three known resistances, one unknown resistance, and a galvanometer. The indication shown by the galvanometer is used to determine the value of the unknown resistance.

yaw: Rotation of an airplane or missile about its vertical axis; turning to the right or left.

zener diode: A diode rectifier designed to prevent the flow of current in a reverse direction until the voltage in that direction reaches a predetermined value. At this time the diode permits a reverse current to flow.

INDEX